STUDENT SOLUTIONS MANUAL

Laurel Technical Services
Gloria Langer

ALGEBRA

FOR COLLEGE STUDENTS

THIRD EDITION

ROBERT BLITZER

PRENTICE HALL, Upper Saddle River, NJ 07458

Acquisitions Editor: ***Karin Wagner***
Supplement Editor: ***Audra Walsh***
Production Editor: ***Dawn Blayer***
Special Projects Manager: ***Barbara A. Murray***
Supplement Cover Manager: ***Paul Gourhan***
Production Coordinator: ***Alan Fisher***

Printed in the United States of America

10 9 8 7 6 5 4 3

ISBN 0-13-752289-4

Prentice-Hall International (UK) Limited, *London*
Prentice-Hall of Australia Pty. Limited, *Sydney*
Prentice-Hall Canada, Inc., *Toronto*
Prentice-Hall Hispanoamericana, S.A., *Mexico*
Prentice-Hall of India Private Limited, *New Delhi*
Prentice-Hall of Japan, Inc., *Tokyo*
Simon & Schuster Asia Pte. Ltd., *Singapore*
Editora Prentice-Hall do Brasil, Ltda., *Rio de Janeiro*

CONTENTS

Chapter

Chapter 1

Problem Set 1.1

1. $\{x|x$ is an even natural number between 14 and 20, inclusively$\}$
$\{14, 16, 18, 20\}$

3. $\{x|x$ is a natural number that is divisible by 5$\}$
$\{5, 10, 15, 20, \ldots\}$

5. $\{x|x$ is a whole number but is not a natural number$\}$
$\{0\}$

7. $\{x|x$ is an integer and $x^2 = 1\}$
$\{-1, 1\}$

9. $\{x|x$ is a positive real number and $x^2 = 5\}$
$\left\{\sqrt{5}\right\}$

11. $\{x|x$ is an integer but not a natural number$\}$
$\{\ldots, -4, -3, -2, -1, 0\}$

13. $\left\{x|\sqrt{x}\right.$ is a natural number less than or equal to 2$\}$
$\{1, 4\}$

15. $\left\{x|x \text{ is the fractional form of } 0.\overline{6}\right\}$
$\left\{\frac{2}{3}\right\}$

17. Natural numbers: $\left\{\sqrt{4},\ 7,\ \frac{18}{2},\ 100\right\}$

19. Integers: $\left\{-10,\ 0,\ \sqrt{4},\ 7,\ \frac{18}{2},\ 100\right\}$

21. Irrational numbers: $\left\{-\sqrt{2},\ \sqrt{3},\ \pi\right\}$

23. -13 is less than or equal to -2; true

25. -6 is greater than 2; false

27. 4 is greater than or equal to -7; true

29. -13 is less than -5; true

31. $-\pi$ is greater than or equal to $-\pi$; true

33. $-\sqrt{2}$ is less than $-\sqrt{2}$; false

35. $\{x|0 < x < 3\}$
(0, 3)

37. $\{x|-2 \le x < 1\}$
[−2, 1)

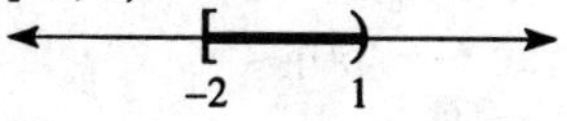

39. $\{x|-2 \le x \le -1\}$
[−2, −1]

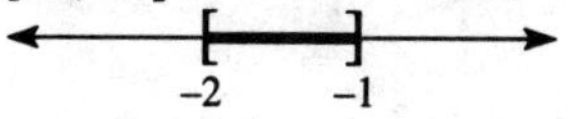

41. $\{x|x \le 2\}$
$(-\infty, 2]$

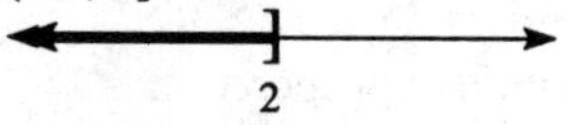

43. $\{x|x > -3\}$
$(-3, \infty)$

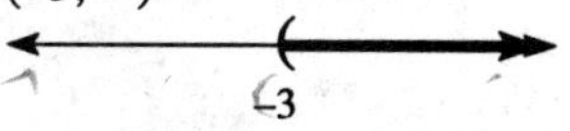

45. $\{x|x < -1\}$
$(-\infty, -1)$

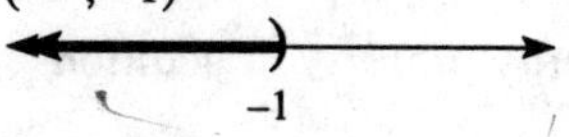

47. $\{x|x \le 0\}$
$(-\infty, 0]$

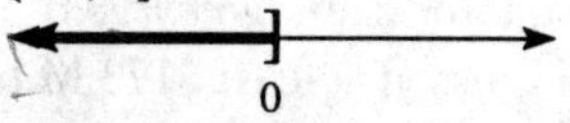

49. x lies between 5 and 12, excluding 5 and 12
set-builder: $\{x|5 < x < 12\}$
interval: (5, 12)

51. x lies between 2 and 13, excluding 2 and including 13
set-builder: $\{x|2 < x \le 13\}$
interval: (2, 13]

53. x is at most 6
set-builder: $\{x|x \le 6\}$
interval: $(-\infty, 6]$

55. x is at least 2 and at most 5
set-building: $\{x|2 \le x \le 5\}$
interval: [2, 5]

57. x is not more than 60
set-builder: $\{x|x \le 60\}$
interval: $(-\infty, 60]$

59. x is negative and at least –2
set-builder: $\{x|-2 \le x < 0\}$
interval: [–2, 0)

61. $3 + 7 = 7 + 3$;
Commutative property of addition

63. $2(7 + 4) = 2 \cdot 7 + 2 \cdot 4$;
Distributive property

65. $6 + (3 + 8) = 6 + (8 + 3)$;
Commutative property of addition

67. $2(3 \cdot 4) = (2 \cdot 3) \cdot 4$;
Associative property of multiplication

69. $(15 + 6) + 0 = 15 + 6$;
Identity property of addition

71. *Home Alone 2: Lost in New York, The Fugitive, Dances with Wolves,* and *Terminator 2: Judgment Day*
each grossed at least \$173 M but less than \$206 M.

73. *Aladdin, Ghost, Home Alone, Lion King,* and *Jurassic Park* each grossed at least \$206 M.

75. *Terminator 2: Judgment Day, Aladdin* and *Ghost* each grossed more than \$184 M but not more than \$217 M.

77. *Robin Hood, Prince of Thieves* grossed at most \$165 M.

79. None of the top ten grossing films from 1990–1995 grossed less than \$165 M.

81. **a.** a is not true.
$\{x|x < 3\} \rightarrow (-\infty, 3)$ *not* $[-\infty, 3)$

b. False; $2 < x < 5$ says 2 is less than x, or, equivalently, x is greater than 2.

c. False; infinity is not considered a real number; it is a compact symbolic notation.

d. d is true

83. Answers may vary.

85. Answers may vary.

87. Explanations may vary.
3.141592653589793238462643383279

89. For example, [250, 300] billion dollars

91. For example, [0, 150] billion dollars or [300, 350]

93. $a \circ b = ab + a$
$b \circ a = ba + b$
Since $ab + a \ne ba + b$, the operation is not commutative.

Problem Set 1.2

1. $7 + 6 \cdot 3 = 7 + 18 = 25$

3. $4(-5) - 6(-3) = -20 + 18 = -2$

5. $6 - 4(-3) - 5$
$= 6 + 12 - 5 = 18 - 5 = 13$

7. $3 - 5(-4 - 2)$
$= 3 - 5(-6) = 3 + 30 = 33$

9. $(2 - 6)(-3 - 5) = (-4)(-8) = 32$

11. $3(-2)^2 - 4(-3)^2 = 3(4) - 4(9)$
$= 12 - 36 = -24$

13. $(2-6)^2 - (3-7)^2$
$= (-4)^2 - (-4)^2 = 16 - 16 = 0$

15. $6(3-5)^3 - 2(1-3)^3$
$= 6(-2)^3 - 2(-2)^3$
$= 6(-8) - 2(-8)$
$= -48 + 16 = -32$

17. $8^2 - 16 \div 2^2 \cdot 4 - 3$
$= 64 - 16 \div 4 \cdot 4 - 3$
$= 64 - 4 \cdot 4 - 3$
$= 64 - 16 - 3$
$= 48 - 3 = 45$

19. $\dfrac{4^2 + 3^3}{5^2 - (-18)}$
$= \dfrac{16+27}{25+18}$
$= \dfrac{43}{43} = 1$

21. $20 - 4\left(\frac{8-2}{3-6}\right) \div \frac{1}{2}$
$= 20 - 4\left(\frac{6}{-3} \div \frac{1}{2}\right)$
$= 20 - 4((-2) \cdot 2)$
$= 20 - 4(-4)$
$= 20 + 16 = 36$

23. $\left(\frac{1}{2}\right)^2 + \left(\frac{6-4}{5}\right)^2 + \left(\frac{5+2}{10}\right)^2$
$= \left(\frac{1}{2}\right)^2 + \left(\frac{2}{5}\right)^2 + \left(\frac{7}{10}\right)^2$
$= \frac{1}{4} + \frac{4}{25} + \frac{49}{100}$
$= \frac{25}{100} + \frac{16}{100} + \frac{49}{100}$
$= \frac{90}{100} = \frac{9}{10}$

25. $-3[8 + (-6)] \div [-4 - (-5)]$
$= -3[2] \div [-4 + 5]$
$= -6 \div 1 = -6$

27. $\left(\frac{1}{2} - \frac{7}{4}\right) \div \left(1 - \frac{3}{8}\right)$
$= \left(\frac{2}{4} - \frac{7}{4}\right) \div \left(\frac{8}{8} - \frac{3}{8}\right)$
$= -\frac{5}{4} \div \frac{5}{8}$
$= -\frac{5}{4} \cdot \frac{8}{5} = -2$

29. $\frac{1}{4} - 6(2+8) \div \left(-\frac{1}{3}\right)\left(-\frac{1}{9}\right)$
$= \frac{1}{4} - 6(10) \cdot (-3)\left(-\frac{1}{9}\right)$
$= \frac{1}{4} - 60 \cdot \frac{1}{3}$
$= \frac{1}{4} - 20$
$= \frac{1}{4} - \frac{80}{4}$
$= -\frac{79}{4}$

31. $6.8 - (0.3)^2 \div 0.09$
$= 6.8 - (0.09) \div 0.09$
$= 6.8 - (0.09) \cdot \frac{1}{0.09}$
$= 6.8 - 1 = 5.8$

33. $\frac{1}{2} - \left(\frac{2}{3} \cdot \frac{9}{5}\right) + \frac{3}{10}$
$= \frac{1}{2} - \frac{6}{5} + \frac{3}{10}$
$= \frac{5}{10} - \frac{12}{10} + \frac{3}{10}$
$= -\frac{4}{10} = -\frac{2}{5}$

35. $8 - 3[-2(2-5) - 4(8-6)]$
$= 8 - 3[-2(-3) - 4(2)]$
$= 8 - 3[6 - 8]$
$= 8 - 3(-2)$
$= 8 + 6 = 14$

37. $\dfrac{2(-2)-4(-3)}{5-8}$
$= \dfrac{-4+12}{-3}$
$= \dfrac{8}{-3} = -\dfrac{8}{3}$

39. $10-(-8)\left[\dfrac{2(-3)-5(6)}{7-(-1)}\right]$
$= 10-(-8)\left[\dfrac{-6-30}{7+1}\right]$
$= 10+8\left[\dfrac{-36}{8}\right]$
$= 10 - 36 = -26$

41. $6-(-12)\left[\dfrac{2-4(3-7)}{-4-5(1-3)}\right]$
$= 6-(-12)\left[\dfrac{2-4(-4)}{-4-5(-2)}\right]$
$= 6+12\left[\dfrac{2+16}{-4+10}\right]$
$= 6+12\left[\dfrac{18}{6}\right]$
$= 6 + 12(3)$
$= 6 + 36 = 42$

43. $2\left[-5-\frac{1}{3}(17+4)\right]$
$= 2\left[-5-\frac{1}{3}(21)\right]$
$= 2[-5 - 7]$
$= 2(-12) = -24$

45. $-3x + 7x - 6x = -2x$

47. $4x^2y - 8x^2y = -4x^2y$

49. $-2x + 7y + 9x = 7x + 7y$

51. $5x^2y - 3xy^2 + 2x^2y = 7x^2y - 3xy^2$

53. $-4x^2 + 5y^2 - 3x^2 - 6y^2 = -7x^2 - y^2$

55. $7x^3y - 3 + 6x^3y - 8 = 13x^3y - 11$

57. $2(x+3) + 3(x+2)$
$= 2x + 6 + 3x + 6$
$= 5x + 12$

59. $-4(b-3) - 2(b+1)$
$= -4b + 12 - 2b - 2$
$= -6b + 10$

61. $5(x^2+3) - 7(x^2-4)$
$= 5x^2 + 15 - 7x^2 + 28$
$= -2x^2 + 43$

63. $-3(y^2-1) - (y^2-7)$
$= -3y^2 + 3 - y^2 + 7$
$= -4y^2 + 10$

65. $-2(x^2-1) - 4(3x^2-5)$
$= -2x^2 + 2 - 12x^2 + 20$
$= -14x^2 + 22$

67. $4(2x+5y) - 6(3x-2y)$
$= 8x + 20y - 18x + 12y$
$= -10x + 32y$

69. $2x - 5(x - 6y)$
$= 2x - 5x + 30y$
$= -3x + 30y$

71. $-3(5x-z) - (z-4y) + 5(x+3y-z)$
$= -15x + 3z - z + 4y + 5x + 15y - 5z$
$= -10x + 19y - 3z$

73. $3[x - 5(5-3x)]$
$= 3[x - 25 + 15x]$
$= 3[16x - 25]$
$= 48x - 75$

75. $4[x-2(x-3y)]$
$=4[x-2x+6y]$
$=4(-x+6y)$
$=-4x+24y$

77. $10x-6x\cdot 3+15y^2\div 5\cdot 3$
$=10x-18x+3y^2\cdot 3$
$=-8x+9y^2$

79. a. $E=155x^2-65x+150$
Substitute 2 for x.
$E=155(2)^2-65(2)+150$
$E=155(4)-130+150$
$E=620+20$
$E=640$
A person walking 2 meters/second uses 640 W of energy.

b. $E=250x+100$
$E=250(4)+100$
$E=1000+100$
$E=1100$
A runner moving at 4 meters/second uses 1100 W of energy.

81. $d=0.042s^2+1.1s$
$s=50$:
$d=0.042(50)^2+1.1(50)$
$d=0.042(2500)+55$
$d=105+55$
$d=160$

$s=30$:
$d=0.042(30)^2+1.1(30)$
$d=0.042(900)+33$
$d=37.8+33$
$d=70.8$
difference in distance:
$160-70.8=89.2$
A car going 50 mph will travel 89.2 ft farther than a car going 30 mph.

83. $R=206.835-(1.015w+0.846s)$
$(w=10,\ s=2.5)$:
$R=206.835-(1.015\cdot 10+0.846\cdot 2.5)$
$R=206.835-(10.15+2.115)$
$R=206.835-12.265$
$R=194.57$

85. $N=(14{,}400+120t+100t^2)\div(144+t^2)$
$(t=2)$:
$N=(14{,}400+120\cdot 2+100\cdot 2^2)\div(144+2^2)$
$N=(14{,}400+240+100\cdot 4)\div(144+4)$
$N=(14{,}640+400)\div(148)$
$N=(15{,}040)\div(148)$
$N\approx 101.62\approx 102$
Approximately 102 bacteria are present after 2 hours.

87. $T=3(A-20)^2\div 50+10$
$A=30$:
$T=3(30-20)^2\div 50+10$
$T=3(10)^2\div 50+10$
$T=3(100)\div 50+10$
$T=300\div 50+10$
$T=6+10$
$T=16$
$A=40$:
$T=3(40-20)^2\div 50+10$
$T=3(20)^2\div 50+10$
$T=3(400)\div 50+10$
$T=1200\div 50+10$
$T=24+10$
$T=34$
percent increase $=\dfrac{34-16}{16}=\dfrac{18}{16}=\dfrac{9}{8}$
$=1.125=112.5\%$

89. $M=2.89h+70.64$
$M=2.89(42)+70.64$
$M=192.02$
The model-based estimate of the body's height is 192.02 cm; the missing person's height is 196 cm. The difference is

approximately 4 cm or roughly $1\frac{3}{4}$ inches. Yes, the body could be that of the missing person.

91. d is true.

93. d is true.

95. Answers may vary.

97. Answers may vary.

99. $8 - 2 \cdot (3 - 4) = 10$

101. $\left(2 \cdot 5 - \frac{1}{2} \cdot 10\right) \cdot 9 = 45$

103.
1. x
2. $2x$
3. $2x + 9$
4. $(2x + 9) + x$
5. $[(2x + 9) + x] \div 3$
6. $[(2x + 9) + x] \div 3 + 4$
7. $[(2x + 9) + x] \div 3 + 4 - x$
$= (3x + 9) \div 3 + 4 - x$
$= x + 3 + 4 - x$
$= 7$

105. Group Activity

Review Problems

107. $\{x | x \le 2\}$, $(-\infty, 2]$

108. $\{..., -6, -4, -2\}$

109. Rewriting $x + 4$ as a, the statement reads $a + (-a) = 0$ which is the inverse property of addition.

Problem Set 1.3

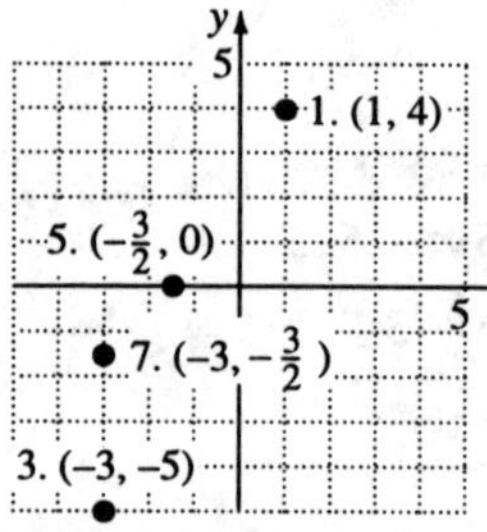

1. (1, 4): I

3. (−3, −5): III

5. $\left(-\frac{3}{2}, 0\right)$: On x-axis

7. $\left(-3, -\frac{3}{2}\right)$: III

9. $A(0, 6)$, $B(2, 4)$, $C(-2, 1)$, $D(2, -4)$, $E(0, -5)$

11. $A(-2, 0)$, $B(-1, 3)$, $C(0, 4)$, $D(1, 3)$, $E(2, 0)$

13.

x	$y = x^2 - 2$
−3	7
−2	2
−1	−1
0	−2
1	−1
2	2
3	7

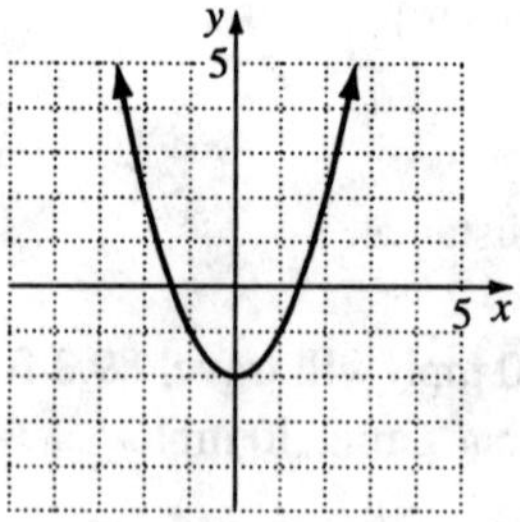

15.

x	$y=2x+1$
–3	–5
–2	–3
–1	–1
0	1
1	3
2	5
3	7

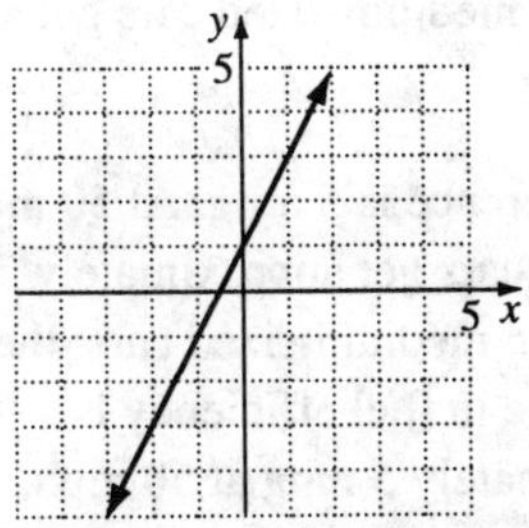

17.

x	$y=-\frac{1}{2}x$
–3	$\frac{3}{2}$
–2	1
–1	$\frac{1}{2}$
0	0
1	$-\frac{1}{2}$
2	–1
3	$-\frac{3}{2}$

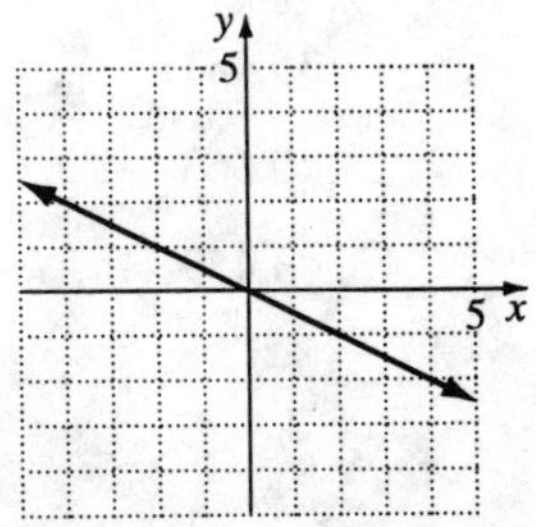

19.

x	$y=\lvert x\rvert+1$
–3	4
–2	3
–1	2
0	1
1	2
2	3
3	4

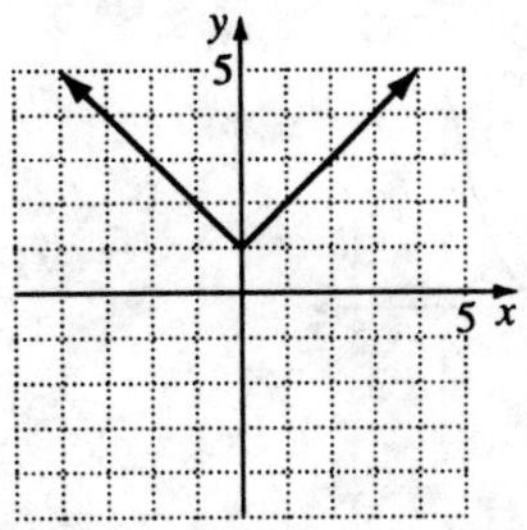

21.

x	$y=\lvert x+1\rvert$
–3	2
–2	1
–1	0
0	1
1	2
2	3
3	4

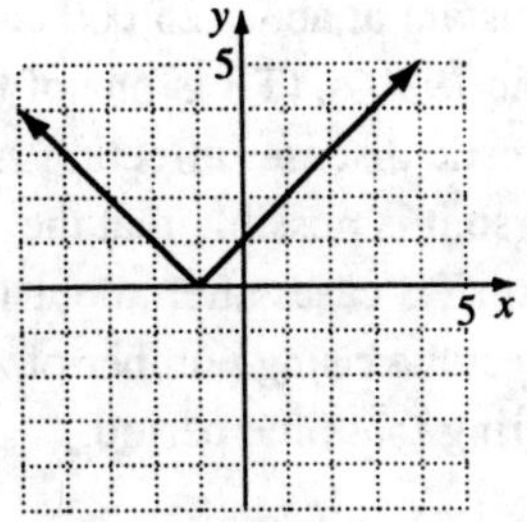

23.

x	$y = x^3$
−3	−27
−2	−8
−1	−1
0	0
1	1
2	8
3	27

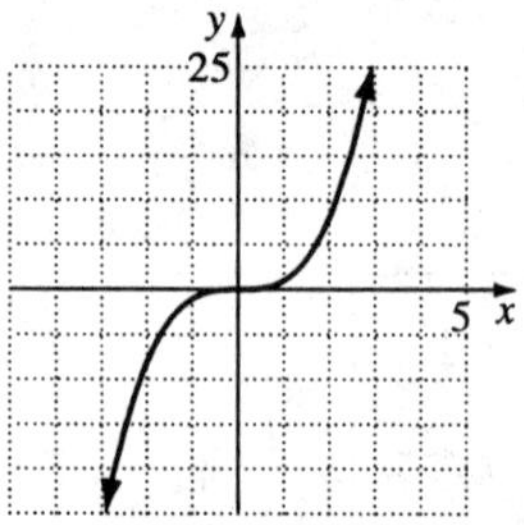

25. **a.** Maximum in approximately 1965; approximately 60,000 people

b. Minimum in approximately 1985; approximately 22,000 people

c. The number of TB cases decreased dramatically from a high in the mid 1960's to a low in the mid 1980's. However, after approximately 1985, the number of cases increased slightly to approximately 25,000 and has remained fairly constant at about 25,000 cases during the 1990's. (TB is one of the opportunistic diseases affecting AIDS patients, so it is possible that the increase in TB cases after about 1985 may reflect the rising number of AIDS cases during that time period.)

27. **a.** 64 ft; after 2 seconds

b. after 4 seconds

29. **a.** Approximately (2070, 19); the percentage of young people and the percentage of senior citizens will be the same in 2070 (approximately 19%).

b. The population will have a greater percentage of people aged 65 or older than people under age 15.

31. **a.** Approximate coordinates: $A(50, 22.5)$, $B(50, 17.5)$
At 50 mph, compact cars get about 22.5 mpg and medium-sized cars get about 17.5 mpg.

b. The difference is 5 mpg. At 50 mph, compact cars get approximately 5 mpg more than medium-sized cars; the difference in fuel efficiency is approximately 5 mpg at 50 mph.

c. compact cars: 35 mph
medium-sized cars: 40 mph

33.

x	$y = -143x^3 + 1810x^2 - 187x + 2331$
0	2331
1	3811
2	8053
3	14,199
4	21,391
5	28,771
6	35,481
7	40,663
8	43,459

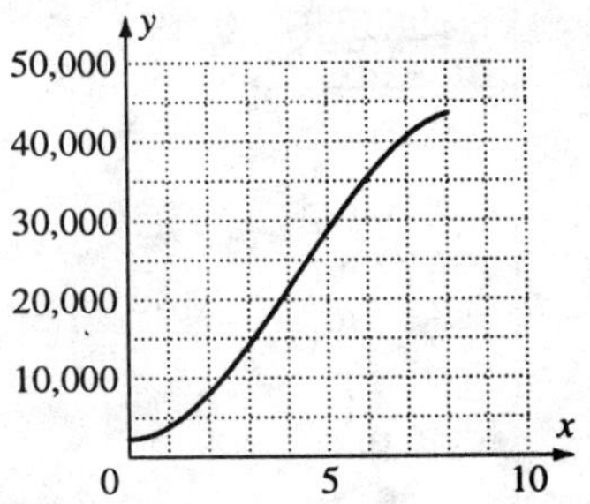

35. a. $40 + 0.35x = 36 + 0.45x$

$4 = 0.1x$

$x = 40$

The total cost for a day's rental is the same for both companies when the truck is driven 40 miles.

b. estimated $y \approx 55$

c. $y = 40 + 0.35x$

$y = 40 + 0.35(40)$

$y = \$54$

$y = 36 + 0.45x$

$y = 36 + 0.45(40)$

$y = \$54$

If a moving truck is driven 40 miles, the daily cost of renting the truck is $54, regardless of which company the truck is rented from.

37. a. $y = -3.1x^2 + 51.4x + 4024.5$

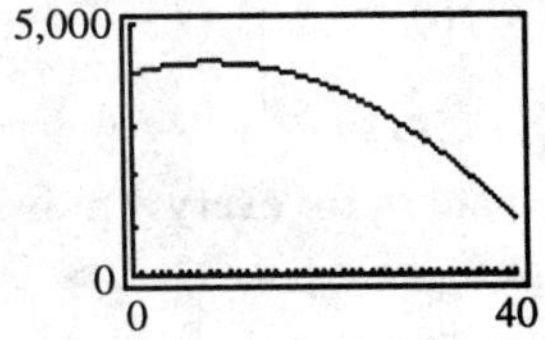

b. $y = -3.1x^2 + 51.4x + 4024.5$

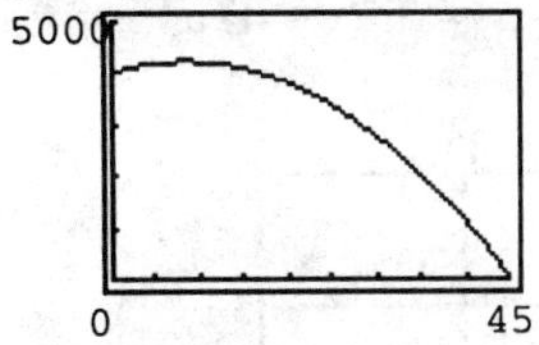

No; formulas that model variables over time tend to be more accurate in the center of the data and less accurate in the tails of the data (near the highest and lowest x-values).

39. a. $y = \dfrac{80,000x}{100 - x}$

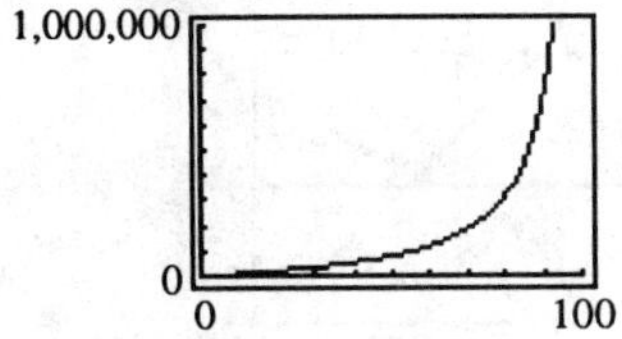

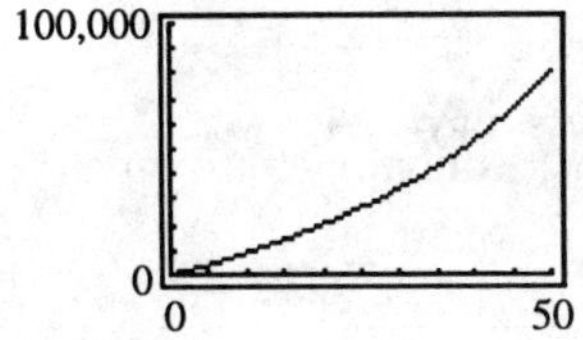

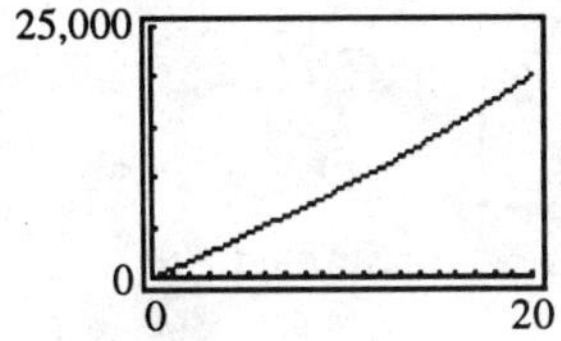

The cleanup costs increase dramatically as x nears 100%.

b. $y = \dfrac{80,000x}{100 - x}$

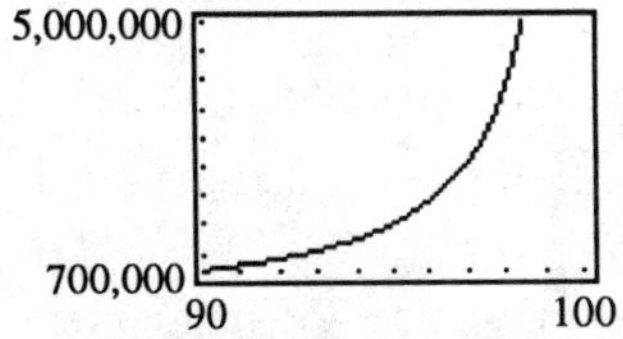

For $90 \le x \le 100$ the graph will show the steep curve approaching an imaginary vertical line at $x = 100$.

41. The simplification is correct; the graphs of $7-(2x-5)-3(2x+4)$ and $-8x$ are the same.

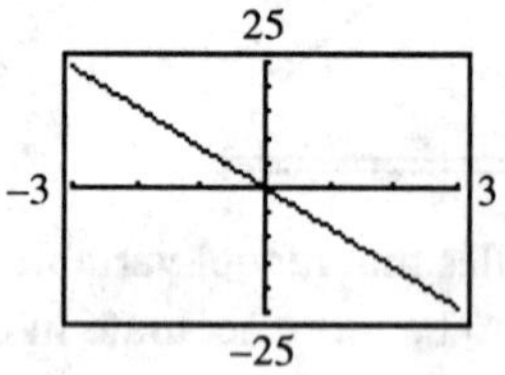

43. The simplification is wrong; the graphs of $4x^2-x^2-3x-3x^2+4=4-3x$ and x^2-3x+4 are not the same. Correct the right side by eliminating the x^2 term.

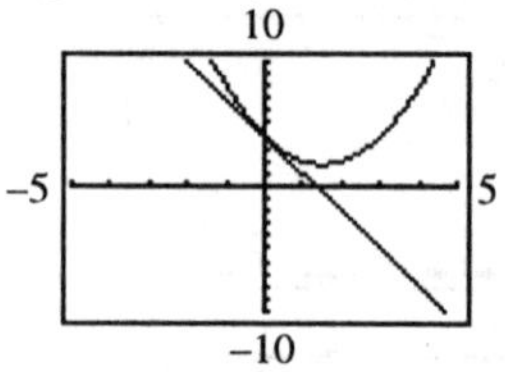

45. Answers may vary.

47.

x	$y=\frac{1}{x+2}$
−4	$-\frac{1}{2}$
−3	−1
−2	undefined
−1	1
0	$\frac{1}{2}$
1	$\frac{1}{3}$
2	$\frac{1}{4}$

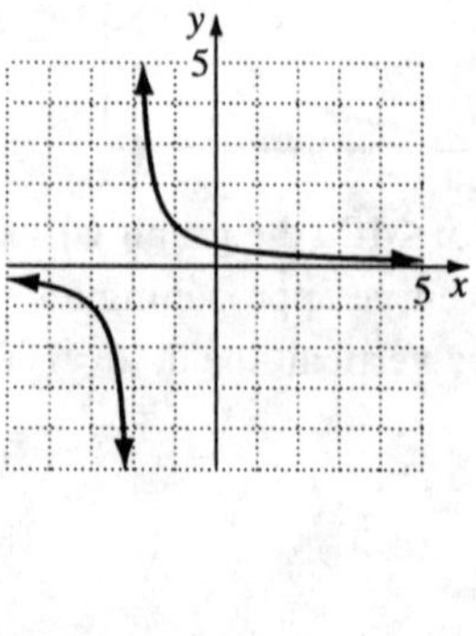

49.

x	$y=\sqrt{x-2}$
2	0
3	1
6	2
11	3
18	4

Graph is undefined for $x < 2$.

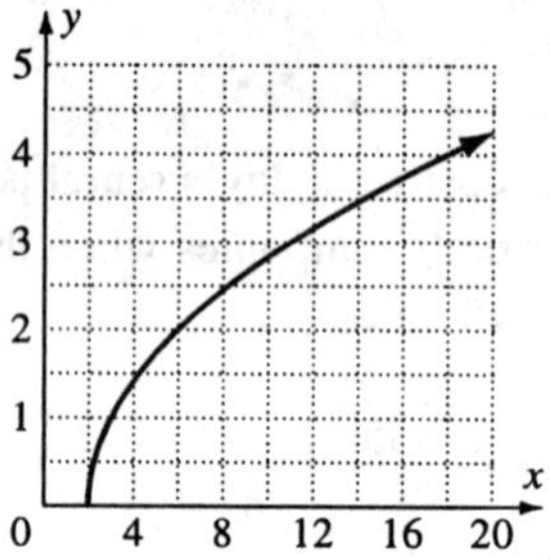

51.

Move	Coordinate	Pattern	
(n)	(0, 0)		
1	(1, −1)		$+n, -n$
2	(3, 1)	(1 + 2, −1 + 2)	$+n, +n$
3	(0, 4)	(3 − 3, 1 + 3)	$-n, +n$
4	(−4, 0)	(0 − 4, 4 − 4)	$-n, -n$
5	(1, −5)	(−4 + 5, 0 − 5)	
6	(7, 1)	(1 + 6, −5 + 6)	
7	(0, 8)	(7 − 7, 1 + 7)	
8	(−8, 0)	(0 − 8, 8 − 8)	

Note the pattern for every 4 moves.
After four moves: (−4, 0)
After eight moves: (−8, 0)

*For the x-coordinate the pattern is: add the move number, add the move number, subtract the move number, subtract the move number (at this point the x-coordinate is the negative value of the move) and then repeat the pattern.

Since $\frac{220}{4} = 55$ (an integer) the x-coordinate for the particle after 220 moves will be in the fourth point of the pattern at the negative value of the move number or –220.

*For the y-coordinate the fourth part of the pattern is on the x-axis.

Thus the coordinate of the particle after 220 moves: (–220, 0)

Review Problems

53. $\left\{-\frac{1}{2}, -\frac{2}{4}, -\frac{3}{6}\right\}$

54. $[12-(13-17)]-[9-(6-10)]$
$= [12-(-4)]-[9-(-4)]$
$= 16-13 = 3$

55. x is positive and at most 7
set-builder: $\{x \mid 0 < x \le 7\}$
interval: $(0, 7]$

Problem Set 1.4

1. $2^4 \cdot 2^2 = 2^{4+2} = 2^6$

3. $5^3 \cdot 5 = 5^{3+1} = 5^4$

5. $x^2 \cdot x^5 = x^{2+5} = x^7$

7. $2x^5 \cdot 3x^8 = (2 \cdot 3)(x^{5+8}) = 6x^{13}$

9. $(-5x^3y^2)(2xy^{17})$
$= (-5)(2)x^{3+1}y^{2+17}$
$= -10x^4y^{19}$

11. $(2^4)^3 = 2^{4 \cdot 3} = 2^{12}$

13. $(x^4)^8 = x^{4 \cdot 8} = x^{32}$

15. $(4x)^3 = 4^3x^3 = 64x^3$

17. $(3xy)^4 = 3^4x^4y^4 = 81x^4y^4$

19. $(2xy^2)^3 = 2^3x^3y^{2 \cdot 3} = 8x^3y^6$

21. $(-3x^2y^5)^2 = (-3)^2x^{2 \cdot 2}y^{5 \cdot 2} = 9x^4y^{10}$

23. $(2xy)(4x)^2$
$= (2xy)(4^2x^2)$
$= (2xy)(16x^2)$
$= 32x^3y$

25. $(4xy)(-2x^2y) + 17x^3y^2$
$= -8x^3y^2 + 17x^3y^2$
$= 9x^3y^2$

27. $(2x)^3(-3xy) + 25x^4y$
$= (2^3x^3)(-3xy) + 25x^4y$
$= (8x^3)(-3xy) + 25x^4y$
$= -24x^4y + 25x^4y = x^4y$

29. $\left(\frac{x}{y}\right)^6 = \frac{x^6}{y^6}$

31. $\left(\frac{-3x}{y}\right)^4 = \frac{(-3)^4x^4}{y^4} = \frac{81x^4}{y^4}$

33. $\left(\frac{x^4}{y^2}\right)^3 = \frac{x^{12}}{y^6}$

35. $\left(\frac{-5x^3}{2y}\right)^3 = \frac{(-5)^3x^9}{2^3y^3} = \frac{-125x^9}{8y^3} = -\frac{125x^9}{8y^3}$

37. $\frac{5^6}{5^3} = 5^{6-3} = 5^3$

39. $\frac{x^{16}}{x^8} = x^{16-8} = x^8$

41. $\frac{8x^7}{2x^4} = 4x^{7-4} = 4x^3$

43. $\frac{-100x^{18}}{25x^{17}} = -4x^{18-17} = -4x$

45. $\frac{-50x^2y^7}{5xy^4} = -10x^{2-1}y^{7-4} = -10xy^3$

47. $\frac{56a^{12}b^{10}c^8}{-7ab^2c^4} = -8a^{12-1}b^{10-2}c^{8-4}$
$= -8a^{11}b^8c^4$

49. $6^0 = 1$

51. $17^0 = 1$

53. $(6x)^0 = 1$

55. $5^{-2} = \frac{1}{5^2} = \frac{1}{25}$

57. $(-4)^{-3} = \frac{1}{(-4)^3} = \frac{1}{-64} = -\frac{1}{64}$

59. $(-4)^{-2} = \frac{1}{(-4)^2} = \frac{1}{16}$

61. $-4^{-2} = -\frac{1}{4^2} = -\frac{1}{16}$

63. $\left(\frac{3}{4}\right)^{-2} = \frac{1}{\left(\frac{3}{4}\right)^2} = \frac{1}{\frac{9}{16}} = \frac{16}{9}$

65. $\frac{1}{5^{-3}} = 5^3 = 125$

67. $\frac{1}{(-3)^{-4}} = (-3)^4 = 81$

69. $\frac{1}{-3^{-4}} = -3^4 = -81$

71. $\frac{20x^4y^3}{5xy^3} = \left(\frac{20}{5}\right)\left(\frac{x^4}{x}\right)\left(\frac{y^3}{y^3}\right)$
$= 4x^{4-1}y^{3-3} = 4x^3y^0 = 4x^3$

73. $\frac{x^3}{x^9} = x^{3-9} = x^{-6} = \frac{1}{x^6}$

75. $\frac{20x^3}{-5x^4} = \left(\frac{20}{-5}\right)\left(\frac{x^3}{x^4}\right)$
$= -4x^{3-4} = -4x^{-1} = \frac{-4}{x} = -\frac{4}{x}$

77. $\frac{16x^3}{8x^{10}} = \left(\frac{16}{8}\right)\left(\frac{x^3}{x^{10}}\right)$
$= 2x^{3-10} = 2x^{-7} = \frac{2}{x^7}$

79. $\frac{20a^3b^8}{2ab^{13}} = \left(\frac{20}{2}\right)(a^{3-1})(b^{8-13})$
$= 10a^2b^{-5} = \frac{10a^2}{b^5}$

81. $\frac{1}{b^{-5}} = b^5$

83. $x^3 \cdot x^{-12} = x^{3-12} = x^{-9} = \frac{1}{x^9}$

85. $(2a^5)(-3a^{-7}) = -6a^{5-7}$
$= -6a^{-2} = \frac{-6}{a^2} = -\frac{6}{a^2}$

87. $(3a^7)(-2a^{-3}) = -6a^{7-3} = -6a^4$

89. $(x^3)^{-6} = x^{3(-6)} = x^{-18} = \frac{1}{x^{18}}$

91. $(x^{-3})^{-6} = x^{(-3)(-6)} = x^{18}$

93. $\dfrac{x^3}{x^{-7}} = x^{3-(-7)} = x^{3+7} = x^{10}$

95. $\dfrac{6y^2}{2y^{-8}} = 3y^{2-(-8)} = 3y^{2+8} = 3y^{10}$

97. $\dfrac{4x^6}{-20x^{-11}} = \dfrac{1}{-5}x^{6-(-11)} = -\dfrac{x^{17}}{5}$

99. $\dfrac{x^{-7}}{x^3} = x^{-7-3} = x^{-10} = \dfrac{1}{x^{10}}$

101. $\dfrac{x^{-7}}{x^{-3}} = x^{-7-(-3)} = x^{-7+3} = x^{-4} = \dfrac{1}{x^4}$

103. $\dfrac{30x^2y^5}{-6x^8y^{-3}} = -5x^{2-8}y^{5-(-3)}$

$= -5x^{-6}y^{5+3} = -5x^{-6}y^8 = \dfrac{-5y^8}{x^6} = -\dfrac{5y^8}{x^6}$

105. $\dfrac{25a^{-8}b^2}{-75a^{-3}b^4} = \dfrac{1}{-3}a^{-8-(-3)}b^{2-4}$

$= \dfrac{1}{-3}a^{-8+3}b^{-2} = \dfrac{1}{-3}a^{-5}b^{-2} = -\dfrac{1}{3a^5b^2}$

107. $\left(\dfrac{x^3}{x^{-5}}\right)^2 = (x^{3-(-5)})^2$

$= (x^{3+5})^2 = (x^8)^2 = x^{16}$

109. $\left(\dfrac{15a^4b^2}{-5a^{10}b^{-3}}\right)^3 = (-3a^{4-10}b^{2-(-3)})^3$

$= (-3a^{-6}b^5)^3$

$= (-3)^3(a^{-6})^3(b^5)^3$

$= -27a^{-18}b^{15} = -\dfrac{27b^{15}}{a^{18}}$

111. $\left(\dfrac{10y^2}{y}\right) + \left(\dfrac{4xy^4}{xy^3}\right)$

$= 10y + 4y$

$= 14y$

113. $T = \left(\dfrac{h}{12.3}\right)^3$

(h = 5 feet 10 inches = 70 inches):

$T = \left(\dfrac{70}{12.3}\right)^3 \approx (5.691)^3 \approx 184$

Threshold weight: 184 pounds

115. d is true; $-3^{-2} = -\dfrac{1}{3^2} = -\dfrac{1}{9}$

117. d is true;

a is *not* true: $2^2 \cdot 2^4 = 2^6$ *not* 2^8

b is *not* true: $5^6 \cdot 5^2 = 5^8$ *not* 25^8

c is *not* true: $2^3 \cdot 3^2 = 8 \cdot 9 = 72 \neq 6^5$

d is true

119. **a.** $y_m = 67.0166(1.00308)^x$,

$y_f = 74.9742(1.00201)^x$

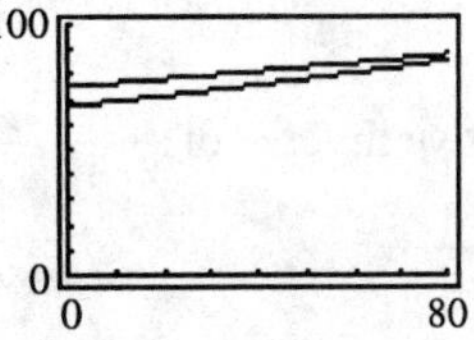

b. When $x = 40$, life expectancy for males is 75.8 years and life expectancy for females is 81.2 years.

c. Both curves increase as x (age) increases.

121. Answers may vary.

123. $(x^{-4n} \cdot x^n)^{-3} = (x^{-4n+n})^{-3}$

$= (x^{-3n})^{-3}$

$= x^{9n}$

125. $\left(\dfrac{x^n y^{3n+1}}{y^n}\right)^3 = (x^n y^{3n+1-n})^3$

$= (x^n y^{2n+1})^3$

$= (x^n)^3 (y^{2n+1})^3$

$= x^{3n} y^{6n+3}$

127. **a.** $\text{megaplex} = 10^{1,000,000} = 10^{10^6}$
$\text{gigaplex} = 10^{1,000,000,000} = 10^{10^9}$
$\text{megaplex}^{1000} = \text{gigaplex}$
$(10^{10^6})^{10^3} = 10^{10^9}$
$10^{10^9} = 10^{10^9}$ True

b. $2^{3,000,000,000} = 2^{3\times10^9}$
$= (2^3)^{10^9} = 8^{10^9}$
gigaplex

Review Problems

128. $[6(xy - 1) - 2xy] - 4[(2xy + 3) - 2(xy + 2)]$
$= [6xy - 6 - 2xy] - 4[2xy + 3 - 2xy - 4]$
$= (4xy - 6) - 4(-1)$
$= 4xy - 6 + 4$
$= 4xy - 2$

129. Multiplicative inverse of $-\frac{3}{4}$
$= \frac{1}{-\frac{3}{4}} = -\frac{4}{3}$

130. $\frac{3}{7^2 + (-7)(-5)} - \left[\left(\frac{2}{3} + \frac{1}{4}\right)\cdot\frac{6}{7}\right]$
$= \frac{3}{49 + 35} - \left(\frac{11}{12}\right)\cdot\frac{6}{7}$
$= \frac{3}{84} - \frac{11}{14}$
$= \frac{1}{28} - \frac{22}{28}$
$= -\frac{21}{28}$
$= -\frac{3}{4}$

Problem Set 1.5

1. $6\times10^{-3} = \frac{6}{1000} = 0.006$

3. $5.93\times10^6 = 5,930,000$

5. $6.284\times10^{-7} = \frac{6,284}{10,000,000}$
$= 0.0000006284$

7. $4.003\times10^{10} = 40,030,000,000$

9. $98,000,000,000 = 9.8\times10^{10}$

11. $746,000,000,000,000,000 = 7.46\times10^{17}$

13. $0.000000023 = 2.3\times10^{-8}$

15. $0.0007924 = 7.924\times10^{-4}$

17. $(2.8\times10^4)(3.2\times10^3)$
$= 2.8(3.2)\times10^{4+3} = 8.96\times10^7$

19. $(4.3\times10^{15})(6.5\times10^{-11})$
$= (4.3)(6.5)\times10^{15+(-11)}$
$= 27.95\times10^4 = 2.795\times10^5$

21. $(6.03\times10^6)(2.01\times10^{-9})$
$= 6.03(2.01)\times10^{6+(-9)}$
$= 12.1203\times10^{-3}$
$= (1.21203\times10^1)\times10^{-3}$
$= 1.21203\times10^{1+(-3)} = 1.21203\times10^{-2}$

23. $(8.04\times10^{-8})(3.01\times10^{-16})$
$= 8.04(3.01)\times10^{-8+(-16)}$
$= 24.2004\times10^{-24}$
$= (2.42004\times10^1)\times10^{-24}$
$= 2.42004\times10^{1+(-24)}$
$= 2.42004\times10^{-23}$

25. $\frac{9.9\times10^8}{1.1\times10^5} = 9\times10^{8-5} = 9\times10^3$

27. $\frac{6.12\times10^7}{3.06\times10^{-4}} = 2\times10^{7-(-4)} = 2\times10^{11}$

29. $\frac{1.5\times10^{-4}}{5.5\times10^7} = 0.\overline{27}\times10^{-4-7}$
$= 0.\overline{27}\times10^{-11}$
$= (2.7\overline{27}\times10^{-1})\times10^{-11}$
$= 2.7\overline{27}\times10^{-1+(-11)} = 2.7\overline{27}\times10^{-12}$

31. $\dfrac{1.21\times10^{-4}}{2.42\times10^{-7}}=0.5\times10^{-4-(-7)}$
$=(5\times10^{-1})\times10^{3}=5\times10^{-1+3}=5\times10^{2}$

33. $(82{,}000{,}000)(3{,}000{,}000{,}000)$
$=(8.2\times10^{7})(3\times10^{9})$
$=24.6\times10^{16}=2.46\times10^{17}$

35. $(0.00037)(8{,}300{,}000)$
$=(3.7\times10^{-4})(8.3\times10^{6})$
$=30.71\times10^{2}=3.071\times10^{3}$

37. $(150{,}000{,}000)(0.00005)(30{,}000)(0.002)$
$=(1.5\times10^{8})(5\times10^{-5})(3\times10^{4})(2\times10^{-3})$
$=45\times10^{4}=4.5\times10^{5}$

39. $\dfrac{95{,}000{,}000{,}000}{5{,}000{,}000}$
$=\dfrac{9.5\times10^{10}}{5\times10^{6}}$
$=1.9\times10^{10-6}$
$=1.9\times10^{4}$

41. $\dfrac{480{,}000{,}000{,}000}{0.00012}$
$=\dfrac{4.8\times10^{11}}{1.2\times10^{-4}}$
$=4\times10^{11-(-4)}$
$=4\times10^{15}$

43. $\dfrac{0.000000096}{16{,}000}$
$=\dfrac{9.6\times10^{-8}}{1.6\times10^{4}}$
$=6\times10^{-8-4}$
$=6\times10^{-12}$

45. $\dfrac{(90{,}000)(0.004)}{(0.0003)(120)}$
$=\dfrac{(9\times10^{4})(4\times10^{-3})}{(3\times10^{-4})(1.2\times10^{2})}$
$=\dfrac{(9)(4)\times10^{4-3}}{(3)(1.2)\times10^{-4+2}}$
$=\dfrac{36\times10}{3.6\times10^{-2}}$
$=10\times10^{1+2}$
$=1\times10\times10^{3}$
$=1\times10^{4}$

47. $\dfrac{(0.000035)(40{,}000)}{(14{,}000)(0.00025)}$
$=\dfrac{(3.5\times10^{-5})(4\times10^{4})}{(1.4\times10^{4})(2.5\times10^{-4})}$
$=\dfrac{(3.5)(4)\times10^{-5+4}}{(1.4)(2.5)\times10^{4-4}}$
$=\dfrac{14\times10^{-1}}{3.5\times10^{0}}$
$=4\times10^{-1}$

49. Approximately 150 billion dollars
$=1.5\times10^{11}$ dollars

51. $\dfrac{\text{average length of human foot}}{\text{length of hydrogen atom}}$
$=\dfrac{200\text{ millimeters}}{0.00000003\text{ millimeters}}$
$=\dfrac{2\times10^{2}}{3\times10^{-8}}$
$=0.\overline{6}\times10^{2-(-8)}$
$=6.\overline{6}\times10^{-1}\times10^{10}$
$=6.\overline{6}\times10^{9}$ times as large

53. $\dfrac{\text{energy released by H-bombs}}{\text{energy released by cricket's chirp}}$

$= \dfrac{10{,}000{,}000{,}000{,}000{,}000{,}000{,}000{,}000{,}000 \text{ ergs}}{9000 \text{ ergs}} = \dfrac{1 \times 10^{28}}{9 \times 10^{3}}$

$= 0.\overline{1} \times 10^{28-3}$

$= 1.\overline{1} \times 10^{-1} \times 10^{25}$

$= 1.\overline{1} \times 10^{24}$ times greater

55. $t = \dfrac{d}{r} = \dfrac{1.2 \times 10^{17}}{1.5 \times 10^{5}}$

$= 0.8 \times 10^{12} = 8 \times 10^{11}$ seconds

$8 \times 10^{11} \text{ sec} \times \dfrac{1 \text{ hour}}{3600 \text{ sec}} \times \dfrac{1 \text{ day}}{24 \text{ hours}} \times \dfrac{1 \text{ year}}{365 \text{ days}} = \dfrac{8 \times 10^{11}}{31536000}$ years

$= \dfrac{8 \times 10^{11}}{3.1536 \times 10^{7}}$ years $\approx 2.5 \times 10^{4}$ years

It would take approximately 2.5×10^{4} years to cross the galaxy moving at half the speed of light.

57. \$194.4 billion = $194.4 \times 10^{9} = 1.944 \times 10^{11}$

So $(\$1.944 \times 10^{11})\left(\dfrac{1 \text{ second}}{\$1000}\right)$

$= 1.944 \times 10^{8}$ seconds and

1.944×10^{8} seconds

$= (1.944 \times 10^{8} \text{ sec})\left(\dfrac{1 \text{ min}}{60 \text{ sec}}\right)\left(\dfrac{1 \text{ hr}}{60 \text{ min}}\right)\left(\dfrac{1 \text{ day}}{24 \text{ hr}}\right)\left(\dfrac{1 \text{ yr}}{365 \text{ days}}\right)$

$= \dfrac{(1.944 \times 10^{8}) \text{ yr}}{(6 \times 10)(6 \times 10)(2.4 \times 10)(3.65 \times 10^{2})}$

$= \dfrac{1.944 \times 10^{8}}{315.36 \times 10^{5}}$ years

$\approx 0.006 \times 10^{3}$ years = 6 years

59. d is true;

$(4 \times 10^{3}) + (3 \times 10^{2})$

$= 4000 + 300 = 4300$

$= 43 \times 10^{2}$

61. $F = \dfrac{(6.67 \times 10^{-11})(5.97 \times 10^{24})(7.35 \times 10^{22})}{(3.84 \times 10^{8})^{2}}$ newtons

$= \dfrac{292.676265 \times 10^{35}}{14.7456 \times 10^{16}} \approx 19.85 \times 10^{19} = 1.985 \times 10^{20}$ newtons

63. After explanation, result is 7.63×10^{-4}

65. Answers may vary.

67. Answers may vary.

69. $(c \times 10^n) \times (d \times 10^m) = (c \times d) \times (10^{n+m}) = cd \times 10^{n+m}$
$(c \times 10^n) \div (d \times 10^m) = \frac{c}{d} \times 10^{n-m}$ or $cd^{-1} \times 10^{n-m}$

Review Problems

71. $y = x^2 + 3$

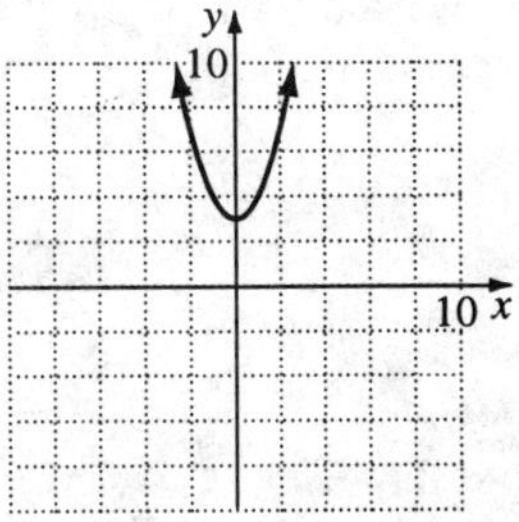

72. $\{x \mid -3 \le x < 5\}$
$[-3, 5)$

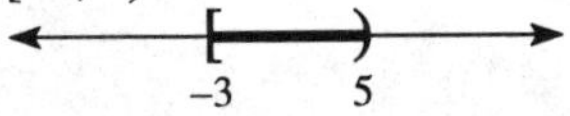

73. $-18 - (-12 + 2 \cdot 3^2)$
$= -18 - (-12 + 2 \cdot 9)$
$= -18 - (-12 + 18)$
$= -18 - (6) = -18 - 6 = -24$

Problem Set 1.6

1. $5x + 3 = 18$
$5x = 15$
$x = 3$
$\{3\}$

3. $6x - 3 = 63$
$6x = 66$
$x = 11$
$\{11\}$

5. $4x - 14 = -82$
$4x = -68$
$x = -17$
$\{-17\}$

7. $14 - 5x = -41$
$-5x = -55$
$x = 11$
$\{11\}$

9. $9(5x - 2) = 45$
$45x - 18 = 45$
$45x = 63$
$x = \frac{7}{5} = 1.4$
$\left\{\frac{7}{5}\right\}$

11. $5x - (2x - 10) = 35$
$5x - 2x + 10 = 35$
$3x = 25$
$x = \frac{25}{3}$
$\left\{\frac{25}{3}\right\}$

13. $3x + 5 = 2x + 13$
$3x + 5 - 2x = 2x + 13 - 2x$
$x + 5 = 13$
$x + 5 - 5 = 13 - 5$
$x = 8$
$\{8\}$

15. $8x - 2 = 7x - 5$
$8x - 7x - 2 = 7x - 5 - 7x$
$x - 2 = -5$
$x - 2 + 2 = -5 + 2$
$x = -3$
$\{-3\}$

17. $7x + 4 = x + 16$
$7x + 4 - x = x + 16 - x$
$6x + 4 = 16$
$6x + 4 - 4 = 16 - 4$
$6x = 12$
$\frac{1}{6}(6x) = \frac{1}{6}(12)$
$x = 2$
$\{2\}$

19. $8y - 3 = 11y + 9$
$8y - 3 - 11y = 11y + 9 - 11y$
$-3y - 3 = 9$
$-3y - 3 + 3 = 9 + 3$
$-3y = 12$
$\left(-\frac{1}{3}\right)(-3y) = \left(-\frac{1}{3}\right)(12)$
$y = -4$
$\{-4\}$

21. $8z + 11.7 = 9z - 15$
$8z + 11.7 - 9z = 9z - 15 - 9z$
$-z + 11.7 = -15$
$-z + 11.7 - 11.7 = -15 - 11.7$
$-z = -26.7$
$z = 26.7$
$\{26.7\}$

23. $\frac{1}{6}y + \frac{1}{2} = \frac{1}{3}$
$6\left(\frac{1}{6}y + \frac{1}{2}\right) = 6\left(\frac{1}{3}\right)$
$y + 3 = 2$
$y = -1$
$\{-1\}$

25. $3(y + 4) + (y - 2) = 2 - (2y - 14)$
$3y + 12 + y - 2 = 2 - 2y + 14$
$4y + 10 = -2y + 16$
$6y = 6$
$y = 1$
$\{1\}$

27. $16 = (3y - 3) - (y - 7)$
$16 = 3y - 3 - y + 7$
$16 = 2y + 4$
$12 = 2y$
$6 = y$
$\{6\}$

29. $2(y + 1) - 3y = 3(3 + 2y)$
$2y + 2 - 3y = 9 + 6y$
$-y + 2 = 9 + 6y$
$-7y = 7$
$y = -1$
$\{-1\}$

31. $5(x - 2) - 2(2x + 1) = 2 + 5x$
$5x - 10 - 4x - 2 = 2 + 5x$
$-4x = 14$
$x = -\frac{7}{2}$
$\left\{-\frac{7}{2}\right\}$

33. $7(x + 1) = 4[x - (3 - x)]$
$7x + 7 = 4[x - 3 + x]$
$7x + 7 = 8x - 12$
$19 = x$
$\{19\}$

35. $\frac{y}{4} = 2 + \frac{y - 3}{3}$
$12\left(\frac{y}{4}\right) = 12\left(2 + \frac{y - 3}{3}\right)$
$3y = 24 + 4(y - 3)$
$3y = 24 + 4y - 12$
$3y = 12 + 4y$
$3y - 4y = 12 + 4y - 4y$
$-y = 12$
$y = -12$
$\{-12\}$

37. $\frac{y-3}{12}+\frac{y-1}{6}=\frac{y+2}{9}-1$

$36\left(\frac{y-3}{12}+\frac{y-1}{6}\right)=36\left(\frac{y+2}{9}-1\right)$

$3(y-3)+6(y-1)=4(y+2)-36$

$3y-9+6y-6=4y+8-36$

$9y-15=4y-28$

$9y-15-4y=4y-28-4y$

$5y-15=-28$

$5y-15+15=-28+15$

$5y=-13$

$\frac{1}{5}(5y)=\frac{1}{5}(-13)$

$y=-\frac{13}{5}$

$\left\{-\frac{13}{5}\right\}$

39. $\frac{x}{5}-6=6+\frac{1}{5}x$

$5\left(\frac{x}{5}-6\right)=5\left(6+\frac{1}{5}x\right)$

$5\left(\frac{x}{5}\right)-5(6)=5(6)+5\left(\frac{1}{5}x\right)$

$x-30=30+x$

$x-30-x=30+x-x$

$-30=30$ contradiction

no solution

$\varnothing$

41. $0.35y-0.1=0.15y+0.2$

$0.35y-0.1-0.15y=0.15y+0.2-0.15y$

$0.20y-0.1=0.2$

$0.20y-0.1+0.1=0.2+0.1$

$0.2y=0.3$

$\left(\frac{1}{0.2}\right)(0.2y)=\left(\frac{1}{0.2}\right)(0.3)$

$y=\frac{3}{2}$

$\left\{\frac{3}{2}\right\}$

43. $0.02(y-100)=62+0.06y$

$0.02y-2=62+0.06y$

$0.02y-2-0.06y=62+0.06y-0.06y$

$-2-0.04y=62$

$-2-0.04y+2=62+2$

$-0.04y=64$

$\left(-\frac{1}{0.04}\right)(-0.04y)=\left(-\frac{1}{0.04}\right)(64)$

$y=-1600$

$\{-1600\}$

45. $0.09(-y+5000)=513-0.12y$

$-0.09y+450=513-0.12y$

$-0.09y+450+0.12y=513-0.12y+0.12y$

$0.03y+450=513$

$0.03y+450-450=513-450$

$0.03y=63$

$\left(\frac{1}{0.03}\right)(0.03y)=\left(\frac{1}{0.03}\right)(63)$

$y=2100$

$\{2100\}$

47. $0.2z=35+0.05(z-100)$

$0.2z=35+0.05z-5$

$0.2z=30+0.05z$

$0.2z-0.05z=30+0.05z-0.05z$

$0.15z=30$

$\left(\frac{1}{0.15}\right)(0.15z)=\left(\frac{1}{0.15}\right)(30)$

$z=200$

$\{200\}$

49. $10x-2(4+5x)=-8$

$10x-8-10x=-8$

$-8=-8$ True

$\{x|x\in R\}$

51. $10x-2(4+5x)=8$

$10x-8-10x=8$

$-8=8$ False

No solution

$\varnothing$

53. $7x-2[3(1-x)]=-(4+2-9x-4x)$

$7x-2(3-3x)=-(6-13x)$

$7x-6+6x=-6+13x$

$13x-6=-6+13x$

$13x-6-13x=-6+13x-13x$

$-6 = -6$ True

$\{x | x \in R\}$

55. $12\left(\frac{y}{3} - \frac{1}{2}\right) = 8\left(\frac{y}{2} - 1\right)$

$4y - 6 = 4y - 8$

$4y - 6 - 4y = 4y - 8 - 4y$

$-6 = -8$ contradiction

no solution

$\varnothing$

57. $3(z - 4) - 5 = -2z + 5z - 9$

$3z - 12 - 5 = 3z - 9$

$-17 = -9$

no solution

$\varnothing$

59. $-9 + 2(1 - 15k) = 7[2 - (3 + 4k)]$

$-9 + 2 - 30k = -7 - 28k$

$-7 - 30k = -7 - 28k$

$0 = 2k$

$0 = k$

$\{0\}$

61. d is true;

$7(y + 1) + 5(-y + 5) = 2(y + 3) - 7$

$7y + 7 - 5y + 25 = 2y + 6 - 7$

$2y + 32 = 2y - 1$

$2y + 32 - 2y = 2y - 1 - 2y$

$32 = -1$ contradiction

no solution

$\varnothing$

63. $y_1 = 6x + 3$

$y_2 = 6x + 2$

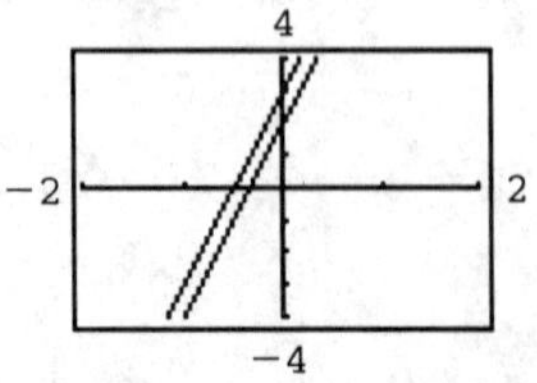

Since $9x + 3 - 3x = 6x + 3$, and $2(3x + 1) = 6x + 2$, the equations are inconsistent (parallel but with different intercepts).

65. $y_1 = \frac{x+1}{2}$

$y_2 = \frac{x-3}{4}$

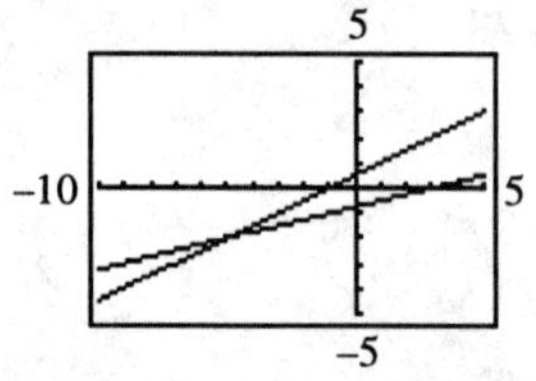

Since $\frac{2x-1}{3} - \frac{x-5}{6} = \frac{4x-2-x+5}{6}$

$= \frac{3x+3}{6} = \frac{x+1}{2}$, and not $\frac{x-3}{4}$, the equations are conditional. $\frac{x+1}{2} = \frac{x-3}{4}$ when $x = -5$, and at $x = -5$, $y = -2$ so the lines intersect at $(-5, -2)$.

67–73. Answers may vary.

Review Problems

75. $y = 2x - 4$

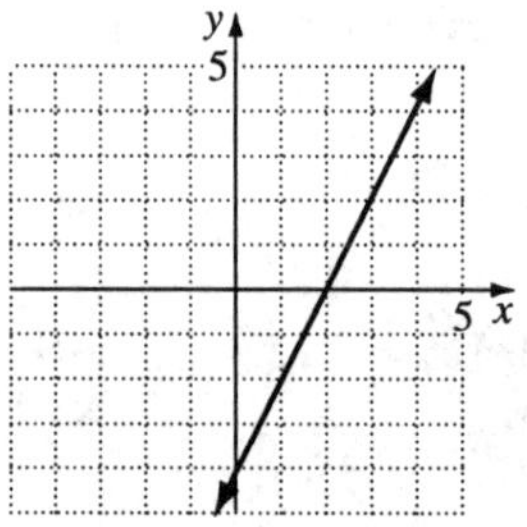

76. $\frac{-30x^2y^8}{10x^{-4}y^{11}} = -3x^{2-(-4)}y^{8-11}$

$= -3x^6y^{-3} = \frac{-3x^6}{y^3} = -\frac{3x^6}{y^3}$

77. $\frac{5 \times 10^{-6}}{20 \times 10^{-3}} = 0.25 \times 10^{-6-(-3)} = 0.25 \times 10^{-3}$

$= (2.5 \times 10^{-1}) \times 10^{-3} = 2.5 \times 10^{-4}$

Problem Set 1.7

1. $D = 0.2F - 1$
$19 = 0.2F - 1$
$20 = 0.2F$
$F = 100$
100 grams of fat per day

3. $d = 5000c - 525{,}000$
$500{,}000 = 5000c - 525{,}000$
$1{,}025{,}000 = 5000c$
$205 = c$
average adult cholesterol level in 1990 was 205 mg/dl
At 180 mg/dl,
$d = 5000(180) - 525{,}000 = 375{,}000.$
So if the average adult cholesterol level could be reduced to 180 mg/dl, 500,000 – 375,000 = 125,000 lives could be saved, compared to the number of deaths in 1990.

5. $D = 2A + V$
$A = \frac{D - V}{2}$

	D	V	A	Rank
Region 1	8	6	$\frac{8-6}{2} = 1$	1
Region 2	23	3	$\frac{23-3}{2} = 10$	3
Region 3	17	7	$\frac{17-7}{2} = 5$	2

Rank of the region in order of average brush age: Region 1, Region 3, Region 2

7. $400 = C + 273$
$127 = C$
$F = \frac{9}{5}(127) + 32 = \frac{1143}{5} + 32 = 228.6 + 32$
$= 260.6 \approx 261$
261°F

9. **a.** $L = 1 + (23 - 1) \cdot 2 = 1 + 22 \cdot 2$
$= 1 + 44 = 45$
The 23rd term is 45.

b. $L = a + (n - 1)d$
$L = a + nd - d$
$L - a + d = nd$
$\frac{1}{d}(L - a + d) = (nd)\left(\frac{1}{d}\right)$
$n = \frac{L - a + d}{d}$

11. $P = C + MC$
$P - C = MC$
$\frac{P - C}{C} = M$
$M = \frac{P - C}{C}$ or $M = \frac{P}{C} - 1$

13. **a.** $7x - 3y = -715$
$-3y = -715 - 7x$
$y = \frac{7}{3}x + 238\frac{1}{3}$

b. 1992 is 7 years after 1985.
$y = \frac{7}{3}(7) + 238\frac{1}{3} = 254\frac{2}{3}$
254,666,667 people

15. $N = \frac{20Ld}{600 + s^2}$
$N\frac{600 + s^2}{20L} = d$
$d = (845)\frac{600 + 20^2}{20(4)} = 10{,}562.5$ feet
≈ 2 miles

17. $3(x - 2) + 3y = 9x$
$3x - 6 + 3y = 9x$
$3y = 6x + 6$
$y = 2x + 2$

19. $\frac{1}{3}x - 2y = 5$
$-2y = -\frac{1}{3}x + 5$
$y = \frac{1}{6}x - \frac{5}{2}$

21. $x(y+3)=4y+2$
$xy+3x=4y+2$
$3x-2=4y-xy$
$3x-2=(4-x)y$
$\frac{3x-2}{4-x}=y$
$x\neq 4$

23. $7y-ay=5$
$y(7-a)=5$
$y=5\cdot\frac{1}{7-a}$
$y=\frac{5}{7-a}$

25. $xr-11=yr+15$
$xr-yr=26$
$r(x-y)=26$
$r=26\cdot\frac{1}{x-y}$
$r=\frac{26}{x-y}$

27. $3A=3\left[\frac{c}{3}(d+b)\right]$
$3A=cd+cb$
$3A-cd=cb$
$b=\frac{3A-cd}{c}$

29. $\frac{3}{4}(x+y)=2(x-z)$
$4\left[\frac{3}{4}(x+y)\right]=4\cdot 2(x-z)$
$3(x+y)=8(x-z)$
$3x+3y=8x-8z$
$3y+8z=5x$
$x=\frac{3y+8z}{5}$

31. $3R=3\left[\frac{1}{3}a(x+c)\right]$
$3R=ax+ac$
$3R-ac=ax$
$x=\frac{3R-ac}{a}$

33. $c\left(\frac{3ab}{c}\right)=c\cdot 7$
$3ab=7c$
$a=\frac{7c}{3b}$

35. $2\left[\frac{1}{2}(a-bx)+b\right]=2\left[\frac{1}{2}(2b-x)\right]$
$a-bx+2b=2b-x$
$a=bx-x$
$a=x(b-1)$
$x=\frac{a}{b-1}$

37. $2S=2\frac{n}{2}(a+1)$
$2S=na+n(1)$
$2S=n(a+1)$
$n=\frac{2S}{a+1}$

39. $14\left(\frac{bcx}{7}+\frac{3b}{2}\right)=14\left(\frac{4bcx}{14}\right)$
$2bcx+21b=4bcx$
$21b=2bcx$
$x=\frac{21b}{2bc}$
$x=\frac{21}{2c}$

41. d is true; None is true.
a is not true: $S=P+Prt$
$S=P(1+rt)$
$P=\frac{S}{1+rt}$
b is not true: $I=Prt$
$t=\frac{I}{Pr}$
c is not true: $S=2LW+2LH+2WH$
Using the distributive property.
$S-2LW=H(2L+2W)$
$H=\frac{S-2LW}{2L+2W}$

43. $y=-0.358709x+256.835$
$224.91=-0.358709x+256.835$
$x\approx 89.0$
1989

45. man: $h = 69.089 + 2.238f$
$f = 50$ cm
$h = 69.089 + 2.238(50) = 180.989$
$180.989 \text{ cm} \approx 180 \text{ cm}$
Yes, this could be a match.

47. $u = \dfrac{fk(k+1)}{n(n+1)}$
$42.16 = f\dfrac{12(12+1)}{36(36+1)}$
$42.16 = f \cdot \dfrac{1}{3} \cdot \dfrac{13}{37}$
$42.16 = f \cdot \dfrac{13}{111}$
$(111)(42.16) = f \cdot \dfrac{13}{111}(111)$
$4679.76 = 13f$
$f = \dfrac{4679.76}{13} \approx 359.98$
Finance charge was approximately \$359.98.

49. a. $y = -0.00949x + 0.479$

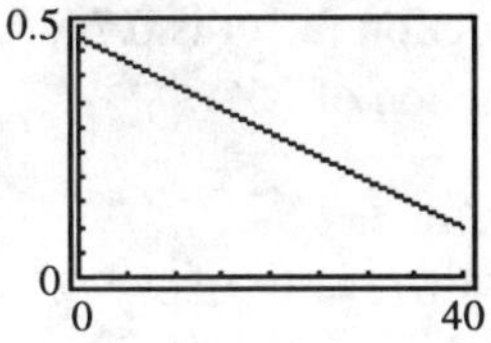

b. The graph indicates the percentage of smokers decreases linearly.

c. $y = -0.00949x + 0.479$
$\dfrac{y - 0.479}{-0.00949} = x$
$x = \dfrac{0.15 - 0.479}{-0.00949} \approx 34.668 \approx 35$
35 years after 1974, or in 2009

d. The graph predicts, unrealistically, that no one will smoke in 2025.

51. a. bank: $y = 8 + 0.05x$
credit union: $y = 2 + 0.08x$

b. $8 + 0.05x = 2 + 0.08x$
$6 = 0.03x$
$200 = x$
200 checks

53. $S = 0.5N + 26$
$1\frac{3}{8} = \frac{11}{8} = \frac{6}{8} + \frac{5}{8} = \frac{3}{4} + 5\left(\frac{1}{8}\right)$
There are five $\frac{1}{8}$-inch increases in nozzle size diameter over $\frac{3}{4}$ of an inch; so 5(5 feet) = 25 feet is *added* to the maximum horizontal range. Thus 94 feet – 25 feet = 69 feet is the maximum horizontal range.
$69 = 0.5N + 26$
$43 = 0.5N$
$N = 86$
The nozzle pressure is 86 pounds.

55. Answers may vary.

Review Problems

56. x is at least 7
$\{x \mid x \ge 7\}$

57. $\frac{2}{5}(x-3) - 4 = \frac{1}{3}x$
$\frac{2}{5}x - \frac{6}{5} - 4 = \frac{1}{3}x$
$\frac{2}{5}x - \frac{1}{3}x = \frac{6}{5} + 4$
$\left(\frac{2}{5} - \frac{1}{3}\right)x = \frac{6}{5} + \frac{20}{5}$
$\left(\frac{6}{15} - \frac{5}{15}\right)x = \frac{26}{5}$
$\frac{1}{15}x = \frac{26}{5}$
$x = \left(\frac{26}{5}\right)15 = 78$

58. $\{x \mid x$ is a natural number less than 17 that is divisible by 3$\}$
$\{3, 6, 9, 12, 15\}$

Problem Set 1.8

1. Let $x =$ the number.
$12x - 6 = 7x + 24$
$12x - 6 - 7x = 24$
$5x - 6 = 24$
$5x - 6 + 6 = 24 + 6$
$5x = 30$
$x = 6$
The number is 6.

3. a. $x + 0.52$ (billion \$)

b. amount in 1993 + amount in 1994 $= 4.32$
$x + (x + 0.52) = 4.32$
$2x = 3.8$
$x = 1.9$
1993: \$1.9 billion, 1994: \$2.42 billion

5. Let $x =$ salary of cabinet member
$VP = 23{,}100 + x$
$S = x - 14{,}800$
$VP + S + x = 2P + 53{,}500$
or $(23{,}100 + x) + (x - 14{,}800) + x$
$= 2(200{,}000) + 53{,}500$
$3x + 8300 = 453{,}500$
$3x = 445{,}200$
$x = \$148{,}400$
cabinet members: \$148,400
Vice President: \$171,500
Senators: \$133,600

7. Let $x =$ annual salary of Pink Floyd (in \$ million).
$S = 2x + 53$
$C = x - 22$
$S + C + x = 10.2(25)$
or $(2x + 53) + (x - 22) + x = 255$
$4x + 31 = 255$
$4x = 224$
$x = 56$
Pink Floyd: \$56 million
Spielberg: \$165 million
Cosby: \$34 million

9. Let $x =$ amount spent on Medicaid in 1995 (billion \$).
$219 = 1.2x$
$x = \$182.5$ billion

11. Let $x =$ original price of VCR.
Price after 30% reduction $= 0.70x$
Price after additional reduction $= 0.60(0.70x)$
Then $372.40 = 0.6(0.7x) = 0.42x$
$x = \$886.67$

13. Let $x =$ original price of item.
After price increase: $x + 0.5x = 1.5x$
After 50% reduction: $0.50(1.5x) = 0.75x$
Then $280 = 0.75x$ or $x = \$373.33$.
The net reduction factor is 0.75, for an overall reduction of 25%.

15. Let $x =$ price of car.
$x + 0.07x = 1.07x = 16{,}050$
$x = \$15{,}000$

17. a. men: $6300 + 1600x$
women: $2100 + 1200x$

b. $6300 + 1600(11) = 2100 + 1200x$
$21{,}800 = 1200x$
$x = 18\frac{1}{6}$
Women need approximately 18 years of education (equivalent to a Master's Degree) to earn the same salary as a man with an eleventh grade education (i.e., who did not finish high school).

c.

x	$y_{male} = 6300 + 1600x$	$y_{female} = 2100 + 1200x$
0	6300	2100
5	14,300	8100
10	22,300	14,100
15	30,300	20,100
20	38,300	26,100

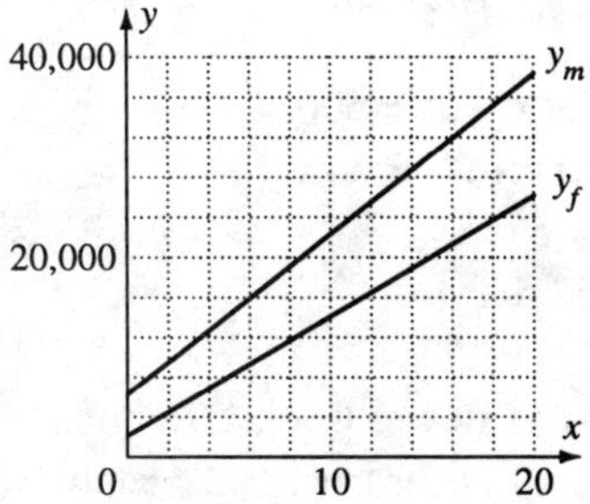

no

d. Answers may vary. For example: Draw a vertical line at $x = 11$ up to the y_m line. This shows that, with 11 years of education, a man's income is about \$24,000. Draw a horizontal line at $y \approx 24{,}000$ until it intersects the y_f line. Draw a vertical line from this point down to the x-axis. The x-value (18 years) shows what education level a woman needs to make \$24,000 a year.

19. Let x = number of years since 1980.
$5.23 + 0.40x = 10.03$
$0.4x = 4.8$
$x = 12$
In 1992

21. Option 1: 7.5% on a 30-year loan is \$7 per thousand, so monthly payment is
$7(50) = \$350$.
Option 2: 7.125% on 30 year loan is \$6.74 per thousand, so monthly payment is
$6.74(50) = \$337$.
When cost for Option 1 = cost for Option 2:
$350x = 337x + 3(0.01)(50{,}000)$
$13x = 1500$
$x \approx 115$ months ≈ 9.62 years
For a 7.125% mortgage, with 3 points, and a 20 year loan, monthly payment is
$7.83(50) = \$391.50$.
Total cost of Option 2 over 30 years (360 months) $= 337(360) + 3(0.01)(50{,}000)$
$= \$122{,}820$
Total cost of 20 year (240 months) loan at 7.125% with 3 points
$= 391.50(240) + 3(0.01)(50{,}000) = \$95{,}460$
So the 20 year loan plan saves
$122{,}820 - 95{,}460 = \$27{,}360$

23. x: Number of minutes talking
$43 + 32(x - 1) + 210 = 573$
$x = 11$
The person talked for 11 minutes.

25. Let w = width and l = length.
Then $l = 4w - 30$.
$P = 2l + 2w$, so $640 = 2(4w - 30) + 2w$
$= 8w - 60 + 2w = 10w - 60$
$10w = 700$
$w = 70$
The dimensions are 70 yards by 250 yards.

27. Let h = height and w = width.
Then $h = w + 3$.
Assembled bookcase has 2 sides of height h and 4 shelves of width w.
$30 = 2(w + 3) + 4w = 2w + 6 + 4w = 6w + 6$
$24 = 6w$
$w = 4$
The bookcase should be 4 feet wide by 7 feet tall.

29. Let l = length of second side.
first side $= 2l - 1$

third side = $2l + 1$
$30 = (2l - 1) + l + (2l + 1) = 5l$
$l = 6$ inches
11 inches, 6 inches, 13 inches

31. Let x = length of shorter side of rectangle.
$x + 2$ = length of longer side
$\left(\frac{1}{2}x+5\right)+2(x+2)=44$
$\frac{1}{2}x+5+2x+4=44$
$\frac{5}{2}x+9=44$
$\frac{5x}{2}=35$
$x = 14$
$x + 2 = 16$
area $= x(x + 2) = 14(16) = 224$
area: 224 cm^2

33. Let x = (common) length of nonparallel sides.
Let b = length of smaller base.
Let B = length of larger base.
Then $x = 3b - 1$
$B = 2 + 5b$
$P = b + x + B + x = 2x + b + B$
$36 = 2(3b - 1) + b + (2 + 5b)$
$= 6b - 2 + b + 2 + 5b = 12b$
so $b = 3$ yards.
Then $x = 3(3) - 1 = 8$ yards and
$B = 2 + 5(3) = 17$ yards.
The trapezoid has bases 3 yards and 17 yards with the 2 nonparallel sides being 8 yards each.

35. Let x = number of hours when the planes are 2500 miles apart.

	Rate, r ·	Time, t =	Distance, d
faster plane	300	x	$300x$
slower plane	200	x	$200x$

$300x + 200x = 2500$
$500x = 2500$
$x = 5$
5 hours

37. Let x = rate of slower truck.
$x + 5$ = rate of faster truck

	Rate, r ·	Time, t =	Distance, d
slower truck	x	5	$5x$
faster truck	$x+5$	5	$5(x + 5)$

$5x + 5(x + 5) = 600$
$5x + 5x + 25 = 600$
$10x = 575$
$x = 57.5$
$x + 5 = 62.5$
speed of slower truck: 57.5 mph;
speed of faster truck: 62.5 mph

39. Let x = time out going.
$10 - x$ = time returning

	Rate, r ·	Time, t =	Distance, d
outgoing	20	x	$20x$
returning	30	$10 - x$	$30(10 - x)$

distance going = distance returning
$20x = 30(10 - x)$
$20x = 300 - 30x$
$50x = 300$
$x = 6$
$10 - x = 4$
outgoing: 6 hr; return: 4 hr
distance $= 20x = 20(6) = 120$
120 miles each way

41. Let x = amount invested at 8%.
$20{,}000 - x$ = amount invested at 5%

	P ·	r ·	t =	Interest
8% stock	x	0.08	2	$0.08(2)x$
5% stock	$20{,}000 - x$	0.05	3	$0.05(3)(20{,}000 - x)$

$0.08(2)x + 0.05(3)(20{,}000 - x) = 3160$
$0.16x + 3000 - 0.15x = 3160$
$0.01x = 160$
$x = 16{,}000$
$20{,}000 - x = 4000$
16,000 at 8% for 2 years; $4000 at 5% for 3 years

43. Let x = amount invested at 9%.
$35{,}000 - x$ = amount invested at 6%

	P ·	r ·	t =	Interest
9% stock	x	0.09	2	$0.09(2)x$
6% stock	$35{,}000 - x$	0.06	4	$0.06(4)(35{,}000 - x)$

$0.09(2)x = 0.06(4)(35{,}000 - x)$
$0.18x = 8400 - 0.24x$
$0.42x = 8400$
$x = 20{,}000$
$35{,}000 - x = 35{,}000 - 20{,}000 = 15{,}000$
$20,000 at 9% for 2 years; $15,000 at 6% for 4 years
Interest: $0.09(2)(20{,}000) = 3600$
$0.06(4)(15{,}000) = 3600$
Total interest: $3600 + $3600 = $7200

45. $-3-5(2-x)+6x=-6-2[-1+4(x-6)]$
$-3-10+5x+6x=-6-2(-1+4x-24)$
$11x-13=-6-2(4x-25)$
$11x-13=-6-8x+50$
$11x-13=-8x+44$
$19x=57$
$x=3$

a	4	8	3
2	b	c	14
1	d	6	13
12	e	11	0

column 4: $3+14+13+0=30$ (total = 30)
row 1: $a+4+8+3=30$
$a+15=30$
$a=15$
row 4: $12+e+11+0=30$
$e+23=30$
$e=7$
column 3: $8+c+6+11=30$
$c+25=30$
$c=5$
row 2: $2+b+5+14=30$
$b+21=30$
$b=9$
row 3: $1+d+6+13=30$
$d+20=30$
$d=10$

15	4	8	3
2	9	5	14
1	10	6	13
12	7	11	0

47.
$$\begin{array}{r} 9567 \\ +\ 1085 \\ \hline 10{,}652 \end{array}$$

49.

Nickel	Dime	Quarter	Half Dollar	Total Value
			1	50¢
		2		50¢
	5			50¢
1	2	1		50¢
3	1	1		50¢
10				50¢
5		1		50¢
2	4			50¢
4	3			50¢
6	2			50¢
8	1			50¢

11 combinations

51. $P=2l+2w=16$, or $l+w=8$ and $w=8-l$.
$A=l\cdot w=l(8-l)=8l-l^2$
Graph this function and observe it is a parabola, opening down, with a maximum at $l=4$. Maximum area is 16 square yards, attained by making the corral a square with side = 4 yards.

53. The company gives a computer to person A (one of the mothers), and to person B (who is A's daughter), and to person C (who is B's daughter). One woman is a daughter and a mother.

55. Three spelling errors plus the false claim that it only contains one mistake.

57. a. Let x = number of years since 1980.
hourly rate: $y=3.82+0.30x$

b. 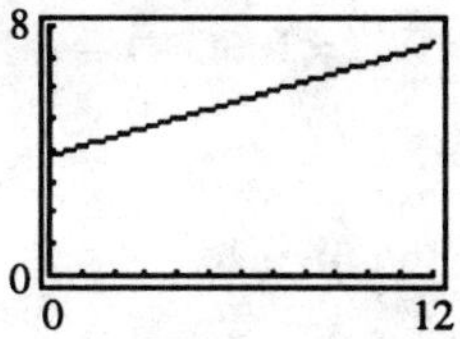

c. Students should use TRACE feature.

d. $6.22 = 3.82 + 0.3x$
$2.4 = 0.3x$
$x = 8$
so hourly rate was \$6.22 in 1988.

59. Answers may vary.

61. Answers may vary.

63. Let x = "my" age.
$x + \frac{1}{2}x + \frac{1}{3}x + 3(3) = 6(20) + 10$
$\frac{11}{6}x + 9 = 120 + 10$
$\frac{11}{6}x = 130 - 9$
$\frac{11}{6}x = 121$
$x = \frac{6}{11}(121)$
$x = 66$
66 years old

65. Let x = Mrs. Ricardo's portion.
$2x$ = boy's portion
$\frac{1}{2}x$ = girl's portion
The sum of the three portions is \$14,000.
$x + 2x + \frac{1}{2}x = 14{,}000$
$\frac{7}{2}x = 14{,}000$
$x = 4000$
Thus, Mrs. Ricardo's portion is \$4000. Her son receives $2x$ or \$8000, and her daughters $\frac{1}{2}x$ or \$2000.

67. Let x = Carla's original amount of money.
$4[3(2x - 30) - 54] - 72 = 48$
$x = 29$
Carla originally had \$29.

69. Let x = amount of money you had originally.
Amount of money spent: $\frac{1}{3}x$
Amount left: $x - \frac{1}{3}x = \frac{2}{3}x$
Amount of money lost: $\frac{2}{3}\left(\frac{2}{3}x\right) = \frac{4}{9}x$
Amount left: $\frac{2}{3}x - \frac{4}{9}x = \frac{6}{9}x - \frac{4}{9}x = \frac{2}{9}x$
$\frac{2}{9}x = 12$
$x = 54$
Original amount of money, \$54

71. Let x = number originally stolen.
To first guard, he gives $\frac{1}{2}x + 2$.
He still has $x - \left(\frac{1}{2}x + 2\right) = \frac{1}{2}x - 2$.
To the second guard, he gives
$\frac{1}{2}\left(\frac{1}{2}x - 2\right) + 2 = \frac{1}{4}x + 1$.
He still has $\frac{1}{2}x - 2 - \left(\frac{1}{4}x + 1\right) = \frac{1}{4}x - 3$.
To the third guard, he gives
$\frac{1}{2}\left(\frac{1}{4}x - 3\right) + 2 = \frac{1}{8}x + \frac{1}{2}$.
He still has $\frac{1}{4}x - 3 - \left(\frac{1}{8}x - \frac{1}{2}\right) = \frac{1}{8}x - \frac{7}{2}$.
Since he leaves with one plant,
$\frac{1}{8}x - \frac{7}{2} = 1$
$8\left(\frac{1}{8}x - \frac{7}{2}\right) = 8(1)$
$x - 28 = 8$
$x = 36$
The thief originally stole 36 plants.

73. Let r = athlete's running speed,
$2r$ = athlete's biking speed and
$0.40r$ = athlete's walking speed.
Let "+" indicate initial direction and
"–" indicate opposite direction.

	Rate, r ·	Time, t =	Distance, d
running	r	8	$r(8)$
biking	$2r$	3	$2r(3)$
walking	$0.40r$	1	$0.40r(1)$

$8r - 3(2r) + 0.40r = 9.6$
$8r - 6r + 0.4r = 9.6$
$2.4r = 9.6$
$r = 4$ (running)
$2r = 8$ (biking)
$0.40r = 0.40(4) = 1.6$ (walking)
running, 4 mph; biking(riding), 8 mph; walking, 1.6 mph

75. Let x = amount invested at 5%.
$50{,}000 - x$ = amount invested at 16% (after 1 year, loses at 8%)
$-0.08(50{,}000 - x) + 0.05(3)x = 5200$
$-4000 + 0.08x + 0.15x = 5200$
$0.23x = 9200$
$x = 40{,}000$
$50{,}000 - x = 10{,}000$
\$40,000 at 5%; \$10,000 at 16%

77. Answers may vary.

Review Problems

78. $P = \dfrac{100(p+q)}{pq}$
$(p = 300,\ q = -50)$
$P = \dfrac{100(300-50)}{300(-50)} = \left(\dfrac{100}{300}\right)\left(\dfrac{250}{-50}\right)$
$= \frac{1}{3}(-5) = -\frac{5}{3}$

79. $D = A(n-1)$
$\dfrac{D}{A} = n - 1$
$\dfrac{D}{A} + 1 = n$
$n = \dfrac{D}{A} + 1$ or $n = \dfrac{D+A}{A}$

80. $10^3 + 10^0 - 10^{-1} = 1000 + 1 - \frac{1}{10}$
$= 1001 - 0.1 = 1000.9$

Chapter 1 Review Problems

1. $\{x \mid x$ is a natural number that is divisible by 4$\}$
$\{4, 8, 12, 16, \ldots\}$

2. $\{x \mid x$ is a whole number but not a natural number$\}$
$\{0\}$

3. $\{x \mid x \le 1\}$
$(-\infty, 1]$

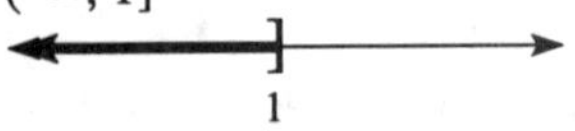

4. $\{x \mid x \ge -2\}$
$[-2, \infty)$

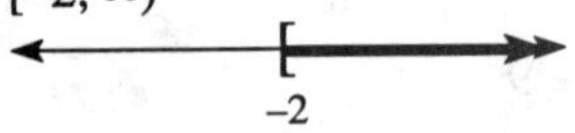

5. $\{x \mid -1 < x \le 2\}$
$(-1, 2]$

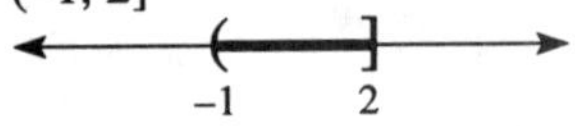

6. x lies between –3 and 6, including –3 and excluding 6.
set-builder: $\{x \mid -3 \le x < 6\}$
interval: $[-3, 6)$

7. x is at most 12.
set-builder: $\{x \mid x \le 12\}$
interval: $(-\infty, 12]$

8. (0, 23,350]; (23,350, 56,550]; (56,550, 117,950]; (117,950, 256,500]; (256,500, ∞)

9. ServiceMaster, Jazzercise, Dairy Queen, Century 21

10. Subway, McDonald's, 7-Eleven

11. Baskin-Robbins

12. McDonald's, 7-Eleven

13. Baskin-Robbins, Jani-King, ServiceMaster

14. Commutative property of addition

15. Associative property of multiplication

16. Distributive property

17. $3 + (-17) + (-25) = -14 - 25 = -39$

18. $16 - (-14) = 16 + 14 = 30$

19. $-11 - [-17 + (-3)] = -11 - (-20) = -11 + 20$
$= 9$

20. $|-17| + |3| - (|10| + |-13|) = 17 + 3 - (10 + 13)$
$= 20 - 23 = -3$

21. $(-0.2)(-0.5) = 0.1$

22. $-\frac{1}{2}(-16)(-3) = -24$

23. $-12 \div \frac{1}{4} = -12 \cdot 4 = -48$

24. $\left(-\frac{1}{2}\right)^3 \cdot 2^4 = \frac{1}{(-2)^3} \cdot 2^4 = -\frac{1}{2^3} \cdot 2^4$
$= -2^{4-3} = -2$

25. $-\frac{2}{7} \div \left(-\frac{3}{7}\right) = -\frac{2}{7} \cdot \left(-\frac{7}{3}\right) = \frac{2}{3}$

26. $-3[4 - (6 - 8)] = -3[4 - (-2)] = -3(4 + 2)$
$= -3(6) = -18$

27. $8^2 - 36 \div 3^2 \cdot 4 - (-7) = 64 - 36 \div 9 \cdot 4 + 7$
$= 64 - 4 \cdot 4 + 7 = 64 - 16 + 7 = 55$

28. $\frac{(-2)^4 + (-3)^2}{2^2 - (-21)} = \frac{16 + 9}{4 + 21} = \frac{25}{25} = 1$

29. $25 \div \left(\frac{8+9}{2^3 - 3}\right) - 6 = 25 \div \left(\frac{17}{8-3}\right) - 6$
$= 25 \div \frac{17}{5} - 6 = 25 \cdot \frac{5}{17} - 6 = \frac{125}{17} - \frac{102}{17}$
$= \frac{23}{17}$

30. $6 - (-20)\left[\frac{6 - 1(6 - 10)}{14 - 3(6 - 8)}\right]$
$= 6 + 20\left[\frac{6 - 1(-4)}{14 - 3(-2)}\right]$
$= 6 + 20\left(\frac{6 + 4}{14 + 6}\right)$
$= 6 + 20\left(\frac{10}{20}\right) = 6 + 10 = 16$

31. $\left(\frac{1}{4} - \frac{3}{8}\right) \div \left(-\frac{3}{5} - \frac{1}{4}\right) = \left(\frac{2}{8} - \frac{3}{8}\right) \div \left(-\frac{12}{20} - \frac{5}{20}\right)$
$= \left(-\frac{1}{8}\right) \div \left(-\frac{17}{20}\right) = \left(-\frac{1}{8}\right) \cdot \left(-\frac{20}{17}\right) = \frac{5}{34}$

32. $\frac{3}{4} - \frac{5}{8} - \left(-\frac{1}{2}\right) = \frac{6}{8} - \frac{5}{8} + \frac{4}{8} = \frac{5}{8}$

33. $\frac{(10 - 6)^2 + (-2)(-3)}{10 - (-4)(3)} = \frac{4^2 + 6}{10 + 12} = \frac{16 + 6}{22}$
$= \frac{22}{22} = 1$

34. $(2.6)(-5.4) \div (1.8) - (-5.7)$
$= -14.04 \div 1.8 + 5.7$
$= -7.8 + 5.7$
$= -2.1$

35. $\frac{9(-1)^3 - 3(-6)^2}{5 - 8} = \frac{9(-1) - 3(36)}{-3}$
$= \frac{-9 - 108}{-3} = \frac{-117}{-3} = 39$

36. $\frac{(7 - 9)^3 - (-4)^2}{2 + 2 \cdot 8 \div 4}$
$= \frac{(-2)^3 - 16}{2 + 16 \div 4}$
$= \frac{-8 - 16}{2 + 4}$
$= -\frac{24}{6} = -4$

37. $2^4(-1)^{50}+|-3|^3=16(1)+3^3=16+27=43$

38. $6-[-3(-2)^3\div(-8)]^2=6-[-3(-8)\div(-8)]^2$
$=6-[24\div(-8)]^2=6-(-3)^2=6-9=-3$

39. $6(2x-3)-5(3x-2)=12x-18-15x+10$
$=-3x-8$

40. $6[b-3(a-6b)]=6[b-3a+18b]$
$=6(19b-3a)=114b-18a$

41. $3x-[y-(2x-3y)]=3x-(y-2x+3y)$
$=3x-(4y-2x)=3x-4y+2x=5x-4y$

42. $-x^2-[3y^2-(2x^2-y^2)]$
$=-x^2-(3y^2-2x^2+y^2)$
$=-x^2-(4y^2-2x^2)=-x^2-4y^2+2x^2$
$=x^2-4y^2$

43. $8x-2x\cdot5+6x^2\div3\cdot2=8x-10x+2x^2\cdot2$
$=-2x+4x^2$

44. $[6(xy-1)-2xy]-[4(2xy+3)-2(xy+2)]$
$=(6xy-6-2xy)-(8xy+12-2xy-4)$
$=(4xy-6)-(6xy+8)$
$=4xy-6-6xy-8$
$=-2xy-14$

45. $-\frac{1}{2}(x+4)-\frac{1}{4}(-3x+8)$
$=-\frac{1}{2}x-2+\frac{3x}{4}-2=\frac{1}{4}x-4$

46. For 1980, $x=60$; for 1930, $x=10$
$y_{1980}-y_{1930}=0.0005(60)^3-0.04(60)^2$
$+2.8(60)+22-[0.0005(10)^3-0.04(10)^2$
$+2.8(10)+22]$
$=108-144+168+22$
$-[0.5-4+28+22]$
$=154-46.5=107.5$ people per square mile

47. $k=293$
$C=k-273$
$C=293-273=20$
$F=\frac{9}{5}C+32$
$F=\frac{9}{5}(20)+32=36+32=68$
68°F

48. $V=C(1-rt)$
$(C=1200,\ r=7\%=0.07,\ t=4)$:
$V=1200[1-0.07(4)]$
$V=1200(1-0.28)$
$V=1200(0.72)$
$V=864$
value; \$864

49. a. $R_m-R_w=0.75(220-30)$
$-[0.65(220-30)]$
$=(0.75-0.65)(220-30)=0.1(190)$
$=19$
The desirable heart rate for men is 19 heartbeats per minute higher (faster) than that for women at age 30. This difference is shown as the vertical difference between the two lines graphed, at age = 30.

b. $R_{20}-R_{70}=0.75(220-20)$
$-[0.75(220-70)]$
$=150-112.5=37.5$
The desirable heart rate for a 20 year old man is 37.5 bpm faster than that of a 70 year old man.

50.

x	−3	−2	−1	0	1	2	3
$y=x^2-1$	8	3	0	−1	0	3	8

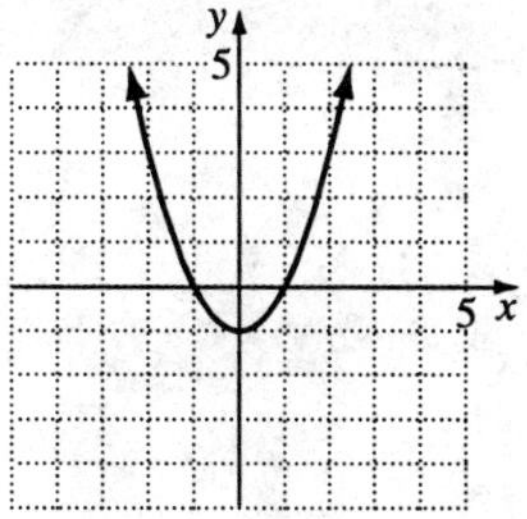

51.

x	−3	−2	−1	0	1	2	3
$y = \lvert 2x \rvert$	6	4	2	0	2	4	6

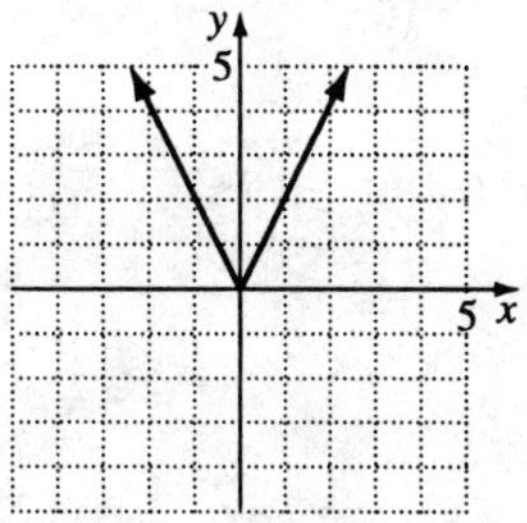

52.

x	−3	−2	−1	0	1	2	3
$y = -2x + 3$	9	7	5	3	1	−1	−3

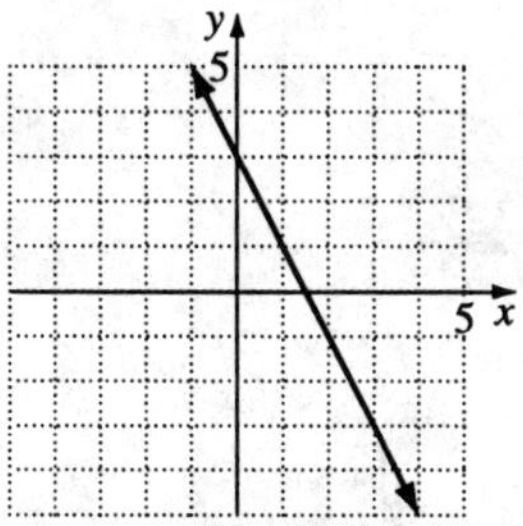

53. Answers may vary.

54. Answers may vary.

55. **a.** Answers may vary.

b. Answers may vary.

c. Answers may vary.

d. Answers may vary.

56. **a.** 5 PM, −4°F

b. 8 PM, 16°F

c. $x = 4, 6$; answers may vary.

d. 12°F; answers may vary.

e. At 7 P.M., 4°F; at 8 P.M., 16°F

increase $= \frac{16-4}{4} = \frac{12}{4} = 3$

300%

57. $y = \frac{10,000x}{100 - x}$

x	0	20	40	50	60
y	100	2500	6666.6	10,000	15,000

x	80	90	99
y	40,000	90,000	990,000

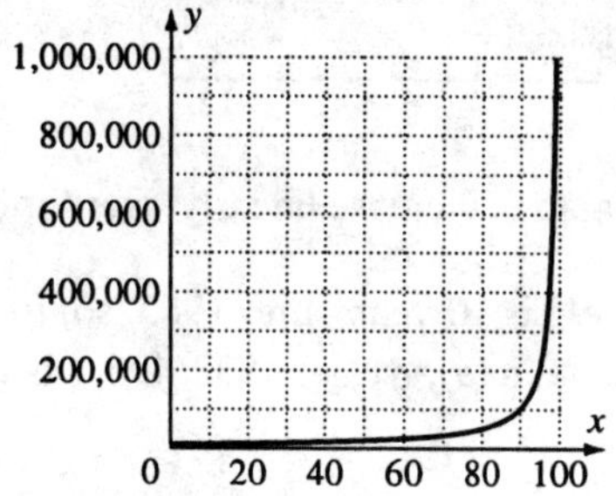

$y = \frac{10,000x}{100 - x}$

The cost of removing the pollutants from the lake increases rapidly as the percent of removed pollutants approaches 100%.

58. a.

x	$y = 553x + 27{,}966$
0	27,966
1	28,519
2	29,072
4	30,178
6	31,284
8	32,390
10	33,496
11	34,049
12	34,602

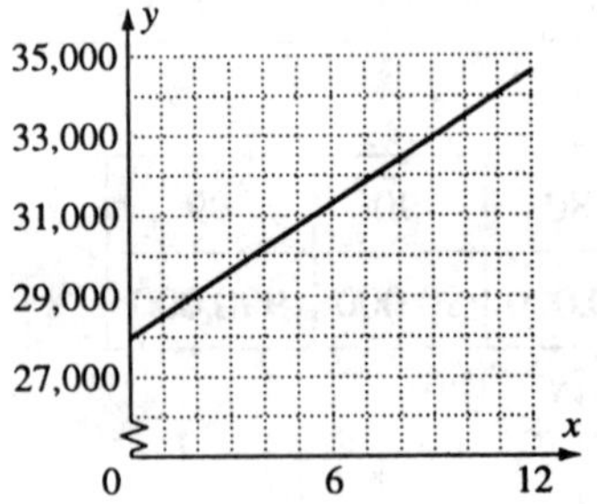

b. Salaries increase linearly over time.

c. 1995 is 10 years after 1985, so, from the table in (a), $y_{1995} = \$33,496$ when $x = 10$.

d. $553x + 27{,}966 = 33{,}000$
$553x = 5034$
$x = 9.1$ years after 1985
So, salary exceeds \$33,000 after 1994.

59. $(-3y^7)(-8y^6) = 24y^{7+6} = 24y^{13}$

60. $(7x^3y)^2 = 7^2x^{3\cdot 2}y^2 = 49x^6y^2$

61. $(-3xy)(2x^2)^3 = (-3xy)(8x^6) = -24x^7y$

62. $\left(\frac{2}{3}\right)^{-2} = \frac{1}{\left(\frac{2}{3}\right)^2} = \frac{1}{\frac{4}{9}} = \frac{9}{4}$

63. $(-6xy)(-3x^2y) - 25x^3y^2$
$= 18x^3y^2 - 25x^3y^2$
$= -7x^3y^2$

64. $\frac{16y^3}{-2y^{10}} = -8y^{3-10} = -8y^{-7} = \frac{-8}{y^7} = -\frac{8}{y^7}$

65. $(-3x^4)(-4x^{-11}) = 12x^{4-11} = 12x^{-7} = \frac{12}{x^7}$

66. $\frac{12x^7}{-4x^{-3}} = -3x^{7-(-3)} = -3x^{7+3} = -3x^{10}$

67. $\frac{-10a^5b^6}{20a^{-3}b^{11}} = -\frac{1}{2}a^{5-(-3)}b^{6-11} = -\frac{1}{2}a^8b^{-5}$
$= -\frac{a^8}{2b^5}$

68. $(-2)^{-3} + 2^{-2} + \frac{1}{2}x^0 = \frac{1}{(-2)^3} + \frac{1}{2^2} + \frac{1}{2}(1)$
$= \frac{1}{-8} + \frac{1}{4} + \frac{1}{2} = -\frac{1}{8} + \frac{2}{8} + \frac{4}{8} = \frac{5}{8}$

69. $\left(\frac{-2}{ab}\right)^5 = \frac{(-2)^5}{a^5b^5} = -\frac{32}{a^5b^5}$

70. $(3x^4y^{-2})(-2x^5y^{-3}) = -6x^{4+5}y^{-2-3}$
$= -6x^9y^{-5} = -\frac{6x^9}{y^5}$

71. $93{,}700{,}000{,}000{,}000 = 9.37 \times 10^{13}$

72. $0.000000409 = 4.09 \times 10^{-7}$

73. $(2.8 \times 10^{13})(4.2 \times 10^{-6})$
$= (2.8)(4.2) \times 10^{13+(-6)}$
$= 11.76 \times 10^7 = 1.176 \times 10^8$

74. $\frac{1.8\times10^{-6}}{4.8\times10^{-8}}=\frac{1.8}{4.8}\times10^{-6-(-8)}=0.375\times10^{2}$
$=3.75\times10^{1}$

75. $(3\times10^{8}\text{ meters / sec})\times(1.97\times10^{4}\text{ sec})$
$=5.91\times10^{12}\text{ meters}$

76. **a.** defense budget $\approx$ \$302 billion
$=\$3.02\times10^{11}$
total: \$1.14 trillion $=\$1.14\times10^{12}$,
$\frac{3.02\times10^{11}}{1.14\times10^{12}}\approx26\%$

b. defense budget $\approx$ \$260 billion
$=\$2.6\times10^{11}$
total budget: $\$1.41\times10^{12}$,
$\frac{2.6\times10^{11}}{1.41\times10^{12}}\approx18\%$

c. No; for example, in 1993 defense was approximately $\frac{260\times10^{9}}{1410\times10^{9}}=18.44\%$ of the budget. Assume in 1994 the defense budget dropped to $\$250\times10^{9}$ whereas the total budget was $\$1300\times10^{9}$. Then the defense percentage would be 19.23%, an increase over 1993.

77. $5x-3=x+5$
$4x=8$
$x=2$
$\{2\}$

78. $8-(4x-5)=x-7$
$8-4x+5=x-7$
$20=5x$
$x=4$
$\{4\}$

79. $2(2x+5)+1=5-2(3-x)$
$4x+10+1=5-6+2x$
$2x=-12$
$x=-6$
$\{-6\}$

80. $2(y-2)=2[y-5(1-y)]$
$2y-4=2(y-5+5y)$
$2y-4=2(6y-5)$
$2y-4=12y-10$
$-10y=-6$
$y=\frac{3}{5}$
$\left\{\frac{3}{5}\right\}$

81. $3(x-1)+7(x-3)=5(2x-5)$
$3x-3+7x-21=10x-25$
$10x-24=10x-25$
$-24=-25$
False, no solution
$\varnothing$

82. $\frac{2y}{3}=6-\frac{y}{4}$
$4(2y)=12(6)-3y$
$8y=72-3y$
$11y=72$
$y=\frac{72}{11}$
$\left\{\frac{72}{11}\right\}$

83. $\frac{3z+1}{3}-\frac{13}{2}=\frac{1-z}{4}$
$4(3z+1)-6(13)=3(1-z)$
$12z+4-78=3-3z$
$15z=77$
$z=\frac{75}{15}$
$\left\{\frac{77}{15}\right\}$

84. $0.07y + 0.06(1400 - y) = 90$
$0.07y + 84 - 0.06y = 90$
$0.01y = 6$
$y = 600$
$\{600\}$

85. $2(x + 6) + 3(x + 1) = 4x + 10 + x + 5$
$2x + 12 + 3x + 3 = 5x + 15$
$5x + 15 = 5x + 15$
$15 = 15$
True
$\{x | x \in R\}$

86. $2x - 3y = 9$
$2x - 9 = 3y$
$y = \frac{2}{3}x - 3$

87. $y = mx + b$
$y - b = mx$
$x = \frac{1}{m}(y - b)$
$m \neq 0$

88. $C = \frac{5}{9}(F - 32)$
$9C = 5(F - 32)$
$\frac{9}{5}C = F - 32$
$F = \frac{9}{5}C + 32$

89. $A = \frac{1}{2}bh$
$2A = bh$
$\frac{2A}{b} = h$
$b \neq 0$

90. $A = 2HW + 2LW + 2LH$
$A - 2LW = 2H(W + L)$
$\frac{A - 2LW}{W + L} = 2H$
$H = \frac{A - 2LW}{2(W + L)}$ when $L + W \neq 0$.

91. $x(y - 2) = 3y + 5$
$xy - 2x = 3y + 5$
$xy - 3y = 2x + 5$
$y(x - 3) = 2x + 5$
$y = \frac{2x + 5}{x - 3}$
$x \neq 3$

92. $\frac{ax}{3} - \frac{bx}{2} = \frac{17}{6}$
Multiply by LCD = 6
$2ax - 3bx = 17$
$x(2a - 3b) = 17$
$x = \frac{17}{2a - 3b}$ for $2a - 3b \neq 0$.

93. $4080 = 420x + 720$
$3360 = 420x$
$x = 8$ years after 1989, or in 1997.

94. $A = 8.34FC$
($A = 2001.6$ lb, $C = 30$ ppm):
$2001.6 = 8.34F(30)$
$F = 8$
8 million gallons/day

95. $F = \frac{1}{4}C + 40$

a. $4F = C + 160$
$4F - 160 = C$
$C = 4F - 160$

b. $C = 200$
$F = \frac{1}{4}C + 40$
$F = \frac{1}{4}(200) + 40$
$F = 50 + 40$
$F = 90$
control temperature at 90°F

96. a. $400 + 2x = 500 + 1.5x$
$0.5x = 100$
$x = 200$ hours

b.

x	Club 1: $y_1 = 400 + 2x$	Club 2: $y_2 = 500 + 1.5x$
0	400	500
50	500	575
150	700	725
200	800	800
250	900	875

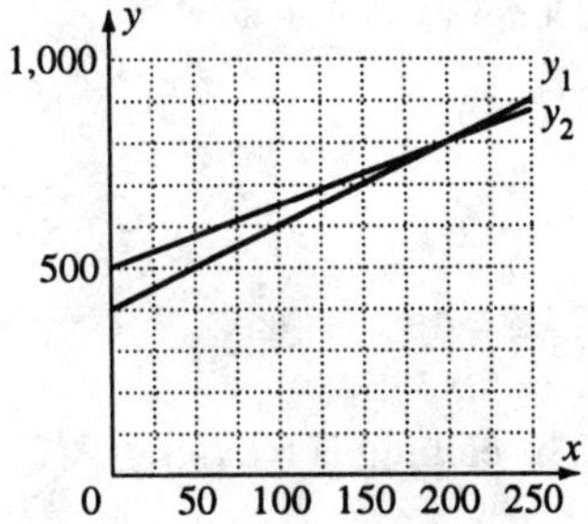

The graphs intersect at (200, 800), where the cost of each club is the same.

97. a. $34x - 10y = -1550$
$34x + 1550 = 10y$
$3.4x + 155 = y$

b. Year 2000 is 15 years after 1985, so $x = 15$. $y = 3.4(15) + 155$
$= 206$ thousand dentists

98. Let $x =$ the number.
$7x - 1 = 9 + 5x$
$2x = 10$
$x = 5$
The number is 5.

99. Let $x =$ number of white female victims (per 1000).
$BM = 2x + 9$
$BF = 2x - 10$
$WM = x + 38$
$BM + BF + WM + WF = 349$
$(2x + 9) + (2x - 10) + (x + 38) + x = 349$
$6x + 37 = 349$
$6x = 312$
$x = 52$
Number of crime victims (per 1000):
black males: 113
black females: 94
white males: 90
white females: 52

100. Let $x =$ original price of computer.
Price after 1st reduction: $0.80x$
Price after 2nd reduction:
$0.90(0.80x) = 0.72x$
$0.72x = 1872$
$x = \$2600$

101. Let $x =$ number of years the system is operating.
Cost for gas system for this period is $12{,}000 + 700x$.
Cost for electric system for this period is $5000 + 1100x$.

a. $12{,}000 + 700x = 5000 + 1100x$
$7{,}000 = 400x$
$x = 17.5$ years
Cost will be
$12{,}000 + 700(17.5) = \$24{,}250$.

b.

x	$y_e = 5000 + 1100x$	$y_g = 12{,}000 + 700x$
0	5000	12,000
5	10,500	15,500
10	16,000	19,000
15	21,500	22,500
20	27,000	26,000

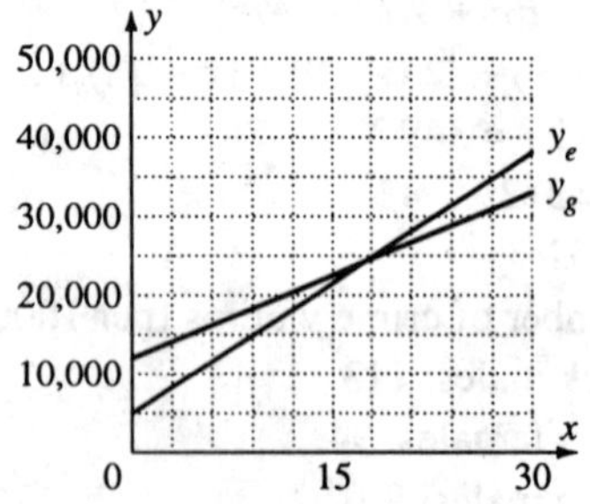

102. From Table 1.11, option 1 results in a payment of \$7.69 per \$1000 of the loan, and option 2 has a payment of \$7.34 per \$1000.
Total cost after x months:
Option 1: $7.69(40)x = 307.60x$
Option 2: $7.34(40)x + 0.03(40{,}000)$
$= 293.60x + 1200$
Then $307.6x = 293.6x + 1200$
$14x = 1200$
$x \approx 85.7$ months

103. Let x be the number of years since 1970.
Budget model: $y = 381 + 43x$ (billion \$)
$381 + 43x = 1370$
$43x = 989$
$x = 23$ years after 1970, or in 1993

104. Let x = sales amount.
Then pay $= 300 + 0.05x$.
$800 = 300 + 0.05x$
$500 = 0.05x$
$x = \$10{,}000$ in sales

105. Let x be the number of bus trips taken.
Cost without coupons $= 1.50x$
Cost with coupons $= 25 + 0.25x$
$1.5x = 25 + 0.25x$
$1.25x = 25$
$x = 20$
If the commuter makes more than 20 trips a month, the coupon book would be worthwhile.

106. Let x = height of bookcase.
Length $= 3x$
The bookcase has 4 horizontal pieces ("length") and 3 vertical pieces ("height")
$4(3x) + 3x = 60$
$15x = 60$
$x = 4$
The bookcase will be 4 ft × 12 ft.

107. Let w = width.
Then length $= 1\frac{1}{4} + w$.
$P = 2l + 2w$

a. $26 = 2\left(1\frac{1}{4} + w\right) + 2w = 2.5 + 4w$
$23.5 = 4w$
$w = 5.875$
The etching is 5.875 in. × 7.125 in.

b. From (a), area $= 5.875 \times 7.125$
$= 5\frac{7}{8} \times 7\frac{1}{8}$
$= \frac{47}{8} \times \frac{57}{8}$
$= \frac{2679}{64}$ sq inches ($\approx$ 41.9 sq in)
Then $75{,}000{,}000 \div 41.9 \approx \$1{,}789{,}976$.
The painting cost approximately \$1,800,000 per square inch.

108. Let x = length of side of the square.
$2x$ = length of the rectangle
$x - 1$ = width of rectangle.
(Perimeter of rectangle)
= 8 + (perimeter of square)
$2(2x) + 2(x - 1) = 8 + (4x)$
$4x + 2x - 2 = 8 + 4x$
$2x = 10$
$x = 5$
Length of side of the square: 5 yards

109. Let t = number of hours when the cars will be 660 miles apart.
$50t + 60t = 660$
$110t = 660$
$t = 6$
6 hours

110. Let x = speed of slower train.
$x + 20$ = speed of faster train
$5x + 5(x + 20) = 500$
$5x + 5x + 100 = 500$
$10x = 400$
$x = 40$ (slower train)
$x + 20 = 60$ (faster train)
Slower train: 40 mph; faster train: 60 mph

111. Let x = amount invested at 6% for 2 years.
$5000 - x$ = amount invested at 7% for 3 years.
$0.06(2)x + 0.07(3)(5000 - x) = 735$
$0.12x + 1050 - 0.21x = 735$
$-0.09x = -315$
$x = 3500$ (at 6%)
$5000 - x = 1500$ (at 7%)
\$3500 at 6% for 2 years; \$1500 at 7% for 3 years

112.

a	3	x
5	b	9
6	c	4

$\frac{x-2}{2} - \frac{x-3}{4} = \frac{7}{4}$
$2(x - 2) - (x - 3) = 7$
$2x - 4 - x + 3 = 7$
$x - 1 = 7$
$x = 8$
Column 3: $8 + 9 + 4 = 21$ (total)
Row 1: $a + 3 + 8 = 21$
$a + 11 = 20$
$a = 10$
Row 2: $5 + b + 9 = 21$
$b + 14 = 21$
$b = 7$
Row 3: $6 + c + 4 = 21$
$c + 10 = 21$
$c = 11$

10	3	8
5	7	9
6	11	4

113. 12 ways
Possible ways:

Dimes	Nickels	Pennies
2	1	0
2	0	5
1	3	0
1	2	5
1	1	10
1	0	15
0	5	0
0	4	5
0	3	10
0	2	15
0	1	20
0	0	25

114. Product = 96
Factors of $96 = 2 \cdot 2 \cdot 2 \cdot 2 \cdot 2 \cdot 3$
One is a teenager: $2 \cdot 2 \cdot 2 \cdot 2 = 16$
Remaining factors: 2 and 3
Ages of three people: 2, 3, and 16

115. $\frac{x}{2x+1}$

Chapter 1 Test

1. $x \geq 6$
   $\{x | x \geq 6\}$
   $[6, \infty)$

2. $2.62 < x \leq 3.13$
   $1980 = 2.76$
   $1990 = 2.63$
   $1993 = 2.63$
   The years are 1980, 1990, and 1993.

3. $24 - 36 \div 4 \cdot 3 = 24 - 9 \cdot 3$
   $= 24 - 27 = -3$

4. $\dfrac{8 - \left[|-4| - (1-3)^2\right]}{5 - (-3)^2 + 6 \div 3} = \dfrac{8 - [4 - (-2)^2]}{5 - 9 + 6 \div 3}$
   $= \dfrac{8 - [4 - 4]}{5 - 9 + 3}$
   $= \dfrac{8 - 0}{-4 + 3}$
   $= \dfrac{8}{-1} = -8$

5. $-7(x-5) - 3(x+6) = -7x + 35 - 3x - 18$
   $= -7x - 3x + 35 - 18 = -10x + 17$

6. $5y^2 - [7y - 3y(9y - 1)]$
   $= 5y^2 - [7y - 27y^2 + 3y]$
   $= 5y^2 - [10y - 27y^2]$
   $= 5y^2 - 10y + 27y^2$
   $= 32y^2 - 10y$

7. $t = 1993 - 1986 = 7$
   $A = -24t + 624$
   $A = -24(7) + 624 = 456$
   $t = 1996 - 1986 = 10$
   $A = -24t + 624$
   $A = -24(10) + 624 = 384$
   $456 - 384 = 72$
   There were 72,000 more on active duty.

8.

x	-3	-2	-1	0	1	2	3
y	-5	0	3	4	3	0	-5

$y = 4 - x^2$

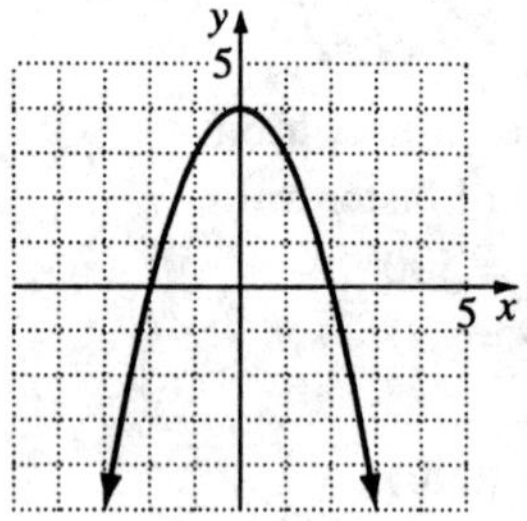

9. $A(6, 90)$; $1978 + 6 = 1984$
   The number of nuclear power plants in the U.S. in 1984 was about 90.

10. Peak point is about (12, 120).
    1978 + 12 =1990
    About 120 plants in 1990.

11. Point (60, 20)
    About age 60

12. As age increase, death rate per 1000 increases.

13. Lines cross at approximately (1989, 5).
    Spending was the same in 1989.
    NASA spent approximately 5 billion dollars.

14. $(-2x^5)(-7x^{-11}) = (-2)(-7)x^{5-11}$
    $= 14x^{-6} = \dfrac{14}{x^6}$

15. $\left(\dfrac{6x^5y^{-6}}{-3xy^{-4}}\right)^2 = \dfrac{6^2 x^{5 \cdot 2} y^{-6 \cdot 2}}{(-3)^2 x^2 y^{-4 \cdot 2}}$
    $= \dfrac{36x^{10}y^{-12}}{9x^2y^{-8}}$
    $= \dfrac{4x^{10-2}}{y^{12-8}} = \dfrac{4x^8}{y^4}$

16. $(3.7\times10^4)(9.0\times10^{13}) = 33.3\times10^{4+13}$
$= 3.33\times10^{18}$

17. $\dfrac{6\times10^{-3}}{1.2\times10^{-8}} = \dfrac{6}{1.2}\times10^{8-3}$
$= 5\times10^5$

18. $3(2x-4) = 9-3(x+1)$
$6x-12 = 9-3x-3$
$6x+3x = 12+9-3$
$9x = 18$
$x = 2$
$\{2\}$

19. $\dfrac{2x-3}{4} = \dfrac{x-4}{2} - \dfrac{x+1}{4}$
$4\left(\dfrac{2x-3}{4}\right) = 4\left(\dfrac{x-4}{2} - \dfrac{x+1}{4}\right)$
$2x-3 = 2(x-4)-(x+1)$
$2x-3 = 2x-8-x-1$
$2x-3 = x-9$
$2x-x = 3-9$
$x = -6$
$\{-6\}$

20. $A = P + Prt$, for r.
$A - P = Prt$
$\dfrac{A-P}{Pt} = r$
$r = \dfrac{A-P}{Pt}$

21. $2s - nf - nl = 0$, for n
$2s = nf + nl$
$2s = n(f+l)$
$\dfrac{2s}{f+l} = n$
$n = \dfrac{2s}{f+l}$

22. $y = 2350x + 22{,}208$
$66{,}858 = 2350x + 22{,}208$
$44{,}650 = 2350x$
$19 = x$
$1984 + 19 = 2003$
It predicts the year will be 2003.

23. Let x = budget for 1994, then
$x - 14$ = budget for 1995,
and $x - 23$ = budget for 1996.
$x + (x-14) + (x-23) = 623$
$3x - 37 = 623$
$3x = 660$
$x = 220$
$x - 14 = 206$
$x - 23 = 197$
The budget for each year was as follows: 1994, $220, billion; 1995, $206 billion; 1996, $197 billion.

24. Let x = annual salary prior to raise, then
$x + 0.09x$ = annual salary after raise.
$x + 0.09x = 34{,}880$
$x(1.09) = 34{,}880$
$x = 32{,}000$
The annual salary before the raise was $32,000.

25. y = population and
x = years since 1980.
Arizona $y = 89x + 2795$
S. Carolina $y = 43x + 3071$
$89x + 2795 = 43x + 3071$
$89x - 43x = 3071 - 2795$
$46x = 276$
$x = 6$
$1980 + 6 = 1986$
The two states had the same population in 1986.

26. y = monthly charge and
x = hours spent on line.
$y = 25$
$y = 0.25x + 10$
$25 = 0.25x + 10$
$15 = 0.25x$
$60 = x$
The rates will be the same with 60 hours on line.

27. Let x = width, then $2x + 1$ = length.
$2x + 2(2x + 1) = 302$
$2x + 2x + 2 = 302$
$4x = 300$
$x = 75$
$2x + 1 = 151$
The width is 75 yards and the length is 151 yards.

28. Let x = amount invested at 6%, then $x + 3000$ = amount invested at 8%.
$0.06x + 0.08(x + 3000) = 520$
$0.06x + 0.08x + 240 = 520$
$0.14x = 280$
$x = 2000$
$x + 3000 = 5000$
The investments were \$2000 at 6% and \$5000 at 8%.

29. Let x = number of rungs.
$\frac{x}{2} = 4 - 6 + 7 + 4$
$\frac{x}{2} = 9$
$x = 18$
The ladder has 18 rungs.

Chapter 2

Problem Set 2.1

1. $\{(1, 3), (1, 7), (1, 10)\}$
Not a function since (1, 3), (1, 7) and (1, 10) have the same first component.
Domain = $\{1\}$
Range = $\{3, 7, 10\}$

3. $\{(2, 3), (2, 4), (3, 3), (3, 4)\}$
Not a function since (2, 3) and (2, 4) have the same first component and (3, 3) and (3, 4) have the same first component.
Domain = $\{2, 3\}$
Range = $\{3, 4\}$

5. $\{(-1, -1), (0, 0), (1, 1), (2, 2)\}$
Function since each first component is assigned exactly one second component.
Domain = $\{-1, 0, 1, 2\}$
Range = $\{-1, 0, 1, 2\}$

7. $\{(1, -1), (2, -2), (3, -3), (4, -4)\}$
Function since each first component is assigned exactly one second component.
Domain = $\{1, 2, 3, 4\}$
Range = $\{-4, -3, -2, -1\}$

9. $f(x) = 3x - 2$

a. $f(0) = 3(0) - 2 = -2$

b. $f(-2) = 3(-2) - 2 = -6 - 2 = -8$

c. $f(7) = 3(7) - 2 = 21 - 2 = 19$

d. $f\left(\frac{2}{3}\right) = 3\left(\frac{2}{3}\right) - 2 = 2 - 2 = 0$

e. $f(2a) = 3(2a) - 2 = 6a - 2$

11. $g(x) = 3x^2 + 5$

a. $g(0) = 3 \cdot 0^2 + 5 = 5$

b. $g(-1) = 3(-1)^2 + 5 = 8$

c. $g(4) = 3(4)^2 + 5 = 53$

d. $g(-3) = 3(-3)^2 + 5 = 32$

e. $g(4b) = 3(4b)^2 + 5 = 3(16b^2) + 5$
$= 48b^2 + 5$

13. $h(x) = 2x^2 + 3x - 1$

a. $h(0) = 2(0)^2 + 3(0) - 1 = -1$

b. $h(3) = 2(3)^2 + 3(3) - 1 = 26$

c. $h(-4) = 2(-4)^2 + 3(-4) - 1 = 19$

d. $h\left(\frac{1}{2}\right) = 2\left(\frac{1}{2}\right)^2 + 3\left(\frac{1}{2}\right) - 1 = 1$

e. $h(5r) = 2(5r)^2 + 3(5r) - 1$
$= 50r^2 + 15r - 1$

15. $f(x) = \frac{2x - 3}{x - 4}$

a. $f(0) = \frac{2(0) - 3}{0 - 4} = \frac{-3}{-4} = \frac{3}{4}$

b. $f(3) = \frac{2(3) - 3}{3 - 4} = \frac{3}{-1} = -3$

c. $f(-4)=\frac{2(-4)-3}{-4-4}=\frac{-11}{-8}=\frac{11}{8}$

d. $f(-5)=\frac{2(-5)-3}{-5-4}=\frac{-13}{-9}=\frac{13}{9}$

e. $f(a+h)=\frac{2(a+h)-3}{(a+h)-4}=\frac{2a+2h-3}{a+h-4}$

f. Denominator = 0 and, therefore, the function is undefined.

17. Yes; each house would map to a single area.

19. No, there could be two people who weigh the same with different number of years of education.

21.

x	$f(x)=x^2$
−2	4
−1	1
0	0
1	1
2	4

x	$g(x)=x^2-2$
−2	2
−1	−1
0	−2
1	−1
2	2

x	$h(x)=x^2+1$
−2	5
−1	2
0	1
1	2
2	5

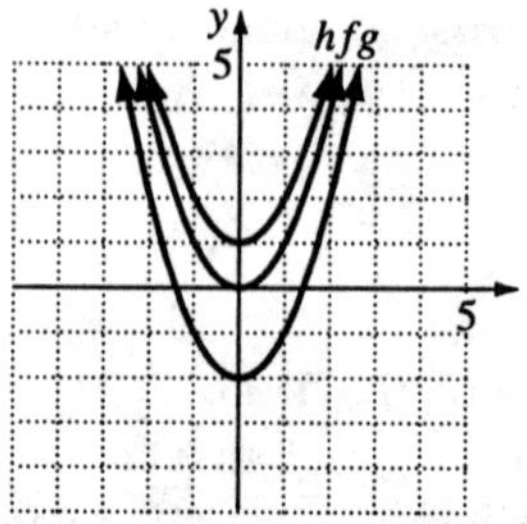

$f(x) \pm c$ is the graph of $f(x)$ shifted vertically up $(+c)$ or down $(-c)$ c units.

23.

x	$f(x)=\sqrt{x}$
0	0
1	1
4	2
9	3

x	$g(x)=\sqrt{x-2}$
2	0
3	1
6	2
11	3

x	$h(x)=\sqrt{x+1}$
–1	0
0	1
3	2
8	3

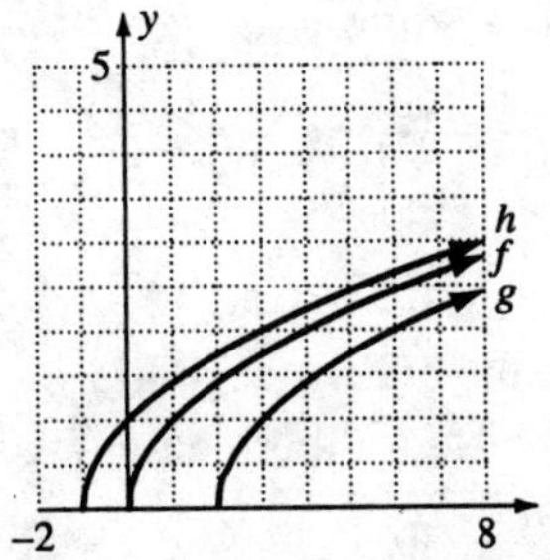

$f(x \pm c)$ is the graph of f shifted horizontally left $(+c)$ or right $(-c)$ c units.

25.

x	$f(x)=x^3$
–2	–8
–1	–1
0	0
1	1
2	8

x	$g(x)=x^3+1$
–2	–7
–1	0
0	1
1	2
2	9

x	$h(x)=(x+1)^3$
–3	$(-2)^3=-8$
–2	$(-1)^3=-1$
–1	$0^3=0$
0	$1^3=1$
1	$2^3=8$

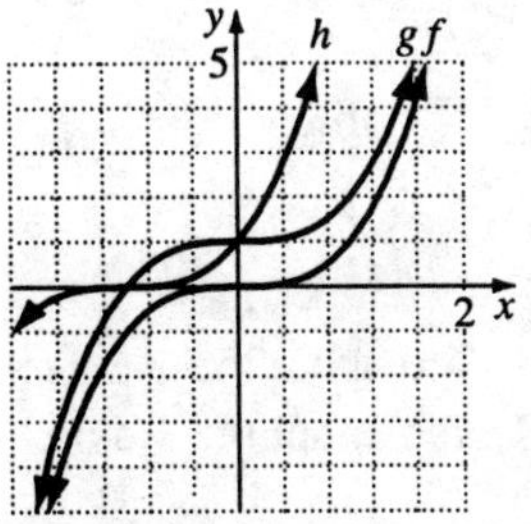

For $c>0$, $f(x)+c$ is $f(x)$ shifted up c units and $f(x+c)$ is $f(x)$ shifted to the left c units.

27.

x	$f(x)=\sqrt{x}$
0	0
1	1
4	2
9	3

x	$g(x)=-\sqrt{x}$
0	0
1	–1
4	–2
9	–3

x	$h(x)=\sqrt{-x}$
0	0
–1	1
–4	2
–9	3

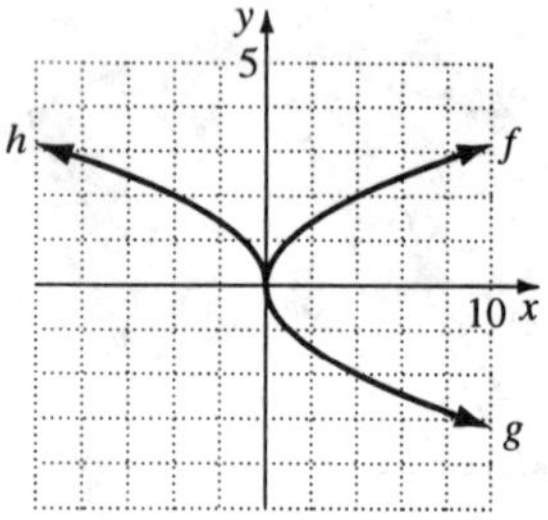

$-f(x)$ is $f(x)$ reflected about the x-axis and $f(-x)$ is $f(x)$ reflected about the y-axis.

29. Not a function

31. Function

33. Function

35. Not a function

37. Function

39. Not a function

41. $f(x)=-0.011x^2+1.22x-8.5$
$f(20)=-0.011(20)^2+1.22(20)-8.5$
$f(20)=11.5$
In 1950, there were approximately 11.5 million union members in the U.S.

43. a. $f(x)=14x^3-17x^2-16x+34$
for $1.5 \le x \le 3.5$
$f(2)=14(2^3)-17(2^2)-16(2)+34$
$f(2) = 112 - 68 - 32 + 34$
$f(2) = 46$
A moth with abdominal width of 2 mm produces 46 eggs.

b. No; 1 is not in the specified domain of $1.5 \le x \le 3.5$.

c. $f(3.2) - f(1.83)$

45. $f(x)=\dfrac{x+66.94}{0.01x+1}$
$f(0)=\dfrac{66.94}{1}=66.94$ years
$f(80)=\dfrac{80+66.94}{0.01(80)+1}=\dfrac{146.94}{1.8}$
≈ 81.63 years
$f(80) - f(0) = 81.63 - 66.94 = 14.69$ years
14.69 years is the difference in life expectancy at birth for those born in 1920 and those born in 2000.

47.

t	$f(t)=-t^4+12t^3-58t^2+132t$
0	0
1	$-1 + 12 - 58 + 132 = 85$
2	$-(2)^4+12(2)^3-58(2)^2+132(2)=112$
3	$-(3)^4+12(3)^3-58(3)^2+132(3)=117$
4	$-(4)^4+12(4)^3-58(4)^2+132(4)=112$
5	$-(5)^4+12(5)^3-58(5)^2+132(5)=85$
6	$-(6)^4+12(6)^3-58(6)^2+132(6)=0$

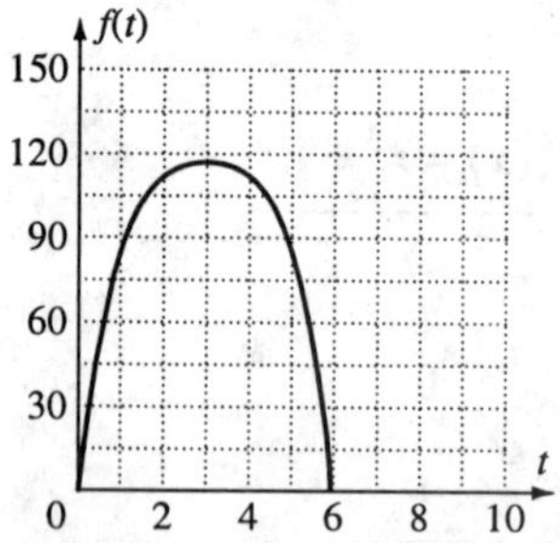

The shape shows the concentration increases, then decreases. The drug is eliminated at $t = 6$ hours after it is administered, which is shown on the graph by the point (6, 0).

49. **a.** Because each year maps onto exactly one consumption figure.

b. Maximum cigarette consumption was about 4.3 thousand per capita in 1965.

c. Minimum cigarette consumption was about 3.0 thousand per capita in 1990.

d. $f(1980) - f(1955) \approx 3.75 - 3.6$
$= 0.15$ thousand per capita
This is the change in consumption per capita over this 25 year period.

e. Approximately 1995

51. d is true.

53. Students should verify.

55. **a.** $P = 2l + 2w$ so $\frac{P}{2} = l + w$ or $w = \frac{P}{2} - l$
here,
$P = 80$, $l = x$, so $w = \frac{80}{2} - x = 40 - x$ so
Area $= l \cdot w = x(40 - x)$.

b.

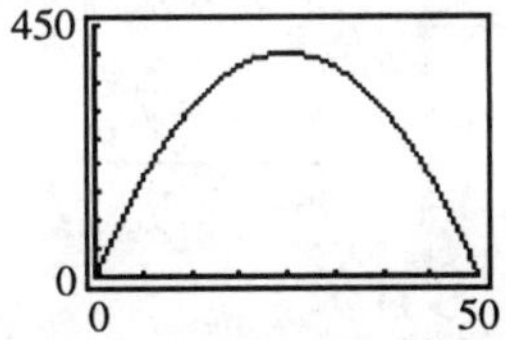

c. Greatest area occurs when the rectangle is a square with sides = 20 meters; answers may vary.

57. Answers may vary.

59. **a.** $f(x)$ if x is even $= -\frac{x}{2}$

b. $f(x)$ if x is odd $= \frac{x+1}{2}$

c. $f(20) + f(40) + f(65)$
$= \frac{-20}{2} + \left(-\frac{40}{2}\right) + \frac{65+1}{2}$
$= -10 + (-20) + 33$
$= 3$

61. Answers may vary.

Review Problems

63. Let x = number of years since 1960.
$f(x) = 179.5 + 2.35x$ (million people)
$297 = 179.5 + 2.35x$
$117.5 = 2.35x$
$x = 50$ years after 1960, or in 2010.

64. $P = \frac{2}{5}w\left(1 + \frac{n}{50}\right)$
$P = 720$
$w = 1000$
$720 = \frac{2}{5}(1000)\left(1 + \frac{n}{50}\right)$
$720 = 400\left(1 + \frac{n}{50}\right)$
$720 = 400 + 8n$
$320 = 8n$
$40 = n$
40 years

65. $(8.6 \times 10^{14})(2.5 \times 10^{-2})$
$= 8.6(2.5) \times 10^{14+(-2)}$
$= 21.5 \times 10^{12} = 2.15 \times 10^{13}$

Problem Set 2.2

1. $m = \frac{1}{2}$

3. $m = -1$

5. $m = -3$

7. $m=-\frac{3}{2}$

9. Let $A=(-3,-3)$, $B=(2,-5)$, $C=(5,-1)$, $D=(0,1)$.

$$M_{AB}=\frac{-3-(-5)}{-3-2}=-\frac{2}{5}$$
$$M_{BC}=\frac{-1-(-5)}{5-2}=\frac{4}{3}$$
$$M_{CD}=\frac{1-(-1)}{0-5}=-\frac{2}{5}$$
$$M_{DA}=\frac{1-(-3)}{0-(-3)}=\frac{4}{3}$$

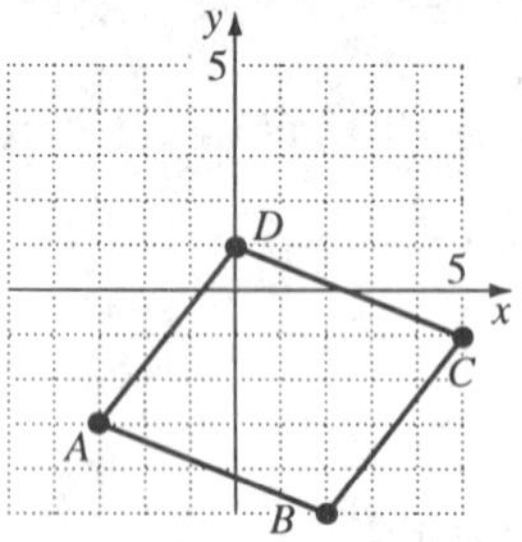

Since both sets of line segments are parallel (same slopes), the vertices of quadrilateral $ABCD$ are those of a parallelogram.

11.

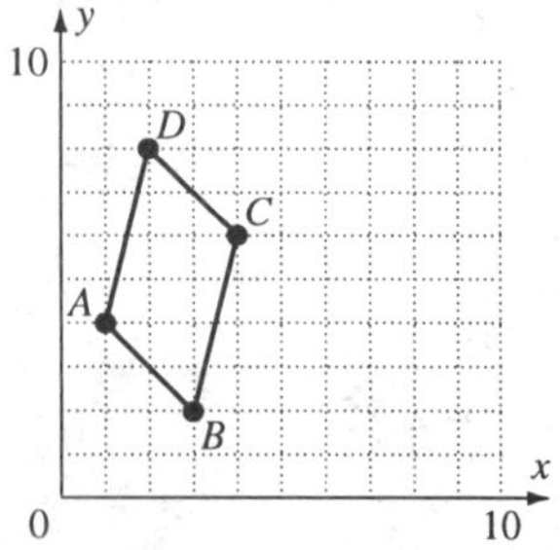

$A(1, 4)$, $B(3, 2)$, $C(4, 6)$, $D(2, 8)$

a. $m_1=\text{slope of } AB=\frac{2-4}{3-1}=\frac{-2}{2}=-1$

$m_2=\text{slope of } CD=\frac{8-6}{2-4}=\frac{2}{-2}=-1$

$m_1=m_2$: AB is parallel to CD.

$m_3=\text{slope of } BC=\frac{6-2}{4-3}=\frac{4}{1}=4$

$m_4=\text{slope of } AD=\frac{8-4}{2-1}=\frac{4}{1}=4$

$m_3=m_4$: AD is parallel to BC.

Opposite sides AB and CD have slope -1, and opposite sides BC and AD have slope 4. Both pairs of opposite sides are parallel; thus $ABCD$ is a parallelogram.

b. $m_1=\text{slope of } AB=-1$

$m_3=\text{slope of } BC=4$

$m_1m_3=-1(4)=-4\neq-1$

The slopes are not negative reciprocals. AB and BC are not perpendicular. Thus, the figure is not a rectangle. Product of slope AB and BC is -4, not -1.

13. $A(1, 3)$, $B(4, 3)$, $C(4, 6)$, $D(1, 6)$

AC: $m=\frac{6-3}{4-1}=1$

BD: $m=\frac{6-3}{1-4}=-1$

Since the slopes of diagonals AC and BD are negative reciprocals, diagonals AC and BD are perpendicular.

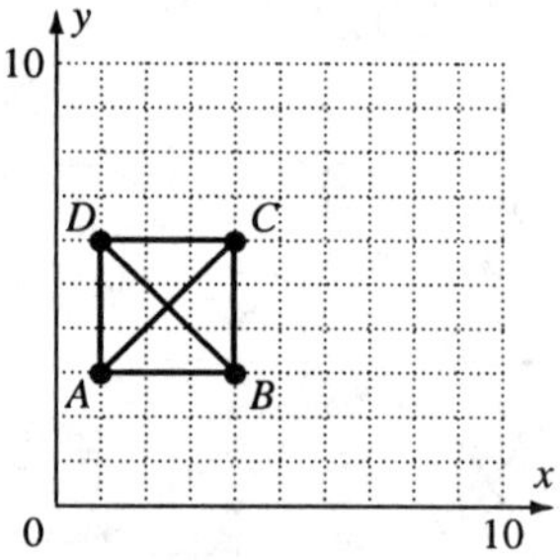

15. $m_1>m_2>m_3>m_4$

17. $y=-2x-1$

$m=-2$; y-intercept: $(0,-1)$

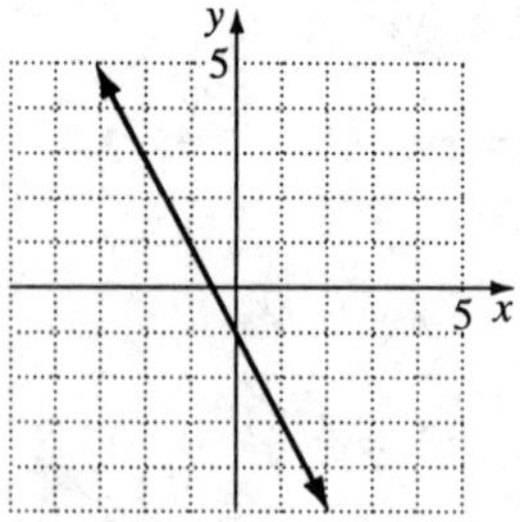

19. $y = \frac{1}{2}x + 1$

$m = \frac{1}{2}$; y-intercept: (0, 1)

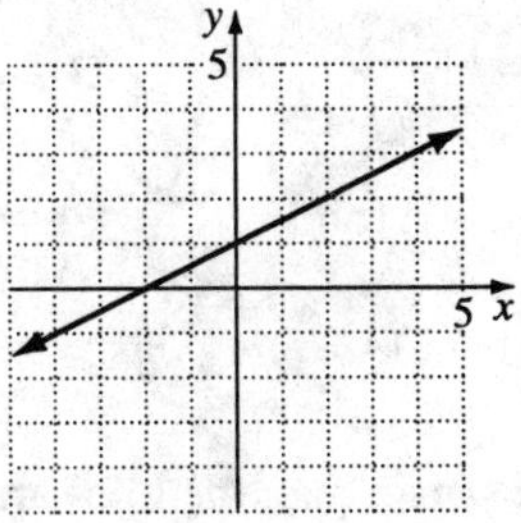

21. $y = \frac{1}{2}x - 1$

$m = \frac{1}{2}$; y-intercept: (0, –1)

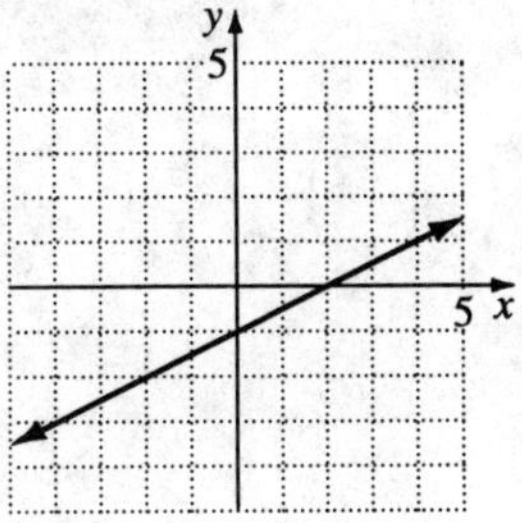

23. $y = -\frac{1}{2}x + 1$

$m = -\frac{1}{2}$; y-intercept: (0, 1)

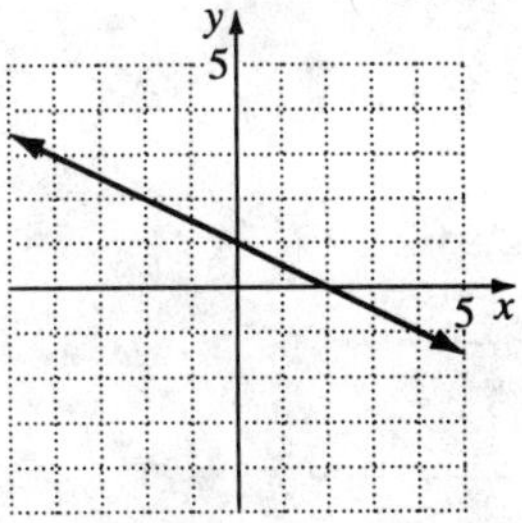

25. $y = -\frac{1}{2}x - 1$

$m = -\frac{1}{2}$; y-intercept: (0, –1)

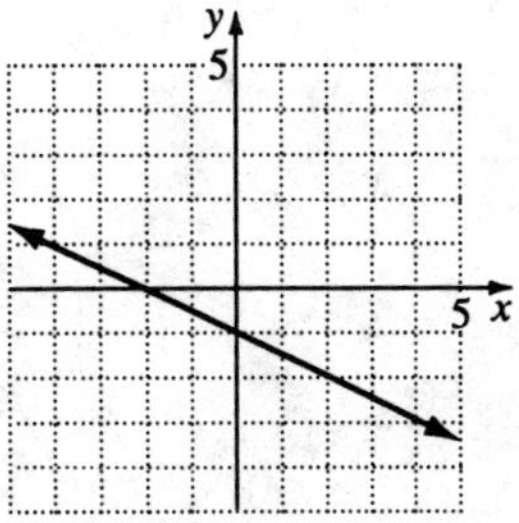

27. $x + y = 3$

$y = -x + 3$

$m = -1$; y-intercept: (0, 3)

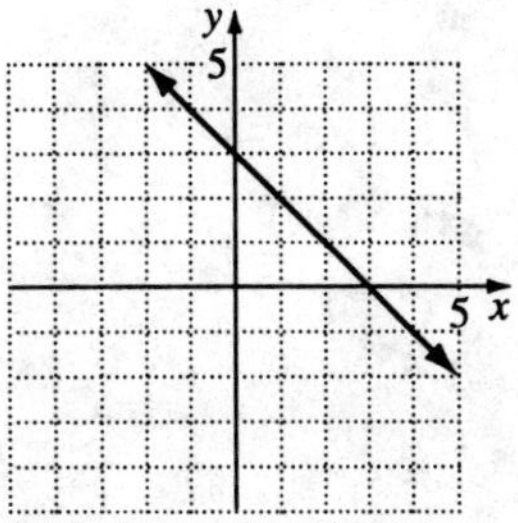

29. $2x + y = -1$

$y = -2x - 1$

$m = -2$; y-intercept: (0, –1)

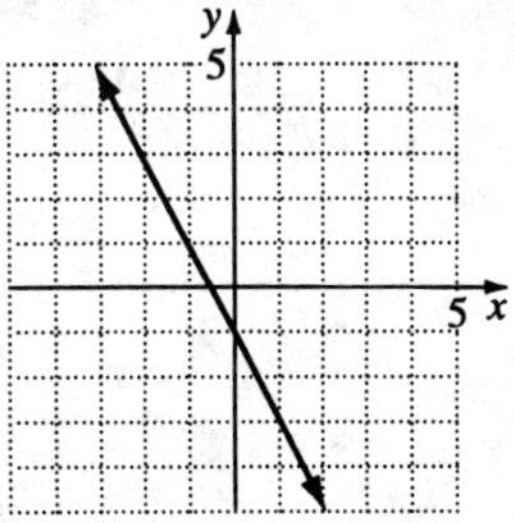

31. $2x+3y=6$
$3y=-2x+6$
$y=-\frac{2}{3}x+2$
$m=-\frac{2}{3}$; *y*-intercept: (0, 2)

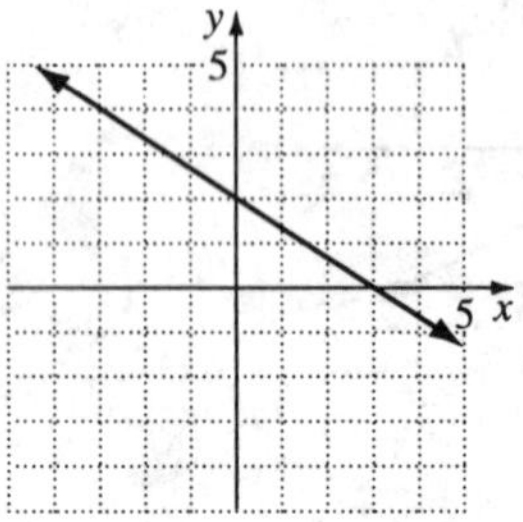

33. $-2x+3y=6$
$3y=2x+6$
$y=\frac{2}{3}x+2$
$m=\frac{2}{3}$; *y*-intercept: (0, 2)

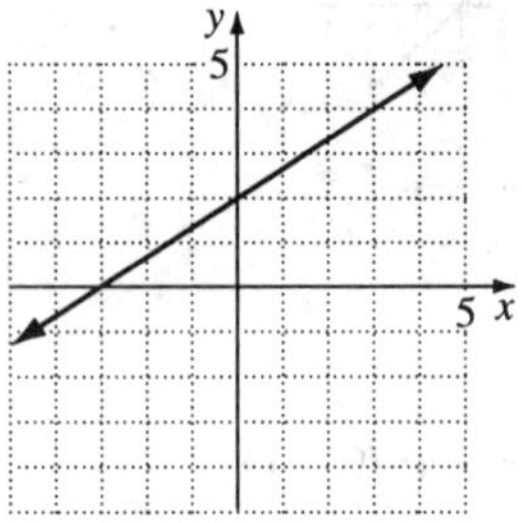

35. $3x+y=5$
$y=-3x+5$
$m=-3,\ b=5$

37. $2x+3y=18$
$3y=-2x+18$
$y=-\frac{2}{3}x+6$
$m=-\frac{2}{3},\ b=6$

39. $8x-4y=12$
$4y=8x-12$
$y=2x-3$
$m=2,\ b=-3$

41. $7x-3y=13$
$3y=7x-13$
$y=\frac{7}{3}x-\frac{13}{3}$
$m=\frac{7}{3},\ b=-\frac{13}{3}$

43. $6x+2y=10$ and $3x+y=7$
$y=-3x+5$ $\quad y=-3x+7$
$m_1=-3$ $\quad m_2=-3$
$m_1=m_2$
Since the slopes are equal, the lines are parallel.

45. $3x-2y=6$ and $2x+3y=-6$
$y=\frac{3}{2}x-3$ $\quad y=-\frac{2}{3}x-2$
$m_1=\frac{3}{2}$ $\quad m_2=-\frac{2}{3}$
Since $m_1m_2=\left(\frac{3}{2}\right)\left(-\frac{2}{3}\right)=-1$, the lines are perpendicular.

47. $2x+y=1$ and $x-y=2$
$y=-2x+1$ $\quad y=x-2$
$m_1=-2$ $\quad m_2=1$
Since $m_1\neq m_2$ and $m_1m_2=(-2)(1)=-2\neq -1$, the lines are neither parallel nor perpendicular.

49. $y=3$

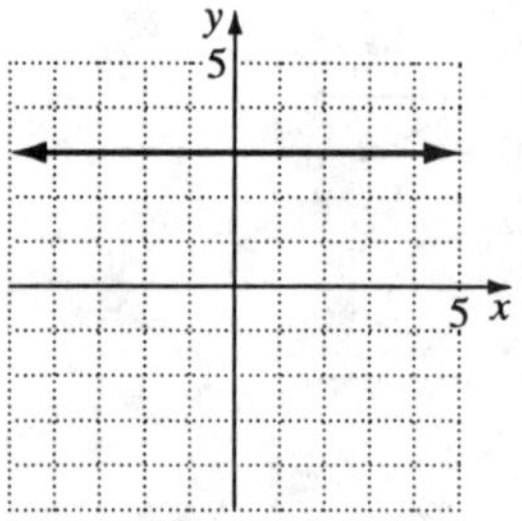

51. $y = -2$

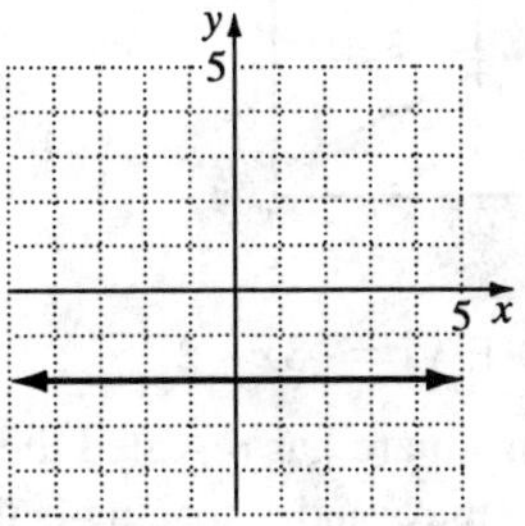

53. $3y = 18$

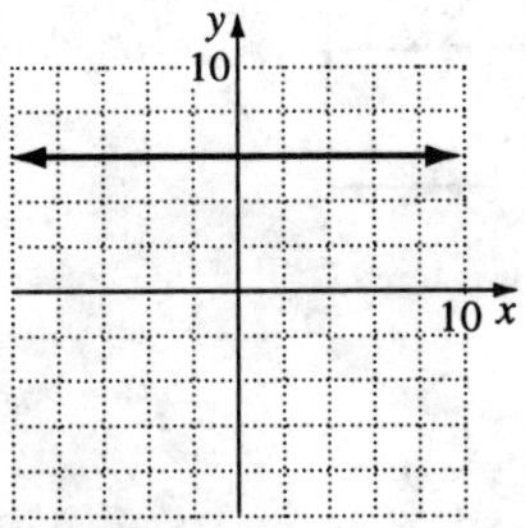

55. $f(x) = 2$

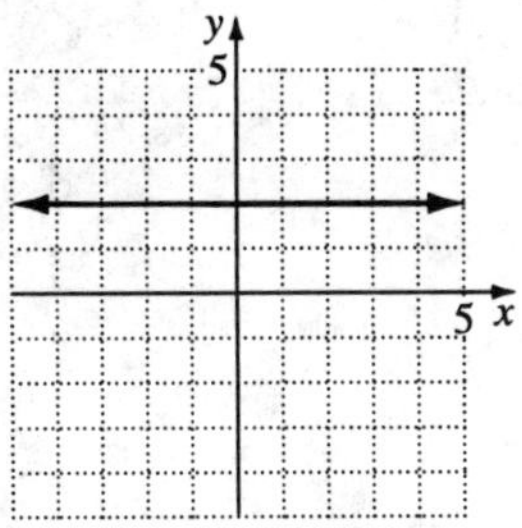

57. $x = -5$

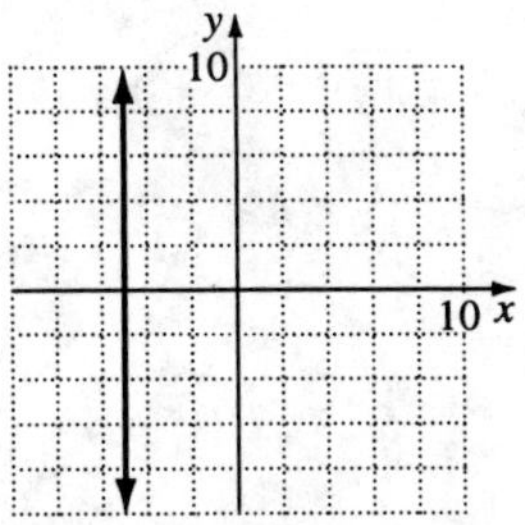

59. $3x - 12 = 0$

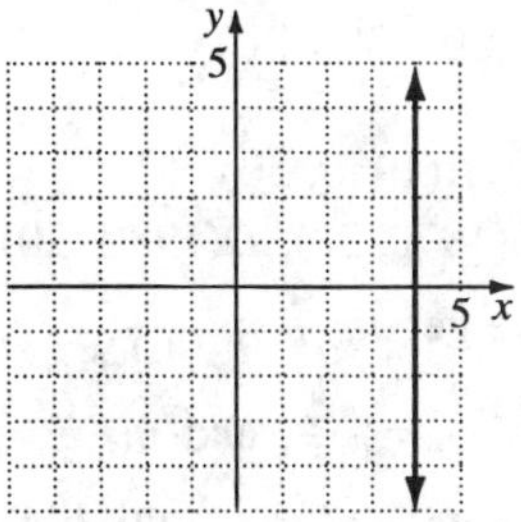

61. $x = 0$

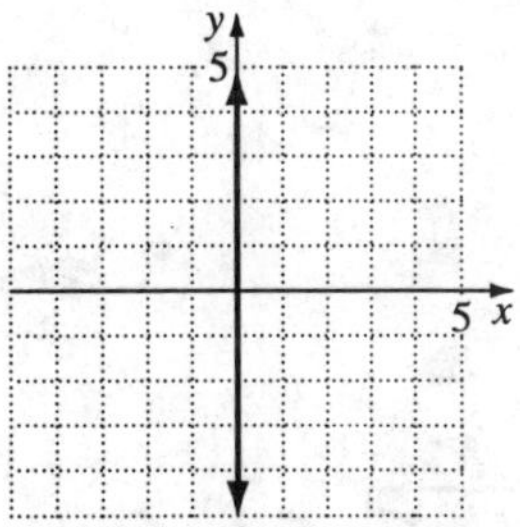

63. $N = 34t + 1549$
$m = 34,\ b = 1549$
There were 1,549,000 nurses in 1985 and the number grew by 34,000 each year thereafter.

65. $p = -0.5d + 100$
$m = -0.5,\ b = 100$
The percentage of hikers found is 100% when the searchers are zero meters apart and decreases by 0.5% for every meter increase the searchers are apart.

67. a. Slope through (1996, 9285) and (1980, 3200)

$$m = \frac{9285 - 3200}{1996 - 1980} = \frac{6085}{16} = 380.3125$$
$$= 380\frac{5}{16}$$

From 1980 to 1996, the cost of education has increased approximately \$380 per year.

b. Using $y = 3200 + 380x$ for x = number of years since 1980,
$y_{2010} = 3200 + 380(30) = \$14,600.$

69. Using estimates $A(1950, 50)$, $B(1960, 43)$, $C(1970, 28)$, $D(1980, 20)$ and $E(1990, 12)$, we get $m_{AB} = -\frac{7}{10}$, $m_{BC} = -\frac{3}{2}$, $m_{CD} = -\frac{4}{5}$, and $m_{DE} = -\frac{4}{5}$.
While the percentage of men with high sperm counts continues to decrease, the rate of decrease is slowing.

71. d is true.

73. a is true.

75. $y = 2x + 4$
$m = 2$

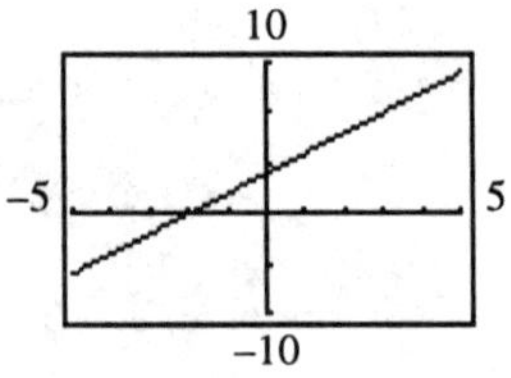

77. $y = -\frac{1}{2}x - 5$
$m = -\frac{1}{2}$

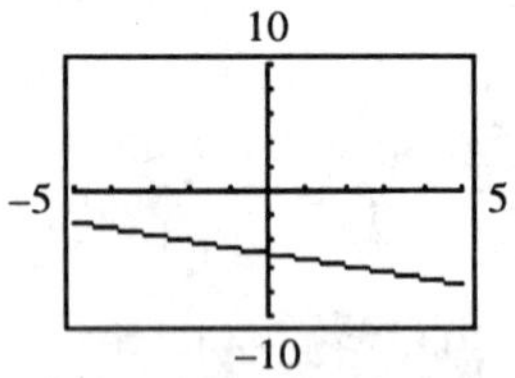

79. a. $y_1 = \frac{1}{3}x + 1,\ y_2 = -3x - 2$
no

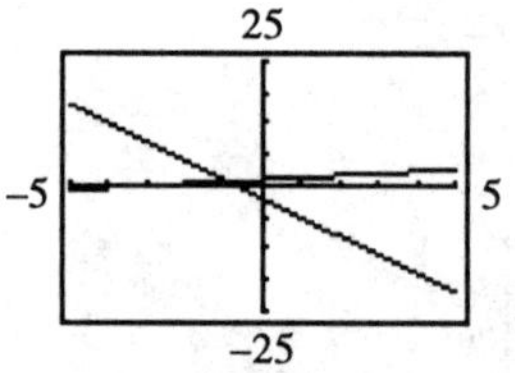

b. $y_1 = \frac{1}{3}x + 1,\ y_2 = -3x - 2$
Yes; zoom square has rescaled the window so the x and y spacing is the same.

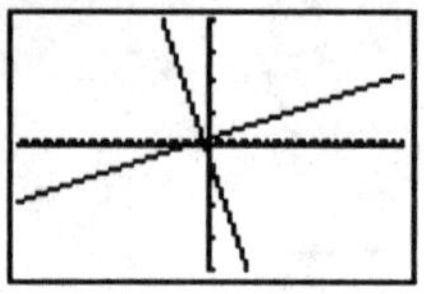

81–83. Answers may vary.

85. $y = mx + b$
x-intercept: let $y = 0$
$0 = mx + b$
$mx = -b$
$x = -\frac{b}{m}$
x-intercept is $-\frac{b}{m}$.

87. $2by = 5x + 9$
$y = \frac{5}{2b}x + \frac{9}{2b}$
Slope:
$\frac{5}{2b} = 1$
$5 = 2b$
$\frac{5}{2} = b$

Review Problems

89. $ay = 3r - by$
$ay + by = 3r$
$y(a + b) = 3r$
$y = \frac{3r}{a+b}$

90. $y = |x| + 2$

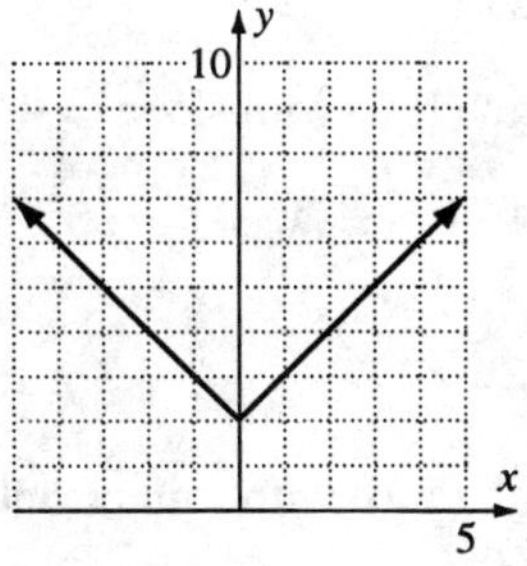

91. $\frac{3x-2}{4} - \frac{x-2}{3} = \frac{13}{4} - \frac{10x-8}{12}$

$12\left(\frac{3x-2}{4} - \frac{x-2}{3}\right) = 12\left(\frac{13}{4} - \frac{10x-8}{12}\right)$

$3(3x-2) - 4(x-2) = 3(13) - (10x-8)$

$9x - 6 - 4x + 8 = 39 - 10x + 8$

$5x + 2 = -10x + 47$

$15x = 45$

$x = 3$

$\{3\}$

Problem Set 2.3

1. $y - 1 = 3(x - 2)$

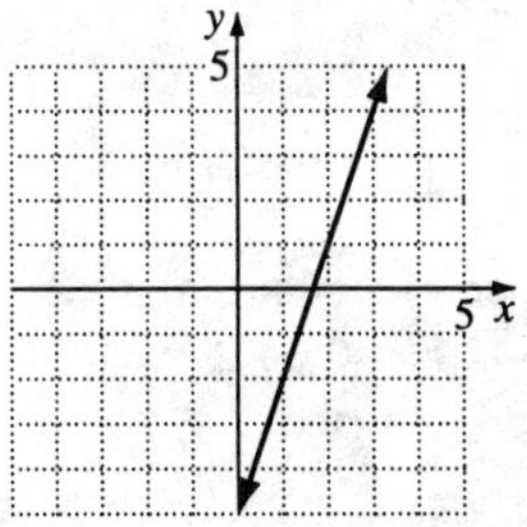

3. $y - 2 = -3(x - 6)$

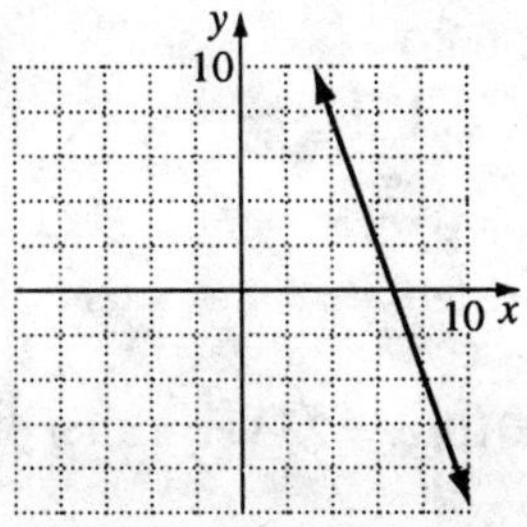

5. $y + 5 = 4(x + 3)$

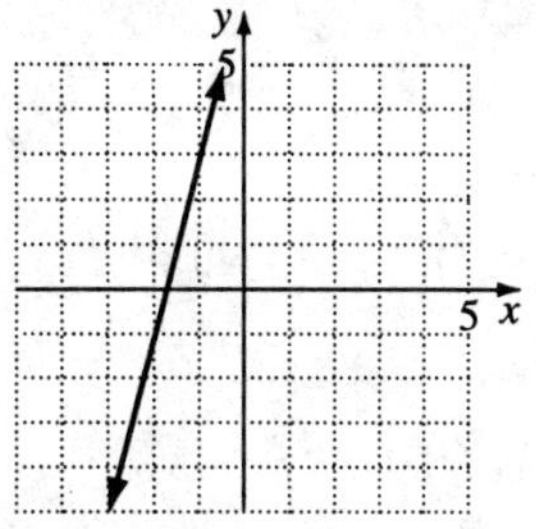

7. $y + 3 = 0(x - 0) = 0$

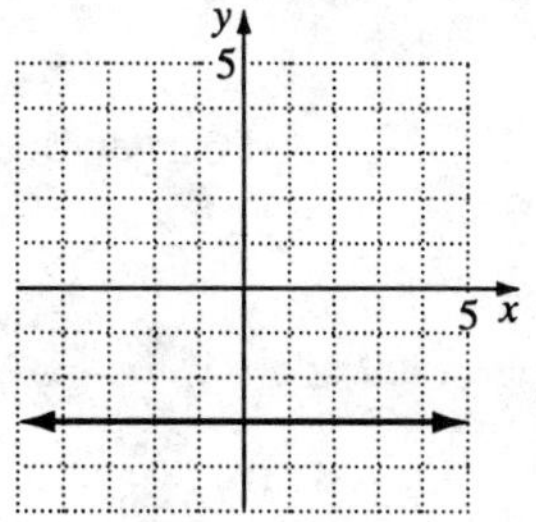

9. $y + 2 = \frac{4}{5}(x - 4)$

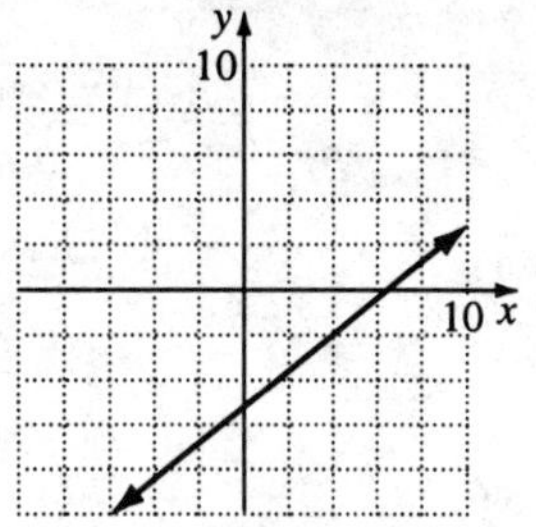

11. $y - 2 = 2(x + 3)$

$y = 2x + 6 + 2$

$y = 2x + 8$

13. $y - 0 = \frac{1}{2}(x - 1)$

$y = \frac{1}{2}x - \frac{1}{2}$

15. $m = \frac{6-4}{1+3} = \frac{2}{4} = \frac{1}{2}$

$y - 4 = \frac{1}{2}(x + 3)$ or $y - 6 = \frac{1}{2}(x - 1)$

$y = \frac{1}{2}x + \frac{11}{2}$

17. $m = \frac{11+2}{9+4} = \frac{13}{13} = 1$
$y + 2 = 1(x + 4)$ or $y - 11 = 1(x - 9)$
$y = x + 2$

19. Passing through (1, 3) and parallel to $y = 2x - 1$
$y = 2x - 1$: $m = 2$
Slope of line parallel = 2
Point-slope: $y - 3 = 2(x - 1)$
$y - 3 = 2x - 2$
Slope-intercept: $y = 2x + 1$

21. Passing through (1, –3) and perpendicular to $y = 2x - 1$.
$y = 2x - 1$: $m = 2$
Slope of line perpendicular: $m = -\frac{1}{2}$
Point-slope: $y + 3 = -\frac{1}{2}(x - 1)$
$y + 3 = -\frac{1}{2}x + \frac{1}{2}$
Slope-intercept: $y = -\frac{1}{2}x - \frac{5}{2}$

23. $m = \frac{5-(-2)}{0-1} = \frac{7}{-1} = -7$
$y + 2 = -7(x - 1)$
$y = -7x + 7 - 2$
$y = -7x + 5$

25. $m = \frac{-8-(-8)}{-2-5} = \frac{-8+8}{-7} = 0$
$y + 8 = 0(x + 2) = 0$
$y + 8 = 0$
$y = -8$

27. $m = \frac{6-5}{-3-2} = \frac{1}{-5} = -\frac{1}{5}$
$y - 5 = -\frac{1}{5}(x - 2)$
$y = -\frac{1}{5}x + \frac{2}{5} + 5$
$y = -\frac{1}{5}x + \frac{27}{5}$

29. Passing through (–3, 2) with slope = 2
Point-slope form: $y - 2 = 2(x + 3)$
$y - 2 = 2x + 6$
Standard form: $2x - y = -8$

31. Passing through (1, 0) with slope $= \frac{1}{2}$
Point-slope: $y - 0 = \frac{1}{2}(x - 1)$
$2y = x - 1$
Standard: $x - 2y = 1$

33. Passing through (5, 6) with x-intercept = 11
(5, 6) and (11, 0)
$m = \frac{0-6}{11-5} = -\frac{6}{6} = -1$
Point-slope: $y - 6 = -1(x - 5)$
$y - 6 = -x + 5$
Standard: $x + y = 11$

35. Passing through (–2, –5) and parallel to $2x + y = 4$
$y = -2x + 4$: $m = -2$
Slope of line parallel: $m = -2$
Point-slope: $y + 5 = -2(x + 2)$
$y + 5 = -2x - 4$
Standard: $2x + y = -9$

37. Containing (1, –3) and perpendicular to $2x - 4y = 12$:
$x - 2y = 6$
$-2y = -x + 6$
$y = \frac{1}{2}x - 3$: $m = \frac{1}{2}$
Slope of line perpendicular: $m = -2$
Point-slope: $y + 3 = -2x + 2$
$y = -2x - 1$
Standard: $2x + y = -1$

39. $(x_1, y_1) = (3, 546)$
$(x_2, y_2) = (5, 666)$
$m = \frac{666-546}{5-3} = \frac{120}{2} = 60$
$y - 546 = 60(x - 3)$ or $y - 666 = 60(x - 5)$
$y = 60x - 180 + 546$
$y = 60x + 366$
For the year 2010, $x = 25$ years after 1985

so $y = 60(25) + 366 = 1866$.
In 2010, the model predicts the total cost of health care expenditures to be 1866 billion dollars, or 1.866 trillion dollars.

41. $m = \frac{55-19}{2000-20,000} = \frac{36}{-18,000} = -\frac{1}{500}$
$y - 55 = -\frac{1}{500}(x - 2000)$
$y - 55 = -\frac{1}{500}x + 4$
$y = -\frac{1}{500}x + 59$
When $y = \$50$,
$50 = -\frac{1}{500}x + 59$
$-9 = -\frac{1}{500}x$
$x = 4500$ shirts can be sold.

43. Estimated values: (0, 58.9) and (18, 59.3)
$m = \frac{59.3-58.9}{18-0} = \frac{0.4}{18} = \frac{1}{45} \approx 0.022$
$y - 58.9 = \frac{1}{45}(x - 0)$ or
$y - 59.3 = \frac{1}{45}(x - 18)$
$y = \frac{1}{45}x + 58.9$
The year 2010 corresponds to $x = 35$.
$y = \frac{1}{45}(35) + 58.9 \approx 59.68$
The average global temperature in 2010 is estimated to be ≈ 60°F.

45–47. Answers may vary.

49. $3x - 2y = 4$
$y = \frac{3}{2}x - 2$
$m = \frac{3}{2}$
Slope of line perpendicular $= -\frac{2}{3}$
y-intercept of $3x - 2y = 4$:
$3(0) - 2y = 4$
$y = -2$
$(x_1, y_1) = (0, -2)$
Point-slope: $y + 2 = -\frac{2}{3}(x - 0)$
Slope-intercept: $y = -\frac{2}{3}x - 2$
$3y = -2x - 6$
Standard: $2x + 3y = -6$

51. $\frac{x}{-3} + \frac{y}{-6} = 1$
$2x + y = -6$
$x = -40$:
$-80 + y = -6$
$y = 74$
$(-40, 74)$
$y = -200$:
$2x - 200 = -6$
$x = 97$
$(97, -200)$

53. Line passing through (0, 4) and $(a - 2, 6)$ and has x-intercept of a: $(a, 0)$
$m = \frac{6-4}{a-2-0} = \frac{2}{a-2}$
$y - 4 = \frac{2}{a-2}(x - 0)$
$y = \frac{2}{a-2}x + 4$
x-intercept: $(a, 0) \rightarrow x = a, y = 0$
$0 = \frac{2a}{a-2} + 4$
$\frac{2a}{a-2} = -4$
$2a = -4a + 8$
$6a = 8$
$a = \frac{4}{3}$

55.

Number	Term		Pattern
first	17	(1, 17)	$8 \cdot 1 + 9 = 17$
second	25	(2, 25)	$8 \cdot 2 + 9 = 25$
third	33	(3, 33)	$8 \cdot 3 + 9 = 33$
fourth	41	(4, 41)	$8 \cdot 4 + 9 = 41$
fifth	49	(5, 49)	$8 \cdot 5 + 9 = 49$
n^{th}	$8n + 9$	$(n, 8n + 9)$	$8n + 9$

57. Answers may vary.

Review Problems

58. $5x - 4 = 2(6x - 1) + 12$
$5x = 12x - 2 + 12 + 4$
$-7x = 14$
$x = -2$
$\{-2\}$

59. $\left(\dfrac{2x^5y^{-2}}{3x^{-4}}\right)^4 = \left(\dfrac{2x^{5-(-4)}}{3y^2}\right)^4$
$= \left(\dfrac{2x^9}{3y^2}\right)^4 = \dfrac{16x^{36}}{81y^8}$

60. $-2^2 - 4(1-3)^5 = -4 - 4(-2)^5$
$= -4 - 4(-32) = 124$

Problem Set 2.4

1. $x + 7 > 10$
$x > 3$
$\{x|x > 3\}$
$(3, \infty)$

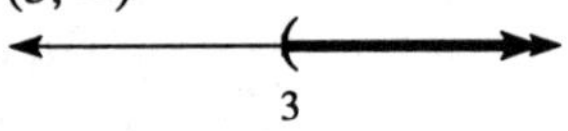

3. $x - 5 \le -7$
$x \le -2$
$\{x|x \le -2\}$
$(-\infty, -2]$

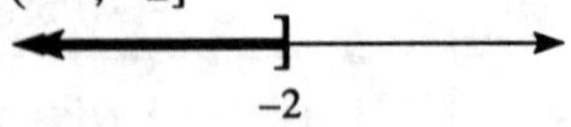

5. $9x < -45$
$x < -5$
$\{x|x < -5\}$
$(-\infty, -5)$

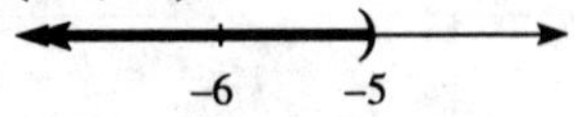

7. $0.5x > 4$
$x > 8$
$\{x|x > 8\}$
$(8, \infty)$

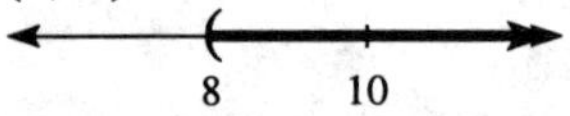

9. $-5x \le 30$
$x \ge -6$
$\{x|x \ge -6\}$
$[-6, \infty)$

11. $5x + 11 > 26$
$5x > 15$
$x > 3$
$\{x|x > 3\}$
$(3, \infty)$

13. $2x + 3 \ge x + 4$
$x + 3 \ge 4$
$x \ge 1$
$\{x|x \ge 1\}$
$[1, \infty)$

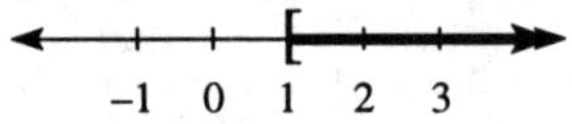

15. $18x + 45 \le 12x - 8$
$6x \le -53$
$x \le \frac{-53}{6} = -8.8\overline{3}$
$\left\{x \middle| x \le -\frac{53}{6}\right\}$
$\left(-\infty, -\frac{53}{6}\right]$

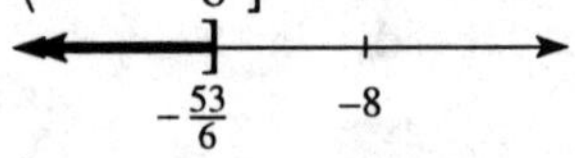

17. $4(x + 1) + 2 \ge 3x + 6$
$4x + 4 + 2 \ge 3x + 6$
$4x + 6 \ge 3x + 6$
$x + 6 \ge 6$

$x \ge 0$
$\{x|x \ge 0\}$
$[0, \infty)$

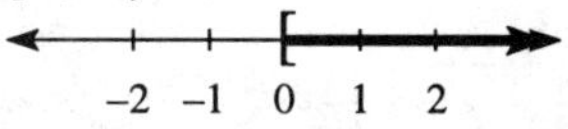

19. $7(2x - 1) < 9x + 11$
$14x - 7 < 9x + 11$
$5x < 18$
$x < \frac{18}{5}$
$\left\{x \middle| x < \frac{18}{5}\right\}$
$\left(-\infty, \frac{18}{5}\right)$

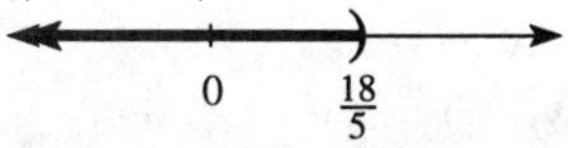

21. $-2(2x - 9) \le 4 - 5(x - 2)$
$-4x + 18 \le 4 - 5x + 10$
$-4x + 18 \le 14 - 5x$
$x \le -4$
$\{x|x \le -4\}$
$(-\infty, -4]$

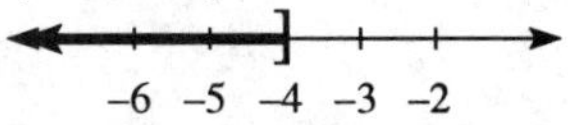

23. $2.6x - 0.2 \ge 1.4x + 2.2$
$1.2x \ge 2.4$
$x \ge 2$
$\{x|x \ge 2\}$
$[2, \infty)$

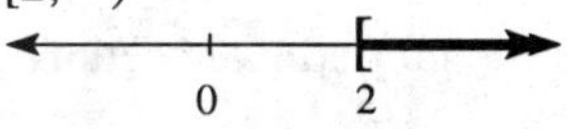

25. $\frac{x}{3} + \frac{2}{5} \le 4$
$5x + 6 \le 60$
$5x \le 54$
$x \le \frac{54}{5} = 10.8$
$\left\{x \middle| x \le \frac{54}{5}\right\}$
$\left(-\infty, \frac{54}{5}\right]$

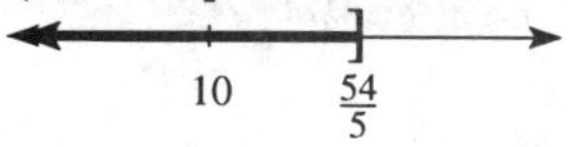

27. $\frac{x+1}{3} - \frac{x-3}{2} < \frac{1}{6}$
$6\left(\frac{x+1}{3} - \frac{x-3}{2}\right) < 6 \cdot \frac{1}{6}$

$2(x + 1) - 3(x - 3) < 1$
$2x + 2 - 3x + 9 < 1$
$-x + 11 < 1$
$-x < -10$
$x > 10$
$\{x|x > 10\}$
$(10, \infty)$

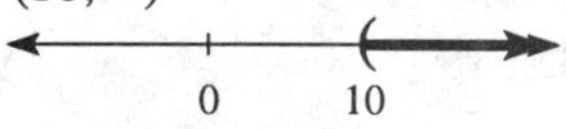

29. $3[3(y + 5) + 8y + 7] + 5[3(y - 6) - 2(3y - 5)]$
$< 2(4y + 3)$
$3(3y + 15 + 8y + 7) + 5(3y - 18 - 6y + 10)$
$< 8y + 6$
$3(11y + 22) + 5(-3y - 8) < 8y + 6$
$33y + 66 - 15y - 40 < 8y + 6$
$18y + 26 < 8y + 6$
$10y < -20$
$y < -2$
$\{y|y < -2\}$
$(-\infty, -2)$

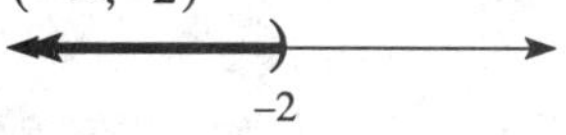

31. $4(3y + 2) - 3y < 3(1 + 3y) - 7$
$12y + 8 - 3y < 3 + 9y - 7$
$9y + 8 < 9y - 4$
$8 < -4$ False
No solution
$\varnothing$

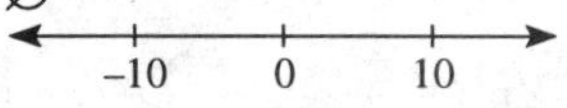

33. Let x = number of years since 1988.
$y = 49{,}391 - 1546x$
$49{,}391 - 1546x < 35{,}477$
$13{,}914 < 1546x$

$9 < x$
After 9 years, or after 1997 (i.e., in 1998, 1999, ...)

35. Men 18–25, Men 26–34

37. $3x + 2 - 5 > -x + 7 + 2x$
$3x - 3 > x + 7$
$2x > 10$
$x > 5$, i.e., $> 5\%$
All groups except men and women age 35 or older

39. Basic rental: weekly cost $y = \$260$
Continental: weekly cost $y = \$80 + \$0.25x$ for driving x miles
$260 < 80 + 0.25x$
$180 < \frac{x}{4}$
$720 < x$
Drive more than 720 miles per week

41. x = home assessment
$1800 + 0.03x < 200 + 0.08x$
$-0.05x < -1600$
$x > 32{,}000$
Homes assessed at more than $32,000.

43. x = number of advertisements
$1600 + 25x \geq 3560$
$x \geq 78.4$
The minimum number of advertisements is 79.

45. b is true;
$\frac{7 - 5x}{-2} \geq -1$
$7 - 5x \leq 2$
$-5x \leq -5$
$x \geq 1$
$4 \in \{x | x \geq 1\}$

47. $y_2 = -3(x - 6) = -3x + 18$
$y_1 = 2x - 2$
$y_2 > y_1$ when
$-3x + 18 > 2x - 2$
$20 > 5x$
$4 > x$, i.e., $x < 4$

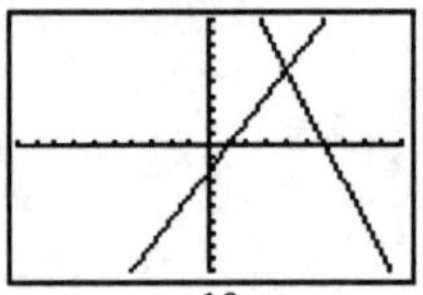

49. $y_2 = 12x - 10$
$y_1 = 2(x - 4) + 10x = 12x - 8$
$y_2 > y_1$ when $12x - 10 > 12x - 8$
$-10 > -8$
No solution; No x–values
Nowhere is $y_2 > y_1$.

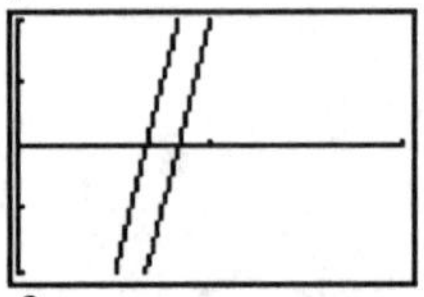

51. a. $y_A = 4 + 0.10x$
$y_B = 2 + 0.15x$

b.

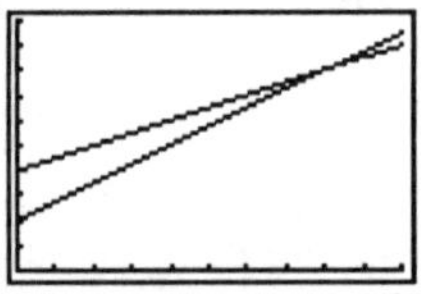

c. Students should find intersection.

d. $y_A < y_B$
$4 + 0.1x < 2 + 0.15x$
$2 < 0.05x$
$40 < x$
Plan A is cheaper (better) when the number of checks written is greater than 40.

53. Answers may vary.

55. Let x = the cost of meat pie.
Each tourist had $x - 7$ and $x - 2$ dollars.
Together then had $(x - 7) + (x - 2)$ or
$2x - 9$ dollars, where
$2x - 9 < x$
$x < 9$
Since the first tourist needed \$7 more, $x > 7$.
Since $x > 7$ and $x < 9$,
$7 < x < 9$
$x = 8$
and the meat pie costs \$8.00.
More than \$7, less than \$9
Cost of meat pie, \$8

Review Problems

57. $y-(3+y)=5+\frac{2(y-2)}{4}$
$y-3-y=5+\frac{1}{2}(y-2)$
$-3=5+\frac{1}{2}y-1$
$-3=4+\frac{1}{2}y$
$-7=\frac{1}{2}y$
$-14=y$
$\{-14\}$

58. $w=\frac{11}{2}h-220$
$w+220=\frac{11}{2}h$
$\frac{2}{11}(w+220)=h$
$h=\frac{2}{11}(w+220)$
(w = 154:)
$h=\frac{2}{11}(154+220)=\frac{2}{11}(374)=2(34)=68$
68 inches

59. $-4x^2-[6y^2-(5x^2-2y^2)]$
$=-4x^2-(6y^2-5x^2+2y^2)$
$=-4x^2-(8y^2-5x^2)$
$=-4x^2-8y^2+5x^2$
$=x^2-8y^2$

Problem Set 2.5

1. $\{1, 2, 3, 4\} \cap \{2, 4, 5\} = \{2, 4\}$

3. $\{1, 3, 5, 7\} \cap \{2, 4, 6, 8, 10\} = \varnothing$

5. $\{7, 9, 10\} \cap \{7, 9, 10\} = \{7, 9, 10\}$

7. $x > 3$ and $x > 6$ is $x > 6$
$\{x|x > 6\}$
$(6, \infty)$

9. $\{x|x \ge 3\} \cap \{x|x \le 6\}$
$\{x|3 \le x \le 6\}$
$[3, 6]$

11. $x < -3$ and $x > 1$: $\varnothing$
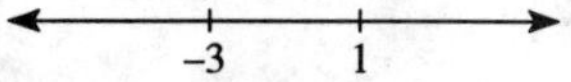

13. $5x < -20$ and $3x > -18$
$x < -4$ and $x > -6$
$\{x|-6 < x < -4\}$
$(-6, -4)$
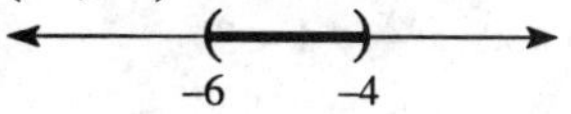

15. $x - 4 \le 2$ and $3x + 1 > -8$
$x \le 6$ and $3x > -9$
$x > -3$
$\{x|-3 < x \le 6\}$
$(-3, 6]$
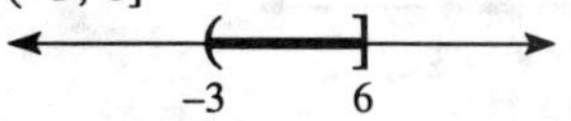

17. $2x > 5x - 15$ and $7x > 2x + 10$
$-3x > -15$ $5x > 10$
$x < 5$ $x > 2$
$\{x|2 < x < 5\}$
$(2, 5)$
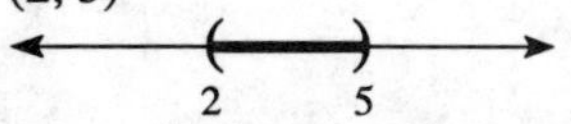

19. $\{y|4(1-y)<-6\}\cap\left\{y\left|\frac{y-7}{5}\le -2\right.\right\}$

$4(1-y)<-6$ $\qquad \frac{y-7}{5}\le -2$

$4-4y<-6$ $\qquad y-7\le -10$

$-4y<-10$ $\qquad y\le -3$

$y>\frac{5}{2}$

$y>\frac{5}{2}$ *and* $y\le -3$ must be satisfied.

No real numbers are less than or equal to –3 and also greater than $\frac{5}{2}$.

The solution is $\varnothing$. No solution

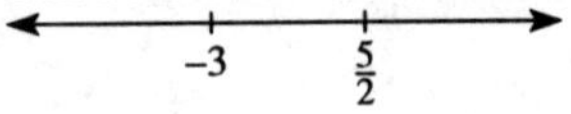

21. $\{x|x-1\le 7x-1\}\cap\{x|4x-7<3-x\}$

$x-1\le 7x-1$ $\qquad 4x-7<3-x$

$-6x\le 0$ $\qquad 5x<10$

$x\ge 0$ $\qquad x<2$

$\{x|0\le x<2\}$

[0, 2)

23. $\left\{x\left|\frac{9+4x}{3}>-5\right.\right\}\cap\left\{x\left|\frac{x}{3}+4<3\right.\right\}$

$\frac{9+4x}{3}>-5$ $\qquad \frac{x}{3}+4<3$

$9+4x>-15$ $\qquad \frac{x}{3}<-1$

$4x>-24$ $\qquad x<-3$

$x>-6$

$\{x|-6<x<-3\}$

(–6, –3)

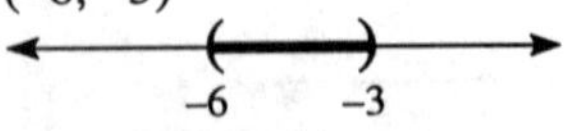

25. $2<x-5\le 7$

$7<x\le 12$

$\{x|7<x\le 12\}$

(7, 12]

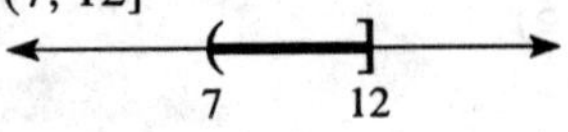

27. $-5\le 2x-7<9$

$2\le 2x<16$

$1\le x<8$

$\{x|1\le x<8\}$

[1, 8)

29. $-6<\frac{1}{2}y-4<-3$

$-2<\frac{1}{2}y<1$

$-4<y<2$

$\{y|-4<y<2\}$

(–4, 2)

31. $-2<-3y\le 3$

$\frac{2}{3}>y\ge -1$

$-1\le y<\frac{2}{3}$

$\left\{y\left|-1\le y<\frac{2}{3}\right.\right\}$

$\left[-1,\frac{2}{3}\right)$

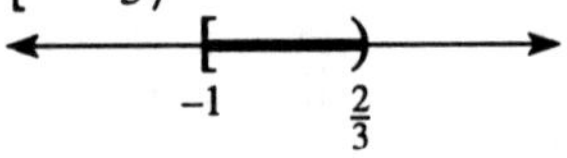

33. $-8<2-3y<10$

$-10<-3y<8$

$\frac{10}{3}>y>-\frac{8}{3}$

$-\frac{8}{3}<y<\frac{10}{3}$

$\left\{y\left|-\frac{8}{3}<y<\frac{10}{3}\right.\right\}$

$\left(-\frac{8}{3},\frac{10}{3}\right)$

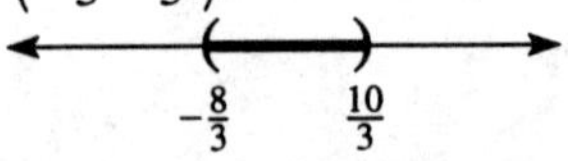

35. $-2\le -\frac{1}{2}y+3\le 6$

$-5\le -\frac{1}{2}y\le 3$

$10\ge y\ge -6$

$-6 \le y \le 10$
$\{y|-6 \le y \le 10\}$
$[-6, 10]$

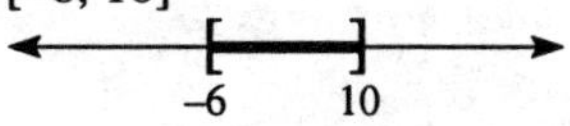

37. $\{1, 2, 3, 4\} \cup \{2, 4, 5\} = \{1, 2, 3, 4, 5\}$

39. $\{1, 3, 5, 7\} \cup \{2, 4, 6, 8, 10\}$
$= \{1, 2, 3, 4, 5, 6, 7, 8, 10\}$

41. $\{7, 9, 10\} \cup \{7, 9, 10\} = \{7, 9, 10\}$

43. $x > 3$ or $x > 6$ is $x > 3$.
$\{x|x > 3\}$
$(3, \infty)$

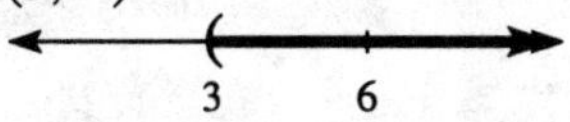

45. $\{x|x \ge 3\} \cup \{x|x \le 6\}$
$= \{x|-\infty < x < \infty\}$ or $(-\infty, \infty)$

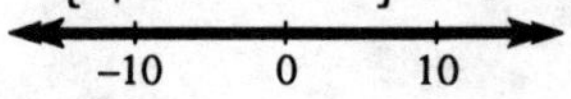

47. $x < -3$ or $x > 1$ is
$\{x|x < -3 \text{ or } x > 1\}$ or $(-\infty, -3) \cup (1, \infty)$

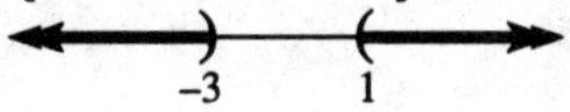

49. $5x < -20$ or $3x > -18$
$x < -4$ or $x > -6$
$\{x|-\infty < x < \infty\}$ or $(-\infty, \infty)$

51. $x - 4 \le 2$ or $3x + 1 > -8$
$x \le 6$ or $3x > -9$
$x > -3$
$\{x|-\infty < x < \infty\}$ or $(-\infty, \infty)$

53. $3x - 12 > 0$ or $2x - 3 \le -6$
$3x > 12$ or $2x \le -3$
$x > 4$ $\quad x \le -\frac{3}{2}$
$\left\{x|x \le -\frac{3}{2} \text{ or } x > 4\right\}$
$\left(-\infty, -\frac{3}{2}\right] \cup (4, \infty)$

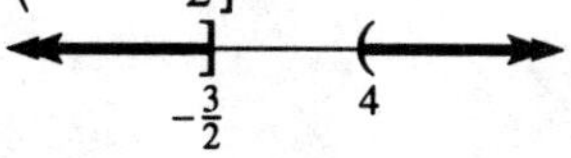

55. $\{x|3x + 6 < 8\} \cup \{x|-2x + 3 > -2\}$
$3x + 6 < 8 \qquad -2x + 3 > -2$
$3x < 2 \qquad -2x > -5$
$x < \frac{2}{3} \qquad x < \frac{5}{2}$
$\left\{x|x < \frac{2}{3}\right\} \cup \left\{x|x < \frac{5}{2}\right\}$
$\left\{x|x < \frac{5}{2}\right\}$
$\left(-\infty, \frac{5}{2}\right)$

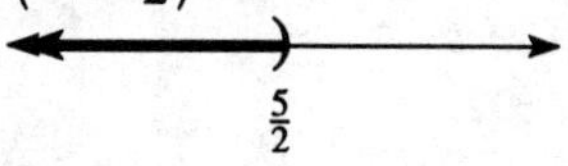

57. $\{x|x - 14 < 26 - 3x\} \cup \{x|31 - 5x > 1 - 8x\}$
$x - 14 < 26 - 3x \qquad 31 - 5x > 1 - 8x$
$4x < 40 \qquad 3x > -30$
$x < 10 \qquad x > -10$
$\{x|x < 10\} \cup \{x|x > -10\}$
$\{x|x \in R\}$
$(-\infty, \infty)$

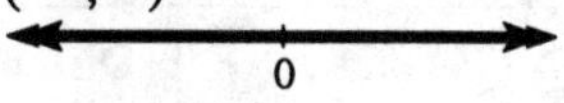

59. $\{y|12y + 6 < 4(5y - 1)\}$
$\cup \{y|9y + 4 \ge 5(3y - 4)\}$
$12y + 6 < 4(5y - 1) \qquad 9y + 4 \ge 5(3y - 4)$
$12y + 6 < 20y - 4 \qquad 9y + 4 \ge 15y - 20$
$-8y < -10 \qquad -6y \ge -24$
$y > \frac{5}{4} \qquad y \le 4$
$\left\{y|y > \frac{5}{4}\right\} \cup \{y|y \le 4\}$
$\{y|y \in R\}$
$(-\infty, \infty)$

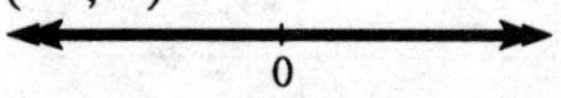

61. $\left\{x \middle| \frac{2x}{9}+5<7\right\} \cup \left\{x \middle| 3-x>\frac{x-3}{2}\right\}$

$\frac{2x}{9}+5<7 \qquad 3-x>\frac{x-3}{2}$

$\frac{2x}{9}<2 \qquad 6-2x>x-3$

$x<9 \qquad -3x>-9$

$x<3$

$\{x|x<9\} \cup \{x|x<3\}$

$\{x|x<9\}$

$(-\infty, 9)$

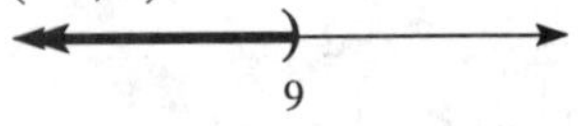

63. $77 \le \frac{9}{5}C+32 \le 104$

$5(77) \le 5\left(\frac{9}{5}C+32\right) \le 5(104)$

$385 \le 9C + 160 \le 520$

$225 \le 9C \le 360$

$25 \le C \le 40$

$25^\circ \le C \le 40^\circ$

65. a. $R_M = -0.03t + 20.86$

$19.78 \le -0.03t + 20.86 \le 20.38$

$-1.08 \le -0.03t \le -0.48$

$36 \ge t \ge 16$ (years after 1948) or between 1964 and 1984

b.

Year	t	Men	R_M	Difference
1948	0	21.1	20.86	0.24
1956	8	20.6	20.62	–0.02
1964	16	20.3	20.38	–0.08
1972	24	20	20.14	–0.14
1980	32	20.19	19.9	0.29
1984	36	19.8	19.78	0.02
1988	40	19.75	19.66	0.09
1992	44	20.01	19.54	0.47

The estimates are off by an average 0.1 seconds overall and by 0.02 seconds on average during 1964 to 1984 (observe that the model is consistently low then high then low, somewhat cyclically). The model is a good fit.

67. $y = 29{,}750 - 132x$ where x is the number of years since 1984.

$28{,}694 \le 29{,}750 - 132x \le 29{,}222$

$-1056 \le -132x \le -528$

$8 \ge x \ge 4$ or between 1988 and 1992

69. $x \ge 9$ and $x \ge 21$ means $x \ge 21$.

U.S. total, U.S. Latino, U.S. African American

71. $120 \le 3x \le 45$

$40 \le x \le 15$

None of the 11 countries and subpopulations meet these criteria.

73. $-\frac{x}{2} \ge -7$ or $2x - 1 < 41$

$-x \ge -14$ or $2x < 42$

$x \le 14$ or $x < 21$ means $x < 21$

U.S. whites, Canada, Australia, United Kingdom, France, Netherlands, Germany, Sweden

75. Let s be the length of the side of the square.

Observe perimeter $= 7s$.

Then $126 < 7s \le 140$

$18 < s \le 20$

The square's sides must be greater than 18 meters but not more than 20 meters.

77. Let

t = time for cars to meet (in hours)

distance $= 40t + 50t = 90t$

$90 <$ distance < 120

$90 < 90t < 120$

$1 < t < \frac{4}{3} \quad \left(\frac{4}{3} \text{ hr} = 1\frac{1}{3} \text{ hr} = 1 \text{ hr } 20 \text{ min}\right)$

Between 1 hour and $1\frac{1}{3}$ hr (or 1 hr 20 min)

79. a is true.

81. Answers may vary.

83. $-7 \le 8 - 3y \le 20$ and $-7 < 6y - 1 < 41$

$-15 \le -3y \le 12 \qquad -6 < 6y < 42$

$5 \ge y \ge -4 \qquad -1 < y < 7$

$-4 \le y \le 5$

$\{y \mid -1 < y \le 5\}$

$(-1, 5]$

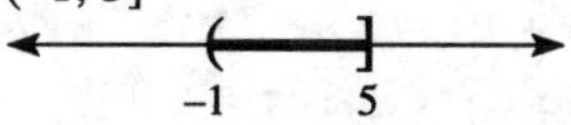

85. $7 < 5x + 12 < 17$ or $-2 < -3x + 4 \le 7$

$-1 < x < 1 \qquad -1 \le x < 2$

$\{x \mid -1 \le x < 2\} = [-1, 2)$

Review Problems

87. Let x = the number.

$6(x - 6) = 9(x - 9)$

$6x - 36 = 9x - 81$

$-3x = -45$

$x = 15$

The number is 15.

88. $p = \frac{5}{11}d + 15$

$(p = 60)$:

$60 = \frac{5}{11}d + 15$

$45 = \frac{5}{11}d$

$99 = d$

The depth is 99 feet.

89.

x	−2	−1	0	1	2
$y = -x^3 + 1$	9	2	1	0	−7

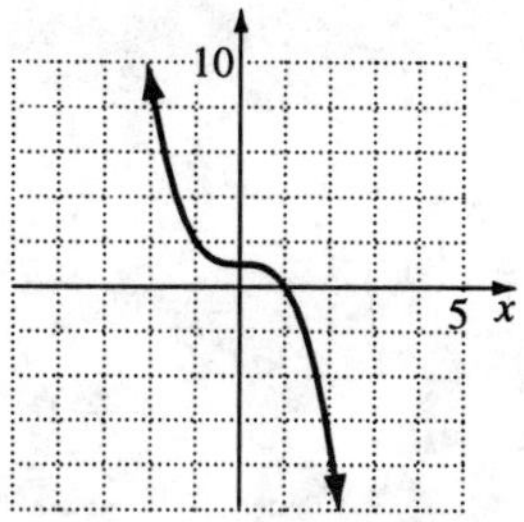

Problem Set 2.6

1. $|x| = 8$

$x = 8$ or -8

$\{-8, 8\}$

3. $|x| = -8$

$\varnothing$

5. $|x - 2| = 7$

$x - 2 = 7$ or $x - 2 = -7$

$x = 9$ or $x = -5$

$\{-5, 9\}$

7. $|x + 3| = 0$

$x + 3 = 0$

$x = -3$

$\{-3\}$

9. $|2x - 1| = 5$

$2x - 1 = 5$ or $2x - 1 = -5$

$2x = 6 \qquad 2x = -4$

$x = 3 \qquad x = -2$

$\{-2, 3\}$

11. $\left|\frac{4x - 2}{3}\right| = 2$

$\frac{4x - 2}{3} = 2$ or $\frac{4x - 2}{3} = -2$

$4x - 2 = 6 \qquad 4x - 2 = -6$

$4x = 8 \qquad 4x = -4$

$x = 2 \qquad x = -1$

$\{-1, 2\}$

13. $2|x+3|-1=13$
$2|x+3|=14$
$|x+3|=7$
$x+3=7$ or $x+3=-7$
$x=4 \qquad x=-10$
$\{-10, 4\}$

15. $|2x-4|+5=2$
$|2x-4|=-3$
$\varnothing$

17. $3|2x-5|-8=-2$
$3|2x-5|=6$
$|2x-5|=2$
$2x-5=2$ or $2x-5=-2$
$2x=7 \qquad 2x=3$
$x=\frac{7}{2} \qquad x=\frac{3}{2}$
$\left\{\frac{3}{2}, \frac{7}{2}\right\}$

19. $|2x+2|=|x+2|$
$2x+2=x+2$ or $2x+2=-(x+2)$
$2x=x \qquad 2x+2=-x-2$
$x=0 \qquad 3x=-4$
$x=-\frac{4}{3}$
$\left\{-\frac{4}{3}, 0\right\}$

21. $|3x-3|=|x+4|$
$3x-3=x+4$ or $3x-3=-(x+4)$
$2x=7 \qquad 3x-3=-x-4$
$x=\frac{7}{2} \qquad 4x=-1$
$x=-\frac{1}{4}$
$\left\{-\frac{1}{4}, \frac{7}{2}\right\}$

23. $|2x-4|=|2x+3|$
$2x-4=2x+3$ or $2x-4=-(2x+3)$
$-4=3$ False $\qquad 2x-4=-2x-3$
$4x=1$
$x=\frac{1}{4}$
$\left\{\frac{1}{4}\right\}$

25. $\left|\frac{2}{3}x-2\right|=\left|\frac{1}{3}x+3\right|$
$\frac{2}{3}x-2=\frac{1}{3}x+3$ or $\frac{2}{3}x-2=-\left(\frac{1}{3}x+3\right)$
$\frac{1}{3}x=5 \qquad \frac{2}{3}x-2=-\frac{1}{3}x-3$
$x=15 \qquad x=-1$
$\{-1, 15\}$

27. $|8x+10|=|2(4x+5)|$
$8x+10=2(4x+5)$ or $8x+10=-2(4x+5)$
$8x+10=8x+10 \qquad 8x+10=-8x-10$
$10=10$ True $\qquad 16x=-20$
$x=-\frac{5}{4}$
$\{x|x \in R\}$

29. $|x|<4$
$-4<x<4$
$\{x|-4<x<4\}$
$(-4, 4)$

31. $|x|\ge 4$
$x\ge 4$ or $x\le -4$
$\{x|x\le -4 \text{ or } x\ge 4\}$
$(-\infty, -4]\cup[4, \infty)$

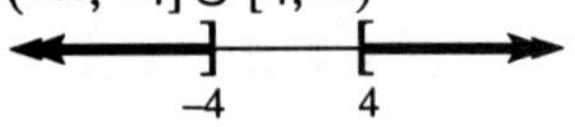

33. $|x-2|<1$
$-1<x-2<1$
$1<x<3$
$\{x|1<x<3\}$
$(1, 3)$

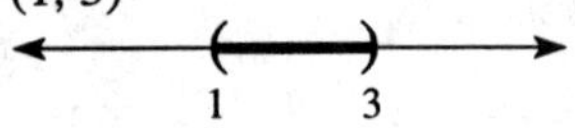

35. $|x+2|\le 1$
$-1\le x+2\le 1$

$-3 \le x \le -1$

$\{x \mid -3 \le x \le -1\}$

$[-3, -1]$

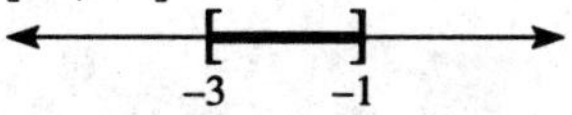

37. $|x+3| > 1$

$x + 3 > 1$ or $x + 3 < -1$

$x > -2 \qquad x < -4$

$\{x \mid x < -4 \text{ or } x > -2\}$

$(-\infty, -4) \cup (-2, \infty)$

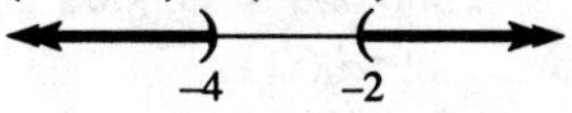

39. $|x-4| \ge 2$

$x - 4 \ge 2$ or $x - 4 \le -2$

$x \ge 6 \qquad x \le 2$

$\{x \mid x \le 2 \text{ or } x \ge 6\}$

$(-\infty, 2] \cup [6, \infty)$

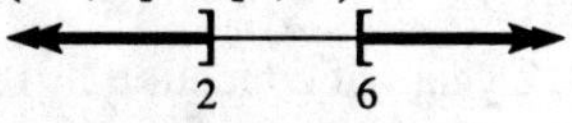

41. $|2y-6| < 8$

$-8 < 2y - 6 < 8$

$-2 < 2y < 14$

$-1 < y < 7$

$\{y \mid -1 < y < 7\}$

$(-1, 7)$

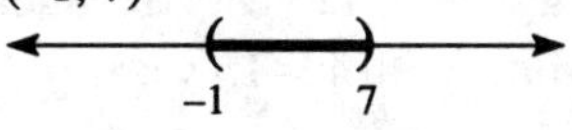

43. $|2(x+1)+3| \le 5$

$-5 \le 2(x+1) + 3 \le 5$

$-5 \le 2x + 2 + 3 \le 5$

$-5 \le 2x + 5 \le 5$

$-10 \le 2x \le 0$

$-5 \le x \le 0$

$\{x \mid -5 \le x \le 0\}$

$[-5, 0]$

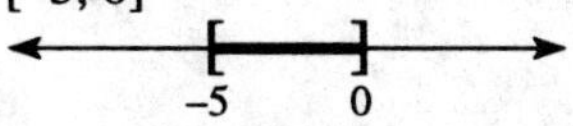

45. $\left|\frac{2(3x-1)}{3}\right| < \frac{1}{6}$

$-\frac{1}{6} < \frac{2(3x-1)}{3} < \frac{1}{6}$

$-1 < 4(3x-1) < 1$

$-1 < 12x - 4 < 1$

$3 < 12x < 5$

$\frac{1}{4} < x < \frac{5}{12}$

$\left\{x \,\middle|\, \frac{1}{4} < x < \frac{5}{12}\right\}$

$\left(\frac{1}{4}, \frac{5}{12}\right)$

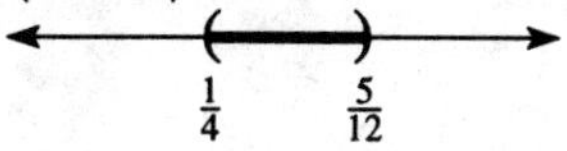

47. $|3x-1| > 13$

$3x - 1 > 13$ or $3x - 1 < -13$

$3x > 14 \qquad 3x < -12$

$x > \frac{14}{3} \qquad x < -4$

$\left\{x \,\middle|\, x < -4 \text{ or } x > \frac{14}{3}\right\}$

$(-\infty, -4) \cup \left(\frac{14}{3}, \infty\right)$

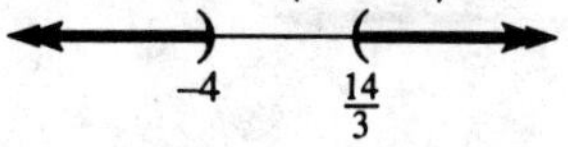

49. $|2(x-3)+3(x+2)-17| \ge 13$

$|2x-6+3x+6-17| \ge 13$

$|5x-17| \ge 13$

$5x - 17 \ge 13$ or $5x - 17 \le -13$

$5x \ge 30 \qquad 5x \le 4$

$x \ge 6 \qquad x \le \frac{4}{5}$

$\left\{x \,\middle|\, x \le \frac{4}{5} \text{ or } x \ge 6\right\}$

$\left(-\infty, \frac{4}{5}\right] \cup [6, \infty)$

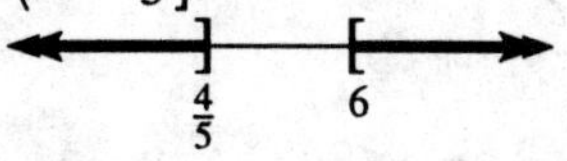

51. $\left|\frac{2y+2}{4}\right| > 2$

$\frac{2y+2}{4} > 2$ or $\frac{2y+2}{4} < -2$

$2y+2 > 8$ $\quad$ $2y+2 < -8$

$2y > 6$ $\quad$ $2y < -10$

$y > 3$ $\quad$ $y < -5$

$\{y|y < -5 \text{ or } y > 3\}$

$(-\infty, -5) \cup (3, \infty)$

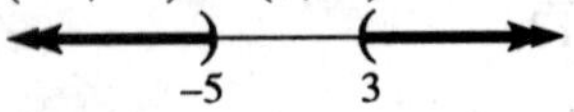

53. $\left|\frac{7x-2}{4}\right| \geq \frac{5}{4}$

$\frac{7x-2}{4} \geq \frac{5}{4}$ or $\frac{7x-2}{4} \leq -\frac{5}{4}$

$7x-2 \geq 5$ $\quad$ $7x-2 \leq -5$

$7x \geq 7$ $\quad$ $7x \leq -3$

$x \geq 1$ $\quad$ $x \leq -\frac{3}{7}$

$\left\{x \middle| x \leq -\frac{3}{7} \text{ or } x \geq 1\right\}$

$\left(-\infty, -\frac{3}{7}\right] \cup [1, \infty)$

55. $|x+2| + 9 \leq 16$

$|x+2| \leq 7$

$-7 \leq x+2 \leq 7$

$-9 \leq x \leq 5$

$\{x|-9 \leq x \leq 5\}$

$[-9, 5]$

57. $2|2x-3| + 10 > 12$

$2|2x-3| > 2$

$|2x-3| > 1$

$2x-3 > 1$ or $2x-3 < -1$

$2x > 4$ $\quad$ $2x < 2$

$x > 2$ $\quad$ $x < 1$

$\{x|x < 1 \text{ or } x > 2\} = (-\infty, 1) \cup (2, \infty)$

59. $|x-2| < -1$

$\varnothing$

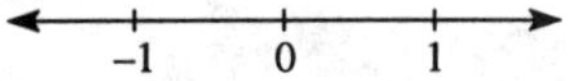

61. $|x-2| > -1$

True for all x

$\{x|-\infty < x < \infty\} = (-\infty, \infty)$

63. $|x - 2{,}560{,}000| \leq 135{,}000$

$-135{,}000 \leq x - 2{,}560{,}000 \leq 135{,}000$

$2{,}425{,}000 \leq x \leq 2{,}695{,}000$

High: 2,695,000 barrels;

Low: 2,425,000 barrels

65. $|2x-30| \leq 30$

$-30 \leq 2x - 30 \leq 30$

$0 \leq 2x \leq 60$

$0 \leq x \leq 30$

Adults with 12 years of education in 1990, adults with 13–15 years of education in 1987 or 1990, all adults with 16 or more years of education

67. $|x - 39.8| = 4$

$x - 39.8 = 4$ or $x - 39.8 = -4$

$x = 43.8$ $\quad$ $x = 35.8$

Adults with less than 12 years of education in 1974 or adults with 13–15 years of education in 1974

69. c is true.

71. Answers may vary.

73. $|x| = x$ when $x \geq 0$

$\{x|x \geq 0\}$

75. $|x+1| = -(x+1)$ when $x + 1 < 0$, $x < -1$

$\{x|x < -1\}$

77. $x-|x|=5$
$x-5=|x|$
$\varnothing$

79. $|x+2|\ge 4x$
$x+2\ge 4x$ or $x+2\le -4x$
$-3x\ge -2$　　$5x\le -2$
$x\le \frac{2}{3}$　　$x\le -\frac{2}{5}$
$\left\{x \middle| x\le \frac{2}{3}\right\}$

Review Problems

81. 5 "width" pieces of size w
2 "height" pieces of size $w+3$
$P=2(w+3)+5w=2w+6+5w=7w+6$
$41=7w+6$
$35=7w$
$w=5$
The bookcase will be
5 feet wide × 8 feet tall.

82. $2(x-3)=4[x-(5x-2)]$
$2x-6=4[x-5x+2]$
$2x-6=4[2-4x]$
$2x-6=8-16x$
$18x=14$
$x=\frac{14}{18}=\frac{7}{9}$
$\left\{\frac{7}{9}\right\}$

83. $-2[3a-(2b-1)-5a]+b$
$=-2[3a-2b+1-5a]+b$
$=-2[-2a-2b+1]+b$
$=4a+4b-2+b$
$=4a+5b-2$

Problem Set 2.7

1. $2x-y>4$; $(2,-4)$
$2(2)-(-4)>4$
$4+4=8>4$　True
Yes, $(2,-4)$ is a solution.

3. $3x-4y\le 12$; $(6,-2)$
$3(6)-4(-2)\le 12$
$18+8=26\le 12$　False
No, $(6,-2)$ is not a solution.

5. $x+y\ge 2$
1. Graph $x+y=2$.
The graph is indicated by a solid line since equality is included ($\ge$).
2. Select a test point not on the line: (0, 0)
$0+0\ge 2$
$0\ge 2$　False
3. Shade the half-plane not containing (0, 0).

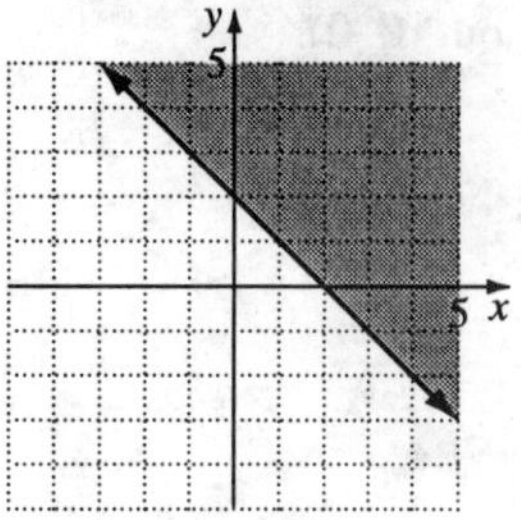

7. $3x-y\ge 6$
1. Graph $3x-y=6$ with a solid line ($\ge$).
2. Test point: (0, 0)
$0-0\ge 6$
$0\ge 6$　False
3. Shade the half-plane not containing (0, 0).

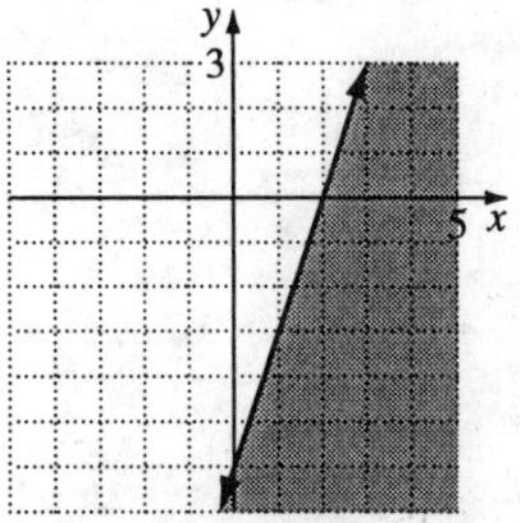

9. $2x+3y>12$
1. Graph $2x+3y=12$ with a dashed line (>).
2. Test point: (0, 0)
$0+0>12$
$0>12$ False
3. Shade the half-plane not containing (0, 0).

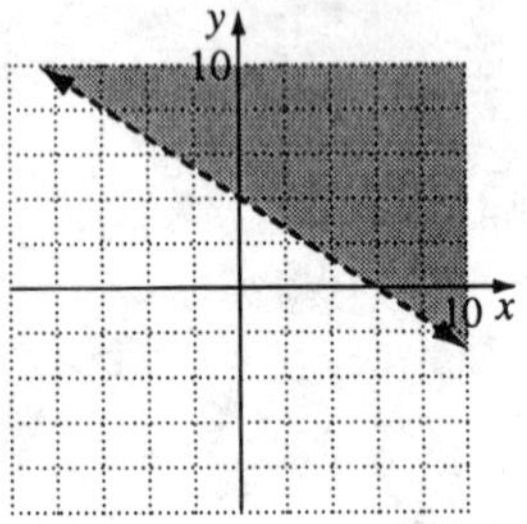

11. $5x+3y\le -15$
Since "≤," graph $5x+3y=-15$ with a solid line.
Test (0, 0):
$5(0)+3(0)=0\le -15$ is false, so shade the half plane without (0, 0).

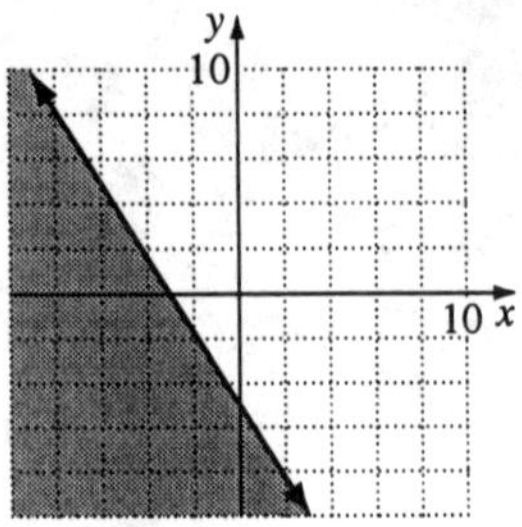

13. $2y-3x>6$
Since ">," use dashed line.
Test (0, 0): $2(0)-3(0)=0>6$ is false, so shade half plane without (0, 0).

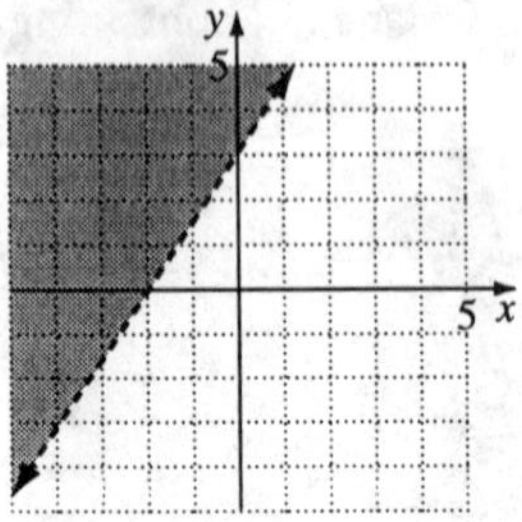

15. $5x+y>3x-2$
Since ">" use dashed line.
Test (0, 0):
$5(0)+0>3(0)-2$
$0>-2$
True, so shade half-plane with (0, 0).

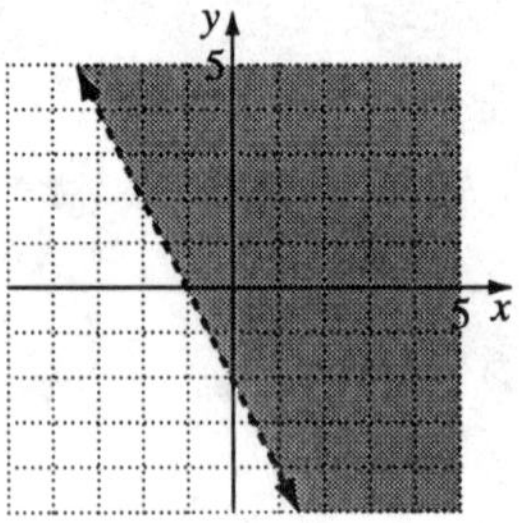

17. $\frac{x}{2}+\frac{y}{3}<1$
Since <, use dashed line.
Test (0, 0):
$\frac{0}{2}+\frac{0}{3}=0<1$ is true, so shade half-plane with (0, 0).

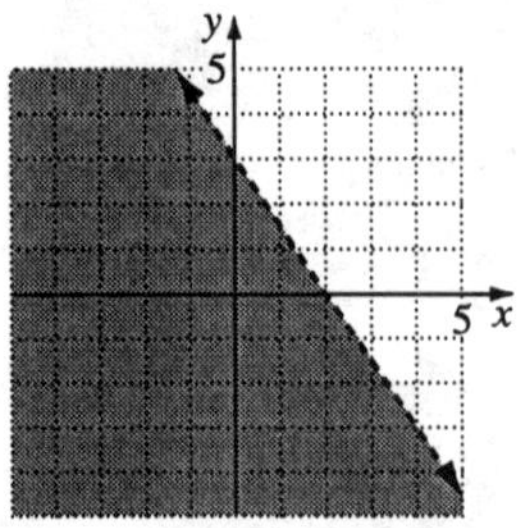

19. $y>2x-1$
Slope $m=2$
y-intercept $b=-1$
Shade above dashed line.

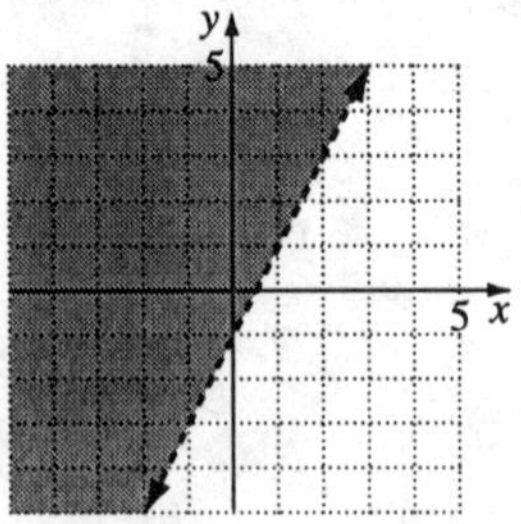

21. $y \geq \frac{2}{3}x - 1$

$m = \frac{2}{3}$, $b = -1$

Shade above solid line.

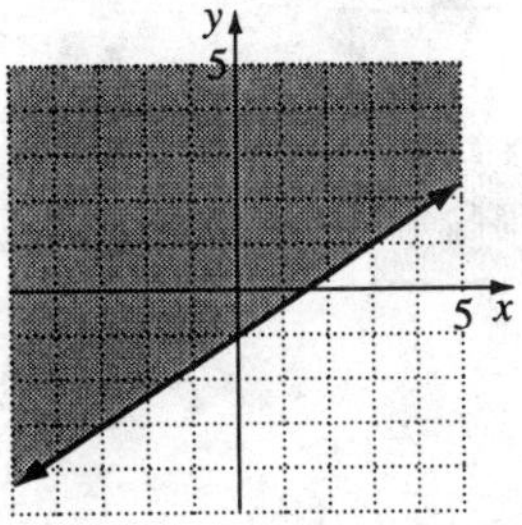

23. $y < x$

1. Graph $y = x$ with a dashed line.
2. Test point, (1, 0): $0 < 1$ True
3. Shade the half-plane containing (1, 0).

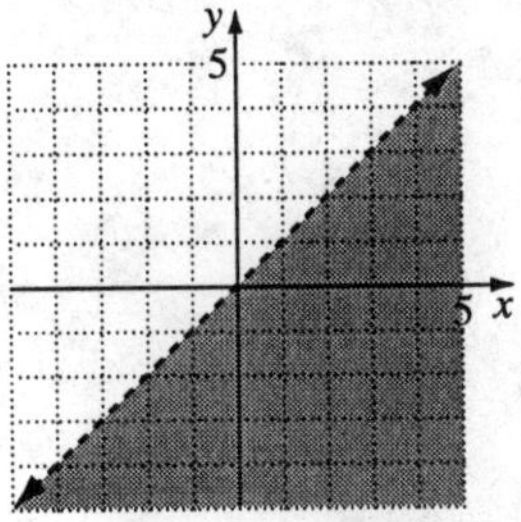

25. $y > -\frac{1}{2}x$

$m = -\frac{1}{2}$, $b = 0$

Shade above dashed line.

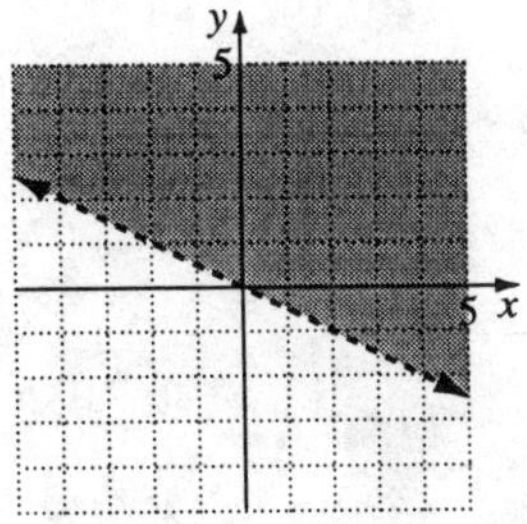

27. $x \leq 1$

1. Graph $x = 1$ with a solid line.
2. Shade half-plane left of $x = 1$.

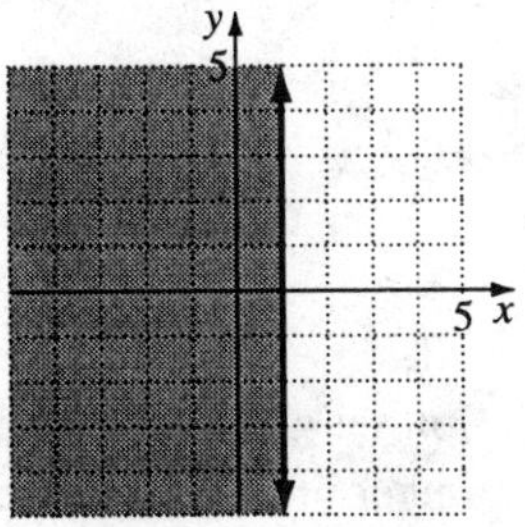

29. $y > 1$

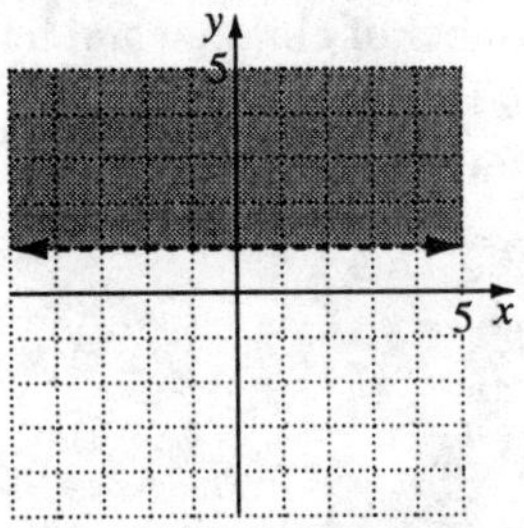

31. $x \geq 0$

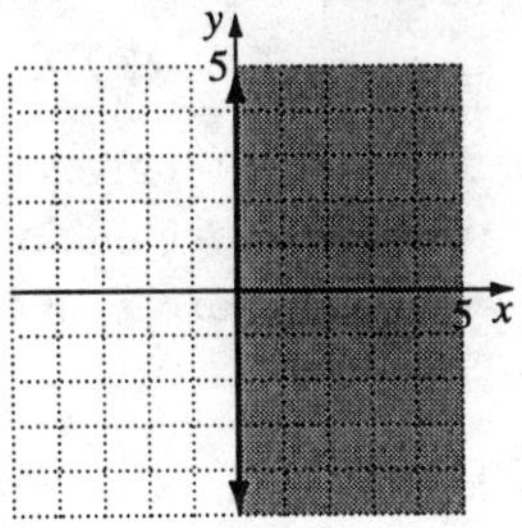

33. $x > -5$

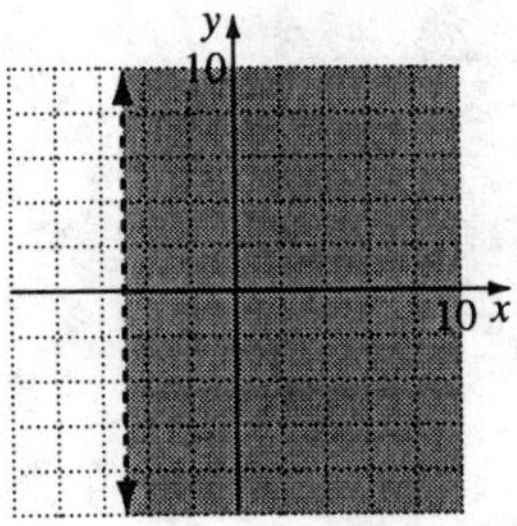

35. $165x + 110y \le 330$ in Quadrant I.

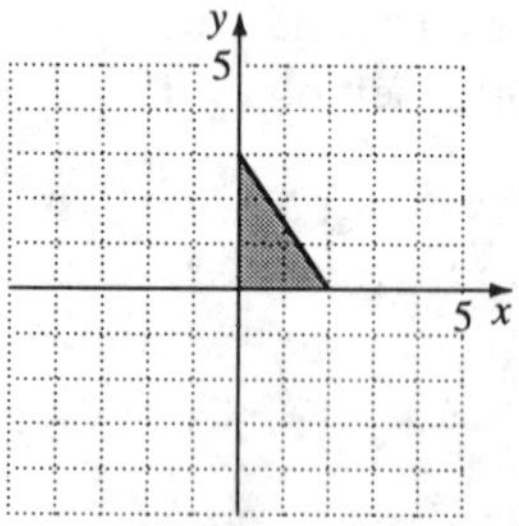

(x, y) in the solution set gives the number of eggs x and the number of ounces of meat y that satisfy the dietary requirements.

37. Let x be the number of children and let y be the number of adults.
$50x + 150y > 2000$ results in an overload.

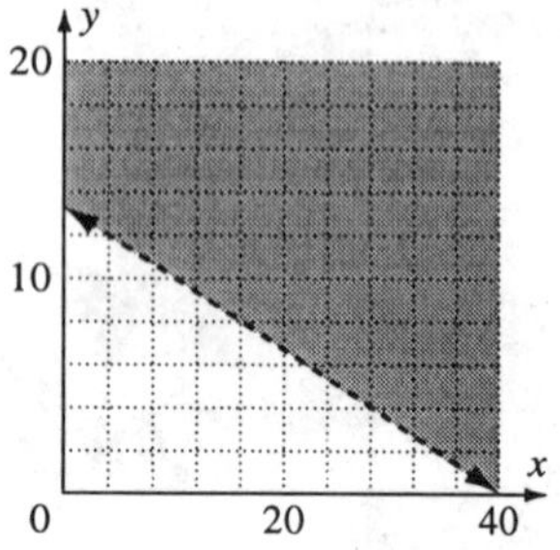

Values (x, y) in the solution set gives a combination of the number of children and adults whose weight would exceed the elevator's stated capacity.

39. b is true.

41. $y \le 6 - 1.5x$

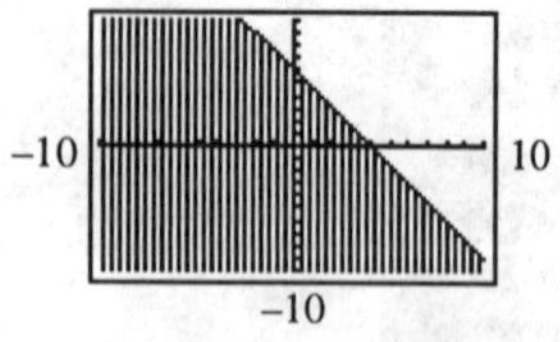

43. $3x + 6y \le 6$
$6y \le -3x + 6$
$y \le -\frac{1}{2}x + 1$

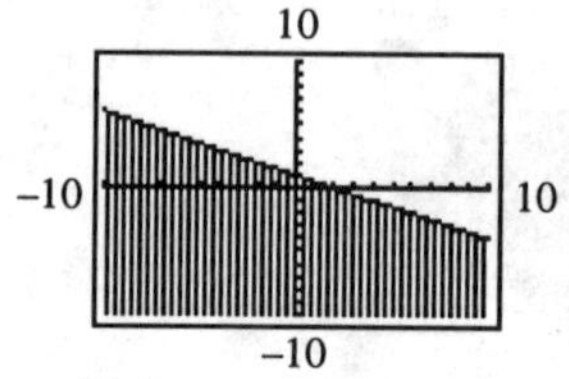

45. Students should verify.

47. Answers may vary.

49. $x - y \ge 3$ and $x + y \le 5$

$x - 3 \ge y$
$y \le x - 3$
Shade plane without (0, 0)

$y \le -x + 5$
Shade half plane with (0, 0)

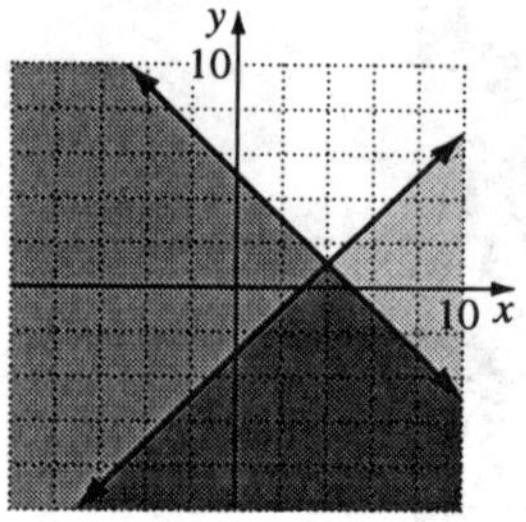

51. $y \ge 2x + 1$ and $y \le 2x + 3$

Shade half-plane without (0, 0).

Shade half-plane with (0, 0).

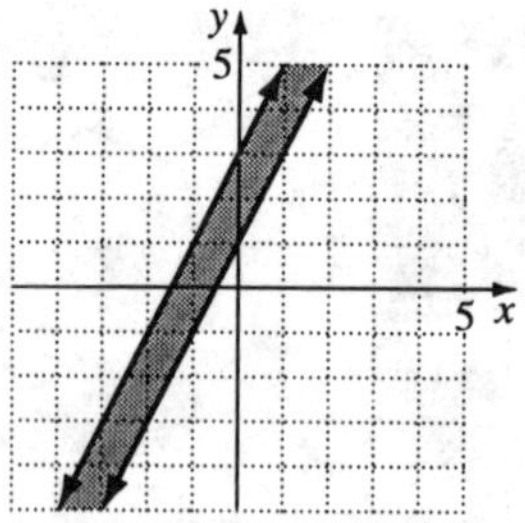

53. $-2 < x \le 5$

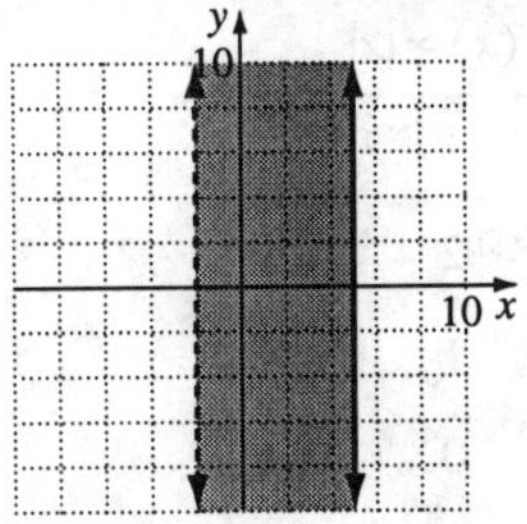

Review Problems

54. $3(x-2) > 5x+8$
$3x-6 > 5x+8$
$-14 > 2x$
$-7 > x$
$x < -7$
$\{x|x < -7\}$

55. Solve $d(f+w) = fl$ for f.
$df + dw = fl$
$df - fl = -dw$
$f(d-l) = -dw$
$f = \frac{-dw}{d-l}$
$f = \frac{dw}{l-d}$

56. $-2[3x-(2y-3)-7x]+4y$
$= -2[3x-2y+3-7x]+4y$
$= -2[-4x-2y+3]+4y$
$= 8x+4y-6+4y$
$= 8x+8y-6$

Chapter 2 Review Problems

1. $\{(2, 7), (3, 7), (5, 7)\}$
Function, since each first component is assigned exactly one second component.
Domain = $\{2, 3, 5\}$
Range = $\{7\}$

2. $\{(1, 10), (2, 500), (13, \pi)\}$
Function, since each first component is assigned exactly one second component.
Domain = $\{1, 2, 13\}$
Range = $\{10, 500, \pi\}$

3. $\{(1, 2), (3, 4), (5, 6)\}$
Function, since each first component is assigned exactly one second component.
Domain = $\{1, 3, 5\}$
Range = $\{2, 4, 6\}$

4. $\{(12, 13), (14, 15), (12, 19)\}$
Not a function, since (12, 13) and (12, 19) have the same first component.
Domain = $\{12, 14\}$
Range = $\{13, 15, 19\}$

5. $f(x) = -7x+5$

a. $f(0) = -7(0) + 5 = 5$

b. $f(-2) = -7(-2) + 5 = 14 + 5 = 19$

c. $f(5) = -7(5) + 5 = -35 + 5 = -30$

d. $f\left(\frac{3}{7}\right) = -7\left(\frac{3}{7}\right) + 5 = -3 + 5 = 2$

e. $f(b+3) = -7(b+3) + 5 = -7b - 21 + 5$
$= -7b - 16$

6. $g(x) = 3x^2 - 5x + 2$

a. $g(0) = 3(0)^2 - 5(0) + 2 = 2$

b. $g(-2) = 3(-2)^2 - 5(-2) + 2$
$= 12 + 10 + 2 = 24$

c. $g(4) = 3(4)^2 - 5(4) + 2$
$= 48 - 20 + 2 = 30$

d. $g\left(\frac{1}{2}\right) = 3\left(\frac{1}{2}\right)^2 - 5\left(\frac{1}{2}\right) + 2 = \frac{1}{4}$

e. $g(3a) = 3(3a)^2 - 5(3a) + 2$
$= 3(9a^2) - 15a + 2 = 27a^2 - 15a + 2$

7.

x	$f(x)=x^2$
–2	4
–1	1
0	0
1	1
2	4

x	$g(x)=x^2+2$
–2	6
–1	3
0	2
1	3
2	6

x	$h(x)=x^2-1$
–2	3
–1	0
0	–1
1	0
2	3

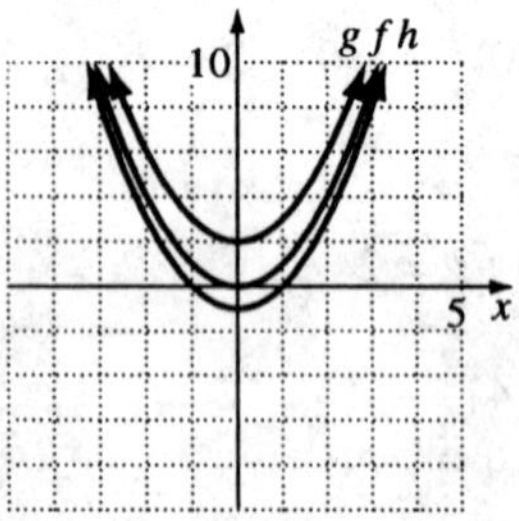

$f(x) \pm c$ is $f(x)$ shifted up (+) or down (–) c units ($c > 0$).

8.

x	$f(x)=\lvert x\rvert$
–3	3
–2	2
–1	1
0	0
1	1
2	2
3	3

$g(x)=|x|+2$
$h(x)=|x|-1$

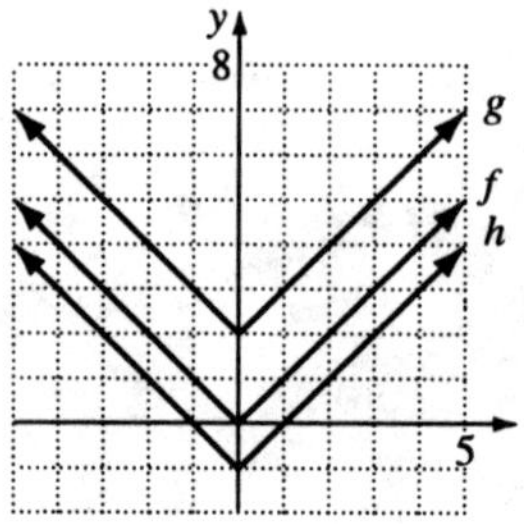

9.

x	$f(x)=\sqrt{x}$
0	0
1	1
4	2
9	3

x	$g(x)=\sqrt{x+2}$
–2	$\sqrt{0}=0$
–1	$\sqrt{1}=1$
2	$\sqrt{4}=2$
7	$\sqrt{9}=3$

x	$h(x)=\sqrt{x-1}$
1	$\sqrt{0}=0$
2	$\sqrt{1}=1$
5	$\sqrt{4}=2$
10	$\sqrt{9}=3$

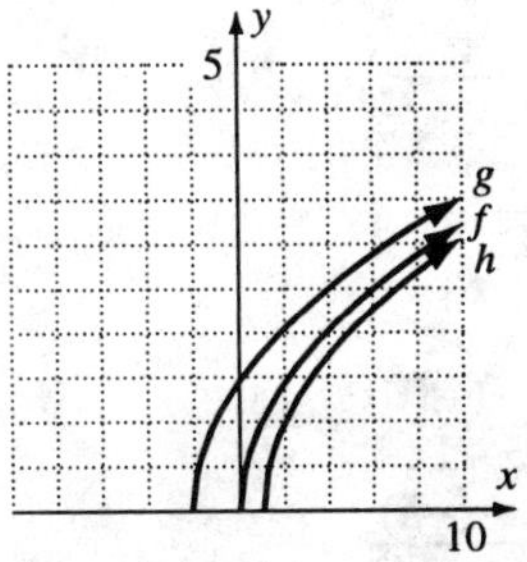

For $c>0$, $f(x+c)$ is $f(x)$ shifted left c units and $f(x-c)$ is $f(x)$ shifted right c units.

10.

x	$f(x)=x^2-3x$
−2	$(-2)^2-3(-2)=4+6=10$
−1	$(-1)^2-3(-1)=1+3=4$
0	$0^2-3(0)=0$
1	$1^2-3(1)=1-3=-2$
2	$2^2-3(2)=4-6=-2$
3	$3^2-3(3)=9-9=0$
4	$4^2-3(4)=16-12=4$
5	$5^2-3(5)=25-15=10$

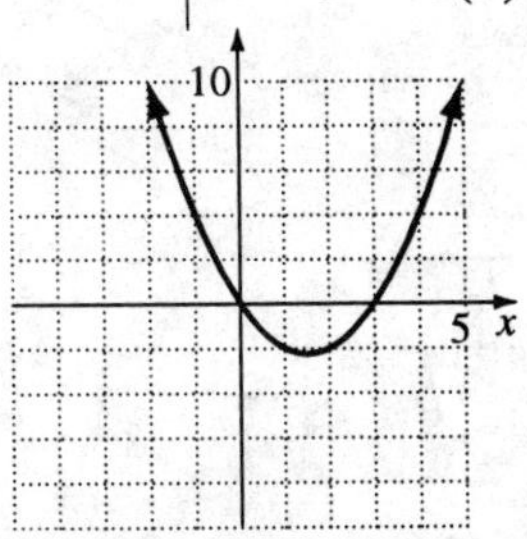

11. Yes, it is a function. Each state has only one number that represents the members of the U.S. Senate.

12. Not a function

13. Function

14. Not a function

15. Function

16. Not a function

17. $f(x)=0.0234x^2-0.5029x+12.5$
$f(10)=0.0234(10^2)-0.5029(10)+12.5$
$=9.811$
$f(10)=9.811$ is the number of infant deaths (per thousand live births) in 1990.

18. a. $f(x)=0.0091x^3+0.1354x^2+2.1336x+83.2653$
$f(39)=\$912.22$ is the weekly salary in 2000.

b. $f(29)-f(19)\approx 480-250$
$=\$230$ is the increase in weekly salary between 1980 and 1990.

19. a. Because each year maps into one and only one violent crime rate.

b. Maximum violent crime rate was 35 (per thousand people) in 1981.

c. Minimum violent crime rate was approximately 28 per thousand in 1986.

d. One cannot expect to accurately predict the value in 2009, 17 years beyond the range of the data and 34 years after 1975 from the information given. Based on 1975 to 1992, one might argue the crime rate is relatively constant, and predict the 2009 rate as between 28 and

38, say. Or one could use the 1985-1992 data and predict a huge crime rate if the linear trend continued, or, one might observe that the data appear to be somewhat cyclical or periodic, and predict the 2009 rate based on this pattern continuing.

20. **a.**

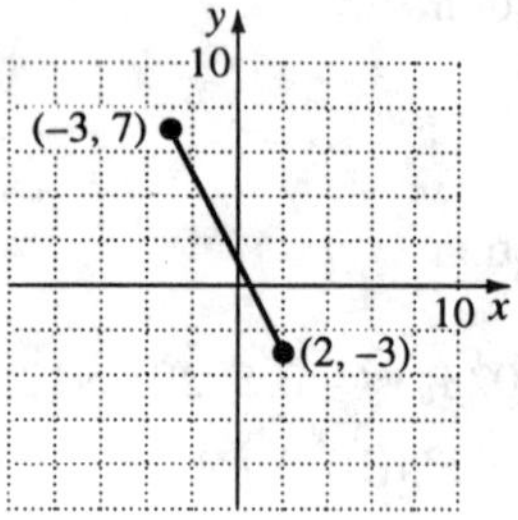

b. $m = \dfrac{7-(-3)}{-3-2} = \dfrac{10}{-5} = -2$

c. $m = \dfrac{-3-7}{2-(-3)} = \dfrac{-10}{5} = -2$
Same slope

d. Choice of points will vary; $m = -2$.

21. **a.** $f(x) = 60$ (percent)

b. Answers may vary.

c. $g(x) = 45$ (percent)

22. Using estimates (1950, 8.3), (1955, 68), (1960, 82), (1990, 98.2), (1995, 98.2)
$m_{1950-1995} = \dfrac{98.2-8.3}{1995-1950} = \dfrac{89.9}{45} \approx 2$
The percentage of households with TV's increased by an average of 2% per year over the 45 years.
$m_{1955-1960} = \dfrac{82-68}{1960-1955} = \dfrac{14}{5} = 2.8$
From 1955 to 1960, the percentage of homes with TV's increased an average of 2.8% per year.
$m_{1990-1995} = \dfrac{0}{5} = 0$
There was virtually no change in the percentage of households with TV's during the period 1990–1995.

23. Label points $A(12, -10)$, $B(3, 4)$, $C(-2, -19)$, $D(-11, -5)$.

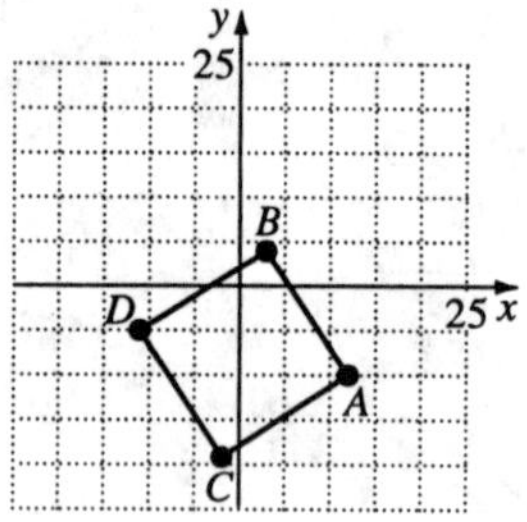

a. $m_{AB} = \dfrac{4-(-10)}{3-12} = \dfrac{14}{-9}$
$m_{BD} = \dfrac{4-(-5)}{3-(-11)} = \dfrac{9}{14}$
$m_{CD} = \dfrac{-5-(-19)}{-11-(-2)} = \dfrac{14}{-9}$
$m_{AC} = \dfrac{-10-(-19)}{12-(-2)} = \dfrac{9}{14}$
Since AB is parallel to CD and BD is parallel to AC, this is a parallelogram. (Their slopes are the same.)

b. This is a rectangle since adjacent sides are perpendicular: AB is perpendicular to BD since
$\left(\dfrac{14}{-9}\right)\left(\dfrac{9}{14}\right) = -1.$

24. $y = \dfrac{1}{2}x - 4$

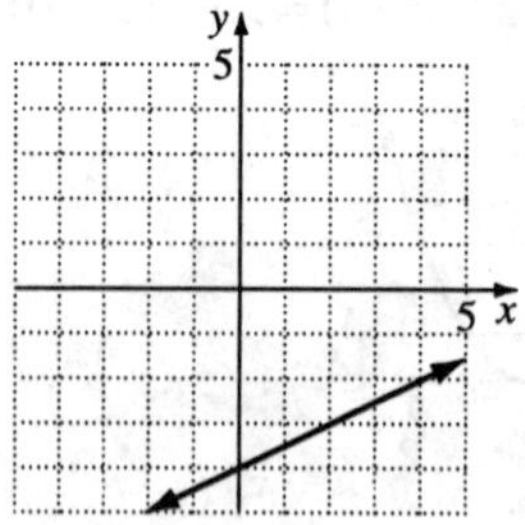

25. $f(x) = -\frac{3}{4}x + 5$

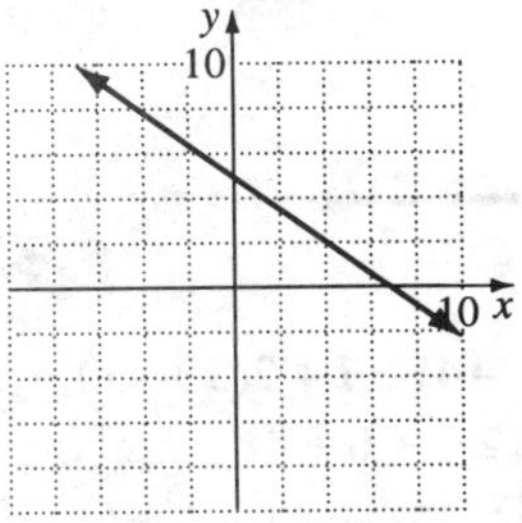

26. $2x - 3y = 6$
$3y = 2x - 6$
$y = \frac{2}{3}x - 2$

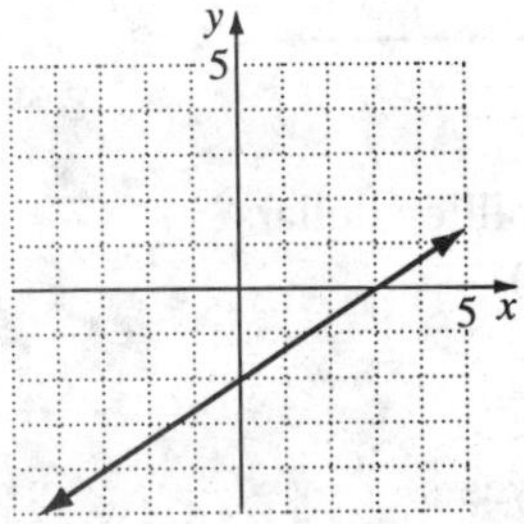

27. $x - 3 = 2$
$x = 5$

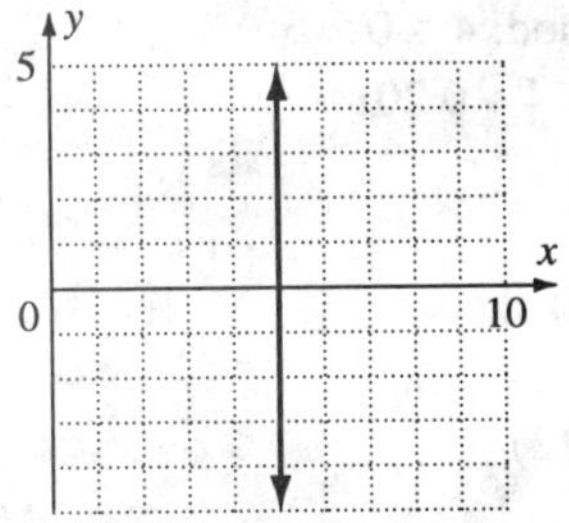

28. $y - (-3) = -2(x - 1)$ or $y + 3 = -2(x - 1)$
$y = -2x + 2 - 3$ or $y = -2x - 1$
$2x + y = -1$

29. $m = \frac{18-3}{-4-1} = \frac{15}{-5} = -3$
$y - 3 = -3(x - 1)$ or $y - 18 = -3(x + 4)$
$y = -3x + 3 + 3$ or $y = -3x + 6$
$3x + y = 6$

30. $2x - 3y = 6$
$3y = 2x - 6$
$y = \frac{2}{3}x - 2$ so $m = \frac{2}{3}$
$y - (-4) = \frac{2}{3}(x - (-1))$ or $y + 4 = \frac{2}{3}(x + 1)$
$y = \frac{2}{3}x + \frac{2}{3} - 4$ or $y = \frac{2}{3}x - \frac{10}{3}$
$\frac{2}{3}x - y = \frac{10}{3}$ or $2x - 3y = 10$

31. $2x - 4y = 8$
$4y = 2x - 8$
$y = \frac{1}{2}x - 2$ so $m = \frac{1}{2}$
A line perpendicular to it would have $m = -2$ so $y - (-3) = -2(x - 2)$ or $y + 3 = -2(x - 2)$.
$y = -2x + 4 - 3$ or $y = -2x + 1$
$2x + y = 1$

32. a. $m = \frac{16-12.1}{1900-1930} = \frac{3.9}{-30} = -0.13$
$y - 16 = -0.13(x - 1900)$
or $y - 12.1 = -0.13(x - 1930)$

b. $y = -0.13x + 247 + 16$
$y = -0.13x + 263$

c. $f(1970) = -0.13(1970) + 263$
$= 6.9$ feet in 1970
$f(1980) = -0.13(1980) + 263$
$= 5.6$ feet in 1980

d. $f(1990) = -0.13(1990) + 263$
$= 4.3$ feet in 1990
No; this is probably unreasonably short. A reasonable range for x would be $1900 \le x \le 1980$.

33. $2x + 4 > 7$
$2x > 3$
$x > \frac{3}{2}$
$\left\{x \middle| x > \frac{3}{2}\right\}$

$\left(\frac{3}{2}, \infty\right)$

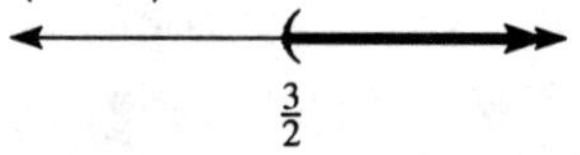

34. $-6x + 3 \le 15$
$-6x \le 12$
$x \ge -2$
$\{x|x \ge -2\}$
$[-2, \infty)$

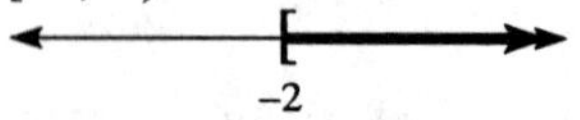

35. $6x - 9 \ge -4x - 3$
$10x \ge 6$
$x \ge \frac{3}{5}$
$\left\{x|x \ge \frac{3}{5}\right\}$
$\left[\frac{3}{5}, \infty\right)$

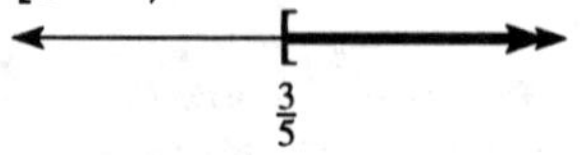

36. $\frac{x}{4} - \frac{1}{2} < \frac{x}{2} + \frac{1}{4}$
$x - 2 < 2x + 1$
$-3 < x$
$x > -3$
$\{x|x > -3\}$
$(-3, \infty)$

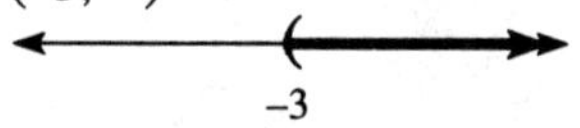

37. $3(x + 5) \le 6(x + 1)$
$3x + 15 \le 6x + 6$
$9 \le 3x$
$x \ge 3$
$\{x|x \ge 3\}$
$[3, \infty)$

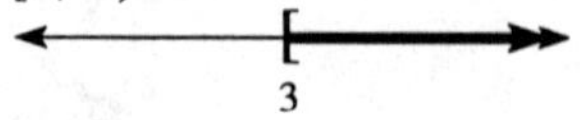

38. $6x + 5 > -2(x - 3) - 25$
$6x + 5 > -2x + 6 - 25$
$8x > 6 - 25 - 5$
$8x > -24$
$x > -3$
$\{x|x > -3\}$
$(-3, \infty)$

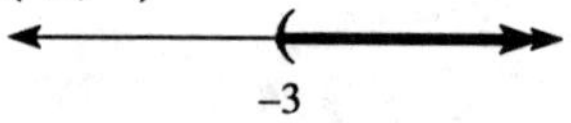

39. $3(2x - 1) - 2(x - 4) \ge 7 + 2(3 + 4x)$
$6x - 3 - 2x + 8 \ge 7 + 6 + 8x$
$4x + 5 \ge 8x + 13$
$-8 \ge 4x$
$x \le -2$
$\{x|x \le -2\}$
$(-\infty, -2]$

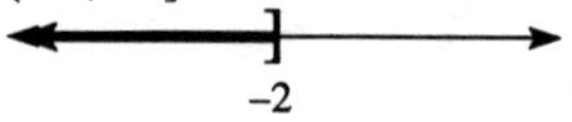

40. $y = 140 + 9x$ million dollars
$140 + 9x > 320$
$9x > 180$
$x > 20$
More than 20 years

41. Let x = number of checks.
first method: $11 + 0.06x$
Second method: $4 + 0.20x$
$11 + 0.06x < 4 + 0.20x$
$-0.14x < -7$
$x > 50$
More than 50 checks

42. $C = \frac{5}{9}(F - 32)$
$15 \le C \le 35$
$15 \le \frac{5}{9}(F - 32) \le 35$
$\frac{9}{5}(15) \le F - 32 \le \frac{9}{5}(35)$
$27 \le F - 32 \le 63$
$59 \le F \le 95$
$59^\circ \le F \le 95^\circ$

43. $A \cap B = \{a, b, c\} \cap \{a, c, d, e\} = \{a, c\}$

44. $A \cap C = \{a, b, c\} \cap \{a, d, f, g\} = \{a\}$

45. $A \cup B = \{a, b, c\} \cup \{a, c, d, e\}$
$= \{a, b, c, d, e\}$

46. $A \cup C = \{a, b, c\} \cup \{a, d, f, g\}$
$= \{a, b, c, d, f, g\}$

47. $x \le 3$ and $x < 6$
$\{x | x \le 3\}$
$(-\infty, 3]$

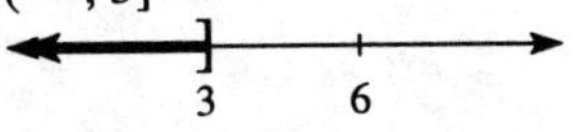

48. $x \le 3$ or $x < 6$
$\{x | x < 6\}$
$(-\infty, 6)$

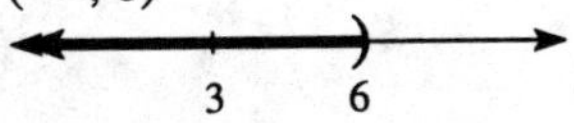

49. $-2x < -12$ and $x - 3 < 5$
$x > 6$ $\quad x < 8$
$\{x | 6 < x < 8\}$
$(6, 8)$

50. $5x + 3 \le 18$ and $2x - 7 \le -5$
$5x \le 15$ $\quad 2x \le 2$
$x \le 5$ $\quad x \le 1$
$\{x | x \le 1\}$
$(-\infty, 1]$

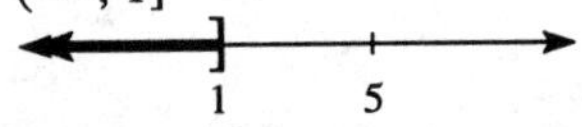

51. $\{x | 2x - 5 > -1\} \cap \{x | 3x < 3\}$
$2x > 4$ $\quad x < 1$
$x > 2$
$\varnothing$

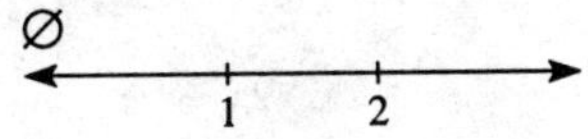

52. $2x - 5 > -1$ or $3x < 3$
$2x > 4$ $\quad x < 1$
$x > 2$
$\{x | x < 1 \text{ or } x > 2\}$
$(-\infty, 1) \cap (2, \infty)$

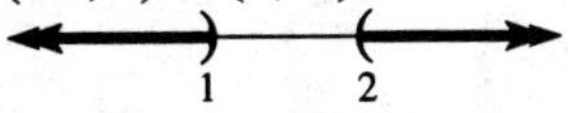

53. $x + 1 \le -3$ or $-4x + 3 < -5$
$x \le -4$ $\quad -4x < -8$
$x > 2$
$\{x | x \le -4 \text{ or } x > 2\}$
$(-\infty, -4] \cup (2, \infty)$

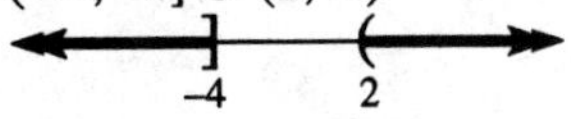

54. $\{x | 5x - 2 \le -22\} \cup \{x | -3x - 2 > 4\}$
$5x \le -20$ $\quad -3x > 6$
$x \le -4$ $\quad x < -2$
$\{x | x < -2\}$
$(-\infty, -2)$

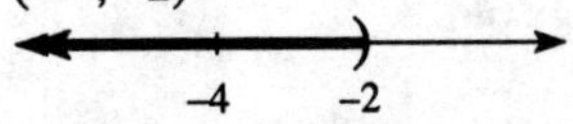

55. $\{x | 5x + 4 \ge -11\} \cup \{x | 1 - 4x \ge 9\}$
$5x \ge -15$ $\quad -8 \ge 4x$
$x \ge -3$ $\quad x \le -2$
$\{x | -\infty < x < \infty\}$
$(-\infty, \infty)$

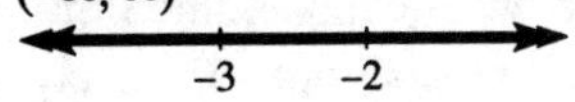

56. $-1 < 4x + 2 \le 6$
$-3 < 4x \le 4$
$-\frac{3}{4} < x \le 1$
$\left\{x \middle| -\frac{3}{4} < x \le 1\right\}$
$\left(-\frac{3}{4}, 1\right]$

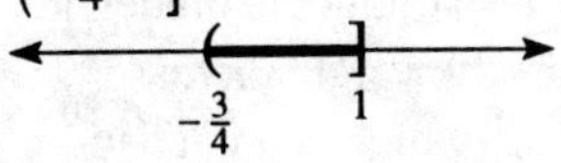

57. $-13 \le 3 - 4x < 13$
$-16 \le -4x < 10$
$4 \ge x > \frac{-10}{4}$
$-\frac{5}{2} < x \le 4$
$\left\{x \middle| -\frac{5}{2} < x \le 4\right\}$
$\left(-\frac{5}{2}, 4\right]$

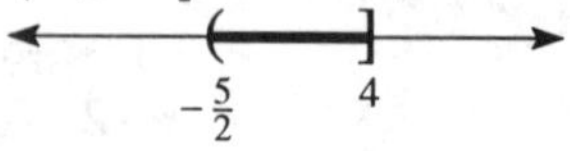

58. $29 \le w \le 53$
$29 \le 6t + 11 \le 53$
$18 \le 6t \le 42$
$3 \le t \le 7$ years old

59. $w \le 65$ and $w > 41$

$6t + 11 \le 65$	$6t + 11 > 41$
$6t \le 54$	$6t > 30$
$t \le 9$	$t > 5$

$5 < t \le 9$ years old

60. $w \le 71$ or $w \le 47$ is $w \le 71$.
$6t + 11 \le 71$
$6t \le 60$
$t \le 10$
$t \le 10$ years old

61. The graph shows the model line fit through the data. The line does not adequately represent the data, which show a curvilinear pattern.

62. $w(t) = 6t + 11$
$w(30) = 6(30) + 11 = 191$
A male 30 years old would weigh 191 pounds. This may give a somewhat realistic response for $t = 30$, but the model indicates continued, steady weight increase throughout life, which is unrealistic.

63. Answers may vary.

64. $|2x + 1| = 7$

$2x + 1 = 7$	or	$2x + 1 = -7$
$2x = 6$	or	$2x = -8$
$x = 3$	or	$x = -4$

$\{-4, 3\}$

65. $|3x + 2| = -5$
$\varnothing$

66. $2|x - 3| - 7 = 10$
$2|x - 3| = 17$
$|x - 3| = \frac{17}{2}$

$x - 3 = \frac{17}{2}$	or	$x - 3 = -\frac{17}{2}$
$x = 11.5$	or	$x = -5.5$

$\{-5.5, 11.5\}$

67. $|4x - 3| = |7x + 9|$

$4x - 3 = 7x + 9$	or	$4x - 3 = -(7x + 9)$
$-12 = 3x$	or	$4x - 3 = -7x - 9$
$-4 = x$	or	$11x = -6$
$x = -4$	or	$x = -\frac{6}{11}$

$\left\{-4, -\frac{6}{11}\right\}$

68. $|2x + 3| \le 15$
$-15 \le 2x + 3 \le 15$
$-18 \le 2x \le 12$
$-9 \le x \le 6$
$\{x | -9 \le x \le 6\}$
$[-9, 6]$

69. $\left|\frac{2x + 6}{3}\right| > 2$

$\frac{2x + 6}{3} > 2$	or	$\frac{2x + 6}{3} < -2$
$2x + 6 > 6$		$2x + 6 < -6$
$2x > 0$		$2x < -12$
$x > 0$		$x < -6$

$\{x | x < -6 \text{ or } x > 0\}$
$(-\infty, -6) \cup (0, \infty)$

70. $|3-2x|<7$
$-7<3-2x<7$
$-10<-2x<4$
$5>x>-2$
$\{x|-2<x<5\}$
$(-2, 5)$

71. $|2x+5|-7\geq -6$
$|2x+5|\geq 1$
$2x+5\geq 1$ or $2x+5\leq -1$
$2x\geq -4$ $\quad$ $2x\leq -6$
$x\geq -2$ $\quad$ $x\leq -3$
$\{x|x\leq -3 \text{ or } x\geq -2\}$
$(-\infty, -3]\cup[-2, \infty)$

72. $|h-6.5|\leq 1$
$-1\leq h-6.5\leq 1$
$5.5\leq h\leq 7.5$
90% of the population sleeps between $5\frac{1}{2}$ and $7\frac{1}{2}$ hours per day (inclusive).

73. $x-3y\leq 6$

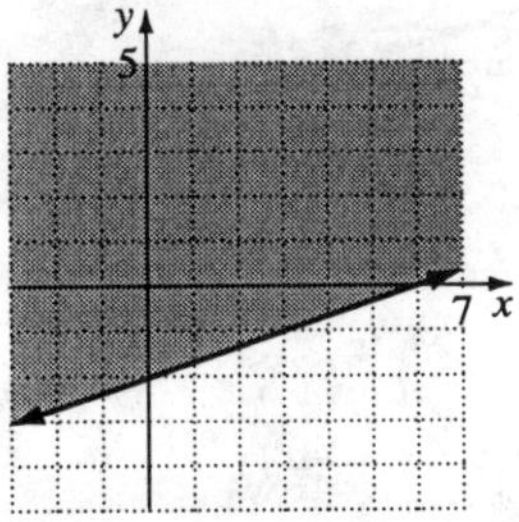

74. $y>-\frac{1}{2}x+3$
shade above line

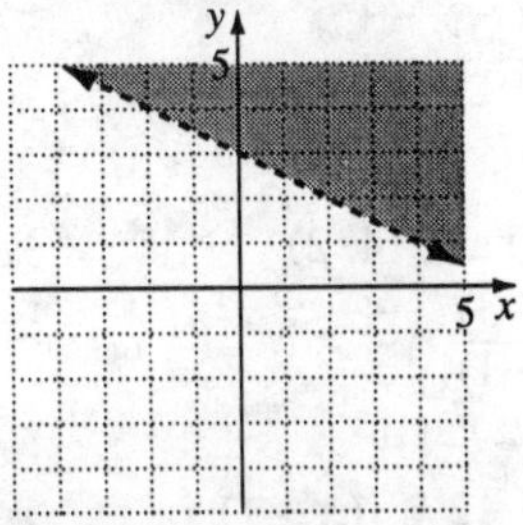

75. $x\leq 2$

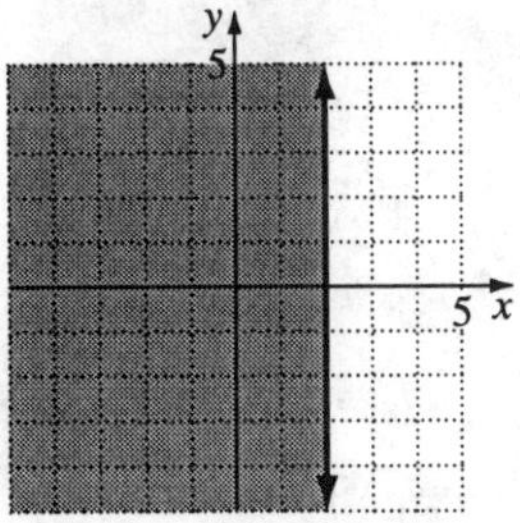

76. $y>-3$

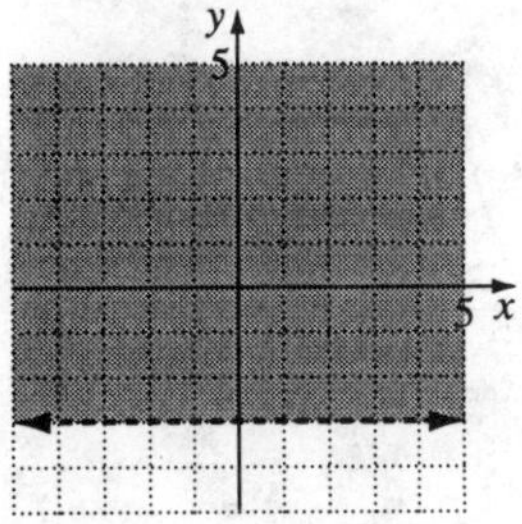

Chapter 2 Test

1. $f(x)=3x^2-7x+3$
$f(2a)=3(2a)^2-7(2a)+3$
$=3(4a^2)-14a+3=12a^2-14a+3$

2. $f(x)=x^2$

x	$f(x)=x^2$
–2	4
–1	1
0	0
1	1
2	4

$g(x) = (x-1)^2$

x	$g(x) = (x-1)^2$
−1	4
0	1
1	0
2	1
3	4

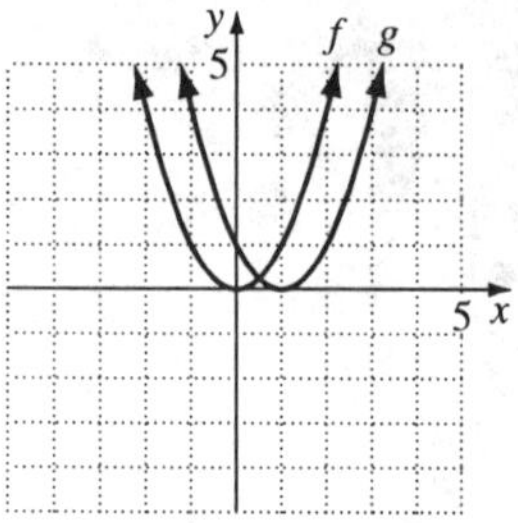

g is f shifted one unit to the right.

3.

x	$f(x) = x^2 + 2x + 1$
−3	4
−2	1
−1	0
0	1
1	4

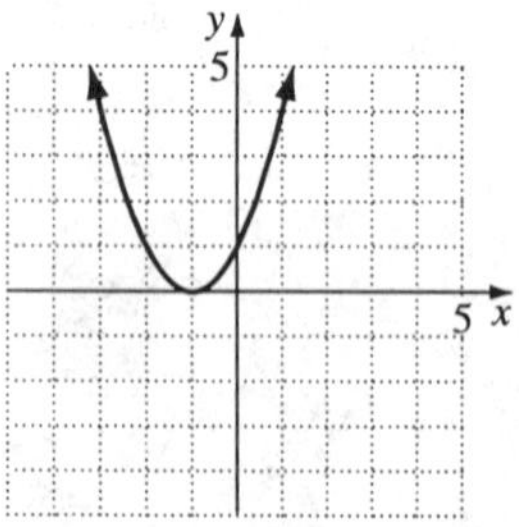

4. b, c
b. has 3 y-coordinates for the same x-coordinate; A vertical line through c passes through more than one point.

5. $f(x) = 0.79x^2 - 2x - 4$
$f(10) = 0.79(10)^2 - 2(10) - 4$
$f(10) = 79 - 20 - 4$
$f(10) = 55$
There is 55 board feet in a 16-foot log whose diameter is 10 inches.

6. d is true. Normal breathing rate in volume varies from about 180 to 150 or 30 cubic inches

7. $A(1976, 4)$, $B(2016, 8)$
$\text{slope} = \dfrac{8-4}{2016-1976} = \dfrac{4}{40} = \dfrac{1}{10}$
The average change per year in world population from 1976 to 2016 is $\dfrac{1}{10}$ billion increase.

8. $y = -\frac{1}{3}x + 2$

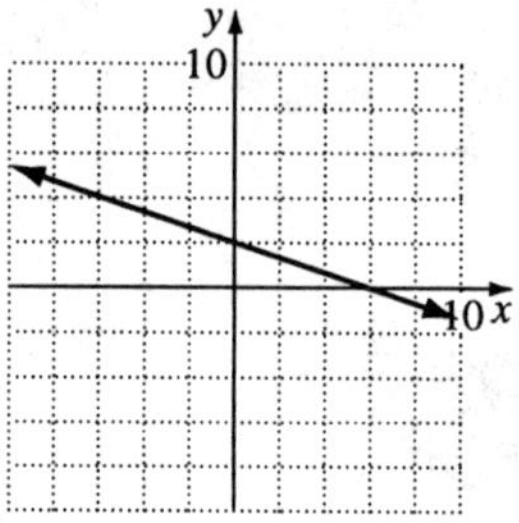

9. $4x - 3y = 12$

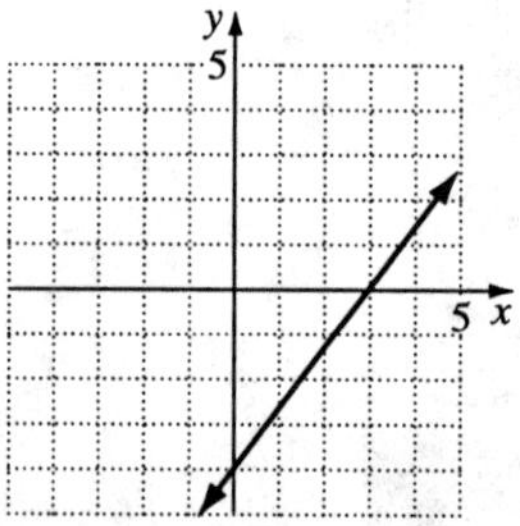

10. Points $(-1, -3)$ and $(4, 2)$
$\text{Slope} = \dfrac{2-(-3)}{4-(-1)} = \dfrac{5}{5} = 1$
$y - (-3) = 1[x - (-1)]$
Point-slope: $y + 3 = 1(x + 1)$ or

$y-2=1(x-4)$
$y+3=x+1$
Slope-intercept: $y=x-2$
Standard: $x-y=2$

11. Point (–2, 3)
Perpendicular to $y=-\frac{1}{2}x-4$
Slope = 2
$y-3=2[x-(-2)]$
Point-slope: $y-3=2(x+2)$
$y-3=2x+4$
Slope-intercept: $y=2x+7$
Standard: $2x-y=-7$

12.

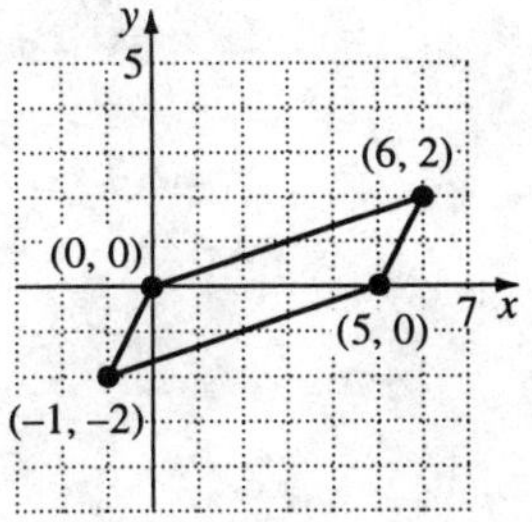

Is the line connecting (0, 0) and (6, 2) parallel to the line connecting (–1, –2) and (5, 0)?
Slope $=\frac{2-0}{6-0}=\frac{2}{6}=\frac{1}{3}$
Slope $=\frac{0-(-2)}{5-(-1)}=\frac{2}{6}=\frac{1}{3}$
Yes, they are parallel.
Is the line connecting (0, 0) and (–1, –2) parallel to the line connecting (6, 2) and (5, 0)?
Slope $=\frac{-2-0}{-1-0}=\frac{-2}{-1}=2$
Slope $=\frac{0-2}{5-6}=\frac{-2}{-1}=2$
Yes, they are parallel.
Yes, these points are the vertices of a parallelogram.

13. Slope $=\frac{7.98-1.98}{3-0}=2$
$y-1.98=2(x-0)$ or
$y-7.98=2(x-3)$
$y=2x+1.98$
Year 2010, $x=2010-1993=17$
$y=2(17)+1.98=35.98$
The prediction is 35.98 million.

14. $-2x+5\geq 4-(x-3)$
$-2x+5>4-x+3$
$-2x+x\geq 4+3-5$
$-x\geq 2$
$x\leq -2$
$\{x|x\leq -2\}$

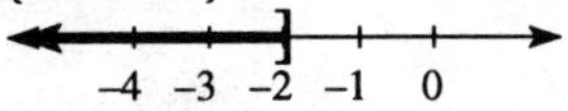

15. $\frac{x}{7}-\frac{1}{5}\leq\frac{x}{5}-\frac{5}{7}$
$35\left(\frac{x}{7}-\frac{1}{5}\right)\leq 35\left(\frac{x}{5}-\frac{5}{7}\right)$
$5x-7\leq 7x-25$
$5x-7x\leq 7-25$
$-2x\leq -18$
$x\geq 9$
$\{x|x\geq 9\}$

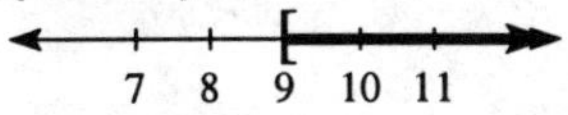

16. $y=ax+b$
where y = monthly rate and
x = hours on line
$y=20$
$y>0.15x+5$
$20>0.15x+5$
$15>0.15x$
$100>x$
It will be a better deal at less than 100 hours of on line service.

17. $\{2, 4, 6, 7, 10\}\cap\{2, 4, 5, 9, 10\}$
$=\{2, 4, 10\}$

18. $\{2, 4, 6, 7, 10\}\cup\{2, 4, 5, 9, 10\}$
$=\{2, 4, 5, 6, 7, 9, 10\}$

19. $-2 < x - 5 \le 7$

$-2 + 5 < x - 5 + 5 \le 7 + 5$

$3 < x \le 12$

$\{x | 3 < x \le 12\}$ or $(3, 12]$

20. $-11 < 4 - 3x < 40$

$-11 - 4 < 4 - 4 - 3x < 40 - 4$

$-15 < -3x < 36$

$\dfrac{-15}{-3} > \dfrac{-3x}{-3} > \dfrac{36}{-3}$

$5 > x > -12$

$\{x | -12 < x < 5\}$ or $(-12, 5)$

21. $x + 3 \le -1$ or $-4x + 3 < -5$

$x + 3 - 3 \le -1 - 3$ $\quad$ $-4x + 3 - 3 < -5 - 3$

$x \le -4$ $\quad$ $-4x < -8$

$x > 2$

$\{x | x \le -4 \text{ or } x > 2\}$ or $(-\infty, -4] \cup (2, \infty)$

22. $x + 6 \ge 4$ and $-2x - 3 \le 2$

$x + 6 - 6 \ge 4 - 6$ $\quad$ $-2x - 3 + 3 \le 2 + 3$

$x \ge -2$ $\quad$ $-2x \le 5$

$x \ge -\dfrac{5}{2}$

Since $-\dfrac{5}{2} < -2$, $x \ge -2$.

$\{x | x \ge -2\}$ or $[-2, \infty)$

23. $2x - 3 < 5$ or $3x - 6 \le 4$

$2x - 3 + 3 < 5 + 3$ $\quad$ $3x - 6 + 6 \le 4 + 6$

$2x < 8$ $\quad$ $3x \le 10$

$x < 4$ $\quad$ $x \le \dfrac{10}{3}$

Since $\dfrac{10}{3} < 4$, $x < 4$.

$\{x | x < 4\}$ or $(-\infty, 4)$

24. $-2x - 4 > -2$ and $x - 3 > -5$

$-2x - 4 + 4 > -2 + 4$ $\quad$ $x - 3 + 3 > -5 + 3$

$-2x > 2$ $\quad$ $x > -2$

$x < -1$ $\quad$ $x > -2$

$\{x | -2 < x < -1\}$ or $(-2, -1)$

25. $\left|\dfrac{2}{3}x - 6\right| = 2$

$\dfrac{2}{3}x - 6 = 2$ or $\dfrac{2}{3}x - 6 = -2$

$\dfrac{2}{3}x = 8$ $\quad$ $\dfrac{2}{3}x = 4$

$x = 12$ $\quad$ $x = 6$

$\{6, 12\}$

26. $\left|\dfrac{2}{3}x - 1\right| = \left|\dfrac{1}{3}x + 3\right|$

$\dfrac{2}{3}x - 1 = \dfrac{1}{3}x + 3$ or $-\left(\dfrac{2}{3}x - 1\right) = \dfrac{1}{3}x + 3$

$3\left(\dfrac{2}{3}x - 1\right) = 3\left(\dfrac{1}{3}x + 3\right)$

or $-3\left(\dfrac{2}{3}x - 1\right) = 3\left(\dfrac{1}{3}x + 3\right)$

$2x - 3 = x + 9$ or $-2x + 3 = x + 9$

$2x - x = 9 + 3$ $\quad$ $-2x - x = 9 - 3$

$x = 12$ $\quad$ $-3x = 6$

$x = -2$

$\{-2, 12\}$

27. $|2x - 1| < 7$

$-7 < 2x - 1 < 7$

$-7 + 1 < 2x - 1 + 1 < 7 + 1$

$-6 < 2x < 8$

$-3 < x < 4$

$\{x | -3 < x < 4\}$ or $(-3, 4)$

28. $|2x - 3| \ge 5$

$2x - 3 \ge 5$ or $2x - 3 \le -5$

$2x \ge 8$ $\quad$ $2x \le -2$

$x \ge 4$ $\quad$ $x \le -1$

$\{x | x \le -1 \text{ or } x \ge 4\}$ or $(-\infty, -1] \cup [4, \infty)$

29. $3x - 2y < 6$

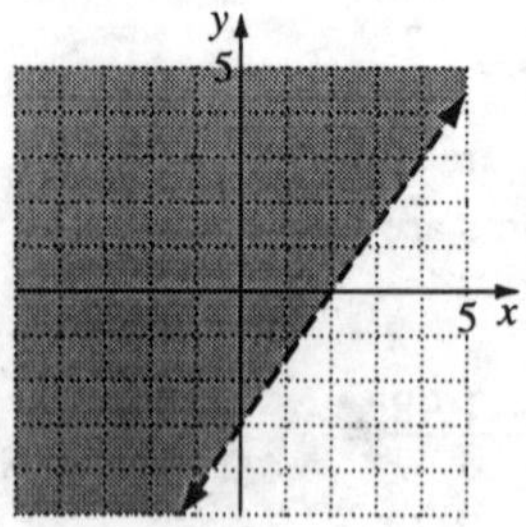

30. $y \geq \frac{1}{2}x - 1$

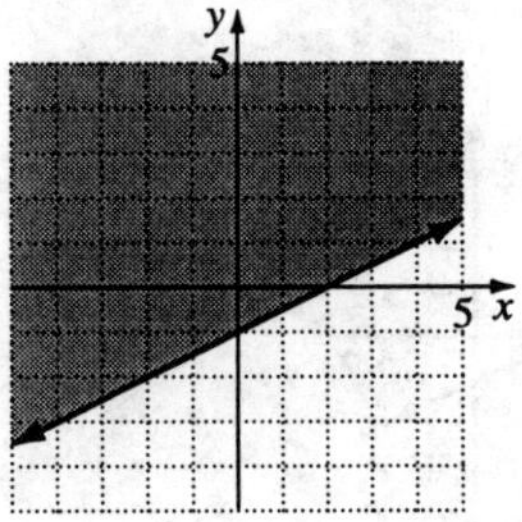

Chapter 3

Problem Set 3.1

1. $x + y = 4$
$x - y = 2$

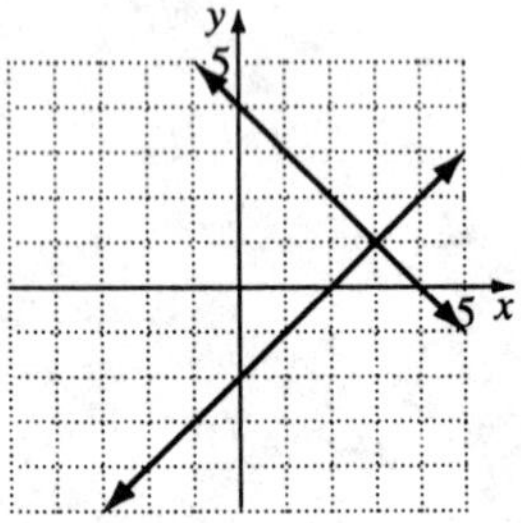

$\{(3, 1)\}$

3. $x - 3y = 2$
$y = 6 - x$

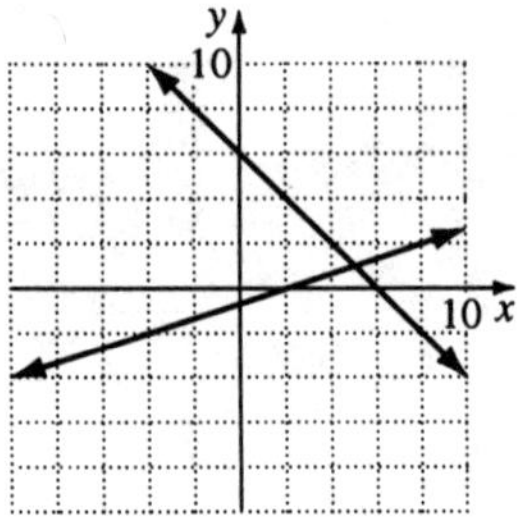

$\{(5, 1)\}$

5. $2x = 3y - 6$
$-3x + y = -5$

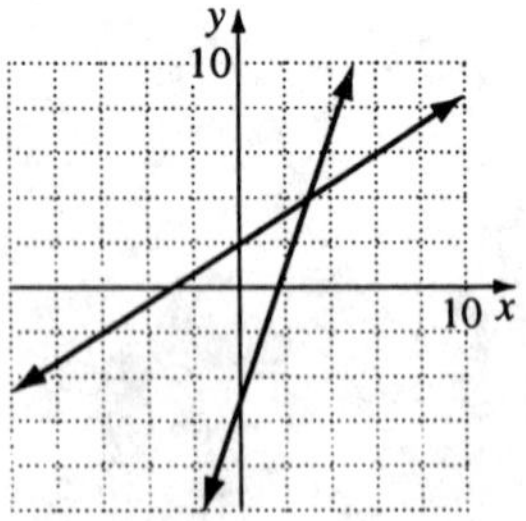

$\{(3, 4)\}$

7. $x + 2y = 1$
$x = 3$

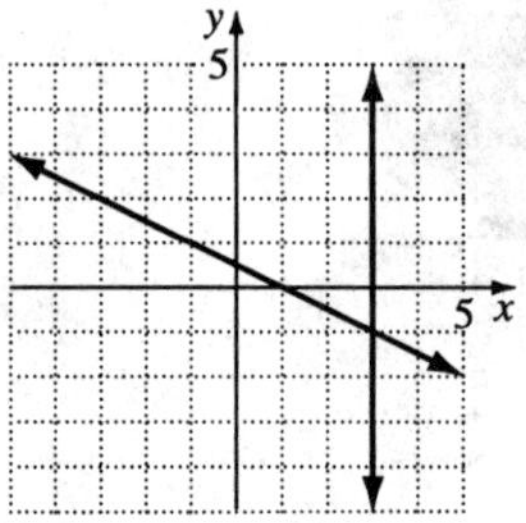

$\{(3, -1)\}$

9. $3x + y = 13$
$6x + 2y = 12$

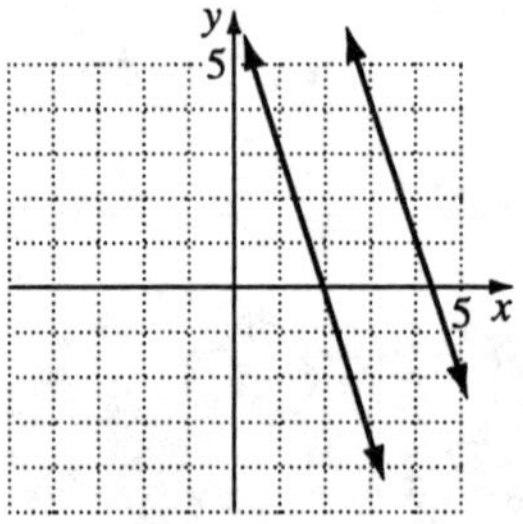

Parallel lines
$\varnothing$ (inconsistent)

11. $\frac{x}{3} - \frac{y}{4} = 1$
$\frac{2x}{3} - \frac{y}{2} = 1$

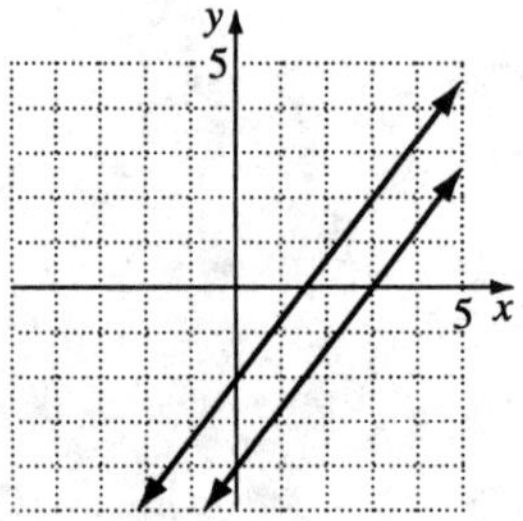

Parallel lines
$\varnothing$ (inconsistent)

13. $2x + 2y = 1$
$4x + 4y = 2$

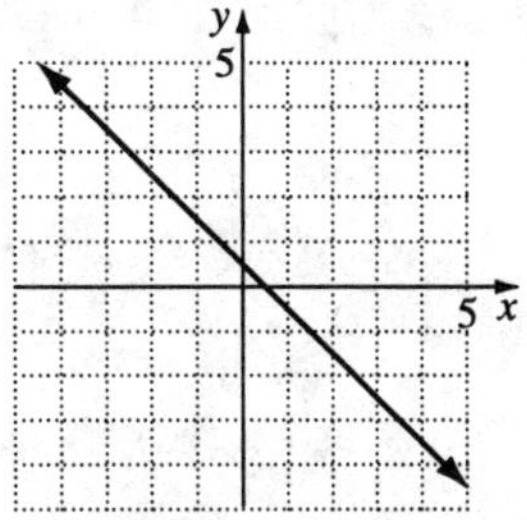

Same line
Dependent

15. $3x - 3y = 6$
$2x - 2y = 4$

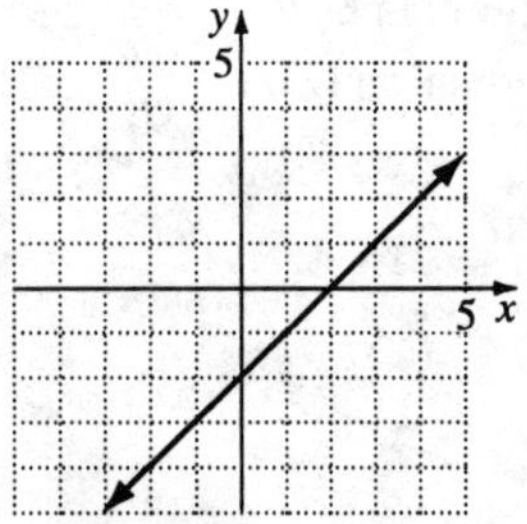

Same line
Dependent

17. $x = 2y - 5$ (1)
$x - 3y = 8$ (2)
Substitute for x in (2).
$(2y - 5) - 3y = 8$
$-y - 5 = 8$
$-y = 13$
$y = -13$

$x = 2y - 5$ (Replace y with -13)
$x = 2(-13) - 5 = -26 - 5 = -31$
$\{(-31, -13)\}$

19. $4x + y = 5$ (1)
$2x - 3y = 13$ (2)
Solve (1) for y and substitute into (2).
$y = 5 - 4x$
$2x - 3(5 - 4x) = 13$
$2x - 15 + 12x = 13$
$14x = 28$
$x = 2$

$y = 5 - 4x$ (replace x with 2)
$y = 5 - 4(2)$
$y = 5 - 8$
$y = -3$
$\{(2, -3)\}$

21. $x + y = 0$ (1)
$3x + 2y = 5$ (2)
Solve (1) for y and substitute into (2).
$y = -x$
$3x + 2(-x) = 5$
$3x - 2x = 5$
$x = 5$

$y = -x$ (replace x with 5)
$y = -5$
$\{(5, -5)\}$

23. $7x - 3y = 23$ (1)
$x + 2y = 13$ (2)
Solve (2) for x and substitute for x in (1).
$x = 13 - 2y$
$7(13 - 2y) - 3y = 23$
$91 - 14y - 3y = 23$
$91 - 17y = 23$
$-17y = -68$
$y = 4$

$x = 13 - 2y$ (Replace y with 4)
$x = 13 - 2(4)$
$x = 5$
$\{(5, 4)\}$

25.
$$\begin{array}{r} x + y = 7 \\ \underline{x - y = 3} \\ 2x = 10 \\ x = 5 \end{array}$$
(Replace x with 5 in first equation.)
$5 + y = 7$
$y = 2$
$\{(5, 2)\}$

27. $12x+3y=15$
$\underline{2x-3y=13}$
$14x=28$
$x=2$

(Replace x with 2 in second equation.)
$2(2)-3y=13$
$4-3y=13$
$-3y=9$
$y=-3$
$\{(2,-3)\}$

29. $x-2y=5$ (1)
$5x-y=-2$ (2)
No change to (1).
Multiply equation (2) by −2.
$x-2y=5$
$\underline{-10x+2y=4}$
$-9x=9$
$x=-1$

$5x-y=-2$
$5(-1)-y=-2$
$y=-5+2$
$y=-3$
$\{(-1,-3)\}$

31. $2x-9y=5$ (1)
$3x-3y=11$ (2)
No change to (1).
Multiply equation (2) by −3.
$2x-9y=5$
$\underline{-9x+9y=-33}$
$-7x=-28$
$x=4$

$3x-3y=11$
$3(4)-3y=11$
$12-3y=11$
$-3y=-1$
$y=\frac{1}{3}$
$\left\{\left(4,\frac{1}{3}\right)\right\}$

33. $3x-7y=1$ (1)
$2x-3y=-1$ (2)
Multiply equation (1) by 2.
Multiply equation (2) by −3.

$6x-14y=2$
$\underline{-6x+9y=3}$
$-5y=5$
$y=-1$

$2x-3y=-1$
$2x-3(-1)=-1$
$2x+3=-1$
$2x=-4$
$x=-2$
$\{(-2,-1)\}$

35. $4x+y=2$ (1)
$2x-3y=8$ (2)
Multiply equation (1) by 3.
No change to equation (2).
$12x+3y=6$
$\underline{2x-3y=8}$
$14x=14$
$x=1$

$4x+y=2$
$4(1)+y=2$
$y=2-4$
$y=-2$
$\{(1,-2)\}$

37. $2y=5-5x$
$9x-15=-3y$
Rearrange
$5x+2y=5$ (1)
$9x+3y=15$ (2)
Multiply equation (1) by 3.
Multiply equation (2) by −2.
$15x+6y=15$
$\underline{-18x-6y=-30}$
$-3x=-15$
$x=5$

$5x+2y=5$
$5(5)+2y=5$
$2y=-20$
$y=-10$
$\{(5,-10)\}$

39. $9x+\frac{4}{3}y=5$ (1)
$4x-\frac{1}{3}y=5$ (2)

Multiply equation (1) by 3.
Multiply equation (2) by 12.

$$\begin{array}{r} 27x + 4y = 15 \\ 48x - 4y = 60 \\ \hline 75x = 75 \\ x = \ \ 1 \end{array}$$

Multiply equation 2 by 3.

$12x - y = 15$

Substitute 1 for x

$12 - y = 15$

$y = -3$

$\{(1, -3)\}$

41. $y - 3x = 2$ (1)

$x = \frac{1}{4}y$ (2)

Substitute for x in equation (1).

$y - 3\left(\frac{1}{4}y\right) = 2$

$y - \frac{3}{4}y = 2$

Multiply by 4.

$4y - 3y = 8$

$y = 8$

$x = \frac{1}{4}y$

Replace y with 8.

$x = \frac{1}{4}(8)$

$x = 2$

$\{(2, 8)\}$

43. $x + 2y - 3 = 0$

$12 = 8y + 4x$

$x + 2y = 3$

$4x + 8y = 12$

$\frac{1}{4} = \frac{2}{8} = \frac{3}{12}$

Dependent; $\{(x, y) | x + 2y - 3 = 0\}$ or $\{(x, y) | 12 = 8y + 4x\}$

45. $2x - y - 5 = 0$

$4x - 2y - 10 = 0$

$2x - y = 5$

$4x - 2y = 10$

Dependent; $\{(x, y) | 2x - y - 5 = 0\}$ or $\{(x, y) | 4x - 2y - 10 = 0\}$

47. $2x - 3y = 7$

$4x - 6y = 3$

Parallel lines, $\varnothing$

49. $x - 3y = 5$

$y = \frac{2}{3}x$

$x - 3\left(\frac{2}{3}x\right) = 5$

$-x = 5$

$x = -5$

$y = \frac{2}{3}(-5) = -\frac{10}{3}$

$\left\{\left(-5, -\frac{10}{3}\right)\right\}$

51. $y = 2x + 5$

$y = -4x + 2$

$2x + 5 = -4x + 2$

$6x = -3$

$x = -\frac{1}{2}$

$y = 2\left(-\frac{1}{2}\right) + 5$

$y = 4$

$\left\{\left(-\frac{1}{2}, 4\right)\right\}$

53. $3(x + y) = 6$ (1)

$3(x - y) = -36$ (2)

Divide equation (1) by 3.
Divide equation (2) by 3.

$$\begin{array}{r} x + y = \ \ 2 \\ x - y = -12 \\ \hline 2x = -10 \\ x = \ -5 \end{array}$$

$x + y = 2$

$-5 + y = 2$

$y = 7$

$\{(-5, 7)\}$

55. $\left.\begin{array}{l} 3(x-3)-2y=0 \\ 2(x-y)=-x-3 \end{array}\right\}$(Rearrange and simplify)

$\left.\begin{array}{l} 3x-9-2y=0 \\ 2x-2y=-x-3 \end{array}\right\}$(Rearrange)

$$\begin{array}{rl} 3x-2y = 9 & \\ \underline{3x-2y=-3} & \text{(Subtract)} \\ 0 = 6 & \text{False} \end{array}$$

Contradiction

No solution, $\varnothing$

57. Costs: $y = 800{,}000 + 45x$

Revenue: $y = 65x$

$$\begin{aligned} \text{Cost} &= \text{Revenue} \\ 800{,}000 + 45x &= 65x \\ 800{,}000 &= 20x \\ 40{,}000 &= x \end{aligned}$$

40,000 units

59. $y = 500x$

$y = 50x + 1000$

a.

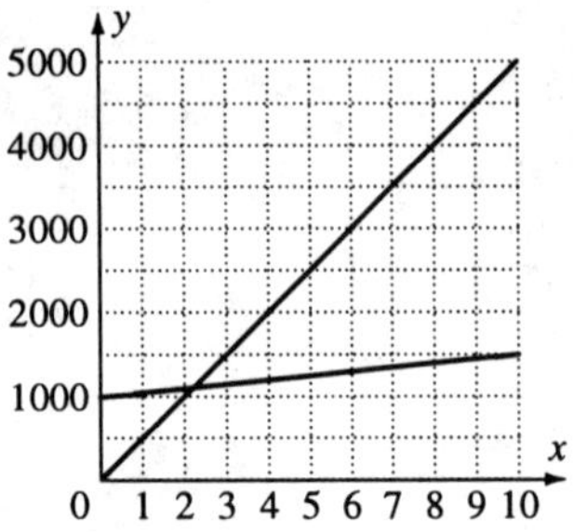

b. Achilles overtakes the tortoise in 2.2 minutes. This occurs approximately 1100 yards from the starting point.

c. $y = 500x$

$y = 50x + 1000$

$500x = 50x + 1000$

$450x = 1000$

$x = 2.\overline{2}$

$y = 500(2.\overline{2})$

$y = 1111.\overline{1}$

$\{(2.\overline{2}, 1111.\overline{1})\}$

61. A

63. Students should verify problems.

65. Answers may vary.

67. Answers may vary.

69. Answers may vary.

71. $3x - 2y = 4$

$6x - 4y = 6$

$3x - 2y = 3$ is a line parallel to $3x - 2y = 4$

Thus, the system is inconsistent.

D

73. $5x - 10y = 40$ (1)

$2x + ky = -30$ (2)

Multiply equation (1) by $\frac{2}{5}$.

$2x - 4y = 16$

$2x + ky = -30$

The system of linear equations is inconsistent when $k = -4$.

75. Parallel lines; inconsistent; no solutions, $\varnothing$.

77. $a_1x + b_1y = c_1$ (1)

$a_2x + b_2y = c_2$ (2)

Multiply equation (1) by a_2.

Multiply equation (2) by $-a_1$.

$$\begin{aligned} a_1a_2x + b_1a_2y &= c_1a_2 \\ \underline{-a_1a_2x - b_2a_1y} &\underline{= -c_2a_1} \\ (b_1a_2 - b_2a_1)y &= c_1a_2 - c_2a_1 \\ y &= \frac{c_1a_2 - c_2a_1}{b_1a_2 - b_2a_1} \end{aligned}$$

$a_1x + b_1y = c_1$ (3)

$a_2x + b_2y = c_2$ (4)

Multiply equation (3) by b_2.

Multiply equation (4) by $-b_1$.

$$\begin{aligned} a_1b_2x + b_1b_2y &= c_1b_2 \\ -a_2b_1x - b_1b_2y &= -c_2b_1 \\ \hline (a_1b_2 - a_2b_1)x &= c_1b_2 - c_2b_1 \\ x &= \frac{c_1b_2 - c_2b_1}{a_1b_2 - a_2b_1} \end{aligned}$$

$$\left\{\left(\frac{c_1b_2 - c_2b_1}{a_1b_2 - a_2b_1}, \frac{c_1a_2 - c_2a_1}{b_1a_2 - b_2a_1}\right)\right\}$$

Review Problems

79. $\frac{x}{3} - \frac{x-5}{5} = 3$

$15\left(\frac{x}{3} - \frac{x-5}{5}\right) = 15(3)$

$5x - 3(x-5) = 45$

$5x - 3x + 15 = 45$

$2x + 15 = 45$

$2x = 30$

$x = 15$

y	10	15
12	a	b
13	c	d

Diagonal:

$15 + a + 13 = 28 + a$ (total)

Row 2: $12 + a + b$

$12 + a + b = 28 + a$

$b = 16$

Column 2:

$10 + a + c$

$10 + a + c = 28 + a$

$c = 18$

y	10	15
12	a	16
13	18	d

All columns, rows and diagonals have the same sum.

Row 1: $y + 10 + 15 = 25 + y$

Row 2: $12 + a + 16 = 28 + a$

Row 3: $13 + 18 + d = 31 + d$

Column 1: $y + 12 + 13 = 25 + y$

Column 2: $10 + a + 18 = 28 + a$

Column 3: $15 + 16 + d = 31 + d$

Diagonal (top left): $y + a + d$

Diagonal (top right): $13 + a + 15 = 28 + a$

Let total $= y + a + d$

Then

$y + a + d = 25 + y \rightarrow a + d = 25$ (1)

$y + a + d = 28 + a \rightarrow y + d = 28$ (2)

$y + a + d = 31 + d \rightarrow y + a = 31$ (3)

Subtract equation (2) from equation (3).

$$\left.\begin{aligned} y + a &= 31 \\ y + d &= 28 \\ \hline a - d &= 3 \end{aligned}\right\}$$

$$\begin{aligned} a - d &= 3 \\ a + d &= 25 \\ \hline 2a &= 28 \\ a &= 14 \end{aligned}$$

$a + d = 25 \rightarrow 14 + d = 25$

$d = 11$

$y + a = 31 \rightarrow y + 14 = 31$

$y = 17$

17	10	15
12	14	16
13	18	11

80. Let x = price before the increase.

$x + 0.16x = 3.19$

$1.16x = 3.19$

$x = 2.75$

$2.75

81. $|2x + 3| = |4x - 5|$

$2x + 3 = 4x - 5$

$-2x = -8$

$x = 4$

or

$2x + 3 = -(4x - 5)$

$2x + 3 = -4x + 5$

$6x = 2$

$x = \frac{1}{3}$

$\left\{\frac{1}{3}, 4\right\}$

Problem Set 3.2

1. Let x = first number and
y = second number.
$2x - y = 9$ (1)
$2x + 3y = -3$ (2)
Multiply equation (1) by 3.
No change to equation (2).

$$\begin{array}{r} 6x - 3y = 27 \\ \underline{2x + 3y = -3} \\ 8x = 24 \\ x = 3 \end{array}$$

$$\begin{aligned} 2x - y &= 9 \\ 2(3) - y &= 9 \\ -y &= 3 \\ y &= -3 \end{aligned}$$

The numbers are 3 and –3.

3. Let x = mg of cholesterol in sponge cake and y = mg of cholesterol in pound cake.
$4x + 2y = 784$
$x + 3y = 366$
$x = 366 - 3y$
$4(366 - 3y) + 2y = 784$
$1464 - 12y + 2y = 784$
$-10y = -680$
$y = 68$
$x = 162$
Sponge cake has 162 mg of cholesterol and pound cake has 68 mg of cholesterol.

5. Let t = the tens' digit and
u = the units' digit.
original number = $10t + u$
number with digits reversed = $10u + t$

$$\begin{array}{r} t + u = 9 \\ \underline{10u + t = (10t + u) + 45} \end{array}$$

(Simplify)
$t + u = 9$ (1)
$-9t + 9u = 45$ (2)
No change to equation (1).
Divide equation (2) by 9.

$$\begin{array}{r} t + u = 9 \\ \underline{-t + u = 5} \\ 2u = 14 \\ u = 7 \end{array}$$

$t + u = 9$
$t + 7 = 9$
$t = 2$
The number is 27.

7. Let x = hourly wages for tutoring and
y = hourly wages for grading.
$70x + 50y = 622.50$ (1)
$90x + 40y = 719$ (2)
Multiply equation (1) by –4.
Multiply equation (2) by 5.

$$\begin{array}{r} -280x - 200y = -2490 \\ \underline{450x + 200y = 3595} \\ 170x = 1105 \\ x = 6.5 \text{ (tutors)} \end{array}$$

$$\begin{aligned} 70x + 50y &= 622.50 \\ 70(6.5) + 50y &= 622.50 \\ 455 + 50y &= 622.50 \\ 50y &= 167.50 \\ y &= 3.35 \text{ (graders)} \end{aligned}$$

Tutors: \$6.50; graders: \$3.35

9. Let x = weekly salary and
y = commission rate.
$x + 2000y = 900$
$x + 5000y = 1350$
$900 - 2000y = 1350 - 5000y$
$3000y = 450$
$y = 0.15$
$x = 600$
\$600 weekly salary and 15% commission

11. Let x = number of day-care centers and y = miles of road repairs.
$25{,}000x + 5000y = 1{,}300{,}000$
$\frac{y}{x} = \frac{50}{3}$
$y = \frac{50x}{3}$

$25,000x + 5000\left(\frac{50x}{3}\right) = 1,300,000$

$25,000x + \frac{250,000x}{3} = 1,300,000$

$325,000x = 3,900,000$

$x = 12$

$y = 200$

12 day-care centers and 200 miles of road repair.

13. Let l = length of lot
and w = width of lot.

$2l + 2w = 360$

$20l + 8w + 8w = 3280$

$l = 180 - w$

$20(180 - w) + 16w = 3280$

$3600 - 20w + 16w = 3280$

$-4w = -320$

$w = 80$

$l = 100$

The lot is 80 ft by 100 ft.

15. Let x = length of EB, x = length of BC,
x = length of AD,
y = length of AE, y = length of EC and
$x + y$ = length of DC.

perimeter of parallelogram $ABCD = 50$

$AE + EB + BC + DC + AD = 50$

$y + x + x + (x + y) + x = 50$

$4x + 2y = 50$

perimeter of trapezoid $AECD = 39$

$AE + EC + DC + AD = 39$

$y + y + (x + y) + x = 39$

$2x + 3y = 39$

$4x + 2y = 50 \quad (1)$

$2x + 3y = 39 \quad (2)$

Divide equation (1) by –2.
No change to equation (2).

$$\begin{array}{r} -2x - y = -25 \\ 2x + 3y = 39 \\ \hline 2y = 14 \end{array}$$

$y = 7\ (AE)$

$2x + y = 25$

$2x + 7 = 25$

$2x = 18$

$x = 9\ (EB)$

$x + y = 9 + 7 = 16\ (DC)$

length AE, 7 m; length EB, 9 m;
length of DC, 16 m

17. Sum of $\angle A$, $\angle B$, and $\angle C = 180°$.

$(x + 8y - 1) + (3y + 4) + (7x + 5) = 180$

$8x + 11y = 172$

Angles B and C are equal.

$$\begin{array}{r} 3y + 4 = 7x + 5 \\ 7x - 3y = \quad -1 \\ 56x + 77y = \quad 1204 \\ -56x + 24y = \quad 8 \\ \hline 101y = \quad 1212 \\ y = \quad 12 \\ x = \quad 5 \end{array}$$

$\angle A = x + 8y - 1 = 5 + 8(12) - 1 = 100°$

$\angle B = 3y + 4 = 3(12) + 4 = 40°$

$\angle C = 7x + 5 = 7(5) + 5 = 40°$

19.
$$\begin{array}{r} x + y + 61 = \quad 180 \\ y = x - 39 \\ x + y = \quad 119 \\ x - y = \quad 39 \\ \hline 2x = \quad 158 \\ x = \quad 79 \\ y = \quad 40 \end{array}$$

The angles are 79° and 40°.

21. $m\angle A + m\angle B = 180$
(angles A and B are supplementary)

$m\angle A = m\angle C$
(corresponding angles are equal in measure)

(Substitute):

$(4x - 2y + 4) + (12x + 6y + 12) = 180$

$4x - 2y + 4 = 6x - 24$

(Simplify):

$16x + 4y = 164 \quad (1)$

$-2x - 2y = -28 \quad (2)$

Divide equation (1) by 4.
Divide equation (2) by 2.

$$\begin{aligned} 4x+y &= 41 \\ -x-y &= -14 \\ \hline 3x &= 27 \\ x &= 9 \\ x+y &= 14 \\ 9+y &= 14 \\ y &= 5 \end{aligned}$$

$x=9,\ y=5$

$m\angle A = (4x-2y+4)°$
$= (4\cdot 9-2\cdot 5+4)° = 30°$

$m\angle B = (12x+6y+12)°$
$= (12\cdot 9+6\cdot 5+12)° = 150°$

$m\angle C = (6x-24)° = (6\cdot 9-24)° = 30°$

$m\angle A = 30°,\ m\angle B = 150°,\ m\angle C = 30°$

23. $[5x-(2y-80)]+2y=180$
(consecutive angles are supplementary)
$5x-(2y-80)=3x$
(opposite angles are equal in measure)
(Simplify):
$5x-2y+80+2y=180$
$5x=100$
$x=20$

$5x-2y+80=3x$
$2x-2y=-80$
$x-y=-40$
$20-y=-40$
$-y=-60$
$y=60$

$(3x)° = (3\cdot 20)° = 60°$
$(2y)° = (2\cdot 60)° = 120°$
$5x-(2y-80)° = [5\cdot 20-(2\cdot 60-80)]°$
$= [100-(120-80)]°$
$= (100-40)° = 60°$

Since opposite angles are equal in measure, the remaining angle is $(2y)° = 120°$.

25.

	Rate, r	· Time, t	= Distance, d
With wind	$x+y$	4	800
Against wind	$x-y$	5	800

x = speed of plane
y = speed of wind
$4(x+y)=800$
$5(x-y)=800$

$$\begin{aligned} x+y &= 200 \\ x-y &= 160 \\ \hline 2x &= 360 \\ x &= 180 \\ y &= 20 \end{aligned}$$

Speed of plane is 180 mph and speed of wind is 20 mph

27. Let x = number of gallons of solution A (30% acid).
$60 - x$ = number of gallons of solution B (60% acid)

	Number of gallons	· Percentage of acid	= Amount of acid
solution A (30% acid)	x	30% (0.30)	$0.30x$
solution B (60% acid)	$60 - x$	60% (0.60)	$0.60(60 - x)$
50% acid solution	60	50% (0.50)	0.50(60)

$0.30x + 0.60(60 - x) = 0.50(60)$
$0.30x + 36 - 0.60x = 30$
$-0.30x = -6$
$x = 20$ (Solution A)
$60 - x = 60 - 20 = 40$ (Solution B)
20 gallons solution A; 40 gallons solution B

29. Let x = number of students at second school.

	Students	Percent	Hispanic
School 1	900	0.40	0.40 (900)
School 2	x	0.75	$0.75x$
Merge	$900 + x$	0.525	$0.525\,(900 + x)$

$0.40(900) + 0.75x = 0.525\,(900 + x)$
$360 + 0.75x = 472.5 + 0.525x$
$0.225x = 112.5$
$x = 500$
There are 500 students at the second school.

31. a. $90{,}597 = 8m + b$
$110{,}428 = 15m + b$

$$\begin{aligned} -90{,}597 &= -8m - b \\ \underline{110{,}428} &\underline{= 15m + b} \\ 19{,}831 &= 7m \\ 2833 &= m \\ 90{,}597 &= 8(2833) + b \\ b &= 67{,}933 \end{aligned}$$

$m = 2833,\ b = 67{,}933$

b. Each year the value of a house increases by $2833.

c. $y = mx + b$
$y = 2833(20) + 67{,}933$
$y = 124{,}593$
The house will be $124,593 in the year 2000.

33. a. $6500 = 40m + b$
$6000 = 60m + b$
$$\begin{array}{r} 6500 = 40m + b \\ \underline{-6000 = -60m - b} \\ 500 = -20m \\ m = -25 \\ b = 7500 \end{array}$$

b. $6200 = 40m + b$
$6300 = 60m + b$
$$\begin{array}{r} -6200 = -40m - b \\ \underline{6300 = 60m + b} \\ 100 = 20m \\ m = 5 \\ b = 6000 \end{array}$$

c. $y = -25x + 7500$
$y = 5x + 6000$
$$\begin{array}{r} -25x + 7500 = 5x + 6000 \\ -30x = -1500 \\ x = 50 \\ y = 6250 \end{array}$$
($50, 6250)
When fans are priced at $50, the supply and demand both equal 6250 fans.

d. $y = -25x + 7500$
$y = 5x + 6000$

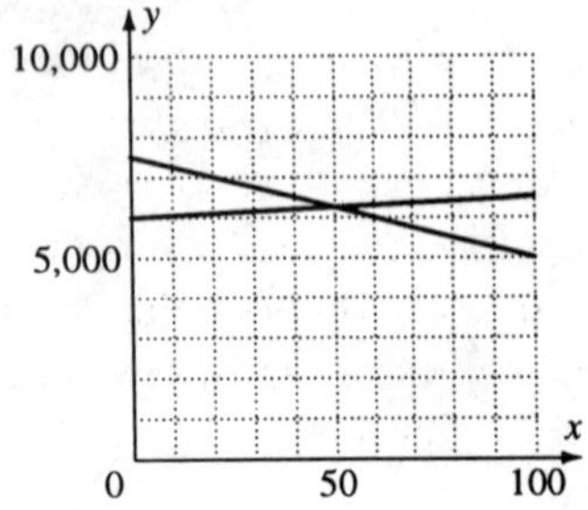

35. $D(p) = 410 - p$
$S(p) = p^2 + 3p - 70$

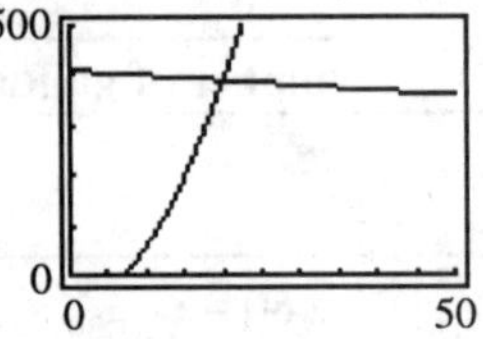

$20; 390 units

37. Let x = cost of mangos (sold at profit of 20%) and y = cost of avocados (sold at loss of 2%).
$x + y = 67$ (solve for y) $\rightarrow$ $y = 67 - x$
(mangos at profit of 20%)
+ (avocados at loss of 2%)
= (total profit of 8.56)
$0.20x + (-0.02y) = 8.56$
$0.20x - 0.02y = 8.56$
(substitute for y)
$0.20x - 0.02(67 - x) = 8.56$
$0.20x - 1.34 + 0.02x = 8.56$
$0.22x = 9.90$
$x = 45$ (mangos)
$x + y = 67$
$45 + y = 67$
$y = 22$ (avocados)
Mangos: $45; Avocados: $22

39. Let x = the assistant's federal tax and y = the assistant's state tax.
$x = 0.2(9800 - y)$
(These equations are translated from the given conditions).
$y = 0.1(9800 - x)$
Multiply both sides of each equation by 10 and simplifying, we obtain:
$10x + 2y = 19{,}600$ (1)
$x + 10y = 9800$ (2)
No change to equation (1).
Multiply equation (2) by –10.

$$\begin{array}{r} 10x+2y=19,600 \\ -10x-100y=-98,000 \\ \hline \text{(Add):}\ -98y=-78,400 \\ y=800 \end{array}$$

Since $x+10y=9800$

$x+10(800)=9800$

$x=1800$

The teacher's assistant paid \$1800 in federal tax and \$800 in state tax.

State: \$800; Federal: \$1800

41. From the given information, we have

$1.5A+1.5B=120$

$2A+B=120$

Solving results in $A=40$ and $B=40$.

Each person is traveling at 40 miles per hour.

Review Problems

43. $|3x+5|\geq 2$

$3x+5\geq 2$ or $3x+5\leq -2$

$3x\geq -3$ or $3x\leq -7$

$x\geq -1$ or $x\leq -\frac{7}{3}$

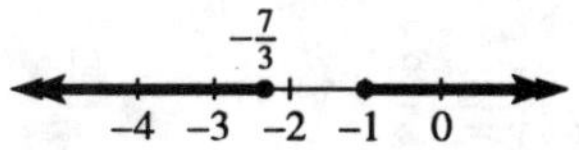

44. $\frac{x}{2}-\frac{x-4}{5}=\frac{23}{10}$

$10\left(\frac{x}{2}-\frac{x-4}{5}\right)=\frac{23}{10}\cdot 10$

$5x-2(x-4)=23$

$5x-2x+8=23$

$3x=15$

$x=5$

45. $f(x)=-x^2-x+2$

x	−3	−2	−1	$-\frac{1}{2}$	0	1	2
$f(x)$	−4	0	2	$2\frac{1}{4}$	2	0	−4

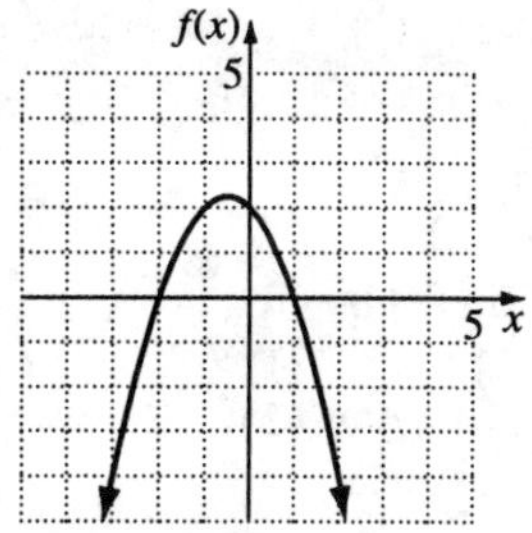

Problem Set 3.3

1. $x+y+z=4$

$2-1+3=4$

$4=4$ True

$x-2y-z=1$

$2-2(-1)-3=1$

$1=1$ True

$2x-y-2z=-1$

$2(2)-(-1)-2(3)=-1$

$-1=-1$ True

The ordered triple (2, −1, 3) makes all three equations true, so it is a solution.

3. $x-2y=2$

$4-2(1)=2$

$2=2$ True

$2x+3y=11$

$2(4)+3(1)=11$

$11=11$ True

$y-4z=-7$

$1-4(2)=-7$

$-7=-7$ True

The ordered triple (4, 1, 2) makes all three equations true, so it is a solution.

(Note: For all systems of equations, even though the equations are not numbered, they will be identified as Equations (1) or (2) or Equations (1), (2) or (3) in order. Problem 1 shows the equation numbering sequence. However the remaining problems will not, except when new equations result from these equations.)

5. $x + y + 2z = 11$ (1)
$x + y + 3z = 14$ (2)
$x + 2y - z = 5$ (3)
(Equations 1 and 2):
Multiply equation (2) by –1.

$$\begin{array}{rl} x+y+2z = 11 & (1) \\ \underline{-x-y-3z = -14} & (2) \\ -z = -3 & \\ z = 3 & \end{array}$$

(Substitute $z = 3$ into Equations 1 and 3):
$x + y + 2(3) = 11$ (1)
$x + 2y - 3 = 5$ (3)
(Simplify) Multiply equation (1) by –1.
Simplify equation (2).

$$\begin{array}{r} -x-y=-5 \\ \underline{x+2y=8} \\ y=3 \end{array}$$

(Substitute $z = 3$ and $y = 3$ into Equation 1):
$x + 3 + 6 = 11$
$x = 2$
$\{(2, 3, 3)\}$

7. $4x - y + 2z = 11$ (1)
$x + 2y - z = -1$ (2)
$2x + 2y - 3z = -1$ (3)
(Equations 1 and 2):
Multiply equation (2) by 2.

$$\begin{array}{rl} 4x-y+2z = 11 & (1) \\ \underline{2x+4y-2z = -2} & (2) \\ 6x+3y = 9 & \end{array}$$

(Equations 2 and 3):
Multiply equation (2) by –3.

$$\begin{array}{rl} -3x-6y+3z = 3 & (2) \\ \underline{2x+2y-3z = -1} & (3) \\ -x-4y = 2 & \end{array}$$

$$\begin{array}{rl} 6x+3y=9 & (4) \\ \underline{-x-4y=2} & (5) \end{array}$$

Divide equation (4) by 3.
Multiply equation (5) by 2.

$$\begin{array}{r} 2x+y=3 \\ \underline{-2x-8y=4} \\ -7y=7 \\ y=-1 \end{array}$$

$2x + y = 3$
$2x + (-1) = 3$
$2x = 4$
$x = 2$
(Substitute $x = 2$ and $y = -1$ into Equation 2):
$x + 2y - z = -1$
$2 + 2(-1) - z = -1$
$-z = -1$
$z = 1$
$\{(2, -1, 1)\}$

9. $3x + 5y + 2z = 0$ (1)
$12x - 15y + 4z = 12$ (2)
$6x - 25y - 8z = 8$ (3)
Multiply equation (1) by 4.
Multiply equation (2) by 2.
No change to equation (3).
$12x + 20y + 8z = 0$
$24x - 30y + 8z = 24$
$6x - 25y - 8z = 8$
(Equations 1 and 3):

$$\begin{array}{r} 12x+20y+8z=0 \\ \underline{6x-25y-8z=8} \\ 18x-5y=8 \end{array}$$

(Equations 2 and 3):

$$\begin{array}{r} 24x-30y+8z=24 \\ \underline{6x-25y-8z=8} \\ 30x-55y=32 \end{array}$$

$$\begin{array}{rl} 18x-5y=8 & (4) \\ \underline{30x-55y=32} & (5) \end{array}$$

Multiply equation (4) by –11.

$$\begin{array}{r} -198x+55y=-88 \\ \underline{30x-55y=32} \\ -168x=-56 \\ x=\frac{1}{3} \end{array}$$

$18x - 5y = 8$
$\left(\text{Substitute for } x = \frac{1}{3}.\right)$
$18\left(\frac{1}{3}\right) - 5y = 8$
$6 - 5y = 8$
$-5y = 2$
$y = -\frac{2}{5}$

$3x + 5y + 2z = 0$

$\left(\text{Substitute for } x = \frac{1}{3} \text{ and } y = -\frac{2}{5}.\right)$

$3\left(\frac{1}{3}\right) + 5\left(-\frac{2}{5}\right) + 2z = 0$

$1 - 2 + 2z = 0$

$2z = 1$

$z = \frac{1}{2}$

$\left\{\left(\frac{1}{3}, -\frac{2}{5}, \frac{1}{2}\right)\right\}$

11. $2x - 4y + 3z = 17$ (1)

$x + 2y - z = 0$ (2)

$4x - y - z = 6$ (3)

No change to equation (1).

Multiply equation (2) by 3.

Multiply equation (3) by 3.

$2x - 4y + 3z = 17$

$3x + 6y - 3z = 0$

$12x - 3y - 3z = 18$

(Equations 1 and 2):

$$\begin{array}{r} 2x - 4y + 3z = 17 \\ \underline{3x + 6y - 3z = \ \ 0} \\ 5x + 2y = 17 \end{array}$$

(Equations 1 and 3):

$$\begin{array}{r} 2x - 4y + 3z = 17 \\ \underline{12x - 3y - 3z = 18} \\ 14x - 7y = 35 \\ 2x - y = \ \ 5 \end{array}$$

$5x + 2y = 17$ (4)

$2x - y = 5$ (5)

No change to equation (1).

Multiply equation (5) by 2.

$$\begin{array}{r} 5x + 2y = 17 \\ \underline{4x - 2y = 10} \\ 9x = 27 \\ x = 3 \end{array}$$

$2x - y = 5$

$2(3) - y = 5$

$-y = -1$

$y = 1$

$x + 2y - z = 0$

$3 + 2(1) - z = 0$

$-z = -5$

$z = 5$

$\{(3, 1, 5)\}$

13. $2x + y = 2$ (1)

$x + y - z = 4$ (2)

$3x + 2y + z = 0$ (3)

Multiply equation (1) by –3.

Add equations (2) and (3) and add the sum to (1).

$$\begin{array}{r} -6x - 3y = -6 \\ \underline{4x + 3y = \ \ 4} \\ -2x = -2 \\ x = \ \ 1 \end{array}$$

$2x + y = 2$ (Equation 1)

$2(1) + y = 2$ $(x = 1)$

$y = 0$

$x + y - z = 4$ (Equation 2)

$1 + 0 - z = 4$ $(x = 1, y = 0)$

$-z = 3$

$z = -3$

$\{(1, 0, -3)\}$

15. $x + y = -4$ (1)

$y - z = 1$ (2)

$2x + y + 3z = -21$ (3)

Multiply equation (2) by 3.

$$\begin{array}{r} 3y - 3z = \ \ 3 \\ \underline{2x + y + 3z = -21} \\ 2x + 4y = -18 \\ x + 2y = \ -9 \end{array}$$ (4)

$x + y = -4$ (1)

$x + 2y = -9$ (4)

Multiply equation (1) by –2.

$$\begin{array}{r} -2x - 2y = \ \ 8 \\ \underline{x + 2y = -9} \\ -x = -1 \\ x = \ \ 1 \end{array}$$

$x + y = -4$

$1 + y = -4$

$y = -5$

$y - z = 1$

$-5 - z = 1$

$-z = 6$

$z = -6$

$\{(1, -5, -6)\}$

17. $x + y = 4$ (1)
$x + z = 4$ (2)
$y + z = 4$ (3)
Multiply equation (3) by –1.
$$\begin{array}{r} x+z=\ \ 4 \\ \underline{-y-z=-4} \\ x-y=\ \ 0 \end{array}$$
$$\begin{array}{rl} x+y=4 & (1) \\ \underline{x-y=0} & (4) \\ 2x=4 & \\ x=2 & \end{array}$$
$$\begin{array}{r} x+y=4 \\ 2+y=4 \\ y=2 \end{array}$$
$$\begin{array}{r} x+z=4 \\ 2+z=4 \\ z=2 \end{array}$$
$\{(2, 2, 2)\}$

19. $3x + 4y + 5z = 8$
$x - 2y + 3z = -6$
$2x - 4y + 6z = 8$
Add –2 times equation (2) to equation 3.
$$\begin{array}{r} -2x+4y-6z=12 \\ \underline{2x-4y+6z=\ \ 8} \\ 0=20 \end{array}\quad \text{False}$$
System is inconsistent, $\varnothing$.

21. $6x - y + 3z = 9$ (1)
$\frac{1}{4}x - \frac{1}{2}y - \frac{1}{3}z = -1$ (2)
$-x + \frac{1}{6}y - \frac{2}{3}z = 0$ (3)
Multiply equation (2) by –12.
Multiply equation (3) by 6.
$$\begin{array}{r} 6x-y+3z=\ \ 9 \\ -3x+6y+4z=12 \\ -6x+y-4z=\ \ 0 \end{array}$$
Equations (1) and (3):
$$\begin{array}{r} 6x-y+3z=\ \ 9 \\ \underline{-6x+y-4z=\ \ 0} \\ -z=\ \ 9 \\ z=-9 \end{array}$$
Equations (2) and (3):
$$\begin{array}{rl} -3x+6y+4z=12 & \\ \underline{-6x+y-4z=\ \ 0} & \\ -9x+7y=12 & (5) \end{array}$$

$$\begin{array}{rl} 6x-y+3z=\ \ 9 & (1) \\ 6x-y-27=\ \ 9 & \\ 6x-y=36 & (4) \end{array}$$

$$\begin{array}{rl} 6x-y=36 & (4) \\ -9x+7y=12 & (5) \end{array}$$
Multiply equation (4) by 7.
$$\begin{array}{r} 42x-7y=\ 252 \\ \underline{-9x+7y=\ \ \ 12} \\ 33x=264 \\ x=\ \ \ \ 8 \end{array}$$
$$\begin{array}{r} 6x-y=\ \ 36 \\ 6(8)-y=\ \ 36 \\ -y=-12 \\ y=\ \ 12 \end{array}$$
$\{(8, 12, -9)\}$

23. $2x + y + 4z = 4$ (1)
$x - y + z = 6$ (2)
$x + 2y + 3z = 5$ (3)
Multiply equation (2) by 2.
$$\begin{array}{rl} 2x-2y+2z=12 & \\ \underline{x+2y+3z=\ \ 5} & \\ 3x+5z=17 & (4) \end{array}$$
Add equation (1) and (2) to give
$3x + 5z = 10.$
Multiply (4) by –1 to give $-3x - 5z = -17$.
$$\begin{array}{r} \underline{-3x-5z=-17} \\ 0=-\ 7 \end{array}\quad \text{False}$$
No solution.
Inconsistent, $\varnothing$.

25. $3(2x + y) + 5z = -1$
$2(x - 3y + 4z) = -9$
$4(1 + x) = -3(z - 3y)$
$$\begin{array}{rl} 6x+3y+5z=-1 & (1) \\ 2x-6y+8z=-9 & (2) \\ 4x-9y+3z=-4 & (3) \end{array}$$
Eliminate y:
$$\begin{array}{lr} 2(1)+(2): & 14x+18z=-11 \\ 3(1)+(3): & \underline{22x+18z=-\ 7} \\ & -8x=-\ 4 \\ & x=\ \frac{1}{2} \end{array}$$

$14x + 18z = -11$
$14\left(\frac{1}{2}\right) + 18z = -11$
$18z = -18$
$z = -1$

$2\left(\frac{1}{2}\right) - 6y + 8(-1) = -9$
$1 - 6y - 8 = -9$
$-6y = -2$
$y = \frac{1}{3}$

$\left\{\left(\frac{1}{2}, \frac{1}{3}, -1\right)\right\}$

27. $3x - 3y - z = 0$
$x - y + z = 0$
$-x + y + z = 0$
Add equations (2) and (3).
$x - y + z = 0$
$-x + y + z = 0$
$2z = 0$
$z = 0$
Input $z = 0$
$3x - 3y = 0$
$x - y = 0$
$-x + y = 0$
Multiply equation (1) by $\frac{1}{3}$.
Multiply equation (3) by –1.
$x - y = 0$
$x - y = 0$
$x - y = 0$
Each equation represents the same line where $x = y$.
Dependent
$\{(x, y, z) | x = y, z = 0\}$

29. $\frac{x}{2} - \frac{y}{2} + \frac{z}{4} = 1$
$\frac{x}{2} + \frac{y}{3} - \frac{z}{4} = 2$
$\frac{x}{4} - \frac{y}{2} + \frac{z}{2} = 2$
or
$2x - 2y + z = 4$
$6x + 4y - 3z = 24$
$x - 2y + 2z = 8$

Eliminate y:
Add –1 times (1) to (3).
$-2x + 2y - z = -4$
$x - 2y + 2z = 8$
$-x + z = 4$ (4)
Add 2 times (1) to (2).
$4x - 4y + 2z = 8$
$6x + 4y - 3z = 24$
$10x - z = 32$ (5)
Add equations (4) and (5).
$-x + z = 4$
$10x - z = 32$
$9x = 36$
$x = 4$
$z = x + 4$
$z = 4 + 4 = 8$
$y = \frac{2x + z - 4}{2}$
$y = \frac{2(4) + 8 - 4}{2} = 6$
$\{(4, 6, 8)\}$

31. Let x = first number, y = second number, and z = third number.
$x + y + z = 16$ (1)
$2x + 3y + 4z = 46$ (2)
$5x - y = 31$ (3)
Multiply equation (1) by –4.
$-4x - 4y - 4z = -64$
$2x + 3y + 4z = 46$
$-2x - y = -18$ (4)
Multiply equation (4) by –1.
Equations (3) and (4):
$5x - y = 31$
$2x + y = 18$
$7x = 49$
$x = 7$
$5x - y = 31$
$5(7) - y = 31$
$-y = -4$
$y = 4$
$x + y + z = 16$
$7 + 4 + z = 16$
$z = 5$
The numbers are 7, 4, and 5.

33. $x + y + z = 6860$

$x = 560 + z$

$x = 2y - 2140$

Eliminate x.

Let (2) equal (3).

$560 + z = 2y - 2140$

$z - 2y = -2700 \quad (4)$

Add (1) and –1 times (2).

$$\begin{array}{r} x + y + z = 6860 \\ -x + z = -560 \\ \hline y + 2z = 6300 \end{array} \quad (5)$$

Add (4) and 2 times (5).

$$\begin{array}{r} -2y + z = -2700 \\ 2y + 4z = 12,600 \\ \hline 5z = 9900 \\ z = 1980 \end{array}$$

$y = 6300 - 2z$

$y = 6300 - 2(1980)$

$y = 2340$

$x = 6860 - y - z$

$x = 6860 - 2340 - 1980$

$x = 2540$

The Missouri is 2540 miles, Mississippi is 2340 miles, and Yukon is 1980 miles.

35. Let x = the largest angle, y = next largest angle, and z = smallest angle.

$x = 80 + z$

$x = 2y - 20$

$x + y + z = 180$

Eliminate x:

Let (1) equal (2).

$80 + z = 2y - 20$

$-2y + z = -100 \quad (4)$

Add (1) and –1 times (3).

$$\begin{array}{r} x - z = 80 \\ -x - y - z = -180 \\ \hline -y - 2z = -100 \end{array} \quad (5)$$

Add (4) and –2 times (5).

$$\begin{array}{r} -2y + z = -100 \\ 2y + 4z = 200 \\ \hline 5z = 100 \\ z = 20 \end{array}$$

$x = 80 + z$

$x = 80 + 20$

$x = 100$

$y = 180 - x - z$

$y = 180 - 100 - 20$

$y = 60$

The angles are 100°, 60°, and 20°.

37. Let x = number of triangles, y = number of rectangles, and z = number of pentagons.

$x + y + z = 40 \quad (1)$

$3x + 4y + 5z = 153 \quad (2)$

$2y + 5z = 72 \quad (3)$

Multiply equation (3) by –1.

$$\begin{array}{r} 3x + 4y + 5z = 153 \\ -2y - 5z = -72 \\ \hline 3x + 2y = 81 \end{array} \quad (4)$$

Equations (1) and (2):

$$\begin{array}{r} -5x - 5y - 5z = -200 \\ 3x + 4y + 5z = 153 \\ \hline -2x - y = -47 \end{array} \quad (5)$$

$$\begin{array}{r} -4x - 2y = -94 \quad (5) \\ 3x + 2y = 81 \quad (4) \\ \hline -x = -13 \\ x = 13 \\ 2x + y = 47 \\ 2(13) + y = 47 \\ y = 21 \end{array}$$

$$\begin{array}{r} x + y + z = 40 \\ 13 + 21 + z = 40 \\ z = 6 \end{array}$$

13 triangles, 21 rectangles, 6 pentagons

39. Let L = length of rectangular solid

W = width of rectangular solid

H = height of rectangular solid

$2L + 2W = 16 \quad (1)$

$2L + 2H = 18 \quad (2)$

$2W + 2H = 14 \quad (3)$

Divide equations (1) and (2) by 2, and (3) by –2.

$$\begin{aligned} L+W &= 8 \\ L+H &= 9 \\ -W-H &= -7 \\ \hline L-W &= 2 \quad (4) \end{aligned}$$

$$\begin{aligned} L+W &= 8 \quad (1) \\ L-W &= 2 \quad (4) \\ \hline 2L &= 10 \\ L &= 5 \end{aligned}$$

$$\begin{aligned} L+W &= 8 \\ 5+W &= 8 \\ W &= 3 \end{aligned}$$

$$\begin{aligned} L+H &= 9 \\ 5+H &= 9 \\ H &= 4 \end{aligned}$$

Length, 5 cm; width, 3 cm; height, 4 cm

41. Let A be the number of servings of Food A, B be the number of servings of Food B, and C be the number of servings of Food C.

$2A + B + 2C = 16$

$A + 2B + 4C = 23$

$3A + 4B + 3C = 29$

Eliminate B and C:

Add –2 times (1) to (2).

$$\begin{aligned} -4A-2B-4C &= -32 \\ A+2B+4C &= 23 \\ \hline -3A &= -9 \\ A &= 3 \end{aligned}$$

$B+2C = 16-2A = 16-6 = 10$

$4B+3C = 29-3A = 29-9 = 20$

$$\begin{aligned} -4B-8C &= -40 \\ 4B+3C &= 20 \\ \hline -5C &= -20 \\ C &= 4 \end{aligned}$$

$B = 10-2C$

$B = 10-2(4)$

$B = 2$

3 servings of A, 2 servings of B, and 4 servings of C

43. Let x = amount invested at 8%, y = amount invested at 10%, and z = amount invested at 12%.

$$x+y+z = 6700 \quad (1)$$

$$0.08x + 0.10y + 0.12z = 716 \quad (2)$$

$$z = (x+y) + 300 \quad (3)$$

Multiply equation (2) by 100.

$8x + 10y + 12z = 71{,}600$

Equations (1) and (3):

$$\begin{aligned} x+y+z &= 6700 \\ -x-y+z &= 300 \\ \hline 2z &= 7000 \\ z &= 3500 \end{aligned}$$

Equations (1) and (2), replace 3500 for z:

$$x+y+3500 = 6700 \quad (5)$$

$$4x+5y+6(3500) = 35{,}800 \quad (4)$$

Multiply equation (5) by –5.

$$\begin{aligned} -5x-5y &= -16{,}000 \\ 4x+5y &= 14{,}800 \\ \hline -x &= -1200 \\ x &= 1200 \end{aligned}$$

$$\begin{aligned} x+y &= 3200 \\ 1200+y &= 3200 \\ y &= 2000 \end{aligned}$$

\$1200 at 8%; \$2000 at 10%; \$3500 at 12%

45. Let x = number of \$2 packages, y = number of \$3 packages, and z = number of \$4 packages.

$$x+y+z = 12$$

$$6x+12y+24z = 162$$

$$2x+3y+4z = 35$$

Add equation (1) to $-\frac{1}{6}$ times equation (2).

$$\begin{aligned} x+y+z &= 12 \\ -x-2y-4z &= -27 \\ \hline -y-3z &= -15 \quad (4) \end{aligned}$$

Add –2 times (1) to (3).

$$\begin{aligned} -2x-2y-2z &= -24 \\ 2x+3y+4z &= 35 \\ \hline y+2z &= 11 \quad (5) \end{aligned}$$

Add equations (4) and (5).

$$\begin{aligned} -y-3z &= -15 \\ y+2z &= 11 \\ \hline -z &= -4 \\ z &= 4 \end{aligned}$$

$y = 15-3z$

$y = 15-3(4)$

$y = 3$

$x = 12 - y - z$
$x = 12 - 3 - 4$
$x = 5$
5 \$2 packages, 3 \$3 packages, and 4 \$4 packages

47. Truck I: x cubic yards
Truck II: y cubic yards
Truck III: z cubic yards
Day 1: $4x + 3y + 5z = 78$ (1)
Day 2: $5x + 4y + 4z = 81$ (2)
Day 3: $3x + 5y + 3z = 69$ (3)
Eliminate z.

$$\begin{array}{lr} 3(2)+-4(3): & 3x - 8y = -33 \\ 4(1)+-5(2): & -9x - 8y = -93 \\ \hline & 12x = 60 \\ & x = 5 \end{array}$$

$3y + 5z = 58$
$4y + 4z = 56$
$5y + 3z = 54$
Add 4(1) to −3(2):

$$\begin{array}{r} 12y + 20z = 232 \\ -12y - 12z = -168 \\ \hline 8z = 64 \\ z = 8 \end{array}$$

$$\begin{aligned} 4(5) + 3y + 5(8) &= 78 \\ 3y &= 18 \\ y &= 6 \end{aligned}$$

Truck I: 5 cu yd
Truck II: 6 cu yd
Truck III: 8 cu yd

49. C is true.

51. Students should verify results.

53. Answers may vary.

55. Two planes could be parallel, one equation in one plane and 2 equations in the other plane.

57. $x - y + 2z - 2w = -1$ (1)
$x - y - z + w = -4$ (2)
$-x + 2y - 2z - w = -7$ (3)
$2x + y + 3z - w = 6$ (4)
Add equations (2) and (3).
$y - 3z = -11$ (5)
Add equations (2) and (4).
$3x + 2z = 2$ (6)
Eliminate w first:

$$\begin{array}{r} x - y + 2z - 2w = -1 \\ 2x - 2y - 2z + 2w = -8 \\ \hline 3x - 3y = -9 \\ x - y = -3 \quad (7) \end{array}$$

Now we have a system of 3 equations:
$y - 3z = -11$ (5)
$3x + 2z = 2$ (6)
$x - y = -3$ (7)

$$\begin{array}{r} y - 3z = -11 \\ x - y = -3 \\ \hline x - 3z = -14 \quad (8) \end{array}$$

System of 2 equations:

$$\begin{array}{rl} 3x + 2z = 2 & (6) \\ -3x + 9z = 42 & (8) \\ \hline 11z = 44 & \\ z = 4 & \end{array}$$

$3x + 2z = 2$ (6)
$3x + 2(4) = 2$
$3x = -6$
$x = -2$

$x - y = -3$ (7)
$-2 - y = -3$
$-y = -1$
$y = 1$

$x - y - z + w = -4$ (2)
$-2 - 1 - 4 + w = -4$
$w = 3$
$\{(-2, 1, 4, 3)\}$

59. x = number of adult tickets
y = number tickets for adult with 1 child
z = number of tickets for adult with 2 children
\$415 was collected:

$20x + 30y + 35z = 415 \quad \left(\text{multiply by } \frac{1}{5}\right)$

28 passenger total: $x + 2y + 3z = 28$

15 adults: $x + y + z = 15$

$4x + 6y + 7z = 83 \quad (1)$

$x + 2y + 3z = 28 \quad (2)$

$x + y + z = 15 \quad (3)$

Eliminate x.

(2) – (3): $y + 2z = 13$ (multiply by –2)

–4(3) + (1): $2y + 3z = 23$

$$\begin{array}{r} -2y - 4z = -26 \\ 2y + 3z = 23 \\ \hline -z = -3 \\ z = 3 \end{array} \quad \text{(add)}$$

$$\begin{aligned} y + 2z &= 13 \\ y + 2(3) &= 13 \\ y &= 7 \end{aligned}$$

$$\begin{aligned} x + y + z &= 15 \\ x + 7 + 3 &= 15 \\ x &= 5 \end{aligned}$$

5 adult tickets, 7 adult with one child, 3 adult with 2 children

61. $x + y = a \quad (1)$

$y + z = b \quad (2)$

$x + z = c \quad (3)$

Multiply equation (2) by –1.

$$\begin{array}{r} x + y = \quad a \\ -y - z = \quad -b \\ \hline x - z = a - b \end{array} \quad (4)$$

Equations 4 and 3:

$$\begin{array}{r} x - z = a - b \\ x + z = c \\ \hline 2x = a - b + c \\ x = \frac{a - b + c}{2} \end{array}$$

Equation 1:

$$\frac{a - b + c}{2} + y = a$$
$$y = \frac{a + b - c}{2}$$

Equation 3:

$$\frac{a - b + c}{2} + z = c$$
$$z = \frac{-a + b + c}{2}$$

$$\left\{\left(\frac{a - b + c}{2}, \frac{a + b - c}{2}, \frac{-a + b + c}{2}\right)\right\}$$

63. Group activity.

Review Problems

65. Line passing through (3, 5) and (4, 2):

$m = \frac{2 - 5}{4 - 3} = -3$

Point-slope: $y - 5 = -3(x - 3)$

or $y - 2 = -3(x - 4)$

$y - 5 = -3x + 9$

Slope-intercept: $y = -3x + 14$

Standard: $3x + y = 14$

66. $y = -2x + 4$

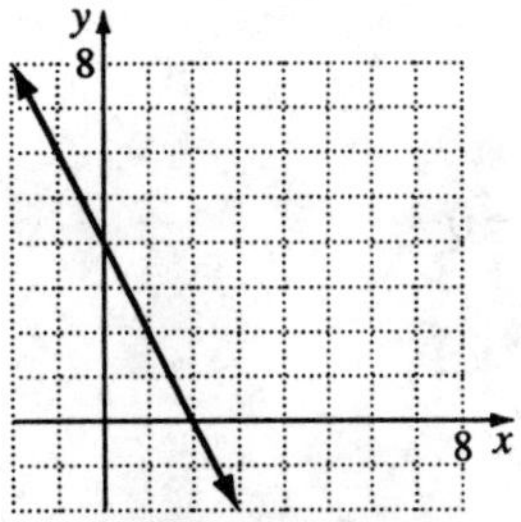

x-intercept: (2, 0)

y-intercept: (0, 4)

area $= \frac{1}{2}(2)(4) = 4$ square units

67. **a.** y is not a function of x.

b. y is a function of x.

c. y is not a function of x.

d. y is a function of x.

Problem Set 3.4

1. $\left[\begin{array}{cc|c} 1 & -3 & 11 \\ 0 & 1 & -3 \end{array}\right]$

$x - 3y = 11$

$y = -3$

$x - 3(-3) = 11$

$x + 9 = 11$

$x = 2$

$\{(2, -3)\}$

3. $\left[\begin{array}{cc|c} 1 & -3 & 1 \\ 0 & 1 & -1 \end{array}\right]$

$x - 3y = 1$
$y = -1$
$x - 3(-1) = 1$
$x + 3 = 1$
$x = -2$
$\{(-2, -1)\}$

5. $\left[\begin{array}{ccc|c} 1 & 2 & 1 & 0 \\ 0 & 1 & 0 & -2 \\ 0 & 0 & 1 & 3 \end{array}\right]$

$x + 2y + z = 0$
$y = -2$
$z = 3$
$x + 2(-2) + 3 = 0$
$x - 4 + 3 = 0$
$x - 1 = 0$
$x = 1$
$\{(1, -2, 3)\}$

7. $\left[\begin{array}{ccc|c} 1 & 1 & 0 & 3 \\ 0 & 1 & \frac{3}{2} & -2 \\ 0 & 0 & 1 & 0 \end{array}\right]$

$x + y = 3$
$y + \frac{3}{2}z = -2$
$z = 0$
$y + \frac{3}{2}(0) = -2$
$y = -2$
$x - 2 = 3$
$x = 5$
$\{(5, -2, 0)\}$

9. $\left[\begin{array}{cc|c} 2 & 2 & 5 \\ 1 & -\frac{3}{2} & 5 \end{array}\right] R_1 \leftrightarrow R_2 \rightarrow \left[\begin{array}{cc|c} 1 & -\frac{3}{2} & 5 \\ 2 & 2 & 5 \end{array}\right]$

11. $\left[\begin{array}{cc|c} -6 & 8 & -12 \\ 3 & 5 & -2 \end{array}\right] -\frac{1}{6}R_1 \rightarrow \left[\begin{array}{cc|c} 1 & -\frac{4}{3} & 2 \\ 3 & 5 & -2 \end{array}\right]$

13. $\left[\begin{array}{cc|c} 1 & -3 & 5 \\ 2 & 6 & 4 \end{array}\right] -2R_1 + R_2 \rightarrow \left[\begin{array}{cc|c} 1 & -3 & 5 \\ 0 & 12 & -6 \end{array}\right]$

15. $\left[\begin{array}{cc|c} 1 & -\frac{3}{2} & \frac{7}{2} \\ 3 & 4 & 2 \end{array}\right] -3R_1 + R_2 \rightarrow \left[\begin{array}{cc|c} 1 & -\frac{3}{2} & \frac{7}{2} \\ 0 & \frac{17}{2} & -\frac{17}{2} \end{array}\right]$

17. $\left[\begin{array}{ccc|c} 1 & -1 & 5 & -6 \\ 3 & 3 & -1 & 10 \\ 1 & 3 & 2 & 5 \end{array}\right] \begin{array}{l} \\ -3R_1 + R_2 \rightarrow \\ -1R_1 + R_3 \rightarrow \end{array} \left[\begin{array}{ccc|c} 1 & -1 & 5 & -6 \\ 0 & 6 & -16 & 28 \\ 0 & 4 & -3 & 11 \end{array}\right]$

19. $\left[\begin{array}{rrr|r} 1 & 1 & -1 & 6 \\ 2 & -1 & 1 & -3 \\ 3 & -1 & -1 & 4 \end{array}\right] \begin{array}{l} \\ -2R_1 + R_2 \rightarrow \\ -3R_1 + R_3 \rightarrow \end{array} \left[\begin{array}{rrr|r} 1 & 1 & -1 & 6 \\ 0 & -3 & 3 & -15 \\ 0 & -4 & 2 & -14 \end{array}\right]$

21. $x + y = 6$
$x - y = 2$

$\left[\begin{array}{rr|r} 1 & 1 & 6 \\ 1 & -1 & 2 \end{array}\right] R_1 - R_2 \rightarrow \left[\begin{array}{rr|r} 1 & 1 & 6 \\ 0 & 2 & 4 \end{array}\right] \tfrac{1}{2}R_2 \rightarrow \left[\begin{array}{rr|r} 1 & 1 & 6 \\ 0 & 1 & 2 \end{array}\right]$

$x + y = 6$
$y = 2$

$x + 2 = 6$ (substitute for y)
$x = 4$
$\{(4, 2)\}$

23. $2x + y = 3$
$x - 3y = 12$

$\left[\begin{array}{rr|r} 2 & 1 & 3 \\ 1 & -3 & 12 \end{array}\right] R_1 - 2R_2 \rightarrow \left[\begin{array}{rr|r} 2 & 1 & 3 \\ 0 & 7 & -21 \end{array}\right] \tfrac{1}{7}R_2 \rightarrow \left[\begin{array}{rr|r} 2 & 1 & 3 \\ 0 & 1 & -3 \end{array}\right]$

$2x + y = 3$
$y = -3$

$2x - 3 = 3$
$2x = 6$
$x = 3$
$\{(3, -3)\}$

25. $5x + 7y = -25$
$11x + 6y = -8$

$\left[\begin{array}{rr|r} 5 & 7 & -25 \\ 11 & 6 & -8 \end{array}\right] 11R_1 - 5R_2 \rightarrow \left[\begin{array}{rr|r} 5 & 7 & -25 \\ 0 & 47 & -235 \end{array}\right] \begin{array}{l} \tfrac{1}{5}R_1 \rightarrow \\ \tfrac{1}{47}R_2 \rightarrow \end{array} \left[\begin{array}{rr|r} 1 & \tfrac{7}{5} & -5 \\ 0 & 1 & -5 \end{array}\right]$

$x + \tfrac{7}{5}y = -5$
$y = -5$

$x + \tfrac{7}{5}(-5) = -5$ (substitute for y)
$x = 2$
$\{(2, -5)\}$

27. $4x-2y=5$
$-2x+y=6$

$$\left[\begin{array}{cc|c} 4 & -2 & 5 \\ -2 & 1 & 6 \end{array}\right] \begin{array}{l} \frac{1}{4}R_1 \rightarrow \\ R_1+2R_2 \rightarrow \end{array} \left[\begin{array}{cc|c} 1 & -\frac{1}{2} & \frac{5}{4} \\ 0 & 0 & 17 \end{array}\right]$$

$x-\frac{1}{2}y=\frac{5}{4}$
$0=17$ (False) No solution
The system is inconsistent.
$\varnothing$

29. $x-2y=1$
$-2x+4y=-2$

$$\left[\begin{array}{cc|c} 1 & -2 & 1 \\ -2 & 4 & -2 \end{array}\right] 2R_1+R_2 \rightarrow \left[\begin{array}{cc|c} 1 & -2 & 1 \\ 0 & 0 & 0 \end{array}\right]$$

$x-2y=1$
$0=0$ (True) Infinitely many solutions
The system is dependent.
$\{(x,y)|x-2y=1\}$

31. $x+y-z=-2$
$2x-y+z=5$
$-x+2y+2z=1$

$$\left[\begin{array}{ccc|c} 1 & 1 & -1 & -2 \\ 2 & -1 & 1 & 5 \\ -1 & 2 & 2 & 1 \end{array}\right] \begin{array}{l} \\ R_2+2R_3 \rightarrow \\ R_1+R_3 \rightarrow \end{array} \left[\begin{array}{ccc|c} 1 & 1 & -1 & -2 \\ 0 & 3 & 5 & 7 \\ 0 & 3 & 1 & -1 \end{array}\right] \begin{array}{l} \\ \\ R_2-R_3 \rightarrow \end{array} \left[\begin{array}{ccc|c} 1 & 1 & -1 & -2 \\ 0 & 3 & 5 & 7 \\ 0 & 0 & 4 & 8 \end{array}\right] \begin{array}{l} \\ \frac{1}{3}R_2 \rightarrow \\ \frac{1}{4}R_3 \rightarrow \end{array}$$

$$\left[\begin{array}{ccc|c} 1 & 1 & -1 & -2 \\ 0 & 1 & \frac{5}{3} & \frac{7}{3} \\ 0 & 0 & 1 & 2 \end{array}\right]$$

$x+y-z=-2$
$y+\frac{5}{3}z=\frac{7}{3}$
$z=2$

$y+\frac{5}{3}(2)=\frac{7}{3}$
$y=-\frac{3}{3}=-1$

$x-1-2=-2$
$x=1$
$\{(1,-1,2)\}$

33. $x+3y=0$
$x+y+z=1$
$3x-y-z=11$

$$\left[\begin{array}{ccc|c} 1 & 3 & 0 & 0 \\ 1 & 1 & 1 & 1 \\ 3 & -1 & -1 & 11 \end{array}\right] \begin{array}{l} \\ R_1-R_2\to \\ 3R_1-R_3\to \end{array} \left[\begin{array}{ccc|c} 1 & 3 & 0 & 0 \\ 0 & 2 & -1 & -1 \\ 0 & 10 & 1 & -11 \end{array}\right] \begin{array}{l} \\ \\ 5R_2-R_3\to \end{array} \left[\begin{array}{ccc|c} 1 & 3 & 0 & 0 \\ 0 & 2 & -1 & -1 \\ 0 & 0 & -6 & 6 \end{array}\right] \begin{array}{l} \\ \frac{1}{2}R_2\to \\ -\frac{1}{6}R_3\to \end{array}$$

$$\left[\begin{array}{ccc|c} 1 & 3 & 0 & 0 \\ 0 & 1 & -\frac{1}{2} & -\frac{1}{2} \\ 0 & 0 & 1 & -1 \end{array}\right]$$

$x+3y=0$
$y-\frac{1}{2}z=-\frac{1}{2}$
$z=-1$

$y-\frac{1}{2}(-1)=-\frac{1}{2}$
$y=-1$

$x+3(-1)=0$
$x=3$
$\{(3,-1,-1)\}$

35. $2x+2y+7z=-1$
$2x+y+2z=2$
$4x+6y+z=15$

$$\left[\begin{array}{ccc|c} 2 & 2 & 7 & -1 \\ 2 & 1 & 2 & 2 \\ 4 & 6 & 1 & 15 \end{array}\right] \begin{array}{l} \\ R_1-R_2\to \\ 2R_1-R_3\to \end{array} \left[\begin{array}{ccc|c} 2 & 2 & 7 & -1 \\ 0 & 1 & 5 & -3 \\ 0 & -2 & 13 & -17 \end{array}\right] \begin{array}{l} \\ \\ 2R_2+R_3\to \end{array} \left[\begin{array}{ccc|c} 2 & 2 & 7 & -1 \\ 0 & 1 & 5 & -3 \\ 0 & 0 & 23 & -23 \end{array}\right] \begin{array}{l} \frac{1}{2}R_1\to \\ \\ \frac{1}{23}R_3\to \end{array}$$

$$\left[\begin{array}{ccc|c} 1 & 1 & \frac{7}{2} & -\frac{1}{2} \\ 0 & 1 & 5 & -3 \\ 0 & 0 & 1 & -1 \end{array}\right]$$

$x+y+\frac{7}{2}z=-\frac{1}{2}$
$y+5z=-3$
$z=-1$

$y+5(-1)=-3$
$y=2$

$x+2+\frac{7}{2}(-1)=-\frac{1}{2}$
$x=1$
$\{(1,2,-1)\}$

37. $x+y+z=6$
$x-z=-2$
$y+3z=11$

$$\left[\begin{array}{ccc|c} 1 & 1 & 1 & 6 \\ 1 & 0 & -1 & -2 \\ 0 & 1 & 3 & 11 \end{array}\right] R_1-R_2 \to \left[\begin{array}{ccc|c} 1 & 1 & 1 & 6 \\ 0 & 1 & 2 & 8 \\ 0 & 1 & 3 & 11 \end{array}\right] R_3-R_2 \to \left[\begin{array}{ccc|c} 1 & 1 & 1 & 6 \\ 0 & 1 & 2 & 8 \\ 0 & 0 & 1 & 3 \end{array}\right]$$

$x+y+z=6$
$y+2z=8$
$z=3$

$y+2(3)=8$
$y=2$

$x+2+3=6$
$x=1$

$\{(1, 2, 3)\}$

39. $x-y+3z=4$
$x+2y-2z=10$
$3x-y+5z=14$

$$\left[\begin{array}{ccc|c} 1 & -1 & 3 & 4 \\ 1 & 2 & -2 & 10 \\ 3 & -1 & 5 & 14 \end{array}\right] \begin{array}{l} \\ R_2-R_1 \to \\ R_3-3R_1 \to \end{array} \left[\begin{array}{ccc|c} 1 & -1 & 3 & 4 \\ 0 & 3 & -5 & 6 \\ 0 & 2 & -4 & 2 \end{array}\right] \begin{array}{l} \\ \frac{1}{3}R_2 \to \\ \frac{1}{2}R_3 \to \end{array} \left[\begin{array}{ccc|c} 1 & -1 & 3 & 4 \\ 0 & 1 & -\frac{5}{3} & 2 \\ 0 & 1 & -2 & 1 \end{array}\right] R_2-R_3 \to$$

$$\left[\begin{array}{ccc|c} 1 & -1 & 3 & 4 \\ 0 & 1 & -\frac{5}{3} & 2 \\ 0 & 0 & \frac{1}{3} & 1 \end{array}\right] 3R_3 \to \left[\begin{array}{ccc|c} 1 & -1 & 3 & 4 \\ 0 & 1 & -\frac{5}{3} & 2 \\ 0 & 0 & 1 & 3 \end{array}\right]$$

$x-y+3z=4$
$y-\frac{5}{3}z=2$
$z=3$

$y-\frac{5}{3}(3)=2$
$y=7$

$x-7+3(3)=4$
$x=2$

$\{(2, 7, 3)\}$

41. $x - 2y + z = 4$
$5x - 10y + 5z = 20$
$-2x + 4y - 2z = -8$

$$\left[\begin{array}{ccc|c} 1 & -2 & 1 & 4 \\ 5 & -10 & 5 & 20 \\ -2 & 4 & -2 & -8 \end{array}\right] \begin{array}{l} \\ 5R_1 - R_2 \rightarrow \\ 2R_1 + R_3 \rightarrow \end{array} \left[\begin{array}{ccc|c} 1 & -2 & 1 & 4 \\ 0 & 0 & 0 & 0 \\ 0 & 0 & 0 & 0 \end{array}\right]$$

$x - 2y + z = 4$
$0 = 0$
$0 = 0$ (True) Infinitely many solutions.

The system is dependent.

$\{(x, y, z) | x - 2y + z = 4\}$

43. $x + y = 1$
$y + 2z = -2$
$2x - z = 0$

$$\left[\begin{array}{ccc|c} 1 & 1 & 0 & 1 \\ 0 & 1 & 2 & -2 \\ 2 & 0 & -1 & 0 \end{array}\right] \begin{array}{l} \\ \\ 2R_1 - R_3 \rightarrow \end{array} \left[\begin{array}{ccc|c} 1 & 1 & 0 & 1 \\ 0 & 1 & 2 & -2 \\ 0 & 2 & 1 & 2 \end{array}\right] \begin{array}{l} \\ \\ 2R_2 - R_3 \rightarrow \end{array} \left[\begin{array}{ccc|c} 1 & 1 & 0 & 1 \\ 0 & 1 & 2 & -2 \\ 0 & 0 & 3 & -6 \end{array}\right] \begin{array}{l} \\ \\ \frac{1}{3}R_3 \rightarrow \end{array} \left[\begin{array}{ccc|c} 1 & 1 & 0 & 1 \\ 0 & 1 & 2 & -2 \\ 0 & 0 & 1 & -2 \end{array}\right]$$

$x + y = 1$
$y + 2z = -2$
$z = -2$

$y + 2(-2) = -2$
$y = 2$

$x + 2 = 1$
$x = -1$

$\{(-1, 2, -2)\}$

45. a. Let x = measure of smaller angle and y = measure of larger angle.

$x + y = 90$
$y = 2x + 15$

or

$x + y = 90$
$-2x + y = 15$

b. $\left[\begin{array}{cc|c} 1 & 1 & 90 \\ -2 & 1 & 15 \end{array}\right] 2R_1 + R_2 \rightarrow \left[\begin{array}{cc|c} 1 & 1 & 90 \\ 0 & 3 & 195 \end{array}\right] \frac{1}{3}R_2 \rightarrow \left[\begin{array}{cc|c} 1 & 1 & 90 \\ 0 & 1 & 65 \end{array}\right]$

$x + y = 90$
$y = 65$

$x + 65 = 90$
$x = 25$

The angles are 25° and 65°.

47. $x + y + z = 180$
$x = y + z - 10$
$y = x + z - 50$

or

$x + y + z = 180$
$x - y - z = -10$
$-x + y - z = -50$

$$\left[\begin{array}{ccc|c} 1 & 1 & 1 & 180 \\ 1 & -1 & -1 & -10 \\ -1 & 1 & -1 & -50 \end{array}\right] \begin{array}{c} \\ -R_1 + R_2 \rightarrow \\ R_1 + R_3 \rightarrow \end{array} \left[\begin{array}{ccc|c} 1 & 1 & 1 & 180 \\ 0 & -2 & -2 & -190 \\ 0 & 2 & 0 & 130 \end{array}\right] R_2 + R_3 \rightarrow \left[\begin{array}{ccc|c} 1 & 1 & 1 & 180 \\ 0 & -2 & -2 & -190 \\ 0 & 0 & -2 & -60 \end{array}\right]$$

$$\begin{array}{c} \\ -\frac{1}{2}R_2 \rightarrow \\ -\frac{1}{2}R_3 \rightarrow \end{array} \left[\begin{array}{ccc|c} 1 & 1 & 1 & 180 \\ 0 & 1 & 1 & 95 \\ 0 & 0 & 1 & 30 \end{array}\right]$$

$x + y + z = 180$
$y + z = 95$
$z = 30$

$y + 30 = 95$
$y = 65$

$x + 65 + 30 = 180$
$x = 85$

The angles are 85°, 65°, and 30°.

49. D is true;
$x - \frac{3}{2}y = 5$
$0 = 6$ False
The system is inconsistent. $\varnothing$

51. Students should verify results.

53. Answers may vary.

55.

$$\begin{aligned} x+y+z+w&=4\\ 2x+y-2z-w&=0\\ x-2y-z-2w&=-2\\ 3x+2y+z+3w&=4 \end{aligned}$$

$$\left[\begin{array}{cccc|c} 1 & 1 & 1 & 1 & 4\\ 2 & 1 & -2 & -1 & 0\\ 1 & -2 & -1 & -2 & -2\\ 3 & 2 & 1 & 3 & 4 \end{array}\right] \begin{array}{r} \\ -2R_1+R_2\to\\ -R_1+R_3\to\\ -3R_1+R_4\to \end{array} \left[\begin{array}{cccc|c} 1 & 1 & 1 & 1 & 4\\ 0 & -1 & -4 & -3 & -8\\ 0 & -3 & -2 & -3 & -6\\ 0 & -1 & -2 & 0 & -8 \end{array}\right] \begin{array}{r} \\ \\ -3R_2+R_3\to\\ -R_2+R_4\to \end{array} \left[\begin{array}{cccc|c} 1 & 1 & 1 & 1 & 4\\ 0 & -1 & -4 & -3 & -8\\ 0 & 0 & 10 & 6 & 18\\ 0 & 0 & 2 & 3 & 0 \end{array}\right]$$

$$\begin{array}{r} \\ \\ \\ R_3-5R_4\to \end{array} \left[\begin{array}{cccc|c} 1 & 1 & 1 & 1 & 4\\ 0 & -1 & -4 & -3 & -8\\ 0 & 0 & 10 & 6 & 18\\ 0 & 0 & 0 & -9 & 18 \end{array}\right] \begin{array}{r} \\ -R_2\to\\ \frac{1}{10}R_3\to\\ -\frac{1}{9}R_4\to \end{array} \left[\begin{array}{cccc|c} 1 & 1 & 1 & 1 & 4\\ 0 & 1 & 4 & 3 & 8\\ 0 & 0 & 1 & \frac{3}{5} & \frac{9}{5}\\ 0 & 0 & 0 & 1 & -2 \end{array}\right]$$

$$\begin{aligned} x+y+z+w&=4\\ y+4z+3w&=8\\ z+\tfrac{3}{5}w&=\tfrac{9}{5}\\ w&=-2 \end{aligned}$$

$z+\frac{3}{5}(-2)=\frac{9}{5}$

$z=3$

$y+4(3)+3(-2)=8$

$y=2$

$x+2+3-2=4$

$x=1$

$\{(1, 2, 3, -2)\}$

57. a.

$$\begin{aligned} y&=ax^3+bx^2+cx+d\\ 14{,}685&=a(5)^3+b(5)^2+c(5)+d\\ 13{,}140&=a(10)^3+b(10)^2+c(10)+d\\ 15{,}095&=a(15)^3+b(15)^2+c(15)+d\\ 15{,}150&=a(20)^3+b(20)^2+c(20)+d \end{aligned}$$

or

$$\begin{aligned} 125a+25b+5c+d&=14{,}685\\ 1000a+100b+10c+d&=13{,}140\\ 3375a+225b+15c+d&=15{,}095\\ 8000a+400b+20c+d&=15{,}150 \end{aligned}$$

$$\left[\begin{array}{cccc|c} 125 & 25 & 5 & 1 & 14{,}685\\ 1000 & 100 & 10 & 1 & 13{,}140\\ 3375 & 225 & 15 & 1 & 15{,}095\\ 8000 & 400 & 20 & 1 & 15{,}150 \end{array}\right] \begin{array}{r} \\ -8R_1+R_2\to\\ -27R_1+R_3\to\\ -64R_1+R_4\to \end{array} \left[\begin{array}{cccc|c} 125 & 25 & 5 & 1 & 14{,}685\\ 0 & -100 & -30 & -7 & -104{,}340\\ 0 & -450 & -120 & -26 & -381{,}400\\ 0 & -1200 & -300 & -63 & -924{,}690 \end{array}\right]$$

$$\begin{array}{r} \\ \\ -\frac{9}{2}R_2+R_3\to\\ -12R_2+R_4\to \end{array} \left[\begin{array}{cccc|c} 125 & 25 & 5 & 1 & 14{,}685\\ 0 & -100 & -30 & -7 & -104{,}340\\ 0 & 0 & 15 & \frac{11}{2} & 88{,}130\\ 0 & 0 & 60 & 21 & 327{,}390 \end{array}\right] -4R_3+R_4\to$$

$$\left[\begin{array}{cccc|c} 125 & 25 & 5 & 1 & 14{,}685 \\ 0 & -100 & -30 & -7 & -104{,}340 \\ 0 & 0 & 15 & \frac{11}{2} & 88{,}130 \\ 0 & 0 & 0 & -1 & -25{,}130 \end{array}\right] \begin{array}{r} \frac{1}{125}R_1 \to \\ -\frac{1}{100}R_2 \to \\ \frac{1}{15}R_3 \to \\ -1R_4 \to \end{array} \left[\begin{array}{cccc|c} 1 & \frac{1}{5} & \frac{1}{25} & \frac{1}{125} & \frac{2937}{25} \\ 0 & 1 & \frac{3}{10} & \frac{7}{100} & \frac{5217}{5} \\ 0 & 0 & 1 & \frac{11}{30} & \frac{17{,}626}{3} \\ 0 & 0 & 0 & 1 & 25{,}130 \end{array}\right]$$

$$a+\frac{1}{5}b+\frac{1}{25}c+\frac{1}{125}d=\frac{2937}{25}$$
$$b+\frac{3}{10}c+\frac{7}{100}d=\frac{5217}{5}$$
$$c+\frac{11}{30}d=\frac{17{,}626}{3}$$
$$d=25{,}130$$

$$c+\frac{11}{30}(25{,}130)=\frac{17{,}626}{3}$$
$$c=-3339$$

$$b+\frac{3}{10}(-3339)+\frac{7}{100}(25{,}130)=\frac{5217}{5}$$
$$b=286$$
$$a+\frac{1}{5}(286)+\frac{1}{25}(-3339)+\frac{1}{125}(25{,}130)=\frac{2937}{25}$$
$$a=-\frac{36}{5}$$
$$y=-\frac{36}{5}x^3+286x^2-3339x+25{,}130$$

b.

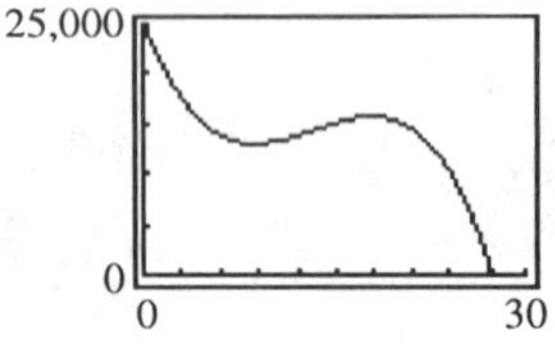

It starts to decrease rapidly.

Review Problems

59. $A=\{1, 2, 8, 9\}$, $B=\{7, 8, 9, 11\}$

a. $A\cup B=\{1, 2, 7, 8, 9, 11\}$

b. $A\cap B=\{8, 9\}$

60. $-3x\ge 6$ or $3x-1<-10$

$x\le -2$ $\quad$ $3x<-9$

$x<-3$

$\{x|x\le -2\}$

$(-\infty, -2]$

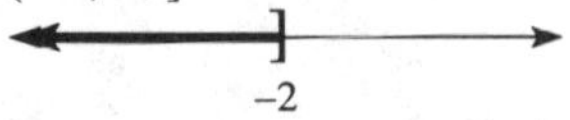

61. Let x = number of miles and y = total cost.

$y=0.14x+30$

$y=0.24x+16$

$0.14x+30=0.24x+16$

$(0.14-0.24)x=16-30$

$-0.1x=-14$

$x=140$

The costs are the same when one drives 140 miles in a day.

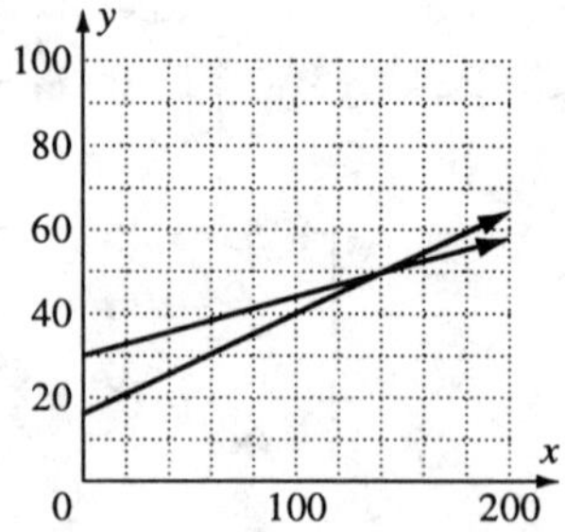

Problem Set 3.5

1. $\begin{vmatrix} 5 & 7 \\ 2 & 3 \end{vmatrix} = 5(3) - 2(7) = 15 - 14 = 1$

3. $\begin{vmatrix} -4 & 1 \\ 5 & 6 \end{vmatrix} = -4(6) - 5(1) = -24 - 5 = -29$

5. $\begin{vmatrix} -7 & 14 \\ 2 & -4 \end{vmatrix} = (-7)(-4) - (2)(14) = 28 - 28 = 0$

7. $\begin{vmatrix} 0 & -1 & 2 \\ -3 & 1 & 3 \\ 1 & -2 & -1 \end{vmatrix} = 3\begin{vmatrix} -1 & 2 \\ -2 & -1 \end{vmatrix} + 1\begin{vmatrix} -1 & 2 \\ 1 & 3 \end{vmatrix}$

$= 3(1 + 4) + 1(-3 - 2)$

$= 15 - 5$

$= 10$

9. $\begin{vmatrix} 2 & 4 & 0 \\ 1 & -2 & -3 \\ 3 & 1 & 0 \end{vmatrix} = 3\begin{vmatrix} 2 & 4 \\ 3 & 1 \end{vmatrix} = 3(-10) = -30$

11. $x + y = 7$

$x - y = 3$

$D = \begin{vmatrix} 1 & 1 \\ 1 & -1 \end{vmatrix} = -1 - 1 = -2$

$D_x = \begin{vmatrix} 7 & 1 \\ 3 & -1 \end{vmatrix} = -7 - 3 = -10$

$D_y = \begin{vmatrix} 1 & 7 \\ 1 & 3 \end{vmatrix} = 3 - 7 = -4$

$x = \frac{D_x}{D} = \frac{-10}{-2} = 5$

$y = \frac{D_y}{D} = \frac{-4}{-2} = 2$

$\{(5, 2)\}$

13. $12x + 3y = 15$

$2x - 3y = 13$

$D = \begin{vmatrix} 12 & 3 \\ 2 & -3 \end{vmatrix} = -36 - 6 = -42$

$D_x = \begin{vmatrix} 15 & 3 \\ 13 & -3 \end{vmatrix} = -45 - 39 = -84$

$D_y = \begin{vmatrix} 12 & 15 \\ 2 & 13 \end{vmatrix} = 156 - 30 = 126$

$x = \frac{D_x}{D} = \frac{-84}{-42} = 2$

$y = \frac{D_y}{D} = \frac{126}{-42} = -3$

$\{(2, -3)\}$

15. $4x - 5y = 17$

$2x + 3y = 3$

$D = \begin{vmatrix} 4 & -5 \\ 2 & 3 \end{vmatrix} = 12 - (-10) = 22$

$D_x = \begin{vmatrix} 17 & -5 \\ 3 & 3 \end{vmatrix} = 51 - (-15) = 66$

$D_y = \begin{vmatrix} 4 & 17 \\ 2 & 3 \end{vmatrix} = 12 - 34 = -22$

$x = \frac{D_x}{D} = \frac{66}{22} = 3$

$y = \frac{D_y}{D} = \frac{-22}{22} = -1$

$\{(3, -1)\}$

17. $x + 2y = 3$

$5x + 10y = 15$

$D = \begin{vmatrix} 1 & 2 \\ 5 & 10 \end{vmatrix} = 10 - 10 = 0$

$D_y = \begin{vmatrix} 3 & 2 \\ 15 & 10 \end{vmatrix} = 30 - 30 = 0$

$D_y = \begin{vmatrix} 1 & 3 \\ 5 & 15 \end{vmatrix} = 15 - 15 = 0$

The system is dependent.

$\{(x, y) | x + 2y = 3\}$

19. $3x+3y-z=10$
$x+9y+2z=16$
$x-y+6z=14$

$$D=\begin{vmatrix}3&3&-1\\1&9&2\\1&-1&6\end{vmatrix}=(3)\begin{vmatrix}9&2\\-1&6\end{vmatrix}-(1)\begin{vmatrix}3&-1\\-1&6\end{vmatrix}+(1)\begin{vmatrix}3&-1\\9&2\end{vmatrix} \quad \text{(column 1)}$$
$$=3(54+2)-(18-1)+(6+9)$$
$$=3(56)-(17)+(15)$$
$$=166$$

$$D_x=\begin{vmatrix}10&3&-1\\16&9&2\\14&-1&6\end{vmatrix}=10\begin{vmatrix}9&2\\-1&6\end{vmatrix}-16\begin{vmatrix}3&-1\\-1&6\end{vmatrix}+14\begin{vmatrix}3&-1\\9&2\end{vmatrix} \quad \text{(column 1)}$$
$$=10(56)-16(17)+14(15)$$
$$=560-272+210$$
$$=498$$

$$D_y=\begin{vmatrix}3&10&-1\\1&16&2\\1&14&6\end{vmatrix}=3\begin{vmatrix}16&2\\14&6\end{vmatrix}-1\begin{vmatrix}10&-1\\14&6\end{vmatrix}+1\begin{vmatrix}10&-1\\16&2\end{vmatrix} \quad \text{(column 1)}$$
$$=3(96-28)-(60+14)+(20+16)$$
$$=3(68)-74+36$$
$$=166$$

$$D_z=\begin{vmatrix}3&3&10\\1&9&16\\1&-1&14\end{vmatrix}=3\begin{vmatrix}9&16\\-1&14\end{vmatrix}-1\begin{vmatrix}3&10\\-1&14\end{vmatrix}+1\begin{vmatrix}3&10\\9&16\end{vmatrix} \quad \text{(column 1)}$$
$$=3(126+16)-(42+10)+(48-90)$$
$$=3(142)-52-42$$
$$=426-52-42$$
$$=332$$

$$x=\frac{D_x}{D}=\frac{498}{166}=3$$
$$y=\frac{D_y}{D}=\frac{166}{166}=1$$
$$z=\frac{D_z}{D}=\frac{332}{166}=2$$
{(3, 1, 2)}

21. $x-y+2z=4$
$3x+2y-4z=2$
$x+y+z=3$

$$D=\begin{vmatrix}1&-1&2\\3&2&-4\\1&1&1\end{vmatrix}=1\begin{vmatrix}-1&2\\2&-4\end{vmatrix}-1\begin{vmatrix}1&2\\3&-4\end{vmatrix}+1\begin{vmatrix}1&-1\\3&2\end{vmatrix} \quad \text{(row 3)}$$

$= (4-4) - (-4-6) + (2+3)$
$= 0 + 10 + 5 = 15$

$D_x = \begin{vmatrix} 4 & -1 & 2 \\ 2 & 2 & -4 \\ 3 & 1 & 1 \end{vmatrix} = 3\begin{vmatrix} -1 & 2 \\ 2 & -4 \end{vmatrix} - 1\begin{vmatrix} 4 & 2 \\ 2 & -4 \end{vmatrix} + 1\begin{vmatrix} 4 & -1 \\ 2 & 2 \end{vmatrix}$ (row 3)

$= 3(4-4) - 1(-16-4) + 1(8+2)$
$= 0 + 20 + 10 = 30$

$D_y = \begin{vmatrix} 1 & 4 & 2 \\ 3 & 2 & -4 \\ 1 & 3 & 1 \end{vmatrix} = 1\begin{vmatrix} 4 & 2 \\ 2 & -4 \end{vmatrix} - 3\begin{vmatrix} 1 & 2 \\ 3 & -4 \end{vmatrix} + 1\begin{vmatrix} 1 & 4 \\ 3 & 2 \end{vmatrix}$ (row 3)

$= (-16-4) - 3(-4-6) + (2-12)$
$= -20 + 30 - 10 = 0$

$D_z = \begin{vmatrix} 1 & -1 & 4 \\ 3 & 2 & 2 \\ 1 & 1 & 3 \end{vmatrix} = 1\begin{vmatrix} -1 & 4 \\ 2 & 2 \end{vmatrix} - 1\begin{vmatrix} 1 & 4 \\ 3 & 2 \end{vmatrix} + 3\begin{vmatrix} 1 & -1 \\ 3 & 2 \end{vmatrix}$ (row 3)

$= (-2-8) - (2-12) + 3(2+3)$
$= -10 + 10 + 15 = 15$

$x = \frac{D_x}{D} = \frac{30}{15} = 2$

$y = \frac{D_y}{D} = \frac{0}{15} = 0$

$z = \frac{D_z}{D} = \frac{15}{15} = 1$

$\{(2, 0, 1)\}$

23. $x - y = 8$
$x + y - z = 1$
$x - 2z = 0$

$D = \begin{vmatrix} 1 & -1 & 0 \\ 1 & 1 & -1 \\ 1 & 0 & -2 \end{vmatrix} = -(-1)\begin{vmatrix} 1 & -1 \\ 1 & -2 \end{vmatrix} + 1\begin{vmatrix} 1 & 0 \\ 1 & -2 \end{vmatrix} - 0\begin{vmatrix} 1 & 0 \\ 1 & -1 \end{vmatrix}$ (column 2)

$= (-2+1) + (-2-0) - 0$
$= -1 - 2 = -3$

$D_x = \begin{vmatrix} 8 & -1 & 0 \\ 1 & 1 & -1 \\ 0 & 0 & -2 \end{vmatrix} = 0\begin{vmatrix} -1 & 0 \\ 1 & -1 \end{vmatrix} - 0\begin{vmatrix} 8 & 0 \\ 1 & -1 \end{vmatrix} - 2\begin{vmatrix} 8 & -1 \\ 1 & 1 \end{vmatrix}$ (row 3)

$= 0 - 0 - 2(8+1)$
$= -18$

$D_y = \begin{vmatrix} 1 & 8 & 0 \\ 1 & 1 & -1 \\ 1 & 0 & -2 \end{vmatrix} = -8\begin{vmatrix} 1 & -1 \\ 1 & -2 \end{vmatrix} + 1\begin{vmatrix} 1 & 0 \\ 1 & -2 \end{vmatrix} - 0\begin{vmatrix} 1 & 0 \\ 1 & -1 \end{vmatrix}$ (col 2)

$= -8(-2+1) + 1(-2-0) - 0$
$= 8 - 2 = 6$

$$D_z = \begin{vmatrix} 1 & -1 & 8 \\ 1 & 1 & 1 \\ 1 & 0 & 0 \end{vmatrix} = 1\begin{vmatrix} -1 & 8 \\ 1 & 1 \end{vmatrix} - 0\begin{vmatrix} 1 & 8 \\ 1 & 1 \end{vmatrix} + 0\begin{vmatrix} 1 & -1 \\ 1 & 1 \end{vmatrix} \text{ (row 3)}$$
$$= (-1 - 8) - 0 + 0 = -9$$
$$x = \frac{D_x}{D} = \frac{-18}{-3} = 6$$
$$y = \frac{D_y}{D} = \frac{6}{-3} = -2$$
$$z = \frac{D_z}{D} = \frac{-9}{-3} = 3$$
$\{(6, -2, 3)\}$

25. $3x + y = -5$
$2x + z = -1$
$x - 2y = -4$

$$D = \begin{vmatrix} 3 & 1 & 0 \\ 2 & 0 & 1 \\ 1 & -2 & 0 \end{vmatrix} = 0\begin{vmatrix} 2 & 0 \\ 1 & -2 \end{vmatrix} - 1\begin{vmatrix} 3 & 1 \\ 1 & -2 \end{vmatrix} + 0\begin{vmatrix} 3 & 1 \\ 2 & 0 \end{vmatrix} \text{ (column 3)}$$
$$= 0 - 1(-6 - 1) + 0$$
$$= 7$$
$$D_x = \begin{vmatrix} -5 & 1 & 0 \\ -1 & 0 & 1 \\ -4 & -2 & 0 \end{vmatrix} = 0\begin{vmatrix} -1 & 0 \\ -4 & -2 \end{vmatrix} - 1\begin{vmatrix} -5 & 1 \\ -4 & -2 \end{vmatrix} + 0\begin{vmatrix} -5 & 1 \\ -1 & 0 \end{vmatrix} \text{ (column 3)}$$
$$= 0 - (10 + 4) + 0$$
$$= -14$$
$$D_y = \begin{vmatrix} 3 & -5 & 0 \\ 2 & -1 & 1 \\ 1 & -4 & 0 \end{vmatrix} = 0\begin{vmatrix} 2 & -1 \\ 1 & -4 \end{vmatrix} - 1\begin{vmatrix} 3 & -5 \\ 1 & -4 \end{vmatrix} + 0\begin{vmatrix} 3 & -5 \\ 2 & -1 \end{vmatrix} \text{ (column 3)}$$
$$= 0 - (-12 + 5) + 0$$
$$= 7$$
$$D_z = \begin{vmatrix} 3 & 1 & -5 \\ 2 & 0 & -1 \\ 1 & -2 & -4 \end{vmatrix} = 3\begin{vmatrix} 0 & -1 \\ -2 & -4 \end{vmatrix} - 2\begin{vmatrix} 1 & -5 \\ -2 & -4 \end{vmatrix} + 1\begin{vmatrix} 1 & -5 \\ 0 & -1 \end{vmatrix} \text{ (column 1)}$$
$$= 3(0 - 2) - 2(-4 - 10) + (-1 - 0)$$
$$= -6 + 28 - 1 = 21$$
$$x = \frac{D_x}{D} = \frac{-14}{7} = -2$$
$$y = \frac{D_y}{D} = \frac{7}{7} = 1$$
$$z = \frac{D_z}{D} = \frac{21}{7} = 3$$
$\{(-2, 1, 3)\}$

27. a. Vertices (3, –5), (2, 6), (–3, 5)

$$\text{Area} = \frac{1}{2}\begin{vmatrix} 3 & -5 & 1 \\ 2 & 6 & 1 \\ -3 & 5 & 1 \end{vmatrix} = \frac{1}{2}\left[3\begin{vmatrix} 6 & 1 \\ 5 & 1 \end{vmatrix} + 5\begin{vmatrix} 2 & 1 \\ -3 & 1 \end{vmatrix} + 1\begin{vmatrix} 2 & 6 \\ -3 & 5 \end{vmatrix}\right]$$

$$= \frac{1}{2}[3(6-5)+5(2+3)+(10+18)]$$

$$= \frac{1}{2}(56) = 28$$

The area is 28 square units.

b.

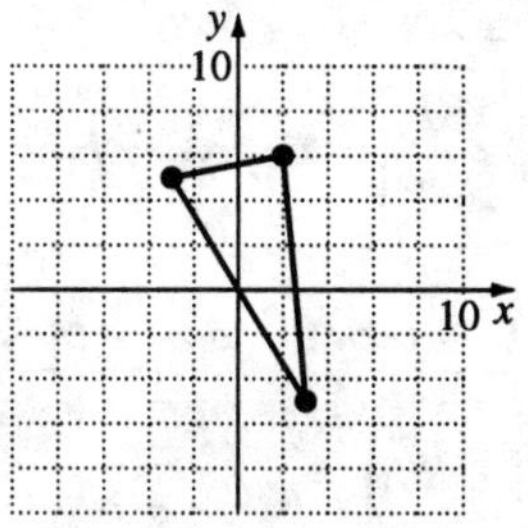

$$A = \frac{1}{2}bh = 28$$

29. (3, –1), (0, –3), (12, 5)

$$\begin{vmatrix} 3 & -1 & 1 \\ 0 & -3 & 1 \\ 12 & 5 & 1 \end{vmatrix} = 3(-3-5)+12(-1+3) = -24+24 = 0\text{; Yes}$$

31. (3, –5) and (–2, 6)

$$\begin{vmatrix} x & y & 1 \\ 3 & -5 & 1 \\ -2 & 6 & 1 \end{vmatrix} = x(-5-6) - y(3+2) + 1(18-10) = -11x - 5y + 8$$

$$-11x - 5y + 8 = 0$$

$$-11x - 5y = -8$$

33. D is true; $\begin{vmatrix} \frac{1}{2} & \frac{1}{4} \\ \frac{1}{3} & -\frac{1}{2} \end{vmatrix} = \frac{1}{2}\left(-\frac{1}{2}\right) - \frac{1}{3}\left(\frac{1}{4}\right) = -\frac{1}{4} - \frac{1}{12} = -\frac{4}{12} = -\frac{1}{3}$

35. Students should verify results.

37. Answers may vary.

39. Answers may vary.

41. Answers may vary.

43. $\begin{vmatrix} 2x+1 & 5 \\ x-1 & -3 \end{vmatrix} = 3x-1$

$-6x-3-5x+5=3x-1$

$-14x=-3$

$x=\frac{3}{14}$

45. Original system:

$a_1x+b_1y=c_1$
$a_2x+b_2y=c_2$

$D=\begin{vmatrix} a_1 & b_1 \\ a_2 & b_2 \end{vmatrix}=a_1b_2-a_2b_1$

$D_x=\begin{vmatrix} c_1 & b_1 \\ c_2 & b_2 \end{vmatrix}=c_1b_2-c_2b_1$

$x=\frac{c_1b_2-c_2b_1}{a_1b_2-a_2b_1}$

$D_y=\begin{vmatrix} a_1 & c_1 \\ a_2 & c_2 \end{vmatrix}=a_1c_2-a_2c_1$

$y=\frac{a_1c_2-a_2c_1}{a_1b_2-a_2b_1}$

New system: Multiply first equation by K.

$Ka_1x+Kb_1y=Kc_1$
$a_2x+b_2y=c_2$

$D=\begin{vmatrix} Ka_1 & Kb_1 \\ a_2 & b_2 \end{vmatrix}=Ka_1b_2-Ka_2b_1$

$=K(a_1b_2-a_2b_1)$

$D_x=\begin{vmatrix} Kc_1 & Kb_1 \\ c_2 & b_2 \end{vmatrix}=Kc_1b_2-Kb_1c_2$

$=K(c_1b_2-b_1c_2)$

$x=\frac{K(c_1b_2-b_1c_2)}{K(a_1b_2-a_2b_1)}=\frac{c_1b_2-b_1c_2}{a_1b_2-a_2b_1}$

$D_y=\begin{vmatrix} Ka_1 & Kc_1 \\ a_2 & c_2 \end{vmatrix}=Ka_1c_2-Ka_2c_1$

$=K(a_1c_2-a_2c_1)$

$y=\frac{K(a_1c_2-a_2c_1)}{K(a_1b_2-a_2b_1)}=\frac{a_1c_2-a_2c_1}{a_1b_2-a_2b_1}$

The values of x and y remain the same.

47. The equation of the line is $y-y_1=\frac{y_2-y_1}{x_2-x_1}(x-x_1)$.

Multiply both sides by (x_2-x_1).

$$(x_2-x_1)(y-y_1)=(y_2-y_1)(x-x_1)$$
$$x_2y-x_2y_1-x_1y+x_1y_1=y_2x-y_2x_1-xy_1+x_1y_1$$
$$x_2y-x_2y_1-x_1y=y_2x-y_2x_1-xy_1 \quad \text{(i)}$$

We must now show that the equation involving the determinant is equivalent to the above equation, namely equation (i).

$$\begin{vmatrix} x & y & 1 \\ x_1 & y_1 & 1 \\ x_2 & y_2 & 1 \end{vmatrix}=0$$

Expanding about the third column:

$$1\begin{vmatrix} x_1 & y_1 \\ x_2 & y_2 \end{vmatrix}-1\begin{vmatrix} x & y \\ x_2 & y_2 \end{vmatrix}+1\begin{vmatrix} x & y \\ x_1 & y_1 \end{vmatrix}=0$$

$$x_1y_2-x_2y_1-x_1y=y_2x-x_2y-xy_1$$

49. **a.** $\begin{vmatrix} a & a \\ 0 & a \end{vmatrix} = a^2 - 0 = a^2$

b. $\begin{vmatrix} a & a & a \\ 0 & a & a \\ 0 & 0 & a \end{vmatrix} = a\begin{vmatrix} a & a \\ 0 & a \end{vmatrix} = a(a^2) = a^3$

c. $\begin{vmatrix} a & a & a & a \\ 0 & a & a & a \\ 0 & 0 & a & a \\ 0 & 0 & 0 & a \end{vmatrix} = a\begin{vmatrix} a & a & a \\ 0 & a & a \\ 0 & 0 & a \end{vmatrix}$

$= a(a^3) = a^4$

d. The diagonal is made up of a's and below the a's are zeros.

e. The number of a's in diagonal is exponent on a in evaluation of determinant.

51. $x + 2y + 6z = 5$
$-3x - 6y + 5z = 8$
$2x + 6y + 9z = 7$

$-1 + 2(0) + 6(1) = 5$
$5 = 5$

$-3(-1) - 6(0) + 5(1) = 8$
$8 = 8$

$2(-1) + 6(0) + 9(1) = 7$
$7 = 7$
Explanations may vary.

Review Problems

52. Let x = number of miles.
$1.40 + 1.60x = 35$
$1.60x = 33.60$
$x = 21$
21 miles

53. $y < 3x + 2$

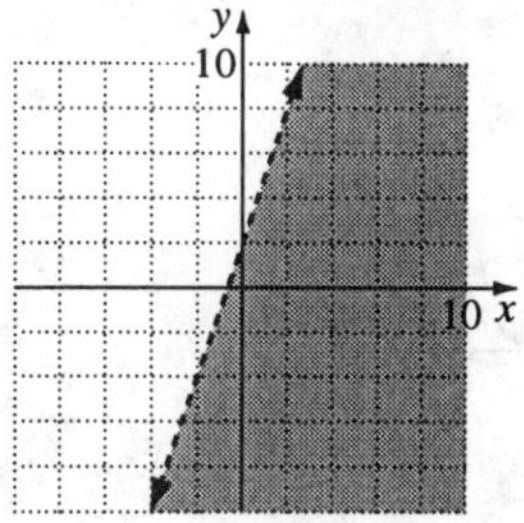

54. $2[5 - 4(x + 2)] = 2(x - 3) + 5$
$2(5 - 4x - 8) = 2x - 6 + 5$
$10 - 8x - 16 = 2x - 1$
$-10x = 5$
$x = -\frac{1}{2}$

Problem Set 3.6

1. $3x + 6y \le 6$
$2x + y \le 8$

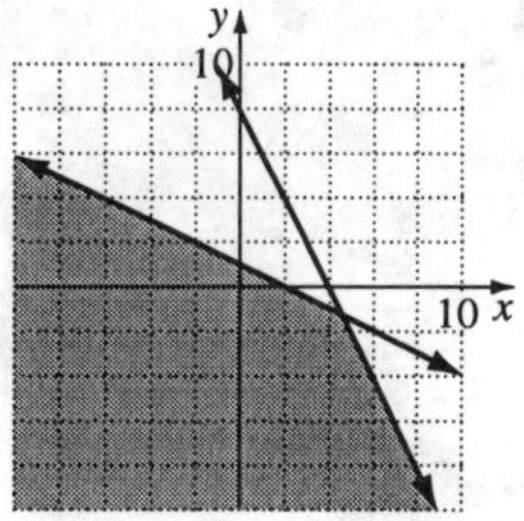

$\left(\frac{14}{3}, -\frac{4}{3}\right)$

3. $y < -2x + 3$
$x - y > 2$

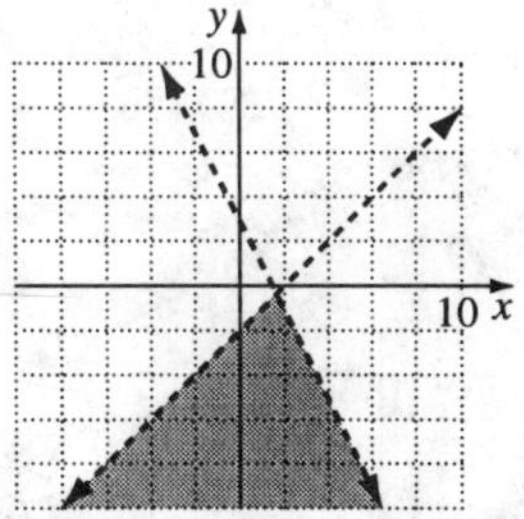

$\left(\frac{5}{3}, -\frac{1}{3}\right)$

5. $y < -2x + 4$
$y < x - 4$

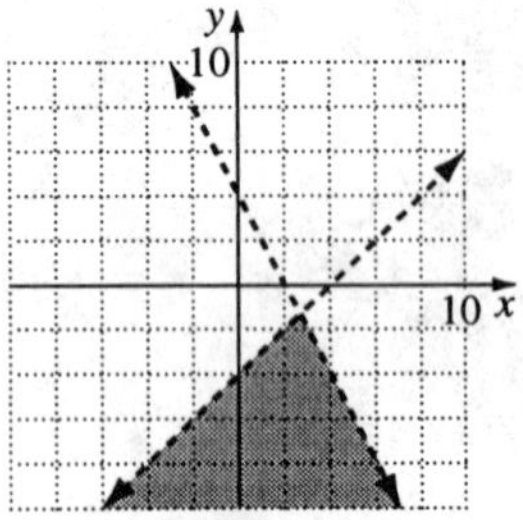

$\left(\frac{8}{3}, -\frac{4}{3}\right)$

7. $-2 < x \le 4$

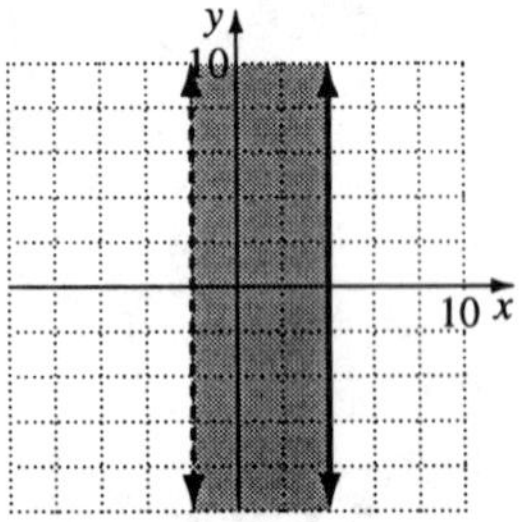

9. $-4 \le y < 2$

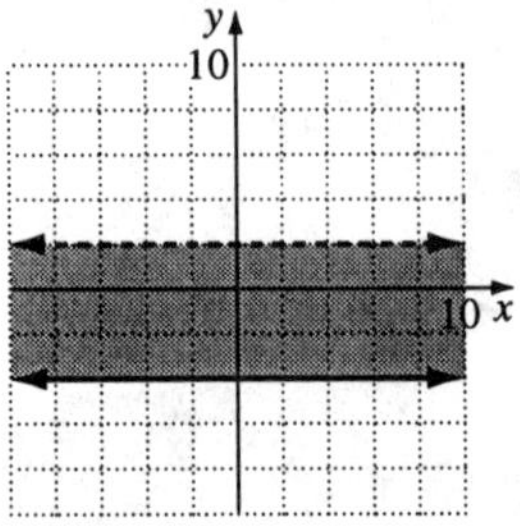

11. $4x - 5y > -20$
$x > -3$

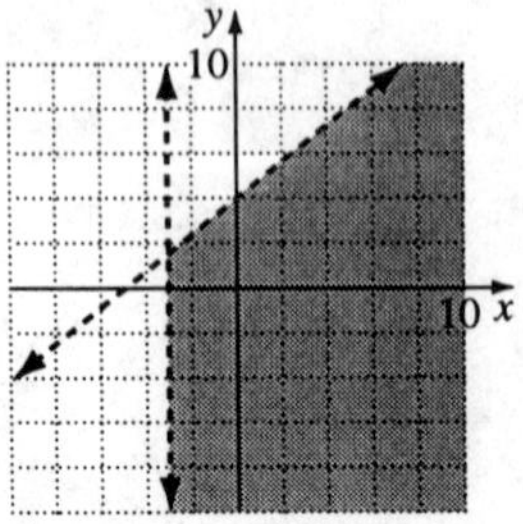

$\left(-3, \frac{8}{5}\right)$

13. $x + y \ge 0$
$y \le 3$
$3x - y \le 6$

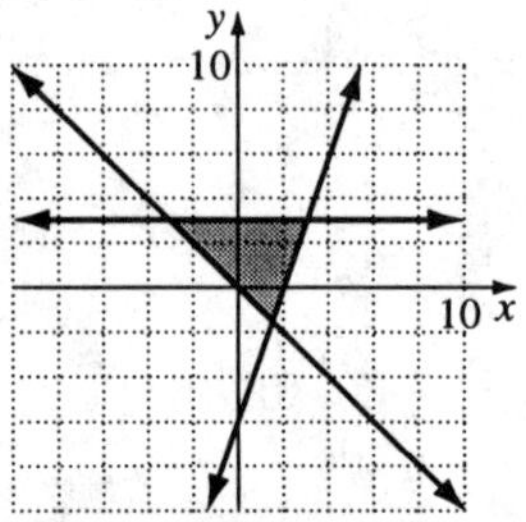

$(-3, 3)$, $\left(\frac{3}{2}, -\frac{3}{2}\right)$, $(3, 3)$

15. $x + 2y \le 20$
$x + y \le 16$
$x \ge 0$
$y \ge 0$

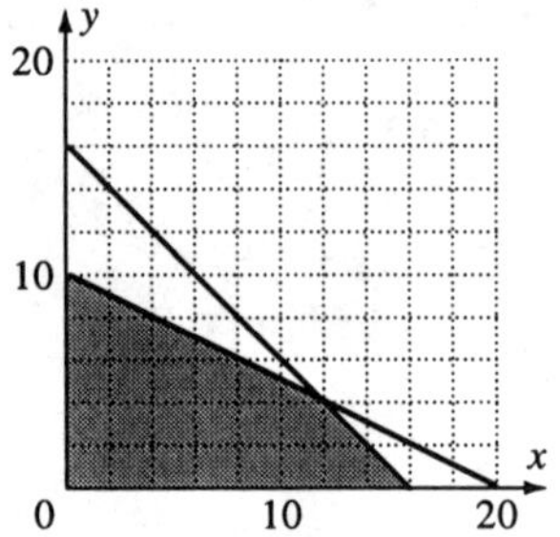

$(12, 4)$, $(0, 10)$, $(16, 0)$, $(0, 0)$

17. $x + 2y \le 8$
$y \ge 2x$
$y \le 3x$

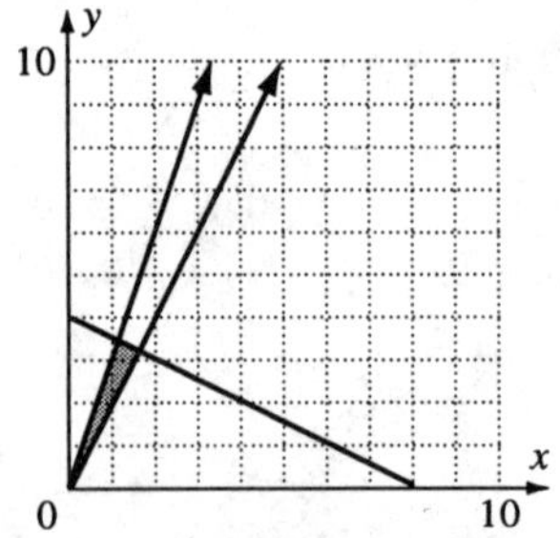

$(0, 0)$, $\left(\frac{8}{5}, \frac{16}{5}\right)$, $\left(\frac{8}{7}, \frac{24}{7}\right)$

19.

Corner	Objective Function $z = 5x + 6y$
(1, 2)	$z = 5(1) + 6(2) = 17$
(2, 10)	$z = 5(2) + 6(10) = 70$
(7, 5)	$z = 5(7) + 6(5) = 65$
(8, 3)	$z = 5(8) + 6(3) = 58$

The maximum value is 70 and the minimum value is 17.

21.

Corner	Objective Function $z = 40x + 50y$
(0, 0)	$z = 40(0) + 50(0) = 0$
(0, 8)	$z = 40(0) + 50(8) = 400$
(4, 9)	$z = 40(4) + 50(9) = 610$
(8, 0)	$z = 40(8) + 50(0) = 320$

The maximum value is 610 and the minimum value is 0.

23.

Corner	Objective Function $z = x + y$
(0, 1)	$z = 0 + 1 = 1$
(6, 13)	$z = 6 + 13 = 19$
(6, 1)	$z = 6 + 1 = 7$

The maximum is 19 at (6, 13) and the minimum is 1 at (0, 1).

25.

Corner	Objective Function $z = -x + 3y$
(2, 2)	$z = -2 + 3(2) = 4$
(2, 8)	$z = -2 + 3(8) = 22$
(6, 12)	$z = -6 + 3(12) = 30$
(6, –6)	$z = -6 + 3(-6) = -24$

The maximum is 30 at (6, 12) and the minimum is –24 at (6, –6).

27. a. $z = 125x + 200y$

b. $x \le 450$
$y \le 200$
$600x + 900y \le 360{,}000$
or
$$\begin{aligned} 2x + 3y &\le 1200 \\ x &\le 450 \\ y &\le 200 \\ x &\ge 0 \\ y &\ge 0 \end{aligned}$$

c. 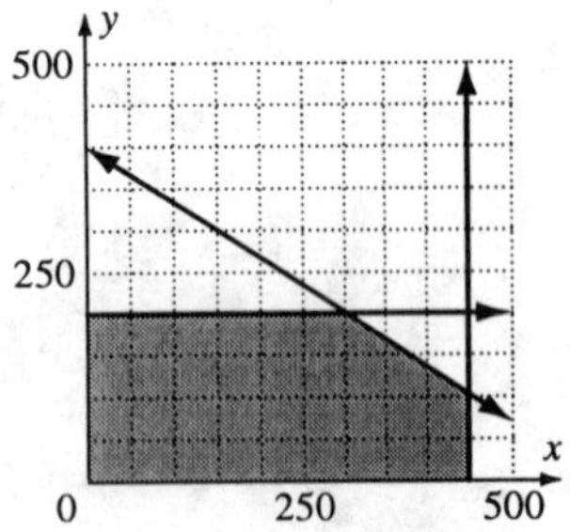

d.

Corner	Objective Function
	$z = 125x + 200y$
(0, 0)	$z = 125(0) + 200(0) = 0$
(0, 200)	$z = 125(0) + 200(200)$ $= 40,000$
(300, 200)	$z = 125(300) + 200(200)$ $= 77,500$
(450, 100)	$z = 125(450) + 200(100)$ $= 76,250$
(450, 0)	$z = 125(450) + 200(0)$ $= 56,250$

e. 300; 200; $77,500

29. Let x = cartons of food and y = cartons of clothing.

$$30x + 20y < 12,000$$
$$50x + 5y \le 18,000$$
$$x \ge 0$$
$$y \ge 0$$
$$z = 5x + 6y$$

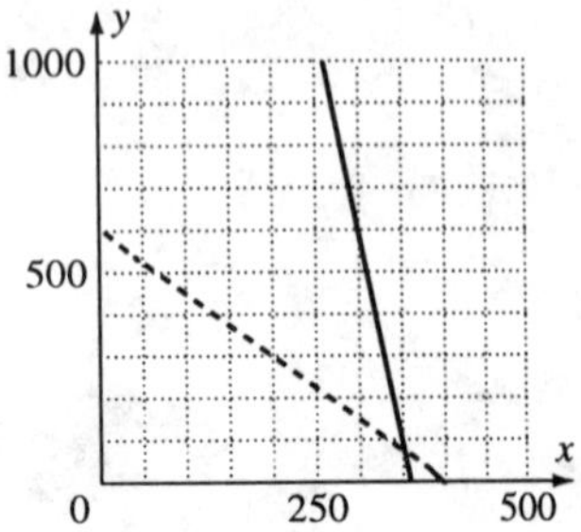

Corner	Objective Function
	$z = 5x + 6y$
(0, 0)	$z = 5(0) + 6(0) = 0$
(0, 600)	$z = 5(0) + 6(600) = 3600$
$\left(\frac{6000}{17}, \frac{1200}{17}\right)$	$z = 5\left(\frac{6000}{17}\right) + 6\left(\frac{1200}{17}\right)$ ≈ 2188
(360, 0)	$z = 5(360) + 6(0) = 1800$

The maximum is at (0, 600) which means 600 boxes of clothing. However, the total volume must be less than 12,000 cubic feet. So 599 boxes of clothing should be sent to help a maximum of 3594 people.

31. Let x = number of parents and y = number of students.

$$x + y \le 150$$
$$2y \ge x$$
$$x \ge 0$$
$$y \ge 0$$
$$z = 2x + y$$

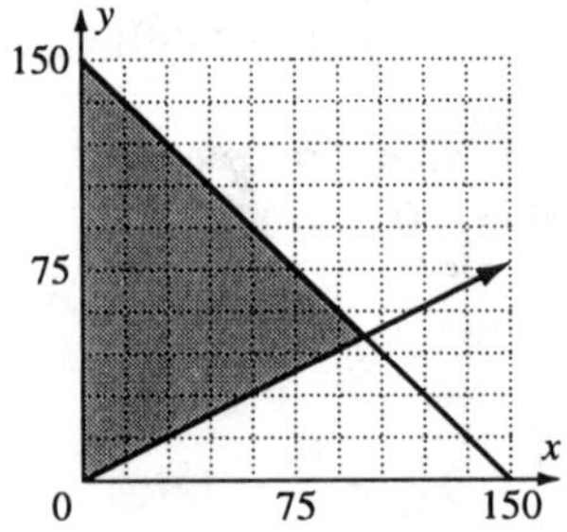

Corner	Objective Function
	$z = 2x + y$
(0, 0)	$z = 2(0) + 0 = 0$
(100, 50)	$z = 2(100) + 50 = 250$
(0, 150)	$z = 2(0) + 150 = 150$

100 parents and 50 students should attend to raise the maximum amount of money, \$250.

33. B is true.

35. Answers may vary.

37. $|x| \le 3$

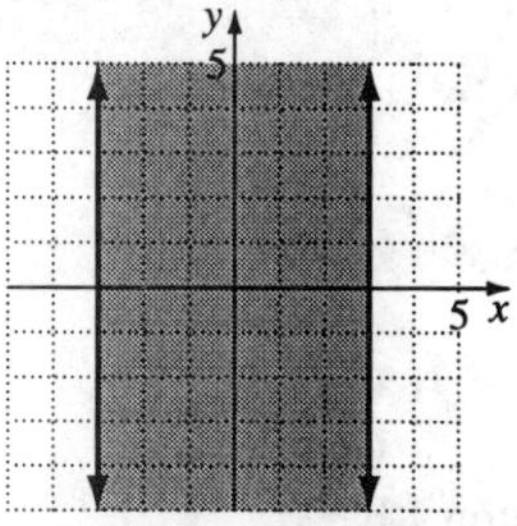

39. $|x| \le 3$
$|y+1| \ge 3$

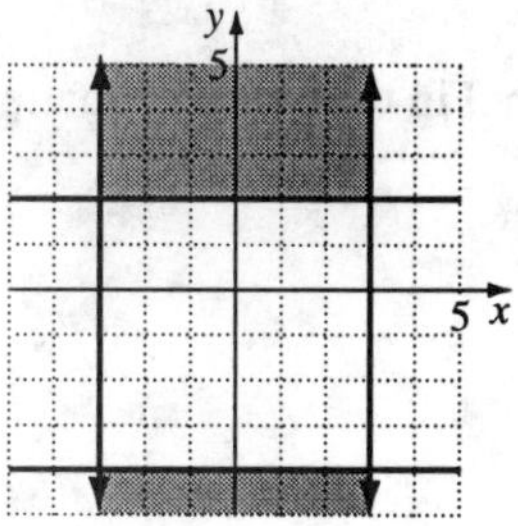

41. $2x + 3y \le 9$
$x - y \le 2$
$x \ge 0$
$y \ge 0$
Objective function
$z = ax + by,\ a = \frac{2}{3}b$
$z = \frac{2}{3}bx + by$
$z = b\left(\frac{2}{3}x + y\right)$

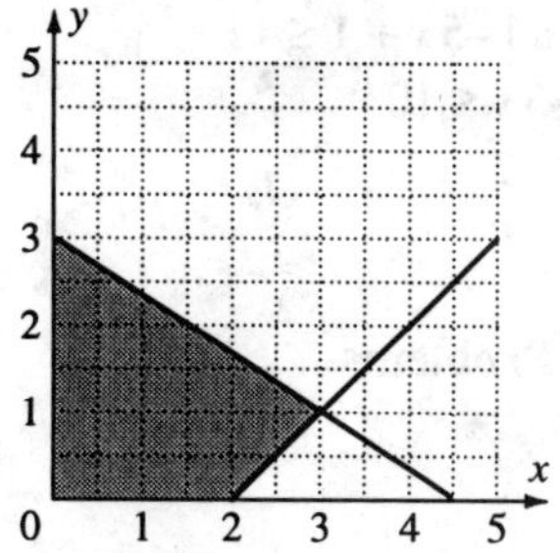

Corner	Objective Function
	$z = b\left(\frac{2}{3}x + y\right)$
(0, 0)	$z = b\left(\frac{2}{3}(0) + 0\right) = 0$
(0, 3)	$z = b\left(\frac{2}{3}(0) + 3\right) = 3b$
(3, 1)	$z = b\left(\frac{2}{3}(3) + 1\right) = 3b$
(2, 0)	$z = b\left(\frac{2}{3}(2) + 0\right) = \frac{4}{3}b$

The objective function has equal values at (0, 3) and (3, 1).

Review Problems

43. $-14\left(\frac{2}{7}\right) - 6 \div (-3) = -4 - (-2)$
$= -4 + 2 = -2$

44. $\frac{x+2}{2} + \frac{x-2}{4} = 8$
$4\left(\frac{x+2}{2} + \frac{x-2}{4}\right) = 8 \cdot 4$
$2(x + 2) + x - 2 = 32$
$2x + 4 + x - 2 = 32$
$3x + 2 = 32$
$3x = 30$
$x = 10$

45. $3x + 5 \geq 26$ and $-5x + 1 \leq 11$
$3x \geq 21$ and $-5x \leq 10$
$x \geq 7$ and $x \geq -2$
$x \geq 7$

Chapter 3 Review Problems

1. $x + y = 5$
$3x - y = 3$

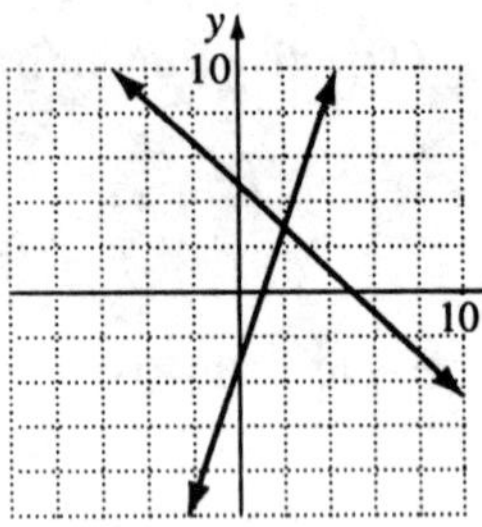

$\{(2, 3)\}$

2. $3x - 2y = 6$
$6x - 4y = 12$

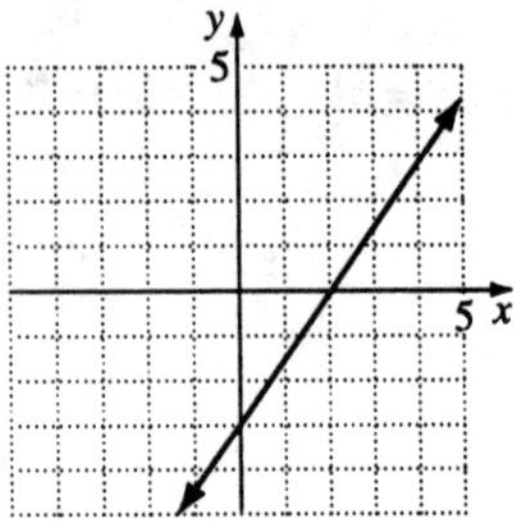

same line
dependent

3. $y = \frac{3}{5}x - 3$
$2x - y = -4$

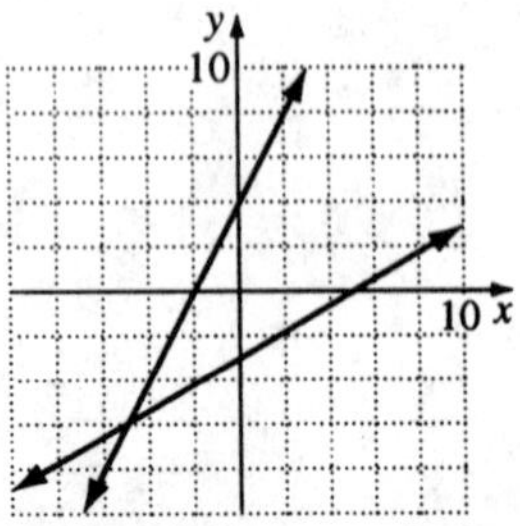

$\{(-5, -6)\}$

4. $y = -x + 4$
$3x + 3y = -6$

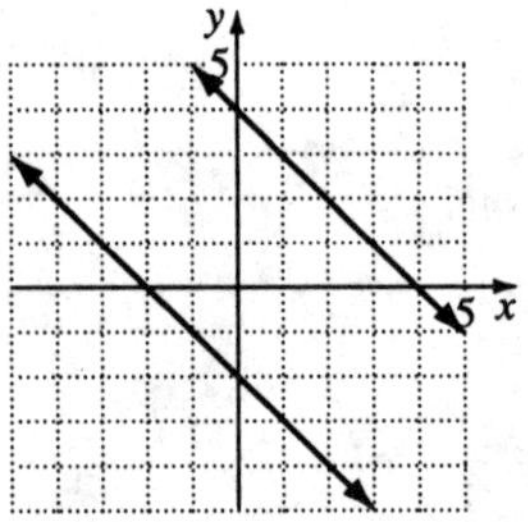

parallel lines
inconsistent

5. $2x - y = 2$ (1)
$x + 2y = 11$ (2)
Multiply equation 1 by 2.

$$\begin{array}{r} 4x - 2y = 4 \\ x + 2y = 11 \\ \hline 5x = 15 \\ x = 3 \end{array}$$

Replace x with 3 in the second equation.
$x + 2y = 11$
$3 + 2y = 11$
$2y = 8$
$y = 4$
$\{(3, 4)\}$

6. $y = 4 - x$ (1)
$3x + 3y = 12$ (2)
Divide equation 2 by −3.

$$\begin{array}{r} x + y = 4 \\ -x - y = -4 \\ \hline 0 = 0 \end{array}$$

True, infinitely many solutions
The system is dependent.
$\{(x, y) | y = 4 - x\}$

7. $5x + 3y = -3$ (1)
$2x + 7y = -7$ (2)
Multiply equation 1 by 7 and equation 2 by −3.

$$\begin{array}{r} 35x + 21y = -21 \\ -6x - 21y = 21 \\ \hline 29x = 0 \\ x = 0 \end{array}$$

Substitute 0 for x in the second equation.
$2x + 7y = -7$
$2(0) + 7y = -7$
$y = -1$
$\{(0, -1)\}$

8. $x - 2y + 3 = 0$ (1)
$2x - 4y + 7 = 0$ (2)
Multiply equation 1 by –2.
$$\begin{array}{r} -2x + 4y = 6 \\ 2x - 4y = -7 \\ \hline 0 = -1 \end{array}$$
False (contradiction)
The system is inconsistent. ∅

9. $\frac{1}{8}x + \frac{3}{4}y = \frac{19}{8}$ (1)
$-\frac{1}{2}x + \frac{3}{4}y = \frac{1}{2}$ (2)
Multiply equation 1 by 16 and equation 2 by 4.
$$\begin{array}{r} 2x + 12y = 38 \\ -2x + 3y = 2 \\ \hline 15y = 40 \\ y = \frac{8}{3} \end{array}$$
Divide equation 1 by 2 and substitute $\frac{8}{3}$ for y.
$x + 6y = 19$
$x + 6\left(\frac{8}{3}\right) = 19$
$x = 3$
$\left\{\left(3, \frac{8}{3}\right)\right\}$

10. Let x = skaters in 1992 (in millions) and y = skaters in 1993 (in millions).
$x + y = 21.8$
$2y - x = 6.2 + 9.2$
$x = 21.8 - y$
$2y - (21.8 - y) = 6.2 + 9.2$
$2y - 21.8 + y = 15.4$
$3y = 37.2$
$y = 12.4$
$x = 21.8 - 12.4 = 9.4$
The number of skaters was 9.4 million in 1992 and 12.4 million in 1993.

11. Let x = mg of cholesterol in an ounce of shrimp and y = mg of cholesterol in an ounce of scallops.
$3x + 2y = 156$
$5x + 3y = 300 - 45$
Add –3 times (2) to 5 times (1).
$$\begin{array}{r} 15x + 10y = 780 \\ -15x - 9y = -765 \\ \hline y = 15 \end{array}$$
$3x + 2y = 156$
$3x + 2(15) = 156$
$3x = 126$
$x = 42$
The amount of cholesterol in one ounce is 42 mg in shrimp and 15 mg in scallops.

12. Let x = cost of 1 pen and y = cost of 1 pad.
$8x + 6y = 16.10$ (1)
$3x + 2y = 5.85$ (2)
Multiply equation 2 by –3.
$$\begin{array}{r} 8x + 6y = 16.10 \\ -9x - 6y = -17.55 \\ \hline -x = -1.45 \\ x = 1.45 \end{array}$$
(pen)
Substitute 1.45 for x in equation 2.
$3x + 2y = 5.85$
$3(1.45) + 2y = 5.85$
$2y = 1.50$
$y = 0.75$ (pad)
cost of one pen: \$1.45; cost of one pad: 75¢

13. perimeter $= A + B + A + 2B + A + A + 2B + B$

$310 = 4A + 6B$

cost $= 30(2A) + 8B$

$2600 = 60A + 8B$

Solve equations:

$2A + 3B = 155 \quad (1)$

$15A + 2B = 650 \quad (2)$

Add 15 times (1) to –2 times (2).

$$\begin{array}{r} 30A + 45B = 2325 \\ -30A - 4B = -1300 \\ \hline 41B = 1025 \\ B = 25 \end{array}$$

Substitute 25 for B in equation 1.

$2A + 3(25) = 155$

$2A = 80$

$A = 40$

$A = 40$ feet

$B = 25$ feet

The lot is two rectangles 40 ft by 50 ft sitting next to each other (can also be described as one long rectangle 80 ft by 50 ft) with another rectangle, 40 ft by 25 ft above the first rectangle.

14. $7x + 14 = 4y + 4$ (Base angles of an isosceles triangle are equal.)

$(4x + 3y + 5) + (4y + 4) + (7x + 14) = 180$

Simplify:

$7x - 4y = -10 \quad (1)$

$11x + 7y = 157 \quad (2)$

Multiply equation 1 by 7 and equation 2 by 4.

$$\begin{array}{r} 49x - 28y = -70 \\ 44x + 28y = 628 \\ \hline 93x = 558 \\ x = 6 \end{array}$$

Substitute 6 for x in equation 1.

$7x - 4y = -10$

$7(6) - 4y = -10$

$y = 13$

$m\angle A = (4x + 3y + 5)° = [4(6) + 3(13) + 5]°$

$= 68°$

$m\angle B = (4y + 4)° = [4(13) + 4]° = 56°$

$m\angle C = (7x + 14)° = [7(6) + 14]° = 56°$

15. Let x = speed of canoeist in calm water and y = rate of current.

	Rate	· Time	= Distance
With current	$x + y$	2	12
Against current	$x - y$	4	12

$\begin{array}{l} 2(x+y) = 12 \\ 4(x-y) = 12 \end{array}$ or $\begin{array}{l} x + y = 6 \\ x - y = 3 \end{array}$

Add equations.

$2x = 9$

$x = 4.5$

$y = 6 - x$

$y = 6 - 4.5$

$y = 1.5$

The canoeist's speed in calm water is 4.5 mph and the rate of the current is 1.5 mph.

16. Let x = liters of 90% H_2SO_4
Let y = liters of 75% H_2SO_4

	percent	liters	total
A	0.90	x	$0.90x$
B	0.75	$20-x$	$0.75(20-x)$
mixture	0.78	20	0.78(20)

$0.90x + 0.75(20 - x) = 0.78(20)$
$0.90x + 15 - 0.75x = 15.6$
$0.15x = 0.6$
$x = 4$
$20 - 4 = 16$
4 liters of 90% sulfuric acid solution and 16 liters of 75% sulfuric acid solution

17. Let x = number of rabbits and y = number of pheasants.
$x + y = 35$ (1)
$4x + 2y = 94$ (2)
Multiply equation 1 by –1 and divide equation 2 by 2.
$-x - y = -35$
$2x + y = 47$
$x = 12$
12 rabbits
$x + y = 35$
$12 + y = 35$
$y = 23$ pheasants
12 rabbits, 23 pheasants

18. a. $400 = 10m + b$
$160 = 14m + b$
Solve each for b and set equal to solve for m.
$400 - 10m = 160 - 14m$
$4m = -240$
$m = -60$
$b = 400 - 10(-60) = 1000$
$y = -60x + 1000$

b. $240 = 10m + b$
$256 = 14m + b$
Solve for m.
$240 - 10m = 256 - 14m$
$4m = 16$
$m = 4$
$b = 240 - 10(4) = 200$
$y = 4x + 200$

c. $y = -60x + 1000$
$y = 4x + 200$

$-60x + 1000 = 4x + 200$
$-64x = -800$
$x = 12.5$

$y = 4(12.5) + 200 = 250$
(12.5, 250)
At a video price of \$12.50, the number sold equals the number supplied, 250.

d. $y = -60x + 1000$
$y = 4x + 200$

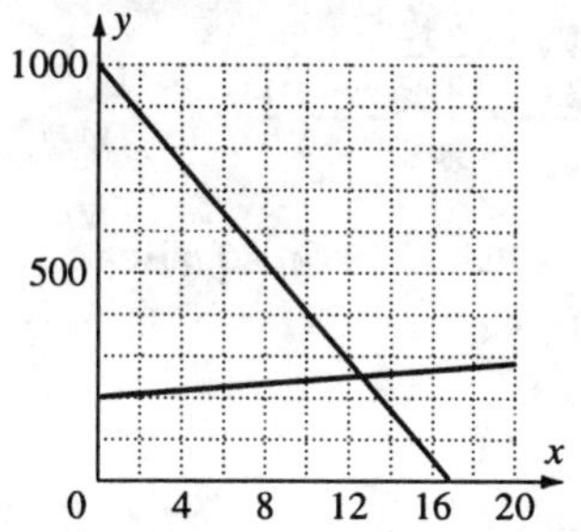

19. Cost: $C(x) = 1.10x + 1080$
Revenue: $R(x) = 1.3x$

a. From the graph, 5400 gallons

b. From the graph, $x < 5400$ gallons

c. Sales cannot be negative.

d. Profit = Revenue – Cost
$= 1.3x - (1.10x + 1080)$
$= 0.2x - 1080$

$x = 8000$ gallons
Profit $= 0.2(8000) - 1080$
$= 1600 - 1080$
$= 520$
\$520 per week
On the graph this is the vertical distance between R(8,000) and C(8,000)

20. $3x - y + 4z = 4$ (1)
$4x + 4y - 3z = 3$ (2)
$2x + 3y + 2z = -4$ (3)
Multiply equation 1 by 4 and add to equation 2.
$$\begin{array}{r} 12x - 4y + 16z = 16 \\ 4x + 4y - 3z = 3 \\ \hline 16x + 13z = 19 \quad (4) \end{array}$$
Multiply equation 1 by 3 and add to equation 3.
$$\begin{array}{r} 9x - 3y + 12z = 12 \\ 2x + 3y + 2z = -4 \\ \hline 11x + 14z = 8 \quad (5) \end{array}$$
Multiply equation 4 by 14 and equation 5 by -13.
$$\begin{array}{r} 224x + 182z = 266 \\ -143x - 182z = -104 \\ \hline 81x = 162 \\ x = 2 \end{array}$$
Substitute 2 for x in equation 5.
$11x + 14z = 8$
$11(2) + 14z = 8$
$14z = -14$
$z = -1$
Substitute -1 for z in equation 1.
$3x - y + 4z = 4$
$3(2) - y + 4(-1) = 4$
$-y = 2$
$y = -2$
$\{(2, -2, -1)\}$

21. $2x - 5y + 2z = -4$ (1)
$5x - 3y - 4z = -18$ (2)
$3x + 2y + 3z = 13$ (3)
Multiply equation 1 by 2 and add to equation 2.
$$\begin{array}{r} 4x - 10y + 4z = -8 \\ 5x - 3y - 4z = -18 \\ \hline 9x - 13y = -26 \quad (4) \end{array}$$
Multiply equation 1 by -3 and equation 3 by 2 and add.
$$\begin{array}{r} -6x + 15y - 6z = 12 \\ 6x + 4y + 6z = 26 \\ \hline 19y = 38 \\ y = 2 \end{array}$$
Substitute 2 for y in equation 4.
$9x - 13(2) = -26$
$9x = 0$
$x = 0$
Substitute 0 for x in equation 1.
$2(0) - 5(2) + 2z = -4$
$0 - 10 + 2z = -4$
$2z = 6$
$z = 3$
$\{(0, 2, 3)\}$

22. $x - z + 2 = 0$ (1)
$y + 3z = 11$ (2)
$x + y + z = 6$ (3)
Multiply equation 1 by -1 and add to 3.
$$\begin{array}{r} -x + z = 2 \\ x + y + z = 6 \\ \hline y + 2z = 8 \quad (4) \end{array}$$
Multiply equation 4 by -1 and add to 2.
$$\begin{array}{r} y + 3z = 11 \\ -y - 2z = -8 \\ \hline z = 3 \end{array}$$
Substitute 3 for z in equation 4.
$y + 2z = 8$
$y + 2(3) = 8$
$y = 2$
Substitute 3 for z in equation 1.
$x - z = -2$
$x - 3 = -2$
$x = 1$
$\{(1, 2, 3)\}$

23. $x + y + z = 50{,}760$
$x = z + 22{,}900$
$x = 2y + 5335$

or

$x + y + z = 50{,}760$ (1)

$x - z = 22{,}900$ (2)

$x - 2y = 5335$ (3)

Add equations 1 and 2.

$$\begin{array}{r} x+y+z=50{,}760 \\ x-z=22{,}900 \\ \hline 2x+y=73{,}660\ (4) \end{array}$$

Multiply equation 3 by –2 and add to equation 4.

$$\begin{array}{r} -2x+4y=-10{,}670 \\ 2x+y=73{,}660 \\ \hline 5y=62{,}990 \\ y=12{,}598 \end{array}$$

Substitute 12,598 for y in equation 4.

$2x + 12{,}598 = 73{,}660$

$2x = 61{,}062$

$x = 30{,}531$

Substitute 30,531 for x in equation 2.

$30{,}531 - z = 22{,}900$

$z = 7631$

$\{(30{,}531, 12{,}598, 7631)\}$

New York City has 30,531 officers, Chicago has 12,598 officers, and Los Angeles has 7631 officers.

24. Let x = smallest angle, y = 2nd smallest angle and z = largest angle.

$z = x + 66$

$z = 3y - 17$

$x + y + z = 180$

or

$-x + z = 66$ (1)

$-3y + z = -17$ (2)

$x + y + z = 180$ (3)

Add equations 1 and 3.

$$\begin{array}{r} -x+z=66 \\ x+y+z=180 \\ \hline y+2z=246\ (4) \end{array}$$

Multiply equation 4 by 3 and add to equation 2.

$$\begin{array}{r} -3y+z=-17 \\ 3y+6z=738 \\ \hline 7z=721 \\ z=103 \end{array}$$

Substitute 103 for z in equation 4.

$y + 2(103) = 246$

$y = 40$

Substitute 103 for z in equation 1.

$-x + 103 = 66$

$x = 37$

$\{(37, 40, 103)\}$

The angles are 37°, 40°, and 103°.

25. Let x = ounces of Food A, y = ounces of Food B and z = ounces of Food C.

$200x + 50y + 10z = 740$

$0.2x + 3y + z = 10.6$

$10y + 30z = 140$

or

$20x + 5y + z = 74$ (1)

$0.2x + 3y + z = 10.6$ (2)

$y + 3z = 14$ (3)

Multiply equation 2 by –100 and add to equation 1.

$$\begin{array}{r} 20x+5y+z=74 \\ -20x-300y-100z=-1060 \\ \hline -295y-99z=-986\ (4) \end{array}$$

Multiply equation 3 by 295 and add to equation 4.

$$\begin{array}{r} 295y+885z=4130 \\ -295y-99z=-986 \\ \hline 786z=3144 \\ z=4 \end{array}$$

Substitute 4 for z in equation 3.

$y + 3(4) = 14$

$y = 2$

Substitute 4 for z in equation 1.

$20x + 5(2) + 4 = 74$

$20x = 60$

$x = 3$

The solution is $\{(3, 2, 4)\}$

3 oz of Food A,

2 oz of Food B,

4 oz of Food C

26. $x+4y=7$
$3x+5y=0$

$$\left[\begin{array}{cc|c}1&4&7\\3&5&0\end{array}\right]\begin{array}{l}\\3R_1-R_2\rightarrow\end{array}\left[\begin{array}{cc|c}1&4&7\\0&7&21\end{array}\right]\begin{array}{l}\\\frac{1}{7}R_2\rightarrow\end{array}\left[\begin{array}{cc|c}1&4&7\\0&1&3\end{array}\right]$$

$x+4y=7$
$y=3$

$x+4(3)=7$
$x=-5$
$\{(-5,3)\}$

27. $6x-3y=1$
$5x+6y=15$

$$\left[\begin{array}{cc|c}6&-3&1\\5&6&15\end{array}\right]\begin{array}{l}\\6R_2-5R_1\rightarrow\end{array}\left[\begin{array}{cc|c}6&-3&1\\0&51&85\end{array}\right]\begin{array}{l}\frac{1}{6}R_1\rightarrow\\\frac{1}{51}R_2\rightarrow\end{array}\left[\begin{array}{cc|c}1&-\frac{1}{2}&\frac{1}{6}\\0&1&\frac{5}{3}\end{array}\right]$$

$x-\frac{1}{2}y=\frac{1}{6}$
$y=\frac{5}{3}$

$x-\frac{1}{2}\left(\frac{5}{3}\right)=\frac{1}{6}$
$x=1$
$\left\{\left(1,\frac{5}{3}\right)\right\}$

28. $x+y+2x=0$
$2x-y-z=1$
$x+2y+3z=1$

$$\left[\begin{array}{ccc|c}1&1&2&0\\2&-1&-1&1\\1&2&3&1\end{array}\right]\begin{array}{l}\\R_2-2R_1\rightarrow\\R_3-R_1\rightarrow\end{array}\left[\begin{array}{ccc|c}1&1&2&0\\0&-3&-5&1\\0&1&1&1\end{array}\right]\begin{array}{l}\\\\R_2+3R_3\rightarrow\end{array}\left[\begin{array}{ccc|c}1&1&2&0\\0&-3&-5&1\\0&0&-2&4\end{array}\right]\begin{array}{l}\\-\frac{1}{3}R_2\rightarrow\\-\frac{1}{2}R_3\rightarrow\end{array}\left[\begin{array}{ccc|c}1&1&2&0\\0&1&\frac{5}{3}&-\frac{1}{3}\\0&0&1&-2\end{array}\right]$$

$x+y+2z=0$
$y+\frac{5}{3}z=-\frac{1}{3}$
$z=-2$

$y+\frac{5}{3}(-2)=-\frac{1}{3}$
$y=3$

$x+3+2(-2)=0$
$x=1$
$\{(1,3,-2)\}$

29. $3x - y + 2z = 2$
$x + 4z = -1$
$3x - 2y = -1$

$$\left[\begin{array}{ccc|c} 3 & -1 & 2 & 2 \\ 1 & 0 & 4 & -1 \\ 3 & -2 & 0 & -1 \end{array}\right] \begin{array}{l} \\ 3R_2 - R_1 \to \\ R_3 - R_1 \to \end{array} \left[\begin{array}{ccc|c} 3 & -1 & 2 & 2 \\ 0 & 1 & 10 & -5 \\ 0 & -1 & -2 & -3 \end{array}\right] \begin{array}{l} \\ \\ R_3 + R_2 \to \end{array} \left[\begin{array}{ccc|c} 3 & -1 & 2 & 2 \\ 0 & 1 & 10 & -5 \\ 0 & 0 & 8 & -8 \end{array}\right] \begin{array}{l} \frac{1}{3}R_1 \to \\ \\ \frac{1}{8}R_3 \to \end{array} \left[\begin{array}{ccc|c} 1 & -\frac{1}{3} & \frac{2}{3} & \frac{2}{3} \\ 0 & 1 & 10 & -5 \\ 0 & 0 & 1 & -1 \end{array}\right]$$

$x - \frac{1}{3}y + \frac{2}{3}z = \frac{2}{3}$
$y + 10z = -5$
$z = -1$

$y + 10(-1) = -5$
$y = 5$

$x - \frac{1}{3}(5) + \frac{2}{3}(-1) = \frac{2}{3}$
$x = 3$
$\{(3, 5, -1)\}$

30. $2x - y - 3z = 1$
$6x - 3y - 9z = 3$
$4x - 2y - 6z = 2$

$$\left[\begin{array}{ccc|c} 2 & -1 & -3 & 1 \\ 6 & -3 & -9 & 3 \\ 4 & -2 & -6 & 2 \end{array}\right] \begin{array}{l} \frac{1}{2}R_1 \to \\ R_2 - 3R_1 \to \\ R_3 - 2R_1 \to \end{array} \left[\begin{array}{ccc|c} 1 & -\frac{1}{2} & -\frac{3}{2} & \frac{1}{2} \\ 0 & 0 & 0 & 0 \\ 0 & 0 & 0 & 0 \end{array}\right]$$

$x - \frac{1}{2}y - \frac{3}{2}z = \frac{1}{2}$
$0 = 0$
True; infinitely many solutions
The system is dependent.
$\{(x, y, z) | 2x - y - 3z = 1\}$

31. $3x + 2y + z = 7$
$x + y - z = 2$
$6x + 4y + 2z = 10$

$$\left[\begin{array}{ccc|c} 3 & 2 & 1 & 7 \\ 1 & 1 & -1 & 2 \\ 6 & 4 & 2 & 10 \end{array}\right] \begin{array}{l} \\ 3R_2 - R_1 \to \\ R_3 - 2R_1 \to \end{array} \left[\begin{array}{ccc|c} 3 & 2 & 1 & 7 \\ 0 & 1 & -4 & -1 \\ 0 & 0 & 0 & -4 \end{array}\right]$$

$3x + 2y + z = 7$
$y - 4z = -1$
$0 = -4$
False; contradiction

The system is inconsistent.
There is no solution; ∅.

32. $\begin{vmatrix} 3 & 2 \\ -1 & 5 \end{vmatrix} = 15-(-2) = 17$

33. $\begin{vmatrix} -2 & -3 \\ -4 & -8 \end{vmatrix} = 16-12 = 4$

34. $\begin{vmatrix} 2 & 4 & -3 \\ -1 & 7 & -4 \\ 1 & -6 & -2 \end{vmatrix} = 2\begin{vmatrix} 7 & -4 \\ -6 & -2 \end{vmatrix} - (-1)\begin{vmatrix} 4 & -3 \\ -6 & -2 \end{vmatrix} + 1\begin{vmatrix} 4 & -3 \\ 7 & -4 \end{vmatrix}$

$= 2(-14-24) + 1(-8-18) + (-16+21)$
$= 2(-38) + (-26) + (5)$
$= -97$

35. $\begin{vmatrix} 4 & 7 & 0 \\ -5 & 6 & 0 \\ 3 & 2 & -4 \end{vmatrix} = 0\begin{vmatrix} -5 & 6 \\ 3 & 2 \end{vmatrix} - 0\begin{vmatrix} 4 & 7 \\ 3 & 2 \end{vmatrix} + (-4)\begin{vmatrix} 4 & 7 \\ -5 & 6 \end{vmatrix}$

$= 0 - 0 - 4[24-(-35)] = -4(59) = -236$

36. $2x - y = 2$
$x + 2y = 11$

$D = \begin{vmatrix} 2 & -1 \\ 1 & 2 \end{vmatrix} = 4+1 = 5$

$D_x = \begin{vmatrix} 2 & -1 \\ 11 & 2 \end{vmatrix} = 4+11 = 15$

$D_y = \begin{vmatrix} 2 & 2 \\ 1 & 11 \end{vmatrix} = 22-2 = 20$

$x = \dfrac{D_x}{D} = \dfrac{15}{5} = 3$

$y = \dfrac{D_y}{D} = \dfrac{20}{5} = 4$

{(3, 4)}

37. $4x = 12 - 3y$
$2x - 6 = 5y$
or
$4x + 3y = 12$
$2x - 5y = 6$

$D = \begin{vmatrix} 4 & 3 \\ 2 & -5 \end{vmatrix} = -20-6 = -26$

$D_x = \begin{vmatrix} 12 & 3 \\ 6 & -5 \end{vmatrix} = -60-18 = -78$

$D_y = \begin{vmatrix} 4 & 12 \\ 2 & 6 \end{vmatrix} = 24-24 = 0$

$x = \dfrac{D_x}{D} = \dfrac{-78}{-26} = 3$

$y = \dfrac{D_y}{D} = \dfrac{0}{-26} = 0$

{(3, 0)}

38. $4x + 2y + 3z = 9$
$3x + 5y + 4z = 19$
$9x + 3y + 2z = 3$

$D = \begin{vmatrix} 4 & 2 & 3 \\ 3 & 5 & 4 \\ 9 & 3 & 2 \end{vmatrix}$

$= 4\begin{vmatrix} 5 & 4 \\ 3 & 2 \end{vmatrix} - 3\begin{vmatrix} 2 & 3 \\ 3 & 2 \end{vmatrix} + 9\begin{vmatrix} 2 & 3 \\ 5 & 4 \end{vmatrix}$

$= 4(10 - 12) - 3(4 - 9) + 9(8 - 15)$
$= 4(-2) - 3(-5) + 9(-7)$
$= -8 + 15 - 63 = -56$

$$D_x = \begin{vmatrix} 9 & 2 & 3 \\ 19 & 5 & 4 \\ 3 & 3 & 2 \end{vmatrix}$$

$$= 9\begin{vmatrix} 5 & 4 \\ 3 & 2 \end{vmatrix} - 19\begin{vmatrix} 2 & 3 \\ 3 & 2 \end{vmatrix} + 3\begin{vmatrix} 2 & 3 \\ 5 & 4 \end{vmatrix}$$

$= 9(10 - 12) - 19(4 - 9) + 3(8 - 15)$
$= 9(-2) - 19(-5) + 3(-7)$
$= -18 + 95 - 21 = 56$

$$D_y = \begin{vmatrix} 4 & 9 & 3 \\ 3 & 19 & 4 \\ 9 & 3 & 2 \end{vmatrix}$$

$$= 4\begin{vmatrix} 19 & 4 \\ 3 & 2 \end{vmatrix} - 3\begin{vmatrix} 9 & 3 \\ 3 & 2 \end{vmatrix} + 9\begin{vmatrix} 9 & 3 \\ 19 & 4 \end{vmatrix}$$

$= 4(38 - 12) - 3(18 - 9) + 9(36 - 57)$
$= 4(26) - 3(9) + 9(-21)$
$= 104 - 27 - 189 = -112$

$$D_z = \begin{vmatrix} 4 & 2 & 9 \\ 3 & 5 & 19 \\ 9 & 3 & 3 \end{vmatrix}$$

$$= 4\begin{vmatrix} 5 & 19 \\ 3 & 3 \end{vmatrix} - 3\begin{vmatrix} 2 & 9 \\ 3 & 3 \end{vmatrix} + 9\begin{vmatrix} 2 & 9 \\ 5 & 19 \end{vmatrix}$$

$= 4(15 - 57) - 3(6 - 27) + 9(38 - 45)$
$= 4(-42) - 3(-21) + 9(-7)$
$= -168 + 63 - 63 = -168$

$$x = \frac{D_x}{D} = \frac{56}{-56} = -1$$

$$y = \frac{D_y}{D} = \frac{-112}{-56} = 2$$

$$z = \frac{D_z}{D} = \frac{-168}{-56} = 3$$

$\{(-1, 2, 3)\}$

39. $2x + z = -4$
$-2y + z = -4$
$2x - 4y + z = -20$

$$D = \begin{vmatrix} 2 & 0 & 1 \\ 0 & -2 & 1 \\ 2 & -4 & 1 \end{vmatrix}$$

$$= 2\begin{vmatrix} -2 & 1 \\ -4 & 1 \end{vmatrix} - 0\begin{vmatrix} 0 & 1 \\ -4 & 1 \end{vmatrix} + 2\begin{vmatrix} 0 & 1 \\ -2 & 1 \end{vmatrix}$$

$= 2(-2 + 4) - 0 + 2(0 + 2) = 4 + 4 = 8$

$$D_x = \begin{vmatrix} -4 & 0 & 1 \\ -4 & -2 & 1 \\ -20 & -4 & 1 \end{vmatrix}$$

$$= -0\begin{vmatrix} -4 & 1 \\ -20 & 1 \end{vmatrix} + (-2)\begin{vmatrix} -4 & 1 \\ -20 & 1 \end{vmatrix} - (-4)\begin{vmatrix} -4 & 1 \\ -4 & 1 \end{vmatrix}$$

$= -0 - 2(-4 + 20) + 4(-4 + 4) = -32$

$$D_y = \begin{vmatrix} 2 & -4 & 1 \\ 0 & -4 & 1 \\ 2 & -20 & 1 \end{vmatrix}$$

$$= 2\begin{vmatrix} -4 & 1 \\ -20 & 1 \end{vmatrix} - 0\begin{vmatrix} -4 & 1 \\ -20 & 1 \end{vmatrix} + 2\begin{vmatrix} -4 & 1 \\ -4 & 1 \end{vmatrix}$$

$= 2(-4 + 20) - 0 + 2(-4 + 4) = 32$

$$D_z = \begin{vmatrix} 2 & 0 & -4 \\ 0 & -2 & -4 \\ 2 & -4 & -20 \end{vmatrix}$$

$$= 2\begin{vmatrix} -2 & -4 \\ -4 & -20 \end{vmatrix} - 0\begin{vmatrix} 0 & -4 \\ -4 & -20 \end{vmatrix} + 2\begin{vmatrix} 0 & -4 \\ -2 & -4 \end{vmatrix}$$

$= 2(40 - 16) - 0 + 2(0 - 8)$
$= 48 - 16 = 32$

$$x = \frac{D_x}{D} = \frac{-32}{8} = -4$$

$$y = \frac{D_y}{D} = \frac{32}{8} = 4$$

$$z = \frac{D_z}{D} = \frac{32}{8} = 4$$

$\{(-4, 4, 4)\}$

40. $3x - 2y = 16$
$12x - 8y = -5$

$$D = \begin{vmatrix} 3 & -2 \\ 12 & -8 \end{vmatrix} = -24 + 24 = 0$$

$$D_x = \begin{vmatrix} 16 & -2 \\ -5 & -8 \end{vmatrix} = -128 - 10 = -138$$

$$D_y = \begin{vmatrix} 3 & 16 \\ 12 & -5 \end{vmatrix} = -15 - 192 = -207$$

$D = 0,\ D_x = -138 \neq 0,\ D_y = -207 \neq 0$

Thus, the system is inconsistent.
There is no solution; $\varnothing$.

41. $x - y - 2z = 1$
$2x - 2y - 4z = 2$
$-x + y + 2z = -1$

$$D = \begin{vmatrix} 1 & -1 & -2 \\ 2 & -2 & -4 \\ -1 & 1 & 2 \end{vmatrix}$$

$$= 1\begin{vmatrix} -2 & -4 \\ 1 & 2 \end{vmatrix} - 2\begin{vmatrix} -1 & -2 \\ 1 & 2 \end{vmatrix} + (-1)\begin{vmatrix} -1 & -2 \\ -2 & -4 \end{vmatrix}$$

$$= (-4 + 4) - 2(-2 + 2) - 1(4 - 4) = 0$$

$$D_x = \begin{vmatrix} 1 & -1 & -2 \\ 2 & -2 & -4 \\ -1 & 1 & 2 \end{vmatrix} = 0$$

$$D_y = \begin{vmatrix} 1 & 1 & -2 \\ 2 & 2 & -4 \\ -1 & -1 & 2 \end{vmatrix} = 0$$

(columns 1 and 2 are identical)

$$D_z = \begin{vmatrix} 1 & -1 & 1 \\ 2 & -2 & 2 \\ -1 & 1 & -1 \end{vmatrix} = 0$$

(columns 1 and 3 are identical)

$D = 0,\ D_x = 0,\ D_y = 0,\ D_z = 0$

Thus, the system is dependent.

$\{(x, y, z) \mid x - y - 2z = 1\}$

42. $3x - y \le 6$
$x + y \ge 2$

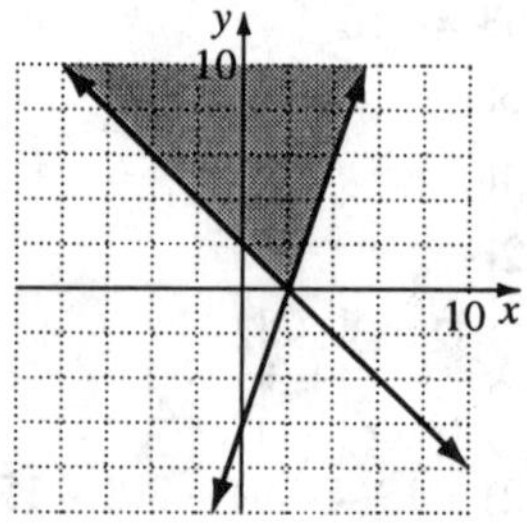

(2, 0)

43. $y < -x + 4$
$y > x - 4$

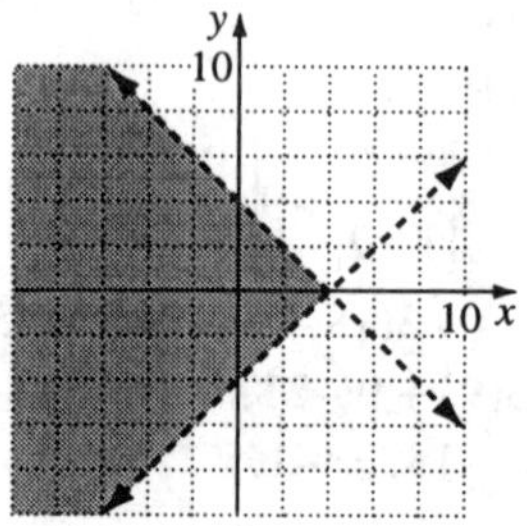

(4, 0)

44. $-3 \le x < 5$

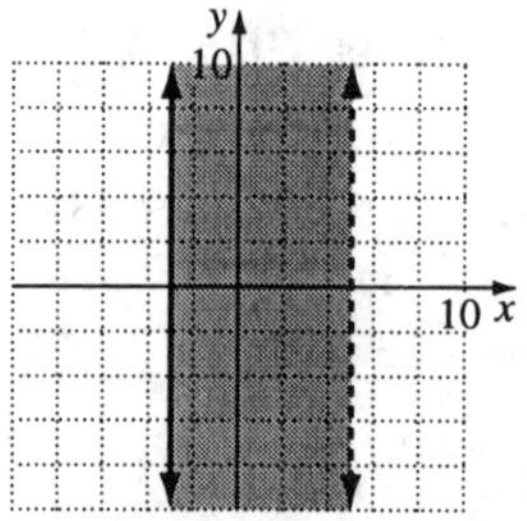

45. $-2 < y \le 6$

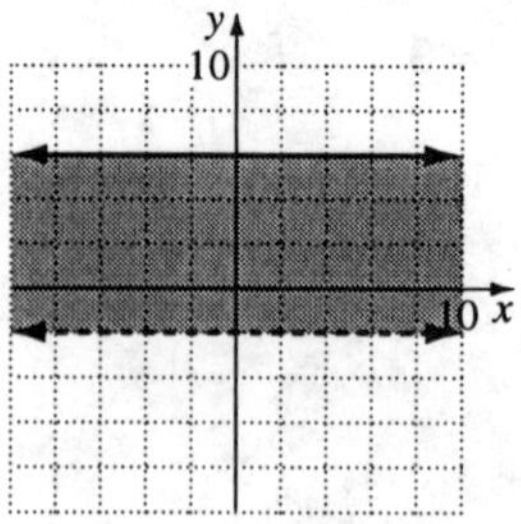

46. $x \geq 3$
$y \leq 0$

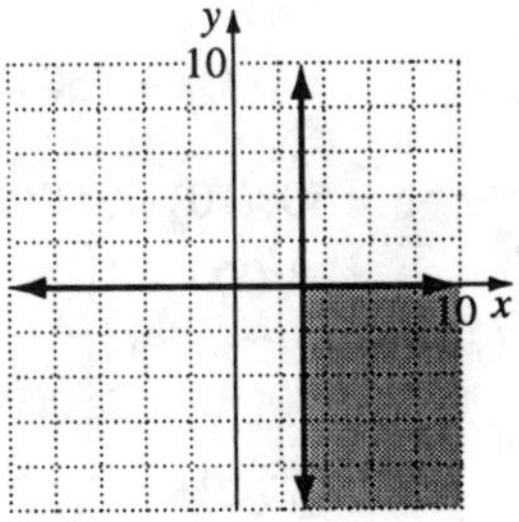

(3, 0)

47. $2x - y > -4$
$x \geq 0$

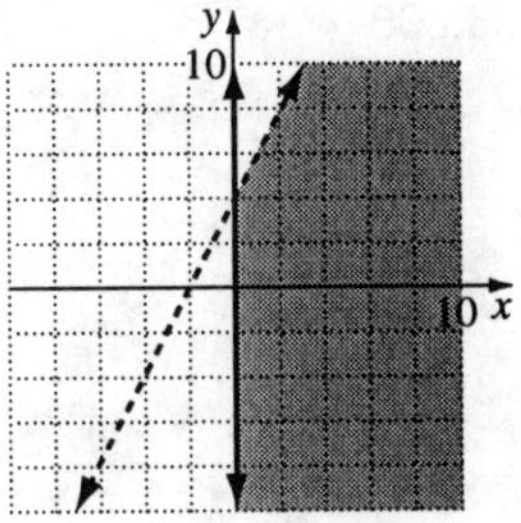

(0, 4)

48. $x + y \leq 6$
$y \geq 2x - 3$

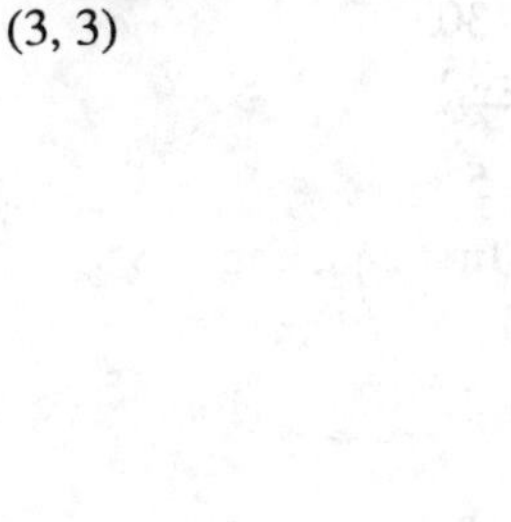

(3, 3)

49. $3x + y \leq 6$
$4x + 6y \leq 24$
$x \geq 0$
$y \geq 0$

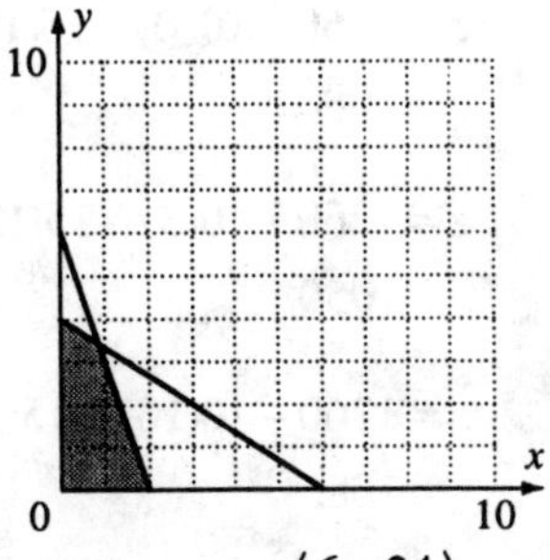

(0, 0), (0, 4), $\left(\frac{6}{7}, \frac{24}{7}\right)$, (2, 0)

50.

Corner	Objective Function $z = 2x + 3y$
$\left(\frac{1}{2}, \frac{1}{2}\right)$	$z = 2\left(\frac{1}{2}\right) + 3\left(\frac{1}{2}\right) = \frac{5}{2}$
(2, 2)	$z = 2(2) + 3(2) = 10$
(4, 0)	$z = 2(4) + 3(0) = 8$
(1, 0)	$z = 2(1) + 3(0) = 2$

The maximum value is 10 and the minimum value is 2.

51. $x + y \geq 12$
$x + y \leq 15$
$0 \leq x \leq 10$
$0 \leq y \leq 12$

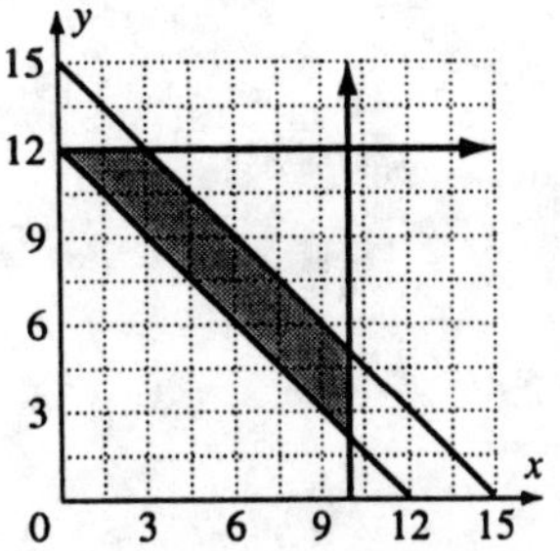

Corner	Objective Function
	$z = 1260 - 10x - 15y$
(0, 12)	$z = 1260 - 10(0) - 15(12)$ $= 1080$
(3, 12)	$z = 1260 - 10(3) - 15(12)$ $= 1050$
(10, 5)	$z = 1260 - 10(10) - 15(5)$ $= 1085$
(10, 2)	$z = 1260 - 10(10) - 15(2)$ $= 1130$

The maximum value is 1130 and the minimum value is 1050.

52. **a.** $z = 500x + 350y$

b. $x + y \le 200$
$x \ge 10$
$y \ge 80$

c.

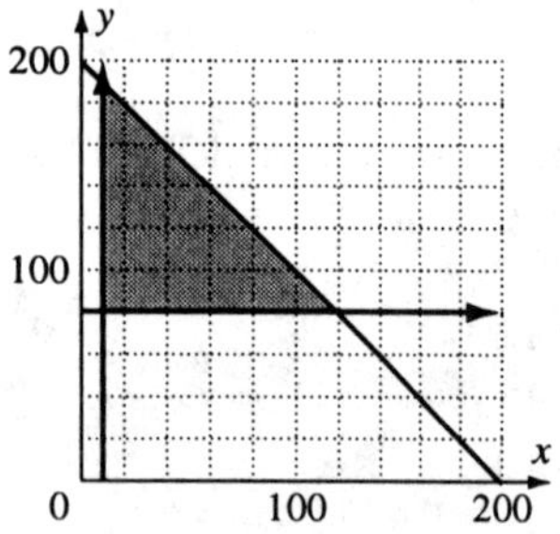

d.

Corner	Objective Function
	$z = 500x + 350y$
(10, 80)	$z = 500(10) + 350(80)$ $= 33,000$
(10, 190)	$z = 500(10) + 350(190)$ $= 71,500$
(120, 80)	$z = 500(120) + 350(80)$ $= 88,000$

e. 120; 80; $88,000

Chapter 3 Test

1. $x + y = 6$
$4x - y = 4$

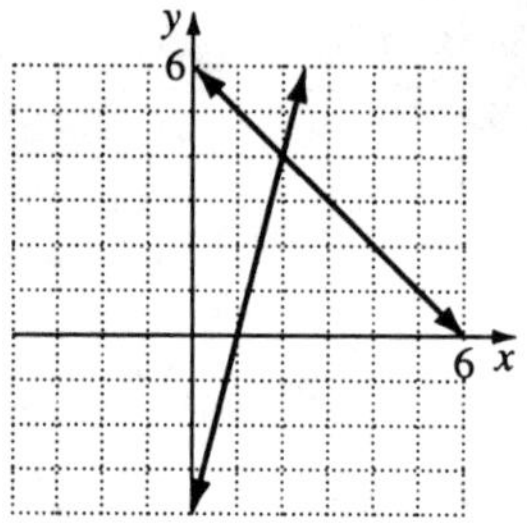

{(2, 4)}

2. $5x + 4y = 10$
$3x + 5y = -7$
Add –3 times first equation to 5 times second equation.

$$\begin{array}{r} -15x - 12y = -30 \\ 15x + 25y = -35 \\ \hline 13y = -65 \\ y = -5 \end{array}$$

Substitute into first equation.
$5x + 4(-5) = 10$
$5x - 20 = 10$
$5x = 30$
$x = 6$
$\{(6, -5)\}$

3. $x = y + 4$
$-3x - 7y = 18$
Substitute $x = y + 4$ into the second equation for x.
$-3(y + 4) - 7y = 18$
$-3y - 12 - 7y = 18$
$-10y = 30$
$y = -3$
Substitute into first equation.
$x = -3 + 4$
$x = 1$
$\{(1, -3)\}$

4. $4x = 2y + 6$
$y = 2x - 3$
Substitute $y = 2x - 3$ into the first equation for y.
$4x = 2(2x - 3) + 6$
$4x = 4x - 6 + 6$
$4x - 4x = -6 + 6$
$0 = 0$
Dependent equations
Infinite number of solutions

5. Let x = teacher's salary and
y = attorney's salary.
$y = 2x - 1783$
$x + y = 52{,}784 + 38{,}192 + 9322$
Substitute $y = 2x - 1783$ into the second equation for y.
$x + (2x - 1783) = 100{,}298$
$3x = 102{,}081$
$x = 34{,}027$
$y = 2(34{,}027) - 1783 = 66{,}271$
The salary for teachers is $34,027 and for attorneys is $66,271.

6. $y = 2x - 39$
$x + y = 90$
Substitute $y = 2x - 39$ into the second equation for y.
$x + (2x - 39) = 90$
$3x = 129$
$x = 43$
$y = 2(43) - 39 = 47$
The angles are 43°, 47° and 90°.

7. Let x = price for one orchestra ticket and
y = price for one mezzanine ticket.
$4x + 3y = 134$
$5x + 2y = 143$
Add –5 times the first equation to 4 times the second equation.
$$\begin{array}{r} -20x - 15y = -670 \\ 20x + 8y = 572 \\ \hline -7y = -98 \\ y = 14 \end{array}$$
Substitute into the first equation.
$4x + 3(14) = 134$
$4x = 92$
$x = 23$
The ticket prices are as follows:
Orchestra, $23
Mezzanine, $14

8. $D(p) = 1000 - 20p$
$S(p) = 250 + 5p$
$1000 - 20p = 250 + 5p$
$-20p - 5p = -1000 + 250$
$-25p = -750$
$p = 30$
Supply will equal demand at a price of $30.
$D(30) = 1000 - 20(30) = 400$
400 units will be supplied and sold.

9. $x + 3y - z = 1$ (1)
$x - y + 2z = -4$ (2)
$2x + y + 3z = 2$ (3)
Add (1) to –1 times (2).
$$\begin{array}{r} x + 3y - z = 1 \\ -x + y - 2z = 4 \\ \hline 4y - 3z = 5 \ (4) \end{array}$$
Add –2 times (1) to (3).
$$\begin{array}{r} -2x - 6y + 2z = -2 \\ 2x + y + 3z = 2 \\ \hline -5y + 5z = 0 \ (5) \end{array}$$
Add 5 times (4) to 4 times (5).

$$\begin{array}{r} 20y-15z=25 \\ -20y+20z=0 \\ \hline 5z=25 \\ z=5 \end{array}$$

Substitute into equation (4).

$4y-3(5)=5$

$4y=20$

$y=5$

Substitute into equation (1).

$x+3(5)-5=1$

$x=-9$

$\{(-9, 5, 5)\}$

10. $x+2y+z=5$ (1)

$y+5z=6$ (2)

$4x+2z=10$ (3)

Add –4 times equation (1) to equation (3).

$$\begin{array}{r} -4x-8y-4z=-20 \\ 4x \qquad +2z=10 \\ \hline -8y-2z=-10 \end{array} \quad (4)$$

Add 8 times equation (2) to equation (4).

$$\begin{array}{r} 8y+40z=48 \\ -8y-2z=-10 \\ \hline 38z=38 \\ z=1 \end{array}$$

Substitute into equation (2).

$y+5(1)=6$

$y=1$

Substitute into equation (1).

$x+2(1)+1=5$

$x=2$

$\{(2, 1, 1)\}$

11. Let x = price of model A,
y = price of model B, and
z = price of model C.

$x+2y+z=300$ (1)

$2x+y+3z=465$ (2)

$2x+y+4z=555$ (3)

Add –1 times equation (2) to equation (3).

$$\begin{array}{r} -2x-y-3z=-465 \\ 2x+y+4z=555 \\ \hline z=90 \end{array}$$

Substitute into equations (1) and (2).

$x+2y=210$ (4)

$2x+y=195$ (5)

Add –2 times equation (4) to equation (5).

$$\begin{array}{r} -2x-4y=-420 \\ 2x+y=195 \\ \hline -3y=-225 \\ y=75 \end{array}$$

Substitute into equation (1).

$x+2(75)+90=300$

$x=60$

Solution: (60, 75, 90)

The prices are as follows:

Model A, \$60

Model B, \$75

Model C, \$90

12. $2x + y = 6$
$3x - 2y = 16$

$$\left[\begin{array}{cc|c} 2 & 1 & 6 \\ 3 & -2 & 16 \end{array}\right] \begin{array}{l} \\ -3R_1 + 2R_2 \rightarrow \end{array} \left[\begin{array}{cc|c} 2 & 1 & 6 \\ 0 & -7 & 14 \end{array}\right] \begin{array}{l} \frac{1}{2}R_1 \rightarrow \\ -\frac{1}{7}R_2 \end{array} \left[\begin{array}{cc|c} 1 & \frac{1}{2} & 3 \\ 0 & 1 & -2 \end{array}\right]$$

$y = -2$
$x + \frac{1}{2}(-2) = 3$
$x = 4$
$\{(4, -2)\}$

13. $x - 4y + 4z = -1$
$2x - y + 5z = 6$
$-x + 3y - z = 5$

$$\left[\begin{array}{ccc|c} 1 & -4 & 4 & -1 \\ 2 & -1 & 5 & 6 \\ -1 & 3 & -1 & 5 \end{array}\right] \begin{array}{l} \\ -2R_1 + R_2 \rightarrow \\ R_1 + R_3 \rightarrow \end{array} \left[\begin{array}{ccc|c} 1 & -4 & 4 & -1 \\ 0 & 7 & -3 & 8 \\ 0 & -1 & 3 & 4 \end{array}\right] \begin{array}{l} \\ \\ R_2 + 7R_3 \rightarrow \end{array} \left[\begin{array}{ccc|c} 1 & -4 & 4 & -1 \\ 0 & 7 & -3 & 8 \\ 0 & 0 & 18 & 36 \end{array}\right] \begin{array}{l} \\ \frac{1}{7}R_2 \rightarrow \\ \frac{1}{18}R_3 \rightarrow \end{array}$$

$$\left[\begin{array}{ccc|c} 1 & -4 & 4 & -1 \\ 0 & 1 & -\frac{3}{7} & \frac{8}{7} \\ 0 & 0 & 1 & 2 \end{array}\right]$$

$x - 4y + 4z = -1$
$y - \frac{3}{7}z = \frac{8}{7}$
$z = 2$
$y - \frac{3}{7}(2) = \frac{8}{7}$
$y = 2$
$x - 4(2) + 4(2) = -1$
$x = -1$
$\{(-1, 2, 2)\}$

14. $\begin{vmatrix} -1 & -3 \\ 7 & 4 \end{vmatrix} = (-1)(4) - (-3)(7) = -4 + 21 = 17$

15. $\begin{vmatrix} 3 & 4 & 5 \\ -1 & -6 & -3 \\ 4 & 9 & 5 \end{vmatrix}$ Across row 1 $= 3\begin{vmatrix} -6 & -3 \\ 9 & 5 \end{vmatrix} - 4\begin{vmatrix} -1 & -3 \\ 4 & 5 \end{vmatrix} + 5\begin{vmatrix} -1 & -6 \\ 4 & 9 \end{vmatrix}$

$= 3(-30 + 27) - 4(-5 + 12) + 5(-9 + 24)$
$= -9 - 28 + 75 = 38$

16. $4x - 3y = 14$
$3x - y = 3$

$$D = \begin{vmatrix} 4 & -3 \\ 3 & -1 \end{vmatrix} = -4 + 9 = 5$$

$$D_x = \begin{vmatrix} 14 & -3 \\ 3 & -1 \end{vmatrix} = -14 + 9 = -5$$

$$D_y = \begin{vmatrix} 4 & 14 \\ 3 & 3 \end{vmatrix} = 12 - 42 = -30$$

$$x = \frac{D_x}{D} = \frac{-5}{5} = -1$$

$$y = \frac{D_y}{D} = \frac{-30}{5} = -6$$

$\{(-1, -6)\}$

17. $2x + 3y + z = 2$
$3x + 3y - z = 0$
$x - 2y - 3z = 1$

$$D = \begin{vmatrix} 2 & 3 & 1 \\ 3 & 3 & -1 \\ 1 & -2 & -3 \end{vmatrix}$$

$$= 2\begin{vmatrix} 3 & -1 \\ -2 & -3 \end{vmatrix} - 3\begin{vmatrix} 3 & -1 \\ 1 & -3 \end{vmatrix} + 1\begin{vmatrix} 3 & 3 \\ 1 & -2 \end{vmatrix}$$

$$= 2(-9 - 2) - 3(-9 + 1) + 1(-6 - 3)$$

$$= -7$$

$$D_x = \begin{vmatrix} 2 & 3 & 1 \\ 0 & 3 & -1 \\ 1 & -2 & -3 \end{vmatrix}$$

$$= 2\begin{vmatrix} 3 & -1 \\ -2 & -3 \end{vmatrix} + 1\begin{vmatrix} 3 & 1 \\ 3 & -1 \end{vmatrix}$$

$$= 2(-9 - 2) + 1(-3 - 3) = -28$$

$$x = \frac{D_x}{D} = \frac{-28}{-7} = 4$$

18. $x + y \geq 2$
$x - y \geq 5$
$\left(\frac{7}{2}, -\frac{3}{2}\right)$

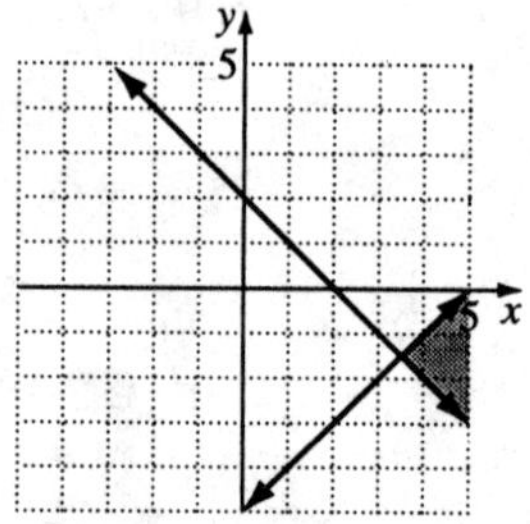

19. $x + 2y \leq 6$
$5x + 4y \leq 20$
$x \geq 0$
$y \geq 0$

$(0, 0)$, $(0, 3)$, $(4, 0)$, $\left(\frac{8}{3}, \frac{5}{3}\right)$

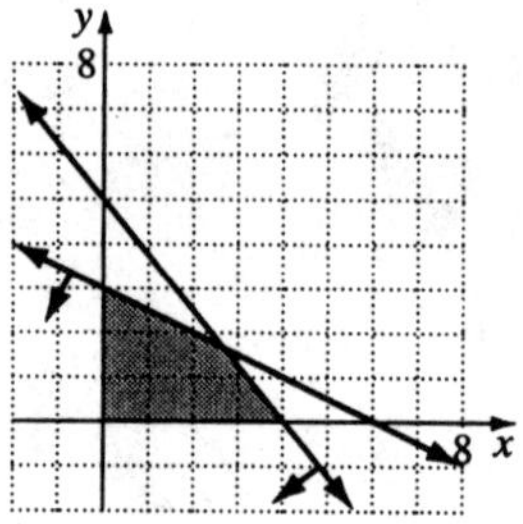

20. $y \leq 2x + 1$
$1 \leq y \leq 3$
$y \leq -\frac{1}{2}x + 6$
$(0, 1)$, $(1, 3)$, $(6, 3)$, $(10, 1)$

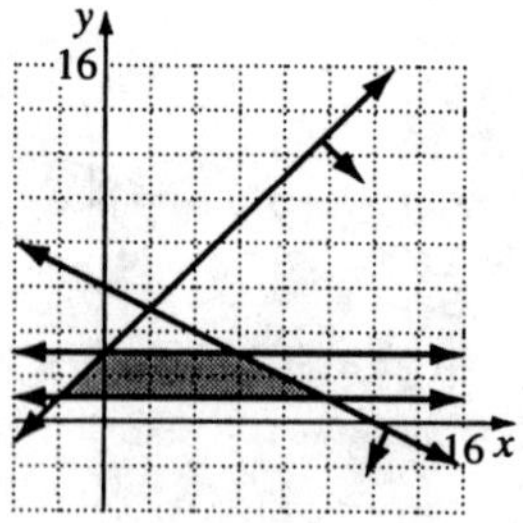

Chapters 1–3 Cumulative Review Problems

1. $\frac{3x}{5}+4=\frac{x}{3}$
$15\left(\frac{3x}{5}+4\right)=15\cdot\frac{x}{3}$
$9x+60=5x$
$4x=-60$
$x=-15$

2. $-2(x-3)-9=3x-4$
$-2x+6-9=3x-4$
$-2x-3=3x-4$
$-5x=-1$
$x=\frac{1}{5}$

3. $\frac{-10x^2y^4}{15x^7y^{-3}}=-\frac{2y^{4+3}}{3x^{7-2}}=-\frac{2y^7}{3x^5}$

4. $(9.3\times10^{-3})(5.1\times10^8)$
$=(9.3)(5.1)\times10^{-3+8}$
$=47.43\times10^5$
$=4.743\times10^6$

5. $2x+5<1$ or $7-2x\le1$
$2x<-4$ or $-2x\le-6$
$x<-2$ or $x\ge3$
$\{x|x<-2 \text{ or } x\ge3\}$
$(-\infty,-2)\cup[3,\infty)$

6. Let x = years since 1989 and
y = millions of dollars lost.
$y=420x+720$
$7860=420x+720$
$7140=420x$
$17=x$
$1989+17=2006$
In year 2006

7. $f(x)=0.000002x^4-0.0002x^3+0.006x^2-0.06x+3.76$
$f(10)=0.000002(10)^4-0.0002(10)^3+0.006(10)^2-0.06(10)+3.76$
$f(10)=3.58$
The average number of people per family in 1950 was 3.58.

8.

terms	Number of dots	Pattern
1	8	$\frac{1^2+7(1)+8}{2}=8$
2	13	$\frac{2^2+7(2)+8}{2}=13$
3	19	$\frac{3^2+7(3)+8}{2}=19$
4	26	$\frac{4^2+7(4)+8}{2}=26$
etc.		
10	89	$\frac{10^2+7(10)+8}{2}=89$
n	$\frac{n^2+7n+8}{2}$	

9. $f(x) = x^2 - 1$

x	–2	–1	0	1	2
$f(x)$	3	0	–1	0	3

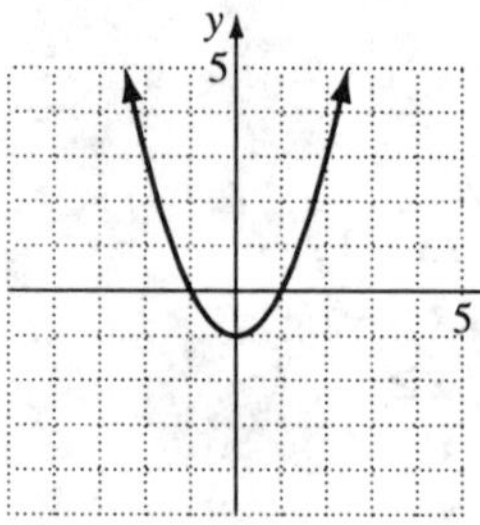

The graph defines y as a function of x, since for each value of x there is exactly one value of y.

10. $|4x - 3| + 2 = 6$
$|4x - 3| = 4$
$4x - 3 = 4$ or $4x - 3 = -4$
$4x = 7$ or $4x = -1$
$x = \frac{7}{4}$ or $x = -\frac{1}{4}$
$\left\{-\frac{1}{4}, \frac{7}{4}\right\}$

11. a. The graphs represent functions since for each value of x there is exactly one value for y.

b. $\left(\frac{1}{2}, 119\right)$; approximately $\frac{1}{2}$ hr after eating a candy bar, the glucose level is at a maximum of about 119 mg per deciliter.

c. $\left(2\frac{1}{2}, 104\right)$; approximately $2\frac{1}{2}$ hours after eating an apple, the glucose level is at a maximum of about 104 mg per deciliter.

d. $\left(1\frac{1}{3}, 96\right)$; approximately $1\frac{1}{3}$ hours after eating either food the glucose is equal at about 96 mg per deciliter.

e. Apple; the graph of the candy bar rises quickly, but drops off quickly.

12. Points (2, 4) and (4, –2)
slope $= \frac{-2-4}{4-2} = \frac{-6}{2} = -3$
Point-slope form:
$y - 4 = -3(x - 2)$
or
$y + 2 = -3(x - 4)$
Slope-intercept form: $y = -3x + 10$
Standard form: $3x + y = 10$

13. Point (2, 3)
parallel to $2x - 3y = 8$ or $y = \frac{2}{3}x - \frac{8}{3}$
Point-slope form: $y - 3 = \frac{2}{3}(x - 2)$
Slope-intercept form: $y = \frac{2}{3}x + \frac{5}{3}$
Standard form: $2x - 3y = -5$

14. $2x - 4y = -8$

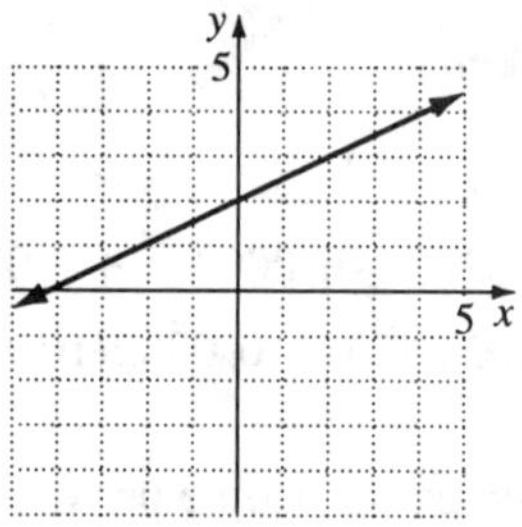

15. $f(x) = \frac{3}{4}x - 6$

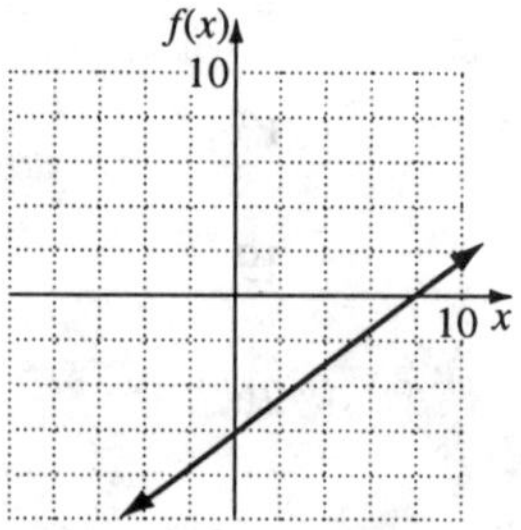

16. $y-3=1$ or $y=4$

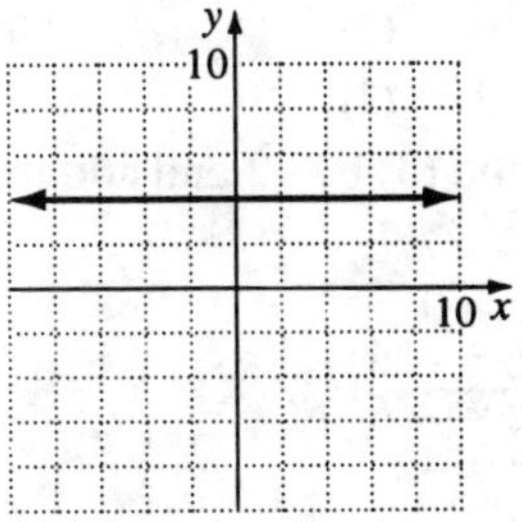

17. $6x-3y<12$ and $x<2$

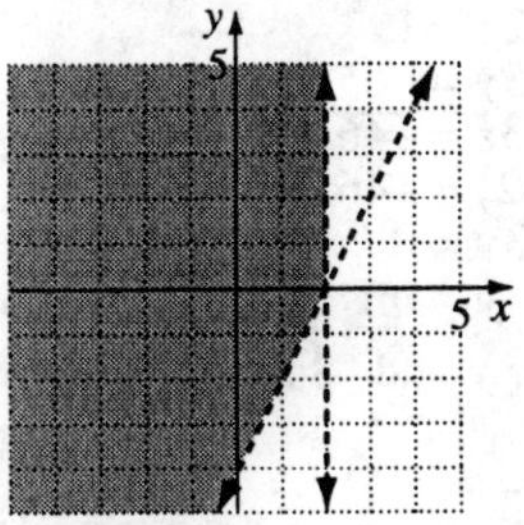

18. $24 \div 4[3-(2-4)]^2-9=6[3-(-2)]^2-9$
$=6[3+2]^2-9$
$=6(5)^2-9$
$=6(25)-9$
$=150-9$
$=141$

19. $|2x-1|>11$
$2x-1>11$ or $2x-1<-11$
$2x>12$ or $2x<-10$
$x>6$ or $x<-5$
$\{x|x<-5 \text{ or } x>6\}$
$(-\infty,-5)\cup(6,\infty)$

20. $R=2(a-b)$
$R=2a-2b$
$R+2b=2a$
$\frac{R+2b}{2}=a$
$a=\frac{R+2b}{2}=\frac{R}{2}+b$

21. a. $A(1900, 1{,}850{,}093)$
$B(1910, 2{,}331{,}542)$
$\text{slope}_{AB}=\frac{2{,}331{,}542-1{,}850{,}093}{1910-1900}$
$=48{,}144.9$
The average rate of change per year from 1900 to 1910 is an increase of 48,144.9 people.

b. $B(1910, 2{,}331{,}542)$
$C(1920, 2{,}284{,}103)$
$\text{slope}_{BC}=\frac{2{,}284{,}103-2{,}331{,}542}{1920-1910}$
$=-4743.9$
The average rate of change from 1910 to 1920 is a decrease of 4743.9 people.

22.

Corner	Objective Function
	$z=3x+2y$
(0, 0)	$z=3(0)+2(0)=0$
(0, 4)	$z=3(0)+2(4)=8$
(3, 3)	$z=3(3)+2(3)=15$
(5, 0)	$z=3(5)+2(0)=15$

The maximum value is 15.

23. $x-y=7$
$2x+y=2$

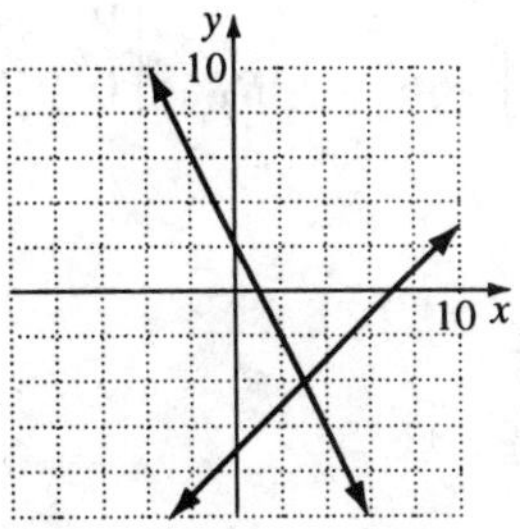

$\{(3,-4)\}$

24. $5x + 4y = 10$ (1)
$3x + 5y = -7$ (2)
Multiply equation (1) by 3 and equation (2) by –5 and add.

$$\begin{array}{r} 15x + 12y = 30 \\ \underline{-15x - 25y = 35} \\ -13y = 65 \\ y = -5 \end{array}$$

$5x + 4(-5) = 10$
$5x - 20 = 10$
$5x = 30$
$x = 6$
$\{(6, -5)\}$

25. $2x - 4y + 3z = 0$ (1)
$5x + 3y - 2z = 19$ (2)
$x - 2y - 5z = 13$ (3)
Multiply equation (3) by –2 and add to equation (1).

$$\begin{array}{r} 2x - 4y + 3z = 0 \\ \underline{-2x + 4y + 10z = -26} \\ 13z = -26 \\ z = -2 \end{array}$$

Multiply equation (3) by –5 and add to equation (2).

$$\begin{array}{r} 5x + 3y - 2z = 19 \\ \underline{-5x + 10y + 25y = -65} \\ 13y + 23z = -46 \\ 13y + 23(-2) = -46 \\ 13y = 0 \\ y = 0 \end{array}$$

Substitute 0 for y and –2 for z in equation (3).
$x - 2y - 5z = 13$
$x - 2(0) - 5(-2) = 13$
$x = 3$
$\{(3, 0, -2)\}$

26. $x - 2y - z = 6$
$y + 4z = 5$
$4x + 2y + 3z = 8$

$$\left[\begin{array}{rrr|r} 1 & -2 & -1 & 6 \\ 0 & 1 & 4 & 5 \\ 4 & 2 & 3 & 8 \end{array}\right] -4R_1 + R_3 \rightarrow \left[\begin{array}{rrr|r} 1 & -2 & -1 & 6 \\ 0 & 1 & 4 & 5 \\ 0 & 10 & 7 & -16 \end{array}\right] -10R_2 + R_3 \rightarrow$$

$$\left[\begin{array}{rrr|r} 1 & -2 & -1 & 6 \\ 0 & 1 & 4 & 5 \\ 0 & 0 & -33 & -66 \end{array}\right] -\tfrac{1}{33}R_3 \rightarrow \left[\begin{array}{rrr|r} 1 & -2 & -1 & 6 \\ 0 & 1 & 4 & 5 \\ 0 & 0 & 1 & 2 \end{array}\right]$$

$x - 2y - z = 6$
$y + 4z = 5$
$z = 2$

$y + 4(2) = 5$
$y = -3$
$x - 2(-3) - 2 = 6$
$x + 6 - 2 = 6$
$x = 2$
$\{(2, -3, 2)\}$

27. $6x+7y=-9$
$4x+6y=0$

$$D=\begin{vmatrix}6 & 7\\4 & 6\end{vmatrix}=36-28=8$$

$$D_x=\begin{vmatrix}-9 & 7\\0 & 6\end{vmatrix}=-54-0=-54$$

$$D_y=\begin{vmatrix}6 & -9\\4 & 0\end{vmatrix}=0+36=36$$

$$x=\frac{D_x}{D}=\frac{-54}{8}=-\frac{27}{4}$$

$$y=\frac{D_y}{D}=\frac{36}{8}=\frac{9}{2}$$

$$\left\{\left(-\frac{27}{4},\frac{9}{2}\right)\right\}$$

28. $2x-7y-z=35$
$x+y=1$
$2y+z=-5$

$$D=\begin{vmatrix}2 & -7 & -1\\1 & 1 & 0\\0 & 2 & 1\end{vmatrix}$$
$$=2\begin{vmatrix}1 & 0\\2 & 1\end{vmatrix}-1\begin{vmatrix}-7 & -1\\2 & 1\end{vmatrix}+0\begin{vmatrix}-7 & -1\\1 & 0\end{vmatrix}$$
$$=2(1-0)-(-7+2)+0$$
$$=2+5=7$$

$$D_x=\begin{vmatrix}35 & -7 & -1\\1 & 1 & 0\\-5 & 2 & 1\end{vmatrix}$$
$$=-1\begin{vmatrix}1 & 1\\-5 & 2\end{vmatrix}-0\begin{vmatrix}35 & -7\\-5 & 2\end{vmatrix}+1\begin{vmatrix}35 & -7\\1 & 1\end{vmatrix}$$
$$=-(2+5)-0+(35+7)$$
$$=-7+42=35$$

$$D_y=\begin{vmatrix}2 & 35 & -1\\1 & 1 & 0\\0 & -5 & 1\end{vmatrix}$$
$$=2\begin{vmatrix}1 & 0\\-5 & 1\end{vmatrix}-1\begin{vmatrix}35 & -1\\-5 & 1\end{vmatrix}+0\begin{vmatrix}35 & -1\\1 & 0\end{vmatrix}$$
$$=2(1+0)-(35-5)+0$$
$$=2-30=-28$$

$$D_z=\begin{vmatrix}2 & -7 & 35\\1 & 1 & 1\\0 & 2 & -5\end{vmatrix}$$
$$=2\begin{vmatrix}1 & 1\\2 & -5\end{vmatrix}-1\begin{vmatrix}-7 & 35\\2 & -5\end{vmatrix}+0\begin{vmatrix}-7 & 35\\1 & 1\end{vmatrix}$$
$$=2(-5-2)-(35-70)+0$$
$$=-14+35=21$$

$$x=\frac{D_x}{D}=\frac{35}{7}=5$$

$$y=\frac{D_y}{D}=-\frac{28}{7}=-4$$

$$z=\frac{D_z}{D}=\frac{21}{7}=3$$

$\{(5,-4,3)\}$

29. Let x = width and y = length.
$2x+2y=288$
$y=2x-6$

$2x+2(2x-6)=288$
$2x+4x-12=288$
$6x=300$
$x=50$
$y=2(50)-6=94$

width = 50 ft
length = 94 ft
The court is 50 ft by 94 ft.

30. $124=30m+b$
$136=40m+b$
Solve each for b, and set equal to solve for m.
$124-30m=136-40m$
$40m-30m=136-124$
$10m=12$
$m=\frac{6}{5}$
$b=124-30\left(\frac{6}{5}\right)=88$
Point-slope form:
$y-124=\frac{6}{5}(x-30)$ or $y-136=\frac{6}{5}(x-40)$

Slope-intercept form:

$y = \frac{6}{5}x + 88$

$y = \frac{6}{5}(60) + 88$

$y = 160$

After running for 60 minutes, pulse rate will be 160.

Chapter 4

Problem Set 4.1

1. $x^2-4x^3+9x-12x^4-6$
$-12x^4-4x^3+x^2+9x-6$
Degree of each term: 4, 3, 2, 1, 0
Degree of the polynomial: 4

3. $7xy+5x^2+6x^3y^2+x^4-3$
$x^4+6x^3y^2+5x^2+7xy-3$
Degree of each term: 4, 5, 2, 2, 0
Degree of the polynomial: 5

5. $4x^3yz^2+8x^5-12y-3x^6y+5x^2y^4z^3$
$-3x^6y+8x^5+4x^3yz^2+5x^2y^4z^3-12y$
Degree of each term: 7, 5, 6, 9, 1
Degree of the polynomial: 9

7. $y=0.071x^2+0.73x$
Binomial, degree 2

9. $w^3+2p^2-6w+6p-6wp$
Five terms, degree 3

11. $15x+6x^2-x^3$
Trinomial, degree 3

13. $f(x)=-x^2+2x+3$
$x=-\frac{b}{2a}=-\frac{2}{2(-1)}=1$
$y=f(1)=-(1)^2+2(1)+3=4$
Vertex is at (1, 4).
$y=-x^2+2x+3$

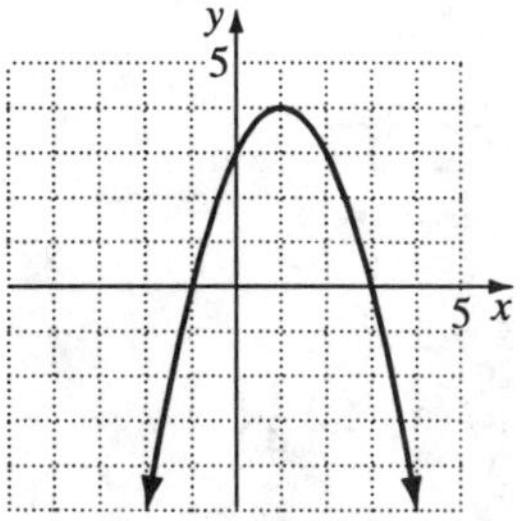

Range is $\{f(x)|f(x)\le 4\}$

15. $f(x)=x^2-x-2$
$x=-\frac{b}{2a}=-\frac{(-1)}{2(1)}=\frac{1}{2}$
$y=f\left(\frac{1}{2}\right)=\left(\frac{1}{2}\right)^2-\frac{1}{2}-2=-2\frac{1}{4}$
Vertex is at $\left(\frac{1}{2},-2\frac{1}{4}\right)$.
$y=x^2-x-2$

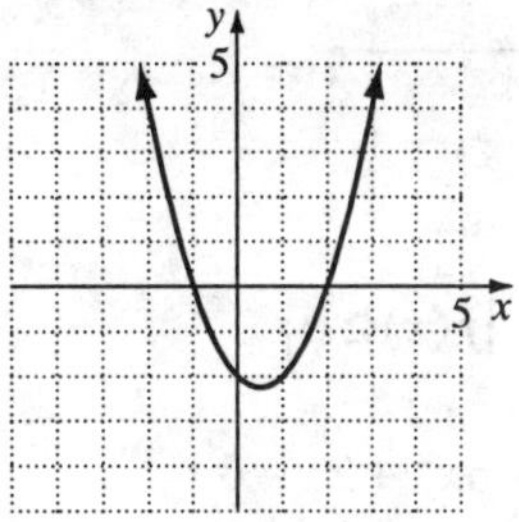

Range is $\left\{f(x)\middle|f(x)\ge -2\frac{1}{4}\right\}$.

17. $f(x)=-x^2-x+2$
$x=-\frac{b}{2a}=-\frac{-1}{2(-1)}=-\frac{1}{2}$
$y=f\left(-\frac{1}{2}\right)=-\left(-\frac{1}{2}\right)^2-\left(-\frac{1}{2}\right)+2=2\frac{1}{4}$
Vertex is at $\left(-\frac{1}{2},2\frac{1}{4}\right)$.
$y=-x^2-x+2$

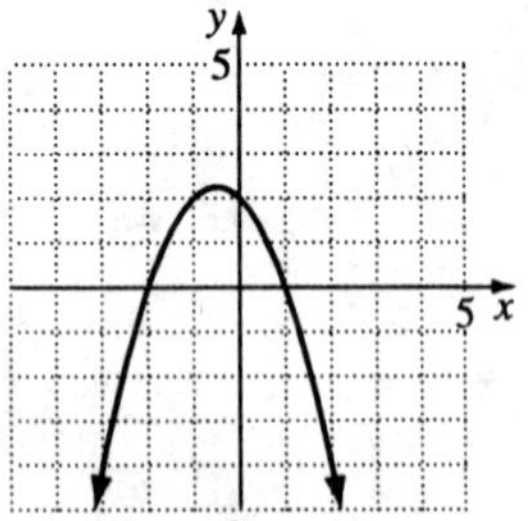

Range is $\left\{f(x)\middle|f(x) \le 2\frac{1}{4}\right\}$

19. $f(x) = 4x^2 + 8x + 4$

$x = -\frac{b}{2a} = -\frac{8}{2(4)} = -1$

$y = f(-1) = 4(-1)^2 + 8(-1) + 4 = 0$

Vertex is at $(-1, 0)$.

$y = 4x^2 + 8x + 4$

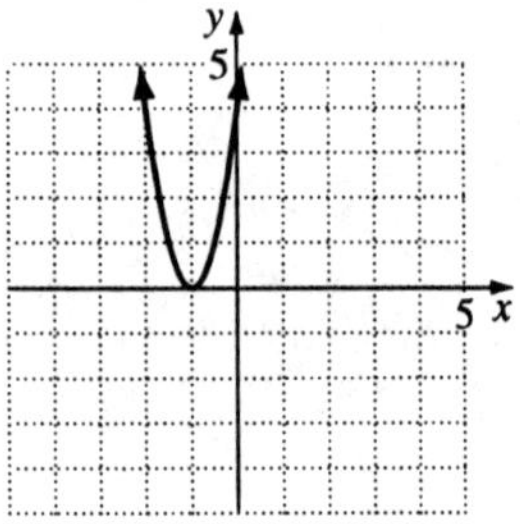

Range is $\{f(x)|f(x) \ge 0\}$.

21. $f(x) = -2x^3 + 6x^2 + 2x - 6$

x	−2	−1	0	1	2	3	4
$f(x)$	30	0	−6	0	6	0	−30

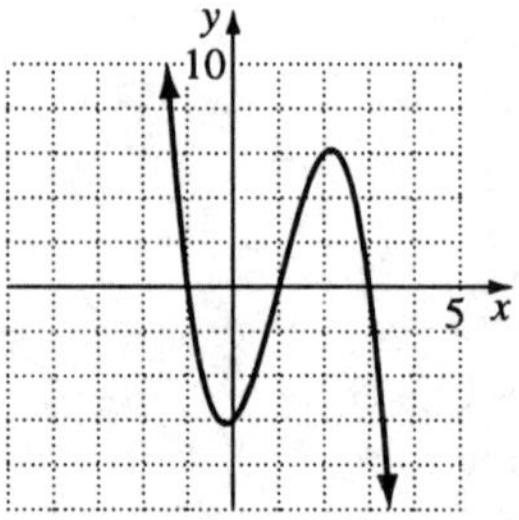

23. $f(x) = x^3 + 2x - 1$

x	−2	−1	0	1	2
$f(x)$	−13	−4	−1	2	11

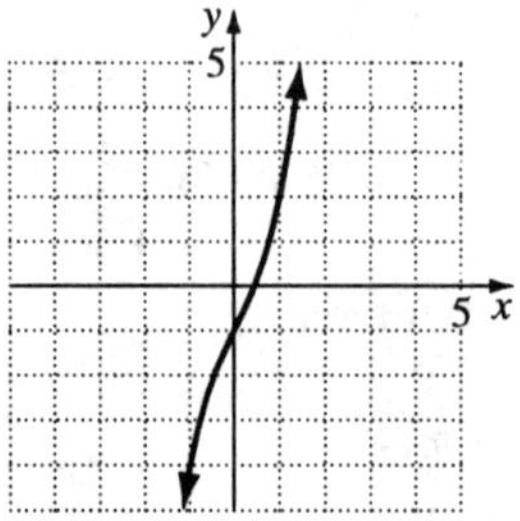

25. $(7x^2 - 3x) + (4x^2 - 7x - 8)$

$= 7x^2 + 4x^2 - 3x - 7x - 8$

$= 11x^2 - 10x - 8$

27. $(-7r^3 + 3r - 2 + 8r^2) + (-3r^2 + 7r + 4)$

$= -7r^3 + 8r^2 - 3r^2 + 3r + 7r - 2 + 4$

$= -7r^3 + 5r^2 + 10r + 2$

29. $(-3x^2y+2xy)+(-5x^3y+4x^2y-7)+(7xy+7)$
$=-5x^3y-3x^2y+4x^2y+2xy+7xy-7+7$
$=-5x^3y+x^2y+9xy$

31. $(17x^3-5x^2+4x-3)-(5x^3-9x^2-8x+11)$
$=(17-5)x^3+(-5+9)x^2+(4+8)x+(-3-11)$
$=12x^3+4x^2+12x-14$

33. $(13r^5+9r^4-5r^2+3r+6)-(-9r^5-7r^3+8r^2+11)$
$=(13+9)r^5+9r^4+7r^3+(-5-8)r^2+3r+(6-11)$
$=22r^5+9r^4+7r^3-13r^2+3r-5$

35. $(-6x^3y^2-8x^2y+11xy-3)-(7x^3y^2-5x^2y+9xy-3)$
$=(-6-7)x^3y^2+(-8+5)x^2y+(11-9)xy+(-3+3)$
$=-13x^3y^2-3x^2y+2xy$

37. $(5x^2-7x-8)+(2x^2-3x+7)-\left(x^2-4x-3\right)$
$=(5+2-1)x^2+(-7-3+4)x+(-8+7+3)$
$=6x^2-6x+2$

39. $(6y^4-5y^3+2y)-(4y^3+3y^2-1)+(y^4-2y^2+7y-3)$
$=(6+1)y^4+(-5-4)y^3+(-3-2)y^2+(2+7)y+(1-3)$
$=7y^4-9y^3-5y^2+9y-2$

41. $f(x)=x^2+2x,\ g(x)=2x+3$
$(f+g)(x)=x^2+2x+2x+3=x^2+4x+3$
$(f-g)(x)=x^2+2x-2x-3=x^2-3$
$f(1)=(1)^2+2(1)=3$
$g(1)=2(1)+3=5$
$(f+g)(1)=(1)^2+4(1)+3=8$

43. $f(x)=-3x^2+5x+7, g(x)=x^2-5x-4$
$(f+g)(x)=-3x^2+5x+7+x^2-5x-4\ =-2x^2+3$
$(f-g)(x)=-3x^2+5x+7-x^2+5x+4\ =-4x^2+10x+11$
$f(1)=-3(1)^2+5(1)+7=9$
$g(1)=(1)^2-5(1)-4=-8$
$(f+g)(1)=-2(1)^2+3=1$

45. $f(x) = 8x^2 - 5x - 4x^3,\ g(x) = 3x^2 - 9x - 7x^3$
$(f+g)(x) = 8x^2 - 5x - 4x^3 + 3x^2 - 9x - 7x^3 = -11x^3 + 11x^2 - 14x$
$(f-g)(x) = 8x^2 - 5x - 4x^3 - 3x^2 + 9x + 7x^3 = 3x^3 + 5x^2 + 4x$
$f(1) = 8(1)^2 - 5(1) - 4(1)^3 = -1$
$g(1) = 3(1)^2 - 9(1) - 7(1)^3 = -13$
$(f+g)(1) = -11(1)^3 + 11(1)^2 - 14(1) = -14$

47. $f(x) = 3x^3 - 7x^2 + 5x - 2,\ g(x) = -7x^3 + 4x^2 - 8x - 3$
$(f+g)(x) = 3x^3 - 7x^2 + 5x - 2 - 7x^3 + 4x^2 - 8x - 3 = -4x^3 - 3x^2 - 3x - 5$
$(f-g)(x) = 3x^3 - 7x^2 + 5x - 2 + 7x^3 - 4x^2 + 8x + 3 = 10x^3 - 11x^2 + 13x + 1$
$f(1) = 3(1)^3 - 7(1)^2 + 5(1) - 2 = -1$
$g(1) = -7(1)^3 + 4(1)^2 - 8(1) - 3 = -14$
$(f+g)(1) = -4(1)^3 - 3(1)^2 - 3(1) - 5 = -15$

49. **a.** $f(x) = x^2 + 2x$
$g(x) = 2x + 3$
$h(x) = (f+g)(x) = x^2 + 2x + 2x + 3 = x^2 + 4x + 3$

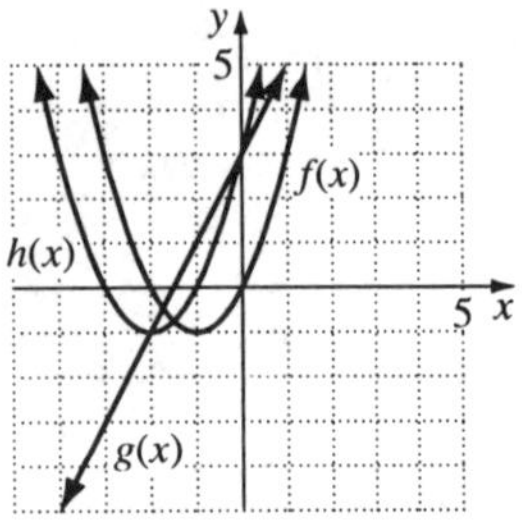

b. At each value of x, the graph of $h(x)$ is the sum of the values of $f(x)$ and $g(x)$.

51. Difference $= h_{\text{moon}} - h_{\text{earth}}$
$= (-2.7t^2 + 48t + 6) - (-16t^2 + 48t + 6)$
$= (-2.7 + 16)t^2 + (48 - 48)t + (6 - 6)$
$= 13.3t^2$

53. $f(x) = 0.0075x^2 - 0.2676x + 14.8$
$x = -\frac{b}{2a} = -\frac{-0.2676}{2(0.0075)} = 17.84$
1940 + 17 = 1957
$f(17.84) = 0.0075(17.84)^2 - 0.2676(17.84) + 14.8 \approx 12.4$
In 1957 the fuel efficiency was poorest at about 12.4 miles per gallon.

55. d is true.

57. Students should verify problems.

59. $y = -0.25x^2 + 40x$

$x = -\frac{b}{2a} = -\frac{40}{2(-0.25)} = 80$

$y = -0.25(80)^2 + 40(80) = 1600$

Vertex is at (80, 1600).

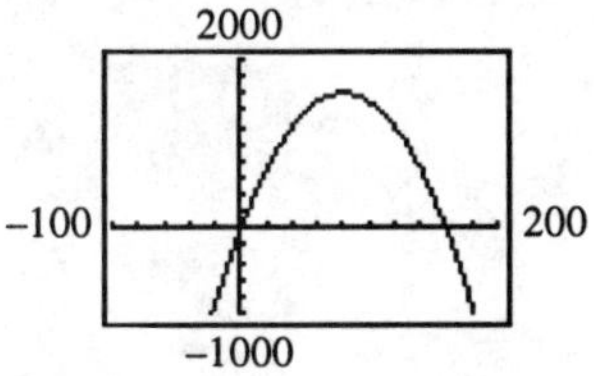

61. $y = 5x^2 + 40x + 600$

$x = -\frac{b}{2a} = -\frac{40}{2(5)} = -4$

$y = 5(-4)^2 + 40(-4) + 600 = 520$

Vertex is at (–4, 520).

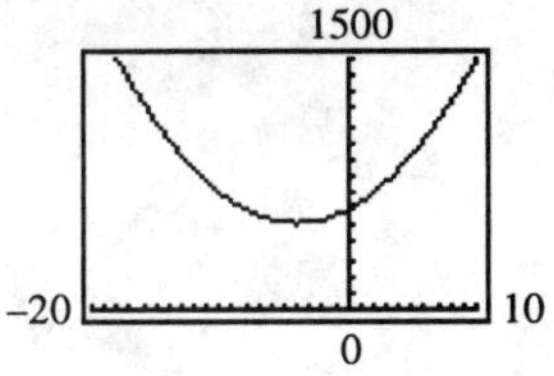

63. **c**. gives a relatively complete picture of the polynomial's graph. **a**. and **b**. give only partial pictures.

65. $y = -x^3 + 8x^2 + 10x - 15$

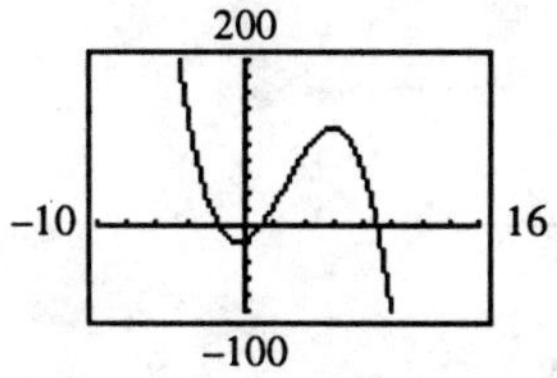

67. $(x^2 - 6x + 3) - (2x + 5) \neq x^2 - 4x - 2$

$(x^2 - 6x + 3) - (2x + 5) = x^2 - 8x - 2$

$y = x^2 - 8x - 2$

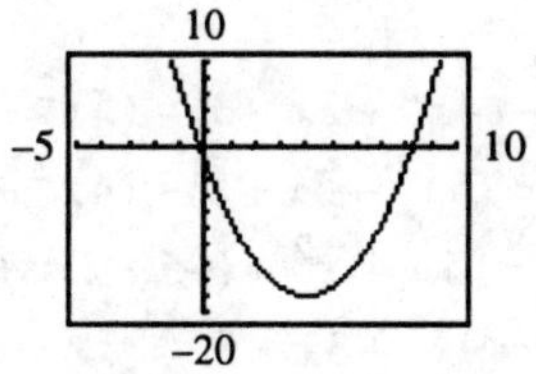

69. $(9x^3 + 5x - 4x^2) - (2x^3 - 5x^2) \neq 7x^3 - 9x^2 + 5x$
$(9x^3 + 5x - 4x^2) - (2x^3 - 5x^2) = 7x^3 + x^2 + 5x$
$y = 7x^3 + x^2 + 5x$

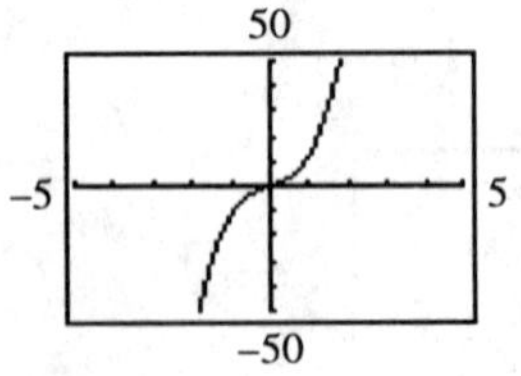

71. Answers may vary.

73. $(x^{2n} - 3x^n + 5) + (4x^{2n} - 3x^n - 4) - (2x^{2n} - 5x^n - 3)$
$= (1 + 4 - 2)x^{2n} + (-3 - 3 + 5)x^n + (5 - 4 + 3)$
$= 3x^{2n} - x^n + 4$

75. $(\text{polynomial}) - (4x^2 + 2x - 3) = (5x^2 - 5x + 8)$
$(\text{polynomial}) = (5x^2 - 5x + 8) + (4x^2 + 2x - 3)$
$= (5 + 4)x^2 + (-5 + 2)x + (8 - 3)$
$= 9x^2 - 3x + 5$

77. $f(0) = -4$ (from the graph)
$g(-6) > g(-5), g(-5) = 6$
$g(-6) > f(0)$
$g(-6)$

79. a. $f(x) = x^4$
$g(x) = (x - 1)^4$
$h(x) = x^4 - 1$

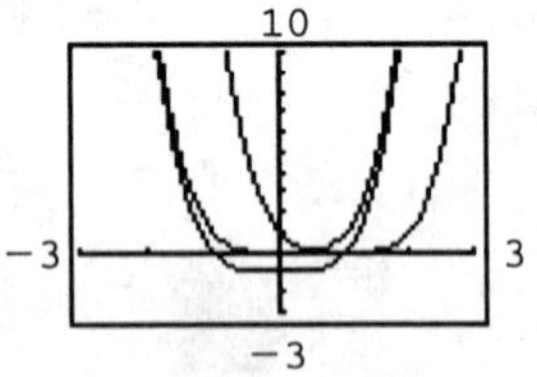

b. They have similar shapes, but different vertex points.

c. Rises, rises

81. $f(x) = x, f\left(\frac{1}{2}\right) = \frac{1}{2}$
$g(x) = x^2, g\left(\frac{1}{2}\right) = \frac{1}{4}$

$h(x) = x^3, h\left(\frac{1}{2}\right) = \frac{1}{8}$

$F(x) = x^4, F\left(\frac{1}{2}\right) = \frac{1}{16}$

$G(x) = x^5, G\left(\frac{1}{2}\right) = \frac{1}{32}$

$H(x) = x^6, H\left(\frac{1}{2}\right) = \frac{1}{64}$

$f\left(\frac{1}{2}\right) > g\left(\frac{1}{2}\right) > h\left(\frac{1}{2}\right) > F\left(\frac{1}{2}\right) > G\left(\frac{1}{2}\right) > H\left(\frac{1}{2}\right)$

From the left to right: $f(x) = x, g(x) = x^2, h(x) = x^3, F(x) = x^4, G(x) = x^5, H(x) = x^6$

83. 1, 8, 27, 64, __, __, __

$1^3, 2^3, 3^3, 4^3, 5^3 = 125, 6^3 = 216, 7^3 = 343$

125, 216, 343; x^3

85. –1, 6, 25, 62, __, __, __

1 – 2, 8 – 2, 27 – 2, 64 – 2, __, __, __

$1^3 - 2,\ 2^3 - 2,\ 3^3 - 2,\ 4^3 - 2,\ 5^3 - 2 = 123,\ 6^3 - 2 = 214,\ 7^3 - 2 = 341$

123, 214, 314; $x^3 - 2$

87. 2, 10, 30, 68, __, __, __

1 + 1, 8 + 2, 27 + 3, 64 + 4, __, __, __

$1^3 + 1,\ 2^3 + 2,\ 3^3 + 3,\ 4^3 + 4,\ 5^3 + 5 = 130;\ 6^3 + 6 = 222,\ 7^3 + 7 = 350$

130, 222, 350; $x^3 + x$

Review Problems

88. $9(x-1) = 1 + 3(x-2)$

$9x - 9 = 1 + 3x - 6$

$9x - 3x = 9 - 5$

$6x = 4$

$x = \frac{4}{6}$

$x = \frac{2}{3}$

89. $2x - 3y < -6$

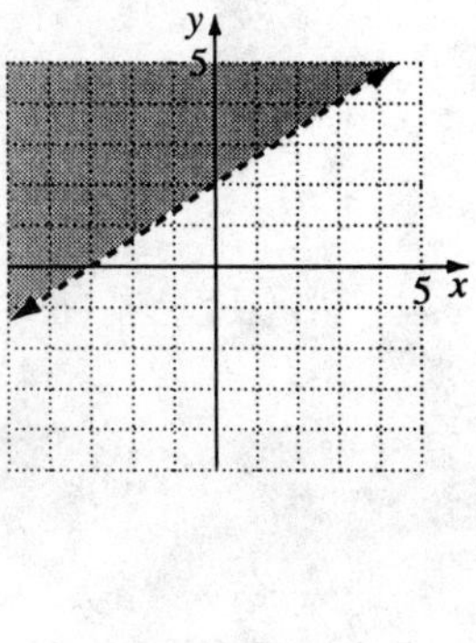

90. Point: (–2, 5)

line: $3x - y = 9$ or $y = 3x - 9$

perpendicular slope $= -\frac{1}{3}$

point-slope form:

$y - 5 = -\frac{1}{3}(x+2)$

slope-intercept form:

$y - 5 = -\frac{1}{3}x - \frac{2}{3}$

$y = -\frac{1}{3}x - \frac{2}{3} + \frac{15}{3}$

$y = -\frac{1}{3}x + \frac{13}{3}$

standard form:

$3y = -x + 13$

$x + 3y = 13$

Problem Set 4.2

1. $y^7 \cdot y^5 = y^{7+5} = y^{12}$

3. $(3x^8)(5x^6) = (3)(5)x^{8+6} = 15x^{14}$

5. $(3x^2y^4)(5xy^7) = 15x^{2+1}y^{4+7} = 15x^3y^{11}$

7. $(-3xy^2z^5)(2xy^7z^4) = -6x^{1+1}y^{2+7}z^{5+4} = -6x^2y^9z^9$

9. $4x^2(3x-2) = 4x^2(3x) + 4x^2(-2) = 12x^3 - 8x^2$

11. $-2x^2(5x^3 - 8x^2 + 7x - 3)$
$= -10x^{2+3} + 16x^{2+2} - 14x^{2+1} + 6x^2$
$= -10x^5 + 16x^4 - 14x^3 + 6x^2$

13. $2xy(4x^2y + 7x - 2y - 3)$
$= 2xy(4x^2y) + 2xy(7x) + 2xy(-2y) + 2xy(-3)$
$= 8x^3y^2 + 14x^2y - 4xy^2 - 6xy$

15. $\frac{1}{3}x^3y^7\left(\frac{1}{2}xy^6 + \frac{2}{5}x^4y^2 + 6\right)$
$= \frac{1}{3}x^3y^7\left(\frac{1}{2}xy^6\right) + \frac{1}{3}x^3y^7\left(\frac{2}{5}x^4y^2\right) + \frac{1}{3}x^3y^7(6)$
$= \frac{1}{6}x^4y^{13} + \frac{2}{15}x^7y^9 + 2x^3y^7$

17. $16x^4y^5z^3\left(-\frac{1}{8}xz + \frac{1}{16}x^4yz^2 - \frac{1}{32}x^6y^2z\right)$
$= 16x^4y^5z^3\left(-\frac{1}{8}xz\right) + 16x^4y^5z^3\left(\frac{1}{16}x^4yz^2\right) + 16x^4y^5z^3\left(-\frac{1}{32}x^6y^2z\right)$
$= -2x^5y^5z^4 + x^8y^6z^5 - \frac{1}{2}x^{10}y^7z^4$

19. $(6uv^3w - 8uv + w^4)(-5u^5v^3w)$
$= (6uv^3w)(-5u^5v^3w) - 8uv(-5u^5v^3w) + w^4(-5u^5v^3w)$
$= -30u^6v^6w^2 + 40u^6v^4w - 5u^5v^3w^5$

21. $(x-3)(x^2 + 2x + 5)$
$= x(x^2 + 2x + 5) - 3(x^2 + 2x + 5)$
$= x^3 + 2x^2 + 5x - 3x^2 - 6x - 15$
$= x^3 - x^2 - x - 15$

23. $(x+5)(x^2-5x+25)$
$= x(x^2-5x+25)+5(x^2-5x+25)$
$= x^3-5x^2+25x+5x^2-25x+125$
$= x^3+125$

25. $(a-b)(a^2+ab+b^2) = a(a^2+ab+b^2)-b(a^2+ab+b^2)$
$= a^3+a^2b+ab^2-a^2b-ab^2-b^3 = a^3-b^3$

27. $(x^2+2x-1)(x^2+3x-4)$
$= x^2(x^2+3x-4)+2x(x^2+3x-4)-1(x^2+3x-4)$
$= x^4+3x^3-4x^2+2x^3+6x^2-8x-x^2-3x+4$
$= x^4+5x^3+x^2-11x+4$

29. $(xy+2)(x^2y^2-2xy+4) = xy(x^2y^2-2xy+4)+2(x^2y^2-2xy+4)$
$= x^3y^3-2x^2y^2+4xy+2x^2y^2-4xy+8$
$= x^3y^3+8$

31. $(5a^2b-3ab+2b^2)(ab-3b+a)$
$= 5a^2b(ab-3b+a)-3ab(ab-3b+a)+2b^2(ab-3b+a)$
$= 5a^3b^2-15a^2b^2+5a^3b-3a^2b^2+9ab^2-3a^2b+2ab^3-6b^3+2ab^2$
$= 5a^3b^2+5a^3b-18a^2b^2-3a^2b+11ab^2+2ab^3-6b^3$

33. $(x+4)(x+7)$
$= x^2+7x+4x+28$
$= x^2+11x+28$

35. $(y+5)(y-6)$
$= y^2-6y+5y-30$
$= y^2-y-30$

37. $(5x+3)(7x+1)$
$= 35x^2+5x+21x+3$
$= 35x^2+26x+3$

39. $(3y-11)(2y-3)$
$= 6y^2-9y-22y+33$
$= 6y^2-31y+33$

41. $(7x-12)(3x+8)$
$= 21x^2+56x-36x-96$
$= 21x^2+20x-96$

43. $(9x^2-4)(3x^2+5)$
$= 27x^4+45x^2-12x^2-20$
$= 27x^4+33x^2-20$

45. $(8x^3-3x)(5x^3-2x)$
$= 40x^6-16x^4-15x^4+6x^2$
$= 40x^6-31x^4+6x^2$

47. $(5z^2+1)(3z-4)$
$= (5z^2)(3z)+(5z^2)(-4)+1(3z)-1(4)$
$= 15z^3-20z^2+3z-4$

49. $(9x-5)(7+4x)$
$=63x+36x^2-35-20x$
$=36x^2+43x-35$

51. $(9x^2+4)(9x^3-4)$
$=81x^5-36x^2+36x^3-16$
$=81x^5+36x^3-36x^2-16$

53. $(3x+7y)(2x-5y)$
$=6x^2-15xy+14xy-35y^2$
$=6x^2-xy-35y^2$

55. $(3x^2yz-2y)(5x^2yz+7y)$
$=15x^4y^2z^2+21x^2y^2z-10x^2y^2z-14y^2$
$=15x^4y^2z^2+11x^2y^2z-14y^2$

57. $(x+3)^2$
$=x^2+2(3x)+9$
$=x^2+6x+9$

59. $(y-5)^2$
$=y^2-2(5y)+25$
$=y^2-10y+25$

61. $(2x+y)^2$
$=(2x)^2+2(2xy)+y^2$
$=4x^2+4xy+y^2$

63. $(5x-3y)^2$
$=(5x)^2-2(5x)(3y)+(3y)^2$
$=25x^2-30xy+9y^2$

65. $(2x^2-3y)^2$
$=(2x^2)^2-2(2x^2)(3y)+(3y)^2$
$=4x^4-12x^2y+9y^2$

67. $(4x^3+2y^5)^2$
$=(4x^3)^2+2(4x^3)(2y^5)+(2y^5)^2$
$=16x^6+16x^3y^5+4y^{10}$

69. $(4ab^2+ab)^2$
$=(4ab^2)^2+2(4ab^2)(ab)+(ab)^2$
$=16a^2b^4+8a^2b^3+a^2b^2$

71. $(y+7)(y-7)$
$=(y)^2-(7)^2$
$=y^2-49$

73. $(5x+3)(5x-3)$
$=(5x)^2-(3)^2$
$=25x^2-9$

75. $(3x+7y)(3x-7y)$
$=(3x)^2-(7y)^2$
$=9x^2-49y^2$

77. $(x^2+yz)(x^2-yz)$
$=(x^2)^2-(yz)^2$
$=x^4-y^2z^2$

79. $(7x^2y-3z)(7x^2y+3z)$
$=(7x^2y)^2-(3z)^2$
$=49x^4y^2-9z^2$

81. $(3x^3y^2z-1)(3x^3y^2z+1)$
$=(3x^3y^2z)^2-1^2=9x^6y^4z^2-1$

83. $(-xy+y^2)(xy+y^2)$
$=(y^2-xy)(y^2+xy)$
$=(y^2)^2-(xy)^2$
$=y^4-x^2y^2$

85. $(3x+7+5y)(3x+7-5y)$
$=(3x+7)^2-(5y)^2$
$=9x^2+42x+49-25y^2$

87. $[5y-(2x+3)][5y+(2x+3)]$
$=(5y)^2-(2x+3)^2$

$= 25y^2 - (4x^2 + 12x + 9)$
$= 25y^2 - 4x^2 - 12x - 9$

89. $(2x + y + 1)^2$
$= [(2x + y) + 1]^2$
$= (2x + y)^2 + 2(2x + y)(1) + 1^2$
$= 4x^2 + 4xy + y^2 + 4x + 2y + 1$

91. $[(3x - 1) + y]^2$
$= (3x - 1)^2 + 2(3x - 1)y + y^2$
$= 9x^2 - 6x + 1 + 6xy - 2y + y^2$

93. $[(3x - 1) + y][(3x - 1) - y]$
$= (3x - 1)^2 - y^2$
$= 9x^2 - 6x + 1 - y^2$

95. $(x + 2)(x - 2)(x^2 + 4)$
$= (x^2 - 4)(x^2 + 4)$
$= x^4 - 16$

97. $(3x - y)(3x + y)(9x^2 + y^2)$
$= (9x^2 - y^2)(9x^2 + y^2)$
$= 81x^4 - y^4$

99. $f(x) = 2x^2 + 3, \ g(x) = x^2 - 4$
$(f + g)(x) = 2x^2 + 3 + x^2 - 4 = 3x^2 - 1$
$(f - g)(x) = 2x^2 + 3 - x^2 + 4 = x^2 + 7$
$(fg)(x) = (2x^2 + 3)(x^2 - 4)$
$= 2x^4 - 8x^2 + 3x^2 - 12$
$= 2x^4 - 5x^2 - 12$

101. $f(x) = x^2 - 2x + 1, \ g(x) = x^3 + x^2 + x + 3$
$(f + g)(x) = x^2 - 2x + 1 + x^3 + x^2 + x + 3 = x^3 + 2x^2 - x + 4$
$(f - g)(x) = x^2 - 2x + 1 - x^3 - x^2 - x - 3 = -x^3 - 3x - 2$
$(fg)(x) = (x^2 - 2x + 1)(x^3 + x^2 + x + 3) = x^5 + x^4 + x^3 + 3x^2 - 2x^4 - 2x^3 - 2x^2 - 6x + x^3 + x^2 + x + 3$
$= x^5 - x^4 + 2x^2 - 5x + 3$

103. $f(x) = x^2 - 2x + 5$
$f(a + 3) = (a + 3)^2 - 2(a + 3) + 5 = a^2 + 6a + 9 - 2a - 6 + 5 = a^2 + 4a + 8$
$f(2a - 1) = (2a - 1)^2 - 2(2a - 1) + 5 = 4a^2 - 4a + 1 - 4a + 2 + 5 = 4a^2 - 8a + 8$
$f(a + h) - f(a) = (a + h)^2 - 2(a + h) + 5 - (a^2 - 2a + 5)$
$= a^2 + 2ah + h^2 - 2a - 2h + 5 - a^2 + 2a - 5$
$= 2ah + h^2 - 2h$

105. $f(x) = 3x^2 + 5x - 2$
$f(a + 3) = 3(a + 3)^2 + 5(a + 3) - 2 = 3(a^2 + 6a + 9) + 5a + 15 - 2 = 3a^2 + 23a + 40$
$f(2a - 1) = 3(2a - 1)^2 + 5(2a - 1) - 2 = 3(4a^2 - 4a + 1) + 10a - 5 - 2 = 12a^2 - 2a - 4$
$f(a + h) - f(a) = 3(a + h)^2 + 5(a + h) - 2 - (3a^2 + 5a - 2)$
$= 3a^2 + 6ah + 3h^2 + 5a + 5h - 2 - 3a^2 - 5a + 2$
$= 6ah + 3h^2 + 5h$

107. Area of shaded region
= area of large rectangle – area of small rectangle
$= 2x(5x+1) - x(4x-1)$
$= 10x^2 + 2x - 4x^2 + x$
$= 6x^2 + 3x$

109. Area of shaded region
= area of triangle – area of square
$= \frac{1}{2} 8x(6x+4) - 2x(2x)$
$= 4x(6x+4) - 4x^2$
$= 24x^2 + 16x - 4x^2$
$= 20x^2 + 16x$

111. $2(x+10)^2 - 2(x+5)^2$
$= 2(x^2 + 20x + 100) - 2(x^2 + 10x + 25)$
$= 2x^2 + 40x + 200 - 2x^2 - 20x - 50$
$= 20x + 150 \text{ cm}^3$

113. $N = \left(\frac{x-4}{4}\right)^2 y$
$= \left(\frac{x^2 - 8x + 16}{16}\right) y$
$= \frac{x^2y - 8xy + 16y}{16}$ or $\frac{x^2y}{16} - \frac{xy}{2} + y$

115. b is true; $(3x+7)(3x-2)$
$= 9x^2 - 6x + 21x - 14$
$= 9x^2 + 15x - 14$
a is *not* true since $(x-5)^2 = x^2 - 10x + 25$.
c is *not* true since $(x^m + 2)(x^m + 4)$
$= x^{2m} + 6x^m + 8$.
d is *not* true since $7x^2 \cdot 4x^3$ is the product of two monomials.

117. $y_1 = (x-2)^2$
$y_2 = x^2 - 4x + 4$
$(x-2)^2 = x^2 - 4x + 4$
$y_1 = y_2$

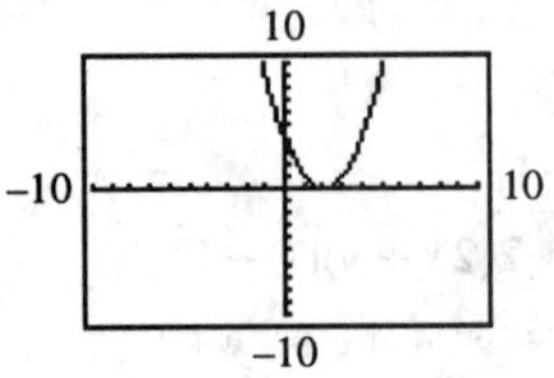

119. $y_1 = (x^2 - 3x + 2)(x - 4)$
$y_2 = x^3 - 7x^2 + 14x - 8$
$(x^2 - 3x + 2)(x - 4) = x^3 - 7x^2 + 14x - 8$
$y_1 = y_2$

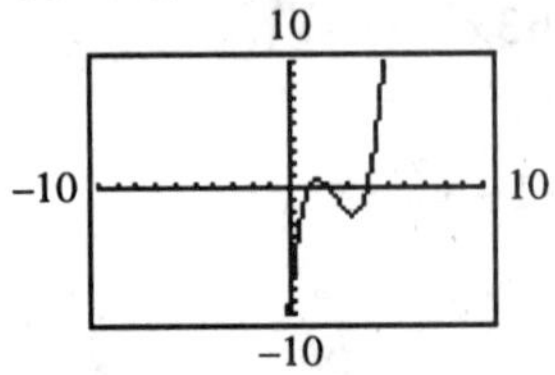

121. Answers may vary.

123. $(y^n + 2)(y^n - 2) - (y^n - 3)^2$
$= y^{2n} - 4 - (y^{2n} - 6y^n + 9)$
$= 6y^n - 13$

125. $P(x)$: desired polynomial
$\frac{P(x)}{4x-5} = 3x + 1$
$P(x) = (4x-5)(3x+1)$
$= 12x^2 - 11x - 5$

127.

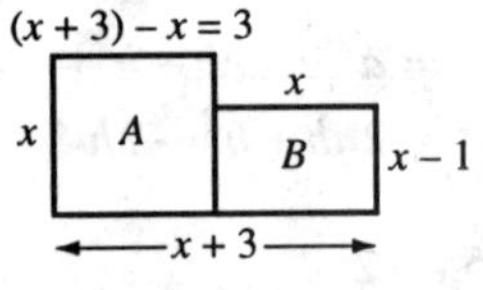

Area of figure
= area A + area B
$= 3x + x(x-1)$
$= 3x + x^2 - x$
$= x^2 + 2x$

129. Area of figure
= area of rectangle + area of small triangle + area of large triangle
$= (x+2)[(x-5)+(x+3)]+\frac{1}{2}x(x-5)+\frac{1}{2}x(x+3)$
$= (x+2)(2x-2)+\frac{1}{2}x^2-\frac{5}{2}x+\frac{1}{2}x^2+\frac{3}{2}x$
$= 2x^2-2x+4x-4+x^2-x$
$= 3x^2+x-4$

131. Volume of figure
= volume of large rectangular solid + volume of small rectangular solid
$= x(x+1)(x+3)+x[(x+3)-x][(2x-1)-(x+1)]$
$= x(x^2+4x+3)+x(3)(x-2)$
$= x^3+4x^2+3x+3x^2-6x$
$= x^3+7x^2-3x$

133. **a.** $(y-1)(y+1)=y^2-1$
b. $(y-1)(y^2+y+1)=y^3+y^2+y-y^2-y-1=y^3-1$
c. $(y-1)(y^3+y^2+y+1)=y^4+y^3+y^2+y-y^3-y^2-y-1=y^4-1$
d. $(y-1)(y^4+y^3+y^2+y+1)=y^5-1$

135. $(5-2)^2=5(5)-5(4)+2(2)$
$(7-3)^2=7(7)-7(6)+3(3)$
$(11-5)^2=11(11)-11(10)+5(5)=121-110+25$

137. Group activity

Review Problems:

138. $|4-3x|\ge 10$
$4-3x\ge 10$ or $4-3x\le -10$
$-3x\ge 6$ $-3x\le -14$
$x\le -2$ $x\ge \frac{14}{3}$
$\left\{x \middle| x\le -2 \text{ or } x\ge \frac{14}{3}\right\}$
$(-\infty,-2]\cup\left[\frac{14}{3},\ \infty\right)$

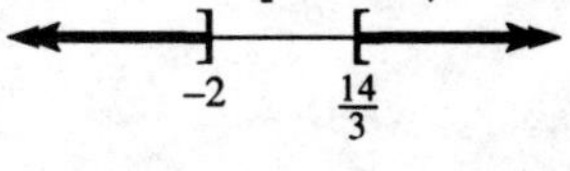

139. $[(2y^2-3y-1)+(3y^2+y-1)]-(y^2+3y-1)$
$=5y^2-2y-2-y^2-3y+1$
$=4y^2-5y-1$

140. $x-y+z=4$
$3x-y+z=6$
$2x+2y-3z=-6$

$$\left[\begin{array}{ccc|c}1&-1&1&4\\3&-1&1&6\\2&2&-3&-6\end{array}\right]\begin{array}{l}\\-3R_1+R_2\to\\-2R_1+R_3\to\end{array}\left[\begin{array}{ccc|c}1&-1&1&4\\0&2&-2&-6\\0&4&-5&-14\end{array}\right]$$

$$-2R_2+R_3\to\left[\begin{array}{ccc|c}1&-1&1&4\\0&2&-2&-6\\0&0&-1&-2\end{array}\right]\begin{array}{l}\\\frac{1}{2}R_2\to\\-1R_3\to\end{array}\left[\begin{array}{ccc|c}1&-1&1&4\\0&1&-1&-3\\0&0&1&2\end{array}\right]$$

$x-y+z=4$
$y-z=-3$
$z=2$
$y-2=-3$
$y=-1$
$x-(-1)+2=4$
$x=1$
$\{(1,-1,2)\}$

Problem Set 4.3

1. $4x-20=4(x-5)$
GCF = 4

3. $18a+27=9(2a+3)$
GCF = 9

5. $12x^2+4x=4x(3x+1)$
GCF = $4x$

7. $9x-18y=9(x-2y)$
GCF = 9

9. $12x^2y-8xy^2=4xy(3x-2y)$
GCF = $4xy$

11. $4xy-7xy^2=xy(4-7y)$
GCF = xy

13. $18x^4+9x^3-27x^2=9x^2(2x^2+x-3)$
GCF $=9x^2$

15. $10y^7-16y^4+8y^3=2y^3(5y^4-8y+4)$
GCF $=2y^3$

17. $15x^5y^3-25x^3y^4=5x^3y^3(3x^2-5y)$
GCF $=5x^3y^3$

19. $2a^2b-5ab^2+7a^2b^2=ab(2a-5b+7ab)$
GCF $=ab$

21. $24xy^3-36x^3y^2+12x^2y^4=12xy^2(2y-3x^2+xy^2)$
GCF $=12xy^2$

23. $26x^5y^3+52x^7y^2-39x^8y^5=13x^5y^2(2y+4x^2-3x^3y^3)$
GCF $=13x^5y^2$

25. $30x^4y+50x^2y^2+20x^3z^3=10x^2(3x^2y+5y^2+2xz^3)$
GCF $=10x^2$

27. $55x^2y^2z^4-77x^3y^2z=11x^2y^2z(5z^3-7x)$
GCF $=11x^2y^2z$

29. $7x^3y^2+14x^2y-42x^5y^3+21xy^4=7xy(x^2y+2x-6x^4y^2+3y^3)$
GCF $=7xy$

31. $70x^3y+42x^2y-28x^2y^2-84xy^3=14xy(5x^2+3x-2xy-6y^2)$
GCF $=14xy$

33. $7(x-3)+y(x-3)=(x-3)(7+y)$
GCF $=x-3$

35. $-2a(x+7y)+4c(x+7y)$
$=(x+7y)(-2a+4c)$
$=-2(x+7y)(a-2c)$ or $2(x+7y)(2c-a)$
GCF $=2(x+7y)$

37. $4x(a+b-c)-2y(a+b-c)$
$=(a+b-c)(4x-2y)$
$=2(a+b-c)(2x-y)$
GCF $=2(a+b-c)$

39. $(x-4y)z^2+(x-4y)z+(x-4y)$
$=(x-4y)(z^2+z+1)$
GCF $=x-4y$

41. $(x-4y)z^2+(x-4y)z+(4y-x)$
$=(x-4y)z^2+(x-4y)z+(x-4y)(-1)$
$=(x-4y)(z^2+z-1)$
$\text{GCF} = x-4y$

43. $5x^3(2a-7b)+15x^2(2a-7b)$
$=5x^2(2a-7b)(x+3)$
$\text{GCF} = 5x^2(2a-7b)$
A second method involves first factoring out $(2a-7b)$:
$=5x^3(2a-7b)+15x^2(2a-7b)$
$=(2a-7b)(5x^3+15x^2)$
$=(2a-7b)(5x^2)(x+3)$
$=5x^2(2a-7b)(x+3)$

45. $77x^3y^2(7a-5b)+11x^2y(7a-5b)$
$=11x^2y(7a-5b)(7xy+1)$
$\text{GCF} = 11x^2y(7a-5b)$
or by first factoring out $(7a-5b)$:
$77x^3y^2(7a-5b)+11x^2y(7a-5b)$
$=(7a-5b)(77x^3y^2+11x^2y)$
$=(7a-5b)(11x^2y)(7xy+1)$
$=11x^2y(7a-5b)(7xy+1)$

47. $-4x+12=4(-x+3)$
$\text{GCF} = 4$
$-4x+12=-4(x-3)$
$\text{GCF} = -4$

49. $-3x^2+27x=3x(-x+9)$
$\text{GCF} = 3x$
$-3x^2+27x=-3x(x-9)$
$\text{GCF} = -3x$

51. $-3x^3+15x^2-21x=3x(-x^2+5x-7)$
$\text{GCF} = 3x$
$-3x^3+15x^2-21x=-3x(x^2-5x+7)$
$\text{GCF} = -3x$

53. $-x^2+7x-5=1(-x^2+7x-5)$
$\text{GCF} = 1$
$-x^2+7x-5=-1(x^2-7x+5)$
$\text{GCF} = -1$

55. $-b^4+3b^3-15b=b(-b^3+3b^2-15)$
$\text{GCF} = b$
$-b^4+3b^3-15b=-b(b^3-3b^2+15)$
$\text{GCF} = -b$

57. $x^3-3x^2+4x-12$
$=x^2(x-3)+4(x-3)$
$=(x-3)(x^2+4)$

59. y^3-y^2+2y-2
$=y^2(y-1)+2(y-1)$
$=(y-1)(y^2+2)$

61. y^3-y^2-5y+5
$=y^2(y-1)-5(y-1)$
$=(y-1)(y^2-5)$

63. $a^2c+5ac+2a+10$
$=ac(a+5)+2(a+5)$
$=(a+5)(ac+2)$

65. $3Y^2+4YZ+24Y+32Z$
$=Y(3Y+4Z)+8(3Y+4Z)$
$=(3Y+4Z)(Y+8)$

67. $2a-6b+ac-3bc$
$=2(a-3b)+c(a-3b)$
$=(a-3b)(2+c)$

69. $3ab+3ac-b-c$
$=3a(b+c)-(b+c)$
$=(b+c)(3a-1)$

71. $5x^2+15-x^2y-3y$
$=5(x^2+3)-y(x^2+3)$
$=(x^2+3)(5-y)$

73. $2ab + ac - 6b - 3c$
$= a(2b + c) - 3(2b + c)$
$= (2b + c)(a - 3)$

75. $a^2 - ab - a + b$
$= a(a - b) - (a - b)$
$= (a - b)(a - 1)$

77. $A = P + Pr + (P + Pr)r$
$= P(1 + r) + P(1 + r)r$
$= P(1 + r)(1 + r)$
$= P(1 + r)^2$

79. $S.A. = 2\pi rh + 2\pi r^2$
$= 2\pi r(h) + 2\pi r(r)$
$= 2\pi r(h + r)$

81. a. $s(t) = -16t^2 + 64t$
$s(t) = -16t(t - 4)$

b. $t = -\frac{b}{2a} = -\frac{64}{2(-16)} = 2$
$s(2) = -16(2)(2 - 4) = 64$
Vertex is at (2, 64).

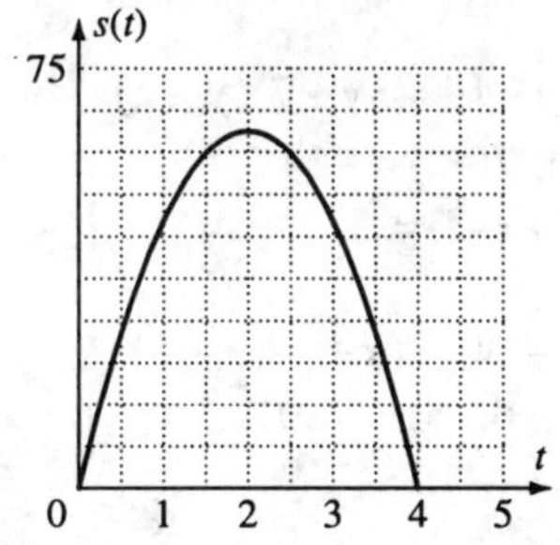

83. c is true.
$x^3 + x^2 + 4x - 4$
$= x^2(x + 1) + 4(x - 1)$
$(x + 1)$ and $(x - 1)$ are different.

85. $x^2 + 3x - 5x - 15 \neq (x + 5)(x - 3)$
$x^2 + 3x - 5x - 15 = x(x + 3) - 5(x + 3)$
$x^2 + 3x - 5x - 15 = (x + 3)(x - 5)$
$y = (x + 3)(x - 5)$

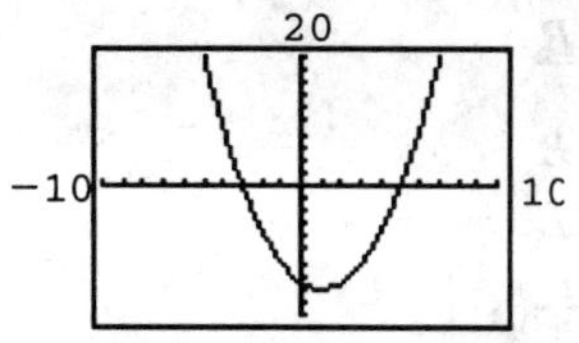

87. $x^3 + x^2 - x - 1 \neq (x^2 + 1)(x - 1)$
$x^3 + x^2 - x - 1 = x^2(x + 1) - 1(x + 1)$
$x^3 + x^2 - x - 1 = (x + 1)(x^2 - 1)$
$y = (x + 1)(x^2 - 1)$

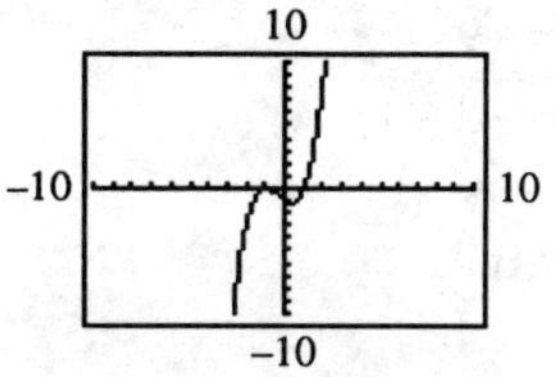

89. Answers may vary.

91. $3x^{3m}y^m + 7x^{2m}y^{2m} = x^{2m}y^m(3x^m + 7y^m)$

93. a. 1st sale price
$= x - 0.05x = x(1 - 0.05) = 0.95x$
2nd sale price $= 0.95x - 0.05(0.95x)$

b. $0.95x - 0.05(0.95x)$
$= 0.95x(1 - 0.05)$
$= 0.95x(0.95) = 0.9025x$
The final price is $0.9025x$.

Review Problems

94. Let l = length and w = width.
$2l + 2w = 22$ or $l + w = 11$
$l = 2w + 2$

$2w + 2 + w = 11$
$3w + 2 = 11$
$3w = 9$
$w = 3$
$l = 2(3) + 2 = 8$
The length is 8 feet and the width is 3 feet.

95. $3A = \frac{2L-B}{4}$
$12A = 2L - B$
$12A + B = 2L$
$L = \frac{12A+B}{2}$

96. $x - y + z = 4$
$2x - y + z = 9$
$x + 2y - z = 5$
Add (1) and –1 times (2).

$$\begin{aligned} x - y + z &= 4 \\ -2x + y - z &= -9 \\ \hline -x \quad &= -5 \\ x \quad &= 5 \end{aligned}$$

Add (1) and (3).

$$\begin{aligned} x - y + z &= 4 \\ x + 2y - z &= 5 \\ \hline 2x + y &= 9 \\ 2(5) + y &= 9 \\ y &= -1 \end{aligned}$$

$x - y + z = 4$
$5 + 1 + z = 4$
$z = -2$
$\{(5, -1, -2)\}$

Problem Set 4.4

Note: Not all possible factorizations are given. Only the pair that satisfies the conditions whose product is *c* and whose sum is *b* are given here.

1. $x^2 + 5x + 6 = (x+3)(x+2)$
product: 3(2) = 6
sum: 3 + 2 = 5
Only the pair 3 and 2 satisfy the conditions.

3. $a^2 + 8a + 15 = (a+5)(a+3)$
product: 5(3) = 15
sum: 5 + 3 = 8

5. $x^2 + 9x + 20 = (x+5)(x+4)$
product = 5(4) = 20
sum: 5 + 4 = 9

7. $d^2 + 10d + 16 = (d+8)(d+2)$
product: 8(2) = 16
sum: 8 + 2 = 10

9. $t^2 + 6t + 9 = (t+3)(t+3) = (t+3)^2$
product: 3(3) = 9
sum: 3 + 3 = 6

11. $Y^2 - 14Y + 49 = (Y-7)(Y-7) = (Y-7)^2$
product: –7(–7) = 49
sum: –7 + (–7) = –14

13. $x^2 - 8x + 15 = (x-5)(x-3)$
product: (–5)(–3) = 15
sum: –5 + (–3) = –8

15. $y^2 - 12y + 20 = (y-10)(y-2)$
product: (–10)(–2) = 20
sum: –10 + (–2) = –12

17. $y^2 + 5y - 14 = (y+7)(y-2)$
product: 7(–2) = –14
sum: 7 + (–2) = 5

19. $x^2 + x - 30 = (x+6)(x-5)$
product: 6(–5) = –30
sum: 6 + (–5) = 1

21. $d^2 - 3d - 28 = (d-7)(d+4)$
product: –7(4) = –28
sum: –7 + 4 = –3

23. $R^2 - 5R - 36 = (R-9)(R+4)$
product: –9(4) = –36
sum: –9 + 4 = –5

25. $-12x + 35 + x^2 = x^2 - 12x + 35$
$= (x-7)(x-5)$

product: $-7(-5) = 35$
sum: $-7 + (-5) = -12$

27. $x^2 - x + 7$
Not factorable
Prime

29. $X^2 + 12XY + 20Y^2 = (X + 10Y)(X + 2Y)$
product: $(10Y)(2Y) = 20Y^2$
sum: $10Y + 2Y = 12Y$

31. $W^2 - 10WZ - 11Z^2 = (W - 11Z)(W + Z)$
product: $(-11Z)(Z) = -11Z^2$
sum: $-11Z + Z = -10Z$

33. $a^2 - ab + b^2$
Not factorable
Prime

35. $2x^3 + 6x^2 + 4x$
$= 2x(x^2 + 3x + 2)$
$= 2x(x + 2)(x + 1)$

37. $2M^2 + 9M + 7 = (2M + 7)(M + 1)$
first term: $2M(M) = 2M^2$
middle term: $7M + 2M = 9M$
last term: $7(1) = 7$

39. $4x^2 + 9x + 2 = (4x + 1)(x + 2)$
first term: $4x(x) = 4x^2$
middle term: $8x + x = 9x$
last term: $1(2) = 2$

41. $5T^2 + 17T + 6 = (5T + 2)(T + 3)$
first term: $5T(T) = 5T^2$
middle term: $3(5T) + 2T = 17T$
last term: $2(3) = 6$

43. $6b^2 + 19b + 15 = (3b + 5)(2b + 3)$
first term: $(3b)(2b) = 6b^2$
middle term: $(3b)(3) + 5(2b) = 19b$
last term: $5(3) = 15$

45. $6x^2 + 17x + 12 = (3x + 4)(2x + 3)$
first term: $3x(2x) = 6x^2$
middle term: $9x + 8x = 17x$
last term: $4(3) = 12$

47. $9y^2 - 30y + 25 = (3y - 5)(3y - 5)$
$= (3y - 5)^2$
first term: $3y(3y) = 9y^2$
middle term: $-15y - 15y = -30y$
last term: $-5(-5) = 25$

49. $4a^2 - 27a + 18 = (4a - 3)(a - 6)$
first term: $4a(a) = 4a^2$
middle term: $-24a - 3a = -27a$
last term: $-3(-6) = 18$

51. $16S^2 - 6S - 27 = (2S - 3)(8S + 9)$
first term: $2S(8S) = 16S^2$
middle term: $18S - 24S = -6S$
last term: $-3(9) = -27$

53. $4y^2 - y - 18 = (4y - 9)(y + 2)$
first term: $4y(y) = 4y^2$
middle term: $8y - 9y = -y$
last term: $-9(2) = -18$

55. $9M^2 - 3MN - 2N^2 = (3M + N)(3M - 2N)$
first term: $3M(3M) = 9M^2$
middle term: $-6MN + 3MN = -3MN$
last term: $N(-2N) = -2N^2$

57. $10x^2 + 29xy - 21y^2 = (5x - 3y)(2x + 7y)$
first term: $5x(2x) = 10x^2$
middle term: $35xy - 6xy = 29xy$
last term: $-3y(7y) = -21y^2$

59. $6a^2 + 14a + 3$
Not factorable
Prime

61. $15w^3 - 25w^2 + 10w$
$= 5w(3w^2 - 5w + 2)$

63. $-x^2 + 6x + 27 = (-x + 9)(x + 3)$
first term: $-x(x) = -x^2$
middle term: $-3x + 9x = 6x$
last term: $9(3) = 27$

65. $-12x^2 + 35x + 3 = (-12x - 1)(x - 3)$
first term: $-12x(x) = -12x^2$
middle term: $36x - x = 35x$
last term: $(-1)(-3) = 3$

67. $4a^4b - 24a^3b - 64a^2b$
$= 4a^2b(a^2 - 6a - 16)$
$= 4a^2b(a + 2)(a - 8)$

69. $36x^3y - 6x^2y - 20xy$
$= 2xy(18x^2 - 3x - 10)$
$= 2xy(6x - 5)(3x + 2)$

71. $4x^2 - 12xy + 9y^2$
$= (2x - 3y)(2x - 3y)$
$= (2x - 3y)^2$

73. $35a^2 - 41ab - 24b^2$
$= (7a + 3b)(5a - 8b)$
first term: $7a(5a) = 35a^2$
middle term: $-56ab + 15ab = -41ab$
last term: $3b(-8b) = -24b^2$

75. $8x^2 - 14xy - 39y^2$
$= (4x - 13y)(2x + 3y)$
first term: $4x(2x) = 8x^2$
middle term: $12xy - 26xy = -14xy$
last term: $(-13y)(3y) = -39y^2$

77. $6x^2y^2 + 13xy + 6$
$= (2xy + 3)(3xy + 2)$
first term: $2xy(3xy) = 6x^2y^2$
middle term: $4xy + 9xy = 13xy$
last term: $3(2) = 6$

79. $13x^3y^3 + 39x^3y^2 - 52x^3y$
$= 13x^3y(y^2 + 3y - 4)$
$= 13x^3y(y + 4)(y - 1)$

81. $y^4 + 5y^2 + 6$
$= x^2 + 5x + 6$ (Let $x = y^2$.)
$= (x + 3)(x + 2)$
$= (y^2 + 3)(y^2 + 2)$ (Substitute y^2 for x)

83. $5m^4 + m^2 - 6$
$= 5x^2 + x - 6$ (Let $x = m^2$.)
$= (5x + 6)(x - 1)$
$= (5m^2 + 6)(m^2 - 1)$ (Substitute m^2 for x)
$= (5m^2 + 6)(m + 1)(m - 1)$

85. $2n^4 + mn^2 - 6m^2$
$= 2x^2 + mx - 6m^2$ (Let $x = n^2$.)
$= (2x - 3m)(x + 2m)$
$= (2n^2 - 3m)(n^2 + 2m)$
(Substitute n^2 for x.)

87. $y^6 - 9y^3 - 36$
$= x^2 - 9x - 36$ (Let $x = y^3$.)
$= (x - 12)(x + 3)$
$= (y^3 - 12)(y^3 + 3)$ (Substitute y^3 for x)

89. $y^8 + 10y^4 - 39$
$= x^2 + 10x - 39$ (Let $x = y^4$.)
$= (x + 13)(x - 3)$
$= (y^4 + 13)(y^4 - 3)$ (Substitute y^4 for x)

91. $(a - 3b)^2 - 5(a - 3b) - 36$
$= x^2 - 5x - 36$ (Let $x = a - 3b$.)
$= (x + 4)(x - 9)$
$= (a - 3b + 4)(a - 3b - 9)$
(Substitute $a - 3b$ for x.)

93. $5(a+b)^2+12(a+b)+7$
$= 5x^2+12x+7$ (Let $x = a+b$.)
$= (5x+7)(x+1)$
$= [5(a+b)+7](a+b+1)$
(Substitute $a+b$ for x)
$= (5a+5b+7)(a+b+1)$

95. $18(x+y)^2-3(x+y)b-28b^2$
$= 18a^2-3ab-28b^2$ (Let $a = x+y$.)
$= (6a+7b)(3a-4b)$
$= [6(x+y)+7b][3(x+y)-4b]$
(Substitute $x+y$ for a.)
$= (6x+6y+7b)(3x+3y-4b)$

97. $6a^2(a-b)^2-13ab(a-b)+6b^2$
$= 6a^2x^2-13axb+6b^2$ (Let $x = a-b$.)
$= (3ax-2b)(2ax-3b)$
$= [3a(a-b)-2b][2a(a-b)-3b]$
(Substitute $a-b$ for x.)
$= (3a^2-3ab-2b)(2a^2-2ab-3b)$

99. **a.** $s(t) = -16t^2+16t+32$
$s(t) = -16(t^2-t-2)$
$s(t) = -16(t-2)(t+1)$

b. $t = -\frac{b}{2a} = -\frac{16}{2(-16)} = \frac{1}{2}$
$s\left(\frac{1}{2}\right) = -16\left(\frac{1}{2}-2\right)\left(\frac{1}{2}+1\right) = 36$
Vertex is at $\left(\frac{1}{2}, 36\right)$.

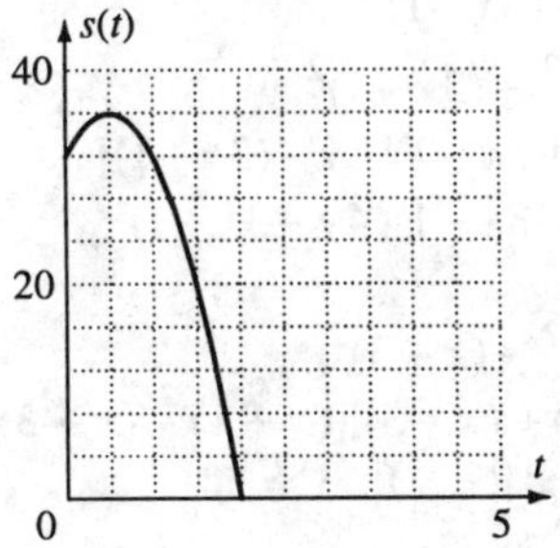

c. The diver rises to 36 feet at $\frac{1}{2}$ second then drops to the water at 2 seconds.

101. b is true.
$8y^2-51y+18 = (8y-3)(y-6)$
First term: $(8y)y = 8y^2$
Middle term: $-48y-3y = -51y$
Last term: $(-3)(-6) = 18$

103. $x^2+7x+12 = (x+4)(x+3)$
$y = (x+4)(x+3)$

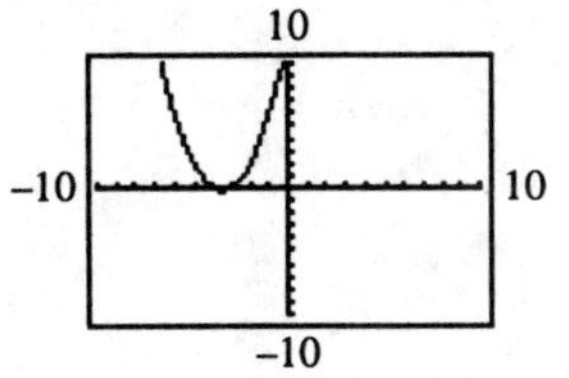

105. $-3x^2+16x+12 \neq -(3x-2)(x+6)$
$-3x^2+16x+12 = -(3x+2)(x-6)$
$y = -(3x+2)(x-6)$

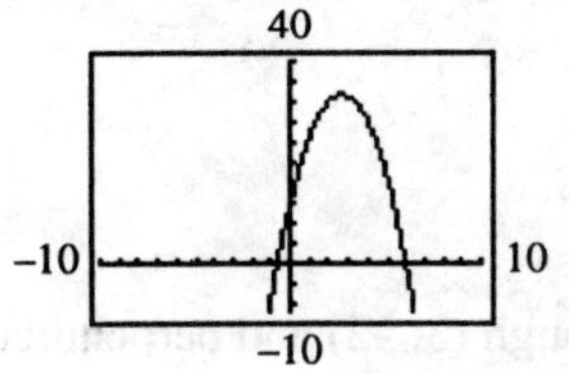

107. $6x^3+5x^2-4x = x(3x+4)(2x-1)$
$y = x(3x+4)(2x-1)$

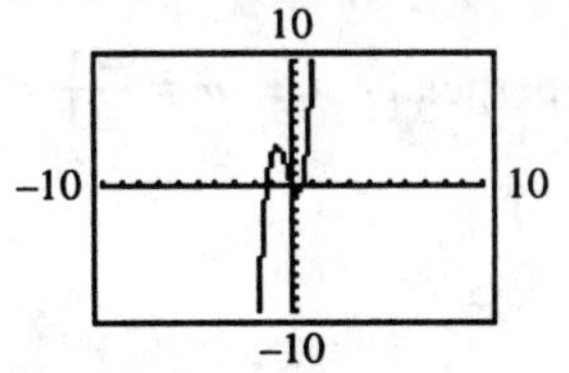

109. Answers may vary.

111. $x^{2n}-8x^n+15 = (x^n-3)(x^n-5)$

113. $4y^{2m}-9y^m+5 = (4y^m-5)(y^m-1)$

115. y^2+Kx+8;
$(y+4)(y+2)$; $K = 6$
$(y-4)(y-2)$; $K = -6$

$(y+8)(y+1)$; $K=9$
$(y-8)(y-1)$; $K=-9$

117. $3x^2+Kx+5$;
$(3x+5)(x+1)$; $K=8$
$(3x-5)(x-1)$; $K=-8$
$(3x+1)(x+5)$; $K=16$
$(3x-1)(x-5)$; $K=-16$

Review Problems

119. $I=Prt+P$
$I=P(rt+1)$
$\frac{I}{rt+1}=P$
$P=\frac{I}{1+rt}$

120. $-2x\le 6$ and $-2x+3<-7$
$x\ge -3$ and $-2x<-10$
$x>5$
$\{x|x>5\}$
$(5,\infty)$

121. The line through (3, –2) and perpendicular to the line whose equation is $x+2y=3$:
slope: $2y=-x+3$
$y=-\frac{1}{2}x+3,\ m=-\frac{1}{2}$
slope of line perpendicular: $m=\frac{-1}{-\frac{1}{2}}=2$

$y+2=2(x-3)$
$y+2=2(x)-6$
$y=2x-8$
$2x-y=8$

Problem Set 4.5

1. $B^2-1=B^2-1^2=(B+1)(B-1)$

3. $25-a^2=5^2-a^2=(5+a)(5-a)$

5. $36x^2-49=(6x)^2-7^2=(6x+7)(6x-7)$

7. $36x^2-49y^2$
$=(6x)^2-(7y)^2$
$=(6x+7y)(6x-7y)$

9. $x^2y^2-a^2b^2$
$=(xy)^2-(ab)^2=(xy+ab)(xy-ab)$

11. $x^2y^6-a^4b^2$
$=(xy^3)^2-(a^2b)^2$
$=(xy^3+a^2b)(xy^3-a^2b)$

13. $4x^2y^6-25a^4b^2$
$=(2xy^3)^2-(5a^2b)^2$
$=(2xy^3+5a^2b)(2xy^3-5a^2b)$

15. $81a^2b^4c^6-49x^8y^2$
$=(9ab^2c^3)^2-(7x^4y)^2$
$=(9ab^2c^3+7x^4y)(9ab^2c^3-7x^4y)$

17. $(x+3)^2-y^2$
$=(x+3+y)(x+3-y)$

19. $(x+y)^2-36$
$=(x+y)^2-6^2$
$=(x+y+6)(x+y-6)$

21. x^2+4 is prime, not factorable

23. $16y^2-(3x-1)^2$
$=(4y)^2-(3x-1)^2$
$=(4y+3x-1)[4y-(3x-1)]$
$=(4y+3x-1)(4y-3x+1)$

25. $(x+1)^2-(x+3)^2$
$[(x+1)+(x+3)][(x+1)-(x+3)]$
$=(2x+4)(x+1-x-3)$
$=2(x+2)(-2)$
$=-4(x+2)$

27. $(2x-1)^2-(3x+2)^2$
$[(2x-1)+(3x+2)][(2x-1)-(3x+2)]$
$=(5x+1)(2x-1-3x-2)$
$=(5x+1)(-x-3)$
$=(5x+1)(-1)(x+3)$
$=-(5x+1)(x+3)$

29. $a^{14}-9=(a^7)^2-3^2=(a^7+3)(a^7-3)$

31. $25x^2+36y^2$ is prime, not factorable

33. $-16+x^2=x^2-16$
$=x^2-4^2=(x+4)(x-4)$

35. $-25A^2+x^2y^2$
$=x^2y^2-25A^2$
$=(xy)^2-(5A)^2$
$=(xy+5A)(xy-5A)$

37. y^4-1
$=(y^2+1)(y^2-1)$
$=(y^2+1)(y+1)(y-1)$

39. $1-81b^2=1^2-(9b)^2=(1+9b)(1-9b)$

41. x^2y^3-16y
$=y(x^2y^2-16)$
$=y[(xy)^2-4^2]$
$=y[(xy+4)(xy-4)]$

43. $3x^3-3x=3x(x^2-1)=3x(x+1)(x-1)$

45. $3x^3y-12xy$
$=3xy(x^2-4)=3xy(x+2)(x-2)$

47. $p^3+q^3=(p+q)(p^2-pq+q^2)$

49. $27x^3-64y^3$
$=(3x)^3-(4y)^3$
$=(3x-4y)[(3x)^2+(3x)(4y)+(4y)^2]$
$=(3x-4y)(9x^2+12xy+16y^2)$

51. $27R^3-1$
$=(3R)^3-1^3$
$=(3R-1)[(3R)^2+3R+1]$
$=(3R-1)(9R^2+3R+1)$

53. $125b^3+64$
$=(5b)^3+4^3$
$=(5b+4)[(5b)^2-(5b)(4)+4^2]$
$=(5b+4)(25b^2-20b+16)$

55. $125+64d^3$
$=5^3+(4d)^3$
$=(5+4d)[5^2-5(4d)+(4d)^2]$
$=(5+4d)(25-20d+16d^2)$

57. $8a^3b^3-27$
$=(2ab)^3-3^3$
$=(2ab-3)[(2ab)^2+(2ab)(3)+3^2]$
$=(2ab-3)(4a^2b^2+6ab+9)$

59. $64-27Y^3Z^3$
$=4^3-(3YZ)^3$
$=(4-3YZ)[4^2+4(3YZ)+(3YZ)^2]$
$=(4-3YZ)(16+12YZ+9Y^2Z^2)$

61. $8x^3+27y^{12}$
$=(2x)^3+(3y^4)^3$
$=(2x+3y^4)[(2x)^2-(2x)(3y^4)+(3y^4)^2]$
$=(2x+3y^4)(4x^2-6xy^4+9y^8)$

63. $a^3b^6 - c^6d^{12}$
$= (ab^2)^3 - (c^2d^4)^3$
$= (ab^2 - c^2d^4)[(ab^2)^2 + (ab^2)(c^2d^4) + (c^2d^4)^2]$
$= (ab^2 - c^2d^4)(a^2b^4 + ab^2c^2d^4 + c^4d^8)$

65. $125x^2 + 27y^3$
$= 5(5x)^2 + (3y)^3$ is prime, not factorable

67. $(x+y)^3 + (x-y)^3$
$[(x+y)+(x-y)][(x+y)^2 - (x+y)(x-y) + (x-y)^2]$
$= (x+y+x-y)(x^2 + 2xy + y^2 - x^2 + y^2 + x^2 - 2xy + y^2)$
$= 2x(x^2 + 3y^2)$

69. $(2x-y)^3 - (2x+y)^3$
$= [(2x-y)-(2x+y)][(2x-y)^2 + (2x-y)(2x+y) + (2x+y)^2]$
$= (2x-y-2x-y)(4x^2 - 4xy + y^2 + 4x^2 - y^2 + 4x^2 + 4xy + y^2)$
$= -2y(12x^2 + y^2)$

71. $y^2 + 4y + 4 = y^2 + 2\cdot y\cdot 2 + 2^2 = (y+2)^2$

73. $z^2 - 10z + 25 = z^2 - 2\cdot z\cdot 5 + 5^2 = (z-5)^2$

75. $9x^2 - 12xy + 4y^2 = (3x)^2 - 2\cdot 3x\cdot 2y + (2y)^2 = (3x-2y)^2$

77. $(x+y)^2 + 2(x+y) + 1 = [(x+y)+1]^2 = (x+y+1)^2$

79. $(v-w)^2 + 4(v-w) + 4$
$= (v-w)^2 + 2\cdot(v-w)\cdot 2 + 2^2$
$= [(v-w)+2]^2$
$= (v-w+2)^2$

81. $x^2 - 6x + 9 - y^2$
$= (x-3)^2 - y^2$
$= (x-3+y)(x-3-y)$

83. $9x^2 - 30x + 25 - 36y^2$
$= (3x-5)^2 - (6y)^2$
$= (3x-5+6y)(3x-5-6y)$

85. $r^2-(16s^2-24s+9)$
$=r^2-(4s-3)^2$
$=(r+4s-3)[r-(4s-3)]$
$=(r+4s-3)(r-4s+3)$

87. y^2-x^2-4x-4
$=y^2-(x^2+4x+4)$
$=y^2-(x+2)^2$
$=(y+x+2)[y-(x+2)]$
$=(y+x+2)(y-x-2)$

89. $z^2-x^2+4xy-4y^2$
$=z^2-(x^2-4xy+4y^2)$
$=z^2-(x-2y)^2$
$=(z+x-2y)[z-(x-2y)]$
$=(z+x-2y)(z-x+2y)$

91. $9a^2-b^2=(3a+b)(3a-b)$

93. a. area = big square – little square
$=A(A)-B(B)$
$=A^2-B^2$

b. area = length × width
$=(A+B)(A-B)$

c. $A^2-B^2=(A+B)(A-B)$

95. $10^2-9^2=(10+9)(10-9)$
$=(19)(1)=19$ True

97. $39^2=(40-1)^2$
$=(40)^2-2(40)+1^2$
$=1600-80+1$
$=1521$

99. d is true; a is *not* true since $9x^2+15x+25$ is prime and $(3x+5)^2=9x^2+30x+25$
b is *not* true since $x^3-27=(x-3)(x^2+3x+9)\neq(x-3)(x^2+6x+9)$
c is *not* true since $x^3-64=(x-4)(x^2+4x+16)\neq(x-4)^3$

101. $x^2+4x+4 \neq (x+4)^2$
$x^2+4x+4=(x+2)^2$
$y=(x+2)^2$

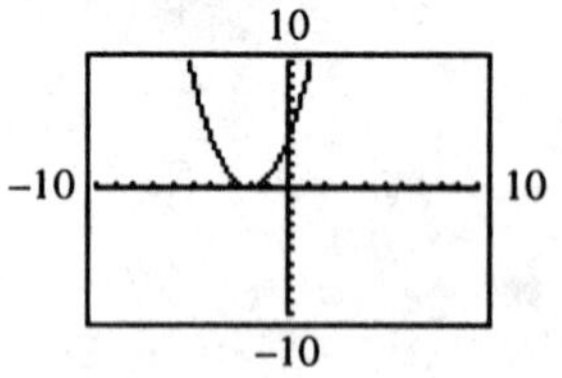

103. $25-(x^2+4x+4) \neq (x+7)(x-3)$
$25-(x^2+4x+4)=-(x+7)(x-3)$
$y=-(x+7)(x-3)$

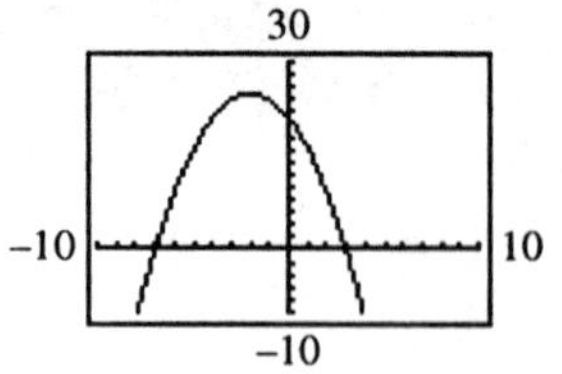

105. $(x-3)^2+8(x-3)+16 \neq (x-1)^2$
$(x-3)^2+8(x-3)+16=(x+1)^2$
$y=(x+1)^2$

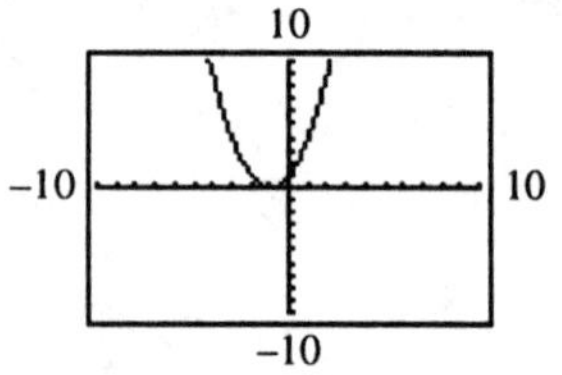

107. $(x+1)^3+1 \neq (x+1)(x^2+x+1)$
$(x+1)^3+1=(x+2)(x^2+x+1)$
$y=(x+2)(x^2+x+1)$

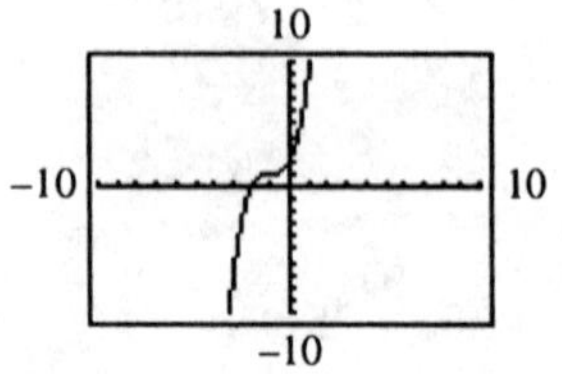

109. The problem occurs in the line that reads: Divide both sides by $(a-b)$. Since $a=b$, $a-b=0$ and division by 0 is undefined.

111. $4x^{2n}+20x^ny^m+25y^{2m}$
$=(2x^n)^2+2\cdot 2x^n\cdot 5y^m+(5y^m)^2$
$=(2x^n+5y^m)^2$

113. $x^2-6xy+ky^2$: $(x-3y)(x-3y)$ gives x^2 and $-6xy$.
$x^2-6xy+ky^2=(x-3y)^2=x^2-6xy+9y^2$
$k=9$

115. $kx^2-112xy+64y^2$: $(7x-8y)(7x-8y)$ gives $-112xy$ and $64y^2$.
$kx^2-112xy+64y^2$
$=(7x-8y)^2$
$=49x^2-112xy+64y^2$
$k=49$

117. y^3+x+x^3+y
$=x^3+y^3+x+y$
$(x+y)(x^2-xy+y^2)+(x+y)$
$(x+y)(x^2-xy+y^2+1)$

119. Group Activity

Review Problems

121. $y > x-2$ and $x+y > -2$

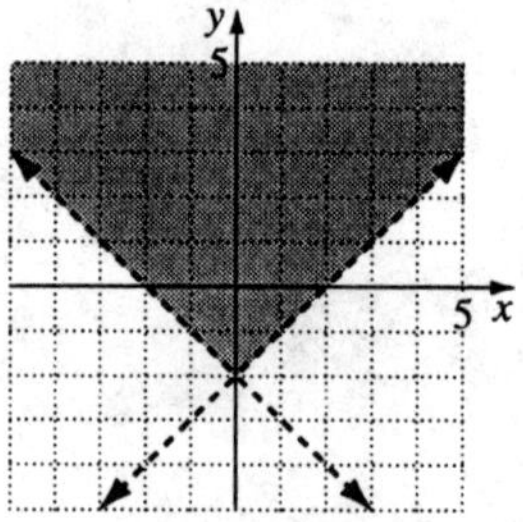

122. Let t = time (in hours) when cars will be 400 miles apart
$60t+40t=400$
$100t=400$
$t=4$
4 hours

123. $3-2(x+7) \geq 3-x$
$3-2x-14 \geq 3-x$
$-2x-11 \geq 3-x$
$-2x+x \geq 11+3$
$-x \geq 14$
$x \leq -14$

Problem Set 4.6

1. $c^3-16c = c(c^2-16) = c(c+4)(c-4)$

3. $3x^2+18x+27$
$= 3(x^2+6x+9)$
$= 3(x+3)^2$

5. $81x^3-3$
$= 3(27x^3-1)$
$= 3(3x-1)(9x^2+3x+1)$

7. $B^2C-16C+32-2B^2$
$= C(B^2-16)-2(B^2-16)$
$= (B^2-16)(C-2)$
$= (B+4)(B-4)(C-2)$

9. $-x^2+12x-27$
$= -(x^2-12x+27)$
$= -(x-9)(x-3)$

11. $4a^2b-2ab-30b$
$= 2b(2a^2-a-15)$
$= 2b(2a+5)(a-3)$

13. $a(y^2-4)-4(y^2-4)$
$= (y^2-4)(a-4)$
$= (y+2)(y-2)(a-4)$

15. $11x^5-11xy^2$
$= 11x(x^4-y^2)$
$= 11x(x^2+y)(x^2-y)$

17. $3x^2+3x+3y-3y^2$
$= 3x^2-3y^2+3x+3y$
$= 3(x^2-y^2)+3(x+y)$
$= 3[(x-y)(x+y)+(x+y)]$
$= 3(x+y)(x-y+1)$

19. $25x^2-xy+36y^2$ is prime, not factorable

21. ax^3+8a
$= a(x^3+8) = a(x+2)(x^2-2x+4)$

23. $s^2-12s+36-49t^2$
$= (s-6)^2-(7t)^2$
$= (s-6+7t)(s-6-7t)$

25. $4m^{10}+12m^5n^3+9n^6$
$= (2m^5)^2+2(2m^5)(3n^3)+(3n^3)^2$
$= (2m^5+3n^3)^2$

27. $9s^2t^2-36t^2$
$= 9t^2(s^2-4)$
$= 9t^2(s+2)(s-2)$

29. $ax+ay+bx+by$
$= a(x+y)+b(x+y)$
$= (x+y)(a+b)$

31. $5x^2yz-5y^3z$
$= 5yz(x^2-y^2)$
$= 5yz(x+y)(x-y)$

33. $20a^3b-245ab^3$
$= 5ab(4a^2-49b^2)$
$= 5ab(2a+7b)(2a-7b)$

35. $63y^2+30y-72$
$= 3(21y^2+10y-24)$
$= 3(7y-6)(3y+4)$

37. $r^6+4r^3s+4s^2$
$= (r^3)^2+2\cdot r^3\cdot 2s+(2s)^2$
$= (r^3+2s)^2$

39. $4ax^3 - 32a$
$= 4a(x^3 - 8)$
$= 4a(x-2)(x^2 + 2x + 4)$

41. $100x^4 + 120x^3y + 36x^2y^2$
$= 4x^2(25x^2 + 30xy + 9y^2)$
$= 4x^2(5x + 3y)^2$

43. $49x^2 + 126xy + 81y^2$
$= (7x)^2 + 2 \cdot (7x)(9y) + (9y)^2$
$= (7x + 9y)^2$

45. $71bx^4 - 71b$
$= 71b(x^4 - 1)$
$= 71b(x^2 + 1)(x^2 - 1)$
$= 71b(x^2 + 1)(x + 1)(x - 1)$

47. $x^2 + 25$ is prime, not factorable

49. $r^3 - s^3 + r - s$
$= (r - s)(r^2 + rs + s^2) + (r - s)$
$= (r - s)(r^2 + rs + s^2 + 1)$

51. $(x^2 - 3)^2 - 4(x^2 - 3) - 12$
$= [(x^2 - 3) - 6][(x^2 - 3) + 2]$
$= (x^2 - 9)(x^2 - 1)$
$= (x + 3)(x - 3)(x + 1)(x - 1)$

53. $a^2 + 4a + 4 - 16b^2$
$= (a + 2)^2 - (4b)^2$
$= (a + 2 + 4b)(a + 2 - 4b)$

55. $27r^3s + 72r^2s^2 + 48rs^3$
$= 3rs(9r^2 + 24rs + 16s^2)$
$= 3rs(3r + 4s)^2$

57. $-6by^3 + 24by$
$= -6by(y^2 - 4)$
$= -6by(y + 2)(y - 2)$

59. $16x^2 + 49y^2$ is prime, not factorable

61. $(3x - y)^2 - 100a^2$
$= (3x - y)^2 - (10a)^2$
$= (3x - y + 10a)(3x - y - 10a)$

63. $(5x + 3y)^2 - 6(5x + 3y) + 9$
$= a^2 - 6a + 9$ (Let $a = 5x + 3y$.)
$= (a - 3)^2$
$= (5x + 3y - 3)^2$ (Substitute $5x + 3y$ for a)

65. $6y^4 - 11y^2 - 10$
$= (2y^2 - 5)(3y^2 + 2)$

67. $48y^4 - 243$
$= 3(16y^4 - 81)$
$= 3(4y^2 + 9)(4y^2 - 9)$
$= 3(4y^2 + 9)(2y + 3)(2y - 3)$

69. $20bx^4 + 220bx^2y + 605by^2$
$= 5b(4x^4 + 44x^2y + 121y^2)$
$= 5b(2x^2 + 11y)^2$

71. $18x^3 + 63x^2 - 36x$
$= 9x(2x^2 + 7x - 4)$
$= 9x(2x - 1)(x + 4)$

73. $4x^7 + 32xy^3$
$= 4x(x^6 + 8y^3)$
$= 4x[(x^2)^3 + (2y)^3]$
$= 4x(x^2 + 2y)(x^4 - 2x^2y + 4y^2)$

75. $x^4 - (x - 6)^2$
$= [x^2 - (x - 6)][x^2 + (x - 6)]$
$= (x^2 - x - 6)(x^2 + x - 6)$
$= (x - 3)(x + 2)(x + 3)(x - 2)$

77. $x^{16} + 1$ is prime

79. $r^4s+3-3r^4-s$
$=r^4s-s-3r^4+3$
$=s(r^4-1)-3(r^4-1)$
$=(r^4-1)(s-3)$
$=(r^2+1)(r^2-1)(s-3)$
$=(r^2+1)(r+1)(r-1)(s-3)$

81. y^3-2y^2-4y+8
$=y^2(y-2)-4(y-2)$
$=(y-2)(y^2-4)$
$=(y-2)(y+2)(y-2)$
$=(y-2)^2(y+2)$

83. ay^3+b-by^3-a
$=ay^3-a-by^3+b$
$=a(y^3-1)-b(y^3-1)$
$=(y^3-1)(a-b)$
$=(y-1)(y^2+y+1)(a-b)$

85. $4x^2+4x-1$ is prime

87. $16x-2x^4$
$=2x(8-x^3)$
$=2x(2-x)(4+2x+x^2)$

89. y^3-2y^2-y+2
$=y^2(y-2)-1(y-2)$
$=(y-2)(y^2-1)=(y-2)(y+1)(y-1)$

91. $a^6b^6-a^3b^3$
$=a^3b^3(a^3b^3-1)$
$=a^3b^3(ab-1)(a^2b^2+ab+1)$

93. $10x^3-6x^2-21x$
$=x(10x^2-6x-21)$

95. $x(x+y)-y(x+y)$
$=(x-y)(x+y)$
$=x^2-y^2$

97. $\pi R^2h-\pi r^2h$
$=\pi h(R^2-r^2)$
$=\pi h(R-r)(R+r)$

99. d is true.

101. Substitute various values of a into each side of the equation. Graph each to see if they are identical.

103. Values of a and the graphs will vary. Students should use a graphing utility to verify the factorization.
$2x^2+18a^2-12ax\neq 2(x-6a)^2$
$2x^2+18a^2-12ax=2(x-3a)^2$

105. $x^3a-16xa^3=xa(x+4a)(x-4a)$

107–109. Answers may vary.

111. $5r^3s^2-5rs^4-5r^2s^2+5rs^3$
$=5rs^2(r^2-s^2-r+s)$
$=5rs^2[(r^2-s^2)-(r-s)]$
$=5rs^2[(r+s)(r-s)-(r-s)]$
$=5rs^2[(r-s)(r+s-1)]$
$=5rs^2(r-s)(r+s-1)$

113. $(25x^2-10xy+y^2)+(10xz-2yz)-24z^2$
$=(5x-y)^2+2z(5x-y)-24z^2$
$=(5x-y+6z)(5x-y-4z)$

115. $x^4-y^4-2x^3y+2xy^3$
$=(x^2+y^2)(x^2-y^2)-2xy(x^2-y^2)$
$=(x^2-y^2)(x^2+y^2-2xy)$
$=(x^2-y^2)(x^2-2xy+y^2)$
$=(x^2-y^2)(x-y)^2$
$=(x+y)(x-y)(x-y)^2$
$=(x+y)(x-y)^3$

117. $x^3+x+2x^4+4x^2+2$
$=x(x^2+1)+2(x^4+2x^2+1)$
$=x(x^2+1)+2(x^2+1)^2$
$=(x^2+1)[x+2(x^2+1)]$
$=(x^2+1)(2x^2+x+2)$

Review Problems

119. $\left[\dfrac{7+(-16)}{|7-10|}\right]\left[\dfrac{12+(-2)}{3+(-2)^3}\right]$
$=\left(\dfrac{-9}{|-3|}\right)\left(\dfrac{10}{3-8}\right)$
$=\left(\dfrac{-9}{3}\right)\left(\dfrac{10}{-5}\right)$
$=(-3)(-2)=6$

120. $\dfrac{3x-1}{5}+\dfrac{x+2}{2}=-\dfrac{3}{10}$
$10\left(\dfrac{3x-1}{5}+\dfrac{x+2}{2}\right)=-\dfrac{3}{10}\cdot 10$
$2(3x-1)+5(x+2)=-3$
$6x-2+5x+10=-3$
$11x+8=-3$
$11x=-11$
$x=-1$

121. $(4x^3y^{-1})^2(2x^{-3}y)^{-1}$
$=4^2x^6y^{-2}(2^{-1}x^3y^{-1})$
$=\left(\dfrac{16x^6}{y^2}\right)\left(\dfrac{x^3}{2y}\right)$
$=\dfrac{16x^{6+3}}{2y^{2+1}}$
$=\dfrac{8x^9}{y^3}$

Problem Set 4.7

For problems 1-56, the check is left to the reader.

1. $(x-7)(x+3)=0$
$x-7=0$ or $x+3=0$
$x=7$ or $x=-3$
A check of both values in the original equation confirms that the solution set is $\{-3, 7\}$.

3. $(2x+3)(5x-1)=0$
$2x+3=0$ or $5x-1=0$
$2x=-3$ or $5x=1$
$x=-\dfrac{3}{2}$ or $x=\dfrac{1}{5}$
$\left\{-\dfrac{3}{2}, \dfrac{1}{5}\right\}$

5. $x^2+x-12=0$
$(x+4)(x-3)=0$
$x+4=0$ or $x-3=0$
$x=-4$ or $x=3$
$\{-4, 3\}$

7. $x^2+6x-7=0$
$(x+7)(x-1)=0$
$x+7=0$ or $x-1=0$
$x=-7$ or $x=1$
$\{-7, 1\}$

9. $3x^2+10x-8=0$
$(x+4)(3x-2)=0$
$x+4=0$ or $3x-2=0$
$x+4=0$ or $3x=2$
$x=-4$ or $x=\dfrac{2}{3}$
$\left\{-4, \dfrac{2}{3}\right\}$

11. $5x^2 - 8x + 3 = 0$
$(5x - 3)(x - 1) = 0$
$5x - 3 = 0$ or $x - 1 = 0$
$5x = 3$ or $x = 1$
$x = \frac{3}{5}$
$\left\{\frac{3}{5}, 1\right\}$

13. $6x^2 - x - 35 = 0$
$(3x + 7)(2x - 5) = 0$
$3x + 7 = 0$ or $2x - 5 = 0$
$3x = -7$ or $2x = 5$
$x = -\frac{7}{3}$ or $x = \frac{5}{2}$
$\left\{-\frac{7}{3}, \frac{5}{2}\right\}$

15. $5x^2 + 26x + 5 = 0$
$(5x + 1)(x + 5) = 0$
$5x + 1 = 0$ or $x + 5 = 0$
$5x = -1$ or $x = -5$
$x = -\frac{1}{5}$
$\left\{-\frac{1}{5}, -5\right\}$

17. $5x^2 - 3x - 2 = 0$
$(5x + 2)(x - 1) = 0$
$5x + 2 = 0$ or $x - 1 = 0$
$5x = -2$ or $x = 1$
$x = -\frac{2}{5}$
$\left\{-\frac{2}{5}, 1\right\}$

19. $3x^2 + 5x - 2 = 0$
$(x + 2)(3x - 1) = 0$
$x + 2 = 0$ or $3x - 1 = 0$
$x = -2$ $\quad$ $3x = 1$
$x = \frac{1}{3}$
$\left\{-2, \frac{1}{3}\right\}$

21. $5x^2 - 8x - 21 = 0$
$(5x + 7)(x - 3) = 0$
$5x + 7 = 0$ or $x - 3 = 0$
$5x = -7$ or $x = 3$
$x = -\frac{7}{5}$
$\left\{-\frac{7}{5}, 3\right\}$

23. $x^2 - x = 2$
$x^2 - x - 2 = 0$
$(x - 2)(x + 1) = 0$
$x - 2 = 0$ or $x + 1 = 0$
$x = 2$ or $x = -1$
$\{-1, 2\}$

25. $3x^2 - 17x = -10$
$3x^2 - 17x + 10 = 0$
$(3x - 2)(x - 5) = 0$
$3x - 2 = 0$ or $x - 5 = 0$
$3x = 2$ or $x = 5$
$x = \frac{2}{3}$
$\left\{\frac{2}{3}, 5\right\}$

27. $x(x - 3) = 54$
$x^2 - 3x - 54 = 0$
$(x - 9)(x + 6) = 0$
$x - 9 = 0$ or $x + 6 = 0$
$x = 9$ or $x = -6$
$\{-6, 9\}$

29. $x(2x + 1) = 3$
$2x^2 + x - 3 = 0$
$(2x + 3)(x - 1) = 0$
$2x + 3 = 0$ or $x - 1 = 0$
$2x = -3$ or $x = 1$
$x = -\frac{3}{2}$
$\left\{-\frac{3}{2}, 1\right\}$

31. $x^2 = \frac{5}{6}x + \frac{2}{3}$
$6x^2 = 5x + 4$
$6x^2 - 5x - 4 = 0$
$(2x + 1)(3x - 4) = 0$
$2x + 1 = 0$ or $3x - 4 = 0$
$2x = -1$ or $3x = 4$
$x = -\frac{1}{2}$ or $x = \frac{4}{3}$
$\left\{-\frac{1}{2}, \frac{4}{3}\right\}$

33. $(x+1)^2 - 5(x+2) = 3x + 7$
$x^2 + 2x + 1 - 5x - 10 = 3x + 7$
$x^2 - 6x - 16 = 0$
$(x - 8)(x + 2) = 0$
$x - 8 = 0$ or $x + 2 = 0$
$x = 8$ or $x = -2$
$\{-2, 8\}$

35. $\frac{1}{6}x^2 + x - \frac{1}{2} = -2$
$x^2 + 6x - 3 = -12$
$x^2 + 6x + 9 = 0$
$(x+3)^2 = 0$
$x + 3 = 0$
$x = -3$
$\{-3\}$

37. $x + (x+2)^2 = 130$
$x + x^2 + 4x + 4 = 130$
$x^2 + 5x - 126 = 0$
$(x + 14)(x - 9) = 0$
$x + 14 = 0$ or $x - 9 = 0$
$x = -14$ or $x = 9$
$\{-14, 9\}$

39. $3(x^2 - 4x - 1) = 2(x + 1)$
$3x^2 - 12x - 3 = 2x + 2$
$3x^2 - 14x - 5 = 0$
$(3x + 1)(x - 5) = 0$
$3x + 1 = 0$ or $x - 5 = 0$
$3x = -1$ or $x = 5$
$x = -\frac{1}{3}$
$\left\{-\frac{1}{3}, 5\right\}$

41. $9x^2 + 6x = -1$
$9x^2 + 6x + 1 = 0$
$(3x+1)^2 = 0$
$3x + 1 = 0$
$3x = -1$
$x = -\frac{1}{3}$
$\left\{-\frac{1}{3}\right\}$

43. $25 = 30x - 9x^2$
$9x^2 - 30x + 25 = 0$
$(3x-5)^2 = 0$
$3x - 5 = 0$
$3x = 5$
$x = \frac{5}{3}$
$\left\{\frac{5}{3}\right\}$

45. $(x + 2)(x - 5) = 8$
$x^2 - 3x - 10 = 8$
$x^2 - 3x - 18 = 0$
$(x + 3)(x - 6) = 0$
$x + 3 = 0$ or $x - 6 = 0$
$x = -3$ or $x = 6$
$\{-3, 6\}$

47. $2x - [(x + 2)(x - 3) + 8] = 0$
$2x - (x^2 - x - 6 + 8) = 0$
$2x - (x^2 - x + 2) = 0$
$2x - x^2 + x - 2 = 0$
$-x^2 + 3x - 2 = 0$
$x^2 - 3x + 2 = 0$
$(x - 1)(x - 2) = 0$
$x - 1 = 0$ or $x - 2 = 0$
$x = 1$ or $x = 2$
$\{1, 2\}$

49. $3[(x+2)^2 - 4x] = 15$
$3(x^2 + 4x + 4 - 4x) = 15$
$3(x^2 + 4) = 15$
$3x^2 + 12 = 15$
$3x^2 - 3 = 0$
$3(x^2 - 1) = 0$
$3(x + 1)(x - 1) = 0$
$x + 1 = 0$ or $x - 1 = 0$
$x = -1$ or $x = 1$
$\{-1, 1\}$

51. $x^3 + 4x^2 - 25x - 100 = 0$
$x^2(x+4) - 25(x+4) = 0$
$(x+4)(x^2 - 25) = 0$
$(x + 4)(x + 5)(x - 5) = 0$
$x + 4 = 0$ or $x + 5 = 0$ or $x - 5 = 0$
$x = -4$ or $x = -5$ or $x = 5$
$\{-5, -4, 5\}$

53. $x^3 - x^2 = 25x - 25$
$x^3 - x^2 - 25x + 25 = 0$
$x^2(x-1) - 25(x-1) = 0$
$(x-1)(x^2 - 25) = 0$
$(x - 1)(x + 5)(x - 5) = 0$
$x - 1 = 0$ or $x + 5 = 0$ or $x - 5 = 0$
$x = 1$ or $x = -5$ or $x = 5$
$\{-5, 1, 5\}$

55. $x^4 + x^3 = 4x^2 + 4x$
$x^4 + x^3 - 4x^2 - 4x = 0$
$x(x^3 + x^2 - 4x - 4) = 0$
$x[x^2(x+1) - 4(x+1)] = 0$
$x(x+1)(x^2 - 4) = 0$
$x(x + 1)(x - 2)(x + 2) = 0$
$x = 0$ or $x + 1 = 0$ or $x - 2 = 0$ or $x + 2 = 0$
$x = 0$ or $x = -1$ or $x = 2$ or $x = -2$
$\{-2, -1, 0, 2\}$

57. $f(x) = x^2 + 6x + 5$
x-intercepts:
$x^2 + 6x + 5 = 0$
$(x + 5)(x + 1) = 0$
$x = -5$ or $x = -1$
$(-5, 0)$ and $(-1, 0)$
y-intercept:
$y = (0)^2 + 6(0) + 5$
$y = 5$
$(0, 5)$
vertex:
$x = -\frac{b}{2a} = -\frac{6}{2(1)} = -3$
$y = f(-3) = (-3)^2 + 6(-3) + 5 = -4$
$(-3, -4)$
$y = x^2 + 6x + 5$

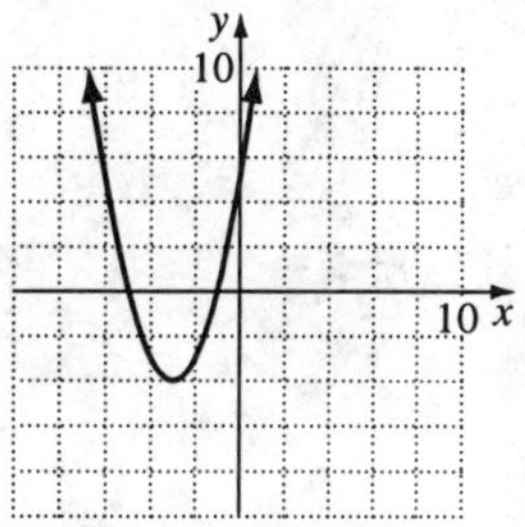

59. $f(x) = x^2 + 4x + 3$
x-intercepts:
$x^2 + 4x + 3 = 0$
$(x + 3)(x + 1) = 0$
$x = -3$ or $x = -1$
$(-3, 0)$ and $(-1, 0)$
y-intercept:
$y = 0^2 + 4(0) + 3$
$y = 3$
$(0, 3)$
vertex:
$x = -\frac{4}{2} = -2$
$y = (-2)^2 + (4)(-2) + 3 = -1$
$(-2, -1)$
$y = x^2 + 4x + 3$

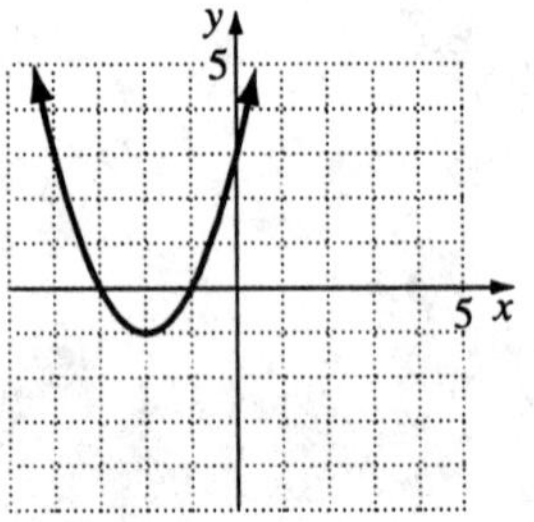

61. $y = -x^2 - 4x - 5$

x-intercepts:

$-x^2 - 4x - 5 = 0$

$-1(x^2 + 4x + 5) = 0$

no x -intercepts

y-intercept:

$y = -0^2 - 4(0) - 5 = -5$

$(0, -5)$

vertex:

$x = -\frac{(-4)}{2(-1)} = -2$

$y = -(-2)^2 - 4(-2) - 5 = -1$

$(-2, -1)$

$y = -x^2 - 4x - 5$

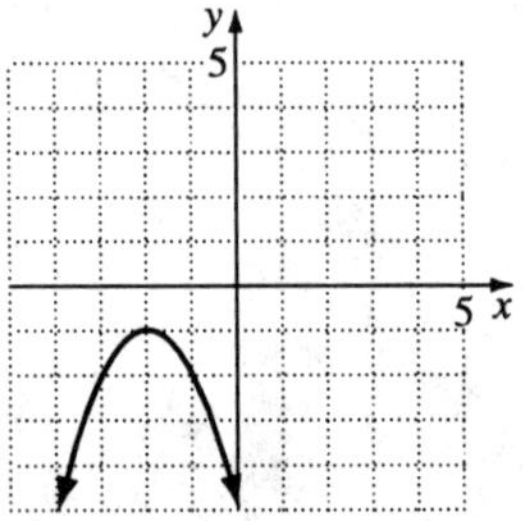

63. $f(x) = -x^2 - 4x - 3$

x-intercepts:

$-x^2 - 4x - 3 = 0$

$-1(x + 3)(x + 1) = 0$

$x = -3$ or $x = -1$

$(-3, 0)$ and $(-1, 0)$

y-intercept:

$y = -0^2 - 4(0) - 3 = -3$

$(0, -3)$

vertex:

$x = -\frac{(-4)}{2(-1)} = -2$

$y = -(-2)^2 - 4(-2) - 3 = 1$

$(-2, 1)$

$y = -x^2 - 4x - 3$

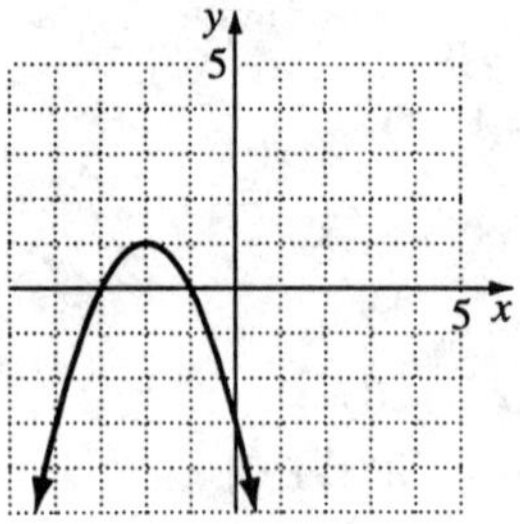

65. $s(t) = -16t^2 + 64t + 80$

a. The ball reaches the ground at $s(t) = 0$. It takes the ball 5 seconds to reach the ground.

b. It reaches a maximum height of 144 ft after 2 seconds.

c.

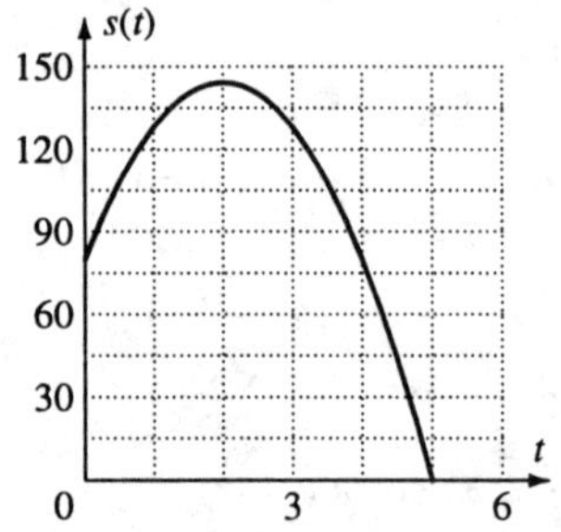

d. $s(t) = 80$

$80 = -16t^2 + 64t + 80$

$16t^2 - 64t = 0$

$16t(t - 4) = 0$

$t - 4 = 0$ or $t = 0$ (Reject $t = 0$ since $t = 0$ represents the initial time when the ball was thrown upward.)

$t = 4$

The ball passes the edge of the top of the building after 4 seconds.

The point (4, 80) is on the graph of the parabola at the same height as (0, 80).

e. Students should verify.

67. a. $P = -\frac{1}{50}A^2 + 2A + 22$

$(P = 72)$:

$72 = -\frac{1}{50}A^2 + 2A + 22$

$\frac{1}{50}A^2 - 2A + 50 = 0$

$A^2 - 100A + 2500 = 0$

$(A - 50)^2 = 0$

$A - 50 = 0$

$A = 50$

Arousal level should be 50.

b. $A = -\frac{2}{2\left(-\frac{1}{50}\right)} = 50$

$P = -\frac{1}{50}(50)^2 + 2(50) + 22 = 72$

Vertex: (50, 72)

The arousal level is 50.

69. $20 = 10 + \frac{7}{6}t - \frac{1}{60}t^2$

$60(20) = 60\left(10 + \frac{7}{6}t - \frac{1}{60}t^2\right)$

$1200 = 600 + 70t - t^2$

$t^2 - 70t + 600 = 0$

$(t - 60)(t - 10) = 0$

$t = 60$ or $t = 10$

Since $0 \le t \le 35$, after 10 days

71. $S(p) = p^2 + 3p - 70$

$D(p) = 410 - p$

$p^2 + 3p - 70 = 410 - p$

$p^2 + 4p - 480 = 0$

$(p + 24)(p - 20) = 0$

$p = -24$ or $p = 20$

The price cannot be negative, so the equilibrium price is \$20.

73. total areas of pool and path = 600 square meters

$(10 + 2x)(20 + 2x) = 600$

$200 + 20x + 40x + 4x^2 = 600$

$4x^2 + 60x - 400 = 0$

$4(x^2 + 15x - 100) = 0$

$4(x - 5)(x + 20) = 0$

$x - 5 = 0$ or $x + 20 = 0$

$x = 5$ or $x = -20$ (Reject since not possible)

The width of the path is 5 meters.

75. Volume = 128 cubic inches

length × width × height = 128

$(x - 4)(x - 4)(2) = 128$

$(x - 4)^2 = 64$ (Dividing by 2)

$x^2 - 8x + 16 - 64 = 0$

$x^2 - 8x - 48 = 0$

$(x - 12)(x + 4) = 0$

$x - 12 = 0$ or $x + 4 = 0$

$x = 12$ or $x = -4$ (not possible)

The dimensions of the piece of square cardboard are 12 inches × 12 inches.

77. Let x = distance up the wall, then

$x + 2$ = length of ladder

$(x + 2)^2 = x^2 + 10^2$

$x^2 + 4x + 4 = x^2 + 100$

$4x - 96 = 0$

$x = 24$

The distance up the wall is 24 m and the length of the ladder is 26 m.

79. Let x = the length of the shorter leg

$2x - 1$ = the length of the longer leg

$2x + 1$ = the length of the hypotenuse

$x^2 + (2x - 1)^2 = (2x + 1)^2$

$x^2 + 4x^2 - 4x + 1 = 4x^2 + 4x + 1$

$x^2 - 8x = 0$

$x(x - 8) = 0$

$x = 0$ (reject) or $\quad x - 8 = 0$

$x = 8$

$2x - 1 = 2(8) - 1 = 15$

$2x + 1 = 2(8) + 1 = 17$

The lengths of the sides of the triangle are 8 inches, 15 inches and 17 inches.

81. a. $y = ax^2 + bx + c$

$4025 = a(0)^2 + b(0) + c$

$4235 = a(10)^2 + b(10) + c$

$3845 = a(20)^2 + b(20) + c$

or

$c = 4025 \quad (1)$

$100a + 10b + c = 4235 \quad (2)$

$400a + 20b + c = 3845 \quad (3)$

Add equation (2) to –1 times equation (3).

$$\begin{array}{r} 100a + 10b + c = 4235 \\ -400a - 20b - c = -3845 \\ \hline -300a - 10b \quad = 390 \end{array}$$

Simplify $-30a - b = 39 \quad (4)$

Substitute $c = 4025$ into equation (2).

$100a + 10b + 4025 = 4235$

$100a + 10b = 210$

$10a + b = 21 \quad (5)$

Add equations (4) and (5).

$$\begin{array}{r} -30a - b = 39 \\ 10a + b = 21 \\ \hline -20a = 60 \end{array}$$

$a = -3$

$b = 21 - 10(-3) = 51$

$y = -3x^2 + 51x + 4025$

b. Vertex:

$x = -\dfrac{51}{2(-3)} = 8.5$

$y = -3(8.5)^2 + 51(8.5) + 4025 = 4241.75$

$(8.5, 4241.75)$

The consumption was at a maximum during 1968.

The consumption was about 4242 per capita.

c.

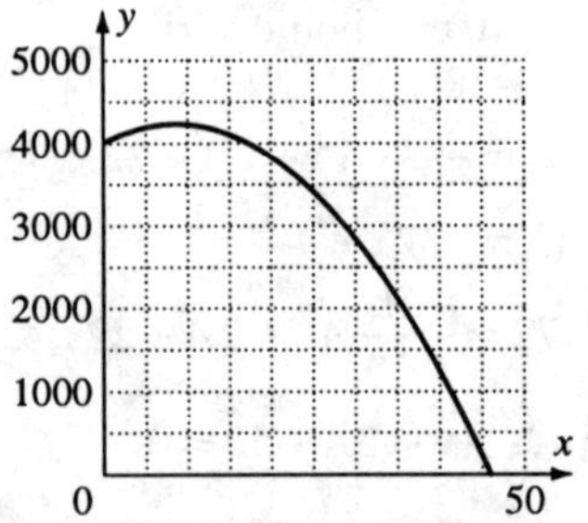

d. Cigarette consumption decreases from 1969 to the present.

83. c is true;

area of large square – area of picture = area of matting

$(x+6)^2 - x^2 = 60$ True

85. $x^2 - 7x + 6 = 0$

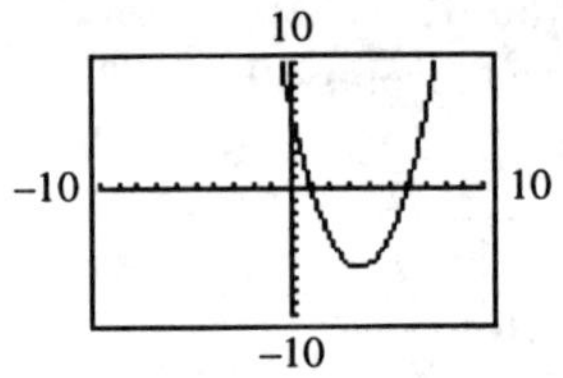

x-intercepts:

(1, 0), (6, 0)

{1, 6}

87. $x^3 - 9x = 0$

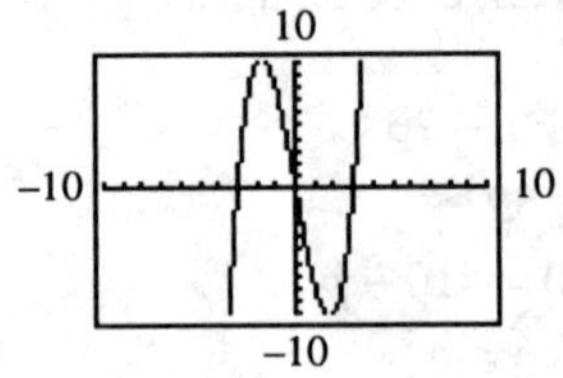

x-intercepts:

(–3, 0), (0, 0), (3, 0)

{–3, 0, 3}

89. $x^4 - 8x^3 + 7x^2 + 72x - 144 = 0$

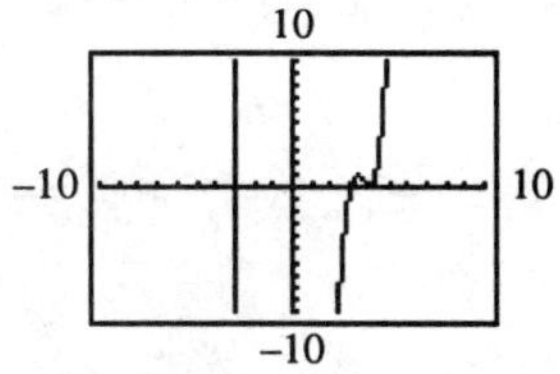

x-intercepts:
(−3, 0), (3, 0), (4, 0)
{−3, 3, 4}

91. Answers may vary.

93. $\left|x^2 + 2x - 36\right| = 12$

$x^2 + 2x - 36 = 12$ or $x^2 + 2x - 36 = -12$
$x^2 + 2x - 48 = 0$ or $x^2 + 2x - 24 = 0$
$(x + 8)(x - 6) = 0$ or $(x + 6)(x - 4) = 0$
$x = -8$ or $x = 6$ or $x = -6$ or $x = 4$
{−8, −6, 4, 6}

95. $\begin{vmatrix} 2x & x \\ -1 & x \end{vmatrix} = 1$

$2x(x) - (-1)(x) = 1$
$2x^2 + x = 1$
$2x^2 + x - 1 = 0$
$(2x - 1)(x + 1) = 0$
$2x - 1 = 0$ or $x + 1 = 0$
$x = \frac{1}{2}$ or $x = -1$
$\left\{-1, \frac{1}{2}\right\}$

97. One dimension: x
Other dimension: $100 - 2x$
Area = 1200
$x(100 - 2x) = 1200$
$100x - 2x^2 = 1200$
$x^2 - 50x + 600 = 0$
$(x - 30)(x - 20) = 0$
$x = 30$ or $x = 20$
$100 - 2x = 40$ or $100 - 2x = 60$
Dimensions:
30 hectometers × 40 hectometers or
20 hectometers × 60 hectometers

99.

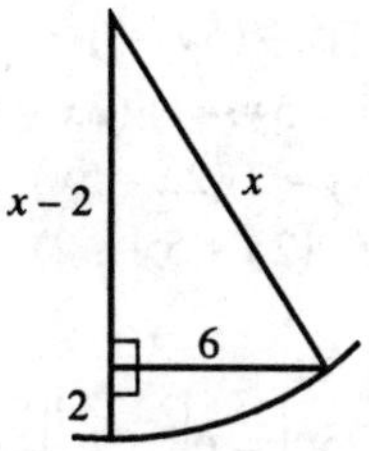

Let x = length of pendulum
$x^2 = (x - 2)^2 + 6^2$
$x^2 = x^2 - 4x + 4 + 36$
$4x = 40$
$x = 10$
length: 10 meters

101. $\dfrac{x^3 - x^2 - x + 1}{x^3 - x^2 + x - 1} = 0$

The only way that the algebraic fraction can equal 0 is if its numerator equals 0.
$x^3 - x^2 - x + 1 = 0$
$x^2(x - 1) - (x - 1) = 0$
$(x - 1)(x^2 - 1) = 0$
$(x - 1)(x + 1)(x - 1) = 0$
$(x + 1)(x - 1)^2 = 0$
$x + 1 = 0$ or $(x - 1)^2 = 0$
$x = -1$ or $x - 1 = 0$
$x = 1$
However, $x = 1$ causes the denominator $(x^3 - x^2 + x - 1)$ to equal 0. Thus, $x = -1$.
{−1}

Review Problems

102. $4x^2-25y^2-4x+10y=4x^2-4x-25y^2+10y$
$=4x^2-4x+1-1-25y^2+10y$ (add $1-1=0$)
$=(4x^2-4x+1)-(25y^2-10y+1)$
$=(2x-1)^2-(5y-1)^2$
$=(2x-1+5y-1)[(2x-1)-(5y-1)]$
$=(2x+5y-2)(2x-5y)$
$=(2x-5y)(2x+5y-2)$

103. $\begin{vmatrix} 5 & 2 & 34 \\ -1 & 3 & 22 \\ 0 & 0 & 4 \end{vmatrix} = 0\begin{vmatrix} 2 & 34 \\ 3 & 22 \end{vmatrix} - 0\begin{vmatrix} 5 & 34 \\ -1 & 22 \end{vmatrix} + 4\begin{vmatrix} 5 & 2 \\ -1 & 3 \end{vmatrix}$
$=0-0+4(15+2)=4(17)=68$

104. $2x-3y+z=0$ (1)
$3x+y+2z=-2$ (2)
$x-2y+z=-2$ (3)

Multiply equation (2) by 3 and add to equation 1.

$$\begin{array}{l} 2x-3y+z=0 \\ 9x+3y+6z=-6 \\ \hline 11x \quad +7z=-6 \quad (4) \end{array}$$

Multiply equation (2) by 2 and add to equation 3.

$$\begin{array}{l} 6x+2y+4z=-4 \\ x-2y+z=-2 \\ \hline 7x \quad +5z=-6 \quad (5) \end{array}$$

Multiply equations (4) and (5) by –5 and 7 respectively, and add together.

$$\begin{array}{l} -55x-35z=30 \\ 49x+35z=-42 \\ \hline -6x \quad =-12 \\ x \quad = 2 \end{array}$$

$7x+5z=-6$ (Equation 5)
$7(2)+5z=-6$
$5z=-20$
$z=-4$

$x - 2y + z = -2$ (Equation 3)
$2 - 2y + (-4) = -2$
$-2y = 0$
$y = 0$
$\{(2, 0, -4)\}$

Chapter 4 Review Problems

1. a. $4y^2 - 8y^3 + 9y$
Trinomial; degree 3

b. $12x^4y^3z \to 4 + 3 + 1 = 8$
Monomial; degree 8

c. $8y^4y^2 - 7xy^6 \to 1 + 6 = 7$
Binomial; degree 7

d. $7x^5 + 3x^3 - 2x^2 + x - 3$
5 terms; degree 5

2. $(-8x^3 + 5x^2 - 7x + 4) + (9x^3 - 11x^2 + 6x - 13) = (-8x^3 + 9x^3) + (5x^2 - 11x^2) + (-7x + 6x) + (4 - 13)$
$= x^3 - 6x^2 - x - 9$

3.
$$\begin{array}{rr} & 7x^3y - 13x^2y - 6y \\ + & \underline{5x^3y + 11x^2y - 8y - 17} \\ & 12x^3y - 2x^2y - 14y - 17 \end{array}$$

4. $(5x^2 - 7x + 3) - (-6x^2 + 4x - 5) = 5x^2 - 7x + 3 + 6x^2 - 4x + 5$
$= 11x^2 - 11x + 8$

5. $(7y^3 - 6y^2 + 5y - 11) - (-8y^3 + 4y^2 - 6y - 17) = 7y^3 - 6y^2 + 5y - 11 + 8y^3 - 4y^2 + 6y + 17$
$= 15y^3 - 10y^2 + 11y + 6$

6. $(4x^5y^2 - 7x^3y - 4) - (-8x^5y^2 - 3x^3y + 4)$
$$\begin{array}{r} 4x^5y^2 - 7x^3y - 4 \\ \underline{8x^5y^2 + 3x^3y - 4} \\ 12x^5y^2 - 4x^3y - 8 \end{array}$$

7. $[(6 + 3a^4b^3 + 4ab^3) + (-5 - 2a^4b^3 - 3ab^3)] - (10 - 6a^4b^3 - 2ab^3)$
$= 1 + a^4b^3 + ab^3 - 10 + 6a^4b^3 + 2ab^3$
$= 7a^4b^3 + 3ab^3 - 9$

8. $f(x) = x^2 - 8x + 15$
Vertex:
$x = -\frac{b}{2a} = -\frac{(-8)}{2(1)} = 4$
$y = f(4) = 4^2 - 8(4) + 15 = -1$
$(4, -1)$

9. $f(x) = 0.337x^2 - 2.265x + 3.962$
$f(10) = 0.337(10)^2 - 2.265(10) + 3.962$
$f(10) = 15.012$
In 1990, the number of mountain bike owners in the U.S. was 15.012 million.

10. $f(x) = -3.1x^2 + 51.4x + 4024.5$
Vertex: $x = -\frac{b}{2a} = -\frac{51.4}{2(-3.1)} \approx 8.3$
$y \approx f(8.3) \approx 4238$
In 1968, the average annual per capita consumption of cigarettes was about 4238.

11. b., c., and d. are all slightly different from the basic graph of a.
b. is a., but moved one unit to the right.
c. is a., but flipped over the y-axis.
d. is c. moved one unit up.
As x goes to $-\infty$, y goes to $-\infty$ or ∞.
As x goes to ∞, y goes to $-\infty$ or ∞.

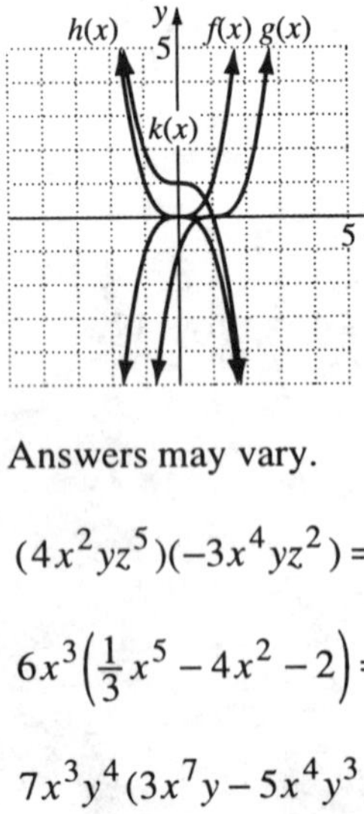

12. Answers may vary.

13. $(4x^2yz^5)(-3x^4yz^2) = -12x^6y^2z^7$

14. $6x^3\left(\frac{1}{3}x^5 - 4x^2 - 2\right) = \frac{6}{3}x^{3+5} + 6(-4)x^{3+2} + 6(-2)x^3 = 2x^8 - 24x^5 - 12x^3$

15. $7x^3y^4(3x^7y - 5x^4y^3 - 6) = 7(3)x^{3+7}y^{4+1} + 7(-5)x^{3+4}y^{4+3} + 7(-6)x^3y^4$
$= 21x^{10}y^5 - 35x^7y^7 - 42x^3y^4$

16. $(2x+5)(3x^2+7x-4) = 2x(3x^2+7x-4)+5(3x^2+7x-4)$
$= 6x^3+14x^2-8x+15x^2+35x-20$
$= 6x^3+29x^2+27x-20$

17. $(3x^2-4x+2)(5x^2+7x-8) = 3x^2(5x^2+7x-8)-4x(5x^2+7x-8)+2(5x^2+7x-8)$
$= 15x^4+21x^3-24x^2-20x^3-28x^2+32x+10x^2+14x-16$
$= 15x^4+x^3-42x^2+46x-16$

18. $(2xy+2)(x^2y-3y+4) = 2xy(x^2y-3y+4)+2(x^2y-3y+4)$
$= 2x^3y^2-6xy^2+8xy+2x^2y-6y+8$
$= 2x^3y^2+2x^2y-6xy^2+8xy-6y+8$

19. $(4x-2)(3x-5) = 4x(3x-5)-2(3x-5)$
$= 12x^2-20x-6x+10$
$= 12x^2-26x+10$

20. $(4xy+5z)(3xy-z) = 4xy(3xy-z)+5z(3xy-z)$
$= 12x^2y^2-4xyz+15xyz-5z^2$
$= 12x^2y^2+11xyz-5z^2$

21. $(8x^3-3x)(7x^3+9x) = 8x^3(7x^3+9x)-3x(7x^3+9x)$
$= 56x^6+72x^4-21x^4-27x^2$
$= 56x^6+51x^4-27x^2$

22. $(3x+7y)^2 = (3x+7y)(3x+7y)$
$= 3x(3x+7y)+7y(3x+7y)$
$= 9x^2+21xy+21xy+49y^2$
$= 9x^2+42xy+49y^2$

23. $(2x^2y-3z)^2 = (2x^2y)^2+2(2x^2y)(-3z)+(-3z)^2$
$= 4x^4y^2-12x^2yz+9z^2$

24. $[(3x-5)+8y]^2 = (3x-5)^2+2(3x-5)(8y)+(8y)^2$
$= 9x^2-30x+25+48xy-80y+64y^2$

25. $(3x+y-2)^2 = (3x+y-2)(3x+y-2)$
$= 3x(3x+y-2)+y(3x+y-2)-2(3x+y-2)$
$= 9x^2+3xy-6x+3xy+y^2-2y-6x-2y+4$
$= 9x^2+6xy-12x-4y+y^2+4$

26. $(2x+7y)(2x-7y)=(2x)^2-(7y)^2=4x^2-49y^2$

27. $(7a^2b+3b)(7a^2b-3b)=(7a^2b)^2-(3b)^2=49a^4b^2-9b^2$

28. $[5y-(2x+7)][5y+(2x+7)]=(5y)^2-(2x+7)^2$
$=25y^2-(4x^2+28x+49)$
$=25y^2-4x^2-28x-49$

29. $f(x)=5x-3$ and $g(x)=8x^2-7x+9$
$(f+g)(x)=(5x-3)+(8x^2-7x+9)$
$=8x^2-2x+6$
$(f-g)(x)=(5x-3)-(8x^2-7x+9)$
$=5x-3-8x^2+7x-9$
$=-8x^2+12x-12$
$(fg)(x)=(5x-3)(8x^2-7x+9)$
$=5x(8x^2-7x+9)-3(8x^2-7x+9)$
$=40x^3-35x^2+45x-24x^2+21x-27$
$=40x^3-59x^2+66x-27$

30. $V(x)=x(4-2x)(5-2x)=(4x-2x^2)(5-2x)$
$=4x(5-2x)-2x^2(5-2x)$
$=20x-8x^2-10x^2+4x^3$
$V(x)=4x^3-18x^2+20x$

31. Let $18-2x$ = width of shaded rectangle
$26-2x$ = length of shaded rectangle
area of shaded rectangle
$=(26-2x)(18-2x)$
$=468-52x-36x+4x^2$
$=468-88x+4x^2$
$=4x^2-88x+468$

32. $x(x+4)-(x+1)(x-2)=17$
$x^2+4x-x^2+x+2=17$
$5x=15$
$x=3$
Dimensions: $x=3$ m by $x+4=3+4=7$ m

33. $f(x)=4x^2-5x+2$

a. $f(a+6)=4(a+6)^2-5(a+6)+2$
$=4(a^2+12a+36)-5a-30+2$
$=4a^2+48a+144-5a-28$
$=4a^2+43a+116$

b. $f(a+h)-f(a)$
$=4(a+h)^2-5(a+h)+2-(4a^2-5a+2)$
$=4(a^2+2ah+h^2)-5a-5h+2-4a^2+5a-2$
$=4a^2+8ah+4h^2-5a-5h+2-4a^2+5a-2$
$=4h^2+8ah-5h$

34. $f(x)=63+20\left(\frac{x-120}{40}\right)-0.4\left(\frac{x-120}{40}\right)^2-1.2\left(\frac{x-120}{40}\right)^3$ where $0\le x\le 240$

$f(140)=63+20\frac{(140-120)}{40}-0.4\left(\frac{140-120}{40}\right)^2-1.2\left(\frac{140-120}{40}\right)^3$
$f(140)=63+20\left(\frac{20}{40}\right)-0.4\left(\frac{20}{40}\right)^2-1.2\left(\frac{20}{40}\right)^3$
$f(140)=63+20\left(\frac{1}{2}\right)-0.4\left(\frac{1}{4}\right)-1.2\left(\frac{1}{8}\right)$
$f(140)=63+10-0.1-0.15$
$f(140)=72.75$

The height of a 140-year old tree is 72.75 feet.

35. $15x^2+3x=3x(5x+1)$

36. $5x^4y^2-20x^3y^3+15x^6y^2=5x^3y^2(x-4y+3x^3)$

37. $x^3+5x^2-2x-10=x^2(x+5)-2(x+5)=(x+5)(x^2-2)$

38. $x^2y^2-36-4y^2+9x^2=(x^2y^2-4y^2)+(9x^2-36)$
$=y^2(x^2-4)+9(x^2-4)$
$=(y^2+9)(x^2-4)$
$=(y^2+9)(x-2)(x+2)$

39. $bc-d-bd+c=bc-bd+c-d=b(c-d)+1(c-d)=(c-d)(b+1)$

40. $x^2+37x+36=(x+36)(x+1)$

41. $x^3-15x^2+26x=x(x^2-15x+26)=x(x-13)(x-2)$

42. $-2x^3+36x^2-64x=-2x(x^2-18x+32)$
$=-2x(x-2)(x-16)$

43. $8y^4-14y^2-15=(4y^2+3)(2y^2-5)$

44. $6x^2-11x-35=(3x+5)(2x-7)$

45. $-2a^4+24a^3-54a^2=-2a^2(a^2-12a+27)=-2a^2(a-9)(a-3)$

46. $6y^6+13y^3-5=(3y^3-1)(2y^3+5)$

47. $3(x+2)^2+14(x+2)+8=[3(x+2)+2][(x+2)+4]$
$=(3x+6+2)(x+2+4)=(3x+8)(x+6)$

48. $4x^2-16=4(x^2-4)=4(x-2)(x+2)$

49. $81x^4-100y^4=(9x^2+10y^2)(9x^2-10y^2)$

50. $(x+4)^2-(3x-1)^2=[(x+4)-(3x-1)][(x+4)+(3x-1)]$
$=(x+4-3x+1)(x+4+3x-1)$
$=(-2x+5)(4x+3)$

51. $x^2(b^2-9)-25(b^2-9)=(b^2-9)(x^2-25)$
$=(b+3)(b-3)(x+5)(x-5)$

52. $x^4-16=(x^2-4)(x^2+4)$
$=(x-2)(x+2)(x^2+4)$

53. $4a^3+32=4(a^3+8)=4(a+2)(a^2-2a+4)$

54. $1-64y^3=(1-4y)(1+4y+16y^2)$

55. $4x^2+12x+9=(2x+3)(2x+3)=(2x+3)^2$

56. x^4+49 is prime.

57. $y^5-y=y(y^4-1)=y(y^2-1)(y^2+1)$
$=y(y+1)(y-1)(y^2+1)$

58. $9x^2-30x+25=(3x-5)(3x-5)=(3x-5)^2$

59. $x^2+6x+9-4a^2=(x+3)^2-4a^2$
$=[(x+3)-2a][(x+3)+2a]$
$=(x+3-2a)(x+3+2a)$

60. $9x^2-21xy+10y^2=(3x-5y)(3x-2y)$

61. $a^2-b^2+4b-4=(a-b)(a+b)+4b-4$
$=[(a+b)-2][(a-b)+2]$
$=(a+b-2)(a-b+2)$

62. $9x^2+30xy+25y^2=(3x+5y)(3x+5y)$
$=(3x+5y)^2$

63. $2a^3+12a^2+18a=2a(a^2+6a+9)=2a(a+3)(a+3)$
$=2a(a+3)^2$

64. $-x^2+4x+21=-(x^2-4x-21)$
$=-(x-7)(x+3)$

65. $5x^4-40x=5x(x^3-8)=5x(x-2)(x^2+2x+4)$

66. $2x^3-x^2-18x+9=x^2(2x-1)-9(2x-1)$
$=(x^2-9)(2x-1)=(x-3)(x+3)(2x-1)$

67. $(x^2-1)^2-11(x^2-1)+24$
$[(x^2-1)-8][(x^2-1)-3]=(x^2-9)(x^2-4)$
$=(x+3)(x-3)(x+2)(x-2)$

68. $x^4-6x^2+9=(x^2-3)(x^2-3)=(x^2-3)^2$

69. $x^3+y+y^3+x=(x+y)(x^2-xy+y^2+1)$

70. $27b^3-125c^3=(3b-5c)(9b^2+15bc+25c^2)$

71. $10x^3y+22x^2y-24xy=2xy(5x^2+11x-12)=2xy(5x-4)(x+3)$

72. $x^4-x^3-x+1=x^3(x-1)-1(x-1)=(x-1)(x^3-1)$
$=(x-1)(x-1)(x^2+x+1)=(x-1)^2(x^2+x+1)$

73. $(x+y)^3-(2x+y)^3=[(x+y)-(2x+y)][(x+y)^2+(x+y)(2x+y)+(2x+y)^2]$
$=(x+y-2x-y)(x^2+2xy+y^2+2x^2+3xy+y^2+4x^2+4xy+y^2)$
$=-x(7x^2+9xy+3y^2)$

74. $x^2+6x+5=0$
$(x+5)(x+1)=0$
$x+5=0$ or $x+1=0$
$x=-5$ or $x=-1$
$\{-5,-1\}$

75. $x(x-12)=-20$
$x^2-12x+20=0$
$(x-10)(x-2)=0$
$x-10=0$ or $x-2=0$
$x=10$ or $x=2$
$\{2, 10\}$

76. $(y-2)(2y+1)=-3$
$2y^2-3y+1=0$
$(2y-1)(y-1)=0$
$2y-1=0$ or $y-1=0$
$y=\frac{1}{2}$ or $y=1$
$\left\{\frac{1}{2},1\right\}$

77. $3x^2=12x$
$3x^2-12x=0$
$3x(x-4)=0$
$3x=0$ or $x-4=0$
$x=0$ or $x=4$
$\{0, 4\}$

78. $x^2+2=6x-7$
$x^2-6x+9=0$
$(x-3)(x-3)=0$
$x-3=0$
$x=3$
$\{3\}$

79. $x^3+5x^2=9x+45$
$x^3+5x^2-9x-45=0$
$x^2(x+5)-9(x+5)=0$
$(x^2-9)(x+5)=0$
$(x-3)(x+3)(x+5)=0$
$x-3=0$ or $x+3=0$ or $x+5=0$.
$x=3$ or $x=-3$ or $x=-5$
$\{-5,-3, 3\}$

80. $y=x^2+5x+4$
x-intercepts:
$x^2+5x+4=0$
$(x+4)(x+1)=0$
$x=-4$ or $x=-1$
$(-4, 0), (-1, 0)$
y-intercepts:
$y=0^2+5(0)+4=4$
$(0, 4)$
vertex:
$x=-\frac{b}{2a}=-\frac{5}{2(1)}=-\frac{5}{2}$
$y=\left(-\frac{5}{2}\right)^2+5\left(-\frac{5}{2}\right)+4=-\frac{9}{4}$
$\left(-\frac{5}{2},-\frac{9}{4}\right)$
$y=x^2+5x+4$

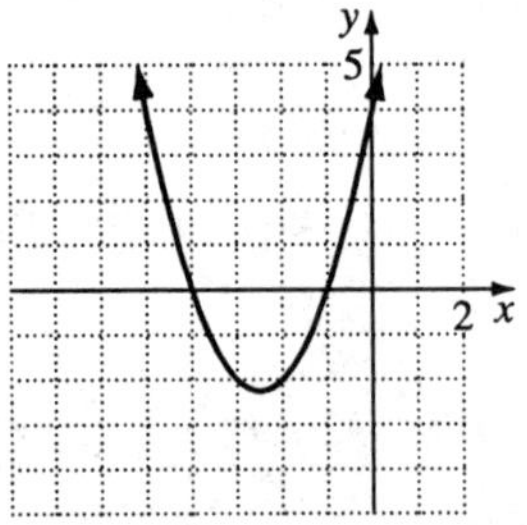

81. $f(x)=-2x^2+12x-10$
x-intercepts:
$-2x^2+12x-10=0$
$-2(x^2-6x+5)=0$
$-2(x-5)(x-1)=0$
$x=5$ or $x=1$
$(1, 0), (5, 0)$
y-intercept:
$y=-2(0)^2+12(0)-10=-10$
$(0,-10)$

vertex:

$x = -\frac{b}{2a} = -\frac{12}{2(-2)} = 3$

$y = -2(3)^2 + 12(3) - 10 = 8$

(3, 8)

$y = -2x^2 + 12x - 10$

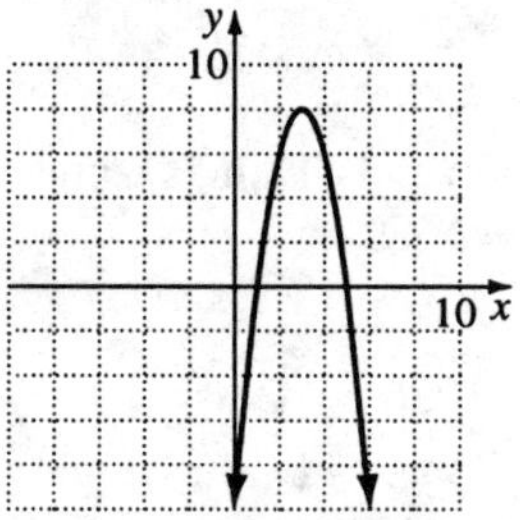

82. $s(t) = -16t^2 - 48t + 640$

a. When $s(t) = 0$,

$-16t^2 - 48t + 640 = 0$

$-16(t^2 + 3t - 40) = 0$

$-16(t + 8)(t - 5) = 0$

$t = -8$ or $t = 5$

Disregard a negative time.

It will reach the ground in 5 seconds.

b. Vertex:

$t = -\frac{b}{2a} = -\frac{(-48)}{2(-16)} = -1.5$

A negative time is not reasonable.

Therefore, it is at a maximum height of 640 feet at 0 seconds.

c.

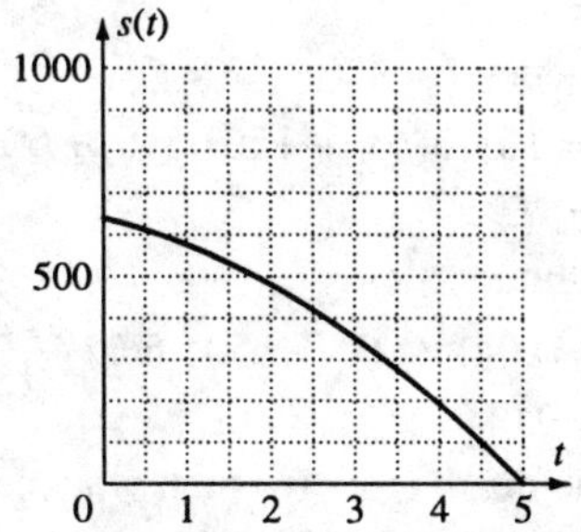

83. $P = 100 + 25x - 5x^2$

(P = 120 millimeters):

$120 = 100 + 25x - 5x^2$

$5x^2 - 25x + 20 = 0$

$5(x^2 - 5x + 4) = 0$

$5(x - 4)(x - 1) = 0$

$x - 4 = 0$ or $x - 1 = 0$

$x = 4$ or $x = 1$ (reject since $4 > 1$)

at most 4 milligrams

84. Let

x = length of the shorter leg

$x + 7$ = length of the longer leg

hypotenuse = 13 meters

$x^2 + (x+7)^2 = 13^2$

$x^2 + x^2 + 14x + 49 = 169$

$2x^2 + 14x - 120 = 0$

$2(x^2 + 7x - 60) = 0$

$2(x + 12)(x - 5) = 0$

$x + 12 = 0$ or $x - 5 = 0$

$x = -12$ (Reject) $x = 5$ (shorter leg)

$x + 7 = 12$ (longer leg)

The length of the longer leg is 12 meters.

85. Let x = width of room, then

$x + 10$ = length of room

$x(x + 10) = 1200$

$x^2 + 10x - 1200 = 0$

$(x - 30)(x + 40) = 0$

$x = 30$ or $x = -40$

Disregard the negative length.

The room is 30 feet by 40 feet.

86. Let

x = width of frame

$x + 10 + x = 10 + 2x$
= width of picture plus frame

$x + 16 + x = 16 + 2x$
= length of picture plus frame

$(10 + 2x)(16 + 2x) = 280$

$160 + 20x + 32x + 4x^2 = 280$

$4x^2 + 52x - 120 = 0$

$4(x^2+13x-30)=0$
$4(x+15)(x-2)=0$
$x+15=0$ or $x-2=0$
$x=-15$ (Reject) or $x=2$
The width of the frame is 2 centimeters.

87. Let
x = width of frame
$4+2x$ = width of frame plus picture
$7+2x$ = length of frame plus picture
area of frame = 26 square centimeters
total area = area of picture + area of frame
$(4+2x)(7+2x)=4(7)+26$
$28+22x+4x^2=28+26$
$4x^2+22x-26=0$
$2(2x^2+11x-13)=0$
$2(x-1)(2x+13)=0$
$x-1=0$ or $2x+13=0$
$x=1$ or $x=-\frac{13}{2}$
The width of the frame is 1 centimeter.

88. Let
x = width of rectangular piece
$2x$ = length of rectangular piece
$x-4$ = width of box
$2x-4$ = length of box
2 = height of box
volume = 480 cubic inches
$(x-4)(2x-4)(2)=480$
$4(x-4)(x-2)=480$
$(\div 4)\quad (x-4)(x-2)=120$
$x^2-6x+8=120$
$(x+8)(x-14)=0$
$x+8=0$ or $x-14=0$
$x=-8$ (Reject) or $x=14$ (width)
$2x=28$ (length)
dimensions of piece of tin:
14 inches × 28 inches

89. $y=ax^2+bx+c$
$1682=a(4)^2+b(4)+c$
$626=a(7)^2+b(7)+c$
$967=a(9)^2+b(9)+c$
or
$16a+4b+c=1682$ (1)
$49a+7b+c=626$ (2)
$81a+9b+c=967$ (3)
Add equation (1) and –1 times equation (2).
$$\begin{aligned}16a+4b+c&=1682\\ -49a-7b-c&=-626\\ \hline -33a-3b&=1056 \quad (4)\end{aligned}$$
Add equation (2) and –1 times equation (3).
$$\begin{aligned}49a+7b+c&=626\\ -81a-9b-c&=-967\\ \hline -32a-2b&=-341 \quad (5)\end{aligned}$$
Add 2 times equation (4) to –3 times equation (5).
$$\begin{aligned}-66a-6b&=2112\\ 96a+6b&=1023\\ \hline 30a&=3135\\ a&=104.5\end{aligned}$$
$-32a-2b=-341$
$b=\frac{-341+32(104.5)}{-2}$
$b=-1501.5$
$c=-16(104.5)-4(-1501.5)+1682$
$c=6016$
$y=104.5x^2-1501.5x+6016$
5 hours:
$y=104.5(5)^2-1501.5(5)+6016$
$y=1121$
6 hours:
$y=104.5(6)^2-1501.5(6)+6016$
$y=769$
11 hours:
$y=104.5(11)^2-1501.5(11)+6016$
$y=2144$
The death rates for men who sleep 5, 6, or 11 hours are 1121, 769, or 2144 deaths, respectively, per 100,000 males per year.

Chapter 4 Test

1. $(4x^3y-19x^2y-7y)$
$+(3x^3y+x^2y+6y-9)$
$=4x^3y+3x^3y-19x^2y$
$+x^2y-7y+6y-9$
$=7x^3y-18x^2y-y-9$

2. $(6x^2-7x-9)-(-5x^2+6x-3)$
$=6x^2-7x-9+5x^2-6x+3$
$=6x^2+5x^2-7x-6x-9+3$
$=11x^2-13x-6$

3. $(-7x^3y)(-12x^4y^2)=(-7)(-12)x^{3+4}y^{1+2}$
$=84x^7y^3$

4. $(7x-9y)(3x+y)=7x(3x+y)-9y(3x+y)$
$=21x^2+7xy-27xy-9y^2$
$=21x^2-20xy-9y^2$

5. $(5ab+4c)(3ab-c)=5ab(3ab-c)$
$+4c(3ab-c)$
$=15a^2b^2-5abc+12abc-4c^2$
$=15a^2b^2+7abc-4c^2$

6. $(x-y)(x^2-3xy-y^2)$
$=x(x^2-3xy-y^2)-y(x^2-3xy-y^2)$
$=x^3-3x^2y-xy^2-x^2y+3xy^2+y^3$
$=x^3-4x^2y+2xy^2+y^3$

7. $(5x+9y)^2=(5x)^2+2(5x)(9y)+(9y)^2$
$=25x^2+90xy+81y^2$

8. $[(2x-7)-3y]^2$
$=(2x-7)^2+2[(2x-7)(-3y)]+(-3y)^2$
$=(2x)^2+2(2x)(-7)+(-7)^2$
$+2(-6xy+21y)+9y^2$
$=4x^2-28x+49-12xy+42y+9y^2$

9. $(7x-3y)(7x+3y)=(7x)^2-(3y)^2$
$=49x^2-9y^2$

10. $x(x+3)(x+1)=x(x^2+4x+3)$
$=x^3+4x^2+3x$

11. $\frac{1}{2}(4x+2)^2=\frac{1}{2}(16x^2+16x+4)$
$=8x^2+8x+2$

12. $f(x)=-1.14x^3-4.82x^2+16.28x-87.49$
The cubic model indicates that the U.S. trade balance is getting more and more negative over time, or the deficit is increasing.

13. $f(5)$
$=-1.14(5)^3-4.82(5)^2+16.28(5)-87.49$
$f(5)=-142.5-120.5+81.4-87.49$
$f(5)=-269.09$
The model estimate was –\$269.09 billion while the actual was –\$159.6 billion. The model estimate is not very good for 1995.

14. $14x^3-15x^2=x^2(14x-15)$

15. $81y^2-25=(9y-5)(9y+5)$

16. $x^3+3x^2-25x-75=x^2(x+3)-25(x+3)$
$=(x^2-25)(x+3)$
$=(x-5)(x+5)(x+3)$

17. $25x^2+9-30x=25x^2-30x+9$
$=(5x-3)(5x-3)$
$=(5x-3)^2$

18. $x^2+10x+25-9y^2=(x+5)^2-(3y)^2$
$=(x+5-3y)(x+5+3y)$

19. x^4+1 is prime.

20. $y^2-16y-36=(y-18)(y+2)$

21. $14x^2+41x+15=(7x+3)(2x+5)$

22. $5p^3-5=5(p^3-1)$
$=5(p-1)(p^2+p+1)$

23. $12x^2-3y^2=3(4x^2-y^2)$
$=3(2x-y)(2x+y)$

24. $12x^2-34x+10=2(6x^2-17x+5)$
$=2(3x-1)(2x-5)$

25. $3x^4-3=3(x^4-1)$
$=3(x^2-1)(x^2+1)$
$=3(x-1)(x+1)(x^2+1)$

26. $27a^3b^6-8b^6=b^6(27a^3-8)$
$=b^6(3a-2)(9a^2+6a+4)$

27. $3x^2-5x-2=0$
$(3x+1)(x-2)=0$
$3x+1=0$ or $x-2=0$
$x=-\frac{1}{3}$ or $x=2$
$\left\{-\frac{1}{3}, 2\right\}$

28. $x(x+6)=-5$
$x^2+6x+5=0$
$(x+5)(x+1)=0$
$x+5=0$ or $x+1=0$
$x=-5$ or $x=-1$
$\{-5, -1\}$

29. $x^3-x=0$
$x(x^2-1)=0$
$x(x-1)(x+1)=0$
$x=0$ or $x-1=0$ or $x+1=0$
$x=1$ or $x=-1$
$\{-1, 0, 1\}$

30. $f(x)=x^2-6x+8$
x-intercepts:
$x^2-6x+8=0$
$(x-4)(x-2)=0$
$x=4$ or $x=2$
(4, 0) and (2, 0)
y-intercept:
$y=0^2-6(0)+8=8$
(0, 8)
Vertex:
$x=-\frac{b}{2a}=-\frac{(-6)}{2(1)}=3$
$y=3^2-6(3)+8=-1$
(3, −1)
$y=x^2-6x+8$

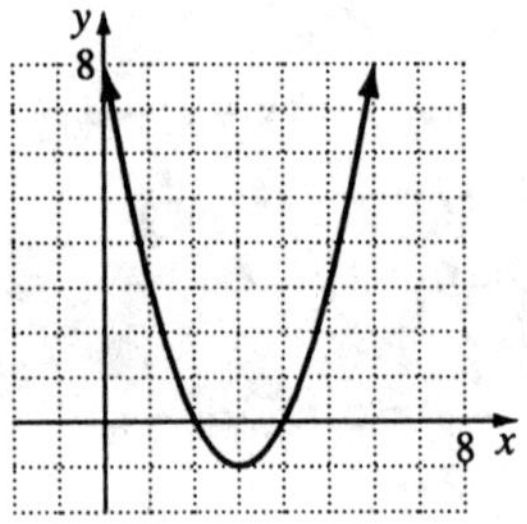

31. $s(t)=-16t^2+16t+32$
Hit water when $s(t)=0$
$0=-16t^2+16t+32$
$0=-16(t^2-t-2)$
$0=-16(t-2)(t+1)$
$t-2=0$ or $t+1=0$
$t=2$ or $t=-1$
Disregard a negative time.
The diver hits the water at 2 seconds.

32. $s(t)=-16t^2+160t$
Maximum height at vertex
$t=-\frac{b}{2a}=-\frac{160}{2(-16)}=5$
$s(5)=-16(5)^2+160(5)=400$
Vertex (5, 400)
At 5 seconds the rock reaches a maximum height of 400 feet.

33. Let x = width of a rectangle then
$3x + 2$ = length of rectangle.
$x(3x+2) = 56$
$3x^2 + 2x - 56 = 0$
$(3x+14)(x-4) = 0$
$3x + 14 = 0$ or $x - 4 = 0$
$x = -\frac{14}{3}$ or $x = 4$
Disregard a negative length.
$x = 4$
$3x + 2 = 3(4) + 2 = 14$
The width is 4 m and the length is 14 m.

Chapters 1–4 Cumulative Review Problems

1. $(x+1) + y = x + (1+y)$
Associative property of addition

2. $\frac{242{,}000}{0.0605} = \frac{2.42\times10^5}{6.05\times10^{-2}}$
$= 0.4\times10^7 = 4\times10^6$ or $4{,}000{,}000$

3. $\frac{6-2^2-(-2)^3}{8-3(2)-(-3)}$
$= \frac{6-4+8}{8-6+3}$
$= \frac{10}{5}$
$= 2$

4. $|2x-8| = |6x-7|$
$2x - 8 = 6x - 7$ or $2x - 8 = -(6x-7)$
$-4x = 1$ $2x - 8 = -6x + 7$
$x = -\frac{1}{4}$ $8x = 15$
$x = \frac{15}{8}$
$\left\{-\frac{1}{4}, \frac{15}{8}\right\}$

5. $8(y+2) - 3(2-y) = 4(2y+6) - 2$
$8y + 16 - 6 + 3y = 8y + 24 - 2$
$3y + 10 = 22$
$3y = 12$
$y = 4$
$\{4\}$

6. $2x + 4 < 10$ and $3x - 1 > 5$
$2x < 6$ and $3x > 6$
$x < 3$ and $x > 2$
$\{x | 2 < x < 3\}$,
$(2, 3)$

7. $x = \frac{ax+b}{c}$
$cx = ax + b$
$cx - ax = b$
$x(c-a) = b$
$x = \frac{b}{c-a}, c \neq a$

8. $2k - k = k$
There are 8 segments with length $\frac{k}{2}$ and 8 segments with length k.
$8\left(\frac{k}{2}\right) + 8(k) = 24$
$4k + 8k = 24$
$12k = 24$
$k = 2$

9. $y = 41{,}000x + 700{,}000$
where x = years after 1996
and y = number infected.
$1{,}479{,}000 = 41{,}000x + 700{,}000$
$779{,}000 = 41{,}000x$
$19 = x$
$1996 + 19 = 2015$
In year 2015; answers may vary.

10. Line passing through (–2, –3) and (2, 5):
$m = \frac{5+3}{2+2} = \frac{8}{4} = 2$
$y + 3 = 2(x+2)$
$y + 3 = 2x + 4$
$y = 2x + 1$

11. From the graph,

$f(5)=3, f(2)=-3, f(-3)=3, f(-2)=-2$

$f(5)-f(2)-|f(-3)-f(-2)|$

$=3-(-3)-|3-(-2)|$

$=3+3-5=1$

12. $x=2(y-5)$

$4x+40=y-7$

(substitute for x)

$4[2(y-5)]+40=y-7$

$8y-40+40=y-7$

$7y=-7$

$y=-1$

$x=2(y-5)$

$x=2(-1-5)$

$x=-12$

$\{(-12,-1)\}$

13. $6x+4y+4z=2$

$7x+5y+z=14$

$5x+4y+3z=4$

(Equations 1 and 2):

$$\begin{array}{lr} (\div 2) & 3x+2y+2z=1 \\ \times -2 & \underline{-14x-10y-2z=-28} \\ & -11x-8y=-27 \quad (4) \end{array}$$

(Equations 2 and 3):

$$\begin{array}{lr} (\times -3) & -21x-15y-3z=-42 \\ & \underline{5x+4y+3z=4} \\ & -16x-11y=-38 \quad (5) \end{array}$$

(Equations 4 and 5):

$$\begin{array}{lr} (\times 11) & -121x-88y=-297 \\ (\times -8) & \underline{128x+88y=304} \\ & 7x=7 \\ & x=1 \end{array}$$

Equation 4 $(\times -1)$: $11x+8y=27$

$11+8y=27$

$8y=16$

$y=2$

(Equation 2): $7x+5y+z=14$

$7(1)+5(2)+z=14$

$z=-13$

$\{(1, 2, -3)\}$

14. $4x-5y=2 \quad \rightarrow \quad 4x-5y=2$

$6x+2y+1=0 \quad \rightarrow \quad 6x+2y=-1$

$D=\begin{vmatrix} 4 & -5 \\ 6 & 2 \end{vmatrix}=8+30=38$

$D_x=\begin{vmatrix} 2 & -5 \\ -1 & 2 \end{vmatrix}=4-5=-1$

$D_y=\begin{vmatrix} 4 & 2 \\ 6 & -1 \end{vmatrix}=-4-12=-16$

$x=\frac{D_x}{D}=\frac{-1}{38}$

$y=\frac{D_y}{D}=-\frac{16}{38}=-\frac{8}{19}$

$\left\{\left(-\frac{1}{38},-\frac{8}{19}\right)\right\}$

15. $6x - y - 3z = 2$
$-3x + y - 3z = 1$
$-2x + 3y + z = -6$

$$\left[\begin{array}{rrr|r} 6 & -1 & -3 & 2 \\ -3 & 1 & -3 & 1 \\ -2 & 3 & 1 & -6 \end{array}\right] \begin{array}{l} \\ 2R_2 + R_1 \rightarrow \\ 3R_3 + R_1 \rightarrow \end{array} \left[\begin{array}{rrr|r} 6 & -1 & -3 & 2 \\ 0 & 1 & -9 & 4 \\ 0 & 8 & 0 & -16 \end{array}\right] R_3 - 8R_2 \rightarrow \left[\begin{array}{rrr|r} 6 & -1 & -3 & 2 \\ 0 & 1 & -9 & 4 \\ 0 & 0 & 72 & -48 \end{array}\right]$$

$$\begin{array}{l} \frac{1}{6}R_1 \rightarrow \\ \\ \frac{1}{72}R_3 \rightarrow \end{array} \left[\begin{array}{rrr|r} 1 & -\frac{1}{6} & -\frac{1}{2} & \frac{1}{3} \\ 0 & 1 & -9 & 4 \\ 0 & 0 & 1 & -\frac{2}{3} \end{array}\right]$$

$x - \frac{1}{6}y - \frac{1}{2}z = \frac{1}{3}$
$y - 9z = 4$
$z = -\frac{2}{3}$

$y - 9\left(-\frac{2}{3}\right) = 4$
$y + 6 = 4$
$y = -2$

$x - \frac{1}{6}(-2) - \frac{1}{2}\left(-\frac{2}{3}\right) = \frac{1}{3}$
$x + \frac{1}{3} + \frac{1}{3} = \frac{1}{3}$
$x = -\frac{1}{3}$
$\left\{\left(-\frac{1}{3}, -2, -\frac{2}{3}\right)\right\}$

16. Cost = 400,000 + 10x
Revenue = 30x
400,000 + 10x = 30x
$-20x = -400{,}000$
$x = 20{,}000$
The break-even point is 20,000 books.

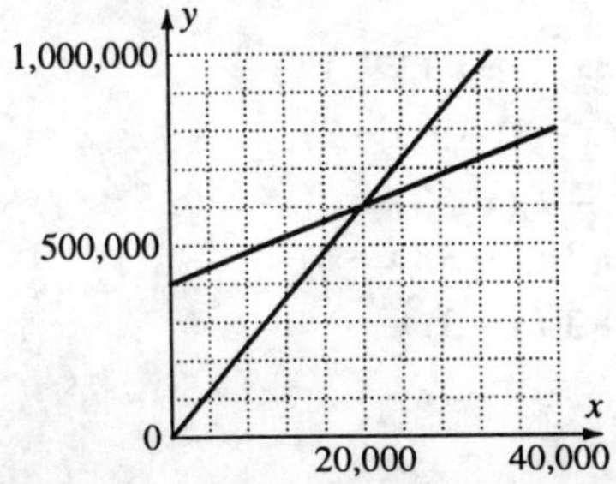

17. Let c = cost of can of paint and p = cost of a paintbrush

$2c + 3p = 53 \quad (1)$

$3c + 1p = 55 \quad (2)$

Solve (2) for p: $p = 55 - 3c$

Substitute into (1):

$2c + 3(55 - 3c) = 53$

$2c + 165 - 9c = 53$

$112 = 7c$

$c = 16$

$p = 55 - 3c$

$p = 55 - 3(16) = 7$

\$16 per can of paint, \$7 per paintbrush

18. Mentally slide all shaded boxes to the top row. The shaded area is $\frac{1}{4}$ the area of the figure.

19. Let l = number of votes for loser.

Then 160 + l = number of votes for winner.

$l + (160 + l) = 2800$

$2l = 2640$

$l = 1320$

$160 + l = 1480$

1320 votes for the losing candidate and 1480 votes for the winning candidate.

20. $c = 1.44t + 318.1$

where t = number of years after 1965

$2(280) = 1.44t + 318.1$

$560 - 318.1 = 1.44t$

$241.9 = 1.44t$

$t = 167.986 \approx 168$ years

1965 + 168 = 2133

In the year 2133.

Explanations may vary.

21. $(6x+1)(2x^2+2x-7)$

$= 6x(2x^2+2x-7) + 1(2x^2+2x-7)$

$= 12x^3 + 12x^2 - 42x + 2x^2 + 2x - 7$

$= 12x^3 + 14x^2 - 40x - 7$

22. Area of trapezoid [shaded area]

= area of rectangle – area of triangle

$= (x+4)(2x+4x) - \frac{1}{2}(2x)(x+4)$

$= (x+4)(6x) - x(x+4)$

$= 6x^2 + 24x - x^2 - 4x$

$= 5x^2 + 20x$

23. $|2x-5| \geq 9$

$2x - 5 \geq 9$ or	$-(2x-5) \geq 9$
$2x \geq 14$	$2x - 5 \leq -9$
$x \geq 7$	$2x \leq -4$
	$x \leq -2$

$\{x \mid x \leq -2 \text{ or } x \geq 7\}$

$(-\infty, -2] \cup [7, \infty)$

24. $\dfrac{-8x^3y^6}{16x^9y^{-4}} = \dfrac{-y^{6+4}}{2x^{9-3}} = \dfrac{-y^{10}}{2x^6}$

25. $2x - y < -4$

$2x + 4 < y$

Graph $x = -2$ and $y = 2x + 4$ with dotted lines. Shade the region which is both to the left of $x = -2$ and above $y = 2x + 4$

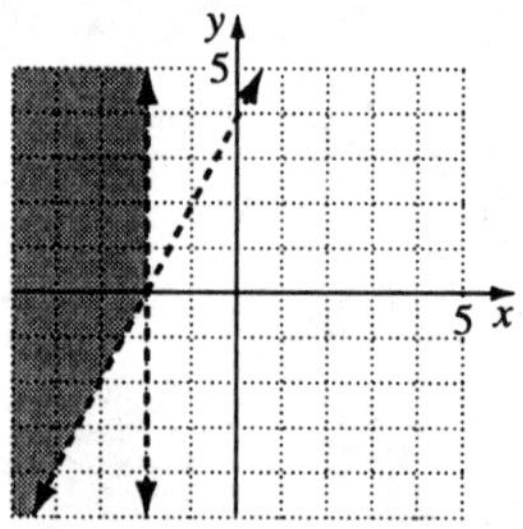

26. $x^4 - x^2 - 12$

$= (x^2 - 4)(x^2 + 3)$

$= (x+2)(x-2)(x^2+3)$

27. $x^3 - 3x^2 - 9x + 27$

$= x^2(x-3) - 9(x-3)$

$= (x^2 - 9)(x-3)$

$= (x+3)(x-3)(x-3)$

$= (x+3)(x-3)^2$

28. $(a+b+c)^2$
$= a^2 + ab + ac + ab + b^2 + bc + ac + bc + c^2$
$= a^2 + 2ab + 2ac + b^2 + 2bc + c^2$

29. Let p = original price
$p - 0.35p = 2080$
$0.65p = 2080$
$p = 3200$
\$3200

30. **a.** $\{1, \sqrt{9}, 6\}$

b. $\{0, 1, \sqrt{9}, 6\}$

c. $\{-5, 0, 1, \sqrt{9}, 6\}$

d. $\{-5, -3.72, 0, 1, \frac{1}{4}, \sqrt{9}, 6\}$

e. $\{\sqrt{10}\}$

f. [all] $\{-5, -3.72, 0, 1, \frac{1}{4}, \sqrt{9}, \sqrt{10}, 6\}$

Chapter 5

Problem Set 5.1

1. $f(x)=\dfrac{3}{x-4}$

$f(x)$ is undefined when $x-4=0$.

$x-4=0$

$x=4$

Domain of $f=\{x \mid x\neq 4\}$ or $(-\infty, 4)\cup(4, \infty)$

3. $h(x)=\dfrac{x+7}{x^2-3x}$

$h(x)$ is undefined when $x^2-3x=0$

$x^2-3x=0$

$x(x-3)=0$

$x=0$ or $x=3$

Domain of $h=$

$\{x \mid x\neq 0, 3\}$ or $(-\infty, 0)\cup(0, 3)\cup(3, \infty)$

5. $f(x)=\dfrac{3}{x^2-1}$

$f(x)$ is undefined when

$x^2-1=0$

$(x-1)(x+1)=0$

$x=1$ or $x=-1$

Domain of f

$=\{x \mid x\neq -1, 1\}$

or $(-\infty, -1)\cup(-1, 1)\cup(1, \infty)$

7. $f(x)=\dfrac{x+8}{x^2-x-12}$

$f(x)$ is undefined when

$x^2-x-12=0$

$(x-4)(x+3)=0$

$x=4$ or $x=-3$

Domain of f

$=\{x \mid x\neq -3, 4\}$

or $(-\infty, -3)\cup(-3, 4)\cup(4, \infty)$

9. a. $h(x)=\dfrac{3}{x^2-1}$

$h(0)=-3$

$h(\pm 2)=\dfrac{3}{4-1}=1$

b. $g(x)=\dfrac{3}{x^2+1}$

$g(0)=3$

$g(\pm 1)=1.5$

c. $f(x)=\dfrac{2}{(x-1)^3}$

$f(0)=-2$

$f(2)=2$

11. a. $f(x)=\dfrac{2}{x-2}$

$f(x)$ is undefined when

$x-2=0$

$x=2$

Domain of f

$=\{x \mid x\neq 2\}$ or $(-\infty, 2)\cup(2, \infty)$

b.

x	1	1.5	1.9	1.99	1.999
$f(x)=\frac{2}{x-2}$	−2	−4	−20	−200	−2000

x	2.001	2.01	2.1	2.5	3
$f(x)=\frac{2}{x-2}$	2000	200	20	4	2

c. $x=2$

$f(x)=\dfrac{2}{x-2}$

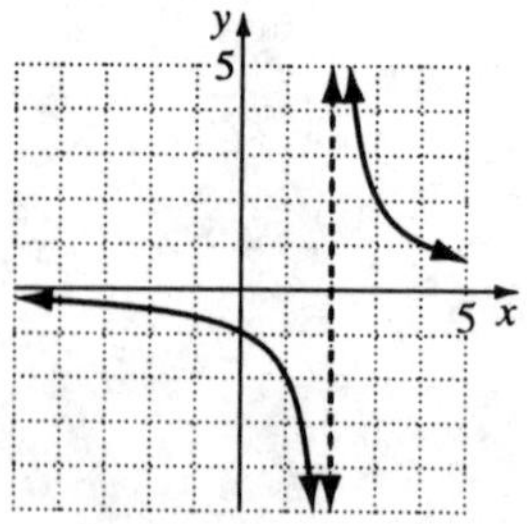

13. $\frac{y-1}{y^2-1}=\frac{y-1}{(y+1)(y-1)}=\frac{1}{y+1}$

15. $\frac{3a+9}{a^2+6a+9}=\frac{3(a+3)}{(a+3)^2}=\frac{3}{a+3}$

17. $\frac{c^2-12c+36}{4c-24}=\frac{(c-6)^2}{4(c-6)}=\frac{c-6}{4}$

19. $\frac{a^2-1}{a^2+2a+1}=\frac{(a+1)(a-1)}{(a+1)^2}=\frac{a-1}{a+1}$

21. $\frac{x^4-y^4}{x^4-2x^2y^2+y^4}$
$=\frac{\left(x^2+y^2\right)\left(x^2-y^2\right)}{\left(x^2+y^2\right)^2}$
$=\frac{x^2-y^2}{x^2+y^2}$

23. $\frac{a^2+7a-18}{a^2-3a+2}=\frac{(a+9)(a-2)}{(a-1)(a-2)}=\frac{a+9}{a-1}$

25. $\frac{2x^2-5x+2}{2x^2-7x+3}=\frac{(2x-1)(x-2)}{(2x-1)(x-3)}=\frac{x-2}{x-3}$

27. $\frac{c^2-14x+49}{c^2-49}=\frac{(c-7)^2}{(c+7)(c-7)}=\frac{c-7}{c+7}$

29. $\frac{x^2-9}{x^3-27}$
$=\frac{(x+3)(x-3)}{(x-3)(x^2+3x+9)}$
$=\frac{x+3}{x^2+3x+9}$

31. $\frac{y^2-9y+18}{y^3-27}$
$=\frac{(y-6)(y-3)}{(y-3)\left(y^2+3y+9\right)}$
$=\frac{y-6}{y^2+3y+9}$

33. $\frac{x^3+12x^2+35x}{x^3+4x^2-21x}$
$=\frac{x\left(x^2+12x+35\right)}{x\left(x^2+4x-21\right)}$
$=\frac{(x+5)(x+7)}{(x+7)(x-3)}$
$=\frac{x+5}{x-3}$

35. $\frac{x^3-9x}{x^3+5x^2+6x}$
$=\frac{x\left(x^2-9\right)}{x\left(x^2+5x+6\right)}$
$=\frac{(x+3)(x-3)}{(x+3)(x+2)}$
$=\frac{x-3}{x+2}$

37. $\frac{x-4}{4x-1}$ cannot be reduced.

39. $\frac{x^2-4}{x^2-9}$ cannot be reduced (although numerator and denominator can be factored).

41. $\frac{x^2+5x+2xy+10y}{x^2-25}$
$=\frac{x(x+5)+2y(x+5)}{(x+5)(x-5)}$
$=\frac{(x+5)(x+2y)}{(x+5)(x-5)}$
$=\frac{x+2y}{x-5}$

43. $\frac{a^2-4}{2-a}=\frac{(a+2)(a-2)}{-(a-2)}=-(a+2)=-a-2$

45. $f(x)=\frac{x^2-x-2}{x+1}$
$=\frac{(x-2)(x+1)}{x+1}$
$=x-2\ (x\neq -1)$

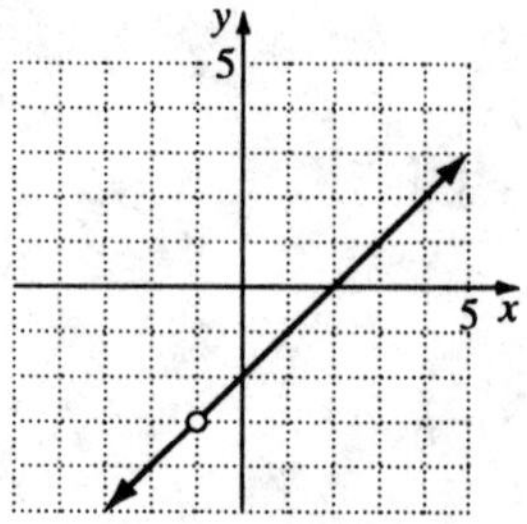

47. $f(x)=\dfrac{x^2-5x-14}{x+2}$

$\dfrac{(x-7)(x+2)}{x+2}$

$x-7\ (x\neq -2)$

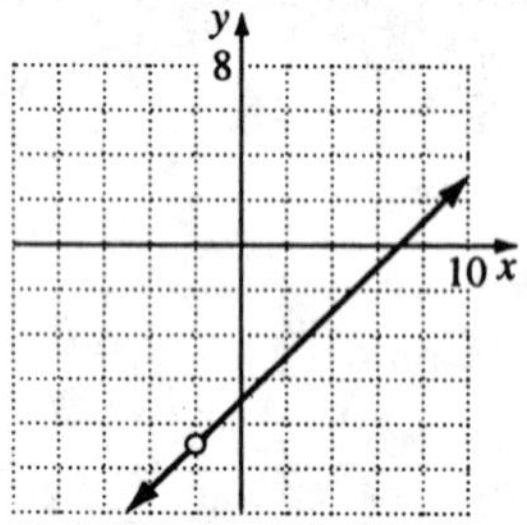

49. $\dfrac{y^2-4}{y-2}\cdot\dfrac{y-2}{y^2+y-6}$

$=\dfrac{(y+2)(y-2)}{y-2}\cdot\dfrac{y-2}{(y+3)(y-2)}$

$=(y+2)\cdot\dfrac{1}{(y+3)}$

$=\dfrac{y+2}{y+3}$

51. $\dfrac{5x+5y}{x-y}\cdot\dfrac{3x-3y}{10}$

$=\dfrac{5(x+y)}{x-y}\cdot\dfrac{3(x-y)}{10}$

$=\dfrac{15(x+y)(x-y)}{10(x-y)}$

$=\dfrac{3(x+y)}{2}$

53. $\dfrac{b^2+b}{b^2-4}\cdot\dfrac{b^2+5b+6}{b^2-1}$

$=\dfrac{b(b+1)(b+3)(b+2)}{(b+2)(b-2)(b+1)(b-1)}$

$=\dfrac{b(b+3)}{(b-2)(b-1)}$

55. $\dfrac{m^2-4}{m^2+4m+4}\cdot\dfrac{2m+4}{m^2+m-6}$

$=\dfrac{(m+2)(m-2)(2)(m+2)}{(m+2)^2(m+3)(m-2)}$

$=\dfrac{2}{m+3}$

57. $\dfrac{b+2}{b^2+7b+6}\cdot(b+1)$

$=\dfrac{(b+2)(b+1)}{(b+6)(b+1)}$

$=\dfrac{b+2}{b+6}$

59. $\dfrac{x^3-8}{x^2-4}\cdot\dfrac{x+2}{3x}$

$=\dfrac{(x-2)\left(x^2+2x+4\right)(x+2)}{(x+2)(x-2)(3x)}$

$=\dfrac{x^2+2x+4}{3x}$

61. $\dfrac{2a^2-13a-7}{a^2-6a-7}\cdot\dfrac{a^2-a-2}{2a^2-5a-3}$

$=\dfrac{(2a+1)(a-7)(a-2)(a+1)}{(a-7)(a+1)(2a+1)(a-3)}$

$=\dfrac{a-2}{a-3}$

63. $\dfrac{2y^2+9y-35}{6y^2-13y-5}\cdot\dfrac{3y^2+10y+3}{y^2+10y+21}$

$=\dfrac{(2y-5)(y+7)(3y+1)(y+3)}{(2y-5)(3y+1)(y+7)(y+3)}$

$=1$

65. $\dfrac{2a+4b}{3ab}\div\dfrac{6a+12b}{6a^2b}$

$=\dfrac{2(a+2b)}{3ab}\cdot\dfrac{6a^2b}{6(a+2b)}$

$=\dfrac{2(6)a^2b(a+2b)}{3(6)ab(a+2b)}$

$=\dfrac{2a}{3}$

67. $\frac{x+y}{x^2-xy}\div\frac{3x+3y}{x-y}$
$=\frac{x+y}{x(x-y)}\cdot\frac{x-y}{3(x+y)}$
$=\frac{(x+y)(x-y)}{3x(x-y)(x+y)}$
$=\frac{1}{3x}$

69. $\frac{x^2-y^2}{(x+y)^2}\div\frac{x-y}{4x+4y}$
$=\frac{x^2-y^2}{(x+y)^2}\cdot\frac{4x+4y}{x-y}$
$=\frac{(x+y)(x-y)(4)(x+y)}{(x+y)^2(x-y)}$
$=4$

71. $\frac{x^3-27}{a^3+8}\div\frac{x-3}{a+2}$
$=\frac{x^3-27}{a^3+8}\cdot\frac{a+2}{x-3}$
$=\frac{(x-3)(x^2+3x+9)(a+2)}{(a+2)(a^2-2a+4)(x-3)}$
$=\frac{x^2+3x+9}{a^2-2a+4}$

73. $\frac{4b^2+20b+25}{5b}\div\frac{4b^2-25}{4b^2}$
$=\frac{(2b+5)^2}{5b}\cdot\frac{4b^2}{4b^2-25}$
$=\frac{(2b+5)^2(4b^2)}{5b(2b+5)(2b-5)}$
$=\frac{4b(2b+5)}{5(2b-5)}$

75. $\frac{a^2-4a-21}{a^2-10a+25}\div\frac{a^2+2a-3}{a^2-6a+5}$
$=\frac{a^2-4a-21}{a^2-10a+25}\cdot\frac{a^2-6a+5}{a^2+2a-3}$
$=\frac{(a-7)(a+3)(a-5)(a-1)}{(a-5)^2(a+3)(a-1)}$
$=\frac{a-7}{a-5}$

77. $\frac{9x^2-12x+4}{2x^2+3x-5}\div\frac{3x^2-8x+4}{2x^2+7x+5}$
$=\frac{9x^2-12x+4}{2x^2+3x-5}\cdot\frac{2x^2+7x+5}{3x^2-8x+4}$
$=\frac{(3x-2)^2(2x+5)(x+1)}{(2x+5)(x-1)(3x-2)(x-2)}$
$=\frac{(3x-2)(x+1)}{(x-1)(x-2)}$

79. $(bx+by+3x+3y)\div\frac{x-y^2}{b+3}$
$=b(x+y)+3(x+y)\cdot\frac{b+3}{x^2-y^2}$
$=\frac{(x+y)(b+3)(b+3)}{(x+y)(x-y)}$
$=\frac{(b+3)^2}{x-y}$

81. $f(x)=x^2-9x+14,\ g(x)=x-2$

a. $\left(\frac{f}{g}\right)(x)=\frac{x^2-9x+14}{x-2}$
$=\frac{(x-2)(x-7)}{x-2}$
$=x-7$

b. $\left(\frac{f}{g}\right)(3)=3-7=-4$

c. $\frac{f(3)}{g(3)}=-4$

83. $f(x)=x^3-16x,\ g(x)=x^2-2x-8$

a. $\left(\frac{f}{g}\right)(x)=\frac{x^3-16x}{x^2-2x-8}$
$=\frac{x(x+4)(x-4)}{(x-4)(x+2)}$
$=\frac{x(x+4)}{x+2}$

b. $\left(\frac{f}{g}\right)(8)=\frac{8(8+4)}{8+2}=9.6$

c. $\frac{f(8)}{g(8)}=9.6$

85. **a.** $C(x) = \dfrac{4x}{100-x}$

$C(80) = \dfrac{4(80)}{100-80} = 16$

cost is $16,000.

b. $C(95) = \dfrac{4(95)}{100-95} = 76$

The cost to remove 95% of pollutants is $76,000.

c. $\{x \mid 0 \le x < 100\}$

It is impossible to remove 100% of the pollutants.

d. It goes to infinity.

87. $P = -0.14t^2 + 0.51t + 31.6$

$C = 0.54t^2 + 12.64t + 107.1$

average cost $= \dfrac{C}{P}$

$= \dfrac{0.54t^2 + 12.64t + 107.1 \text{ (billions)}}{-0.14t^2 + 0.51t + 31.6 \text{ (millions)}}$

1994: $t = 4$

average cost $= \dfrac{166.3 \times 10^9}{31.4 \times 10^6}$

$\approx 5.296 \times 10^3$

$\approx$ \$5294/person

1995: $t = 5$

average cost $= \dfrac{183.8 \times 10^9}{30.65 \times 10^6}$

$\approx 5.997 \times 10^3$

$\approx$ \$5997/person

1996: $t = 6$

average cost $= \dfrac{202.4 \times 10^9}{29.62 \times 10^6}$

$\approx 6.833 \times 10^3$

$\approx$ \$6833/person

1997: $t = 7$

average cost $= \dfrac{222.04 \times 10^9}{28.31 \times 10^6}$

$\approx 7.843 \times 10^3$

$\approx$ \$7843/person

average cost for 1994 through 1997:

$\dfrac{(1994) + (1995) + (1996) + (1997)}{4}$

$\approx 6.492 \times 10^3$

$\approx$ \$6492/person

89. **a.** Let x = speed on outgoing trip then $x - 10$ = speed on return trip

$T(x) = \dfrac{600}{x} + \dfrac{600}{x-10}$

b. $T(50) = \dfrac{600}{50} + \dfrac{600}{50-10}$

$T(50) = 27$

The total time for a trip traveling 600 miles at 50 miles per hour and 600 miles at 40 miles per hour is 27 hours.

91. **a.** 35 words, 11 words, 7 words

b. $N(t) = \dfrac{5t+30}{t}$

$N(1) = \dfrac{5(1)+30}{1} = 35$

$N(5) = \dfrac{5(5)+30}{5} = 11$

$N(15) = \dfrac{5(15)+30}{15} = 7$

c. The numbers of words remembered decreases as time goes on.

d. No matter how much time goes by, the number of words remembered will always be 5.

93. Unshaded area = $30(x + 2)$

Total area = $2x(5x + 10)$

$R(x) = \dfrac{30(x+2)}{2x(5x+10)} = \dfrac{30(x+2)}{2x(5)(x+2)} = \dfrac{3}{x}$

95. c is true.

97. d is true;

a is *not* true:

All rational expressions can be multiplied.

b is *not* true:

$x \div y = \frac{x}{y} \neq \frac{1}{x} \cdot y$

c is *not* true:

$\frac{x}{y} \div \frac{y}{x} = \frac{x}{y} \cdot \frac{x}{y} = \frac{x^2}{y^2} \neq 1$

d is true.

99.

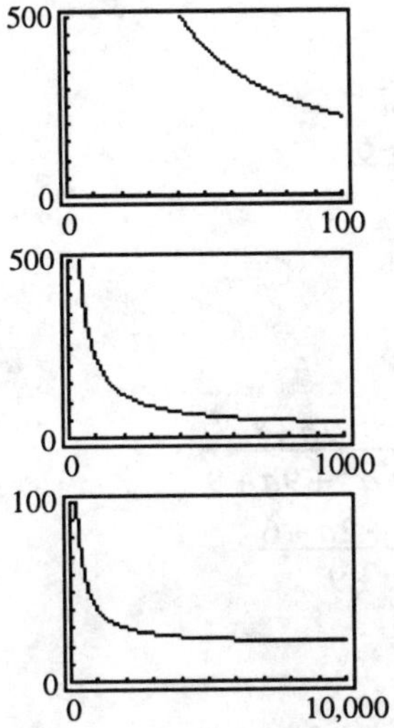

The graph show how the price approaches \$20 as the number of canoes produced increases.

101. a. $f(x) = \frac{x^2 - x - 2}{x - 2}$, $g(x) = x + 1$

They look exactly the same.

b. $f(x) = \frac{x^2 - x - 2}{x - 2}$

$= \frac{(x-2)(x+1)}{x-2}$

$= x + 1, \ x \neq 2$

$f(x) = x + 1, \ x \neq 2$

No; They do not represent the same function because $f(x)$ is undefined at $x = 2$.

c. $g(x) = 3$ at $x = 2$,

$f(x) =$ at $x = 2$, i.e., is undefined at $x = 2$.

103.–107. Answers may vary.

109. $\frac{6x^{3n} + 6x^{2n}y^n}{x^{2n} - y^{2n}}$

$= \frac{6x^{2n}(x^n + y^n)}{(x^n + y^n)(x^n - y^n)}$

$= \frac{6x^{2n}}{x^n - y^n}$

111. $\frac{4y^3 - 12y^2}{4y^n + 8} \div \frac{y^{n+1} - 2y}{y^{2n} - 4}$

$= \frac{4y^3 - 12y^2}{4y^n + 8} \cdot \frac{y^{2n} - 4}{y^{n+1} - 2y}$

$= \frac{4y^2(y-3)(y^n + 2)(y^n - 2)}{4(y^n + 2)(y)(y^n - 2)}$

$= y(y - 3)$

113. $\left(\frac{a-b}{4c} \div \frac{b-a}{c}\right) \div \frac{a-b}{c^2}$

$= \left(\frac{a-b}{4c} \cdot \frac{c}{(-1)(a-b)}\right) \div \frac{a-b}{c^2}$

$= \left(-\frac{1}{4}\right) \cdot \frac{c^2}{a-b}$

$= \frac{-c^2}{4(a-b)}$

115. $\frac{c^4 + 64d^4}{c^2 + 4cd + 8d^2}$

$= \frac{c^4 + 16c^2d^2 + 64d^4 - 16c^2d^2}{c^2 + 4cd + 8d^2}$

$= \frac{(c^2 + 8d^2)^2 - (4cd)^2}{c^2 + 4cd + 8d^2}$

$= \frac{(c^2 + 8d^2 + 4cd)(c^2 + 8d^2 - 4cd)}{(c^2 + 4cd + 8d^2)}$

$= c^2 - 4cd + 8d^2$

117. Area of trapezoid

$= \frac{1}{2} \cdot \frac{1}{8x + 4}(3x + x)$

$= \frac{4x}{2 \cdot 4(2x + 1)}$

$= \frac{x}{2(2x+1)}$

Area of rectangle

$= 3 \cdot \frac{1}{14x+7} = \frac{3}{7(2x+1)}$

$\frac{\text{Area of trapezoid}}{\text{Area of rectangle}}$

$= \frac{x}{2(2x+1)} \cdot \frac{7(2x+1)}{3} = \frac{7x}{6}$

119. Group activity

Review Problems

120. $(2x-5)(3x^2-5x+4)$

$= 2x(3x^2-5x+4) - 5(3x^2-5x+4)$

$= 6x^3 - 10x^2 + 8x - 15x^2 + 25x - 20$

$= 6x^3 - 25x^2 + 33x - 20$

121. $\frac{(0.000012)(400{,}000)}{0.000006}$

$= \frac{(1.2\times10^{-5})(4\times10^{5})}{6\times10^{-6}}$

$= \frac{(1.2)(4)}{6}\times10^{-5+5+6}$

$= 0.8\times10^{6}$

$= 8\times10^{5}$ or 800,000

122. $y \ge 3x-2$

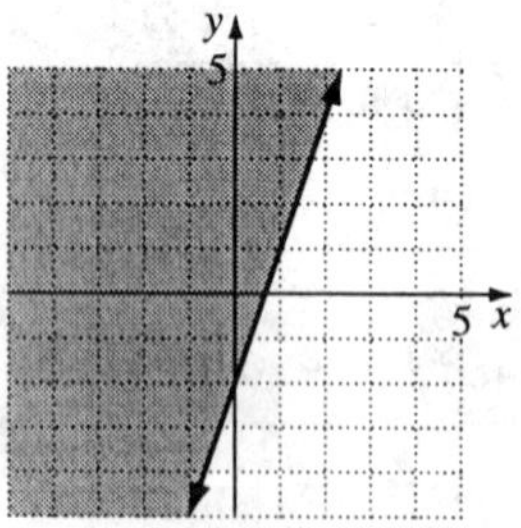

Problem Set 5.2

1. $\frac{3}{2x} + \frac{7}{2x} = \frac{3+7}{2x} = \frac{10}{2x} = \frac{5}{x}$

3. $\frac{8}{5xy^2} - \frac{3}{5xy^2} = \frac{8-3}{5xy^2} = \frac{5}{5xy^2} = \frac{1}{xy^2}$

5. $\frac{x+5y}{x+y} + \frac{x-3y}{x+y}$

$= \frac{x+5y+x-3y}{x+y}$

$= \frac{2x+2y}{x+y}$

$= \frac{2(x+y)}{x+y}$

$= 2$

7. $\frac{3x+2}{x-4} - \frac{x-6}{x-4}$

$= \frac{3x+2-x+6}{x-4}$

$= \frac{2x+8}{x-4}$

$= \frac{2(x+4)}{x-4}$

9. $\frac{a^2+7a+3}{a^2+9a+9} + \frac{2a+6}{a^2+9a+9}$

$= \frac{a^2+7a+3+2a+6}{a^2+9a+9}$

$= \frac{a^2+9a+9}{a^2+9a+9}$

$= 1$

11. $\frac{a^2-4a}{a^2-a-6} - \frac{a-6}{a^2-a-6}$

$= \frac{a^2-4a-(a-6)}{a^2-a-6}$

$= \frac{a^2-4a-a+6}{a^2-a-6}$

$= \frac{a^2-5a+6}{a^2-a-6}$

$= \frac{(a-3)(a-2)}{(a-3)(a+2)}$

$= \frac{a-2}{a+2}$

13. $\frac{9x}{10} - \frac{7x}{10} + \frac{3x}{10}$

$= \frac{9x-7x+3x}{10}$

$= \frac{5x}{10}$

$= \frac{x}{2}$

15. $\frac{3a+b}{a+b}+\frac{2a+3b}{a+b}-\frac{4a+3b}{a+b}$
$=\frac{(3a+b)+(2a+3b)-(4a+3b)}{a+b}$
$=\frac{3a+b+2a+3b-4a-3b}{a+b}$
$=\frac{a+b}{a+b}$
$=1$

17. $\frac{3a^3+4b^3}{a^2-b^2}-\frac{5b^3+2a^3}{a^2-b^2}$
$=\frac{\left(3a^3+4b^3\right)-\left(5b^3+2a^3\right)}{a^2-b^2}$
$=\frac{3a^3+4b^3-5b^3-2a^3}{a^2-b^2}$
$=\frac{a^3-b^3}{a^2-b^2}$
$=\frac{(a-b)\left(a^2+ab+b^2\right)}{(a-b)(a+b)}$
$=\frac{a^2+ab+b^2}{a+b}$

19. $\frac{3}{2x}+\frac{4}{3x}$ (LCD is $6x$)
$=\frac{3}{2x}\cdot\frac{3}{3}+\frac{4}{3x}\cdot\frac{2}{2}$
$=\frac{9}{6x}+\frac{8}{6x}$
$=\frac{9+8}{6x}$
$=\frac{17}{6x}$

21. $\frac{5}{2x^2}-\frac{2}{7x}$ $\left(\text{LCD is } 14x^2\right)$
$=\frac{5}{2x^2}\cdot\frac{7}{7}-\frac{2}{7x}\cdot\frac{2x}{2x}$
$=\frac{35}{14x^2}-\frac{4x}{14x^2}$
$=\frac{35-4x}{14x^2}$

23. $\frac{3}{6x^3}-\frac{2}{9x^2}$ $\left(\text{LCD is } 18x^3\right)$
$=\frac{3}{6x^3}\cdot\frac{3}{3}-\frac{2}{9x^2}\cdot\frac{2x}{2x}$
$=\frac{9}{18x^3}-\frac{4x}{18x^3}$
$=\frac{9-4x}{18x^3}$

25. $\frac{2a}{3c^2}-\frac{3b}{4cd}$ $\left(\text{LCD is } 12c^2d\right)$
$=\frac{2a}{3c^2}\cdot\frac{4d}{4d}-\frac{3b}{4cd}\cdot\frac{3c}{3c}$
$=\frac{8ad}{12c^2d}-\frac{9bc}{12c^2d}$
$=\frac{8ad-9bc}{12c^2d}$

27. $\frac{3a}{2c^2}-\frac{2a}{3cd}+\frac{a}{6d^2}$ $\left(\text{LCD is } 6c^2d^2\right)$
$=\frac{3a}{2c^2}\cdot\frac{3d^2}{3d^2}-\frac{2a}{3cd}\cdot\frac{2cd}{2cd}+\frac{a}{6d^2}\cdot\frac{c^2}{c^2}$
$=\frac{9ad^2}{6c^2d^2}-\frac{4acd}{6c^2d^2}+\frac{ac^2}{6c^2d^2}$
$=\frac{9ad^2-4acd+ac^2}{6c^2d^2}$

29. $\frac{2b-2c}{b^2c}+\frac{b-c}{bc^2}$ $\left(\text{LCD is } b^2c^2\right)$
$=\frac{2b-2c}{b^2c}\cdot\frac{c}{c}+\frac{b-c}{bc^2}\cdot\frac{b}{b}$
$=\frac{(2b-2c)c}{b^2c^2}+\frac{(b-c)b}{b^2c^2}$
$=\frac{2bc-2c^2+b^2-bc}{b^2c^2}$
$=\frac{b^2+bc-2c^2}{b^2c^2}$

31. $\frac{b-2y}{4b^2y}+\frac{2b+y}{6by^2}$ $\left(\text{LCD is } 12b^2y^2\right)$
$=\frac{b-2y}{4b^2y}\cdot\frac{3y}{3y}+\frac{2b+y}{6by^2}\cdot\frac{2b}{2b}$
$=\frac{3y(b-2y)}{12b^2y^2}+\frac{2b(2b+y)}{12b^2y^2}$
$=\frac{3by-6y^2+4b^2+2by}{12b^2y^2}$
$=\frac{5by-6y^2+4b^2}{12b^2y^2}$

33. $\frac{4x-3y}{6xy}-\frac{x-4z}{8xz}-\frac{3y-z}{4yz}$ (LCD is $24xyz$)

$=\frac{4x-3y}{6xy}\cdot\frac{4z}{4z}-\frac{x-4z}{8xz}\cdot\frac{3y}{3y}-\frac{3y-z}{4yz}\cdot\frac{6x}{6x}$

$=\frac{4z(4x-3y)-3y(x-4z)-6x(3y-z)}{24xyz}$

$=\frac{16xz-12yz-3xy+12yz-18xy+6xz}{24xy}$

$=\frac{22xz-21xy}{24xyz}$

$=\frac{x(22z-21y)}{24xyz}$

$=\frac{22z-21y}{24yz}$

35. $\frac{10}{x+4}-\frac{2}{x-6}$ (LCD is $(x+4)(x-6)$)

$=\frac{10}{x+4}\cdot\frac{x-6}{x-6}-\frac{2}{x-6}\cdot\frac{x+4}{x+4}$

$=\frac{10(x-6)-2(x+4)}{(x+4)(x-6)}$

$=\frac{10x-60-2x-8}{(x+4)(x-6)}$

$=\frac{8x-68}{(x+4)(x-6)}$

37. $\frac{b}{b-c}-\frac{c}{b+c}$ (LCD is $(b-c)(b+c)$)

$=\frac{b}{b-c}\cdot\frac{b+c}{b+c}-\frac{c}{b+c}\cdot\frac{b-c}{b-c}$

$=\frac{b(b+c)-c(b-c)}{(b-c)(b+c)}$

$=\frac{b^2+bc-bc+c^2}{(b-c)(b+c)}$

$=\frac{b^2+c^2}{(b-c)(b+c)}$

39. $\frac{3}{a+1}-\frac{3}{a}$ (LCD is $a(a+1)$)

$=\frac{3}{a+1}\cdot\frac{a}{a}-\frac{3}{a}\cdot\frac{a+1}{a+1}$

$=\frac{3a-3(a+1)}{a(a+1)}$

$=\frac{3a-3a-3}{a(a+1)}$

$=\frac{-3}{a(a+1)}$

41. $\frac{5x}{x-2}-\frac{x-1}{x+2}$ (LCD is $(x-2)(x+2)$)

$=\frac{5x}{x-2}\cdot\frac{x+2}{x+2}-\frac{x-1}{x+2}\cdot\frac{x-2}{x-2}$

$=\frac{5x(x+2)-(x-1)(x-2)}{(x-2)(x+2)}$

$=\frac{5x^2+10x-x^2+3x-2}{(x-2)(x+2)}$

$=\frac{4x^2+13x-2}{(x-2)(x+2)}$

43. $\frac{3x}{x-3}-\frac{x+4}{x+2}$ (LCD is $(x-3)(x+2)$)

$=\frac{3x}{x-3}\cdot\frac{x+2}{x+2}-\frac{x+4}{x+2}\cdot\frac{x-3}{x-3}$

$=\frac{3x(x+2)-(x+4)(x-3)}{(x-3)(x+2)}$

$=\frac{3x^2+6x-x^2-x+12}{(x-3)(x+2)}$

$=\frac{2x^2+5x+12}{(x-3)(x+2)}$

45. $\frac{a-b}{a+b}-\frac{a+b}{a-b}$ (LCD is $(a+b)(a-b)$)

$=\frac{a-b}{a+b}\cdot\frac{a-b}{a-b}-\frac{a+b}{a-b}\cdot\frac{a+b}{a+b}$

$=\frac{(a-b)(a-b)-(a+b)(a+b)}{(a+b)(a-b)}$

$=\frac{a^2-2ab+b^2-a^2-2ab-b^2}{(a+b)(a-b)}$

$=\frac{-4ab}{(a+b)(a-b)}$

47. $\frac{4}{x+2}-\frac{3}{x+1}+\frac{2}{x}$ (LCD is $x(x+1)(x+2)$)

$=\frac{4}{x+2}\cdot\frac{x}{x}\cdot\frac{x+1}{x+1}-\frac{3}{x+1}\cdot\frac{x}{x}\cdot\frac{x+2}{x+2}+\frac{2}{x}\cdot\frac{x+1}{x+1}\cdot\frac{x+2}{x+2}$

$=\frac{4x(x+1)-3x(x+2)+2(x+1)(x+2)}{x(x+1)(x+2)}$

$=\frac{4x^2+4x-3x^2-6x+2x^2+6x+4}{x(x+1)(x+2)}$

$=\frac{3x^2+4x+4}{x(x+1)(x+2)}$

49. $\frac{5}{2b-8}+\frac{3}{4b-2}$

$=\frac{5}{2(b-4)}+\frac{3}{2(2b-1)}$

(LCD is $2(b-4)(2b-1)$)

$=\frac{5}{2(b-4)}\cdot\frac{2b-1}{2b-1}+\frac{3}{2(2b-1)}\cdot\frac{b-4}{b-4}$

$=\frac{5(2b-1)+3(b-4)}{2(b-4)(2b-1)}$

$=\frac{10b-5+3b-12}{2(b-4)(2b-1)}$

$=\frac{13b-17}{2(b-4)(2b-1)}$

51. $\frac{4}{x^2+6x+9}+\frac{4}{x+3}$

$=\frac{4}{(x+3)^2}+\frac{4}{(x+3)}$

$=\frac{4}{(x+3)^2}+\frac{4(x+3)}{(x+3)^2}$

$=\frac{4+4x+12}{(x+3)^2}=\frac{4x+16}{(x+3)^2}$

53. $4+\frac{x+3}{x-5}$ (LCD is $x-5$)

$=4\cdot\frac{x-5}{x-5}+\frac{x+3}{x-5}$

$=\frac{4x-20}{x-5}+\frac{x+3}{x-5}$

$=\frac{4x-20+x+3}{x-5}$

$=\frac{5x-17}{x-5}$

55. $3+\frac{1}{x+2}-\frac{2}{x^2-4}$ (LCD is x^2-4)

$=3\cdot\frac{x^2-4}{x^2-4}+\frac{1}{x+2}\cdot\frac{x-2}{x-2}-\frac{2}{x^2-4}$

$=\frac{3x^2-12+x-2-2}{x^2-4}$

$=\frac{3x^2+x-16}{x^2-4}$

57. $\frac{c}{c^2-10c+25}-\frac{c-4}{2c-10}$

$=\frac{c}{(c-5)^2}-\frac{c-4}{2(c-5)}$ (LCD is $2(c-5)^2$)

$=\frac{c}{(c-5)^2}\cdot\frac{2}{2}-\frac{c-4}{2(c-5)}\cdot\frac{c-5}{c-5}$

$=\frac{2c-(c-4)(c-5)}{2(c-5)^2}$

$=\frac{2c-c^2+9c-20}{2(c-5)^2}$

$=\frac{-c^2+11c-20}{2(c-5)^2}$

59. $\frac{a-b}{3a+3b}+\frac{a+b}{2a-2b}$

$=\frac{a-b}{3(a+b)}+\frac{a+b}{2(a-b)}$

(LCD is $6(a+b)(a-b)$)

$=\frac{a-b}{3(a+b)}\cdot\frac{2(a-b)}{2(a-b)}+\frac{a+b}{2(a-b)}\cdot\frac{3(a+b)}{3(a+b)}$

$=\frac{2(a-b)(a-b)+3(a+b)(a+b)}{6(a+b)(a-b)}$

$$= \frac{2a^2-4ab+2b^2+3a^2+6ab+3b^2}{6(a+b)(a-b)}$$
$$= \frac{5a^2+2ab+5b^2}{6(a+b)(a-b)}$$

61. $\frac{b+2}{b^2+b-2}+\frac{2}{b^2-1}$
$$= \frac{b+2}{(b+2)(b-1)}+\frac{2}{(b+1)(b-1}$$
$$= \frac{1}{b-1}+\frac{2}{(b+1)(b-1)}$$
(LCD is $(b+1)(b-1)$)
$$= \frac{1}{b-1}\cdot\frac{b+1}{b+1}+\frac{2}{(b+1)(b-1)}$$
$$= \frac{b+3}{(b+1)(b-1)}$$

63. $\frac{y+3}{y^2-y-2}-\frac{y-1}{y^2+2y+1}$
$$= \frac{y+3}{(y+1)(y-2)}-\frac{y-1}{(y+1)^2}$$
(LCD is $(y+1)^2(y-2)$)
$$= \frac{y+3}{(y+1)(y-2)}\cdot\frac{y+1}{y+1}-\frac{y-1}{(y+1)^2}\cdot\frac{y-2}{y-2}$$
$$= \frac{(y+3)(y+1)}{(y+1)^2(y-2)}-\frac{(y-1)(y-2)}{(y+1)^2(y-2)}$$
$$= \frac{y^2+4y+3-\left(y^2-3y+2\right)}{(y+1)^2(y-2)}$$
$$= \frac{7y+1}{(y+1)^2(y-2)}$$

65. $\frac{x^2+x+2}{x^3-1}-\frac{1}{x-1}$
$$= \frac{x^2+x+2}{(x-1)\left(x^2+x+1\right)}-\frac{1}{x-1}$$
(LCD is $(x-1)\left(x^2+x+1\right)$)
$$= \frac{x^2+x+2}{(x-1)\left(x^2+x+1\right)}-\frac{1}{x-1}\cdot\frac{x^2+x+1}{x^2+x+1}$$
$$= \frac{x^2+x+2-\left(x^2+x+1\right)}{(x-1)\left(x^2+x+1\right)}$$
$$= \frac{1}{(x-1)\left(x^2+x+1\right)}$$

67. $\frac{1}{y-x}+\frac{1}{x-y}$
$$= \frac{1}{y-x}+\frac{1}{x-y}\cdot\frac{-1}{-1}$$
$$= \frac{1}{y-x}+\frac{-1}{-x+y}$$
$$= \frac{1}{y-x}-\frac{1}{y-x}$$
$$= \frac{1-1}{y-x}$$
$$= \frac{0}{y-x}$$
$$= 0$$

69. $\frac{y^2}{y-7}+\frac{6y+7}{7-y}$
$$= \frac{y^2}{y-7}+\frac{6y+7}{7-y}\cdot\frac{-1}{-1}$$
$$= \frac{y^2}{y-7}+\frac{(-1)(6y+7)}{y-7}$$
$$= \frac{y^2-6y-7}{y-7}$$
$$= \frac{(y-7)(y+1)}{y-7}$$
$$= y+1$$

71. $\frac{x}{1}-\frac{3}{x-2}$
$$= \frac{x}{1}\cdot\frac{x-2}{x-2}-\frac{3}{x-2}$$
$$= \frac{x(x-2)-3}{x-2}$$
$$= \frac{x^2-2x-3}{x-2}$$

73. $\frac{x+3y}{x^2-7xy+12y^2}-\frac{x-3y}{x^2-xy-12y^2}$
$$= \frac{x+3y}{(x-4y)(x-3y)}-\frac{x-3y}{(x-4y)(x+3y)}$$
(LCD is $(x-4y)(x-3y)(x+3y)$)

$$= \frac{x+3y}{(x-4y)(x-3y)} \cdot \frac{(x+3y)}{(x+3y)} - \frac{x-3y}{(x-4y)(x+3y)} \cdot \frac{(x-3y)}{(x-3y)}$$

$$= \frac{(x+3y)(x+3y)-(x-3y)(x-3y)}{(x-4y)(x-3y)(x+3y)}$$

$$= \frac{x^2+6xy+9y^2-\left(x^2-6xy+9y^2\right)}{(x-4y)(x-3y)(x+3y)}$$

$$= \frac{12xy}{(x-4y)(x-3y)(x+3y)}$$

75. $\dfrac{\frac{3}{y}}{y-\frac{1}{y}}$

$$= \frac{\frac{3}{y}}{y-\frac{1}{y}} \cdot \frac{y}{y}$$

$$= \frac{\frac{3}{y} \cdot y}{y \cdot y - \frac{1}{y} \cdot y}$$

$$= \frac{3}{y^2-1}$$

77. $\dfrac{3-\frac{2}{b}}{\frac{2}{b}+\frac{3}{b}}$

$$= \frac{3-\frac{2}{b}}{\frac{2}{b}+\frac{3}{b}} \cdot \frac{b}{b}$$

$$= \frac{3b-2}{2+3}$$

$$= \frac{3b-2}{5}$$

79. $\dfrac{\frac{1}{y}+\frac{1}{y^2}}{1+\frac{1}{y}} \cdot \dfrac{y^2}{y^2}$

$$= \frac{\frac{1}{y} \cdot y^2 + \frac{1}{y^2} \cdot y^2}{1 \cdot y^2 + \frac{1}{y} \cdot y^2}$$

$$= \frac{y+1}{y^2+y}$$

$$= \frac{y+1}{y(y+1)}$$

$$= \frac{1}{y}$$

81. $\dfrac{\frac{x}{y}+\frac{y}{x}}{\frac{1}{y}+\frac{1}{x}} \cdot \dfrac{xy}{xy}$

$$= \frac{\frac{x}{y} \cdot xy + \frac{y}{x} \cdot xy}{\frac{1}{y} \cdot xy + \frac{1}{x} \cdot xy}$$

$$= \frac{x^2+y^2}{x+y}$$

83. $\dfrac{\frac{b^2-c^2}{b}}{\frac{b-c}{b^2}}$

$$= \frac{\frac{b^2-c^2}{b}}{\frac{b-c}{b^2}} \cdot \frac{b^2}{b^2}$$

$$= \frac{b\left(b^2-c^2\right)}{b-c}$$

$$= \frac{b(b+c)(b-c)}{b-c}$$

$$= b(b+c)$$

85. $\dfrac{\frac{x}{y}-\frac{y}{x}}{\frac{x^2}{y}-y}$

$$= \frac{\frac{x}{y}-\frac{y}{x}}{\frac{x^2}{y}-y} \cdot \frac{xy}{xy}$$

$$= \frac{x^2-y^2}{x^3-xy^2}$$

$$= \frac{x^2-y^2}{x\left(x^2-y^2\right)}$$

$$= \frac{1}{x}$$

87. $\dfrac{y+5+\frac{6}{y}}{y-\frac{9}{y}}$

$$= \frac{y+5+\frac{6}{y}}{y-\frac{9}{y}} \cdot \frac{y}{y}$$

$$= \frac{y^2+5y+6}{y^2-9}$$

$$= \frac{(y+3)(y+2)}{(y+3)(y-3)}$$
$$= \frac{y+2}{y-3}$$

89. $\dfrac{1-\frac{1}{a}}{\frac{a+1}{a}}$
$$= \frac{1-\frac{1}{a}}{\frac{a+1}{a}} \cdot \frac{a}{a}$$
$$= \frac{a-1}{a+1}$$

91. $\dfrac{\frac{1}{x}+\frac{1}{y}}{x+y}$
$$= \frac{\frac{1}{x}+\frac{1}{y}}{x+y} \cdot \frac{xy}{xy}$$
$$= \frac{y+x}{(x+y)xy}$$
$$= \frac{1}{xy}$$

93. $\dfrac{b^2-c^2}{\frac{1}{b}+\frac{1}{c}}$
$$= \frac{b^2-c^2}{\frac{1}{b}+\frac{1}{c}} \cdot \frac{bc}{bc}$$
$$= \frac{bc\left(b^2-c^2\right)}{c+b}$$
$$= \frac{bc(b+c)(b-c)}{c+b}$$
$$= bc(b-c)$$

95. $\dfrac{3-\frac{1}{c}}{3c-1}$
$$= \frac{3-\frac{1}{c}}{3c-1} \cdot \frac{c}{c}$$
$$= \frac{3c-1}{(3c-1)c}$$
$$= \frac{1}{c}$$

97. $\dfrac{b-3}{b-\frac{3}{b-2}}$
$$= \frac{b-3}{b-\frac{3}{b-2}} \cdot \frac{b-2}{b-2}$$
$$= \frac{(b-3)(b-2)}{b(b-2)-3}$$
$$= \frac{(b-3)(b-2)}{b^2-2b-3}$$
$$= \frac{(b-3)(b-2)}{(b+1)(b-3)}$$
$$= \frac{b-2}{b+1}$$

99. $\dfrac{y+\frac{12}{y-7}}{y-3}$
$$= \frac{y+\frac{12}{y-7}}{y-3} \cdot \frac{y-7}{y-7}$$
$$= \frac{y(y-7)+12}{(y-3)(y-7)}$$
$$= \frac{y^2-7y+12}{(y-3)(y-7)}$$
$$= \frac{(y-4)(y-3)}{(y-3)(y-7)}$$
$$= \frac{y-4}{y-7}$$

101. $\dfrac{\frac{3}{y-2}-\frac{4}{y+2}}{\frac{7}{y^2-4}}$
$$= \frac{\frac{3}{y-2}-\frac{4}{y+2}}{\frac{7}{(y+2)(y-2)}} \cdot \frac{(y+2)(y-2)}{(y+2)(y-2)}$$
$$= \frac{3(y+2)-4(y-2)}{7}$$
$$= \frac{-y+14}{7}$$

103. $\dfrac{\frac{4}{b-2}+1}{\frac{3}{b^2-4}+1}$
$$= \frac{\frac{4}{b-2}+1}{\frac{3}{(b+2)(b-2)}+1} \cdot \frac{(b+2)(b-2)}{(b+2)(b-2)}$$
$$= \frac{4(b+2)+(b+2)(b-2)}{3+(b+2)(b-2)}$$

$$= \frac{(b+2)[4+(b-2)]}{3+b^2-4}$$
$$= \frac{(b+2)(b+2)}{b^2-1}$$
$$= \frac{(b+2)^2}{b^2-1}$$

105. $$\frac{\frac{6}{x^2+2x-15}-\frac{1}{x-3}}{\frac{1}{x+5}+1}$$
$$= \frac{\frac{6}{(x+5)(x-3)}-\frac{1}{x-3}}{\frac{1}{x+5}+1} \cdot \frac{(x+5)(x-3)}{(x+5)(x-3)}$$
$$= \frac{6-(x+5)}{x-3+(x+5)(x-3)}$$
$$= \frac{6-x-5}{(x-3)[1+(x+5)]}$$
$$= \frac{-x+1}{(x-3)(x+6)}$$

107. $$\frac{\frac{3}{x+2y}-\frac{2y}{x^2+2xy}}{\frac{3y}{x^2+2xy}+\frac{5}{x}}$$
$$= \frac{\frac{3}{x+2y}-\frac{2y}{x(x+2y)}}{\frac{3y}{x(x+2y)}+\frac{5}{x}} \cdot \frac{x(x+2y)}{x(x+2y)}$$
$$= \frac{3x-2y}{3y+5(x+2y)}$$
$$= \frac{3x-2y}{3y+5x+10y}$$
$$= \frac{3x-2y}{5x+13y}$$

109. a. $t = \frac{24}{10-x} + \frac{24}{10+x}$

b. LCD is $100-x^2$
$$t = \frac{24}{10-x} \cdot \frac{10+x}{10+x} + \frac{24}{10+x} \cdot \frac{10-x}{10-x}$$
$$t = \frac{24(10+x)+24(10-x)}{(10-x)(10+x)}$$
$$t = \frac{240+24x+240-24x}{100-x^2}$$
$$t = \frac{480}{100-x^2}$$

c. $t(2) = \frac{480}{100-(2)^2}$

$t(2) = 5$

If the current is 2 miles per hour, then the total trip will take 5 hours.

111. $\frac{2d}{\frac{d}{r_1}+\frac{d}{r_2}}$ (LCD is r_1r_2)
$$= \frac{2d}{\frac{dr_2}{r_1r_2}+\frac{dr_1}{r_1r_2}}$$
$$= \frac{2d}{\frac{dr_2+dr_1}{r_1r_2}}$$
$$= \frac{2dr_1r_2}{dr_2+dr_1}$$
$$= \frac{d(2r_1r_2)}{d(r_2+r_1)}$$
$$= \frac{2r_1r_2}{r_2+r_1}$$
$$\frac{2(30)(20)}{20+30}$$
$= 24$

The average speed is 24 miles per hour. Explanations may vary.

113. d is true.

115. a. $2x + 2\left(\frac{2500}{x}\right)$
$$= \frac{2x^2}{x} + \frac{5000}{x}$$
$$= \frac{2x^2+5000}{x}$$

b.

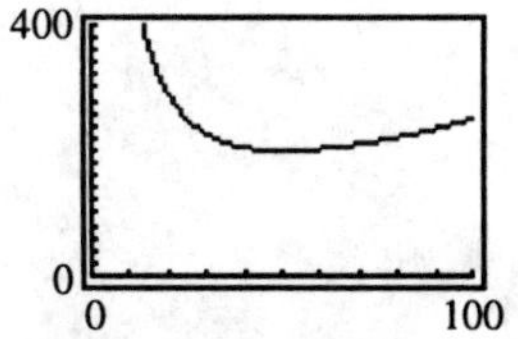

c. $(50, 200)$

d. $\frac{2500}{50} = 50$ feet square

117–123. Answers may vary.

125. **a.** $x = 1,\ y = 2,\ z = 3$

$$\frac{y-z}{1+yz} = \frac{2-3}{1+6} = -\frac{1}{7}$$

$$\frac{z-x}{1+xz} = \frac{3-1}{1+3} = \frac{2}{4} = \frac{1}{2}$$

$$\frac{x-y}{1+xy} = \frac{1-2}{1+2} = -\frac{1}{3}$$

$$-\frac{1}{7} + \frac{1}{2} - \frac{1}{3} = \frac{1}{42}$$

$$\left(-\frac{1}{7}\right)\left(\frac{1}{2}\right)\left(-\frac{1}{3}\right) = \frac{1}{42}$$

They are equal.

b. $$\frac{y-z}{1+yz} + \frac{z-x}{1+xz} + \frac{x-y}{1+xy}$$

$$= \frac{(y-z)(1+xz)(1+xy) + (z-x)(1+yz)(1+xy) + (x-y)(1+yz)(1+xz)}{(1+yz)(1+xz)(1+xy)}$$

$$= \frac{xy^2 - xz^2 + yz^2 - x^2y + x^2z - y^2z}{(1+yz)(1+xz)(1+xy)}$$

$$= \frac{(y-z)(z-x)(x-y)}{(1+yz)(1+xz)(1+xy)}$$

$$= \left(\frac{y-z}{1+yz}\right)\left(\frac{z-x}{1+xz}\right)\left(\frac{x-y}{1+xy}\right)$$

127. $(x-y)^{-1} + (x+y)^{-1}$

$$= \frac{1}{x-y} + \frac{1}{x+y}$$

$$= \frac{1}{x-y} \cdot \frac{x+y}{x+y} + \frac{1}{x+y} \cdot \frac{x-y}{x-y}$$

$$= \frac{x+y+x-y}{x^2-y^2}$$

$$= \frac{2x}{x^2-y^2}$$

129. Area of first shaded region = area of rectangle – area of semicircle

$$= (x+2) \cdot \frac{2}{x} - \frac{1}{2}\pi\left(\frac{2}{x}\right)^2$$

$$= \frac{2(x+2)}{x} - \frac{2\pi}{x^2}$$

Area of second shaded region = area of first shaded region + area of semicircle

$$= \frac{2(x+2)}{x} - \frac{2\pi}{x^2} + \frac{2\pi}{x^2}$$

$$= \frac{2(x+2)}{x}$$

Sum of areas

$$= \frac{2(x+2)}{x} - \frac{2\pi}{x^2} + \frac{2(x+2)}{x}$$

$$= \frac{4(x+2)}{x} \cdot \frac{x}{x} - \frac{2\pi}{x^2}$$

$$= \frac{4x^2 + 8x - 2\pi}{x^2}$$

131. a. Harmonic mean $= \dfrac{1}{\frac{\frac{1}{x}+\frac{1}{y}}{2}}$

$= \dfrac{2}{\frac{1}{x}+\frac{1}{y}} \cdot \dfrac{xy}{xy}$

$= \dfrac{2xy}{y+x}$

$= \dfrac{2xy}{x+y}$

b.

1260		
$a = 504$		360
315	280	$b = 252$

$a = \dfrac{2(1260)(315)}{1260+315} = 504$

b: Since 280 is the harmonic mean of 315 and b:

$\dfrac{2(315)b}{315+b} = 280$

$630b = 280(315+b)$

$630b = 280b + 88{,}200$

$350b = 88{,}200$

$b = 252$

Using similar methods, the table can be completed giving this square:

1260	840	630
504	420	360
315	280	252

Review Problems

133. $|3x-1| \le 14$

$3x - 1 \le 14$ and $3x - 1 \ge -14$

$3x \le 15$ and $3x \ge -13$

$x \le 5$ and $x \ge -\dfrac{13}{3}$

$\left\{x \middle| -\dfrac{13}{3} \le x \le 5\right\}$

134. $2x^4 - 9x^2 = x^2\left(2x^2 - 9\right)$

135. $y^2 + 27 = 12y$

$y^2 - 12y + 27 = 0$

$(y-9)(y-3) = 0$

$y - 9 = 0$ or $y - 3 = 0$

$y = 9$ or $y = 3$

$\{3, 9\}$

Problem Set 5.3

1. $y^7 \div y^5 = \dfrac{y^7}{y^5} = y^{7-5} = y^2$

3. $\dfrac{15x^8}{5x^6} = 3x^{8-6} = 3x^2$

5. $\dfrac{25x^3y^7}{-5xy^5} = -5x^{3-1}y^{7-5} = -5x^2y^2$

7. $\left(-54x^7y^4z^5\right) \div \left(3x^3yz^2\right)$

$= \dfrac{-54x^7y^4z^5}{3x^3yz^2}$

$= -18x^4y^3z^3$

9. $\dfrac{24x^7 - 15x^4 + 18x^3}{3x}$

$= \dfrac{24x^7}{3x} - \dfrac{15x^4}{3x} + \dfrac{18x^3}{3x}$

$= 8x^6 - 5x^3 + 6x^2$

11. $\dfrac{28x^3 - 14x^2 - 35x}{7x}$

$= \dfrac{28x^3}{7x} - \dfrac{14x^2}{7x} - \dfrac{35x}{7x}$

$= 4x^2 - 2x - 5$

13. $\dfrac{18x^6 - 12x^4 - 36x^2}{-6x}$

$= \dfrac{18x^6}{-6x} - \dfrac{12x^4}{-6x} - \dfrac{36x^2}{-6x}$

$= -3x^5 + 2x^3 + 6x$

15. $\dfrac{6x^7-3x^4+x^2-5x+2}{3x^3}$

$=\dfrac{6x^7}{3x^3}-\dfrac{3x^4}{3x^3}+\dfrac{x^2}{3x^3}-\dfrac{5x}{3x^3}+\dfrac{2}{3x^3}$

$=2x^4-x+\dfrac{1}{3x}-\dfrac{5}{3x^2}+\dfrac{2}{3x^3}$

17. $\dfrac{x^2y-x^3y^3-x^5y^5}{x^2y}$

$=\dfrac{x^2y}{x^2y}-\dfrac{x^3y^3}{x^2y}-\dfrac{x^5y^5}{x^2y}$

$=1-xy^2-x^3y^4$

19. $\dfrac{16x^3y^2-28x^2y^3-20x^2y^5}{4x^2y^2}$

$=\dfrac{16x^3y^2}{4x^2y^2}-\dfrac{28x^2y^3}{4x^2y^2}-\dfrac{20x^2y^5}{4x^2y^2}$

$=4x-7y-5y^3$

21. $\dfrac{36x^4y^3-18x^3y^2-12x^2y}{6x^3y^3}$

$=\dfrac{36x^4y^3}{6x^3y^3}-\dfrac{18x^3y^2}{6x^3y^3}-\dfrac{12x^2y}{6x^3y^3}$

$=6x-\dfrac{3}{y}-\dfrac{2}{xy^2}$

23. $\dfrac{8x^4y^2-6xy^3-14x^3y-12xy}{-4x^2y^2}$

$=\dfrac{8x^4y^2}{-4x^2y^2}-\dfrac{6xy^3}{-4x^2y^2}-\dfrac{14x^3y}{-4x^2y^2}-\dfrac{12xy}{-4x^2y^2}$

$=-2x^2+\dfrac{3y}{2x}+\dfrac{7x}{2y}+\dfrac{3}{xy}$

25. $\left(a^2+8a+15\right)\div(a+5)=(a+3)$

$$\begin{array}{r|l} & a+3 \\ \hline a+5 & a^2+8a+15 \\ & \underline{a^2+5a} \\ & \quad 3a+15 \\ & \quad \underline{3a+15} \\ & \qquad 0 \end{array}$$

27. $\left(b^2-4b-12\right)\div(b-6)=b+2$

$$\begin{array}{r|l} & b+2 \\ \hline b-6 & b^2-4b-12 \\ & \underline{b^2-6b} \\ & \quad 2b-12 \\ & \quad \underline{2b-12} \\ & \qquad 0 \end{array}$$

29. $\left(24-10c-c^2\right)\div(2-c)=c+12$

$$\begin{array}{r|l} & c+12 \\ \hline -c+2 & -c^2-10c+24 \\ & \underline{-c^2+2c} \\ & \quad -12c+24 \\ & \quad \underline{-12c+24} \\ & \qquad 0 \end{array}$$

31. $\left(b^3-2b^2-5b+6\right)\div(b-3)=b^2+b-2$

$$\begin{array}{r|l} & b^2+b-2 \\ \hline b-3 & b^3-2b^2-5b+6 \\ & \underline{b^3-3b^2} \\ & \quad b^2-5b \\ & \quad \underline{b^2-3b} \\ & \qquad -2b+6 \\ & \qquad \underline{-2b+6} \\ & \qquad\quad 0 \end{array}$$

33. $\left(6b^3+17b^2+27b+20\right)\div(3b+4)$

$=2b^2+3b+5$

$$\begin{array}{r|l} & 2b^2+3b+5 \\ \hline 3b+4 & 6b^3+17b^2+27b+20 \\ & \underline{6b^3+8b^2} \\ & \quad 9b^2+27b \\ & \quad \underline{9b^2+12b} \\ & \qquad 15b+20 \\ & \qquad \underline{15b+20} \\ & \qquad\quad 0 \end{array}$$

35. $\left(a^3+3a^2b+2ab^2\right)\div(a+b)=a^2+2ab$

$$\begin{array}{r|l} & a^2+2ab \\ \hline a+b & a^3+3a^2b+2ab^2 \\ & \underline{a^3+a^2b} \\ & 2a^2b+2ab^2 \\ & \underline{2a^2b+2ab^2} \\ & 0 \end{array}$$

37. $\left(12x^2+x-4\right)\div(3x-2)=4x+3+\frac{2}{3x-2}$

$$\begin{array}{r|l} & 4x+3 \\ \hline 3x-2 & 12x^2+x-4 \\ & \underline{12x^2-8x} \\ & 9x-4 \\ & \underline{9x-6} \\ & 2 \end{array}$$

39. $\frac{2y^3+7y^2+9y-20}{y+3}=2y^2+y+6-\frac{38}{y+3}$

$$\begin{array}{r|l} & 2y^2+y+6 \\ \hline y+3 & 2y^3+7y^2+9y-20 \\ & \underline{2y^3+6y^2} \\ & y^2+9y \\ & \underline{y^2+3y} \\ & 6y-20 \\ & \underline{6y+18} \\ & -38 \end{array}$$

41. $\frac{4x^4-4x^2+6x}{x-4}$
$=4x^3+16x^2+60x+246+\frac{984}{x-4}$

$$\begin{array}{r|l} & 4x^3+16x^2+60x+246 \\ \hline x-4 & 4x^4+0x^3-4x^2+6x+0 \\ & \underline{4x^4-16x^3} \\ & 16x^3-4x^2 \\ & \underline{16x^3-64x^2} \\ & 60x^2+6x \\ & \underline{60x^2-240x} \\ & 246x+0 \\ & \underline{246x-984} \\ & 984 \end{array}$$

43. $\frac{x^3-1}{x-1}=x^2+x+1$

$$\begin{array}{r|l} & x^2+x+1 \\ \hline x-1 & x^3+0x^2+0x-1 \\ & \underline{x^3-x^2} \\ & x^2+0x \\ & \underline{x^2-x} \\ & x-1 \\ & \underline{x-1} \\ & 0 \end{array}$$

45. $\frac{6a^3+13a^2-11a-15}{3a^2-a-3}=2a+5$

$$\begin{array}{r|l} & 2a+5 \\ \hline 3a^2-a-3 & 6a^3+13a^2-11a-15 \\ & \underline{6a^3-2a^2-6a} \\ & 15a^2-5a-15 \\ & \underline{15a^2-5a-15} \\ & 0 \end{array}$$

47. $\frac{y^4+y^3-3y^2-y+2}{y^2+3y+2}=y^2-2y+1$

$$\begin{array}{r|l} & y^2-2y+1 \\ \hline y^2+3y+2 & y^4+y^3-3y^2-y+2 \\ & \underline{y^4+3y^3+2y^2} \\ & -2y^3-5y^2-y \\ & \underline{-2y^3-6y^2-4y} \\ & y^2+3y+2 \\ & \underline{y^2+3y+2} \\ & 0 \end{array}$$

49. $\dfrac{18y^4+9y^3+3y^2}{3y^2+1}$

$=6y^2+3y-1+\dfrac{-3y+1}{3y^2+1}$

$$\begin{array}{r}
6y^2+3y-1 \\
3y^2+1\overline{)18y^4+9y^3+3y^2+0y+0} \\
\underline{18y^4\qquad\;\;+6y^2}\qquad\qquad \\
9y^3-3y^2+0y\quad \\
\underline{9y^3\qquad\;\;+3y}\quad \\
-3y^2-3y+0 \\
\underline{-3y^2\qquad\;\;-1} \\
-3y+1
\end{array}$$

51. $\left(2x^2+x-10\right)\div(x-2)=2x+5$

$$\begin{array}{r|rrr}
2 & 2 & 1 & -10 \\
 & & 4 & +10 \\
\hline
 & 2 & 5 & 0
\end{array}$$

53. $\left(3x^2+7x-20\right)\div(x+5)=3x-8+\dfrac{20}{x+5}$

$$\begin{array}{r|rrr}
-5 & 3 & 7 & -20 \\
 & & -15 & 40 \\
\hline
 & 3 & -8 & 20
\end{array}$$

55. $\left(4x^3-3x^2+3x-1\right)\div(x-1)$

$=4x^2+x+4+\dfrac{3}{x-1}$

$$\begin{array}{r|rrrr}
1 & 4 & -3 & 3 & -1 \\
 & & 4 & 1 & 4 \\
\hline
 & 4 & 1 & 4 & 3
\end{array}$$

57. $\left(6y^5-2y^3+4y^2-3y+1\right)\div(y-2)$

$=6y^4+12y^3+22y^2+48y+93+\dfrac{187}{y-2}$

$$\begin{array}{r|rrrrrr}
2 & 6 & 0 & -2 & 4 & -3 & 1 \\
 & & 12 & 24 & 44 & 96 & 186 \\
\hline
 & 6 & 12 & 22 & 48 & 93 & 187
\end{array}$$

59. $\left(x^2-5x-5x^3+x^4\right)\div(5+x)$

$=\left(x^4-5x^3+x^2-5x\right)\div(x+5)$

$=x^3-10x^2+51x-260+\dfrac{1300}{x+5}$

$$\begin{array}{r|rrrrr}
-5 & 1 & -5 & 1 & -5 & 0 \\
 & & -5 & 50 & -255 & 1300 \\
\hline
 & 1 & -10 & 51 & -260 & 1300
\end{array}$$

61. $\dfrac{z^5+z^3-2}{z-1}=z^4+z^3+2z^2+2z+2$

$$\begin{array}{r|rrrrrr}
1 & 1 & 0 & 1 & 0 & 0 & -2 \\
 & & 1 & 1 & 2 & 2 & 2 \\
\hline
 & 1 & 1 & 2 & 2 & 2 & 0
\end{array}$$

63. $\dfrac{y^4-256}{y-4}=y^3+4y^2+16y+64$

$$\begin{array}{r|rrrrr}
4 & 1 & 0 & 0 & 0 & -256 \\
 & & 4 & 16 & 64 & 256 \\
\hline
 & 1 & 4 & 16 & 64 & 0
\end{array}$$

65. $\dfrac{2y^5-3y^4+y^3-y^2+2y-1}{y+2}$

$=2y^4-7y^3+15y^2-31y+64-\dfrac{129}{y+2}$

$$\begin{array}{r|rrrrrr}
-2 & 2 & -3 & 1 & -1 & 2 & -1 \\
 & & -4 & 14 & -30 & 62 & -128 \\
\hline
 & 2 & -7 & 15 & -31 & 64 & -129
\end{array}$$

67. $\dfrac{16t^4+8t^3}{2t^2}=\dfrac{16t^4}{2t^2}+\dfrac{8t^3}{2t^2}=8t^2+4t$

$t=4$: $8\left(4^2\right)+4(4)=128+16=144$

144 gallons

69. Area = length × width

length = $\dfrac{\text{area}}{\text{width}}$

length = $\dfrac{6x^2+11x-35}{2x+7}=3x-5$

71. $V = \frac{1}{3}Bh$

$V = 8y^3 - 12y^2 - 18y - 5$

$B = 12y^2 + 12y + 3$

$8y^3 - 12y^2 - 18y - 5 = \frac{1}{3}\left(12y^2 + 12y + 3\right)h$

$8y^3 - 12y^2 - 18y - 5 = \left(4y^2 + 4y + 1\right)h$

$\frac{8y^3 - 12y^2 - 18y - 5}{4y^2 + 4y + 1} = h$

$$\begin{array}{r} 2y - 5 \\ 4y^2 + 4y + 1 \overline{)\, 8y^3 - 12y^2 - 18y - 5} \\ \underline{8y^3 + 8y^2 + 2y} \\ -20y^2 - 20y - 5 \\ \underline{-20y^2 - 20y - 5} \\ 0 \end{array}$$

$h = 2y - 5$

73. c is true.

75. $\frac{x^4 + 6x^3 + 6x^2 - 10x - 3}{x^2 + 2x - 3}$

$= x^2 + 4x + 1, \ x \neq -3, \ x \neq 1$

$y = x^2 + 4x + 1$

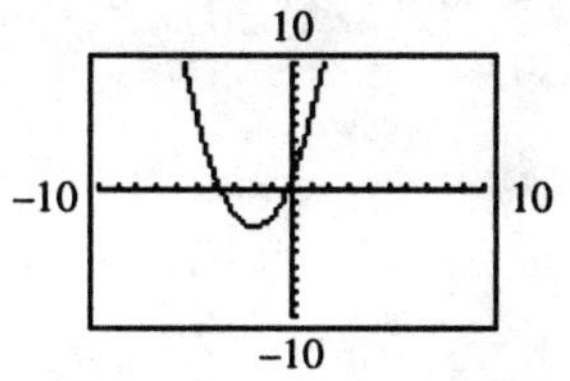

77. $\frac{3x^4 + 4x^3 - 32x^2 - 5x - 20}{x + 4}$

$= 3x^3 - 8x^2 - 5, \ x \neq -4$

$y = 3x^3 - 8x^2 - 5$

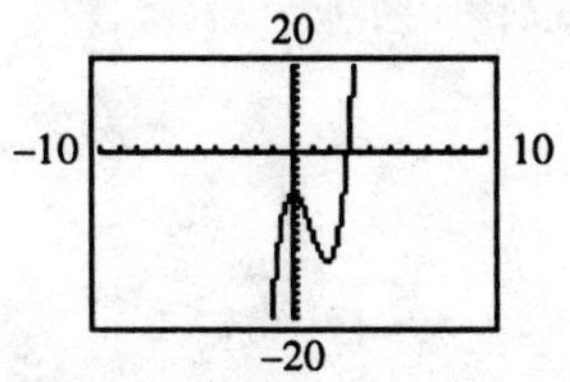

79–81. Answers may vary.

83. $\frac{2x^2 - 7x + 9}{\text{(polynomial)}} = 2x - 3 + \frac{3}{\text{(polynomial)}}$

$2x^2 - 7x + 9 = (2x - 3)\text{(polynomial)} + 3$

$2x^2 - 7x + 6 = (2x - 3)\text{(polynomial)}$

$\frac{2x^2 - 7x + 6}{2x - 3} = \text{(polynomial)}$

$$\begin{array}{r|rrr} \frac{3}{2} & 2 & -7 & 6 \\ & & 3 & -6 \\ \hline & 2 & -4 & 0 \\ (\div 2) & 1 & -2 & 0 \end{array} \quad \text{or} \quad \begin{array}{r} x - 2 \\ 2x - 3 \overline{)\, 2x^2 - 7x + 6} \\ \underline{2x^2 - 3x} \\ -4x + 6 \\ \underline{-4x + 6} \\ 0 \end{array}$$

polynomial $= x - 2$

85. $\frac{27y^{3n} - 1}{3y^n - 1} = 9y^{2n} + 3y^n + 1$

$$\begin{array}{r} 9y^{2n} + 3y^n + 1 \\ 3y^n - 1 \overline{)\, 27y^{3n} + 0y^{2n} + 0y^n - 1} \\ \underline{27y^{3n} - 9y^{2n}} \\ 9y^{2n} + 0y^n \\ \underline{9y^{2n} - 3y^n} \\ 3y^n - 1 \\ \underline{3y^n - 1} \\ 0 \end{array}$$

87. $\left(18x^2 + 27x + K\right) \div (6x + 5)$

$= 3x + 2 + \frac{K - 10}{6x + 5}$

$$\begin{array}{r} 3x + 2 \\ 6x + 5 \overline{)\, 18x^2 + 27x + K} \\ \underline{18x^2 + 15x} \\ 12x + K \\ \underline{12x + 10} \\ K - 10 \end{array}$$

$K - 10 = 0$ for a zero remainder

$K = 10$

89. $\left(-2x^3+3x^2+x+K\right)\div(x+1)$

$=-2x^2+5x-4+\dfrac{K+4}{x+1}$

$$\begin{array}{r|rrrr} -1 & -2 & 3 & 1 & K \\ & & 2 & -5 & 4 \\ \hline & -2 & 5 & -4 & K+4 \end{array}$$

The remainder is 3 when $K+4=3$.

$K=-1$

Review Problems

91. $3x+5y=0$

$x+4y=7$

$D=\begin{vmatrix} 3 & 5 \\ 1 & 4 \end{vmatrix}=12-5=7$

$D_x=\begin{vmatrix} 0 & 5 \\ 7 & 4 \end{vmatrix}=0-35=-35$

$D_y=\begin{vmatrix} 3 & 0 \\ 1 & 7 \end{vmatrix}=21-0=21$

$x=\dfrac{D_x}{D}=-\dfrac{35}{7}=-5$

$y=\dfrac{D_y}{D}=\dfrac{21}{7}=3$

$\{(-5, 3)\}$

92. $16x^3+250$

$=2\left(8x^3+125\right)$

$=2(2x+5)\left(4x^2-10x+25\right)$

93. $3x-y+3z=4$

$x+2y+z=-1$

$2x-3y+z=1$

(Equations 1 and 2):

$$\begin{array}{rr} & 3x-y+3z=4 \\ (\times -3) & -3x-6y-3z=3 \\ \hline & -7y=7 \\ & y=-1 \end{array}$$

(Equations 2 and 3):

$$\begin{array}{rr} (\times -1) & -x-2y-z=1 \\ & 2x-3y+z=1 \\ \hline & x-5y=2 \\ & x-5(-1)=2 \\ & x=-3 \end{array}$$

(Equation 2):

$x+2y+z=-1$

$-3+2(-1)+z=-1$

$z=4$

$\{(-3, -1, 4)\}$

Problem Set 5.4

1. $\dfrac{2}{3x}+\dfrac{1}{4}=\dfrac{11}{6x}-\dfrac{1}{3}$

$12x\left(\dfrac{2}{3x}+\dfrac{1}{4}\right)=12x\left(\dfrac{11}{6x}-\dfrac{1}{3}\right)$

$12x\cdot\dfrac{2}{3x}+12x\cdot\dfrac{1}{4}=12x\cdot\dfrac{11}{6x}-12x\cdot\dfrac{1}{3}$

$8+3x=22-4x$

$7x=14$

$x=2$

$\{2\}$

3. $\dfrac{4}{5}+\dfrac{7}{2a}=\dfrac{13}{2a}-\dfrac{4}{20}$

$20a\left(\dfrac{4}{5}+\dfrac{7}{2a}\right)=20a\left(\dfrac{13}{2a}-\dfrac{4}{20}\right)$

$16a+70=130-4a$

$20a=60$

$a=3$

$\{3\}$

5. $\dfrac{2}{y-3}+\dfrac{3y+1}{y+3}=3$

$(y-3)(y+3)\left[\dfrac{2}{y-3}+\dfrac{3y+1}{y+3}\right]=(y-3)(y+3)\cdot 3$

$2(y+3)+(3y+1)(y-3)=3(y-3)(y+3)$

$2y+6+3y^2-8y-3=3y^2-27$

$-6y+3=-27$

$-6y=-30$

$y=5$

$\{5\}$

7. $\frac{2x-4}{x-4}-2=\frac{20}{x+4}$

$(x-4)(x+4)\left[\frac{2x-4}{x-4}-2\right]=(x-4)(x+4)\frac{20}{x+4}$

$(x+4)(2x-4)-2(x-4)(x+4)=20(x-4)$

$2x^2+4x-16-2x^2+32=20x-80$

$4x+16=20x-80$

$-16x=-96$

$x=6$

$\{6\}$

9. $\frac{3c}{c+1}=\frac{5c}{c-1}-2$

$(c+1)(c-1)\frac{3c}{c+1}=(c+1)(c-1)\left[\frac{5c}{c-1}-2\right]$

$3c(c-1)=5c(c+1)-2(c+1)(c-1)$

$3c^2-3c=5c^2+5c-2c^2+2$

$3c^2-3c=3c^2+5c+2$

$-3c=5c+2$

$-8c=2$

$c=\frac{2}{-8}$

$c=-\frac{1}{4}$

$\left\{-\frac{1}{4}\right\}$

11. $\frac{x+5}{x^2-4}-\frac{3}{2x-4}=\frac{1}{2x+4}$

$\frac{x+5}{(x+2)(x-2)}-\frac{3}{2(x-2)}=\frac{1}{2(x+2)}$

$2(x+2)(x-2)\left[\frac{x+5}{(x+2)(x-2)}-\frac{3}{2(x-2)}\right]=2(x+2)(x-2)\cdot\frac{1}{2(x+2)}$

$2(x+5)+3(x+2)=x-2$

$2x+10-3x-6=x-2$

$-x+4=x-2$

$6=2x$

$3=x$

$\{3\}$

13. $\frac{4}{a+2}+\frac{2}{a-4}=\frac{30}{a^2-2a-8}$

$(a+2)(a-4)\left[\frac{4}{a+2}+\frac{2}{a-4}\right]=(a+2)(a-4)\frac{30}{(a+2)(a-4)}$

$4(a-4)+2(a+2)=30$

$6a-12=30$

$6a=42$

$a=7$

$\{7\}$

15. $\frac{y+5}{y+1}-\frac{y}{y+2}=\frac{4y+1}{y^2+3y+2}$

$\frac{y+5}{y+1}-\frac{y}{y+2}=\frac{4y+1}{(y+1)(y+2)}$

$(y+1)(y+2)\left[\frac{y+5}{y+1}-\frac{y}{y+2}\right]=(y+1)(y+2)\frac{4y+1}{(y+1)(y+2)}$

$(y+2)(y+5)-y(y+1)=4y+1$

$y^2+7y+10-y^2-y=4y+1$

$6y+10=4y+1$

$2y=-9$

$y=\frac{-9}{2}$

$\left\{-\frac{9}{2}\right\}$

17. $\frac{c}{c+4}=\frac{2}{5}-\frac{4}{c+4}$

$5(c+4)\frac{c}{c+4}=5(c+4)\left[\frac{2}{5}-\frac{4}{c+4}\right]$

$5c=2(c+4)-20$

$5c=2c+8-20$

$5c=2c-12$

$3c=-12$

$c=-4$

The value –4 makes two denominators in the original equation equal to zero. Thus, –4 is not a solution. The equation has no solution.

$\varnothing$

19. $\frac{3y}{y-3}-\frac{5y}{y-3}=-2$

$(y-3)\left[\frac{3y}{y-3}-\frac{5y}{y-3}\right]=(y-3)(-2)$

$3y-5y=-2y+6$

$-2y=-2y+6$

$0=6$

This contradiction indicates that the equation has no solution.

$\varnothing$

21. $\frac{6}{x}-\frac{x}{3}=1$

$3x\left(\frac{6}{x}-\frac{x}{3}\right)=3x(1)$

$18-x^2=3x$

$0=x^2+3x-18$

$0=(x+6)(x-3)$

$x+6=0$ or $x-3=0$

$x=-6$ $\quad$ $x=3$

$\{-6, -3\}$

23. $\frac{2}{x}+\frac{9}{x+4}=1$

$x(x+4)\left(\frac{2}{x}+\frac{9}{x+4}\right)=x(x+4)\cdot 1$

$2(x+4)+9x=x(x+4)$

$2x+8+9x=x^2+4x$

$11x+8=x^2+4x$

$0=x^2-7x-8$

$0=(x+1)(x-8)$

$x+1=0 \quad \text{or} \quad x-8=0$

$x=-1 \qquad\qquad x=8$

$\{-1, 8\}$

25. $\frac{1}{x-1}+\frac{1}{x-4}=\frac{5}{4}$

$4(x-1)(x-4)\left[\frac{1}{x-1}+\frac{1}{x-4}\right]=4(x-1)(x-4)\cdot\frac{5}{4}$

$4(x-4)+4(x-1)=5(x-1)(x-4)$

$4x-16+4x-4=5x^2-25x+20$

$8x-20=5x^2-25x+20$

$0=5x^2-33x+40$

$0=(5x-8)(x-5)$

$5x-8=0 \quad \text{or} \quad x-5=0$

$x=\frac{8}{5} \qquad\qquad x=5$

$\left\{\frac{8}{5}, 5\right\}$

27. $\frac{x^2+10}{x-5}=\frac{7x}{x-5}$

$(x-5)\frac{x^2+10}{x-5}=(x-5)\frac{7x}{x-5}$

$x^2+10=7x$

$x^2-7x+10=0$

$(x-5)(x-2)=0$

$x-5=0 \quad \text{or} \quad x-2=0$

$x=5$ (Reject $x=5$ (causes division by 0)) $\qquad x=2$

$\{2\}$

29. $\frac{7}{y+5}-\frac{3}{y-1}=\frac{8}{y-6}$

$(y+5)(y-1)(y-6)\left[\frac{7}{y+5}-\frac{3}{y-1}\right]=(y+5)(y-1)(y-6)\frac{8}{y-6}$

$7(y-1)(y-6)-3(y+5)(y-6)=8(y+5)(y-1)$

$7y^2-49y+42-3y^2+3y+90=8y^2+32y-40$

$4y^2-46y+132=8y^2+32y-40$

$0=4y^2+78y-172$

$\frac{1}{2}\cdot 0=\frac{1}{2}\left(4y^2+78y-172\right)$

$0=2y^2+39y-86$

$0=(2y+43)(y-2)$

$2y+43=0$ or $y-2=0$

$y=-\frac{43}{2}$ $\quad y=2$

$\left\{-\frac{43}{2},\ 2\right\}$

31. $\frac{24}{10+y}+\frac{24}{10-y}=5$

$(10+y)(10-y)\left[\frac{24}{10+y}+\frac{24}{10-y}\right]=(10+y)(10-y)\cdot 5$

$24(10-y)+24(10+y)=5\left(100-y^2\right)$

$240-24y+240+24y=500-5y^2$

$480=500-5y^2$

$5y^2-20=0$

$y^2-4=0$

$(y+2)(y-2)=0$

$y+2=0$ or $y-2=0$

$y=-2$ $\quad y=2$

$\{-2, 2\}$

33. $\frac{x}{x-5}+\frac{17}{25-x^2}=\frac{1}{x+5}$

$(x-5)(x+5)\left[\frac{x}{x-5}+\frac{17}{25-x^2}\right]=(x-5)(x+5)\left[\frac{1}{x+5}\right]$

$x(x+5)+(-1)17=x-5$

$x^2+5x-17=x-5$

$x^2+4x-12=0$

$(x+6)(x-2)=0$

$x+6=0$ or $x-2=0$

$x=-6$ $\quad x=2$

$\{-6, 2\}$

35. $\frac{5}{y-3}=\frac{30}{y^2-9}+1$

$(y+3)(y-3)\left[\frac{5}{y-3}\right]=(y+3)(y-3)\left[\frac{30}{y^2-9}+1\right]$
$5(y+3)=30+(y+3)(y-3)$
$5y+15=30+y^2-9$
$0=y^2-5y+6$
$0=(y-3)(y-2)$

$y-3=0$ or $y-2=0$

$y=3$ Reject 3 (it causes division by zero in the original equation) $y=2$

$\{2\}$

37. $\frac{8}{x+1}+\frac{2}{1-x^2}=\frac{x}{x-1}$

$(x+1)(x-1)\left[\frac{8}{x+1}+\frac{2}{(1-x)(1+x)}\right]=\left(\frac{x}{x-1}\right)(x+1)(x-1)$
$8(x-1)+(-1)\cdot 2=x(x+1)$
$8x-8-2=x^2+x$
$0=(x-5)(x-2)$

$x-5=0$ or $x-2=0$

$x=5$ $x=2$

$\{2, 5\}$

39. $\frac{x}{x+5}+\frac{x}{5-x}=\frac{15+5x}{x^2-25}$

$(x+5)(x-5)\left[\frac{x}{x+5}+\frac{x}{5-x}\right]=\left[\frac{15+5x}{(x+5)(x-5)}\right](x+5)(x-5)$
$x(x-5)+(-1)x(x+5)=15+5x$
$x^2-5x-x^2-5x=15+5x$
$-10x=15+5x$
$-15x=15$
$x=-1$
$\{-1\}$

41. $\frac{x+2}{x^2-x}-\frac{6}{x^2-1}=0$

$x(x-1)(x+1)\left[\frac{x+2}{x(x-1)}-\frac{6}{(x-1)(x+1)}\right]=x(x-1)(x+1)\cdot 0$
$(x+1)(x+2)-6x=0$
$x^2+3x+2-6x=0$
$x^2-3x+2=0$
$(x-1)(x-2)=0$

$x-1=0$ or $x-2=0$

$x=1$ (reject; it causes division by 0) $x=2$

$\{2\}$

43. Even though this equation appears "complicated," it is not quadratic.

$$\frac{1}{x^3-8}-\frac{2}{x^2+2x+4}=\frac{3}{(2-x)\left(x^2+2x+4\right)}$$

$$(x-2)(x^2+2x+4)\left[\frac{1}{(x-2)(x^2+2x+4)}-\frac{2}{x^2+2x+4}\right]=(x-2)(x^2+2x+4)\left[\frac{3}{(2-x)(x^2+2x+4)}\right]$$

$1-2(x-2)=(-1)3$
$1-2x+4=-3$
$-2x=-8$
$x=4$
$\{4\}$

45. $5y^{-2}+1=6y^{-1}$

$\frac{5}{y^2}+1=\frac{6}{y}$

$y^2\left(\frac{5}{y^2}+1\right)=y^2\left(\frac{6}{y}\right)$

$5+y^2=6y$
$y^2-6y+5=0$
$(y-5)(y-1)=0$
$y-5=0$ or $y-1=0$
$y=5$ $\quad y=1$
$\{1, 5\}$

47. $C=\frac{4p}{100-p}$

$196=\frac{4p}{100-p}$

$196(100-p)=4p$
$19,600-196p=4p$
$19,600=200p$
$98=p$

98%

The height of the bar representing 98% is at 196 million dollars.

49. $t(x)=\frac{40}{x}+\frac{40}{x+30}$

$5=\frac{40}{x}+\frac{40}{x+30}$

$5=\frac{40(x+30)+40x}{x(x+30)}$

$5x(x+30)=40x+1200+40x$
$5x^2+150x=80x+1200$
$5x^2+70x-1200=0$
$5\left(x^2+14x-240\right)=0$
$5(x+24)(x-10)=0$
$x=-24$ or $x=10$

Disregard a negative speed. The outgoing speed is 10 miles per hour.

51. $C(t)=\frac{30t}{t^2+2}$

$10=\frac{30t}{t^2+2}$

$10\left(t^2+2\right)=30t$
$10t^2+20=30t$
$10t^2-30t+20=0$
$10\left(t^2-3t+2\right)=0$
$10(t-2)(t-1)=0$
$t=2$ or $t=1$

After 1 minute and after 2 minutes.

53. $P=20-\frac{4}{t+1}$

$P = 19$, since P is given in thousands.

$19=20-\frac{4}{t+1}$

$(t+1)19=(t+1)\left[20-\frac{4}{t+1}\right]$

$19t+19=20(t+1)-4$
$19t + 19 = 20t + 16$
$3=t$

The population will be 19,000 in 3 years after 1990, or in 1993.

55. $y = \dfrac{100}{x}$

$(y = 4 \text{ milligrams})$:

$4 = \dfrac{100}{x}$

$4x = 100$

$x = 25$

25 species per thousand individuals.

57. $C = \dfrac{4p}{100 - p}$

$C(100 - p) = 4p$

$100C - Cp = 4p$

$100C = Cp + 4p$

$100C = p(C + 4)$

$\dfrac{100C}{C+4} = p$

$p = \dfrac{100C}{C+4}$

$p = \dfrac{100(196)}{196+4}$

$p = 98$

59. a. $t = 6 + \dfrac{20}{n+2}$, for n

$t = \dfrac{6(n+2)+20}{n+2}$

$t(n+2) = 6n + 12 + 20$

$nt + 2t = 6n + 32$

$nt - 6n = 32 - 2t$

$n(t-6) = 32 - 2t$

$n = \dfrac{32 - 2t}{t - 6}$

b. $n = \dfrac{32 - 2(7)}{7 - 6}$

$n = \dfrac{18}{1}$

$n = 18$

18 trials

61. $\dfrac{1}{p} + \dfrac{1}{q} = \dfrac{1}{f}$, for q

$\dfrac{q+p}{pq} = \dfrac{1}{f}$

$f(q + p) = pq$

$fq + fp = pq$

$fq - pq = -fp$

$q(f - p) = -fp$

$q = \dfrac{-fp}{f - p}$ or $\dfrac{fp}{p - f}$

63. $F = \dfrac{Gm_1m_2}{d^2}$, for G

$Fd^2 = Gm_1m_2$

$\dfrac{Fd^2}{m_1m_2} = G$

$G = \dfrac{Fd^2}{m_1m_2}$

65. $d = \dfrac{fl}{f + w}$, for f

$d(f + w) = fl$

$df + dw = fl$

$df - fl = -dw$

$f(d - l) = -dw$

$f = \dfrac{-dw}{d - l}$ or $\dfrac{dw}{l - d}$

67. $v = \dfrac{d_2 - d_1}{t_2 - t_1}$, for t_2

$v(t_2 - t_1) = d_2 - d_1$

$vt_2 - vt_1 = d_2 - d_1$

$vt_2 = d_2 - d_1 + vt_1$

$t_2 = \dfrac{d_2 - d_1 + vt_1}{v}$

69. $x = \dfrac{F_G(r + p)}{F_G - F_S}$, for F_G

$x(F_G - F_S) = F_G(r + p)$

$F_Gx - F_Sx = F_Gr + F_Gp$

$F_Gx - F_Gr - F_Gp = F_Sx$

$F_G(x - r - p) = F_Sx$

$F_G = \dfrac{F_Sx}{x - r - p}$

71. $p = \dfrac{t^2dN}{3.78}$, for d

$3.78p = t^2dN$

$d = \dfrac{3.78p}{t^2N}$

73. $\frac{1}{R}=\frac{1}{R_1}+\frac{1}{R_2}$
$R_1=2R_2$
$(R = 60 \text{ ohms})$:
$\frac{1}{60}=\frac{1}{2R_2}+\frac{1}{R_2}$
$60R_2\left(\frac{1}{60}\right)=60R_2\left(\frac{1}{2R_2}+\frac{1}{R_2}\right)$
$R_2=30+60$
$R_2=90$
$R_1=2(90)=180$
$R_1=180$ ohms, $R_2=90$ ohms

75. $\overline{C}(x)=\frac{80,000+20x}{x}$
$22=\frac{80,000+20x}{x}$
$22x = 80,000 + 20x$
$2x = 80,000$
$x = 40,000$
40,000 calculators

77. $\frac{\text{hits}}{\text{times at bat}}=0.3$
$0.3=\frac{45+x}{185+x}$
$0.3(185 + x) = 45 + x$
$55.5 + 0.3x = 45 + x$
$10.5 = 0.7x$
$15 = x$

79. Walking Rate: x
Driving Rate: $12x$
Time spent walking + Time spent driving $= 2\frac{1}{2}$ hours
$\text{time} = \frac{\text{distance}}{\text{rate}}$
$\frac{4}{x}+\frac{72}{12x}=\frac{5}{2}$
$\frac{4}{x}+\frac{6}{x}=\frac{5}{2}$
$x=4$
Walking rate: 4 mph

81. Jogger's rate $= x$
Cyclist's rate $= 2x$
Time it takes jogger – time it takes cyclist $= 4$ hours
$\frac{40}{x}-\frac{40}{2x}=4$
$\frac{40}{x}-\frac{20}{x}=4$
$x=5$
Cyclist's rate: 2(5) = 10 mph

83. Time for Bud working alone: x
Time for Lou working alone: $x + 5$
Part of the job done by Bud in 6 days + Part of the job done by Lou in 6 days = one whole job
$\frac{6}{x}+\frac{6}{x+5}=1$
$x = 10$ or $x = -3$ (reject)
Time for Bud working alone: 10 days

85. Let x = number of minutes needed to fill the sink.

	Fractional part of job completed in 1 minute	Time spent working together	Fractional part of job completed in x minutes
Fill (5 minutes)	$\frac{1}{5}$	x	$\frac{x}{5}$
Drain ($2 \cdot 5 = 10$ minutes)	$\frac{1}{10}$	x	$\frac{x}{10}$

(Fraction of sink filled by faucet) – (fraction of sink emptied by drain) = 1
$\frac{x}{5}-\frac{x}{10}=1$
$10\cdot\frac{x}{5}-10\cdot\frac{x}{10}=10\cdot 1$

$2x - x = 10$
$x = 10$
10 minutes

87. $\frac{3}{4}\cdot$ (perimeter of first square) = (reciprocal of second square's area) + $\frac{1}{2}\cdot$ (perimeter of third square)

$\frac{3}{4}\cdot 4\cdot\frac{1}{4x-8} = \frac{1}{6\cdot 6} + \frac{1}{2}\cdot 4\cdot\frac{1}{3x-6}$

$\frac{3}{4(x-2)} = \frac{1}{36} + \frac{2}{3(x-2)}$

$36(x-2)\cdot\frac{3}{4(x-2)} = 36(x-2)\cdot\frac{1}{36} + 36(x-2)\cdot\frac{2}{3(x-2)}$

$27 = x - 2 + 24$
$-x = -5$
$x = 5$

89. d is true.

91–95. Students should verify results.

97. Answers may vary.

99. $V = C\left(1 - \frac{T}{N}\right)$

$V = C - \frac{TC}{N}$
$VN = CN - TC$
$TC = CN - VN$
$TC = N(C - V)$
$\frac{TC}{C-V} = N$
$N = \frac{TC}{C-V}$

101. $\frac{13w - 6z}{w + 3z} = 3$

$(w+3z)\left(\frac{13w-6z}{w+3z}\right) = (w+3z)3$
$13w - 6z = 3w + 9z$
$10w = 15z$
$w = \frac{3}{2}z$ (solve for w in terms of z)

$\frac{w+z}{w-z} = \frac{\frac{3}{2}z+z}{\frac{3}{2}z-z} = \frac{3z+2z}{3z-2z} = \frac{5z}{z} = 5$
(substitute for w)

103. $\left|\frac{y+1}{y+8}\right|=\frac{2}{3}$

$\frac{y+1}{y+8}=\frac{2}{3}$

$3(y+8)\cdot\frac{y+1}{y+8}=3(y+8)\cdot\frac{2}{3}$

$3(y+1)=2(y+8)$

$3y+3=2y+16$

$y=13$

or

$\frac{y+1}{y+8}=-\frac{2}{3}$

$3(y+1)=-2(y+8)$

$3y+3=-2y-16$

$5y=-19$

$y=-\frac{19}{5}$

$\left\{-\frac{19}{5}, 13\right\}$

105. $x=\frac{1-a}{1+a}$ and $a=\frac{1+y}{1-y}$

$a(1-y)=1+y$

$a-ay=1+y$

$a-1=ay+y$

$a-1=y(a+1)$

$\frac{a-1}{a+1}=y$

$x+y$

$=\frac{1-a}{1+a}+\frac{a-1}{a+1}=\frac{1-a+a-1}{a+1}=\frac{0}{a+1}=0$

107. $\frac{4x-b}{x-5}=3$

The solution set for $\frac{4x-b}{x-5}=3$ is $\varnothing$ if $x=5$.

$4x-b=3(x-5)$

$4(5)-b=3(5-5)$

$20-b=3(0)$

$20-b=0$

$20=b$

$b=20$

109. Let r = low interest rate,

then $r+1$ = higher interest rate

$I=Prt$, where $t=1$, $P=\frac{I}{r}$

$\frac{96}{r} = \frac{108}{1+r}$
$96 + 96r = 108r$
$r = 8$
$r + 1 = 9$
The rates are 8% and 9%.

111. Let x = train's speed on old schedule
$x + 2$ = train's speed on new schedule

	D (miles)	R	$T = \frac{D}{R}$
train on old schedule	351	x	$\frac{351}{x}$
train on new schedule	351	$x+2$	$\frac{351}{x+2}$

Time on new schedule = time on old schedule $- \frac{1}{4}$

$\frac{351}{x+2} = \frac{351}{x} - \frac{1}{4}$
$4x(x+2)\left[\frac{351}{x+2}\right] = 4x(x+2)\left[\frac{351}{x} - \frac{1}{4}\right]$
$1404x = 1404x + 2808 - x^2 - 2x$
$x^2 + 2x - 2808 = 0$
$(x + 54)(x - 52) = 0$
$x + 54 = 0$ or $x - 52 = 0$
$x = -54$ (Reject) $x = 52$ (old)
$x + 2 = 52 + 2 = 54$ (new)
Train's speed on new schedule: 54 mph

Review Problems

113. $f(x) = x^2 - 6x + 8$

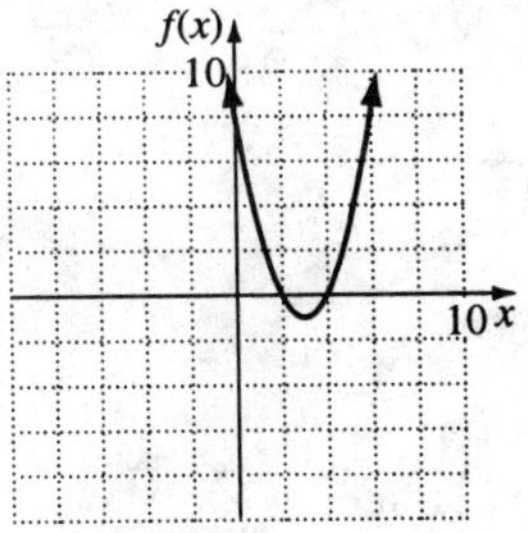

114. $7x^2 - 8x - 9 - (9 - 8x - 7x^2)$
$7x^2 - 8x - 9 - 9 + 8x + 7x^2$
$= 14x^2 - 18$

115. $(4x^2 - y)^2$
$= (4x^2 - y)(4x^2 - y)$
$= (4x^2)^2 + 2(4x^2)(-y) + (-y)^2$
$= 16x^4 - 8x^2y + y^2$

Problem Set 5.5

1. a. $y = kx$
$35 = k(5)$
$k = 7$

b. $y = 7x$

c. $y = 7(12)$
$y = 84$

3. a. $y = kx^2$
$48 = k(3)^2$
$k = 5.\overline{3}$

b. $y = 5.\overline{3}x^2$

c. $y = 5.\overline{3}(7)^2$
$y = 261.\overline{3}$

5. a. $y = \frac{k}{x}$
$10 = \frac{k}{5}$
$k = 50$

b. $y = \frac{50}{x}$

c. $y = \frac{50}{2}$
$y = 25$

7. a. $y = \frac{k}{x^2}$
$3 = \frac{k}{5^2}$
$k = 75$

b. $y = \frac{75}{x^2}$

c. $y = \frac{75}{(10)^2}$
$y = \frac{3}{4}$

9. a. $z = k\left(\frac{x}{y}\right)$
$\frac{1}{3} = k\left(\frac{10}{2}\right)$
$k = \frac{1}{15}$

b. $z = \frac{1}{15}\left(\frac{x}{y}\right)$

c. $z = \frac{1}{15}\left(\frac{30}{5}\right)$
$z = \frac{2}{5}$

11. a. $y = kxz$
$25 = k(2)(5)$
$k = \frac{5}{2}$

b. $y = \frac{5}{2}xz$

c. $y = \frac{5}{2}(8)(12)$
$y = 240$

13. a. $y = k\left(\frac{x^2 z}{R^3}\right)$
$\frac{1}{3} = k\left[\frac{(4)(6)}{8}\right]$
$k = \frac{1}{9}$

b. $y = \frac{1}{9}\left(\frac{x^2 z}{R^3}\right)$

c. $y = \frac{1}{9}\left(\frac{9 \cdot 2}{1}\right)$
$y = 2$

15. Given:
$C = kd$
$d = 2$ feet, $C = 2\pi$ feet
$2\pi = k(2)$
$\pi = k$
$C = \pi d$
At $r = 8$ feet;

$d = 16$ feet, $C = ?$
$C = \pi(16)$
$C = 16\pi$
The circumference is 16π ft.

17. Given:
$d = kr^2$
$d = 200$ feet, $r = 60$ mph
$200 = k(60)^2$
$200 = k(3600)$
$\frac{1}{18} = k$
$d = \frac{1}{18}r^2$
At $r = 100$ mph; $d = ?$
$d = \frac{1}{18}r^2$
$d = \frac{1}{18}(100)^2$
$d = \frac{10,000}{18} = 555\frac{5}{9}$
The stopping distance is $555\frac{5}{9}$ feet.

19. Given:
$V = \frac{k}{p}$
$V = 32\text{ cm}^3,\ p = 8\text{ lb}$
$32 = \frac{k}{8}$
$256 = k$
$V = \frac{256}{p}$
At $V = 40\text{ cm}^3,\ p = ?$
$40 = \frac{256}{p}$
$p = \frac{256}{40} = 6.4$
The pressure is 6.4 lb.

21. Let
I = intensity
d = distance
$I = \frac{k}{d^2}$
Given:
$I = 25$ ft.-candles, $d = 4$ ft.
$25 = \frac{k}{4^2}$
$k = 25(16) = 400$
$I = \frac{400}{d^2}$
At $d = 6$ ft, $I = ?$
$I = \frac{400}{6^2} = \frac{400}{36} = 11\frac{1}{9}$
The illumination is $11\frac{1}{9}$ foot candles.

23. Let
V = volume
T = temperature
P = pressure
$V = \frac{kT}{P}$
Given: $T = 100$ Kelvin, $P = 15\text{ kgm/m}^2$,
$V = 20\text{m}^3$
Find k.
$20 = \frac{k(100)}{15}$
$k = 3$
$V = \frac{3T}{P}$
At $T = 150$ Kelvin,
$P = 30\text{ kgm/m}^2,\ V = ?$
$V = \frac{3(150)}{30} = 15$
The volume is 15 m^3.

25. Let
I = simple interest
r = interest rate
t = time that money is invested
$I = krt$
Given:
$r = 12\% = 0.12$, $t = 2$ years, $I = \$280$
$280 = k(0.12)(2)$
$\frac{3500}{3} = k$
$I = \frac{3500}{3}rt$

At $r = 16\% = 0.16$, $t = 4$ years, $I = ?$

$I = \frac{3500}{3}(0.16)(4) = 746.6\bar{6}$

The investment will yield approximately $746.67.

27. Let

F = force of attraction

m_1 = first mass

m_2 = second mass

d = distance between masses

$F = \frac{km_1m_2}{d^2}$

Given:

$m_1 = 4$ units, $m_2 = 2$ units, $d = 3$ feet, $F = 16$ units

$16 = \frac{k4(2)}{3^2}$

$18 = k$

$F = \frac{18m_1m_2}{d^2}$

At $m_1 = 5$ units, $m_2 = 3$ units, $d = 2$ feet, $F = ?$

$F = \frac{18(5)(3)}{2^2}$

$F = 67.5$

The force is 67.5 units.

29. $I = \frac{kw}{h}$

$21 = \frac{k(150)}{70}$

$k = 9.8$

$I = \frac{9.8w}{h}$

$I = \frac{9.8(240)}{74}$

$I \approx 31.8$

Index of 32, which is not in the desirable range.

31. Let

R = electrical resistance

L = length

D = diameter

$R = \frac{kL}{D^2}$

Given:

$L = 720$ feet, $D = \frac{1}{4}$- inch, $R = 1\frac{1}{2}$ ohms

$\frac{3}{2} = \frac{k(720)}{\left(\frac{1}{4}\right)^2}$

$\frac{\frac{3}{2}\left(\frac{1}{16}\right)}{720} = k$

$\frac{1}{7680} = k$

$R = \frac{L}{7680D^2}$

for $L = 960$ and $D = 2\left(\frac{1}{4}\right) = \frac{1}{2}$

$R = \frac{960}{\left(\frac{1}{2}\right)^2} \cdot \frac{1}{7680}$

$R = 0.5$ ohms

33. b is true.

35. a is not true.

37–41. Answers may vary.

43. $y = \frac{K_1}{x}, \quad x = \frac{K_2}{z}$

$y = \frac{K_1}{\frac{K_2}{z}} = \frac{K_1 z}{K_2}$

y is directly proportional to z

constant of variation: $\frac{K_1}{K_2}$

45. $F = \frac{18m_1m_2}{d^2}$

$F = \frac{18(2m_1)(2m_2)}{\left(\frac{d}{2}\right)^2}$

$F = \frac{18(2)(2)m_1m_2}{\frac{d^2}{(2)(2)}}$

$F = (2)^4\left[\frac{18m_1m_2}{d^2}\right]$

The force of attraction is 16 times that of the original force.

47. Group activity.

Review Problems

49. $y^4 - 3y^3 + 2y^2 - 6y$
$= y^3(y-3) + 2y(y-3)$
$= \left(y^3 + 2y\right)(y-3)$
$= y\left(y^2 + 2\right)(y-3)$

50. $3x - 6y < 12$

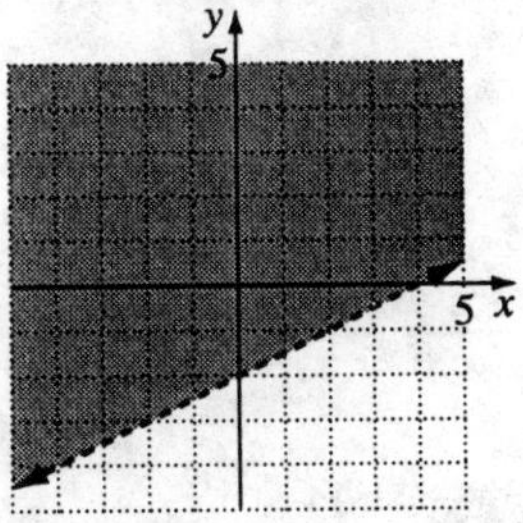

51. $x + 3y + 2z = 5$ (1)
$x + 5y + 5z = 6$ (2)
$3x + 3y - z = 10$ (3)
Add equation (1) to −1 times equation (2).
$x + 3y + 2z = 5$
$-x - 5y - 5z = -6$
$-2y - 3z = -1$ (4)
Add −3 times equation (1) to equation (3).
$-3x - 9y - 6z = -15$
$3x + 3y - z = 10$
$-6y - 7z = -5$ (5)
Add −3 times equation (4) to equation (5).
$6y + 9z = 3$
$-6y - 7z = -5$
$2z = -2$
$z = -1$
Substitute into equation (4)
$-2y - 3(-1) = -1$
$-2y = -4$
$y = 2$
Substitute into equation (1).
$x + 3(2) + 2(-1) = 5$
$x + 4 = 5$
$x = 1$
$x = 1,\ y = 2,\ z = -1$
$\{(1, 2, -1)\}$

Chapter 5 Review Problems

1. $f(x) = \dfrac{7x}{9x - 18}$
$f(x)$ is undefined when
$9x - 18 = 0$
$x = 2$
Domain of $f = \{x \mid x \neq 2\}$ or
$(-\infty,\ 2) \cup (2,\ \infty)$

2. $f(x) = \dfrac{x+3}{(x-1)(x+5)}$
$f(x)$ is undefined when
$(x - 1)(x + 5) = 0$
$x - 1 = 0$ or $x + 5 = 0$
$x = 1$ $\quad x = -5$
Domain of $f = \{x \mid x \neq -5, 1\}$ or
$(-\infty,\ -5) \cup (-5,\ 1) \cup (1,\ \infty)$

3. $f(x) = \dfrac{7x + 14}{2x^2 + 5x - 3}$
$f(x)$ is undefined when
$2x^2 + 5x - 3 = 0$
$(2x - 1)(x + 3) = 0$
$2x - 1 = 0$ or $x + 3 = 0$
$x = \frac{1}{2}$ $\quad x = -3$
Domain of $f = \left\{x \mid x \neq -3,\ \frac{1}{2}\right\}$ or
$(-\infty,\ -3) \cup \left(-3,\ \frac{1}{2}\right) \cup \left(\frac{1}{2},\ \infty\right)$

4. $f(x)=\dfrac{x^2-25}{x^2+4}$

$f(x)$ is undefined when $x^2+4=0$
No real number substituted into the denominator will cause x^2+4 to equal zero. No values of x need to be excluded. The domain of f is the set of all real numbers.
Domain of $f=\{x|x\in R\}$ or $(-\infty,\ \infty)$

5. $f(x)=\dfrac{1}{x+2}$

a. $f(x)$ is undefined when
$x+2=0$
$x=-2$
Domain of $f=\{x\mid x\neq -2\}$ or $(-\infty,\ -2)\cup(-2,\ \infty)$

b. –1, –2, –10, –100, –1000; 1000, 100, 10, 2, 1

c. $x=-2$ (dotted)
$f(x)=\dfrac{1}{x+2}$

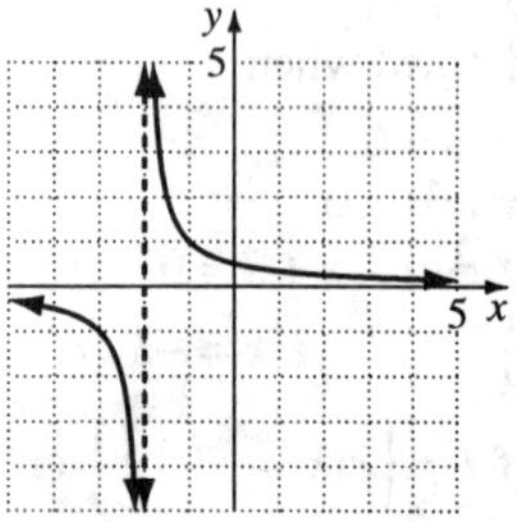

6. c; $f(x)=\dfrac{-2}{x-1}$

$f(0)=\dfrac{-2}{-1}=2$

$f(2)=\dfrac{-2}{2-1}=-2$

7. a. $C(x)=250{,}000+3x$

b. $\overline{C}(x)=\dfrac{250{,}000+3x}{x}$

c. $\overline{C}(1000)=\dfrac{250{,}000+3(1000)}{1000}=253$

$\overline{C}(10{,}000)=\dfrac{250{,}000+3(10{,}000)}{10{,}000}=28$

$\overline{C}(100{,}000)=\dfrac{250{,}000+3(100{,}000)}{100{,}000}$
$=5.5$

The cost per clock drops from \$253 to \$28 to \$5.50 as the number of clocks produced increases from 1000, to 10,000, to 100,000, respectively. Cost drops as number produced increases.

8. $\dfrac{5x^3-35x}{15x^2}=\dfrac{5x^3}{15x^2}-\dfrac{35x}{15x^2}=\dfrac{x}{3}-\dfrac{7}{3x}$

9. $\dfrac{x^2+6x-7}{x^2-49}=\dfrac{(x+7)(x-1)}{(x+7)(x-7)}=\dfrac{x-1}{x-7}$

10. $\dfrac{6m^2+7m+2}{2m^2-9m-5}$
$=\dfrac{(3m+2)(2m+1)}{(2m+1)(m-5)}$
$=\dfrac{3m+2}{m-5}$

11. $\dfrac{y^3-8}{y^2-4}$
$=\dfrac{(y-2)\left(y^2+2y+4\right)}{(y+2)(y-2)}$
$=\dfrac{y^2+2y+4}{y+2}$

12. $\dfrac{3x^2+15xy+12y^2}{3x^3+9x^2y-12xy^2}$
$=\dfrac{3\left(x^2+5xy+4y^2\right)}{3x\left(x^2+3xy-4y^2\right)}$
$=\dfrac{3(x+y)(x+4y)}{3x(x-y)(x+4y)}$
$=\dfrac{x+y}{x(x-y)}$

13. $f(x)=\dfrac{x^2-7x+12}{x-4}$
$=\dfrac{(x-4)(x-3)}{x-4}$
$=x-3\ (x\neq 4)$

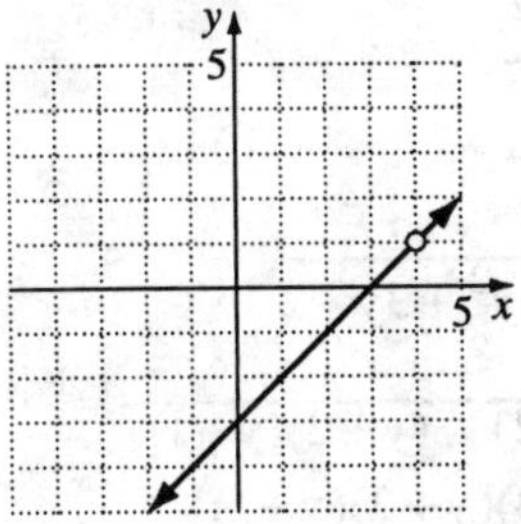

14. $\dfrac{5x^2-5}{3x+12}\cdot\dfrac{x+4}{x-1}$
$=\dfrac{5(x-1)(x+1)}{3(x+4)}\cdot\dfrac{x+4}{x-1}$
$=\dfrac{5(x+1)}{3}$

15. $\dfrac{x^2-9x+14}{x^3+2x^2}\cdot\dfrac{x^2-4}{(x-2)^2}$
$=\dfrac{(x-2)(x-7)}{x^2(x+2)}\cdot\dfrac{(x+2)(x-2)}{(x-2)^2}$
$=\dfrac{x-7}{x^2}$

16. $\dfrac{y^4-81}{y^2+9}\cdot\dfrac{4y-20}{y^2-8y+15}$
$=\dfrac{\left(y^2+9\right)(y+3)(y-3)}{y^2+9}\cdot\dfrac{4(y-5)}{(y-5)(y-3)}$
$=4(y+3)$

17. $\dfrac{5xy-10y^2}{x^2-3xy+2y^2}\cdot\dfrac{x+2y}{x^2}\cdot\dfrac{3x^2-3xy}{xy+2y^2}$
$=\dfrac{5y(x-2y)}{(x-2y)(x-y)}\cdot\dfrac{x+2y}{x^2}\cdot\dfrac{3x(x-y)}{y(x+2y)}$
$=\dfrac{15}{x}$

18. $\dfrac{25x^2-1}{x^2-25}\div\dfrac{5x+1}{x+5}$
$=\dfrac{(5x-1)(5x+1)}{(x-5)(x+5)}\cdot\dfrac{x+5}{5x+1}$
$=\dfrac{5x-1}{x-5}$

19. $\dfrac{x^2+16x+64}{2x^2-128}\div\dfrac{3x^2+30x+48}{x^2-6x-16}$
$=\dfrac{x^2+16x+64}{2x^2-128}\cdot\dfrac{x^2-6x-16}{3x^2+30x+48}$
$=\dfrac{(x+8)^2}{2(x+8)(x-8)}\cdot\dfrac{(x-8)(x+2)}{3(x+8)(x+2)}$
$=\dfrac{1}{6}$

20. $\dfrac{a^3-27}{a^2+3a+9}\div(ab+ac-3b-3c)$
$=\dfrac{a^3-27}{a^2+3a+9}\cdot\dfrac{1}{ab+ac-3b-3c}$
$=\dfrac{(a-3)\left(a^2+3a+9\right)}{a^2+3a+9}\cdot\dfrac{1}{(b+c)(a-3)}$
$=\dfrac{1}{b+c}$

21. $\dfrac{y^3-8}{y^4-16}\div\dfrac{y^2+2y+4}{y^2+4}$
$=\dfrac{y^3-8}{y^4-16}\cdot\dfrac{y^2+4}{y^2+2y+4}$
$=\dfrac{(y-2)\left(y^2+2y+4\right)}{\left(y^2+4\right)(y+2)(y-2)}\cdot\dfrac{y^2+4}{y^2+2y+4}$
$=\dfrac{1}{y+2}$

22. $f(x)=2x^2+x-15,\ \ g(x)=x+3$

a. $\left(\dfrac{f}{g}\right)(x)=\dfrac{2x^2+x-15}{x+3}$
$=\dfrac{(2x-5)(x+3)}{x+3}$
$=2x-5$

b. $\left(\dfrac{f}{g}\right)(6)=2(6)-5=7$

c. $\dfrac{f(6)}{g(6)}=7$

23. a. $t=\dfrac{30}{x}+\dfrac{30}{x+10}$

b. $t=\dfrac{30(x+10)+30x}{x(x+10)}$
$=\dfrac{30x+300+30x}{x^2+10x}$
$=\dfrac{60x+300}{x^2+10x}$

24. $\dfrac{x^3}{x^3+125}+\dfrac{5}{x^3+125}$
$=\dfrac{x^3+5}{x^3+125}$

25. $\dfrac{2x-7}{x^2-9}-\dfrac{x-4}{x^2-9}$
$=\dfrac{2x-7-(x-4)}{x^2-9}$
$=\dfrac{x-3}{x^2-9}$
$=\dfrac{x-3}{(x-3)(x+3)}$
$=\dfrac{1}{x+3}$

26. $\dfrac{1}{x}+\dfrac{2}{x-5}$ (LCD is $x(x-5)$)
$=\dfrac{1}{x}\cdot\dfrac{x-5}{x-5}+\dfrac{2}{x-5}\cdot\dfrac{x}{x}$
$=\dfrac{x-5+2x}{x(x-5)}$
$=\dfrac{3x-5}{x(x-5)}$

27. $\dfrac{3x^2}{9x^2-16}-\dfrac{x}{3x+4}$ $\left(\text{LCD is } 9x^2-16\right)$
$=\dfrac{3x^2-x(3x-4)}{9x^2-16}$
$=\dfrac{3x^2-3x^2+4x}{9x^2-16}$
$=\dfrac{4x}{9x^2-16}$

28. $\dfrac{7}{x^2y^3}-\dfrac{5}{xy^3}+\dfrac{4}{x^2y}$ $\left(\text{LCD is } x^2y^3\right)$
$=\dfrac{7}{x^2y^3}-\dfrac{5}{xy^3}\cdot\dfrac{x}{x}+\dfrac{4}{x^2y}\cdot\dfrac{y^2}{y^2}$
$=\dfrac{7-5x+4y^2}{x^2y^3}$

29. $\dfrac{y}{y^2+5y+6}-\dfrac{2}{y^2+3y+2}$
$=\dfrac{y}{(y+3)(y+2)}-\dfrac{2}{(y+2)(y+1)}$
(LCD is $(y+3)(y+2)(y+1)$)
$=\dfrac{y(y+1)-2(y+3)}{(y+3)(y+2)(y+1)}$
$=\dfrac{y^2+y-2y-6}{(y+3)(y+2)(y+1)}$
$=\dfrac{y^2-y-6}{(y+3)(y+2)(y+1)}$
$=\dfrac{(y-3)(y+2)}{(y+3)(y+2)(y+1)}$
$=\dfrac{y-3}{(y+3)(y+1)}$

30. $\frac{x}{x+3}+\frac{x}{x-3}-\frac{9}{x^2-9}$

$=\frac{x}{x+3}+\frac{x}{x-3}-\frac{9}{(x+3)(x-3)}$

(LCD is $(x+3)(x-3)$)

$=\frac{x}{x+3}\cdot\frac{x-3}{x-3}+\frac{x}{x-3}\cdot\frac{x+3}{x+3}-\frac{9}{(x+3)(x-3)}$

$=\frac{x(x-3)+x(x+3)-9}{(x+3)(x-3)}$

$=\frac{x^2-3x+x^2+3x-9}{(x+3)(x-3)}$

$=\frac{2x^2-9}{(x+3)(x-3)}$

31. $\frac{4}{a^2+a-2}-\frac{2}{a^2-4}+\frac{3}{a^2-4a+4}$

$=\frac{4}{(a+2)(a-1)}-\frac{2}{(a+2)(a-2)}+\frac{3}{(a-2)^2}$ (LCD is $(a-2)^2(a+2)(a-1)$)

$=\frac{4}{(a+2)(a-1)}\cdot\frac{(a-2)^2}{(a-2)^2}-\frac{2}{(a+2)(a-2)}\cdot\frac{(a-2)(a-1)}{(a-2)(a-1)}+\frac{3}{(a-2)^2}\cdot\frac{(a+2)(a-1)}{(a+2)(a-1)}$

$=\frac{4(a-2)^2-2(a-2)(a-1)+3(a+2)(a-1)}{(a-2)^2(a+2)(a-1)}$

$=\frac{4a^2-16a+16-2a^2+6a-4+3a^2+3a-6}{(a-2)^2(a+2)(a-1)}$

$=\frac{5a^2-7a+6}{(a-2)^2(a+2)(a-1)}$

32. $\frac{4+\frac{1}{x}}{x^2}$

$=\frac{4+\frac{1}{x}}{x^2}\cdot\frac{x}{x}$

$=\frac{4x+1}{x^3}$

33. $\frac{4-\frac{1}{y^2}}{4+\frac{4}{y}+\frac{1}{y^2}}$

$=\frac{4-\frac{1}{y^2}}{4+\frac{4}{y}+\frac{1}{y^2}}\cdot\frac{y^2}{y^2}$

$=\frac{4y^2-1}{4y^2+4y+1}$

$=\frac{(2y+1)(2y-1)}{(2y+1)^2}$

$=\frac{2y-1}{2y+1}$

34. $\dfrac{3-\frac{1}{x+3}}{3+\frac{1}{x+3}}$

$=\dfrac{3-\frac{1}{x+3}}{3+\frac{1}{x+3}}\cdot\dfrac{x+3}{x+3}$

$=\dfrac{3(x+3)-1}{3(x+3)+1}$

$=\dfrac{3x+9-1}{3x+9+1}$

$=\dfrac{3x+8}{3x+10}$

35. $\dfrac{\frac{1}{y+5}+1}{\frac{6}{y^2+2y-15}-\frac{1}{y-3}}$

$=\dfrac{\frac{1}{y+5}+1}{\frac{6}{(y+5)(y-3)}-\frac{1}{y-3}}\cdot\dfrac{(y+5)(y-3)}{(y+5)(y-3)}$

$=\dfrac{y-3+(y+5)(y-3)}{6-(y+5)}$

$=\dfrac{y-3+y^2+2y-15}{6-y-5}$

$=\dfrac{y^2+3y-18}{1-y}$

$=\dfrac{(y+6)(y-3)}{1-y}$

36. $\dfrac{35x^2y^3-25x^2y^2-15x^4y^3}{5xy^2}$

$=\dfrac{35x^2y^3}{5xy^2}-\dfrac{25x^2y^2}{5xy^2}-\dfrac{15x^4y^3}{5xy^2}$

$=7xy-5x-3x^3y$

37. $\left(6x^2-5x+5\right)\div(2x+3)=3x-7+\dfrac{26}{2x+3}$

$$\begin{array}{r} 3x-7 \\ 2x+3\overline{\big)6x^2-5x+5} \\ \underline{6x^2+9x} \\ -14x+5 \\ \underline{-14x-21} \\ 26 \end{array}$$

38. $\left(10x^3-26x^2+17x-13\right)\div(5x-3)$

$=2x^2-4x+1-\dfrac{10}{5x-3}$

$$\begin{array}{r} 2x^2-4x+1-\frac{10}{5x-3} \\ 5x-3\overline{\big)10x^3-26x^2+17x-13} \\ \underline{10x^3-\ \ 6x^2} \\ -20x^2+17x \\ \underline{-20x^2+12x} \\ 5x-13 \\ \underline{5x-\ \ 3} \\ -10 \end{array}$$

39. $\left(4x^4+6x^3+3x-1\right)\div\left(2x^2+1\right)$

$=2x^2+3x-1$

$$\begin{array}{r} 2x^2+3x-1 \\ 2x^2+1\overline{\big)4x^4+6x^3+0x^2+3x-1} \\ \underline{4x^4+2x^2} \\ 6x^3-2x^2+3x \\ \underline{6x^3+3x} \\ -2x^2-1 \\ \underline{-2x^2-1} \\ 0 \end{array}$$

40. $\dfrac{4x^3-3x^2-2x+1}{x+1}=4x^2-7x+5-\dfrac{4}{x+1}$

$$\begin{array}{r|rrrr} -1 & 4 & -3 & -2 & 1 \\ & & -4 & 7 & -5 \\ \hline & 4 & -7 & 5 & -4 \end{array}$$

41. $\left(3y^4-2y^2-10y\right)\div(y-2)$

$=3y^3+6y^2+10y+10+\dfrac{20}{y-2}$

$$\begin{array}{r|rrrrr} 2 & 3 & 0 & -2 & -10 & 0 \\ & & 6 & 12 & 20 & 20 \\ \hline & 3 & 6 & 10 & 10 & 20 \end{array}$$

42. $\dfrac{2y}{y-2}-3=\dfrac{4}{y-2}$

$(y-2)\left[\dfrac{2y}{y-2}\right]-3(y-2)=(y-2)\dfrac{4}{y-2}$

$2y-3(y-2)=4$

$2y-3y+6=4$

$-y+6=4$

$2=y$

Since 2 causes a denominator to be 0 in the original equation, 2 is not a solution. The equation has no solution. ∅

43. $\frac{1}{y-5}-\frac{3}{y+5}=\frac{6}{y^2-25}$

$\frac{1}{y-5}-\frac{3}{y+5}=\frac{6}{(y+5)(y-5)}$

$(y+5)(y-5)\left[\frac{1}{y-5}-\frac{3}{y+5}\right]=(y+5)(y-5)\cdot\frac{6}{(y+5)(y-5)}$

$y+5-3(y-5)=6$

$y+5-3y+15=6$

$-2y+20=6$

$-2y=-14$

$y=7$

$\{7\}$

44. $\frac{x+5}{x+1}-\frac{x}{x+2}=\frac{4x+1}{x^2+3x+2}$

$\frac{x+5}{x+1}-\frac{x}{x+2}=\frac{4x+1}{(x+1)(x+2)}$

$(x+1)(x+2)\left[\frac{x+5}{x+1}-\frac{x}{x+2}\right]=(x+1)(x+2)\cdot\frac{4x+1}{(x+1)(x+2)}$

$(x+5)(x+2)-x(x+1)=4x+1$

$x^2+7x+10-x^2-x=4x+1$

$6x+10=4x+1$

$2x=-9$

$x=-\frac{9}{2}$

$\left\{-\frac{9}{2}\right\}$

45. $\frac{2}{3}-\frac{5}{3y}=\frac{1}{y^2}$

$3y^2\cdot\frac{2}{3}-3y^2\cdot\frac{5}{3y}=3y^2\cdot\frac{1}{y^2}$

$2y^2-5y=3$

$2y^2-5y-3=0$

$(2y+1)(y-3)=0$

$2y+1=0$ or $y-3=0$

$y=-\frac{1}{2}$ $\quad$ $y=3$

$\left\{-\frac{1}{2},\ 3\right\}$

46. $\frac{2}{y-1}=\frac{1}{4}+\frac{7}{y+2}$

$4(y-1)(y+2)\cdot\frac{2}{y-1}=4(y-1)(y+2)\left[\frac{1}{4}+\frac{7}{y+2}\right]$

$8y+16=y^2+y-2+28y-28$

$0=y^2+21y-46$

$0=(y+23)(y-2)$

$y+23=0$ or $y-2=0$

$y=-23$ $y=2$

$\{-23, 2\}$

47. $\frac{2y+7}{y+5}-\frac{y-8}{y-4}=\frac{y+18}{y^2+y-20}$

$(y+5)(y-4)\left[\frac{2y+7}{y+5}-\frac{y-8}{y-4}\right]=(y+5)(y-4)\cdot\frac{y+18}{(y+5)(y-4)}$

$(y-4)(2y+7)-(y+5)(y-8)=y+18$

$2y^2-y-28-y^2+3y+40=y+18$

$y^2+y-6=0$

$(y+3)(y-2)=0$

$y+3=0$ or $y-2=0$

$y=-3$ $y=2$

$\{-3, 2\}$

48. $C=\frac{4p}{100-p}$

a. $16=\frac{4p}{100-p}$

$16(100-p)=4p$

$1600-16p=4p$

$1600=20p$

$80=p$

80%

b. $C=\frac{4p}{100-p}$

$C(100-p)=4p$

$100C-pC=4p$

$100C=pC+4p$

$100C=p(C+4)$

$p=\frac{100C}{C+4}$

$p=\frac{100(16)}{16+4}=80$

c. $p\neq 100$

It is impossible to remove 100% of pollutants.

d. Increases rapidly as x approaches 100; cost increases to infinity as percent approaches 100%.

49. $P=30-\frac{9}{t+1}$

$P=29$ (since P is in thousands):

$29=30-\frac{9}{t+1}$

$(t+1)29=(t+1)\left[30-\frac{9}{t+1}\right]$

$29t+29=30t+30-9$

$29t+29=30t+21$

$8=t$

$1985+8=1993$

The population of 29,000 will occur 8 years from 1985, in 1993.

50. $C(t) = \dfrac{5t}{t^2 + 1}$

a. points on graph $\left(\frac{1}{2},\ 2\right)$ and (2, 2)
After $\frac{1}{2}$ hr and 2 hr.

b. point on graph (3, 1.5)
$C(3) = \dfrac{5(3)}{3^2 + 1} = 1.5$
1.5 mg/L

c. It rises rapidly, then decreases slowly.

51. $p = \dfrac{R - C}{n}$, for C
$pn = R - C$
$C = R - pn$

52. $T = \dfrac{A - p}{pr}$, for p
$prT = A - p$
$prT + p = A$
$p(rT + 1) = A$
$p = \dfrac{A}{rT + 1}$

53. $\dfrac{1}{p} + \dfrac{1}{q} = \dfrac{1}{f}$, for p
$\dfrac{q + p}{pq} = \dfrac{1}{f}$
$f(q + p) = pq$
$fq + fp = pq$
$fq = pq - fp$
$fq = p(q - f)$
$p = \dfrac{fq}{q - f}$

54. $I = \dfrac{En}{Rn + r}$, for n
$I(Rn + r) = En$
$IRn + Ir = En$
$Ir = En - IRn$
$Ir = n(E - IR)$
$n = \dfrac{Ir}{E - IR}$

55. $\dfrac{1}{p} + \dfrac{1}{q} = \dfrac{1}{f}$
$\dfrac{1}{4f} + \dfrac{1}{f + 3} = \dfrac{1}{f}$
$4f(f + 3)\left(\dfrac{1}{4f} + \dfrac{1}{f + 3}\right) = 4f(f + 3)\dfrac{1}{f}$
$f + 3 + 4f = 4(f + 3)$
$5f + 3 = 4f + 12$
$f = 9$
focal length of 9 cm.

56. $\overline{C}(x) = \dfrac{100{,}000 + 60x}{x}$
$80 = \dfrac{100{,}000 + 60x}{x}$
$80x = 100{,}000 + 60x$
$20x = 100{,}000$
$x = 5000$
5000 desks

57. Let x = walking rate
$3x$ = cycling rate
total time = 7 hours

	D (miles)	R	$T = \frac{D}{R}$
walking	8	x	$\frac{8}{x}$
cycling	60	$3x$	$\frac{60}{3x}$

Time spent cycling + time spent walking = 7 hours
$\dfrac{60}{3x} + \dfrac{8}{x} = 7$
$\dfrac{20}{x} + \dfrac{8}{x} = 7$
$x\left(\dfrac{20}{x} + \dfrac{8}{x}\right) = x \cdot 7$
$20 + 8 = 7x$
$28 = 7x$
$4 = x$
$x = 4$ (walking)
$3x = 3(4) = 12$ (cycling)
Cycling rate: 12 mph

58. Let x = rate of boat in still water
$x - 3$ = rate of boat against current
$x + 3$ = rate of boat with current
rate of stream: 3 mph
total time = 3 hours

	D (miles)	R	$T = \frac{D}{R}$
against current	12	$x - 3$	$\frac{12}{x-3}$
with current	12	$x + 3$	$\frac{12}{x+3}$

(time upstream) + (time downstream) = 3

$$\frac{12}{x-3} + \frac{12}{x+3} = 3$$

$$(x-3)(x+3)\left[\frac{12}{x-3} + \frac{12}{x+3}\right] = (x-3)(x+3) \cdot 3$$

$$12x + 36 + 12x - 36 = 3x^2 - 27$$

$$-3x^2 + 24x + 27 = 0$$

$$-3\left(x^2 - 8x - 9\right) = 0$$

$$-3(x - 9)(x + 1) = 0$$

$$x - 9 = 0 \quad \text{or} \quad x + 1 = 0$$

$$x = 9 \qquad\qquad x = -1 \text{ (reject)}$$

boat's rate in still water: 9 mph

59. Let x = number of days for Norman's mother to do the job alone
$x - 9$ = number of days for Norman to do the job alone
total time to do the job together: 20 days

	Fractional part of job completed in one day	Time spent working together	Fractional part of job completed
Norman	$\frac{1}{x-9}$	20	$\frac{20}{x-9}$
Mother	$\frac{1}{x}$	20	$\frac{20}{x}$

(Norman) + (mother) = (1 complete job)

$$\frac{20}{x-9} + \frac{20}{x} = 1$$

$$x(x-9)\left[\frac{20}{x-9} + \frac{20}{x}\right] = x(x-9) \cdot 1$$

$$20x + 20x - 180 = x^2 - 9x$$

$$0 = x^2 - 49x + 180$$

$$0 = (x - 45)(x - 4)$$

$x-45=0$ or $x-4=0$

$x=45$ (mother) $x=4$ (reject since $x-9$ is negative)

$x-9=45-9=36$ (Norman)

mother: 45 days; Norman: 36 days

60. Let x = the time to fill the pool

	Fractional part of job completed in 1 hour	Time spent working together	Fractional part of job completed in x hours
Pipe A (8 hours)	$\frac{1}{8}$	x	$\frac{x}{8}$
Pipe B (12 hours)	$\frac{1}{12}$	x	$\frac{x}{12}$

(pipe A-fill) – (pipe B-drain) = $\left(\text{pool } \frac{1}{2} \text{ full}\right)$

$\frac{x}{8}-\frac{x}{12}=\frac{1}{2}$

$24\left[\frac{x}{8}-\frac{x}{12}\right]=24\cdot\frac{1}{2}$

$3x-2x=12$

$x=12$

time elapsed: 12 hours

61. $M=kE$

$17.6=k(110)$

$k=0.16$

$M=0.16E$

$28.8=0.16E$

$E=180$

180 kg

62. $W=\frac{k}{l}$

$500=\frac{k}{10}$

$k=5000$

$W=\frac{5000}{l}$

$W=\frac{5000}{125}$

$W=40$

40 pounds

63. $D=kt^2$

$144=k(3)^2$

$k=16$

$D=16t^2$

$D=16(7)^2$

$D=784$

784 feet

64. $F=kas^2$

$150=k(4\cdot 5)(30)^2$

$k=0.008\overline{3}$

$F=0.008\overline{3}(3\cdot 4)(60)^2$

$F=360$

360 pounds; Yes, hurricane shutters should be put on.

Chapter 5 Test

1. $f(x)=\dfrac{x^2-2x}{x^2-7x+10}$

$x^2-7x+10\neq 0$

$(x-5)(x-2)\neq 0$

$x\neq 5,\ x\neq 2$

Domain of $f=\{x|x\neq 2,\ 5\}$

$f(x)=\dfrac{x^2-2x}{x^2-7x+10}=\dfrac{x(x-2)}{(x-5)(x-2)}$

$=\dfrac{x}{x-5},\ x\neq 2,\ x\neq 5$

2. The domain of $f(x)=\dfrac{1}{x+2}$ does not include $x=-2$ and $f(-2)=\dfrac{1}{4}$.

3. $\dfrac{x^2}{x^2-6}\cdot\dfrac{x^2+7x+12}{x^2+3x}$

$=\dfrac{x^2}{x^2-6}\cdot\dfrac{(x+4)(x+3)}{x(x+3)}$

$=\dfrac{x(x+4)}{x^2-6}$

$=\dfrac{x^2+4x}{x^2-6}$

4. $\dfrac{x+2y}{x+3y}\div\dfrac{3x+6y}{x^2-9y^2}=\dfrac{x+2y}{x+3y}\cdot\dfrac{x^2-9y^2}{3x+6y}$

$=\dfrac{x+2y}{x+3y}\cdot\dfrac{(x-3y)(x+3y)}{3(x+2y)}$

$=\dfrac{x-3y}{3}$

5. $\dfrac{x^3-1}{(x-1)^3}\div\dfrac{x^2+x+1}{x^2-1}=\dfrac{x^3-1}{(x-1)^3}\cdot\dfrac{x^2-1}{x^2+x+1}$

$=\dfrac{(x-1)(x^2+x+1)}{(x-1)(x-1)(x-1)}\cdot\dfrac{(x-1)(x+1)}{x^2+x+1}$

$=\dfrac{x+1}{x-1}$

6. $\dfrac{x}{x+3}+\dfrac{5}{x-3}=\dfrac{x}{x+3}\cdot\dfrac{x-3}{x-3}+\dfrac{5}{x-3}\cdot\dfrac{x+3}{x+3}$

$=\dfrac{x(x-3)+5(x+3)}{(x+3)(x-3)}$

$=\dfrac{x^2-3x+5x+15}{x^2-9}$

$=\dfrac{x^2+2x+15}{x^2-9}$

7. $\dfrac{2}{x^2-4x+3}+\dfrac{3x}{x^2+x-2}$

$=\dfrac{2}{(x-3)(x-1)}+\dfrac{3x}{(x+2)(x-1)}$

$=\dfrac{2}{(x-3)(x-1)}\cdot\dfrac{x+2}{x+2}$

$+\dfrac{3x}{(x+2)(x-1)}\cdot\dfrac{x-3}{x-3}$

$=\dfrac{2(x+2)+3x(x-3)}{(x-3)(x-1)(x+2)}$

$=\dfrac{2x+4+3x^2-9x}{(x-3)(x-1)(x+2)}$

$=\dfrac{3x^2-7x+4}{(x-3)(x-1)(x+2)}$

$=\dfrac{(3x-4)(x-1)}{(x-3)(x-1)(x+2)}$

$=\dfrac{3x-4}{(x-3)(x+2)}$

8. $\dfrac{5x}{x^2-4}-\dfrac{2}{x^2+x-2}$

$=\dfrac{5x}{(x-2)(x+2)}-\dfrac{2}{(x+2)(x-1)}$

$=\dfrac{5x}{(x-2)(x+2)}\cdot\dfrac{x-1}{x-1}$

$-\dfrac{2}{(x+2)(x-1)}\cdot\dfrac{x-2}{x-2}$

$=\dfrac{5x(x-1)-2(x-2)}{(x-2)(x+2)(x-1)}$

$=\dfrac{5x^2-5x-2x+4}{(x-2)(x+2)(x-1)}$

$=\dfrac{5x^2-7x+4}{(x-2)(x+2)(x-1)}$

9. $\frac{xy}{x^2-y^2}-\frac{y}{x+y}$

$=\frac{xy}{(x-y)(x+y)}-\frac{y}{x+y}\cdot\frac{x-y}{x-y}$

$=\frac{xy-y(x-y)}{(x-y)(x+y)}$

$=\frac{xy-xy+y^2}{(x-y)(x+y)}$

$=\frac{y^2}{x^2-y^2}$

10. $\frac{1}{a}+\frac{2a-5}{6a+9}-\frac{4}{2a^2+3a}$

$=\frac{1}{a}+\frac{2a-5}{3(2a+3)}-\frac{4}{a(2a+3)}$

$=\frac{1}{a}\cdot\frac{3(2a+3)}{3(2a+3)}+\frac{2a-5}{3(2a+3)}\cdot\frac{a}{a}$

$-\frac{4}{a(2a+3)}\cdot\frac{3}{3}$

$=\frac{3(2a+3)+a(2a-5)-4\cdot 3}{3a(2a+3)}$

$=\frac{6a+9+2a^2-5a-12}{3a(2a+3)}$

$=\frac{2a^2+a-3}{3a(2a+3)}$

$=\frac{(2a+3)(a-1)}{3a(2a+3)}$

$=\frac{a-1}{3a}$

11. $f(x)=\frac{x^2-2x-15}{x-5}$

$f(x)=\frac{(x-5)(x+3)}{x-5}$

$f(x)=x+3,\ x\neq 5$

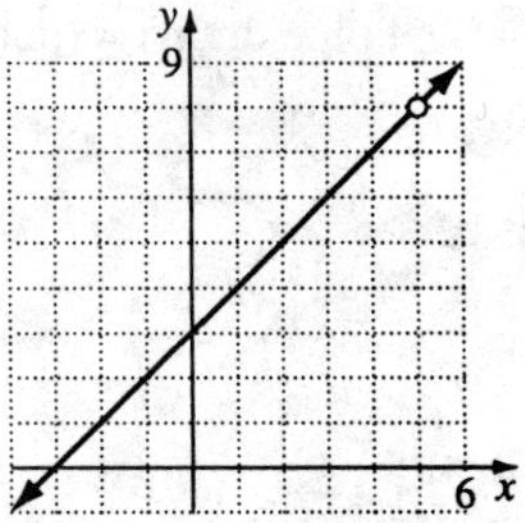

12. $\frac{\frac{x}{4}-\frac{1}{x}}{1+\frac{x+4}{x}}=\frac{\frac{x^2-4}{4x}}{\frac{x+x+4}{x}}$

$=\frac{x^2-4}{4x}\cdot\frac{x}{2x+4}$

$=\frac{(x-2)(x+2)x}{4x(2)(x+2)}$

$=\frac{x-2}{8}$

13. $\frac{\frac{x+2}{x^2-9}}{\frac{4}{x-3}+\frac{x}{x^2-6x+9}}=\frac{\frac{x+2}{x^2-9}}{\frac{4(x-3)+x}{(x-3)(x-3)}}$

$=\frac{x+2}{(x-3)(x+3)}\cdot\frac{(x-3)(x-3)}{5x-12}$

$=\frac{(x+2)(x-3)}{(x+3)(5x-12)}$

14. $(16x^2y^3-10x^2y^2+12x^4y^3)\div 4x^2y$

$=\frac{16x^2y^3}{4x^2y}-\frac{10x^2y^2}{4x^2y}+\frac{12x^4y^3}{4x^2y}$

$=4y^2-\frac{5y}{2}+3x^2y^2$

15. $(9x^3-3x^2-3x+4)\div(3x+2)$

$=3x^2-3x+1+\frac{2}{3x+2}$

$$\begin{array}{r|l} & 3x^2-3x+1+\frac{2}{3x+2} \\ 3x+2 & 9x^3-3x^2-3x+4 \\ & \underline{9x^3+6x^2} \\ & \quad -9x^2-3x \\ & \quad \underline{-9x^2-6x} \\ & \qquad 3x+4 \\ & \qquad \underline{3x+2} \\ & \qquad\quad 2 \end{array}$$

16. $(3x^4+11x^3-20x^2+7x+35)\div(x+5)$

$=3x^3-4x^2+7$

–5	3	11	–20	7	35
		–15	20	0	–35
	3	–4	0	7	0

17. $x+\frac{6}{x}=-5$

$x\left(x+\frac{6}{x}\right)=-5x$

$x^2+6=-5x$

$x^2+5x+6=0$

$(x+3)(x+2)=0$

$x+3=0$ or $x+2=0$

$x=-3$ or $x=-2$

$\{-3, -2\}$

18. $\frac{2}{x-3}-\frac{4}{x+3}=\frac{8}{x^2-9}$

$x^2-9\left(\frac{2}{x-3}-\frac{4}{x+3}\right)=x^2-9\left(\frac{8}{x^2-9}\right)$

$2(x+3)-4(x-3)=8$

$2x+6-4x+12=8$

$-2x=-10$

$x=5$

$\{5\}$

19. $\overline{C}=\frac{6x+3000}{x}$

$10=\frac{6x+3000}{x}$

$10x=6x+3000$

$4x=3000$

$x=750$

750 manuals must be produced.

20. $E=\frac{t+67}{0.01t+1}$

$78=\frac{t+67}{0.01t+1}$

$78(0.01t+1)=t+67$

$0.78t+78=t+67$

$-0.22t=-11$

$t=50$

$1950+50=2000$

The child will be born in year 2000.

21. $R=\frac{as}{a+s}$, for a.

$R(a+s)=as$

$Ra+Rs=as$

$Rs=as-Ra$

$Rs=a(s-R)$

$\frac{Rs}{s-R}=a$

$a=\frac{Rs}{s-R}$

22. $f(x)=3x^2+11x+10$

$g(x)=x+2$

$\left(\frac{f}{g}\right)(x)=\frac{3x^2+11x+10}{x+2}$

$=\frac{(3x+5)(x+2)}{x+2}$

$=3x+5$

$\left(\frac{f}{g}\right)(-3)=3(-3)+5=-4$

23. Let x = the time to fill the pool.

	First Pipe	Second Pipe
Fraction part of job completed in 1 hour	$\frac{1}{3}$	$\frac{1}{4}$
Time spent working together	x	x
Fractional part of job completed in x hours	$\frac{x}{3}$	$\frac{x}{4}$

(1st pipe fill) – (2nd pipe drain) = (pool full)

$\frac{x}{3}-\frac{x}{4}=1$

$12\left(\frac{x}{3}-\frac{x}{4}\right)=1\cdot 12$

$4x-3x=12$

$x=12$

It will take 12 hours.

24. Let x = current's speed.

	With current	Against current
Distance	3	2
Rate	$20+x$	$20-x$
Time $=\frac{D}{R}$	$\frac{3}{20+x}$	$\frac{2}{20-x}$

$\frac{3}{20+x}=\frac{2}{20-x}$

$3(20-x)=2(20+x)$

$60-3x=40+2x$

$-5x=-20$

$x=4$

The speed of the current is 4 mph.

25. Let I = intensity

d = distance

k = constant.

$I=\frac{k}{d^2}$

$20=\frac{k}{(15)^2}$

$k=4500$

$I=\frac{4500}{d^2}$

$I=\frac{4500}{(10)^2}$

$I=45$

The intensity is 45 foot candles.

Chapters 1–5 Cumulative Review Problems

1. $2(x+1)-7(3+2x)=-4(2x+1)$

$2x+2-21-14x=-8x-4$

$-12x-19=-8x-4$

$-15=4x$

$-\frac{15}{4}=x$

$\left\{-\frac{15}{4}\right\}$

2. $\frac{8(-6)-(-7)(4)-(-2)}{6-(3)(4)}$

$=\frac{-48+28+2}{6-12}$

$=\frac{-18}{-6}$

$=3$

3. $\{x|2x+5\le 11\}\cap\{x|-3x>18\}$

$2x\le 6 \qquad x<-6$

$x\le 3$

$\{x\,|\,x<-6\}$ or $(-\infty, -6)$

4. $s(t)=-16t^2+48t+64$

a. $0=-16t^2+48t+64$

$0=-16(t^2-3t-4)$

$0=-16(t-4)(t+1)$

$t-4=0$ or $t+1=0$

$t=4 \qquad t=-1$

Disregard a negative time. The ball will hit the ground in 4 seconds.

b. maximum height at

$x=-\frac{b}{2a}=-\frac{48}{2(-16)}=1.5$

$s(1.5)=-16(1.5)^2+48(1.5)+64$

$s(1.5)=100$

At 1.5 seconds, it reaches a maximum height of 100 feet.

c. $s(t)=-16t^2+48t+64$

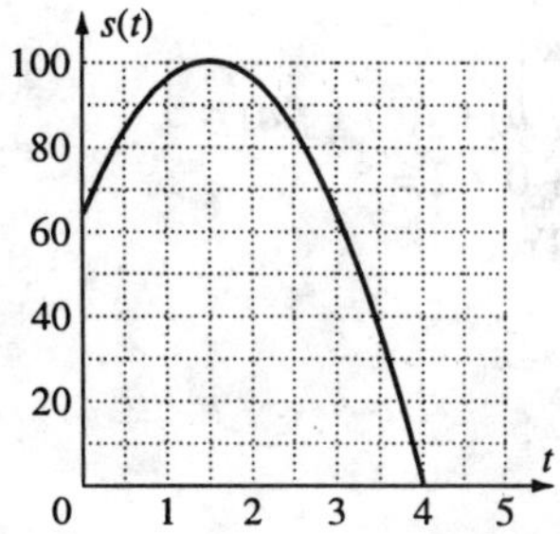

5. points (1976, 6) and (1995, 53.3)

$\text{slope} = \frac{53.3-6}{1995-1976} = \frac{47.3}{19} \approx 2.5$

The average number of people in HMO has risen by 2.5 million each year from 1976 to 1995.

6. Let A = Ana's weight

$230 - A$ = Juan's weight

$274 - A$ = Mike's weight

(Ana's weight) < (Juan's weight)

$A < 230 - A$

$2A < 230$

$A < 115$ (Ana)

$-A > -115$

$230 - A > 115$ (Juan)

$274 - A > 159$ (Mike)

(Ana's weight) < (Juan's weight) < (Mike's weight)

Thus, Mike weighs the most, which is necessarily the case, since

Ana's weight < Juan's weight
< Mike's weight.

7. **a.** No; answers may vary.

b. (1935, 200)
In the mid- to late-1930's, there were about 200 executions.

c. About 2040.

8. d is true;

$f(a) + f(b) + f(c) + f(d) = 5$

$3 + 0 + 0 + 2 = 5$

$5 = 5$ True

9. $2x - y = 4$

$x + y = 5$

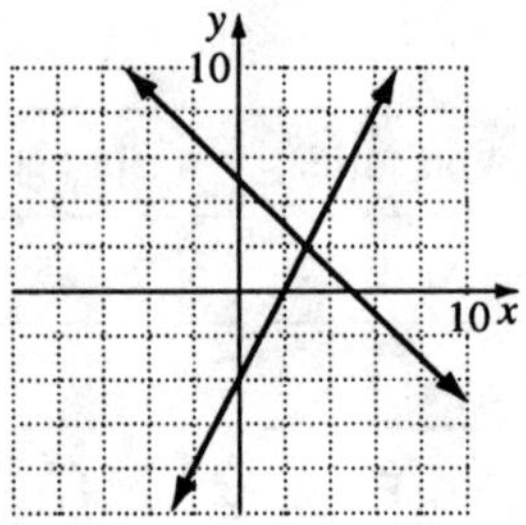

{(3, 2)}

10. $4x + 3y + 3z = 4$ (1)

$3x + 2z = 2$ (2)

$2x - 5y = -4$ (3)

2(eqn. 1) + (−3)(eqn. 2):

$$\begin{array}{rl} 8x + 6y + 6z = & 8 \\ -9x - 6z = & -6 \\ \hline -x + 6y = & 2 \end{array}$$

or $x = 6y - 2$ (4)

Substitute for x in (3):

$2(6y - 2) - 5y = -4$

$12y - 4 - 5y = -4$

$7y = 0$

$y = 0$

Substitute value of y in (4):

$x = 6(0) - 2 = -2$

Substitute x and y into (1):

$4(-2) + 3(0) + 3z = 4$

$-8 + 0 + 3z = 4$

$3z = 12$

$z = 4$

{(−2, 0, 4)}

11. Let x = rate of plane in still air
y = rate of wind

	R	T (hours)	$D = RT$	(1950 km)
against wind	$x-y$	3	$3(x-y)$	
with wind	$x+y$	2	$2(x+y)$	

$$\begin{aligned} 3(x-y) = 1950 &\rightarrow & x-y &= 650 \\ 2(x+y) = 1950 &\rightarrow & x+y &= 975 \\ & & 2x &= 1625 \\ & & x &= 812.5 \text{ (plane)} \\ & & x+y &= 975 \\ & & 812.5+y &= 975 \\ & & y &= 162.5 \text{ (wind)} \end{aligned}$$

speed of plane in still air: 812.5 kph; speed of wind; 162.5 kph

12. $\begin{vmatrix} 5 & 2 & 1 \\ 3 & 0 & -2 \\ -4 & -1 & 2 \end{vmatrix} = -2\begin{vmatrix} 3 & -2 \\ -4 & 2 \end{vmatrix} + 0\begin{vmatrix} 5 & 1 \\ -4 & 2 \end{vmatrix} - (-1)\begin{vmatrix} 5 & 1 \\ 3 & -2 \end{vmatrix}$

$= -2(6 - 8) + 0 + 1(-10 - 3)$
$= 4 - 13$
$= -9$

13. $x - y = 1$
$2y + 3z = 8$
$2x + z = 6$

$$\left[\begin{array}{ccc|c} 1 & -1 & 0 & 1 \\ 0 & 2 & 3 & 8 \\ 2 & 0 & 1 & 6 \end{array}\right] R_3 - 2R_1 \rightarrow \left[\begin{array}{ccc|c} 1 & -1 & 0 & 1 \\ 0 & 2 & 3 & 8 \\ 0 & 2 & 1 & 4 \end{array}\right] R_2 - R_3 \rightarrow \left[\begin{array}{ccc|c} 1 & -1 & 0 & 1 \\ 0 & 2 & 3 & 8 \\ 0 & 0 & 2 & 4 \end{array}\right] \begin{array}{c} \frac{1}{2}R_2 \rightarrow \\ \frac{1}{2}R_3 \rightarrow \end{array} \left[\begin{array}{ccc|c} 1 & -1 & 0 & 1 \\ 0 & 1 & \frac{3}{2} & 4 \\ 0 & 0 & 1 & 2 \end{array}\right]$$

$x - y = 1$
$y + \frac{3}{2}z = 4$
$z = 2$
$y + \frac{3}{2}(2) = 4$
$y = 1$
$x - 1 = 1$
$x = 2$
$\{(2, 1, 2)\}$

14. Let I = illumination
d = distance from source
$I = \frac{K}{d^2}$
Given: $d = 4$ feet, $I = 75$ foot-candles
$75 = \frac{K}{16}$
$1200 = K$
$I = \frac{1200}{d^2}$
At $d = 9$ feet, $I = ?$
$I = \frac{1200}{81} = 14\frac{22}{27}$
$14\frac{22}{27}$ foot-candles, or approximately
14.815 foot-candles

15. $(3x-4)(2x^2-5x+3)$
$= 3x(2x^2-5x+3)-4(2x^2-5x+3)$
$= 6x^3-15x^2+9x-8x^2+20x-12$
$= 6x^3-23x^2+29x-12$

16. $2x^2 = 7x+4$
$2x^2-7x-4=0$
$(2x+1)(x-4)=0$
$2x+1=0$ or $x-4=0$
$x = -\frac{1}{2}$ $x = 4$
$\left\{-\frac{1}{2}, 4\right\}$

17. $3x(x)-3(x-4) = 3x^2-3x+12$

18. Rent = $4500 + 5x$
Purchase = 7000
$7000 = 4500 + 5x$
$2500 = 5x$
$x = 500$
500 copies

19. $6x^2+11x-10 = (3x-2)(2x+5)$

20. $3(x+5)^2-11(x+5)-4$
$= [3(x+5)+1][(x+5)-4]$
$= (3x+15+1)(x+5-4)$
$= (3x+16)(x+1)$

21. Let x = width of rectangle
$2x + 3$ = length of rectangle
area = 65 square yards
$x(2x+3) = 65$
$2x^2+3x-65=0$
$(2x+13)(x-5)=0$
$2x+13=0$ or $x-5=0$
$x = -\frac{13}{2}$ (Reject) $x = 5$ (width)
$2x+3 = 2(5)+3 = 13$ (length)
Rectangle's dimensions: width: 5 yards; length: 13 yards

22. $f(x) = 0.011x^2-0.097x+4.1$
$f(10) = 0.011(10)^2-0.097(10)+4.1 = 4.23$
$f(0) = 0.011(0)^2-0.097(0)+4.1 = 4.1$
$f(10)-f(0) = 4.23-4.1 = 0.13$
130,000 more people hold more than one job in 1980 than in 1970.

23. $(a+b)^2 = a^2 + 2ab + b^2$

24. $\frac{2x-10}{x+2} + \frac{x+4}{x-2}$ $(\text{LCD} = x^2 - 4)$

$= \frac{2x-10}{x+2} \cdot \frac{x-2}{x-2} + \frac{x+4}{x-2} \cdot \frac{x+2}{x+2}$

$= \frac{(2x-10)(x-2)+(x+4)(x+2)}{(x+2)(x-2)}$

$= \frac{2x^2 - 4x - 10x + 20 + x^2 + 2x + 4x + 8}{(x+2)(x-2)}$

$= \frac{3x^2 - 8x + 28}{(x+2)(x-2)}$

25. $\frac{\frac{1}{x} - \frac{1}{3}}{x+3}$

$= \frac{\frac{3-x}{3x}}{x+3}$

$= \frac{3-x}{3x} \cdot \frac{1}{x+3}$

$= \frac{3-x}{3x^2 + 9x}$

26. $(-7x^3y^2 - 5xy^3) - (-5x^2y^2 + x^3y^2) - (2xy^3 - 6x^2y^2)$

$= -7x^3y^2 - 5xy^3 + 5x^2y^2 - x^3y^2 - 2xy^3 + 6x^2y^2$

$= -8x^3y^2 - 7xy^3 + 11x^2y^2$

27. $x + \frac{12}{x} = -7$

$x\left(x + \frac{12}{x}\right) = -7x$

$x^2 + 12 = -7x$

$x^2 + 7x + 12 = 0$

$(x+4)(x+3) = 0$

$x+4=0$ or $x+3=0$

$x=-4$ $\qquad$ $x=-3$

$\{-4, -3\}$

28. $f(x) = \frac{2x^2 + x - 3}{x-1}$

$= \frac{(2x+3)(x-1)}{x-1}$

$= 2x + 3 \ (x \neq 1)$

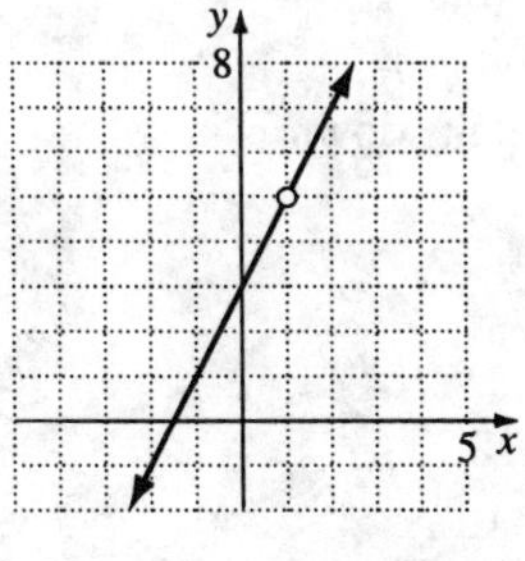

29. $(3x^2 + 10x + 10) \div (x+2) = 3x + 4 + \frac{2}{x+2}$

$$\begin{array}{r} 3x + 4 + \frac{2}{x+2} \\ x+2\overline{)3x^2 + 10x + 10} \\ \underline{3x^2 + \ 6x} \\ 4x + 10 \\ \underline{4x + \ 8} \\ 2 \end{array}$$

30. No; x-coordinate 3 has two different y-coordinates 4 and 11

Chapter 6

Problem Set 6.1

1. $\sqrt{49}=7$ because $7^2=49$.

3. $-\sqrt{49}=-\left(\sqrt{49}\right)=-7$

5. $\sqrt{-49}$ is not a real number because there is no real number whose square is –49.

7. $\sqrt[4]{16}=2$ because $2^4=16$.

9. $\sqrt[5]{-1}=-1$ because $(-1)^5=-1$.

11. $\sqrt{\frac{1}{9}}=\frac{1}{3}$ because $\left(\frac{1}{3}\right)^2=\frac{1}{9}$.

13. $\sqrt[3]{-\frac{1}{64}}=-\frac{1}{4}$ because $\left(-\frac{1}{4}\right)^3=-\frac{1}{64}$.

15. $\sqrt[5]{\frac{1}{32}}=\frac{1}{2}$ because $\left(\frac{1}{2}\right)^5=\frac{1}{32}$.

17. $\sqrt[4]{-16}$ is not a real number.

19. $\left(\sqrt[3]{2}\right)^3=2$

21. $\left(\sqrt[5]{-3}\right)^5=-3$

23. $f(x)=\sqrt{x},\ g(x)=\sqrt{x}+3$

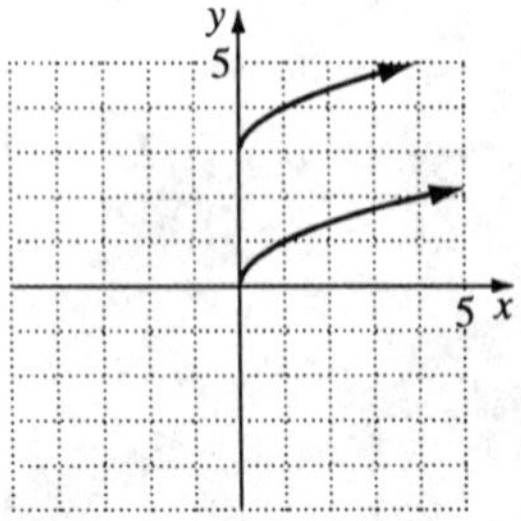

Domain of $f=[0,\infty)$
Range of $f=[0,\infty)$
Domain of $g=[0,\infty)$
Range of $g=[3,\infty)$
g is f shifted 3 units up.

25. $f(x)=\sqrt{x},\ g(x)=\sqrt{x+3}$

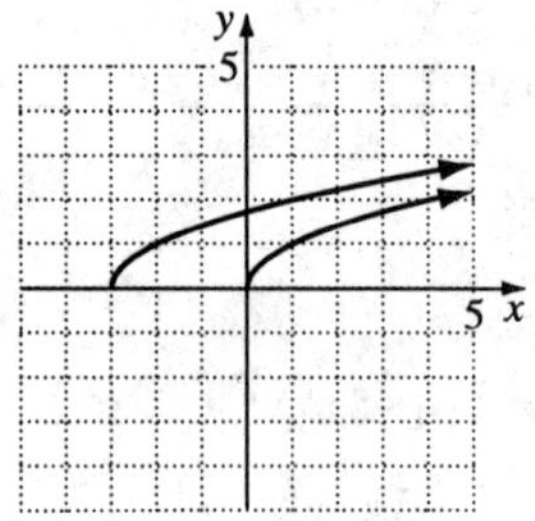

Domain of $f=[0,\infty)$
Range of $f=[0,\infty)$
Domain of $g=[-3,\infty)$
Range of $g=[0,\infty)$
g is f shifted 3 units left.

27. $f(x)=\sqrt{x},\ g(x)=\sqrt{x}-4$

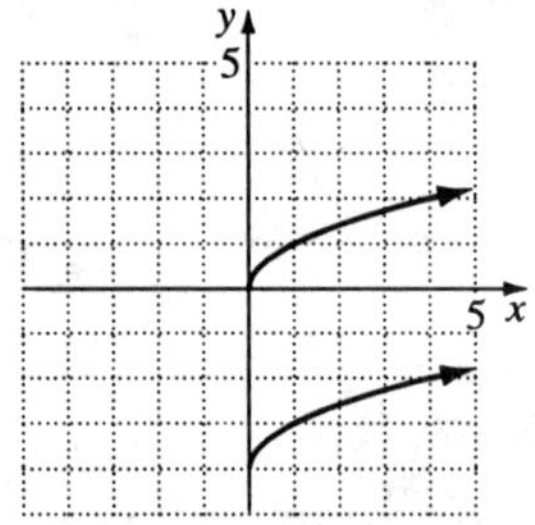

Domain of $f=[0,\infty)$
Range of $f=[0,\infty)$
Domain of $g=[0,\infty)$
Range of $g=[-4,\infty)$
g is f shifted 4 units down.

29. $f(x)=\sqrt{x}$, $g(x)=\sqrt{x-4}$

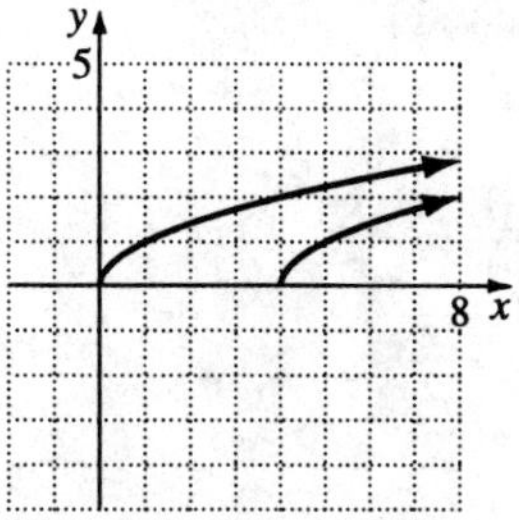

Domain of $f = [0, \infty)$
Range of $f = [0, \infty)$
Domain of $g = [4, \infty)$
Range of $g = [0, \infty)$
g is f shifted 4 units right.

31. $f(x)=\sqrt[3]{x}$, $g(x)=\sqrt[3]{x}-1$

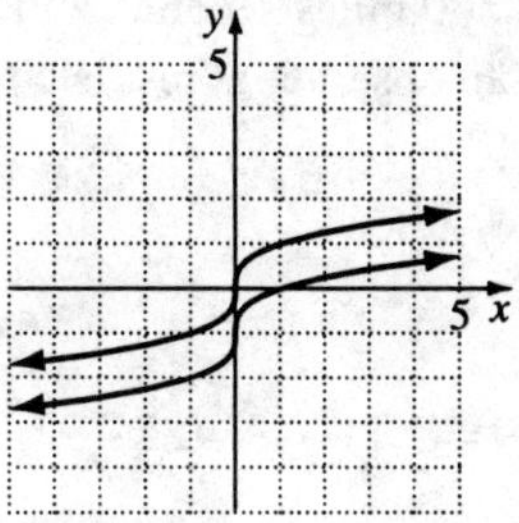

Domain of $f = (-\infty, \infty)$
Range of $f = (-\infty, \infty)$
Domain of $g = (-\infty, \infty)$
Range of $g = (-\infty, \infty)$
g is f shifted 1 unit down.

33. $f(x)=\sqrt[3]{x}$, $g(x)=\sqrt[3]{x-1}$

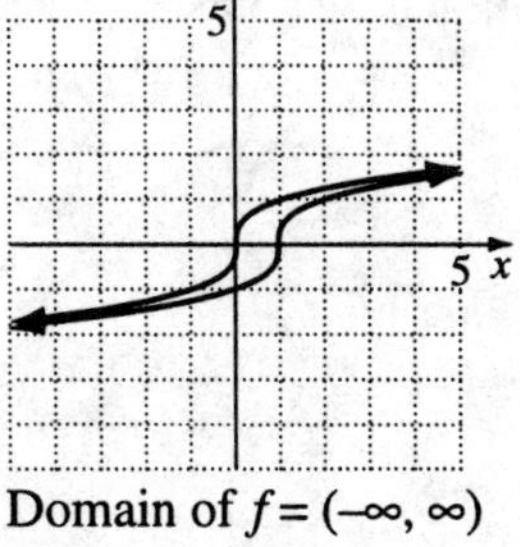

Domain of $f = (-\infty, \infty)$
Range of $f = (-\infty, \infty)$
Domain of $g = (-\infty, \infty)$
Range of $g = (-\infty, \infty)$
g is f shifted 1 unit right.

35. $f(x)=\sqrt[3]{x}$, $g(x)=\sqrt[3]{x+3}$

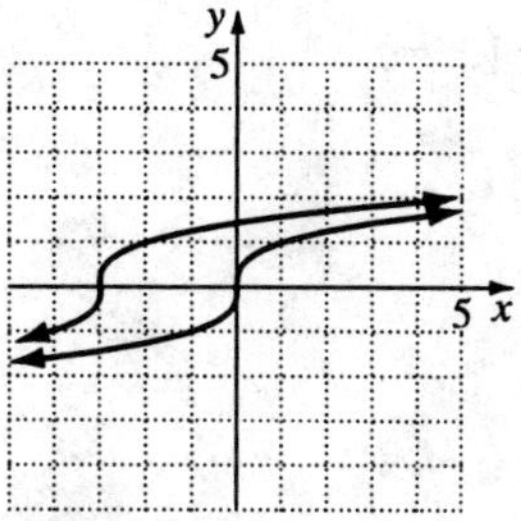

Domain of $f = (-\infty, \infty)$
Range of $f = (-\infty, \infty)$
Domain of $g = (-\infty, \infty)$
Range of $g = (-\infty, \infty)$
g is f shifted 3 units left.

37. $f(x)=\sqrt[4]{x}$, $g(x)=\sqrt[4]{x-1}$

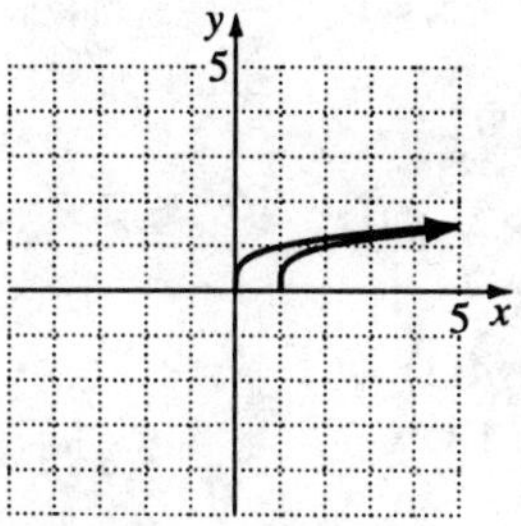

Domain of $f = [0, \infty)$
Range of $f = [0, \infty)$
Domain of $g = [1, \infty)$
Range of $g = [0, \infty)$
g is f shifted 1 unit right.

39. $f(x)=\sqrt{x+5}$
$x+5\ge 0$
$x\ge -5$
Domain of $f = [-5, \infty)$

41. $h(x)=\sqrt{4-x}$
$4-x\ge 0$
$x\le 4$
Domain of $h = (-\infty, 4]$

43. $g(x)=\sqrt{4x-12}$
$4x-12\geq 0$
$4x\geq 12$
$x\geq 3$
Domain of $g=[3,\infty)$

45. $r(x)=\sqrt{12-6x}$
$12-6x\geq 0$
$-6x\geq -12$
$x\leq 2$
Domain of $r=(-\infty, 2]$

47. $f(x)=\sqrt[3]{x+5}$
Domain of $f=(-\infty,\infty)$

49. $f(x)=\sqrt[4]{x+5}$
$x+5\geq 0$
$x\geq -5$
Domain of $f=[-5,\infty)$

51. $2\sqrt{x}-\sqrt[3]{y}+4\sqrt[5]{z}$
$=2\sqrt{36}-\sqrt[3]{-8}+4\sqrt[5]{1}$
$(x=36, y=-8, z=1)$
$=2(6)-(-2)+4(1)$
$=12+2+4=18$

53. $A=\sqrt[3]{-\frac{b}{2}+\sqrt{\frac{b^2}{4}+\frac{a^3}{27}}}$
$=\sqrt[3]{\frac{-2}{2}+\sqrt{\frac{2^2}{4}+\frac{(-3)^3}{27}}}$ $(b=2, a=-3)$
$=\sqrt[3]{-1+\sqrt{1-1}}$
$=\sqrt[3]{-1+\sqrt{0}}$
$=\sqrt[3]{-1}=-1$

55. $\frac{-b\pm\sqrt{b^2-4ac}}{2a}=\frac{-3\pm\sqrt{(3)^2-4(5)(-8)}}{2(5)}$
$(a=5, b=3, c=-8)$
$=\frac{-3\pm\sqrt{9-(-160)}}{10}$
$=\frac{-3\pm\sqrt{169}}{10}$
$=\frac{-3\pm 13}{10}$
$=\frac{-3+13}{10}=\frac{10}{10}=1$
or
$=\frac{-3-13}{10}=\frac{-16}{10}=-\frac{8}{5}$

57. $L(v)=400\sqrt{1-\frac{v^2}{c^2}}$
$v=148{,}800,\ c=186{,}000$
$L(148{,}800)=400\sqrt{1-\frac{(148{,}800)^2}{(186{,}000)^2}}$
$=400\sqrt{1-0.25}$
$=400\sqrt{0.75}\approx 346.4$
The length of a 400-meter tall starship moving at 148,800 miles/second from the perspective of an observer at rest is about 346.4 meters.

59. $N=2\sqrt{Q}-9$
$(Q=121)$:
$N=2\sqrt{121}-9=2(11)-9=22-9=13$
13 syllables

61. $f(x)=4\left(\sqrt{x}\right)^5+17{,}300$
(f in dollars; x in years)
$f(16)=4\left(\sqrt{16}\right)^5+17{,}300=4(4^5)+17{,}300$
$=4096+17{,}300=\$21{,}396$
The yearly income for a person with 16 years of education is \$21,396.

63. $H = \left(10.45 + \sqrt{100 \cdot w} - w\right)(33 - t)$
(H in kilocalorie/square meter/hour,
t in degrees Celsius, w in meters/second)
when $H = 2000$ kilocalories/square meter/hour flesh freezes in 1 minute
If $w = 4$ meters/second, $t = 0°C$, is $H \geq 2000$
$H = \left(10.45 + \sqrt{100(4)} - 4\right)(33 - 0)$
$= \left(10.45 + \sqrt{400} - 4\right)(33)$
$= (10.45 + 20 - 4)(33)$
$= (26.45)(33) = 872.85$
Since $872.45 < 2000$, exposed flesh will not freeze under these conditions.

65. d is true.
a is not true; $f(x) = \sqrt[3]{x} + 1$ is $y = \sqrt[3]{x}$ shifted one unit up, whereas, $g(x) = \sqrt[3]{x+1}$ is $y = \sqrt[3]{x}$ shifted to the left.
b is not true; If n is odd and b is negative, then $\sqrt[n]{b}$ is a real number. c is not true; $\sqrt{4} = 2$, $\sqrt[3]{4} \approx 1.6...$, $\sqrt[6]{4} \approx 1.3$
d is true.

67. Students should verify results.

69. Answers may vary.

71. $\sqrt[3]{\sqrt{\sqrt{169} + \sqrt{9}} + \sqrt{\sqrt[3]{1000} + \sqrt[3]{216}}} = \sqrt[3]{\sqrt{13+3} + \sqrt{10+6}}$
$= \sqrt[3]{\sqrt{16} + \sqrt{16}} = \sqrt[3]{4+4} = \sqrt[3]{8} = 2$

73.

Keystroke	$2\ \sqrt{\ }$	$\sqrt{\ }$	$\sqrt{\ }$	$\sqrt{\ }$	$\sqrt{\ }$	$\sqrt{\ }$
Display	1.4142136	1.1892071	1.0905077	1.0442738	1.0218971	1.0108893

$\sqrt{\ }$	$\sqrt{\ }$	$\sqrt{\ }$
1.0054299	1.0027113	etc.

Pattern: The numbers approach 1. The decimal part is (more or less) halved.

75. $f(x) = \sqrt{x},\ g(x) = \sqrt{-x}$

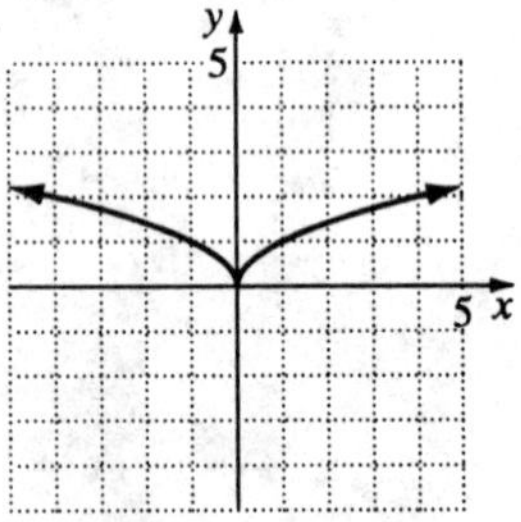

Domain of $f = [0, \infty)$
Range of $f = [0, \infty)$
Domain of $g = (-\infty, 0]$
Range of $g = [0, \infty)$
g is f reflected about the y-axis.

77. h is f shifted 3 units to the right.
$h(x) = \sqrt{x-3}$

Review Problems

79. $\dfrac{x^3+64}{x^2-16} = \dfrac{(x+4)(x^2-4x+16)}{(x+4)(x-4)}$
$= \dfrac{x^2-4x+16}{x-4}\ (x \neq \pm 4)$

80. $125 - 8x^3 = (5)^3 - (2x)^3$
$= (5-2x)[5^2 + 5(2x) + (2x)^2]$
$= (5-2x)(25 + 10x + 4x^2)$

81. $x + 3y - z = 5$
$2x - 5y - z = -8$
$-x + 2y + 3z = 13$
Multiply equation 2 by −1 and add to 1.
$x + 3y - z = 5$
$\underline{-2x + 5y + z = 8}$
$-x + 8y = 13$ (4)
Multiply equation 1 by 3 and add to equation 3.
$3x + 9y - 3z = 15$
$\underline{-x + 2y + 3z = 13}$
$2x + 11y = 28$ (5)
Multiply equation 4 by 2 and add to 5.
$-2x + 16y = 26$
$\underline{2x + 11y = 28}$
$27y = 54$
$y = 2$
Equation 4:
$-x + 8y = 13$
$-x + 8(2) = 13$
$-x = -3$
$x = 3$
Equation 1:
$x + 3y - z = 5$
$3 + 3(2) - z = 5$
$-z = -4$
$z = 4$
$\{(3, 2, 4)\}$

Problem Set 6.2

1. $\sqrt[3]{4^3} = 4$
$\left(\sqrt[n]{x^n} = n \text{ when } n \text{ is odd}\right)$

3. $\sqrt[3]{(-2)^3} = -2$

5. $\sqrt[4]{2^4} = |2| = 2$
$\left(\sqrt[n]{x^n} = |x| \text{ when } n \text{ is even}\right)$

7. $\sqrt[4]{(-2)^4} = |-2| = 2$

9. $\sqrt[5]{(-2)^5} = -2$

11. $\sqrt[17]{(-9)^{17}} = -9$

13. $\sqrt[18]{(-9)^{18}} = |-9| = 9$

15. $\sqrt[n]{(-8)^n}$ (n is odd) $= -8$

17. $\sqrt[n]{(-8)^n}$ (n is even) $= |-8| = 8$

19. $x^{1/4} = \sqrt[4]{x}$

21. $36^{1/2} = \sqrt{36} = 6$

23. $8^{1/3} = \sqrt[3]{8} = 2$

25. $(xy)^{1/6} = \sqrt[6]{xy}$

27. $16^{5/2} = \left(\sqrt{16}\right)^5 = 1024$

29. $(8x)^{2/3} = \left(\sqrt[3]{8x}\right)^2 = \left(\sqrt[3]{8} \cdot \sqrt[3]{x}\right)^2$
$= \left(2\sqrt[3]{x}\right)^2 = 4\sqrt[3]{x^2}$

31. $-27^{4/3} = -\left(\sqrt[3]{27}\right)^4 = -(3)^4 = -81$

33. $64^{-1/2} = \frac{1}{\sqrt{64}} = \frac{1}{8}$

35. $\left(\frac{1}{64}\right)^{-1/3} = 64^{1/3} = \sqrt[3]{64} = 4$

37. $(-64)^{-1/3} = \frac{1}{\sqrt[3]{-64}} = -\frac{1}{4}$

39. $\left(\frac{1}{32}\right)^{-3/5} = 32^{3/5} = \left(\sqrt[5]{32}\right)^3 = 2^3 = 8$

41. $16^{-5/2} = \frac{1}{\left(\sqrt{16}\right)^5} = \frac{1}{4^5} = \frac{1}{1024}$

43. $\sqrt[3]{7} = 7^{1/3}$

45. $\sqrt{x^3} = x^{3/2}$

47. $\sqrt[5]{x^3} = x^{3/5}$

49. $\sqrt[4]{ab} = (ab)^{1/4}$

51. $\sqrt[5]{x^2yz^4} = (x^2yz^4)^{1/5}$

53. $\left(\sqrt{19xy}\right)^3 = (19xy)^{3/2}$

55. $\left(\sqrt[6]{11xy^2}\right)^5 = (11xy^2)^{5/6}$

57. $3^{3/4} \cdot 3^{1/4} = 3^{3/4+1/4} = 3^{4/4} = 3$

59. $\frac{16^{3/4}}{16^{1/4}} = 16^{3/4-1/4} = 16^{1/2} = \sqrt{16} = 4$

61. $x^{3/4} \cdot x^{1/3} = x^{3/4+1/3} = x^{9/12+4/12} = x^{13/12}$

63. $\frac{x^{4/5}}{x^{1/5}} = x^{4/5-1/5} = x^{3/5}$

65. $(x^{2/3})^3 = x^{\frac{2}{3}\cdot 3} = x^2$

67. $\frac{x^{1/3}}{x^{3/4}} = x^{\frac{1}{3}-\frac{3}{4}} = x^{\frac{4}{12}-\frac{9}{12}} = x^{-5/12} = \frac{1}{x^{5/12}}$

69. $(7y^{1/3})(2y^{1/4}) = 14y^{\frac{1}{3}+\frac{1}{4}} = 14y^{\frac{4}{12}+\frac{3}{12}}$
$= 14y^{7/12}$

71. $(3x^{3/4})(-5x^{-1/2}) = -15x^{\frac{3}{4}-\frac{1}{2}} = -15x^{\frac{3}{4}-\frac{2}{4}}$
$= -15x^{1/4}$

73. $\frac{20x^{1/2}}{5x^{1/4}} = 4x^{\frac{1}{2}-\frac{1}{4}} = 4x^{\frac{2}{4}-\frac{1}{4}} = 4x^{1/4}$

75. $\frac{80y^{1/6}}{10y^{1/4}} = 8y^{\frac{1}{6}-\frac{1}{4}} = 8y^{\frac{2}{12}-\frac{3}{12}} = 8y^{-1/12}$
$= \frac{8}{y^{1/12}}$

77. $(2x^{1/5}y^2z^{2/5})^5 = 2^5(x^{1/5})^5(y^2)^5(z^{2/5})^5$
$= 32xy^{10}z^2$

79. $(25x^4y^6)^{1/2} = 25^{1/2}(x^4)^{1/2}(y^6)^{1/2}$
$= 25^{1/2}x^2y^3 = 5x^2y^3$

81. $(16xy^{1/4}z^{2/3})^{1/4} = 16^{1/4}x^{1/4}y^{1/16}z^{1/6}$
$= 2x^{1/4}y^{1/16}z^{1/6}$

83. $\left(\frac{2x^{1/4}}{5y^{1/3}}\right)^3 = \frac{(2x^{1/4})^3}{(5y^{1/3})^3} = \frac{8x^{3/4}}{125y}$

85. $\left(\frac{x^3}{y^5}\right)^{-1/2} = \frac{x^{-3/2}}{y^{-5/2}} = \frac{y^{5/2}}{x^{3/2}}$

87. $\sqrt[9]{a^3} = a^{3/9} = a^{1/3} = \sqrt[3]{a}$

89. $\sqrt[3]{8x^6} = (8x^6)^{1/3} = 8^{1/3}x^{6/3} = 2x^2$

91. $\sqrt[5]{x^{10}y^{15}} = (x^{10}y^{15})^{1/5} = x^{10/5}y^{15/5}$
$= x^2y^3$

93. $\sqrt[9]{2^3x^3y^6} = (2^3x^3y^6)^{1/9} = 2^{1/3}x^{1/3}y^{2/3}$
$= (2xy^2)^{1/3} = \sqrt[3]{2xy^2}$

95. $\sqrt[9]{27x^3y^6} = \sqrt[9]{3^3x^3y^6} = (3^3x^3y^6)^{1/9}$
$= 3^{1/3}x^{1/3}y^{2/3} = (3xy^2)^{1/3} = \sqrt[3]{3xy^2}$

97. $\sqrt[3]{3}\cdot\sqrt{3} = 3^{1/3}\cdot 3^{1/2} = 3^{5/6}$
$= \sqrt[6]{3^5}$ or $\sqrt[6]{243}$

99. $\sqrt[4]{2}\cdot\sqrt{2} = 2^{1/4}\cdot 2^{1/2} = 2^{3/4} = \sqrt[4]{2^3}$
$= \sqrt[4]{8}$

101. $\frac{\sqrt{3}}{\sqrt[3]{3}} = \frac{3^{1/2}}{3^{1/3}} = 3^{1/2-1/3} = 3^{1/6} = \sqrt[6]{3}$

103. $x^{3/2} + x^{1/2} = x^{1/2}(x+1)$

105. $6x^{1/3} + 3x^{4/3} = 3x^{1/3}(2+x)$

107. $x^{2/5} + x^{-3/5} = x^{-3/5}(x+1) = \frac{x+1}{x^{3/5}}$

109. $15x^{-1/2} - 20x^{-5/2} = 5x^{-5/2}(3x^2-4)$
$= \frac{5(3x^2-4)}{x^{5/2}}$

111. $P = 4\cdot 4^{3/2} + 60 = 4\left(\sqrt{4}\right)^3 + 60$
$= 4\cdot 8 + 60 = 92$
$92 - 62 = 30$ tons beyond 62 tons
Fines = (30)(10,000) = 300,000
\$300,000

113. $R = \frac{1}{4}(16^{1/4}+3)16^{-3/4}$
$= \frac{1}{4}\left(\sqrt[4]{16}+3\right)\cdot\frac{1}{\left(\sqrt[4]{16}\right)^3} = \frac{1}{4}(2+3)\cdot\frac{1}{8}$
$= \frac{5}{32}$
$\frac{5}{32}$ units of pollution

115. $r = \left(\frac{A}{P}\right)^{1/t} - 1$
(P = \$80,000, A = \$120,000, t = 4 years):
$r = \left(\frac{120{,}000}{80{,}000}\right)^{1/4} - 1 \approx 0.107 = 10.7\%$

117. $f(d) = 0.07d^{3/2}$
$f(9) = 0.07(9)^{3/2} = 0.07(27) = 1.89$
The duration of a storm whose diameter is 9 miles is 1.89 hours.

119. b is true.

121. $y_1 = \sqrt{x^2}$ and $y_2 = x$

a. From the graph $\sqrt{x^2} = x$ when $x \ge 0$.

b. $\sqrt{x^2} \ne -x$ when $x < 0$.

123. **a.** It only graphs for $x \ge 0$.

b. It graphs for all real numbers x.

125. $N = 13.49(0.967)^t - 1$

a. $N = 13.49(0.967)^{31} - 1 \approx 3.8$
4 O-rings

b.

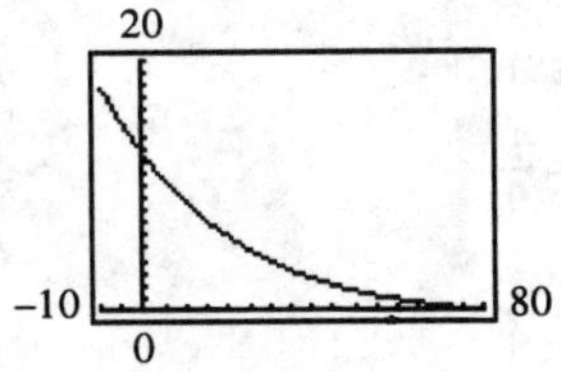

No; as the temperature drops, the number of O-rings that fail increases.

127. Answers may vary.

129. Answers may vary.

131. $\dfrac{8^{-4/3} + 2^{-2}}{16^{-3/4} + 2^{-1}} = \dfrac{\frac{1}{8^{4/3}} + \frac{1}{2^2}}{\frac{1}{16^{3/4}} + \frac{1}{2}} = \dfrac{\frac{1}{(\sqrt[3]{8})^4} + \frac{1}{4}}{\frac{1}{(\sqrt[4]{16})^3} + \frac{1}{2}}$

$= \dfrac{\frac{1}{16} + \frac{1}{4}}{\frac{1}{8} + \frac{1}{2}} \cdot \dfrac{16}{16} = \dfrac{1+4}{2+8} = \dfrac{5}{10} = \dfrac{1}{2}$

The birthday boy ate $\frac{1}{2}$ of the cake. Thus, $\frac{1}{2}$ was left over. The professor ate half of that, or $\frac{1}{2} \cdot \frac{1}{2} = \frac{1}{4}$ of the cake.

133. $6x^{2/3} - 17x^{1/3} + 12 = (3x^{1/3} - 4)(2x^{1/3} - 3)$

135. $[3 + (27^{2/3} + 32^{2/5})]^{3/2} - 9^{1/2} = \left[3 + \left(\sqrt[3]{27}\right)^2 + \left(\sqrt[5]{32}\right)^2\right]^{3/2} - 9^{1/2}$

$= [3 + 3^2 + 2^2]^{3/2} - 9^{1/2} = 16^{3/2} - 9^{1/2}$

$= \left(\sqrt{16}\right)^3 - \sqrt{9} = 4^3 - 3 = 64 - 3 = 61$

137. Line 5:

$$\left[(x+1)-\frac{1}{2}(2x+1)\right]^2=\left[x-\frac{1}{2}(2x+1)\right]^2$$

$$\left[x+1-x-\frac{1}{2}\right]^2=\left(x-x-\frac{1}{2}\right)^2$$

$$\left(\frac{1}{2}\right)^2=\left(-\frac{1}{2}\right)^2$$

In terms of specific numbers, the line with the square brackets says $\left[\frac{1}{2}\right]^2=\left[-\frac{1}{2}\right]^2$, which is correct. The error occurs in the next step, where the writer took the square root of both sides. If $\left[\frac{1}{2}\right]^2=\left[-\frac{1}{2}\right]^2$, it is incorrect to take the square root and assert that $\frac{1}{2}=-\frac{1}{2}$. You cannot take the square root of both sides of an equation without first checking to see if sign problems might occur.

139. $81^{1/4}=\sqrt[4]{81}=3$

$81^{1/2}=\sqrt{81}=9$

$81^{1/2}>81^{1/4}$

In general, if $b>1$ and $n>m$, $b^{1/m}>b^{1/n}$.

141. Let x = number.

$$\frac{x^{-2/3}}{\sqrt[3]{x}}=\frac{1}{4}$$

$$\frac{x^{-2/3}}{x^{1/3}}=\frac{1}{4}$$

$$x^{-1}=\frac{1}{4}$$

$$\frac{1}{x}=\frac{1}{4}$$

$$x=4$$

The number is 4.

Review Problems

143.

$x+4y-2z=-3$

$2x+y+z=3$

$-5x-2y+3z=-14$

$$\left[\begin{array}{ccc|c}1&4&-2&-3\\2&1&1&3\\-5&-2&3&-14\end{array}\right]\begin{array}{l}\\R_2-2R_1\to\\5R_1+R_3\to\end{array}\left[\begin{array}{ccc|c}1&4&-2&-3\\0&-7&5&9\\0&18&-7&-29\end{array}\right]\begin{array}{l}\\\\18R_2+7R_3\to\end{array}\left[\begin{array}{ccc|c}1&4&-2&-3\\0&-7&5&9\\0&0&41&-41\end{array}\right]\begin{array}{l}\\-\frac{1}{7}R_2\to\\\frac{1}{41}R_3\to\end{array}$$

$$\left[\begin{array}{ccc|c}1&4&-2&-3\\0&1&-\frac{5}{7}&-\frac{9}{7}\\0&0&1&-1\end{array}\right]$$

$x+4y-2z=-3$

$y-\frac{5}{7}z=-\frac{9}{7}$

$z=-1$

$y-\frac{5}{7}(-1)=-\frac{9}{7}$

$y=-\frac{14}{7}=-2$

$x + 4(-2) - 2(-1) = -3$
$x - 6 = -3$
$x = 3$
$\{(3, -2, -1)\}$

144. $x^6 - x^2 = x^2(x^4 - 1)$
$= x^2(x^2 - 1)(x^2 + 1) = x^2(x - 1)(x + 1)(x^2 + 1)$

145. $\frac{5}{y+2} - \frac{3}{y+5} = \frac{9}{y^2 + 7y + 10}$

$$\frac{5(y+5)}{(y+2)(y+5)} - \frac{3(y+2)}{(y+5)(y+2)} = \frac{9}{(y+2)(y+5)}$$

$$(y+2)(y+5)\left[\frac{5(y+5)}{(y+2)(y+5)} - \frac{3(y+2)}{(y+5)(y+2)}\right] = (y+2)(y+5)\left[\frac{9}{(y+2)(y+5)}\right]$$

$5(y + 5) - 3(y + 2) = 9$
$5y + 25 - 3y - 6 = 9$
$2y + 19 = 9$
$2y = -10$
$y = -5$
Since –5 causes the denominator to be zero in the original equation, –5 is not a solution. The equation has no solution. ∅

Problem Set 6.3

1. $\sqrt{3}\sqrt{5} = \sqrt{3 \cdot 5} = \sqrt{15}$

3. $\sqrt[3]{2}\sqrt[3]{9} = \sqrt[3]{2 \cdot 9} = \sqrt[3]{18}$

5. $\sqrt[4]{11}\sqrt[4]{3} = \sqrt[4]{11 \cdot 3} = \sqrt[4]{33}$

7. $\sqrt{x+3}\sqrt{x-3} = \sqrt{(x+3)(x-3)}$
$= \sqrt{x^2 - 9}$

9. $\sqrt{7}\sqrt{2xy} = \sqrt{7 \cdot 2xy} = \sqrt{14xy}$

11. $\sqrt[7]{7x^2y}\sqrt[7]{11x^3y^2} = \sqrt[7]{7x^2y \cdot 11x^3y^2}$
$= \sqrt[7]{77x^5y^3}$

13. $\sqrt{20} = \sqrt{4 \cdot 5} = \sqrt{2^2 \cdot 5} = 2\sqrt{5}$

15. $\sqrt{80} = \sqrt{16 \cdot 5} = \sqrt{4^2 \cdot 5} = 4\sqrt{5}$

17. $\sqrt{250} = \sqrt{25 \cdot 10} = \sqrt{5^2 \cdot 10} = 5\sqrt{10}$

19. $7\sqrt{28} = 7\sqrt{4 \cdot 7} = 7\sqrt{2^2 \cdot 7} = 7\left(2\sqrt{7}\right)$
$= 14\sqrt{7}$

21. $2\sqrt{98} = 2\sqrt{49 \cdot 2} = 14\sqrt{2}$

23. $\sqrt[3]{54} = \sqrt[3]{27 \cdot 2} = \sqrt[3]{3^3 \cdot 2} = 3\sqrt[3]{2}$

25. $\sqrt[5]{64} = \sqrt[5]{32 \cdot 2} = \sqrt[5]{2^5 \cdot 2} = 2\sqrt[5]{2}$

27. $6\sqrt[3]{16} = 6\sqrt[3]{8 \cdot 2} = 6\sqrt[3]{2^3 \cdot 2}$
$= 6(2)\left(\sqrt[3]{2}\right) = 12\sqrt[3]{2}$

29. $\sqrt{x^7} = \sqrt{x^6 \cdot x} = \sqrt{(x^3)^2 \cdot x} = x^3\sqrt{x}$

31. $\sqrt[3]{y^8} = \sqrt[3]{y^6 \cdot y^2} = \sqrt[3]{(y^2)^3 \cdot y^2} = y^2\sqrt[3]{y^2}$

33. $\sqrt[5]{z^{16}} = \sqrt[5]{z^{15} \cdot z} = \sqrt[5]{(z^3)^5 \cdot z} = z^3 \sqrt[5]{z}$

35. $\sqrt{x^8 y^9} = \sqrt{x^8 y^8 \cdot y} = \sqrt{(x^4 y^4)^2 \cdot y}$
$= x^4 y^4 \sqrt{y}$

37. $\sqrt[3]{x^{14} y^3 z} = \sqrt[3]{x^{12} x^2 y^3 z} = (x^{12} y^3)^{1/3} \sqrt[3]{x^2 z}$
$= x^4 y \sqrt[3]{x^2 z}$

39. $\sqrt{12y} = \sqrt{4 \cdot 3y} = \sqrt{2^2 \cdot 3y} = 2\sqrt{3y}$

41. $\sqrt{48x^3} = \sqrt{16x^2 \cdot 3x} = \sqrt{(4x)^2 \cdot 3x}$
$= 4x\sqrt{3x}$

43. $\sqrt[3]{32x^{13}} = \sqrt[3]{8x^{12} \cdot 4x} = \sqrt[3]{(2x^4)^3 \cdot 4x}$
$= 2x^4 \sqrt[3]{4x}$

45. $\sqrt[3]{81x^8 y^6} = \sqrt[3]{27x^6 y^6 \cdot 3x^2}$
$= \sqrt[3]{(3x^2 y^2)^3 \cdot 3x^2} = 3x^2 y^2 \sqrt[3]{3x^2}$

47. $\sqrt[4]{80x^{10}} = \sqrt[4]{16x^8 \cdot 5x^2} = \sqrt[4]{(2x^2)^4 \cdot 5x^2}$
$= 2x^2 \sqrt[4]{5x^2}$

49. $\sqrt[5]{64x^6 y^{17}} = \sqrt[5]{32x^5 y^{15} \cdot 2xy^2}$
$= \sqrt[5]{(2xy^3)^5 \cdot 2xy^2} = 2xy^3 \sqrt[5]{2xy^2}$

51. $\sqrt[3]{18x^{15} y^7 z^2} = \sqrt[3]{x^{15} y^6 \cdot 18yz^2}$
$= \sqrt[3]{(x^5 y^2)^3 \cdot 18yz^2} = x^5 y^2 \sqrt[3]{18yz^2}$

53. $6\sqrt[4]{32x^{19} y^8 z^9} = 6\sqrt[4]{16x^{16} y^8 z^8 \cdot 2x^3 z}$
$= 6\sqrt[4]{(2x^4 y^2 z^2)^4 \cdot 2x^3 z} = 12x^4 y^2 z^2 \sqrt[4]{2x^3 z}$

55. $\sqrt[3]{(x+y)^4} = \sqrt[3]{(x+y)^3 \cdot (x+y)}$
$= (x+y) \sqrt[3]{x+y}$

57. $\sqrt{3}\sqrt{6} = \sqrt{18} = \sqrt{9 \cdot 2} = 3\sqrt{2}$

59. $(2\sqrt{5})(3\sqrt{20}) = 6\sqrt{100} = 6(10) = 60$

61. $\sqrt[3]{9} \cdot \sqrt[3]{6} = \sqrt[3]{54} = \sqrt[3]{27 \cdot 2} = 3\sqrt[3]{2}$

63. $\sqrt{5x^3}\sqrt{8x^2} = \sqrt{40x^5} = \sqrt{4x^4 \cdot 10x}$
$= 2x^2\sqrt{10x}$

65. $\sqrt{6xy^4}\sqrt{2x^3 y^7} = \sqrt{12x^4 y^{11}}$
$= \sqrt{4x^4 y^{10} \cdot 3y} = 2x^2 y^5 \sqrt{3y}$

67. $\sqrt[3]{25x^4 y^2}\sqrt[3]{5xy^{12}} = \sqrt[3]{125x^5 y^{14}}$
$= \sqrt[3]{125x^3 y^{12} \cdot x^2 y^2} = 5xy^4 \sqrt[3]{x^2 y^2}$

69. $\sqrt[4]{8x^2 y^3 z^6}\sqrt[4]{2x^4 yz} = \sqrt[4]{16x^6 y^4 z^7}$
$= \sqrt[4]{16x^4 y^4 z^4 \cdot x^2 z^3} = 2xyz\sqrt[4]{x^2 z^3}$

71. $\sqrt[5]{8x^4 y^6}\sqrt[5]{8xy^7} = \sqrt[5]{64x^5 y^{13}}$
$= \sqrt[5]{32x^5 y^{10} \cdot 2y^3} = 2xy^2 \sqrt[5]{2y^3}$

73. $\sqrt[3]{(x+2)^2}\sqrt[3]{(x+2)^5} = \sqrt[3]{(x+2)^7}$
$= \sqrt[3]{(x+2)^6 (x+2)} = (x+2)^2 \sqrt[3]{x+2}$

75. $2\sqrt[3]{x^4 y^5}\sqrt[3]{8x^{15} y} = 2\sqrt[3]{8x^{19} y^6}$
$= 2\sqrt[3]{8x^{18} y^6 \cdot x} = 4x^6 y^2 \sqrt[3]{x}$

77. $r = 2\sqrt{5L}$
$r = 2\sqrt{5(40)} = 2\sqrt{200} = 2\sqrt{100 \cdot 2}$
$= 2 \cdot 10\sqrt{2}$
$= 20\sqrt{2}$
The speed of the car was $20\sqrt{2}$ mph.

79. d is true.
$\sqrt[5]{3^{25}} = \sqrt[5]{(3^5)^5} = 3^5 = 243$

81. $\sqrt{x^4} = x^2 \quad (x \geq 0)$

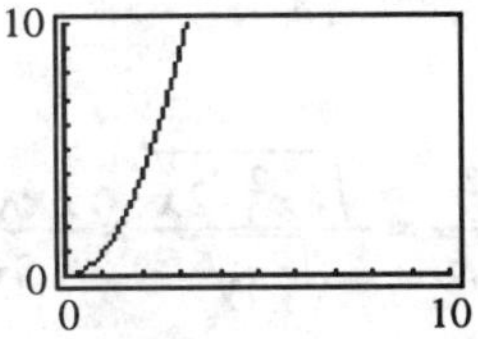

83. $\sqrt{18x^2} \neq 9\sqrt{2}x \quad (x \geq 0)$

$\sqrt{18x^2} = 3x\sqrt{2} \quad (x \geq 0)$

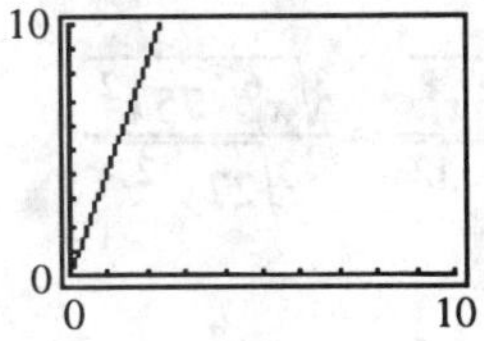

85. Answers may vary.

87. Answers will vary.

$$\sqrt{a+b} = \sqrt{a} + \sqrt{b}$$
$$a+b = a+b+2\sqrt{ab}$$
$$0 = \sqrt{ab}$$
$$0 = ab$$
$$a = 0 \text{ or } b = 0$$
$$a = 0,\ b = 1$$
$$a = 4,\ b = 0$$
$$a = 16,\ b = 0$$

89. Multiplies the square root by $\sqrt{3}$.

91. Multiply the number by 27.

Review Problems

93. $\dfrac{8x^3y^2}{-24x^7y^{-5}} = -\frac{1}{3}x^{-4}y^7 = \dfrac{-y^7}{3x^4}$

94. $(4y^3 - 22y^2 + 44y - 35) \div (2y - 5) = 2y^2 - 6y + 7$

$$\begin{array}{r|l} & 2y^2 - 6y + 7 \\ \hline 2y-5 & 4y^3 - 22y^2 + 44y - 35 \\ & \underline{4y^3 - 10y^2} \\ & -12y^2 + 44y \\ & \underline{-12y^2 + 30y} \\ & 14y - 35 \\ & \underline{14y - 35} \\ & 0 \end{array}$$

95. $H = \dfrac{62.4Ns}{33{,}000}$

$33{,}000H = 62.4Ns$

$\dfrac{33{,}000H}{62.4N} = s$

$s = \dfrac{33{,}000(468)}{62.4(1500)} = \dfrac{15{,}444{,}000}{93{,}600} = 165$

The dam is 165 feet high.

Problem Set 6.4

1. $\frac{\sqrt{40}}{\sqrt{5}} = \sqrt{\frac{40}{5}} = \sqrt{8} = \sqrt{4 \cdot 2} = 2\sqrt{2}$

3. $\frac{\sqrt[3]{48}}{\sqrt[3]{3}} = \sqrt[3]{\frac{48}{3}} = \sqrt[3]{16} = \sqrt[3]{8 \cdot 2} = 2\sqrt[3]{2}$

5. $\frac{\sqrt{54x^3}}{\sqrt{6x}} = \sqrt{\frac{54x^3}{6x}} = \sqrt{9x^2} = 3x$

7. $\frac{-10\sqrt[3]{15}}{-2\sqrt[3]{5}} = \frac{-10}{-2}\sqrt[3]{\frac{15}{5}} = 5\sqrt[3]{3}$

9. $\frac{-12\sqrt[4]{8}}{6\sqrt[4]{2}} = \frac{-12}{6}\sqrt[4]{\frac{8}{2}} = -2\sqrt[4]{4}$

11. $\frac{\sqrt{200x^3}}{\sqrt{10x^{-1}}} = \sqrt{\frac{200x^3}{10x^{-1}}} = \sqrt{20x^4} = \sqrt{4 \cdot 5x^4}$
$= 2x^2\sqrt{5}$

13. $\frac{\sqrt[3]{108x^4y^5}}{\sqrt[3]{2xy^3}} = \sqrt[3]{\frac{108x^4y^5}{2xy^3}} = \sqrt[3]{54x^3y^2}$
$= \sqrt[3]{27 \cdot 2x^3y^2} = 3x\sqrt[3]{2y^2}$

15. $\frac{\sqrt{98x^2y}}{\sqrt{2x^{-3}y}} = \sqrt{\frac{98x^2y}{2x^{-3}y}} = \sqrt{49x^5}$
$= \sqrt{49(x^2)^2x} = 7x^2\sqrt{x}$

17. $\frac{\sqrt{x^2-y^2}}{\sqrt{x-y}} = \sqrt{\frac{x^2-y^2}{x-y}} = \sqrt{\frac{(x+y)(x-y)}{x-y}}$
$= \sqrt{x+y}$

19. $\sqrt{\frac{11}{4}} = \frac{\sqrt{11}}{\sqrt{4}} = \frac{\sqrt{11}}{2}$

21. $\sqrt[3]{\frac{19}{27}} = \frac{\sqrt[3]{19}}{\sqrt[3]{27}} = \frac{\sqrt[3]{19}}{3}$

23. $\sqrt{\frac{x^2}{36y^8}} = \frac{\sqrt{x^2}}{\sqrt{36y^8}} = \frac{x}{6y^4}$

25. $\sqrt{\frac{8x^3}{25y^6}} = \frac{\sqrt{8x^3}}{\sqrt{25y^6}} = \frac{\sqrt{4x^2 \cdot 2x}}{\sqrt{25y^6}} = \frac{2x\sqrt{2x}}{5y^3}$

27. $\sqrt[3]{\frac{x^4}{8y^3}} = \frac{\sqrt[3]{x^4}}{\sqrt[3]{8y^3}} = \frac{\sqrt[3]{x^3 \cdot x}}{\sqrt[3]{8y^3}} = \frac{x\sqrt[3]{x}}{2y}$

29. $\sqrt[3]{\frac{75x^8}{27y^{12}}} = \frac{\sqrt[3]{75x^8}}{\sqrt[3]{27y^{12}}} = \frac{\sqrt[3]{x^6 \cdot 75x^2}}{\sqrt[3]{27y^{12}}}$
$= \frac{x^2\sqrt[3]{75x^2}}{3y^4}$

31. $\sqrt[4]{\frac{9y^6}{x^8}} = \frac{\sqrt[4]{9y^6}}{\sqrt[4]{x^8}} = \frac{\sqrt[4]{y^4 \cdot 9y^2}}{\sqrt[4]{x^8}} = \frac{y\sqrt[4]{9y^2}}{x^2}$

33. $\sqrt[5]{\frac{64x^{13}}{y^{20}}} = \frac{\sqrt[5]{64x^{13}}}{\sqrt[5]{y^{20}}} = \frac{\sqrt[5]{32x^{10} \cdot 2x^3}}{\sqrt[5]{y^{20}}}$
$= \frac{2x^2\sqrt[5]{2x^3}}{y^4}$

35. $\frac{1}{\sqrt{2}} = \frac{1}{\sqrt{2}} \cdot \frac{\sqrt{2}}{\sqrt{2}} = \frac{\sqrt{2}}{2}$

37. $\sqrt{\frac{7}{10}} = \frac{\sqrt{7}}{\sqrt{10}} \cdot \frac{\sqrt{10}}{\sqrt{10}} = \frac{\sqrt{70}}{10}$

39. $\sqrt{\frac{5}{8}} = \frac{\sqrt{5}}{\sqrt{8}} \cdot \frac{\sqrt{8}}{\sqrt{8}} = \frac{\sqrt{40}}{8} = \frac{2\sqrt{10}}{8} = \frac{\sqrt{10}}{4}$

41. $\frac{1}{\sqrt[3]{2}} = \frac{1}{\sqrt[3]{2}} \cdot \frac{\sqrt[3]{4}}{\sqrt[3]{4}} = \frac{\sqrt[3]{4}}{2}$

43. $\frac{6}{\sqrt[3]{4}} = \frac{6}{\sqrt[3]{4}} \cdot \frac{\sqrt[3]{16}}{\sqrt[3]{16}} = \frac{6\sqrt[3]{16}}{4} = \frac{3\sqrt[3]{16}}{2}$
$= \frac{3\sqrt[3]{8 \cdot 2}}{2} = 3\sqrt[3]{2}$

45. $\sqrt[3]{\frac{2}{3}}=\frac{\sqrt[3]{2}}{\sqrt[3]{3}}\cdot\frac{\sqrt[3]{9}}{\sqrt[3]{9}}=\frac{\sqrt[3]{18}}{3}$

47. $\sqrt[3]{\frac{2}{9}}=\frac{\sqrt[3]{2}}{\sqrt[3]{9}}\cdot\frac{\sqrt[3]{81}}{\sqrt[3]{81}}=\frac{\sqrt[3]{162}}{9}=\frac{3\sqrt[3]{6}}{9}=\frac{\sqrt[3]{6}}{3}$

49. $\frac{1}{\sqrt[4]{3}}=\frac{1}{\sqrt[4]{3}}\cdot\frac{\sqrt[4]{27}}{\sqrt[4]{27}}=\frac{\sqrt[4]{27}}{3}$

51. $\sqrt[4]{\frac{5}{2}}=\frac{\sqrt[4]{5}}{\sqrt[4]{2}}\cdot\frac{\sqrt[4]{8}}{\sqrt[4]{8}}=\frac{\sqrt[4]{40}}{2}$

53. $\sqrt{1+\frac{1}{2}}=\sqrt{\frac{3}{2}}=\frac{\sqrt{3}}{\sqrt{2}}\cdot\frac{\sqrt{2}}{\sqrt{2}}=\frac{\sqrt{6}}{2}$

55. $\frac{3}{\sqrt{x}}=\frac{3}{\sqrt{x}}\cdot\frac{\sqrt{x}}{\sqrt{x}}=\frac{3\sqrt{x}}{x}$

57. $\frac{5}{\sqrt{3x}}=\frac{5}{\sqrt{3x}}\cdot\frac{\sqrt{3x}}{\sqrt{3x}}=\frac{5\sqrt{3x}}{3x}$

59. $\sqrt{\frac{3}{5x}}=\frac{\sqrt{3}}{\sqrt{5x}}\cdot\frac{\sqrt{5x}}{\sqrt{5x}}=\frac{\sqrt{15x}}{5x}$

61. $\sqrt{\frac{x}{7y}}=\frac{\sqrt{x}}{\sqrt{7y}}\cdot\frac{\sqrt{7y}}{\sqrt{7y}}=\frac{\sqrt{7xy}}{7y}$

63. $\sqrt{\frac{3x}{7y}}=\frac{\sqrt{3x}}{\sqrt{7y}}\cdot\frac{\sqrt{7y}}{\sqrt{7y}}=\frac{\sqrt{21xy}}{7y}$

65. $\frac{4}{\sqrt[3]{x}}=\frac{4}{\sqrt[3]{x}}\cdot\frac{\sqrt[3]{x^2}}{\sqrt[3]{x^2}}=\frac{4\sqrt[3]{x^2}}{x}$

67. $\sqrt[3]{\frac{2}{x^2}}=\frac{\sqrt[3]{2}}{\sqrt[3]{x^2}}\cdot\frac{\sqrt[3]{x}}{\sqrt[3]{x}}=\frac{\sqrt[3]{2x}}{x}$

69. $\frac{5}{\sqrt{12y}}=\frac{5}{\sqrt{12y}}\cdot\frac{\sqrt{12y}}{\sqrt{12y}}=\frac{5\sqrt{12y}}{12y}=\frac{10\sqrt{3y}}{12y}$

$=\frac{5\sqrt{3y}}{6y}$

71. $\frac{7}{\sqrt[3]{2x^2}}=\frac{7}{\sqrt[3]{2x^2}}\cdot\frac{\sqrt[3]{4x}}{\sqrt[3]{4x}}=\frac{7\sqrt[3]{4x}}{2x}$

73. $\frac{3}{\sqrt[4]{x}}=\frac{3}{\sqrt[4]{x}}\cdot\frac{\sqrt[4]{x^3}}{\sqrt[4]{x^3}}=\frac{3\sqrt[4]{x^3}}{x}$

75. $\frac{7}{\sqrt[5]{y^2}}=\frac{7}{\sqrt[5]{y^2}}\cdot\frac{\sqrt[5]{y^3}}{\sqrt[5]{y^3}}=\frac{7\sqrt[5]{y^3}}{y}$

77. $\sqrt[3]{\frac{7}{2x}}=\frac{\sqrt[3]{7}}{\sqrt[3]{2x}}\cdot\frac{\sqrt[3]{4x^2}}{\sqrt[3]{4x^2}}=\frac{\sqrt[3]{28x^2}}{2x}$

79. $\frac{2}{\sqrt[3]{16x^2y}}=\frac{2}{\sqrt[3]{16x^2y}}\cdot\frac{\sqrt[3]{4xy^2}}{\sqrt[3]{4xy^2}}=\frac{2\sqrt[3]{4xy^2}}{4xy}$

$=\frac{\sqrt[3]{4xy^2}}{2xy}$

81. $\sqrt[5]{\frac{3}{8x^3}}=\frac{\sqrt[5]{3}}{\sqrt[5]{8x^3}}\cdot\frac{\sqrt[5]{4x^2}}{\sqrt[5]{4x^2}}=\frac{\sqrt[5]{12x^2}}{2x}$

83. $\sqrt[4]{\frac{5}{3x}}=\frac{\sqrt[4]{5}}{\sqrt[4]{3x}}\cdot\frac{\sqrt[4]{27x^3}}{\sqrt[4]{27x^3}}=\frac{\sqrt[4]{135x^3}}{3x}$

85. $\sqrt{\frac{3}{32y^2}}=\frac{\sqrt{3}}{\sqrt{32y^2}}=\frac{\sqrt{3}}{4y\sqrt{2}}\cdot\frac{\sqrt{2}}{\sqrt{2}}=\frac{\sqrt{6}}{4y\cdot 2}$

$=\frac{\sqrt{6}}{8y}$

87. $\sqrt{\frac{11}{20y^3}}=\frac{\sqrt{11}}{\sqrt{20y^3}}=\frac{\sqrt{11}}{2y\sqrt{5y}}\cdot\frac{\sqrt{5y}}{\sqrt{5y}}$

$=\frac{\sqrt{55y}}{2y(5y)}=\frac{\sqrt{55y}}{10y^2}$

89. $\sqrt{\frac{3x}{32y^3}}=\frac{\sqrt{3x}}{\sqrt{32y^3}}=\frac{\sqrt{3x}}{4y\sqrt{2y}}\cdot\frac{\sqrt{2y}}{\sqrt{2y}}$

$=\frac{\sqrt{6xy}}{4y(2y)}=\frac{\sqrt{6xy}}{8y^2}$

91. $\dfrac{3x}{\sqrt{8x^3}} = \dfrac{3x}{2x\sqrt{2x}} \cdot \dfrac{\sqrt{2x}}{\sqrt{2x}} = \dfrac{3x\sqrt{2x}}{2x(2x)}$
$= \dfrac{3x\sqrt{2x}}{4x^2}$

93. $\dfrac{7x^2}{\sqrt[3]{2x^5}} = \dfrac{7x^2}{x\sqrt[3]{2x^2}} \cdot \dfrac{\sqrt[3]{4x}}{\sqrt[3]{4x}} = \dfrac{7x^2\sqrt[3]{4x}}{x(2x)}$
$= \dfrac{7\sqrt[3]{4x}}{2}$

95. $\dfrac{7}{\sqrt[5]{32x^7y^4}} = \dfrac{7}{2x\sqrt[5]{x^2y^4}} \cdot \dfrac{\sqrt[5]{x^3y}}{\sqrt[5]{x^3y}} = \dfrac{7\sqrt[5]{x^3y}}{2x(xy)}$
$= \dfrac{7\sqrt[5]{x^3y}}{2x^2y}$

97. $\dfrac{\sqrt{20x^2y}}{\sqrt{5x^3y^3}} = \sqrt{\dfrac{20x^2y}{5x^3y^3}} = \sqrt{\dfrac{4}{xy^2}} = \dfrac{\sqrt{4}}{\sqrt{xy^2}}$
$= \dfrac{2}{y\sqrt{x}} \cdot \dfrac{\sqrt{x}}{\sqrt{x}} = \dfrac{2\sqrt{x}}{xy}$

99. $\sqrt[3]{\dfrac{5}{36x^4y}} = \dfrac{\sqrt[3]{5}}{\sqrt[3]{36x^4y}} = \dfrac{\sqrt[3]{5}}{x\sqrt[3]{36xy}} \cdot \dfrac{\sqrt[3]{6x^2y^2}}{\sqrt[3]{6x^2y^2}}$
$= \dfrac{\sqrt[3]{30x^2y^2}}{x(6xy)} = \dfrac{\sqrt[3]{30x^2y^2}}{6x^2y}$

101. b is true.

$\dfrac{1}{\sqrt[12]{x^7}} = \dfrac{1}{\sqrt[12]{x^7}} \cdot \dfrac{\sqrt[12]{x^5}}{\sqrt[12]{x^5}} = \dfrac{\sqrt[12]{x^5}}{\sqrt[12]{x^{12}}} = \dfrac{\sqrt[12]{x^5}}{x}$

103. $\sqrt{\dfrac{2}{x}} = \dfrac{\sqrt{2x}}{x}$, $x > 0$

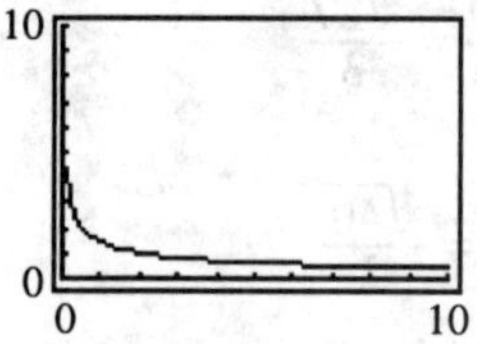

105. $\dfrac{x}{\sqrt[4]{2}} \neq \dfrac{\sqrt[4]{2x}}{2}$

$\dfrac{x}{\sqrt[4]{2}} = \dfrac{x}{\sqrt[4]{2}} \cdot \dfrac{\sqrt[4]{8}}{\sqrt[4]{8}} = \dfrac{x\sqrt[4]{8}}{2}$

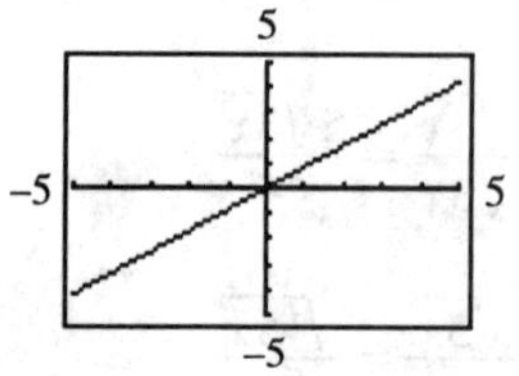

107. Answers may vary.

109. $\dfrac{5}{\sqrt{2x-3y}} = \dfrac{5}{\sqrt{2x-3y}} \cdot \dfrac{\sqrt{2x-3y}}{\sqrt{2x-3y}}$
$= \dfrac{5\sqrt{2x-3y}}{\sqrt{(2x-3y)^2}} = \dfrac{5\sqrt{2x-3y}}{2x-3y}$

Review Problems

111. $\dfrac{\frac{1}{y^2} - \frac{1}{100}}{\frac{1}{y} - \frac{1}{10}} \cdot \dfrac{100y^2}{100y^2} = \dfrac{100 - y^2}{100y - 10y^2}$
$= \dfrac{(10+y)(10-y)}{10y(10-y)} = \dfrac{10+y}{10y}$

112. $(2x - 3(4x^2 - 5x - 1) = 2x(4x^2 - 5x - 1) - 3(4x^2 - 5x - 1)$
$= 8x^3 - 10x^2 - 2x - 12x^2 + 15x + 3$
$= 8x^3 - 22x^2 + 13x + 3$

113. $\frac{4x}{x+1}+\frac{5}{x}=4$

$x(x+1)\left[\frac{4x}{x+1}+\frac{5}{x}\right]=x(x+1)4$

$4x^2+5(x+1)=4x(x+1)$

$4x^2+5x+5=4x^2+4x$

$4x^2-4x^2+5x-4x=-5$

$x=-5$

$\{-5\}$

Problem Set 6.5

1. $7\sqrt{3}+8\sqrt{3}=(7+8)\sqrt{3}=15\sqrt{3}$

3. $8\sqrt{7x}-7\sqrt{7x}-\sqrt{7x}=(8-7-1)\sqrt{7x}=0$

5. $3\sqrt{13}-2\sqrt{5}-2\sqrt{13}+4\sqrt{5}=3\sqrt{13}-2\sqrt{13}-2\sqrt{5}+4\sqrt{5}$
$=\sqrt{13}+2\sqrt{5}$

7. $\sqrt{2}-\sqrt{11}+6\sqrt{2}+4\sqrt{11}=\sqrt{2}+6\sqrt{2}-\sqrt{11}+4\sqrt{11}$
$=(1+6)\sqrt{2}+(-1+4)\sqrt{11}=7\sqrt{2}+3\sqrt{11}$

9. $3\sqrt{15}-2\sqrt{7}-2\sqrt{15}+2\sqrt{7}=3\sqrt{15}-2\sqrt{15}-2\sqrt{7}+2\sqrt{7}$
$=\sqrt{15}+0\sqrt{7}=\sqrt{15}$

11. $\sqrt{50}+\sqrt{18}=\sqrt{25\cdot 2}+\sqrt{9\cdot 2}$
$=5\sqrt{2}+3\sqrt{2}=8\sqrt{2}$

13. $3\sqrt{18}-5\sqrt{50}=3\sqrt{9\cdot 2}-5\sqrt{25\cdot 2}$
$=9\sqrt{2}-25\sqrt{2}=-16\sqrt{2}$

15. $3\sqrt{8}-\sqrt{32}+3\sqrt{72}-\sqrt{75}=3\sqrt{4\cdot 2}-\sqrt{16\cdot 2}+3\sqrt{36\cdot 2}-\sqrt{25\cdot 3}$
$=3(2)\sqrt{2}-4\sqrt{2}+3(6)\sqrt{2}-5\sqrt{3}$
$=6\sqrt{2}-4\sqrt{2}+18\sqrt{2}-5\sqrt{3}=20\sqrt{2}-5\sqrt{3}$

17. $8\sqrt{\frac{1}{2}}-\frac{1}{2}\sqrt{8}=8\cdot\frac{1}{\sqrt{2}}\cdot\frac{\sqrt{2}}{\sqrt{2}}-\frac{1}{2}\sqrt{4\cdot 2}$
$=\frac{8\sqrt{2}}{2}-\frac{1}{2}(2)\sqrt{2}=4\sqrt{2}-\sqrt{2}=3\sqrt{2}$

19. $\frac{\sqrt{63}}{3}+7\sqrt{3}=\frac{\sqrt{9\cdot 7}}{3}+7\sqrt{3}$
$=\frac{3\sqrt{7}}{3}+7\sqrt{3}=\sqrt{7}+7\sqrt{3}$

21. $\sqrt{25x}+\sqrt{16x}=5\sqrt{x}+4\sqrt{x}=9\sqrt{x}$

23. $\sqrt[4]{32}+3\sqrt[4]{1250}=\sqrt[4]{16\cdot 2}+3\sqrt[4]{625\cdot 2}$
$=2\sqrt[4]{2}+3\cdot 5\sqrt[4]{2}=2\sqrt[4]{2}+15\sqrt[4]{2}$
$=17\sqrt[4]{2}$

25. $4\sqrt[3]{40}-3\sqrt[3]{320}+2\sqrt[3]{625}=4\sqrt[3]{8\cdot 5}-3\sqrt[3]{64\cdot 5}+2\sqrt[3]{125\cdot 5}$
$=8\sqrt[3]{5}-12\sqrt[3]{5}+10\sqrt[3]{5}=6\sqrt[3]{5}$

27. $\frac{1}{4}\sqrt{2}+\frac{2}{3}\sqrt{8}=\frac{\sqrt{2}}{4}+\frac{2}{3}\sqrt{4\cdot 2}$
$=\frac{\sqrt{2}}{4}+\frac{2\cdot 2\sqrt{2}}{3}=\frac{\sqrt{2}}{4}+\frac{4\sqrt{2}}{3}$
$=\frac{3\sqrt{2}}{12}+\frac{16\sqrt{2}}{12}=\frac{19\sqrt{2}}{12}$

29. $\frac{\sqrt{45}}{4}-\sqrt{80}+\frac{\sqrt{20}}{3}=\frac{\sqrt{9\cdot 5}}{4}-\sqrt{16\cdot 5}+\frac{\sqrt{4\cdot 5}}{3}$
$=\frac{3\sqrt{5}}{4}-\frac{4\sqrt{5}}{1}+\frac{2\sqrt{5}}{3}=\frac{9\sqrt{5}}{12}-\frac{48\sqrt{5}}{12}+\frac{8\sqrt{5}}{12}$
$=\frac{-31\sqrt{5}}{12}$

31. $7\sqrt[3]{2}+8\sqrt[3]{16}-2\sqrt[3]{54}=7\sqrt[3]{2}+8\sqrt[3]{8\cdot 2}-2\sqrt[3]{27\cdot 2}$
$=7\sqrt[3]{2}+8(2)\sqrt[3]{2}-2(3)\sqrt[3]{2}$
$=7\sqrt[3]{2}+16\sqrt[3]{2}-6\sqrt[3]{2}=17\sqrt[3]{2}$

33. $2\sqrt[3]{2}-\sqrt[3]{16}-\sqrt[3]{54}=2\sqrt[3]{2}-\sqrt[3]{8\cdot 2}-\sqrt[3]{27\cdot 2}$
$=2\sqrt[3]{2}-2\sqrt[3]{2}-3\sqrt[3]{2}=-3\sqrt[3]{2}$

35. $5\sqrt{12x}-2\sqrt{3x}=5\sqrt{4\cdot 3x}-2\sqrt{3x}$
$=10\sqrt{3x}-2\sqrt{3x}=8\sqrt{3x}$

37. $\sqrt[3]{54x^5}+2x\sqrt[3]{16x^2}-7\sqrt[3]{2x^5}=\sqrt[3]{27\cdot 2x^3x^2}+2x\sqrt[3]{8\cdot 2x^2}-7\sqrt[3]{2x^3x^2}$
$=3x\sqrt[3]{2x^2}+4x\sqrt[3]{2x^2}-7x\sqrt[3]{2x^2}=0$

39. $16\sqrt{\frac{5}{8}}+6\sqrt{\frac{5}{2}}=\frac{16\sqrt{5}}{\sqrt{8}}\cdot\frac{\sqrt{2}}{\sqrt{2}}+\frac{6\sqrt{5}}{\sqrt{2}}\cdot\frac{\sqrt{2}}{\sqrt{2}}$
$=\frac{16\sqrt{10}}{4}+\frac{6\sqrt{10}}{2}=4\sqrt{10}+3\sqrt{10}=7\sqrt{10}$

41. $12\sqrt{\frac{2}{3}}+24\sqrt{\frac{1}{6}}=\frac{12\sqrt{2}}{\sqrt{3}}+\frac{24}{\sqrt{6}}$
$=\frac{12\sqrt{2}}{\sqrt{3}}\cdot\frac{\sqrt{3}}{\sqrt{3}}+\frac{24}{\sqrt{6}}\cdot\frac{\sqrt{6}}{\sqrt{6}}$
$=\frac{12\sqrt{6}}{3}+\frac{24\sqrt{6}}{6}=4\sqrt{6}+4\sqrt{6}=8\sqrt{6}$

43. $\sqrt{2}\left(\sqrt{3}+\sqrt{7}\right)=\sqrt{2}\sqrt{3}+\sqrt{2}\sqrt{7}=\sqrt{6}+\sqrt{14}$

45. $4\sqrt{3}\left(2\sqrt{5}+3\sqrt{7}\right)=4\sqrt{3}\left(2\sqrt{5}\right)+4\sqrt{3}\left(3\sqrt{7}\right)$
$=8\sqrt{15}+12\sqrt{21}$

47. $5\sqrt{6}\left(7\sqrt{8}-2\sqrt{12}\right)=35\sqrt{48}-10\sqrt{72}$
$=35\sqrt{16\cdot 3}-10\sqrt{36\cdot 2}$
$=35(4)\sqrt{3}-10(6)\sqrt{2}=140\sqrt{3}-60\sqrt{2}$

49. $4\sqrt{x}\left(7\sqrt{2}-3\sqrt{y}\right)=4\sqrt{x}\left(7\sqrt{2}\right)-4\sqrt{x}\left(3\sqrt{y}\right)$
$=28\sqrt{2x}-12\sqrt{xy}$

51. $\sqrt{2x}\left(\sqrt{6x}-3\sqrt{x}\right)=\sqrt{12x^2}-3\sqrt{2x^2}$
$=\sqrt{4\cdot 3x^2}-3\sqrt{2x^2}=2x\sqrt{3}-3x\sqrt{2}$

53. $\sqrt[3]{2}\left(\sqrt[3]{6}+4\sqrt[3]{5}\right)=\sqrt[3]{2}\sqrt[3]{6}+\sqrt[3]{2}\left(4\sqrt[3]{5}\right)$
$=\sqrt[3]{12}+4\sqrt[3]{10}$

55. $\left(\sqrt{2}+\sqrt{7}\right)\left(\sqrt{3}+\sqrt{5}\right)=\sqrt{2}\sqrt{3}+\sqrt{2}\sqrt{5}+\sqrt{7}\sqrt{3}+\sqrt{7}\sqrt{5}$
$=\sqrt{6}+\sqrt{10}+\sqrt{21}+\sqrt{35}$

57. $\left(\sqrt{2}-\sqrt{7}\right)\left(\sqrt{3}-\sqrt{5}\right)=\sqrt{2}\sqrt{3}-\sqrt{2}\sqrt{5}-\sqrt{7}\sqrt{3}+\sqrt{7}\sqrt{5}$
$=\sqrt{6}-\sqrt{10}-\sqrt{21}+\sqrt{35}$

59. $\left(4\sqrt{2}+5\sqrt{7}\right)\left(2\sqrt{3}+3\sqrt{5}\right)=4\sqrt{2}\left(2\sqrt{3}\right)+4\sqrt{2}\left(3\sqrt{5}\right)+5\sqrt{7}\left(2\sqrt{3}\right)+5\sqrt{7}\left(3\sqrt{5}\right)$
$=8\sqrt{6}+12\sqrt{10}+10\sqrt{21}+15\sqrt{35}$

61. $\left(3\sqrt{2}-2\sqrt{8}\right)\left(2\sqrt{3}-4\sqrt{5}\right)=6\sqrt{6}-12\sqrt{10}-4\sqrt{24}+8\sqrt{40}$
$=6\sqrt{6}-12\sqrt{10}-4\sqrt{4\cdot 6}+8\sqrt{4\cdot 10}$
$=6\sqrt{6}-12\sqrt{10}-4(2)\sqrt{6}+8(2)\sqrt{10}$
$=6\sqrt{6}-12\sqrt{10}-8\sqrt{6}+16\sqrt{10}=4\sqrt{10}-2\sqrt{6}$

63. $\left(\sqrt{5}+7\right)\left(\sqrt{5}-7\right)=\sqrt{25}-49$
$=5-49=-44$

65. $\left(2+5\sqrt{3}\right)\left(2-5\sqrt{3}\right)=4-25\sqrt{9}$
$=4-75=-71$

67. $\left(\sqrt{3}+\sqrt{5}\right)^2=\left(\sqrt{3}\right)^2+2\sqrt{3}\sqrt{5}+\left(\sqrt{5}\right)^2$
$=3+2\sqrt{15}+5=8+2\sqrt{15}$

69. $\left(2\sqrt{3}-4\sqrt{7}\right)^2=\left(2\sqrt{3}\right)^2-2\left(2\sqrt{3}\right)\left(4\sqrt{7}\right)+\left(4\sqrt{7}\right)^2$
$=12-16\sqrt{21}+112=124-16\sqrt{21}$

71. $\left(\sqrt{x}+\sqrt{3}\right)\left(\sqrt{x}+\sqrt{2}\right)=\left(\sqrt{x}\right)^2+\sqrt{2x}+\sqrt{3x}+\sqrt{2\cdot 3}$
$=x+\sqrt{2x}+\sqrt{3x}+\sqrt{6}$

73. $\left(x+\sqrt{y}\right)^2=x^2+2x\sqrt{y}+\left(\sqrt{y}\right)^2=x^2+2x\sqrt{y}+y$

75. $\left(\sqrt{x}+\sqrt{y}\right)^2=\left(\sqrt{x}\right)^2+2\sqrt{x}\sqrt{y}+\left(\sqrt{y}\right)^2$
$=x+2\sqrt{xy}+y$

77. $\left(x+\sqrt[3]{y^2}\right)\left(2x-\sqrt[3]{y^2}\right)=2x^2-x\sqrt[3]{y^2}+2x\sqrt[3]{y^2}-\left(\sqrt[3]{y^2}\right)^2$
$=2x^2+x\sqrt[3]{y^2}-y\sqrt[3]{y}$

79. $\frac{8}{\sqrt{5}-2}\cdot\frac{\sqrt{5}+2}{\sqrt{5}+2}=\frac{8\left(\sqrt{5}+2\right)}{5-4}$
$=8\left(\sqrt{5}+2\right)=8\sqrt{5}+16$

81. $\frac{13}{\sqrt{11}+3}\cdot\frac{\sqrt{11}-3}{\sqrt{11}-3}=\frac{13\left(\sqrt{11}-3\right)}{11-9}$
$=\frac{13\sqrt{11}-39}{2}$

83. $\frac{6}{\sqrt{5}+\sqrt{3}}\cdot\frac{\sqrt{5}-\sqrt{3}}{\sqrt{5}-\sqrt{3}}=\frac{6\left(\sqrt{5}-\sqrt{3}\right)}{5-3}$
$=\frac{6\left(\sqrt{5}-\sqrt{3}\right)}{2}=3\left(\sqrt{5}-\sqrt{3}\right)=3\sqrt{5}-3\sqrt{3}$

85. $\frac{11}{\sqrt{7}-\sqrt{3}}\cdot\frac{\sqrt{7}+\sqrt{3}}{\sqrt{7}+\sqrt{3}}=\frac{11\left(\sqrt{7}+\sqrt{3}\right)}{7-3}$
$=\frac{11\sqrt{7}+11\sqrt{3}}{4}$

87. $\frac{\sqrt{5}}{\sqrt{7}+\sqrt{3}}\cdot\frac{\sqrt{7}-\sqrt{3}}{\sqrt{7}-\sqrt{3}}=\frac{\sqrt{5}\left(\sqrt{7}-\sqrt{3}\right)}{7-3}$
$=\frac{\sqrt{35}-\sqrt{15}}{4}$

89. $\frac{\sqrt{2}}{\sqrt{7}-\sqrt{2}}\cdot\frac{\sqrt{7}+\sqrt{2}}{\sqrt{7}+\sqrt{2}}=\frac{\sqrt{14}+2}{7-2}=\frac{\sqrt{14}+2}{5}$

91. $\frac{8}{3+2\sqrt{2}}\cdot\frac{3-2\sqrt{2}}{3-2\sqrt{2}}=\frac{8\left(3-2\sqrt{2}\right)}{9-4\cdot 2}$
$=\frac{8\left(3-2\sqrt{2}\right)}{1}=24-16\sqrt{2}$

93. $\frac{25}{5\sqrt{2}-3\sqrt{5}}\cdot\frac{5\sqrt{2}+3\sqrt{5}}{5\sqrt{2}+3\sqrt{5}}=\frac{25\left(5\sqrt{2}+3\sqrt{5}\right)}{25\cdot 2-9\cdot 5}$
$=\frac{25\left(5\sqrt{2}+3\sqrt{5}\right)}{5}=5\left(5\sqrt{2}+3\sqrt{5}\right)$
$=25\sqrt{2}+15\sqrt{5}$

95. $\frac{\sqrt{3}}{2\sqrt{3}+3\sqrt{5}}\cdot\frac{2\sqrt{3}-3\sqrt{5}}{2\sqrt{3}-3\sqrt{5}}=\frac{\sqrt{3}\left(2\sqrt{3}-3\sqrt{5}\right)}{4\cdot 3-9\cdot 5}$
$=\frac{2\cdot 3-3\sqrt{15}}{12-45}=\frac{6-3\sqrt{15}}{-33}=\frac{3\left(2-\sqrt{15}\right)}{-33}$
$=\frac{2-\sqrt{15}}{-11}=-\frac{2-\sqrt{15}}{11}=\frac{-2+\sqrt{15}}{11}$

97. $\frac{7}{\sqrt{x}-5}\cdot\frac{\sqrt{x}+5}{\sqrt{x}+5}=\frac{7\left(\sqrt{x}+5\right)}{x-25}=\frac{7\sqrt{x}+35}{x-25}$

99. $\frac{\sqrt{y}}{\sqrt{y}+3}\cdot\frac{\sqrt{y}-3}{\sqrt{y}-3}=\frac{\sqrt{y}(\sqrt{y}-3)}{y-9}=\frac{y-3\sqrt{y}}{y-9}$

101. $\frac{2\sqrt{3}-1}{2\sqrt{3}+1}\cdot\frac{2\sqrt{3}-1}{2\sqrt{3}-1}=\frac{4\cdot 3-2\sqrt{3}-2\sqrt{3}+1}{4\cdot 3-1}$
$=\frac{13-4\sqrt{3}}{11}$

103. $\frac{\sqrt{5}+\sqrt{3}}{\sqrt{5}-\sqrt{3}}\cdot\frac{\sqrt{5}+\sqrt{3}}{\sqrt{5}+\sqrt{3}}=\frac{5+\sqrt{15}+\sqrt{15}+3}{5-3}$
$=\frac{8+2\sqrt{15}}{2}=\frac{2\left(4+\sqrt{15}\right)}{2}=4+\sqrt{15}$

105. $\frac{\sqrt{x}+1}{\sqrt{x}+3}\cdot\frac{\sqrt{x}-3}{\sqrt{x}-3}=\frac{x-3\sqrt{x}+\sqrt{x}-3}{x-9}$
$=\frac{x-2\sqrt{x}-3}{x-9}$

107. $\frac{\sqrt{y}}{\sqrt{x}+\sqrt{y}}\cdot\frac{\sqrt{x}-\sqrt{y}}{\sqrt{x}-\sqrt{y}}=\frac{\sqrt{xy}-y}{x-y}$

109. $\frac{3\sqrt{5}+2\sqrt{2}}{4\sqrt{5}+\sqrt{2}}\cdot\frac{4\sqrt{5}-\sqrt{2}}{4\sqrt{5}-\sqrt{2}}=\frac{12\cdot 5-3\sqrt{10}+8\sqrt{10}-2\cdot 2}{16\cdot 5-2}$

$=\frac{56+5\sqrt{10}}{78}$

111. $\frac{2\sqrt{x}+\sqrt{y}}{4\sqrt{x}+3\sqrt{y}}\cdot\frac{4\sqrt{x}-3\sqrt{y}}{4\sqrt{x}-3\sqrt{y}}=\frac{8x-6\sqrt{xy}+4\sqrt{xy}-3y}{16x-9y}$

$=\frac{8x-2\sqrt{xy}-3y}{16x-9y}$

113. $\frac{x^3-y^3}{\sqrt{x}+\sqrt{y}}\cdot\frac{\sqrt{x}-\sqrt{y}}{\sqrt{x}-\sqrt{y}}=\frac{(x^3-y^3)\left(\sqrt{x}-\sqrt{y}\right)}{x-y}$

$=\frac{(x-y)(x^2+xy+y^2)\left(\sqrt{x}-\sqrt{y}\right)}{x-y}$

$=(x^2+xy+y^2)\left(\sqrt{x}-\sqrt{y}\right)$

115. Perimeter:

$P=\sqrt{45}+\sqrt{125}+\sqrt{80}$

$P=3\sqrt{5}+5\sqrt{5}+4\sqrt{5}$

$P=12\sqrt{5}$

Area:

$A=\frac{1}{2}\left(\sqrt{125}\right)\left(\sqrt{20}\right)$

$A=\frac{1}{2}\left(5\sqrt{5}\right)\left(2\sqrt{5}\right)$

$=\frac{10}{2}\left(\sqrt{5}\right)^2=5\cdot 5=25$

117. $A=\sqrt{s(s-a)(s-b)(s-c)}$

$s=\frac{1}{2}(9+4+12)=\frac{25}{2}$

$a=9,\ b=12,\ c=4$

$A=\sqrt{\frac{25}{2}\left(\frac{25}{2}-9\right)\left(\frac{25}{2}-12\right)\left(\frac{25}{2}-4\right)}$

$A=\sqrt{\frac{25}{2}\left(\frac{7}{2}\right)\left(\frac{1}{2}\right)\left(\frac{17}{2}\right)}$

$A=\frac{5\sqrt{119}}{4}$

$V=A\sqrt{18}=\frac{5\sqrt{119}}{4}\cdot 3\sqrt{2}=\frac{15\sqrt{238}}{4}$

The volume is $\frac{15\sqrt{238}}{4}$ cubic units.

119. c is true.

121. $\sqrt{8}x+\sqrt{2}x\neq\sqrt{10}x$

$x\sqrt{8}+x\sqrt{2}=2x\sqrt{2}+x\sqrt{2}=3x\sqrt{2}$

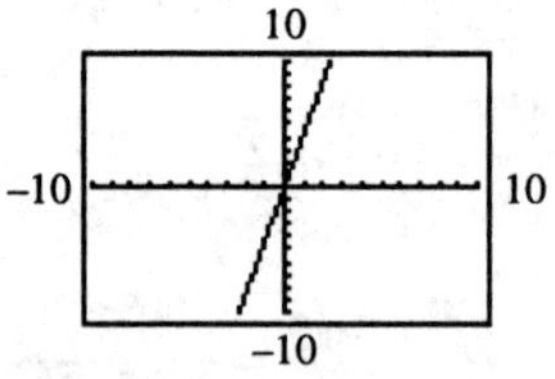

123. $8\sqrt{x}+2\sqrt{x}=10\sqrt{x},\ x\ge 0$

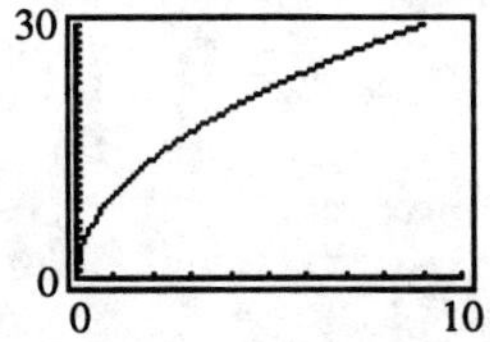

125. $\dfrac{\sqrt{x^7}+\sqrt{x^3}}{\sqrt{x}}=x^3+x,\ x>0$

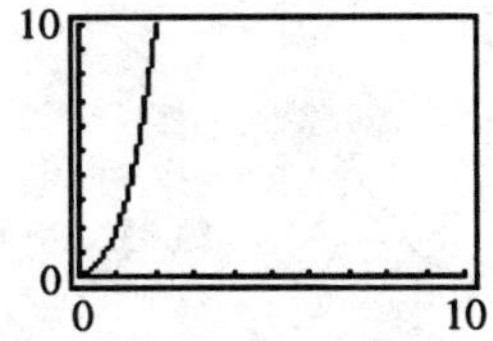

127. Answers may vary.

129. Answers may vary.

131. Answers may vary.

133. Answers may vary.

135. $\left|\dfrac{3x+\sqrt{32}}{2}\right|=\sqrt{50}$

$\left|\dfrac{3x+4\sqrt{2}}{2}\right|=5\sqrt{2}$

$\dfrac{3x+4\sqrt{2}}{2}=5\sqrt{2}$ or $\dfrac{3x+4\sqrt{2}}{2}=-5\sqrt{2}$

$3x+4\sqrt{2}=10\sqrt{2}$ $\quad$ $3x+4\sqrt{2}=-10\sqrt{2}$

$3x=6\sqrt{2}$ $\quad$ $3x=-14\sqrt{2}$

$x=2\sqrt{2}$ $\quad$ $x=\frac{-14}{3}\sqrt{2}$

$\left\{2\sqrt{2},\ \frac{-14}{3}\sqrt{2}\right\}$

137.

via.	4.	$\overline{p}$.	R_x	6.	→	$(4+\sqrt{6})$
	4.	$\overline{m}$.	R_x	6.	→	$(4-\sqrt{6})$
	16.	$\overline{m}$.	6.		→	$16-6$
	Production 10				→	10

or $(4+\sqrt{6})(4-\sqrt{6})=16-6=10$

139. $\sqrt{2\sqrt{15}+8}=\sqrt{3}+\sqrt{5}$

$\left(\sqrt{2\sqrt{15}+8}\right)^2=\left(\sqrt{3}+\sqrt{5}\right)^2$

$2\sqrt{15}+8=\left(\sqrt{3}+\sqrt{5}\right)\left(\sqrt{3}+\sqrt{5}\right)$

$2\sqrt{15}+8=3+\sqrt{15}+\sqrt{15}+5$

$2\sqrt{15}+8=2\sqrt{15}+8$ True

141. $\frac{1}{\sqrt{2}+\sqrt{3}+\sqrt{4}} = \frac{1}{(\sqrt{2}+\sqrt{3})+2} \cdot \frac{(\sqrt{2}+\sqrt{3})-2}{(\sqrt{2}+\sqrt{3})-2}$

$= \frac{\sqrt{2}+\sqrt{3}-2}{(\sqrt{2}+\sqrt{3})^2-4} = \frac{\sqrt{2}+\sqrt{3}-2}{2+2\sqrt{6}+3-4}$

$= \frac{\sqrt{2}+\sqrt{3}-2}{2\sqrt{6}+1} = \frac{\sqrt{2}+\sqrt{3}-2}{2\sqrt{6}+1} \cdot \frac{2\sqrt{6}-1}{2\sqrt{6}-1}$

$= \frac{\sqrt{2}(2\sqrt{6}-1)+\sqrt{3}(2\sqrt{6}-1)-2(2\sqrt{6}-1)}{(2\sqrt{6})^2-1^2}$

$= \frac{2\sqrt{12}-\sqrt{2}+2\sqrt{18}-\sqrt{3}-4\sqrt{6}+2}{4(6)-1}$

$= \frac{2(2\sqrt{3})-\sqrt{2}+2(3\sqrt{2})-\sqrt{3}-4\sqrt{6}+2}{24-1}$

$= \frac{4\sqrt{3}-\sqrt{3}-\sqrt{2}+6\sqrt{2}-4\sqrt{6}+2}{23}$

$= \frac{3\sqrt{3}+5\sqrt{2}-4\sqrt{6}+2}{23}$

143. Given:
area of $BEGC = 2\text{ m}^2$
area of $ADFB = 3\text{ m}^2$
Since the area of $ADFB = 3$, $DF = \sqrt{3}$ and $AD = \sqrt{3}$.
Since the area of $BEGC = 2$, $EG = \sqrt{2}$ and $CG = \sqrt{2}$.
Area of shaded region $= (DH)(HG)$
$= (DF + FH)(CH - CG)$
$= (DF + EG)(AD - CG)$
$= (\sqrt{3}+\sqrt{2})(\sqrt{3}-\sqrt{2}) = 3-2 = 1$
The area of the shaded region is 1 m^2.

145. $\frac{\sqrt{a+h}-\sqrt{a}}{h}$

$= \frac{\sqrt{a+h}-\sqrt{a}}{h}\left(\frac{\sqrt{a+h}+\sqrt{a}}{\sqrt{a+h}+\sqrt{a}}\right)$

$= \frac{(\sqrt{a+h})^2-(\sqrt{a})^2}{h(\sqrt{a+h}+\sqrt{a})}$

$= \frac{a+h-a}{h(\sqrt{a+h}+\sqrt{a})}$

$= \frac{1}{\sqrt{a+h}+\sqrt{a}}$

Review Problems

147. $\sqrt{5y+6}+\sqrt{3y-2} = 6$
$\sqrt{5(2)+6}+\sqrt{3(2)-2} = 6$
$\sqrt{16}+\sqrt{4} = 6$
$4+2 = 6$
$6 = 6$

148. $4x^2 = 3x + 10$

$4x^2 - 3x - 10 = 0$

$(4x + 5)(x - 2) = 0$

$4x + 5 = 0$ or $x - 2 = 0$

$x = -\frac{5}{4}$ or $x = 2$

$\left\{-\frac{5}{4}, 2\right\}$

149. Let x = time it takes both pipes working together

1 complete job = pool is completely emptied

	Fractional part of the job completed in 1 minute	Time spent working together	Fractional part of job completed in x minutes
pipe to fill (30 minutes)	$\frac{1}{30}$	x	$\frac{x}{30}$
pipe to drain (20 minutes)	$\frac{1}{20}$	x	$\frac{x}{20}$

Fraction of the pool emptied by the second pipe in x minutes
– Fraction of the pool filled by the first pipe in x minutes = 1.

$\frac{x}{20} - \frac{x}{30} = 1$

$60\left(\frac{x}{20} - \frac{x}{30}\right) = 60(1)$

$3x - 2x = 60$

$x = 60$

It will take 60 minutes (or 1 hour) before the pool has no water.

Problem Set 6.6

1. $\sqrt{3x-1}=4$
$3x-1=16$
$3x=17$
$x=\frac{17}{3}$
$\left\{\frac{17}{3}\right\}$
Check:
$\sqrt{3\left(\frac{17}{3}\right)-1}=4$
$\sqrt{17-1}=4$
$\sqrt{16}=4$
$4=4$
The solution checks.

3. $\sqrt{2x+4}-6=0$
$\sqrt{2x+4}=6$
$\left(\sqrt{2x+4}\right)^2=6^2$
$2x+4=36$
$2x=32$
$x=16$
Check:
$\sqrt{2(16)+4}-6=0$
$\sqrt{32+4}-6=0$
$\sqrt{36}-6=0$
$6-6=0$
$0=0$
The solution checks.
$\{16\}$

5. $\sqrt{3x-1}=-4$
$\left(\sqrt{3x-1}\right)^2=(-4)^2$
$3x-1=16$
$3x=17$
$x=\frac{17}{3}$
Check:
$\sqrt{3\left(\frac{17}{3}\right)-1}=-4$
$\sqrt{16}=-4$
$4\neq-4$
$\frac{17}{3}$ is an extraneous solution.
$\varnothing$

7. $\sqrt{2x+4}+6=0$
$\sqrt{2x+4}=-6$
$2x+4=36$
$x=16$
Check:
$\sqrt{2(16)+4}+6=0$
$12\neq0$
16 is an extraneous solution.
$\varnothing$

9. $\sqrt{6x+7}=x+2$
$6x+7=(x+2)^2$
$6x+7=x^2+4x+4$
$0=x^2-2x-3$
$0=(x-3)(x+1)$
$x-3=0$ or $x+1=0$
$x=3$ or $x=-1$
$\{-1, 3\}$
Check:
$\sqrt{6(-1)+7}=-1+2$
$\sqrt{1}=1$
$1=1$
The solution checks.
Check:
$\sqrt{6(3)+7}=3+2$
$\sqrt{25}=5$
$5=5$
The solution checks.

11. $\sqrt{3y+1}-3y+11=0$
$\left(\sqrt{3y+1}\right)^2=(3y-11)^2$
$3y+1=9y^2-66y+121$
$0=9y^2-69y+120$
$0=3y^2-23y+40$

$0 = (3y-8)(y-5)$
$3y-8=0$ or $y-5=0$
$y=\frac{8}{3}$ (extraneous) $y=5$ (checks)
$\{5\}$

13. $\sqrt{5x+9}-x+1=0$
$\left(\sqrt{5x+9}\right)^2 = (x-1)^2$
$5x+9 = x^2-2x+1$
$0 = x^2-7x-8$
$0 = (x-8)(x+1)$
$x=8$ (checks) or $x=-1$ (extraneous)
$\{8\}$

15. $\sqrt{z-1}-7=-z$
$\left(\sqrt{z-1}\right)^2 = (7-z)^2$
$z-1 = 49-14z+z^2$
$0 = z^2-15z+50$
$0 = (z-10)(z-5)$
$z=10$ (extraneous) or $z=5$ (checks)
$\{5\}$

17. $z+\sqrt{5z-1}-5=0$
$\sqrt{5z-1} = 5-z$
$5z-1 = 25-10z+z^2$
$0 = z^2-15z+26$
$0 = (z-13)(z-2)$
$z=13$ (extraneous) or $z=2$ (checks)
$\{2\}$

19. $\sqrt[3]{4x^2-3x}=1$
$\left(\sqrt[3]{4x^2-3x}\right)^3 = 1^3$
$4x^2-3x-1=0$
$(x-1)(4x+1)=0$
$x=1$
$x=-\frac{1}{4}$ (Both check)
$\left\{1, -\frac{1}{4}\right\}$

21. $\sqrt[3]{2x^2+3x}=-1$
$2x^2+3x=-1$
$2x^2+3x+1=0$
$(2x+1)(x+1)=0$
$x=-\frac{1}{2}$ or $x=-1$
(Both check)
$\left\{-\frac{1}{2}, -1\right\}$

23. $\sqrt{y-8}=\sqrt{y}-2$
$\left(\sqrt{y-8}\right)^2 = \left(\sqrt{y}-2\right)^2$
$y-8 = y-4\sqrt{y}+4$
$-12 = -4\sqrt{y}$
$3=\sqrt{y}$
$9=y$ (checks)
$\{9\}$

25. $\sqrt{y}+1=\sqrt{y+1}$
$\left(\sqrt{y}+1\right)^2 = \left(\sqrt{y+1}\right)^2$
$y+2\sqrt{y}+1 = y+1$
$2\sqrt{y}=0$
$\sqrt{y}=0$
$y=0$ (checks)
$\{0\}$

27. $\sqrt{y+5}-\sqrt{y-3}=2$
$\left(\sqrt{y+5}\right)^2 = \left(2+\sqrt{y-3}\right)^2$
$y+5 = 4+4\sqrt{y-3}+y-3$
$y+5 = y+1+4\sqrt{y-3}$
$4 = 4\sqrt{y-3}$
$1=\sqrt{y-3}$
$1=y-3$
$4=y$ (checks)
$\{4\}$

29. $\sqrt{y-3}-4=\sqrt{y-3}$
$-4=0$
$\varnothing$

31. $\sqrt{x+2}=1-\sqrt{x-3}$
$x+2=1-2\sqrt{x-3}+x-3$
$x+2=x-2-2\sqrt{x-3}$
$4=-2\sqrt{x-3}$
$-2=\sqrt{x-3}$
$4=x-3$
$7=x$ (does not check)
$\varnothing$

33. $\sqrt{y+2}+\sqrt{y-1}=3$
$\sqrt{y+2}=3-\sqrt{y-1}$
$y+2=9-6\sqrt{y-1}+y-1$
$y+2=8+y-6\sqrt{y-1}$
$-6=-6\sqrt{y-1}$
$1=\sqrt{y-1}$
$1=y-1$
$2=y$ (checks)
$\{2\}$

35. $\sqrt{y}+\sqrt{2}=\sqrt{y+2}$
$\left(\sqrt{y}+\sqrt{2}\right)^2=\left(\sqrt{y+2}\right)^2$
$y+2\sqrt{2y}+2=y+2$
$2\sqrt{2y}=0$
$\sqrt{2y}=0$
$2y=0$
$y=0$ (checks)
$\{0\}$

37. $\sqrt{y}+1=\sqrt{5y-1}$
$\left(\sqrt{y}+1\right)^2=\left(\sqrt{5y-1}\right)^2$
$\left(\sqrt{y}+1\right)\left(\sqrt{y}+1\right)=5y-1$
$y+2\sqrt{y}+1=5y-1$
$2\sqrt{y}=4y-2$
$\frac{1}{2}\left(2\sqrt{y}\right)=\frac{1}{2}(4y-2)$
$\sqrt{y}=2y-1$
$\left(\sqrt{y}\right)^2=(2y-1)^2$
$y=4y^2-4y+1$
$0=4y^2-5y+1$
$0=(4y-1)(y-1)$
$y=\frac{1}{4}$ or $y=1$
Check $\frac{1}{4}$:
$\sqrt{\frac{1}{4}}+1=\sqrt{5\left(\frac{1}{4}\right)-1}$
$\frac{1}{2}+1=\sqrt{\frac{5}{4}-1}$
$\frac{1}{2}+1=\sqrt{\frac{1}{4}}$
$1\frac{1}{2}\neq\frac{1}{2}$
$\frac{1}{4}$ is extraneous.
Check 1:
$\sqrt{1}+1=\sqrt{5(1)-1}$
$1+1=\sqrt{4}$
$2=2$ checks
$\{1\}$

39. $\sqrt{4z-3}-\sqrt{8z+1}+2=0$
$\sqrt{4z-3}=\sqrt{8z+1}-2$
$\left(\sqrt{4z-3}\right)^2=\left(\sqrt{8z+1}-2\right)^2$
$4z-3=\left(\sqrt{8z+1}-2\right)\left(\sqrt{8z+1}-2\right)$
$4z-3=8z+1-4\sqrt{8z+1}+4$
$4z-3=8z+5-4\sqrt{8z+1}$
$-4z-8=-4\sqrt{8z+1}$
$-\frac{1}{4}(-4z-8)=-\frac{1}{4}\left(-4\sqrt{8z+1}\right)$
$z+2=\sqrt{8z+1}$
$(z+2)^2=\left(\sqrt{8z+1}\right)^2$
$z^2+4z+4=8z+1$
$z^2-4z+3=0$

$(z-1)(z-3)=0$
$z=1$ or $z=3$
(Both check.)
$\{1, 3\}$

41. $2\sqrt{3x-2}+\sqrt{2x-3}=5$
$\left(2\sqrt{3x-2}\right)^2=\left(5-\sqrt{2x-3}\right)^2$
$4(3x-2)=25-10\sqrt{2x-3}+2x-3$
$12x-8=2x+22-10\sqrt{2x-3}$
$10x-30=-10\sqrt{2x-3}$
$x-3=-\sqrt{2x-3}$
$(x-3)^2=\left(-\sqrt{2x-3}\right)^2$
$x^2-6x+9=2x-3$
$x^2-8x+12=0$
$(x-6)(x-2)=0$
$x=6$ or $x=2$
Check 6:
$2\sqrt{3(6)-2}+\sqrt{2(6)-3}=5$
$2(4)+3=5$
$11\neq 5$
6 is extraneous.
Check 2:
$2\sqrt{3(2)-2}+\sqrt{2(2)-3}=5$
$2(2)+1=5$
$5=5$ checks
$\{2\}$

43. $\sqrt{2y+3}+\sqrt{y+2}=2$
$\left(\sqrt{2y+3}\right)^2=\left(2-\sqrt{y+2}\right)^2$
$2y+3=4-4\sqrt{y+2}+y+2$
$y-3=-4\sqrt{y+2}$
$(y-3)^2=\left(-4\sqrt{y+2}\right)^2$
$y^2-6y+9=16(y+2)$
$y^2-22y-23=0$
$(y+1)(y-23)=0$
$y=-1$ or $y=23$ (extraneous)
$\{-1\}$

45. $\sqrt{3-y}+\sqrt{2+y}=1$
$\sqrt{3-y}=1-\sqrt{2+y}$
$3-y=1-2\sqrt{2+y}+2+y$
$-2y=-2\sqrt{2+y}$
$y=\sqrt{2+y}$
$y^2=2+y$
$y^2-y-2=0$
$(y+1)(y-2)=0$
$y=-1$ or $y=2$
(Both are extraneous)
$\varnothing$

47. $\left(\sqrt{y+1+\sqrt{7y+4}}\right)^2=3^2$
$y+1+\sqrt{7y+4}=9$
$\sqrt{7y+4}=8-y$
$\left(\sqrt{7y+4}\right)^2=(8-y)^2$
$7y+4=64-16y+y^2$
$0=y^2-23y+60$
$0=(y-20)(y-3)$
$y-20=0$ or $y-3=0$
$y=20$ or $y=3$
Check 20:
$\sqrt{20+1+\sqrt{7(20)+4}}-3=0$
$\sqrt{21+\sqrt{144}}-3=0$
$\sqrt{21+12}-3=0$
$\sqrt{33}-3\neq 0$
20 is extraneous.
Check 3:
$\sqrt{3+1\sqrt{7(3)+4}}-3=0$
$\sqrt{4+\sqrt{25}}-3=0$
$\sqrt{4+5}-3=0$
$\sqrt{9}-3=0$
$0=0$ (checks)
$\{3\}$

49. $40,000 = 5000\sqrt{100-x}$
$8 = \sqrt{100-x}$
$64 = 100 - x$
$x = 36$
to age 36

51. $c = \sqrt{gH}$
$24 = \sqrt{32H}$
$576 = 32H$
$18 = H$
The water depth is 18 feet.

53. $b = \sqrt{\frac{3V}{H}}$
$8 = \sqrt{\frac{3V}{6}}$
$8 = \sqrt{\frac{V}{2}}$
$64 = \frac{V}{2}$
$128 = V$
The volume is 128 ft^3.

55. $N = 1220\sqrt[3]{t+42} + 4900$
$9780 = 1220\sqrt[3]{t+42} + 4900$
$4880 = 1220\sqrt[3]{t+42}$
$4 = \sqrt[3]{t+42}$
$64 = t + 42$
$22 = t$
$1930 + 22 = 1952$
In year 1952

57. $S = 2\sqrt{5L}$
$\frac{S}{2} = \sqrt{5L}$
$\left(\frac{S}{2}\right)^2 = 5L$
$\frac{S^2}{4} + 5 = L$
$L = \frac{S^2}{20}$

59. $r = \sqrt[3]{\frac{GMt^2}{4\pi^2}}$
$r^3 = \frac{GMt^2}{4\pi^2}$
$4\pi^2 r^3 = GMt^2$
$\frac{4\pi^2 r^3}{Gt^2} = M$
$M = \frac{4\pi^2 r^3}{Gt^2}$

61. c is true.
$T = 2\pi\sqrt{\frac{L}{32}}$
$\frac{T}{2\pi} = \sqrt{\frac{L}{32}}$
$\frac{T^2}{(2\pi)^2} = \frac{L}{32}$
$\frac{32T^2}{4\pi^2} = L$
$\frac{8T^2}{\pi^2} = L$

63. Students should verify results.

65. $\sqrt{4-x} - \sqrt{x+6} = 2$

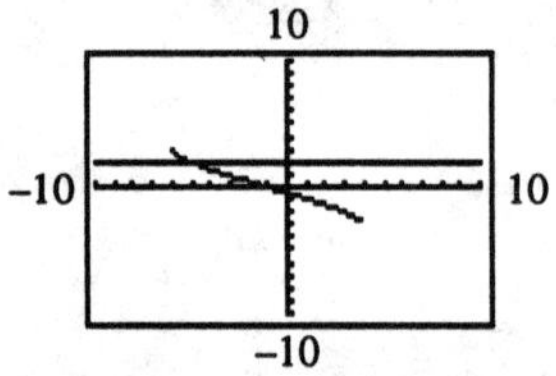

Solution: {–5}
Check:
$\sqrt{4-(-5)} - \sqrt{-5+6} = 2$
$\sqrt{9} - \sqrt{1} = 2$
$3 - 1 = 2$
$2 = 2$ (checks)

67. $\sqrt[3]{6x+9}=-3$

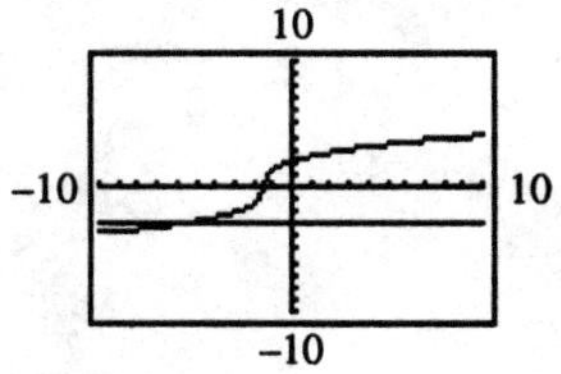

Solution: $\{-6\}$
Check:
$\sqrt[3]{6(-6)+9}=-3$
$\sqrt[3]{-27}=-3$
$-3=-3$ (checks)

69. Answers may vary.

71. Answers may vary.

73. $T=\dfrac{T_0}{\sqrt{1-\frac{v^2}{c^2}}}$

$4=\dfrac{2}{\sqrt{1-\frac{v^2}{(186{,}000)^2}}}$

$\sqrt{1-\dfrac{v^2}{(186{,}000)^2}}=\dfrac{1}{2}$

$1-\dfrac{v^2}{(186{,}000)^2}=\dfrac{1}{4}$

$\dfrac{v^2}{(186{,}000)^2}=\dfrac{3}{4}$

$v^2=\dfrac{3(186{,}000)^2}{4}$

$v=\dfrac{186{,}000\sqrt{3}}{2}$

$v=93{,}000\sqrt{3}$

The starship is traveling $93{,}000\sqrt{3}$ miles/second.

75. $\sqrt[3]{y\sqrt{y}}=9$

$y\sqrt{y}=9^3$

$y^{3/2}=9^3$

$(y^{3/2})^{2/3}=(9^3)^{2/3}$

$y=9^2$

$y=81$

77. Square both sides of the given equation.

$y^2=12+\sqrt{12+\sqrt{12+\ldots}}$

$y^2=12+y$

The second term on the right is given as y.

$y^2-y-12=0$

$(y-4)(y+3)=0$

$y=4$ or $y=-3$

Since y is equal to positive square root, we must reject −3. The exact value of the given expression is 4.

79. Group activity

Review Problems

80. $\dfrac{\sqrt[3]{5}}{\sqrt[4]{5}}=\dfrac{5^{1/3}}{5^{1/4}}=5^{1/3-1/4}=5^{4/12-3/12}$

$=5^{1/12}=\sqrt[12]{5}$

81. $I=\dfrac{k}{R}$

$0.5=\dfrac{k}{240}$

$k=120$

$I=\dfrac{120}{R}$

$I=\dfrac{120}{460}$

$I=\dfrac{6}{23}$

The current is $\dfrac{6}{23}$ amperes.

82. $x^2-6x+9-y^2=(x-3)^2-y^2$

$=[(x-3)+y][(x-3)-y]$

$=(x+y-3)(x-y-3)$

Problem Set 6.7

1. $\sqrt{-4} = \sqrt{4}i = 2i$

3. $\sqrt{-17} = \sqrt{17}i$

5. $\sqrt{-28} = \sqrt{28}i = \sqrt{4 \cdot 7}i = 2\sqrt{7}i$

7. $\sqrt{-45} = \sqrt{45}i = \sqrt{9 \cdot 5}i = 3\sqrt{5}i$

9. $\sqrt{-\frac{4}{9}} = \sqrt{\frac{4}{9}}i = \frac{2}{3}i$

11. $5\sqrt{-12} = 5\sqrt{4 \cdot 3}i = 5(2)\sqrt{3}i = 10\sqrt{3}i$

13. $(3 + 2i) + (5 + i) = (3 + 5) + (2 + 1)i$
$= 8 + 3i$

15. $(7 + 2i) + (1 - 4i) = (7 + 1) + (2 - 4)i$
$= 8 - 2i$

17. $(3 + 2i) - (5 + i) = (3 - 5) + (2 - 1)i$
$= -2 + i$

19. $(7 + 2i) - (1 - 4i) = (7 - 1) + (2 + 4)i$
$= 6 + 6i$

21. $\left(2 + i\sqrt{3}\right) + \left(7 + 4i\sqrt{3}\right) = (2 + 7) + \left(\sqrt{3} + 4\sqrt{3}\right)i$
$= 9 + 5i\sqrt{3}$

23. $\left(5 + 2i\sqrt{32}\right) + \left(11 - 5i\sqrt{8}\right) = (5 + 11) + \left(2\sqrt{32} - 5\sqrt{8}\right)i$
$= 16\left(2\sqrt{16 \cdot 2} - 5\sqrt{4 \cdot 2}\right)i$
$= 16 + \left(8\sqrt{2} - 10\sqrt{2}\right)i$
$= 16 - 2i\sqrt{2}$

25. $\left(5 + 2i\sqrt{32}\right) - \left(11 - 5i\sqrt{8}\right) = \left(5 + 2i\sqrt{32}\right) + \left(-11 + 5i\sqrt{8}\right)$
$= (5 - 11) + \left(2\sqrt{32} + 5\sqrt{8}\right)i$
$= -6 + \left(2\sqrt{16 \cdot 2} + 5\sqrt{4 \cdot 2}\right)i$
$= -6 + \left(8\sqrt{2} + 10\sqrt{2}\right)i = -6 + 18i\sqrt{2}$

27. $\sqrt{-49}+\sqrt{-100}=\sqrt{49}i+\sqrt{100}i=7i+10i$
$=17i$

29. $\sqrt{-64}-\sqrt{-25}=8i-5i=3i$

31. $3\sqrt{-49}+5\sqrt{-100}=3(7i)+5(10i)$
$=21i+50i=71i$

33. $\sqrt{-72}+\sqrt{-50}=\sqrt{36\cdot 2}i+\sqrt{25\cdot 2}i$
$=6\sqrt{2}i+5\sqrt{2}i=11\sqrt{2}i$

35. $5\sqrt{-8}-3\sqrt{-18}=5\sqrt{4\cdot 2}i-3\sqrt{9\cdot 2}i$
$=10\sqrt{2}i-9\sqrt{2}i=\sqrt{2}i$

37. $\frac{3}{5}\sqrt{-50}+\frac{1}{2}\sqrt{-32}=\frac{3}{5}\sqrt{25\cdot 2}i+\frac{1}{2}\sqrt{16\cdot 2}i$
$=\frac{3}{5}(5)\sqrt{2}i+\frac{1}{2}(4)\sqrt{2}i=3\sqrt{2}i+2\sqrt{2}i$
$=5\sqrt{2}i$

39. $(7+3i)(5+2i)=35+14i+15i+6i^2$
$=35+14i+15i+(-6)$
$=29+29i$

41. $(3+4i)(4-7i)=12-21i+16i-28i^2$
$=12-5i-28(-1)$
$=40-5i$

43. $(-5-4i)(3+7i)=-15-47i-28i^2$
$=-15-47i-(-28)=13-47i$

45. $(7-5i)(-2-3i)=-14-21i+10i+15i^2$
$=-14-21i+10i+(-15)=-29-11i$

47. $(3+5i)(3-5i)=9-15i+15i-25i^2$
$=9-25(-1)=9+25=34$

49. $(-5+3i)(-5-3i)=(-5)^2-(3i)^2$
$=25-9i^2=25-9(-1)=34$

51. $(2+3i)^2=(2+3i)(2+3i)$
$=4+12i+9i^2=4+12i+9(-1)$
$=-5+12i$

53. $4i(7-6i)=28i-24i^2$
$=28i-24(-1)=24+28i$

55. $\sqrt{-7}\sqrt{-2}=\left(\sqrt{7}i\right)\left(\sqrt{2}i\right)=\sqrt{14}i^2$
$=\sqrt{14}(-1)=-\sqrt{14}$

57. $\sqrt{-9}\sqrt{-4}=(3i)(2i)=6i^2=6(-1)=-6$

59. $\sqrt{-7}\sqrt{-25}=\left(\sqrt{7}i\right)(5i)=5\sqrt{7}i^2$
$=5\sqrt{7}(-1)=-5\sqrt{7}$

61. $\sqrt{-8}\sqrt{-3}=\left(\sqrt{4\cdot 2}i\right)\left(\sqrt{3}i\right)=2\sqrt{6}i^2$
$=2\sqrt{6}(-1)=-2\sqrt{6}$

63. $\left(2\sqrt{-8}\right)\left(3\sqrt{-6}\right)=\left(2\sqrt{8}i\right)\left(3\sqrt{6}i\right)=6\sqrt{48}i^2$
$=6\sqrt{16\cdot 3}(-1)=6(4)\sqrt{3}(-1)=-24\sqrt{3}$

65. $\left(3\sqrt{-5}\right)\left(-4\sqrt{-12}\right)=\left(3\sqrt{5}i\right)\left(-4\sqrt{12}i\right)$
$=-12\sqrt{60}i^2=-12\sqrt{4\cdot 15}(-1)$
$=-12(2)\sqrt{15}(-1)=24\sqrt{15}$

67. $\sqrt{-2}\left(3-\sqrt{-8}\right)=\sqrt{2}i\left(3-\sqrt{8}i\right)$
$=3\sqrt{2}i-\sqrt{16}i^2=3\sqrt{2}i+4=4+3\sqrt{2}i$

69. $\sqrt{-6}\left(2\sqrt{3}+\sqrt{-6}\right)=\sqrt{6}i\left(2\sqrt{3}+\sqrt{6}i\right)$
$=2\sqrt{18}i+\sqrt{36}i^2=2\sqrt{18}i+6(-1)$
$=2i\sqrt{9\cdot 2}-6=-6+6i\sqrt{2}$

71. $\left(3-\sqrt{-8}\right)\left(4+\sqrt{-2}\right)=\left(3-\sqrt{8}i\right)\left(4+\sqrt{2}i\right)$
$=12+3\sqrt{2}i-4\sqrt{8}i-\sqrt{16}i^2$
$=12+3\sqrt{2}i-8\sqrt{2}i-4(-1)=12-5\sqrt{2}i+4$
$=16-5\sqrt{2}i$

73. $\left(8-\sqrt{-6}\right)\left(2-\sqrt{-3}\right)=\left(8-\sqrt{6}i\right)\left(2-\sqrt{3}i\right)$
$=16-8\sqrt{3}i-2\sqrt{6}i+\sqrt{18}i^2$
$=16-8\sqrt{3}i-2\sqrt{6}i-3\sqrt{2}$
$=16-3\sqrt{2}-\left(8\sqrt{3}+2\sqrt{6}\right)i$

75. $x^2 - 2x + 2 = 0$
$(1+i)^2 - 2(1+i) + 2 \stackrel{?}{=} 0$
$1 + 2i - 1 - 2 - 2i + 2 \stackrel{?}{=} 0$
$0 = 0$
Yes

77. $x^2 - 6x + 13 = 0$
$(3-2i)^2 - 6(3-2i) + 13 \stackrel{?}{=} 0$
$9 - 12i - 4 - 18 + 12i + 13 \stackrel{?}{=} 0$
$0 = 0$
Yes

79. $x^2 - 12x + 40 = 0$
$(5+2i)^2 - 12(5+2i) + 40 \stackrel{?}{=} 0$
$25 + 20i - 4 - 60 - 24i + 40 \stackrel{?}{=} 0$
$1 - 4i \neq 0$
No

81. a. $i^{14} = (i^2)^7 = (-1)^7 = -1$

b. $i^{31} = i^{30} i = (i^2)^{15} i = (-1)^{15} i = -i$

c. $i^{22} = (i^2)^{11} = (-1)^{11} = -1$

d. $i^{37} = i^{36} \cdot i = (i^2)^{18} \cdot i = (-1)^{18} i = 1i = i$

83. $\frac{2i}{3-i} = \frac{2i}{3-i} \cdot \frac{3+i}{3+i}$
$= \frac{2i(3+i)}{9-i^2} = \frac{2i(3+i)}{10} = \frac{i(3+i)}{5}$
$= \frac{3i+i^2}{5} = \frac{3i-1}{5} = -\frac{1}{5} + \frac{3}{5}i$

85. $\frac{3-i}{2i} = \frac{3-i}{2i} \cdot \frac{i}{i} = \frac{3i - i^2}{2i^2} = \frac{3i-(-1)}{2(-1)}$
$= \frac{3i+1}{-2} = -\frac{1}{2} - \frac{3}{2}i$

87. $\frac{1+i}{1-i} = \frac{1+i}{1-i} \cdot \frac{1+i}{1+i} = \frac{1+2i+i^2}{1-i^2}$
$= \frac{1+2i+(-1)}{1-(-1)} = \frac{2i}{2} = i$

89. $\frac{2+3i}{3-i} = \frac{2+3i}{3-i} \cdot \frac{3+i}{3+i} = \frac{6+11i+3i^2}{9-i^2}$
$= \frac{3+11i}{10} = \frac{3}{10} + \frac{11}{10}i$

91. $\frac{3+2i}{2+i} = \frac{3+2i}{2+i} \cdot \frac{2-i}{2-i} = \frac{6+i-2i^2}{4-i^2}$
$= \frac{8+i}{5} = \frac{8}{5} + \frac{1}{5}i$

93. $\frac{-4+7i}{-2-5i} = \frac{-4+7i}{-2-5i} \cdot \frac{-2+5i}{-2+5i} = \frac{8-34i+35i^2}{4-25i^2}$
$= \frac{-27-34i}{29} = -\frac{27}{29} - \frac{34}{29}i$

95. $\frac{8-5i}{2i} \cdot \frac{i}{i} = \frac{8i-5i^2}{2i^2} = \frac{8i+5}{-2} = -\frac{5}{2} - 4i$

97. $\frac{4+7i}{-3i} \cdot \frac{i}{i} = \frac{4i+7i^2}{-3i^2} = \frac{4i-7}{3} = -\frac{7}{3} + \frac{4}{3}i$

99. $\frac{7}{3i} \cdot \frac{i}{i} = \frac{7i}{3i^2} = \frac{7i}{-3} = -\frac{7}{3}i$

101. $\frac{\sqrt{-125}}{\sqrt{5}} = \sqrt{\frac{125}{5}}i = \sqrt{25}i = 5i$

103. $\frac{\sqrt{-24}}{\sqrt{-6}} = \sqrt{\frac{24}{6}} \cdot \frac{i}{i} = \sqrt{4} = 2$

105. $\frac{\sqrt{-200}}{\sqrt{5}} = \sqrt{\frac{200}{5}}i = i\sqrt{40}$
$= i\sqrt{4 \cdot 10} = 2i\sqrt{10}$

107. $E = IR$
$E = (4-5i)(3+7i)$
$= 12 + 28i - 15i - 35i^2$
$= 12 + 13i + 35 = 47 + 13i$
The voltage is $(47 + 13i)$ volts.

109. Sum: $\left(5+\sqrt{15}i\right) + \left(5-\sqrt{15}i\right) = 10$
Product: $\left(5+\sqrt{15}i\right)\left(5-\sqrt{15}i\right) = 25 - 15i^2$
$= 25 - 15(-1) = 40$

111. $X = X_L - X_C$

$22i = X_L - 9i$

$X_L = 22i + 9i$

$X_L = 31i$

The inductive reactance is $31i$ ohms.

113. b is true.

115. Answers may vary.

117. Answers may vary.

119. Answers may vary.

121. Answers may vary.

123. $(8+9i)(2-i)-(1-i)(1+i) = (16+10i-9i^2)-(1-i^2)$

$= (25+10i)-(2) = 23+10i$

125. $\frac{1+i}{1+2i}\cdot\frac{1-2i}{1-2i}+\frac{1-i}{1-2i}\cdot\frac{1+2i}{1+2i} = \frac{1-i-2i^2}{1-4i^2}+\frac{1+i-2i^2}{1-4i^2}$

$= \frac{3-i}{5}+\frac{3+i}{5} = \frac{6}{5}$

127. $\sqrt{-36}\sqrt{-4} = (6i)(2i) = 12i^2 = 12(-1) = -12$

Integers: x, $x+1$, $x+2$

$x + x + 1 + x + 2 = -12$

$3x + 3 = -12$

$3x = -15$

$x = -5$

The integers are –5, –4, and –3.

129. i

opposite of $i = -i$

reciprocal of $i = \frac{1}{i} = \frac{1}{i}\cdot\frac{i}{i} = \frac{i}{i^2} = \frac{i}{-1} = -i$

Thus, the opposite and reciprocal of i are the same number.

131. $a + bi$ and $a - bi$ are a number and its conjugate.

$a + bi = a - bi$ when $b = -b$ or $2b = 0$ or

$b = 0$

Thus a complex number that is its own conjugate is $a + bi$ when $b = 0$.

133. The flaw is in Line 4, since $\sqrt{\frac{a}{b}} = \frac{\sqrt{a}}{\sqrt{b}}$ only if $a \geq 0$ and $b > 0$.

Review Problems

134. $\frac{4x^2+8x+3}{2x^2-x-1} \div \frac{4x^2+12x+9}{x^2-1} = \frac{(2x+1)(2x+3)}{(2x+1)(x-1)} \cdot \frac{(x+1)(x-1)}{(2x+3)(2x+3)}$

$= \frac{x+1}{2x+3}$

135. $V = kr^3$

Given: $V = 36\pi$ cubic meters, $r = 3$ meters

$36\pi = k(3^3)$

$\frac{4}{3}\pi = k$

$V = \frac{4}{3}\pi r^3$

Find v when $r = 5$ meters.

$V = \frac{4}{3}\pi(5^3)$

$V = \frac{4}{3}\pi(125)$

$V = \frac{500}{3}\pi$

The volume is $\frac{500}{3}\pi\ \text{m}^3$.

136. $x - 2y - z = 1$ (1)

$2x + y - z = -1$ (2)

$x + y + z = 2$ (3)

Add equations (1) and (3).

$x - 2y - z = 1$

$x + y + z = 2$

$2x - y = 3$ (4)

Add equations (2) and (3).

$2x + y - z = -1$

$x + y + z = 2$

$3x + 2y = 1$ (5)

Add 2 times equation (4) to (5).

$4x - 2y = 6$

$3x + 2y = 1$

$7x = 7$

$x = 1$

Substitute into equation (4).

$2(1) - y = 3$

$y = -1$

Substitute into equation (3).

$1 - 1 + z = 2$

$z = 2$

$\{(1, -1, 2)\}$

Chapter 6 Review Problems

1. $\sqrt{81} = \sqrt{9^2} = 9$

2. $\sqrt{\frac{4}{49}} = \sqrt{\left(\frac{2}{7}\right)^2} = \frac{2}{7}$

3. $\sqrt[3]{-27} = \sqrt[3]{(-3)^3} = -3$

4. $\sqrt[3]{\frac{64}{125}} = \sqrt[3]{\left(\frac{4}{5}\right)^3} = \frac{4}{5}$

5. $\sqrt[5]{-32} = \sqrt[5]{(-2)^5} = -2$

6. $h(t) = 3\sqrt{t} - 0.23t$

$h(25) = 3\sqrt{25} - 0.23(25)$

$= 3(5) - 5.75 = 9.25$

The height of a bamboo plant 25 weeks after it comes through the soil is 9.25 inches.

7. $f(x) = \sqrt{x-2}$

$x - 2 \geq 0$

$x \geq 2$

Domain of $f = [2, \infty)$

8. $g(x) = \sqrt{100 - 4x}$

$100 - 4x \geq 0$

$100 \geq 4x$

$25 \geq x$

Domain of $g = (-\infty, 25]$

9. $h(x) = \sqrt[3]{x-2}$

Domain of $h = (-\infty, \infty)$

10. $f(x)=\sqrt{x},\ g(x)=\sqrt{x}+2$

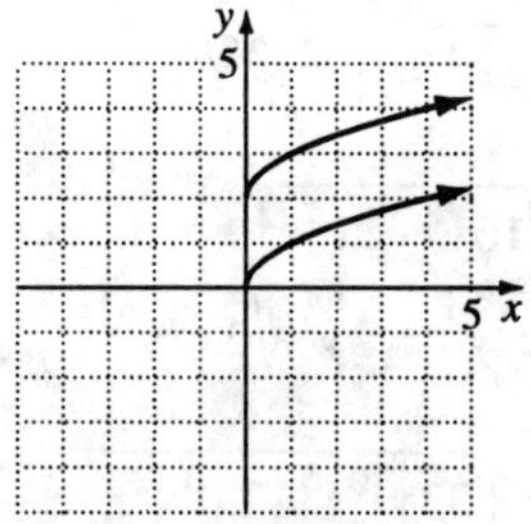

Domain of $f=[0,\infty)$
Range of $f=[0,\infty)$
Domain of $g=[0,\infty)$
Range of $g=[2,\infty)$
g is f shifted 2 units up.

11. $f(x)=\sqrt{x},\ g(x)=\sqrt{x+2}$

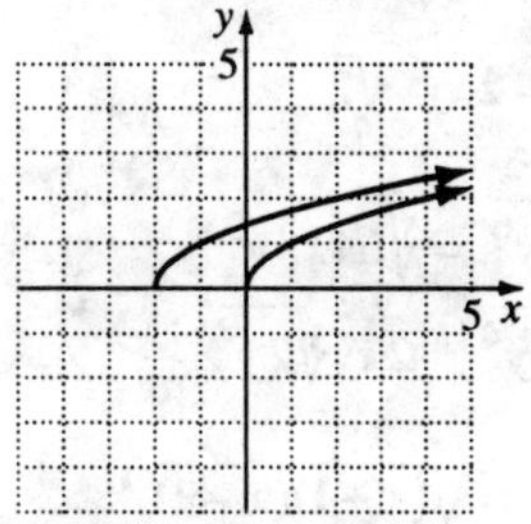

Domain of $f=[0,\infty)$
Range of $f=[0,\infty)$
Domain of $g=[-2,\infty)$
Range of $g=[0,\infty)$
g is f shifted 2 units left.

12. $f(x)=\sqrt[3]{x},\ g(x)=\sqrt[3]{x-1}$

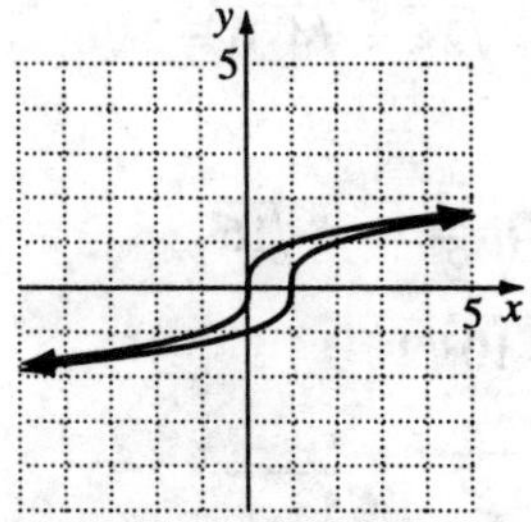

Domain of $f=(-\infty,\infty)$
Range of $f=(-\infty,\infty)$
Domain of $g=(-\infty,\infty)$
Range of $g=(-\infty,\infty)$
g is f shifted one unit right.

13. $L(v)=500\sqrt{1-\frac{v^2}{c^2}}$

$L(167{,}400)=500\sqrt{1-\frac{(167{,}400)^2}{(186{,}000)^2}}$
$=500\sqrt{0.19}$
≈ 217.9

The length of a 500-meter starship moving at 167,400 miles/second from the perspective of an observer at rest is about 217.9 meters.

14. $\sqrt[3]{5^3}=5$

15. $\sqrt[3]{(-5)^3}=-5$

16. $\sqrt{5^2}=5$

17. $\sqrt{(-5)^2}=-5$

18. $\sqrt[6]{(-2)^6}=-2$

19. $\sqrt[7]{(-2)^7}=-2$

20. $16^{3/2}=\left(\sqrt{16}\right)^3=4^3=64$

21. $8^{-2/3}=\frac{1}{\left(\sqrt[3]{8}\right)^2}=\frac{1}{2^2}=\frac{1}{4}$

22. $\left(\frac{1}{32}\right)^{4/5}=\left(\sqrt[5]{\frac{1}{32}}\right)^4=\left(\frac{1}{2}\right)^4=\frac{1}{16}$

23. $(7x^{1/3})(4x^{1/4})=(7)(4)x^{\frac{1}{3}+\frac{1}{4}}=28x^{7/12}$

24. $\frac{80y^{3/4}}{-20y^{1/5}}=\left(\frac{80}{-20}\right)y^{\frac{3}{4}-\frac{1}{5}}=-4y^{11/20}$

25. $(9x^{-2}y^{1/2})^{3/2} = 9^{3/2}x^{-2\cdot 3/2}y^{\frac{1}{2}\cdot\frac{3}{2}}$
$= 27x^{-3}y^{3/4} = \dfrac{27y^{3/4}}{x^3}$

26. $\left(\dfrac{32x^5}{y^{2/3}}\right)^{1/5} = \dfrac{32^{1/5}x^{5\cdot 1/5}}{y^{\frac{2}{3}\cdot\frac{1}{5}}} = \dfrac{2x}{y^{2/15}}$

27. $\sqrt[9]{a^6} = a^{6/9} = a^{2/3}$

28. $\sqrt[8]{x^2y^4} = (x^2y^4)^{1/8} = x^{2/8}y^{4/8} = x^{1/4}y^{1/2}$

29. $\sqrt[6]{2^4x^4y^2} = (2^4x^4y^2)^{1/6} = 2^{4/6}x^{4/6}y^{2/6}$
$= 2^{2/3}x^{2/3}y^{1/3}$

30. $\sqrt{3}\sqrt[3]{3} = 3^{1/2}\cdot 3^{1/3} = 3^{\frac{1}{2}+\frac{1}{3}} = 3^{5/6}$

31. $\dfrac{\sqrt[3]{2}}{\sqrt[4]{2}} = \dfrac{2^{1/3}}{2^{1/4}} = 2^{\frac{1}{3}-\frac{1}{4}} = 2^{1/12}$

32. $x^{1/2} + x^{3/4} = x^{1/2}(1 + x^{1/4})$

33. $4x^{-1/2} - 8x^{1/2} = 4x^{-1/2}(1-2x)$
$= \dfrac{4(1-2x)}{x^{1/2}}$

34. $D = (2H)^{1/2}$
$D = [2(6+794)]^{1/2}$
$= [2(800)]^{1/2} = \sqrt{1600} = 40$
The person can see 40 miles.

35. $f(d) = 0.07d^{3/2}$
$f(16) = 0.07(16)^{3/2}$
$= 0.07(64) = 4.48$
The duration of a storm whose diameter is 16 miles is 4.48 hours.

36. $\sqrt{20x^3} = \sqrt{4x^2\cdot 5x} = \sqrt{(2x)^2\cdot 5x}$
$= 2x\sqrt{5x}$

37. $\sqrt[3]{54x^8y^6} = \sqrt[3]{27x^6y^6\cdot 2x^2}$
$= \sqrt[3]{(3x^2y^2)^3\cdot 2x^2} = 3x^2y^2\sqrt[3]{2x^2}$

38. $\sqrt[4]{32x^3y^{10}} = \sqrt[4]{16y^8\cdot 2x^3y^2}$
$= \sqrt[4]{(2y^2)^4\cdot 2x^3y^2} = 2y^2\sqrt[4]{2x^3y^2}$

39. $\sqrt[5]{64x^3y^{12}z^6} = \sqrt[5]{32y^{10}z^5\cdot 2x^3y^2z}$
$= \sqrt[5]{(2y^2z)^5\cdot 2x^3y^2z} = 2y^2z\sqrt[5]{2x^3y^2z}$

40. $\sqrt{6x^3}\cdot\sqrt{4x^2} = \sqrt{24x^5} = \sqrt{4x^4\cdot 6x}$
$= 2x^2\sqrt{6x}$

41. $\sqrt[3]{4x^2y}\cdot\sqrt[3]{4xy^5} = \sqrt[3]{16x^3y^6}$
$= \sqrt[3]{8x^3y^6\cdot 2} = 2xy^2\sqrt[3]{2}$

42. $\sqrt[5]{8x^4y^3}\cdot\sqrt[5]{8xy^6} = \sqrt[5]{64x^5y^9}$
$= \sqrt[5]{32x^5y^5\cdot 2y^4} = 2xy\sqrt[5]{2y^4}$

43. $\sqrt{x+1}\cdot\sqrt{x-1} = \sqrt{(x+1)(x-1)}$
$= \sqrt{x^2-1}$

44. $v = 3\sqrt{d}$
$v = 3\sqrt{75} = 3\sqrt{25\cdot 3} = 3\cdot 5\sqrt{3}$
$= 15\sqrt{3}$ feet/second

45. $\dfrac{\sqrt{48}}{\sqrt{2}} = \sqrt{\dfrac{48}{2}} = \sqrt{24} = \sqrt{4\cdot 6} = 2\sqrt{6}$

46. $\dfrac{-10\sqrt[3]{32}}{2\sqrt[3]{2}} = -5\sqrt[3]{\dfrac{32}{2}} = -5\sqrt[3]{16}$
$= -5\cdot 2\sqrt[3]{2} = -10\sqrt[3]{2}$

47. $\dfrac{\sqrt[4]{64x^7}}{\sqrt[4]{2x^2}} = \sqrt[4]{\dfrac{64x^7}{2x^2}} = \sqrt[4]{32x^5}$
$= \sqrt[4]{16x^4\cdot 2x} = 2x\sqrt[4]{2x}$

48. $\dfrac{\sqrt{200x^3y^2}}{\sqrt{2x^{-2}y}} = \sqrt{\dfrac{200x^3y^2}{2x^{-2}y}} = \sqrt{100x^5y}$
$= \sqrt{100x^4 \cdot xy} = 10x^2\sqrt{xy}$

49. $\sqrt[3]{\dfrac{16}{125}} = \dfrac{\sqrt[3]{16}}{\sqrt[3]{125}} = \dfrac{\sqrt[3]{8 \cdot 2}}{\sqrt[3]{125}} = \dfrac{2\sqrt[3]{2}}{5}$

50. $\sqrt{\dfrac{x^3}{100y^4}} = \dfrac{\sqrt{x^3}}{\sqrt{100y^4}} = \dfrac{\sqrt{x^2 \cdot x}}{\sqrt{100y^4}}$
$= \dfrac{x\sqrt{x}}{10y^2}$

51. $\sqrt[4]{\dfrac{3y^5}{16x^{20}}} = \dfrac{\sqrt[4]{3y^5}}{\sqrt[4]{16x^{20}}} = \dfrac{\sqrt[4]{y^4 \cdot 3y}}{\sqrt[4]{16x^{20}}} = \dfrac{y\sqrt[4]{3y}}{2x^5}$

52. $\dfrac{4}{\sqrt{6}} = \dfrac{4}{\sqrt{6}} \cdot \dfrac{\sqrt{6}}{\sqrt{6}} = \dfrac{4\sqrt{6}}{6} = \dfrac{2\sqrt{6}}{3}$

53. $\sqrt{\dfrac{2}{7}} = \dfrac{\sqrt{2}}{\sqrt{7}} \cdot \dfrac{\sqrt{7}}{\sqrt{7}} = \dfrac{\sqrt{14}}{7}$

54. $\dfrac{12}{\sqrt[3]{9}} = \dfrac{12}{\sqrt[3]{9}} \cdot \dfrac{\sqrt[3]{3}}{\sqrt[3]{3}} = \dfrac{12\sqrt[3]{3}}{3} = 4\sqrt[3]{3}$

55. $\sqrt{\dfrac{2x}{5y}} = \dfrac{\sqrt{2x}}{\sqrt{5y}} \cdot \dfrac{\sqrt{5y}}{\sqrt{5y}} = \dfrac{\sqrt{10xy}}{5y}$

56. $\dfrac{14}{\sqrt[3]{2x^2}} = \dfrac{14}{\sqrt[3]{2x^2}} \cdot \dfrac{\sqrt[3]{4x}}{\sqrt[3]{4x}} = \dfrac{14\sqrt[3]{4x}}{2x}$
$= \dfrac{7\sqrt[3]{4x}}{x}$

57. $\sqrt[4]{\dfrac{7}{3x}} = \dfrac{\sqrt[4]{7}}{\sqrt[4]{3x}} \cdot \dfrac{\sqrt[4]{27x^3}}{\sqrt[4]{27x^3}} = \dfrac{\sqrt[4]{189x^3}}{3x}$

58. $\dfrac{5}{\sqrt[5]{32x^4y}} = \dfrac{5}{\sqrt[5]{32x^4y}} \cdot \dfrac{\sqrt[5]{xy^4}}{\sqrt[5]{xy^4}} = \dfrac{5\sqrt[5]{xy^4}}{2xy}$

59. $\sqrt{\dfrac{5}{8x^3}} = \dfrac{\sqrt{5}}{\sqrt{8x^3}} = \dfrac{\sqrt{5}}{2x\sqrt{2x}} \cdot \dfrac{\sqrt{2x}}{\sqrt{2x}}$
$= \dfrac{\sqrt{10x}}{2x(2x)} = \dfrac{\sqrt{10x}}{4x^2}$

60. $\sqrt{\dfrac{6x^5}{y^3}} = \dfrac{\sqrt{6x^5}}{\sqrt{y^3}} = \dfrac{x^2\sqrt{6x}}{y\sqrt{y}} \cdot \dfrac{\sqrt{y}}{\sqrt{y}}$
$= \dfrac{x^2\sqrt{6xy}}{y^2}$

61. $\dfrac{4}{\sqrt[3]{2x^5}} = \dfrac{4}{x\sqrt[3]{2x^2}} \cdot \dfrac{\sqrt[3]{4x}}{\sqrt[3]{4x}} = \dfrac{4\sqrt[3]{4x}}{x(2x)}$
$= \dfrac{2\sqrt[3]{4x}}{x^2}$

62. $\dfrac{18}{\sqrt[5]{32x^7y^4}} = \dfrac{18}{2x\sqrt[5]{x^2y^4}} \cdot \dfrac{\sqrt[5]{x^3y}}{\sqrt[5]{x^3y}}$
$= \dfrac{18\sqrt[5]{x^3y}}{2x(xy)} = \dfrac{9\sqrt[5]{x^3y}}{x^2y}$

63. $5\sqrt{18} + 3\sqrt{8} - \sqrt{2} = 5\sqrt{9 \cdot 2} + 3\sqrt{4 \cdot 2} - \sqrt{2}$
$= 5(3)\sqrt{2} + 3(2)\sqrt{2} - \sqrt{2}$
$= 15\sqrt{2} + 6\sqrt{2} - \sqrt{2} = 20\sqrt{2}$

64. $6\sqrt{2x^3} - \sqrt{49x^2} + 7\sqrt{18x^3} = 6\sqrt{x^2 \cdot 2x} - \sqrt{7^2 \cdot x^2} + 7\sqrt{9x^2 \cdot 2x}$
$= 6x\sqrt{2x} - 7x + 21x\sqrt{2x} = 27x\sqrt{2x} - 7x$

65. $2\sqrt[3]{6} + 5\sqrt[3]{48} = 2\sqrt[3]{6} + 5\sqrt[3]{8 \cdot 6} = 2\sqrt[3]{6} + 5 \cdot 2\sqrt[3]{6} = 2\sqrt[3]{6} + 10\sqrt[3]{6} = 12\sqrt[3]{6}$

66. $7\sqrt[3]{16x^4} - 3\sqrt[3]{2x} = 7\sqrt[3]{8x^3 \cdot 2x} - 3\sqrt[3]{2x} = 14x\sqrt[3]{2x} - 3\sqrt[3]{2x} = \sqrt[3]{2x}(14x - 3)$

67. $2x\sqrt[4]{32y^5}-3y\sqrt[4]{162x^4y}+\sqrt[4]{2x^4y^4}=2x\sqrt[4]{2^4y^4(2y)}-3y\sqrt[4]{3^4x^4(2y)}+\sqrt[4]{x^4y^4(2)}$
$=2x\left(2y\sqrt[4]{2y}\right)-3y\left(3x\sqrt[4]{2y}\right)+xy\sqrt[4]{2}$
$=(4xy-9xy)\sqrt[4]{2y}+xy\sqrt[4]{2}$
$=-5xy\sqrt[4]{2y}+xy\sqrt[4]{2}$ or $xy\sqrt[4]{2}\left(1-5\sqrt[4]{y}\right)$

68. $6\sqrt{\frac{1}{3}}+4\sqrt{\frac{1}{2}}=\frac{6}{\sqrt{3}}\cdot\frac{\sqrt{3}}{\sqrt{3}}+\frac{4}{\sqrt{2}}\cdot\frac{\sqrt{2}}{\sqrt{2}}$
$=\frac{6\sqrt{3}}{3}+\frac{4\sqrt{2}}{2}=2\sqrt{3}+2\sqrt{2}$

69. $5\sqrt{3}\left(2\sqrt{6}+4\sqrt{15}\right)=10\sqrt{18}+20\sqrt{45}$
$=10\sqrt{9\cdot 2}+20\sqrt{9\cdot 5}=10(3)\sqrt{2}+20(3)\sqrt{5}$
$=30\sqrt{2}+60\sqrt{5}$

70. $\sqrt{2x}\left(\sqrt{6x}+3\sqrt{x}\right)=\sqrt{12x^2}+3\sqrt{2x^2}$
$=\sqrt{4\cdot 3x^2}+3\sqrt{2x^2}=2x\sqrt{3}+3x\sqrt{2}$

71. $2\sqrt[3]{5}\left(3\sqrt[3]{50}-4\sqrt[3]{2}\right)=6\sqrt[3]{250}-8\sqrt[3]{10}$
$=6\sqrt[3]{125\cdot 2}-8\sqrt[3]{10}=30\sqrt[3]{2}-8\sqrt[3]{10}$

72. $\left(5\sqrt{2}-4\sqrt{3}\right)\left(7\sqrt{2}+3\sqrt{3}\right)=35\sqrt{4}+15\sqrt{6}-28\sqrt{6}-12\sqrt{9}$
$=35\cdot 2-13\sqrt{6}-12\cdot 3=34-13\sqrt{6}$

73. $\left(\sqrt{x}+\sqrt{11}\right)\left(\sqrt{y}+\sqrt{11}\right)=\sqrt{xy}+\sqrt{11x}+\sqrt{11y}+11$

74. $\left(\sqrt{7}+\sqrt{5}\right)^2=\left(\sqrt{7}\right)^2+2\sqrt{7}\sqrt{5}+\left(\sqrt{5}\right)^2$
$=7+2\sqrt{35}+5=12+2\sqrt{35}$

75. $\left(2\sqrt{3}-\sqrt{10}\right)^2=\left(2\sqrt{3}\right)^2-2\left(2\sqrt{3}\right)\left(\sqrt{10}\right)+\left(\sqrt{10}\right)^2$
$=4\cdot 3-4\sqrt{30}+10=22-4\sqrt{30}$

76. $\left(\sqrt{7}+\sqrt{13}\right)\left(\sqrt{7}-\sqrt{13}\right)=\left(\sqrt{7}\right)^2-\left(\sqrt{13}\right)^2$
$=7-13=-6$

77. $\left(7+3\sqrt{5}\right)\left(7-3\sqrt{5}\right)=7^2-\left(3\sqrt{5}\right)^2$
$=49-9(5)=4$

78. $\dfrac{6}{\sqrt{3}-1}=\dfrac{6}{\sqrt{3}-1}\cdot\dfrac{\sqrt{3}+1}{\sqrt{3}+1}=\dfrac{6\left(\sqrt{3}+1\right)}{3-1}$
$=\dfrac{6\left(\sqrt{3}+1\right)}{2}=3\left(\sqrt{3}+1\right)=3\sqrt{3}+3$

79. $\dfrac{\sqrt{7}}{\sqrt{5}+\sqrt{3}}=\dfrac{\sqrt{7}}{\sqrt{5}+\sqrt{3}}\cdot\dfrac{\sqrt{5}-\sqrt{3}}{\sqrt{5}-\sqrt{3}}$
$=\dfrac{\sqrt{35}-\sqrt{21}}{5-3}=\dfrac{\sqrt{35}-\sqrt{21}}{2}$

80. $\dfrac{7}{2\sqrt{5}-3\sqrt{7}}=\dfrac{7}{2\sqrt{5}-3\sqrt{7}}\cdot\dfrac{2\sqrt{5}+3\sqrt{7}}{2\sqrt{5}+3\sqrt{7}}$
$=\dfrac{7\left(2\sqrt{5}+3\sqrt{7}\right)}{4(5)-9(7)}=\dfrac{7\left(2\sqrt{5}+3\sqrt{7}\right)}{-43}$
$=\dfrac{14\sqrt{5}+21\sqrt{7}}{-43}=-\dfrac{14\sqrt{5}+21\sqrt{7}}{43}$
$=\dfrac{-14\sqrt{5}-21\sqrt{7}}{43}$

81. $\dfrac{\sqrt{y}+5}{\sqrt{y}-3}=\dfrac{\sqrt{y}+5}{\sqrt{y}-3}\cdot\dfrac{\sqrt{y}+3}{\sqrt{y}+3}$
$=\dfrac{y+3\sqrt{y}+5\sqrt{y}+15}{y-9}=\dfrac{y+8\sqrt{y}+15}{y-9}$

82. $\dfrac{\sqrt{7}+\sqrt{3}}{\sqrt{7}-\sqrt{3}}=\dfrac{\sqrt{7}+\sqrt{3}}{\sqrt{7}-\sqrt{3}}\cdot\dfrac{\sqrt{7}+\sqrt{3}}{\sqrt{7}+\sqrt{3}}$
$=\dfrac{7+\sqrt{21}+\sqrt{21}+3}{7-3}=\dfrac{10+2\sqrt{21}}{4}$
$=\dfrac{2\left(5+\sqrt{21}\right)}{4}=\dfrac{5+\sqrt{21}}{2}$

83. $\dfrac{2\sqrt{x}}{\sqrt{x}+\sqrt{y}}=\dfrac{2\sqrt{x}}{\sqrt{x}+\sqrt{y}}\cdot\dfrac{\sqrt{x}-\sqrt{y}}{\sqrt{x}-\sqrt{y}}$
$=\dfrac{2x-2\sqrt{xy}}{x-y}$

84. $\dfrac{\sqrt{3a}+\sqrt{b}}{\sqrt{5a}+\sqrt{2b}}=\dfrac{\sqrt{3a}+\sqrt{b}}{\sqrt{5a}+\sqrt{2b}}\cdot\dfrac{\sqrt{5a}-\sqrt{2b}}{\sqrt{5a}-\sqrt{2b}}$
$=\dfrac{\sqrt{15a^2}-\sqrt{6ab}+\sqrt{5ab}-\sqrt{2b^2}}{5a-2b}$
$=\dfrac{a\sqrt{15}+\sqrt{5ab}-\sqrt{6ab}-b\sqrt{2}}{5a-2b}$ or
$\dfrac{a\sqrt{15}-b\sqrt{2}+\left(\sqrt{5}-\sqrt{6}\right)\sqrt{ab}}{5a-2b}$

85. $\dfrac{3\sqrt{7}+1}{\sqrt{3}-4\sqrt{5}}=\dfrac{3\sqrt{7}+1}{\sqrt{3}-4\sqrt{5}}\cdot\dfrac{\sqrt{3}+4\sqrt{5}}{\sqrt{3}+4\sqrt{5}}$
$=\dfrac{3\sqrt{21}+12\sqrt{35}+\sqrt{3}+4\sqrt{5}}{3-16(5)}$
$=\dfrac{3\sqrt{21}+12\sqrt{35}+\sqrt{3}+4\sqrt{5}}{-77}$
$=\dfrac{-3\sqrt{21}-12\sqrt{35}-\sqrt{3}-4\sqrt{5}}{77}$

86. $E=\dfrac{w}{20\sqrt{a}}$

$E=\dfrac{12}{20\sqrt{8}}=\dfrac{12}{20\cdot 2\sqrt{2}}=\dfrac{3}{10\sqrt{2}}\cdot\dfrac{\sqrt{2}}{\sqrt{2}}$
$=\dfrac{3\sqrt{2}}{20}$

The evaporation is $\dfrac{3\sqrt{2}}{20}$ inch.

87. Perimeter:
$P=2\sqrt{54}+2\sqrt{24}$
$P=2\sqrt{9\cdot 6}+2\sqrt{4\cdot 6}$
$P=6\sqrt{6}+4\sqrt{6}$
$P=10\sqrt{6}$
Area:
$A=\sqrt{54}\sqrt{24}$
$A=\sqrt{1296}$
$A=36$

88. Perimeter:

$P = \sqrt{90} + \sqrt{160} + \sqrt{250}$

$P = \sqrt{9 \cdot 10} + \sqrt{16 \cdot 10} + \sqrt{25 \cdot 10}$

$P = 3\sqrt{10} + 4\sqrt{10} + 5\sqrt{10}$

$P = 12\sqrt{10}$

Area:

$A = \frac{1}{2}\sqrt{250}\sqrt{40}$

$A = \frac{1}{2}\left(5\sqrt{10}\right)\left(2\sqrt{10}\right)$

$A = 50$

89. $\sqrt{2x+4} = 6$

$\left(\sqrt{2x+4}\right)^2 = 6^2$

$2x + 4 = 36$

$2x = 32$

$x = 16$

Check:

$\sqrt{2(16)+4} = 6$

$\sqrt{36} = 6$

$6 = 6$

The solution checks.

$\{16\}$

90. $\sqrt{x-5} + 9 = 4$

$\sqrt{x-5} = -5$

$x - 5 = 25$

$x = 30$

Check:

$\sqrt{30-5} + 9 = 4$

$\sqrt{25} + 9 = 4$

$5 + 9 = 4$

$14 = 4$

No

$x = 30$ is extraneous.

No solution

$\varnothing$

91. $\sqrt{2x-3} + x = 3$

$\sqrt{2x-3} = 3 - x$

$2x - 3 = 9 - 6x + x^2$

$x^2 - 8x + 12 = 0$

$(x-6)(x-2) = 0$

$x = 6$ or $x = 2$

Check:

$\sqrt{2(6)-3} + 6 = 3$

$3 + 6 = 3$

$9 = 3$

No

$\sqrt{2(2)-3} + 2 = 3$

$1 + 2 = 3$

$3 = 3$

Yes

$x = 6$ is extraneous.

$\{2\}$

92. $\sqrt[3]{x^2 + 6x} + 2 = 0$

$\sqrt[3]{x^2 + 6x} = -2$

$x^2 + 6x = -8$

$x^2 + 6x + 8 = 0$

$(x+4)(x+2) = 0$

$x = -4$ or $x = -2$

Check:

$\sqrt[3]{(-4)^2 + 6(-4)} + 2 = 0$

$\sqrt[3]{-8} + 2 = 0$

$-2 + 2 = 0$

$0 = 0$

Yes

$\sqrt[3]{(-2)^2 + 6(-2)} + 2 = 0$

$\sqrt[3]{-8} + 2 = 0$

$-2 + 2 = 0$

$0 = 0$

Yes

$\{-4, -2\}$

93. $\sqrt{x-4}+\sqrt{x+1}=5$
$\sqrt{x-4}=5-\sqrt{x+1}$
$x-4=25-10\sqrt{x+1}+x+1$
$\sqrt{x+1}=3$
$x+1=9$
$x=8$
Check:
$\sqrt{8-4}+\sqrt{8+1}=5$
$\sqrt{4}+\sqrt{9}=5$
$2+3=5$
$5=5$ yes
$\{8\}$

94. $2-\sqrt{3y+1}+\sqrt{y-1}=0$
$\sqrt{y-1}=\sqrt{3y+1}-2$
$\left(\sqrt{y-1}\right)^2=\left(\sqrt{3y+1}-2\right)^2$
$y-1=3y+1-4\sqrt{3y+1}+4$
$y-1=3y+5-4\sqrt{3y+1}$
$-2y-6=-4\sqrt{3y+1}$
$y+3=2\sqrt{3y+1}$
$(y+3)^2=\left(2\sqrt{3y+1}\right)^2$
$y^2+6y+9=4(3y+1)$
$y^2+6y+9=12y+4$
$y^2-6y+5=0$
$(y-1)(y-5)=0$
$y-1=0$ or $y-5=0$
$y=1$ $\quad y=5$
Check 1:
$2-\sqrt{3(1)+1}+\sqrt{1-1}=0$
$2-2+0=0$
$0=0$ checks
Check 5:
$2-\sqrt{3(5)+1}+\sqrt{5-1}=0$
$2-4+2=0$
$0=0$ checks
$\{1, 5\}$

95. $r=\sqrt{\dfrac{V}{\pi h}}$

a. $2=\sqrt{\dfrac{V}{\pi\cdot 6}}$
$4=\dfrac{V}{6\pi}$
$24\pi=V$
The volume is 24π ft^3.

b. $\dfrac{650}{8}=81.25$
$24\pi\approx 75.4$
Since $81.25>75.4$, the tank will not do the job.

96. $t=\sqrt{\dfrac{2s}{g}}$ for s
$t^2=\dfrac{2s}{g}$
$gt^2=2s$
$\dfrac{gt^2}{2}=s$
$s=\dfrac{gt^2}{2}$

97. $r=\sqrt[3]{\dfrac{2mM}{c}}$ for c
$r^3=\dfrac{2mM}{c}$
$c=\dfrac{2mM}{r^3}$

98. $v=\sqrt{24L}$
$60=\sqrt{24L}$
$3600=24L$
$150=L$
The skid marks are 150 feet long.

99. $\sqrt{-81}=9i$

100. $\sqrt{-63}=3i\sqrt{7}$

101. $-\sqrt{-8}=-2i\sqrt{2}$

102. $(7+12i)+(5-10i)=12+2i$

103. $(7-12i)-(-3-7i)=(7-12i)+(3+7i)$
$=10-5i$

104. $(7-5i)(2+3i)=14+21i-10i-15i^2$
$=14+21i-10i-15(-1)=29+11i$

105. $(2+5i)^2=(2+5i)(2+5i)$
$=4+10i+10i+25i^2=4+20i+25(-1)$
$=-21+20i$

106. $\frac{3i}{5+i}=\frac{3i}{5+i}\cdot\frac{5-i}{5-i}=\frac{15i-3i^2}{25-i^2}$
$=\frac{15i-3(-1)}{25-(-1)}=\frac{3+15i}{26}=\frac{3}{26}+\frac{15i}{26}$

107. $\frac{3-4i}{4+2i}=\frac{3-4i}{4+2i}\cdot\frac{4-2i}{4-2i}=\frac{12-6i-16i+8i^2}{16-4i^2}$
$=\frac{12-22i+8(-1)}{16-4(-1)}=\frac{4-22i}{20}$
$=\frac{4}{20}-\frac{22}{20}i=\frac{1}{5}-\frac{11}{10}i$

108. $\frac{5+i}{3i}=\frac{5+i}{3i}\cdot\frac{i}{i}=\frac{5i+i^2}{3i^2}=\frac{5i+(-1)}{3(-1)}$
$=\frac{-1+5i}{-3}=\frac{-1}{-3}+\frac{5i}{-3}=\frac{1}{3}-\frac{5i}{3}$

109. $i^{23}=i^{22}i=(i^2)^{11}i=(-1)^{11}i=-i$

110. $2\sqrt{-100}+3\sqrt{-36}=2(10i)+3(6i)$
$=20i+18i=38i$

111. $\sqrt{-5}\cdot\sqrt{-9}=i\sqrt{5}(3i)=-3\sqrt{5}$

112. $\frac{\sqrt{-24}}{\sqrt{-6}}=\frac{2i\sqrt{6}}{i\sqrt{6}}=2$

113. $\left(2+\sqrt{-8}\right)\left(3+\sqrt{-2}\right)=\left(2+2i\sqrt{2}\right)\left(3+i\sqrt{2}\right)$
$=6+2i\sqrt{2}+6i\sqrt{2}+2i^2(2)$
$=2+8i\sqrt{2}$

114. $x^2-2x+5=0$
$(1-2i)^2-2(1-2i)+5=0$
$1-4i-4-2+4i+5=0$
$0=0$

115. $E=IR$
$430-330i=(35-40i)R$
$R=\frac{430-330i}{35-40i}$
$R=\frac{5(86-66i)}{5(7-8i)}$
$R=\frac{86-66i}{7-8i}\cdot\frac{7+8i}{7+8i}$
$R=\frac{602+688i-462i-528i^2}{49+64}$
$R=\frac{1130+226i}{113}$
$R=10+2i$
The impedance is $(10+2i)$ ohms.

Chapter 6 Test

1. $\sqrt[3]{\frac{-8}{125}}=\frac{\sqrt[3]{-8}}{\sqrt[3]{125}}=\frac{-2}{5}=-\frac{2}{5}$

2. $f(x)=\sqrt{8-2x}$
$8-2x\geq 0$
$-2x\geq -8$
$x\leq 4$
Domain of $f=(-\infty, 4]$

3. $f(x)=\sqrt{x}$, $g(x)=\sqrt{x-2}$

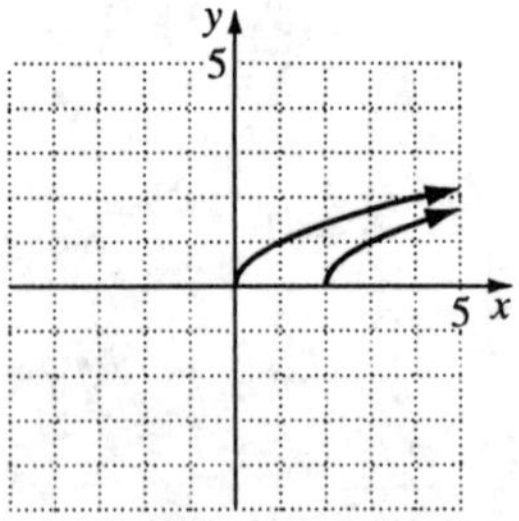

g is f shifted 2 units right.

4. $64^{-2/3}=\frac{1}{64^{2/3}}=\frac{1}{\left(\sqrt[3]{64}\right)^2}=\frac{1}{4^2}=\frac{1}{16}$

5. $(25x^{-4}y^{1/2})^{3/2} = 25^{3/2}x^{-12/2}y^{3/4}$
$= \dfrac{125y^{3/4}}{x^6}$

6. $\sqrt[8]{x^4} = x^{4/8} = x^{1/2}$

7. $\sqrt{2}\cdot\sqrt[3]{2} = 2^{1/2}\cdot 2^{1/3} = 2^{\frac{1}{2}+\frac{1}{3}} = 2^{5/6}$

8. $6x^{1/2} + 12x^{-1/2} = 6x^{-1/2}(x+2)$
$= \dfrac{6(x+2)}{x^{1/2}}$

9. $f(d) = 0.07d^{3/2}$
$f(9) = 0.07(9)^{3/2} = 0.07(27) = 1.89$
The duration of a storm whose diameter is 9 miles is 1.89 hours.

10. $\sqrt[3]{16x^{10}y^{11}} = \sqrt[3]{8x^9y^9\cdot 2xy^2}$
$= \sqrt[3]{(2x^3y^3)^3\cdot 2xy^2} = 2x^3y^3\sqrt[3]{2xy^2}$

11. $\sqrt[3]{4x^4}\cdot\sqrt[3]{2x^7} = \sqrt[3]{(4x^4)(2x^7)}$
$= \sqrt[3]{8x^{11}} = \sqrt[3]{8x^9\cdot x^2} = 2x^3\sqrt[3]{x^2}$

12. $s = \sqrt{30fd}$
$s = \sqrt{30(0.5)(50)}$
$s = \sqrt{750}$
$s = \sqrt{25\cdot 30}$
$s = 5\sqrt{30}$
The speed is approximately $5\sqrt{30}$ mph.

13. $\dfrac{\sqrt[3]{32x^8}}{\sqrt[3]{2x^4}} = \sqrt[3]{\dfrac{32x^8}{2x^4}} = \sqrt[3]{16x^4} = \sqrt[3]{8x^3\cdot 2x}$
$= 2x\sqrt[3]{2x}$

14. $\sqrt{\dfrac{3}{5}} = \dfrac{\sqrt{3}}{\sqrt{5}}\cdot\dfrac{\sqrt{5}}{\sqrt{5}} = \dfrac{\sqrt{15}}{5}$

15. $\dfrac{5}{\sqrt[3]{5x^2}} = \dfrac{5}{\sqrt[3]{5x^2}}\cdot\dfrac{\sqrt[3]{25x}}{\sqrt[3]{25x}} = \dfrac{5\sqrt[3]{25x}}{5x}$
$= \dfrac{\sqrt[3]{25x}}{x}$

16. $\dfrac{3}{\sqrt{18x^5}} = \dfrac{3}{3x^2\sqrt{2x}}\cdot\dfrac{\sqrt{2x}}{\sqrt{2x}} = \dfrac{3\sqrt{2x}}{3x^2(2x)}$
$= \dfrac{\sqrt{2x}}{2x^3}$

17. $3\sqrt{18} + 4\sqrt{32} - 5\sqrt{8} = 3\sqrt{9\cdot 2} + 4\sqrt{16\cdot 2} - 5\sqrt{4\cdot 2}$
$= 3\cdot 3\sqrt{2} + 4\cdot 4\sqrt{2} - 5\cdot 2\sqrt{2}$
$= 9\sqrt{2} + 16\sqrt{2} - 10\sqrt{2} = 15\sqrt{2}$

18. $\left(4\sqrt{3}+\sqrt{2}\right)\left(6\sqrt{3}-5\sqrt{2}\right) = 4\sqrt{3}\left(6\sqrt{3}-5\sqrt{2}\right) + \sqrt{2}\left(6\sqrt{3}-5\sqrt{2}\right)$
$= 24\sqrt{9} - 20\sqrt{6} + 6\sqrt{6} - 5\sqrt{4}$
$= 24(3) - 14\sqrt{6} - 5(2) = 62 - 14\sqrt{6}$

19. $\left(7-\sqrt{3}\right)^2 = 7^2 - 2(7)\left(\sqrt{3}\right)+\left(\sqrt{3}\right)^2$
$= 49 - 14\sqrt{3} + 3 = 52 - 14\sqrt{3}$

20. $\dfrac{6}{\sqrt{5}-\sqrt{2}} = \dfrac{6}{\sqrt{5}-\sqrt{2}} \cdot \dfrac{\sqrt{5}+\sqrt{2}}{\sqrt{5}+\sqrt{2}}$
$= \dfrac{6\left(\sqrt{5}+\sqrt{2}\right)}{5-2} = 2\left(\sqrt{5}+\sqrt{2}\right)$

21. $\sqrt{x-3}+5 = x$
$\sqrt{x-3} = x-5$
$\left(\sqrt{x-3}\right)^2 = (x-5)^2$
$x-3 = x^2-10x+25$
$x^2-11x+28=0$
$(x-7)(x-4)=0$
$x=7$ or $x=4$
Check 7:
$\sqrt{7-3}+5=7$
$\sqrt{4}+5=7$
$2+5=7$
$7=7$ checks
Check 4:
$\sqrt{4-3}+5=4$
$\sqrt{1}+5=4$
$6=4$ extraneous
$\{7\}$

22. $\sqrt{x+4}+\sqrt{x-1}=5$
$\sqrt{x+4}=5-\sqrt{x-1}$
$\left(\sqrt{x+4}\right)^2 = \left(5-\sqrt{x-1}\right)^2$
$x+4 = 25-10\sqrt{x-1}+(x-1)$
$\sqrt{x-1}=2$
$x-1=4$
$x=5$
Check:
$\sqrt{5+4}+\sqrt{5-1}=5$
$\sqrt{9}+\sqrt{4}=5$
$3+2=5$
$5=5$ checks
$\{5\}$

23. Perimeter $= 2\sqrt{8}+2\sqrt{18}$
$= 2\sqrt{4\cdot 2}+2\sqrt{9\cdot 2}$
$= 2\cdot 2\sqrt{2}+2\cdot 3\sqrt{2}$
$= 4\sqrt{2}+6\sqrt{2}$
$= 10\sqrt{2}$
The perimeter is $10\sqrt{2}$ in.

24. $T = 2\pi\sqrt{\dfrac{L}{32}}$
$\dfrac{\pi}{4} = 2\pi\sqrt{\dfrac{L}{32}}$
$\dfrac{\pi}{4(2\pi)} = \sqrt{\dfrac{L}{32}}$
$\left(\dfrac{1}{8}\right)^2 = \left(\sqrt{\dfrac{L}{32}}\right)^2$
$\dfrac{1}{64} = \dfrac{L}{32}$
$\dfrac{32}{64} = L$
$\dfrac{1}{2} = L$
The pendulum's length is $\dfrac{1}{2}$ ft.

25. $\sqrt{-75} = \sqrt{(-1)(25)(3)} = 5i\sqrt{3}$

26. $(5+3i)-(-6-9i) = (5+3i)+(6+9i)$
$= (5+6)+(3+9)i = 11+12i$

27. $(3-4i)(2+5i) = 3(2+5i)-4i(2+5i)$
$= 6+15i-8i-20i^2 = 6+7i-20(-1)$
$= 26+7i$

28. $(7-2i)^2 = 7^2-2(7)(2i)+(2i)^2$
$= 49-28i-4 = 45-28i$

29. $\dfrac{3+i}{1-2i} = \dfrac{3+i}{1-2i}\cdot\dfrac{1+2i}{1+2i} = \dfrac{3+7i+2i^2}{1-4i^2}$
$= \dfrac{3+7i-2}{1+4} = \dfrac{1+7i}{5} = \dfrac{1}{5}+\dfrac{7}{5}i$

30. $\sqrt{-25}\sqrt{-4} = (5i)(2i) = 10i^2 = -10$

Chapters 1–6 Cumulative Review Problems

1. $\frac{(0.00072)(0.003)}{0.00024} = \frac{(7.2\times10^{-4})(3\times10^{-3})}{2.4\times10^{-4}}$

$= \frac{(7.2)(3)}{2.4}\times10^{-4-3+4} = 9\times10^{-3}$

2. $\frac{-60\div(-2)(-3)+5-(-7)+3}{81\div(-9)+3(-5)+2-(-2)} = \frac{30(-3)+5+7+3}{-9-15+2+2}$

$= \frac{-90+15}{-24+4} = \frac{-75}{-20} = \frac{15}{4}$

3. $|6x-8| > 4$

$6x-8 < -4$ or $6x-8 > 4$

$6x < 4$ or $6x > 12$

$x < \frac{2}{3}$ or $x > 2$

$\left\{x \middle| x < \frac{2}{3} \text{ or } x > 2\right\}$, $\left(-\infty, \frac{2}{3}\right)\cup(2, \infty)$

4. $3x-5 = 4(2x+3)-7$

$3x-5 = 8x+12-7$

$3x-5 = 8x+5$

$3x-8x = 5+5$

$-5x = 10$

$x = -2$

$\{-2\}$

5. a. points (1960, 23) and (1980, 46)

slope $= \frac{46-23}{1980-1960} = \frac{23}{20} = 1.15$

The average annual rate of increase from 1960 to 1980 for lung cancer deaths is 1.15 per 100,000 people.

b. points (1950, 10) and (1980, 10)

The rate is constant at $y = 10$ per 100,000 people.

6. a. $V = lwh = (x-8)(x-8)4$

$= 4(x^2-16x+64) = 4x^2-64x+256$

The volume is $4x^2-64x+256$ cubic inches.

b. $4x^2-64x+256 = 256$

$4x(x-16) = 0$

$4x = 0$ or $x-16 = 0$

$x = 0$ or $x = 16$

Disregard a dimension of 0. The box was 16 in. by 16 in.

7. points (–4, –1) and (3, 4)

slope $= \frac{4-(-1)}{3-(-4)} = \frac{5}{7}$

point-slope form:

$y+1 = \frac{5}{7}(x+4)$ or $y-4 = \frac{5}{7}(x-3)$

slope-intercept form:

$y = \frac{5}{7}x + \frac{13}{7}$

8. $2x-y+z = -5$ (1)

$x-2y-3z = 6$ (2)

$x+y-2z = 1$ (3)

Multiply equation 1 by 3 and add to equation 2.

$$\begin{array}{r} 6x-3y+3z = -15 \\ x-2y-3z = 6 \\ \hline 7x-5y = -9 \end{array} \quad (4)$$

Multiply equation 1 by 2 and add to equation 3.

$$\begin{array}{r} 4x-2y+2z = -10 \\ x+y-2z = 1 \\ \hline 5x-y = -9 \end{array} \quad (5)$$

Multiply equation 5 by –5 and add to equation 4.

$$\begin{array}{r} 7x-5y = -9 \\ -25x+5y = 45 \\ \hline -18x = 36 \\ x = -2 \end{array}$$

Substitute –2 for x in equation 5.

$5x - y = -9$
$5(-2) - y = -9$
$-y = 1$
$y = -1$
Substitute -1 for y in equation 1.
$2x - y + z = -5$
$2(-2) + 1 + z = -5$
$z = -2$
$\{(-2, -1, -2)\}$

9. About (80, 700)
At age 80, there are 700 deaths per 1000 population for each, men and women.
To the right of the intersection point, the rate for men is larger than women as they both get older.

10. $x + 2y < 2$ and $2y - x > 4$

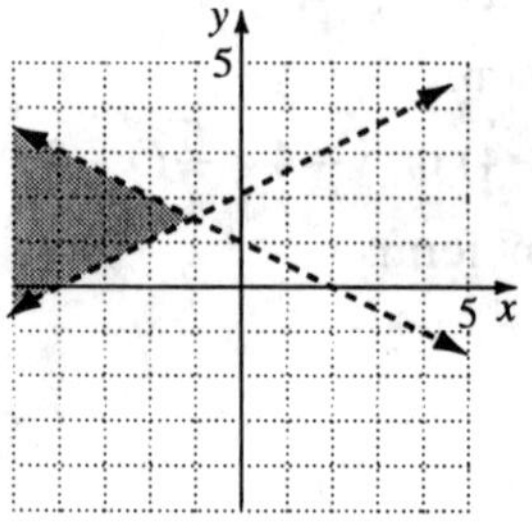

11. $S = 2LW + 2LH + 2WH$, for W
$S - 2LH = 2LW + 2WH$
$S - 2LH = W(2L + 2H)$
$W = \dfrac{S - 2LH}{2L + 2H}$

12. 1-50¢ piece
2 quarters
1 quarter, 2 dimes, 1 nickel
1 quarter, 1 dime, 3 nickels
1 quarter, 5 nickels
5 dimes
4 dimes, 2 nickels
3 dimes, 4 nickels
2 dimes, 6 nickels
1 dime, 8 nickels
10 nickels
11 combinations

13. $3x^2y^4 - 48y^6 = 3y^4(x^2 - 16y^2)$
$= 3y^4(x - 4y)(x + 4y)$

14. $f(t) = 0.002t^2 + 0.41t + 7.34$
$f(10) = 0.002(10)^2 + 0.41(10) + 7.34$
$f(10) = 11.64$
There was 11.64% of the American population that had graduated from college in 1970.

15. $y = 14.60x + 270$ where
y = weekly earnings for workers in U.S. and
x = years since 1980.
$620.40 = 14.60x + 270$
$350.4 = 14.6x$
$24 = x$
$1980 + 24 = 2004$
The earnings will reach $620.40 in year 2004.

16. $I = \dfrac{2V}{R + 2r}$, for R
$I(R + 2r) = 2V$
$IR + 2Ir = 2V$
$IR = 2V - 2Ir$
$R = \dfrac{2V - 2Ir}{I}$

17. Let x = units of corn and
y = units of soybeans.
$0.25x + 0.4y = 14$
$0.4x + 0.2y = 18$
Add the first equation to –2 times the second equation.

$$\begin{array}{r} 0.25x + 0.4y = 14 \\ -0.8x - 0.4y = -36 \\ \hline -0.55x = -22 \\ x = 40 \end{array}$$

Substitute into the first equation.
$0.25(40) + 0.4y = 14$
$0.4y = 4$
$y = 10$
The feed should contain 40 units of corn and 10 units of soybeans.

18. $$\frac{8x^2}{3x^2-12} \div \frac{40}{2-x} = \frac{8x^2}{3(x-2)(x+2)} \cdot \frac{-1(x-2)}{40}$$
$$= -\frac{x^2}{15(x+2)}$$

19. $$\frac{x+\frac{1}{y}}{y+\frac{1}{x}} = \frac{\frac{xy+1}{y}}{\frac{xy+1}{x}} = \frac{xy+1}{y} \cdot \frac{x}{xy+1} = \frac{x}{y}$$

20. $$\frac{x}{x-3} - \frac{3x}{x^2-x-6} = \frac{4x^2-4x-18}{x^2-x-6}$$
$$\frac{x}{x-3} \cdot \frac{x+2}{x+2} - \frac{3x}{(x-3)(x+2)} = \frac{4x^2-4x-18}{(x-3)(x+2)}$$
$$\frac{x^2+2x-3x}{(x-3)(x+2)} = \frac{4x^2-4x-18}{(x-3)(x+2)}$$
$$\frac{x^2-x}{(x-3)(x+2)} - \frac{4x^2-4x-18}{(x-3)(x+2)} = 0$$
$$\frac{-3x^2+3x+18}{(x-3)(x+2)} = 0$$
$$\frac{-3(x^2-x-6)}{(x-3)(x+2)} = 0$$
$$\frac{-3(x-3)(x+2)}{(x-3)(x+2)} = 0$$
$-3 = 0$ False
No solution; $\varnothing$

21. Let x = width, then $2x + 1$ = length.
$2x + 2(2x + 1) = 110$
$2x + 4x + 2 = 110$
$6x = 108$
$x = 18$
$2x + 1 = 37$
The width = 18 in. and the length = 37 in.

22. $(2x-3)(4x^2-5x-2) = 2x(4x^2-5x-2) - 3(4x^2-5x-2)$
$= 8x^3 - 10x^2 - 4x - 12x^2 + 15x + 6$
$= 8x^3 - 22x^2 + 11x + 6$

23. $y = x^2 + 2x - 3$
x-intercepts:
$0 = x^2 + 2x - 3$
$0 = (x+3)(x-1)$
$x = -3$ and $x = 1$
$(-3, 0)$ and $(1, 0)$
y-intercept:
$y = 0^2 + 2(0) - 3 = -3$
$(0, -3)$
vertex:
$x = -\frac{b}{2a} = -\frac{2}{2(1)} = -1$
$y = (-1)^2 + 2(-1) - 3 = -4$
$(-1, -4)$

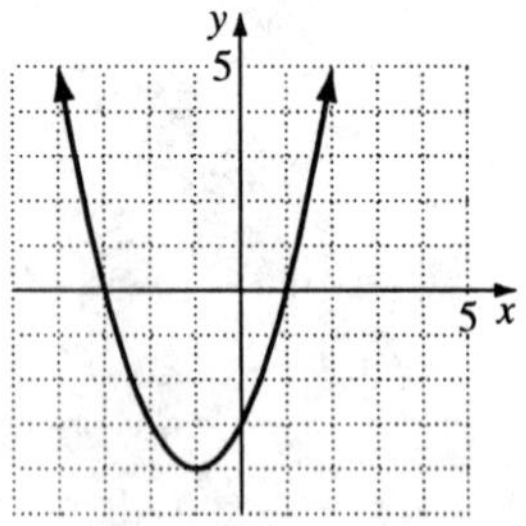

24. **a.** $C(x) = 20x + 50{,}000$

b. $\overline{C}(x) = \frac{20x + 50{,}000}{x}$

c. $\overline{C}(1000) = \frac{20(1000) + 50{,}000}{1000} = 70$
$\overline{C}(10{,}000) = \frac{20(10{,}000) + 50{,}000}{10{,}000}$
$= 25$
$\overline{C}(100{,}000) = \frac{20(100{,}000) + 50{,}000}{100{,}000}$
$= 20.5$
The cost per chair to produce 1000, 10,000, and 100,000 chairs is \$70, \$25, and \$20.50, respectively. As the number of chairs produced increases, the cost per chair approaches \$20.

25. Let L = loudness, d = distance and k = constant.
$L = \frac{k}{d^2}$
$28 = \frac{k}{8^2}$
$k = 1792$

$L = \frac{1792}{d^2}$
$L = \frac{1792}{4^2}$
$L = 112$
The loudness is 112 decibels.

26. Approximate maximum point (1934, 17)
In 1934, the suicide rate was at a maximum (from 1900 to 1990) at 17 per 100,000 people.
Answers may vary.

27. Area of shaded = Area of circle – Area of triangle

$$= \pi r^2 - \frac{1}{2}(2r)(r) = \pi r^2 - r^2$$

$$= r^2(\pi - 1)$$

28. $\dfrac{4x^3 + 12x^2 + x - 12}{x+3} = 4x^2 + 1 - \dfrac{15}{x+3}$

$$\begin{array}{r} 4x^2 + 1 - \frac{15}{x+3} \\ x+3 \overline{\smash{)}\,4x^3 + 12x^2 + x - 12} \\ \underline{4x^3 + 12x^2} \qquad\quad \\ x - 12 \\ \underline{x + 3} \\ -15 \end{array}$$

29. $3\sqrt{2y^3} - y\sqrt{200y} + \sqrt{32y^3} = 3\sqrt{y^2(2y)} - y\sqrt{100(2y)} + \sqrt{16y^2(2y)}$

$$= 3y\sqrt{2y} - 10y\sqrt{2y} + 4y\sqrt{2y}$$

$$= (3y - 10y + 4y)\sqrt{2y}$$

$$= -3y\sqrt{2y}$$

30. $\dfrac{16-15i}{6-i} = \dfrac{16-15i}{6-i} \cdot \dfrac{6+i}{6+i} = \dfrac{96+16i-90i-15i^2}{36-i^2}$

$$= \frac{96-74i+15}{36+1} = \frac{111-74i}{37} = 3-2i$$

Chapter 7

Problem Set 7.1

1. $y = x^2 + 2x - 3$

x-intercepts:

$0 = x^2 + 2x - 3$

$0 = (x + 3)(x - 1)$

$x = -3$ or $x = 1$

$(-3, 0)$ and $(1, 0)$

y-intercept:

$y = 0^2 + 2(0) - 3 = -3$

$(0, -3)$

vertex:

$x = -\frac{b}{2a} = -\frac{2}{2(1)} = -1$

$y = (-1)^2 + 2(-1) - 3 = -4$

$(-1, -4)$

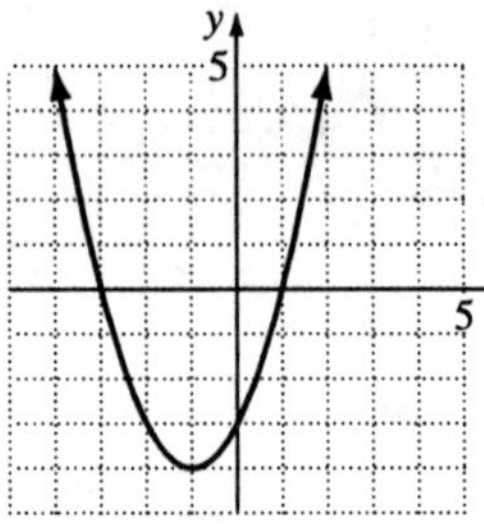

3. $y = x^2 - 9x + 14$

x-intercepts:

$0 = x^2 - 9x + 14$

$0 = (x - 7)(x - 2)$

$x = 7$ or $x = 2$

$(7, 0)$ and $(2, 0)$

y-intercept:

$y = 0^2 - 9(0) + 14 = 14$

$(0, 14)$

vertex:

$x = -\frac{b}{2a} = -\frac{(-9)}{2(1)} = \frac{9}{2}$

$y = \left(\frac{9}{2}\right)^2 - 9\left(\frac{9}{2}\right) + 14 = -\frac{25}{4}$

$\left(\frac{9}{2}, -\frac{25}{4}\right)$

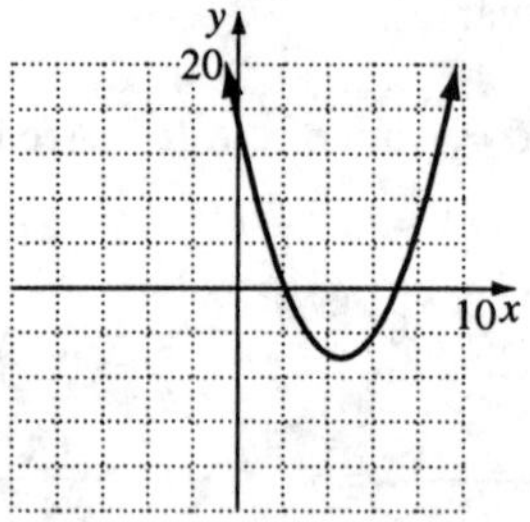

5. $f(x) = 3x^2 - 8x + 4 = -\frac{4}{3}$

x-intercepts:

$0 = 3x^2 - 8x + 4$

$0 = (3x - 2)(x - 2)$

$3x - 2 = 0$ or $x - 2 = 0$

$x = \frac{2}{3}$ or $x = 2$

$\left(\frac{2}{3}, 0\right)$ and $(2, 0)$

y-intercept:

$y = 3(0)^2 - 8(0) + 4 = 4$

$(0, 4)$

vertex:

$x = -\frac{b}{2a} = -\frac{(-8)}{2(3)} = \frac{8}{6} = \frac{4}{3}$

$y = 3\left(\frac{4}{3}\right)^2 - 8\left(\frac{4}{3}\right) + 4 = -\frac{4}{3}$

$\left(\frac{4}{3}, -\frac{4}{3}\right)$

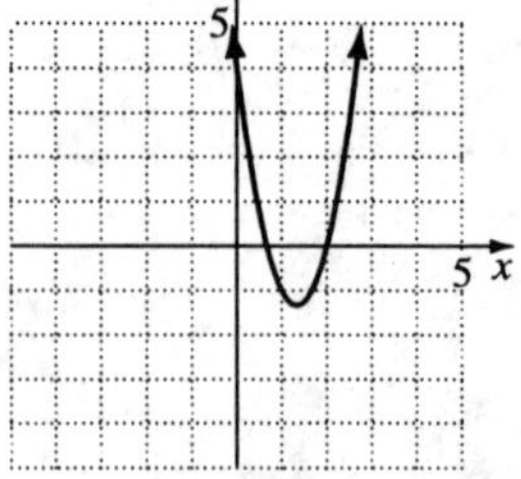

7. $g(x) = 2x^2 + 5x - 7$

x-intercepts:

$0 = 2x^2 + 5x - 7$

$0 = (2x + 7)(x - 1)$

$x = -\frac{7}{2}$ or $x = 1$

$\left(-\frac{7}{2}, 0\right)$ and $(1, 0)$

y-intercept:

$y = 2(0)^2 + 5(0) - 7 = -7$

$(0, -7)$

vertex:

$x = -\frac{b}{2a} = -\frac{5}{2(2)} = -\frac{5}{4}$

$y = 2\left(-\frac{5}{4}\right)^2 + 5\left(-\frac{5}{4}\right) - 7 = -\frac{81}{8}$

$\left(-\frac{5}{4}, -\frac{81}{8}\right)$

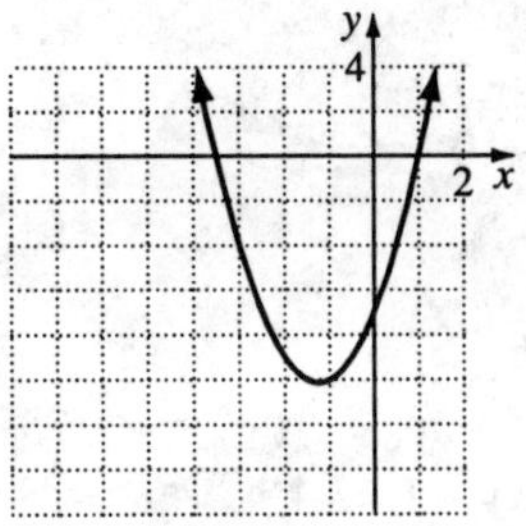

9. $8x^2 = 15 - 14x$

$8x^2 + 14x - 15 = 0$

$(4x - 3)(2x + 5) = 0$

$4x - 3 = 0$ or $2x + 5 = 0$

$x = \frac{3}{4}$ or $x = -\frac{5}{2}$

$\left\{-\frac{5}{2}, \frac{3}{4}\right\}$

11. $7x + 10 = 6x^2$

$0 = 6x^2 - 7x - 10$

$0 = (6x + 5)(x - 2)$

$6x + 5 = 0$ or $x - 2 = 0$

$x = -\frac{5}{6}$ or $x = 2$

$\left\{-\frac{5}{6}, 2\right\}$

13. $x(3x + 1) = 2$

$3x^2 + x = 2$

$3x^2 + x - 2 = 0$

$(3x - 2)(x + 1) = 0$

$3x - 2 = 0$ or $x + 1 = 0$

$x = \frac{2}{3}$ or $x = -1$

$\left\{-1, \frac{2}{3}\right\}$

15. $2(x^2 - 3) = 2(x + 3) + 3x$

$2x^2 - 6 = 2x + 6 + 3x$

$2x^2 - 5x - 12 = 0$

$(2x + 3)(x - 4) = 0$

$2x + 3 = 0$ or $x - 4 = 0$

$x = -\frac{3}{2}$ or $x = 4$

$\left\{-\frac{3}{2}, 4\right\}$

17. $(x + 3)(x - 2) = 2(x - 2)(x + 2) + 2$

$x^2 + x - 6 = 2x^2 - 8 + 2$

$0 = x^2 - x$

$0 = x(x - 1)$

$x = 0$ or $x = 1$

$\{0, 1\}$

For **19–59.**, the check is left to the student.

19. $y^2 = 100$

$y = \pm\sqrt{100}$

$y = \pm 10$

$\{-10, 10\}$

21. $y^2 = 7$

$y = \pm\sqrt{7}$

$\left\{-\sqrt{7}, \sqrt{7}\right\}$

23. $x^2 = 75$

$x = \pm\sqrt{75}$

$= \pm\sqrt{25 \cdot 3}$

$= \pm 5\sqrt{3}$

$\left\{-5\sqrt{3}, 5\sqrt{3}\right\}$

25. $z^2 = -4$
$z = \pm\sqrt{-4}$
$= \pm i\sqrt{4} = \pm 2i$
$\{-2i, 2i\}$

27. $4y^2 = 100$
$y^2 = 25$
$y = \pm\sqrt{25} = \pm 5$
$\{-5, 5\}$

29. $3x^2 = 25$
$x^2 = \frac{25}{3}$
$x = \pm\sqrt{\frac{25}{3}}$
$= \pm\frac{5}{\sqrt{3}} \cdot \frac{\sqrt{3}}{\sqrt{3}}$
$= \pm\frac{5\sqrt{3}}{3}$
$\left\{-\frac{5\sqrt{3}}{3}, \frac{5\sqrt{3}}{3}\right\}$

31. $7x^2 + 2 = 13$
$7x^2 = 11$
$x^2 = \frac{11}{7}$
$x = \pm\sqrt{\frac{11}{7}}$
$= \pm\frac{\sqrt{11}}{\sqrt{7}} \cdot \frac{\sqrt{7}}{\sqrt{7}}$
$= \pm\frac{\sqrt{77}}{7}$
$\left\{-\frac{\sqrt{77}}{7}, \frac{\sqrt{77}}{7}\right\}$

33. $4x^2 + 7 = 3(x^2 + 1)$
$4x^2 + 7 = 3x^2 + 3$
$x^2 = -4$
$x = \pm\sqrt{-4} = \pm 2i$
$\{-2i, 2i\}$

35. $4(x^2 + 2x) + 7 = 3x^2 + 8x + 2$
$4x^2 + 8x + 7 = 3x^2 + 8x + 2$
$x^2 = -5$
$x = \pm\sqrt{-5} = \pm i\sqrt{5}$
$\left\{-i\sqrt{5}, i\sqrt{5}\right\}$

37. $(x+4)(x+1) = 5x - 71$
$x^2 + 5x + 4 = 5x - 71$
$x^2 = -75$
$x = \pm\sqrt{-75}$
$= \pm\sqrt{(25)(3)(-1)}$
$= \pm 5i\sqrt{3}$
$\left\{-5i\sqrt{3}, 5i\sqrt{3}\right\}$

39. $(x+7)^2 = 9$
$x + 7 = \pm 3$
$x + 7 = 3$ or $x + 7 = -3$
$x = -4$ or $x = -10$
$\{-10, -4\}$

41. $(3y-1)^2 = 16$
$3y - 1 = \pm 4$
$3y - 1 = 4$ or $3y - 1 = -4$
$3y = 5$ or $3y = -3$
$y = \frac{5}{3}$ or $y = -1$
$\left\{-1, \frac{5}{3}\right\}$

43. $(2x+7)^2 = 5$
$2x + 7 = \pm\sqrt{5}$
$2x + 7 = \sqrt{5}$ or $2x + 7 = -\sqrt{5}$
$2x = -7 + \sqrt{5}$ or $2x = -7 - \sqrt{5}$
$x = \frac{-7+\sqrt{5}}{2}$ or $x = \frac{-7-\sqrt{5}}{2}$
$\left\{\frac{-7-\sqrt{5}}{2}, \frac{-7+\sqrt{5}}{2}\right\}$

45. $(5y-4)^2 = 24$
$5y-4 = \pm\sqrt{24}$
$5y-4 = \sqrt{24}$ or $5y-4 = -\sqrt{24}$
$5y = 4+2\sqrt{6}$ or $5y = 4-2\sqrt{6}$
$y = \frac{4+2\sqrt{6}}{5}$ or $y = \frac{4-2\sqrt{6}}{5}$
$\left\{\frac{4-2\sqrt{6}}{5}, \frac{4+2\sqrt{6}}{5}\right\}$

47. $(x-3)^2 = -4$
$x-3 = \pm\sqrt{-4}$
$x-3 = \sqrt{-4}$ or $x-3 = -\sqrt{-4}$
$x = 3+2i$ or $x = 3-2i$
$\{3-2i, 3+2i\}$

49. $(2y+5)^2 = -5$
$2y+5 = \pm\sqrt{-5}$
$2y+5 = \sqrt{-5}$ or $2y+5 = -\sqrt{-5}$
$2y = -5+i\sqrt{5}$ or $2y = -5-i\sqrt{5}$
$y = \frac{-5+i\sqrt{5}}{2}$ or $y = \frac{-5-i\sqrt{5}}{2}$
$\left\{\frac{-5-i\sqrt{5}}{2}, \frac{-5+i\sqrt{5}}{2}\right\}$

51. $(3z-2)^2 = -50$
$3z-2 = \pm\sqrt{-50}$
$3z-2 = \sqrt{-50}$ or $3z-2 = -\sqrt{-50}$
$3z = 2+5i\sqrt{2}$ or $3z = 2-5i\sqrt{2}$
$z = \frac{2+5i\sqrt{2}}{3}$ or $z = \frac{2-5i\sqrt{2}}{3}$
$\left\{\frac{2-5i\sqrt{2}}{3}, \frac{2+5i\sqrt{2}}{3}\right\}$

53. $3(5x-4)^2 - 81 = 0$
$3(5x-4)^2 = 81$
$(5x-4)^2 = 27$
$5x-4 = \pm\sqrt{27}$
$5x-4 = \sqrt{27}$ or $5x-4 = -\sqrt{27}$
$5x = 4+3\sqrt{3}$ or $5x = 4-3\sqrt{3}$
$x = \frac{4+3\sqrt{3}}{5}$ or $x = \frac{4-3\sqrt{3}}{5}$
$\left\{\frac{4-3\sqrt{3}}{5}, \frac{4+3\sqrt{3}}{5}\right\}$

55. $4(2x+5)^2 + 100 = 0$
$4(2x+5)^2 = -100$
$(2x+5)^2 = -25$
$2x+5 = \pm\sqrt{-25}$
$2x+5 = \sqrt{-25}$ or $2x+5 = -\sqrt{-25}$
$2x = -5+5i$ or $2x = -5-5i$
$x = \frac{-5+5i}{2}$ or $x = \frac{-5-5i}{2}$
$\left\{\frac{-5-5i}{2}, \frac{-5+5i}{2}\right\}$

57. $4(5x-3)^2 - 1 = 0$
$4(5x-3)^2 = 1$
$(5x-3)^2 = \frac{1}{4}$
$5x-3 = \pm\sqrt{\frac{1}{4}}$
$5x-3 = \pm\frac{1}{2}$
$5x-3 = \frac{1}{2}$ or $5x-3 = -\frac{1}{2}$
$5x = 3+\frac{1}{2}$ or $5x = 3-\frac{1}{2}$
$5x = \frac{7}{2}$ or $5x = \frac{5}{2}$
$x = \frac{7}{10}$ or $x = \frac{1}{2}$
$\left\{\frac{1}{2}, \frac{7}{10}\right\}$

59. $\left(x-\frac{1}{3}\right)^2 = \frac{1}{3}$
$x-\frac{1}{3} = \pm\sqrt{\frac{1}{3}}$
$x-\frac{1}{3} = \sqrt{\frac{1}{3}}$ or $x-\frac{1}{3} = -\sqrt{\frac{1}{3}}$
$x = \frac{1}{3}+\frac{1}{\sqrt{3}}\cdot\frac{\sqrt{3}}{\sqrt{3}}$ or $x = \frac{1}{3}-\frac{1}{\sqrt{3}}\cdot\frac{\sqrt{3}}{\sqrt{3}}$
$x = \frac{1}{3}+\frac{\sqrt{3}}{3}$ or $x = \frac{1}{3}-\frac{\sqrt{3}}{3}$

$x = \dfrac{1+\sqrt{3}}{3}$ or $x = \dfrac{1-\sqrt{3}}{3}$

$\left\{\dfrac{1-\sqrt{3}}{3}, \dfrac{1+\sqrt{3}}{3}\right\}$

61. (4, –3) and (–6, 2)

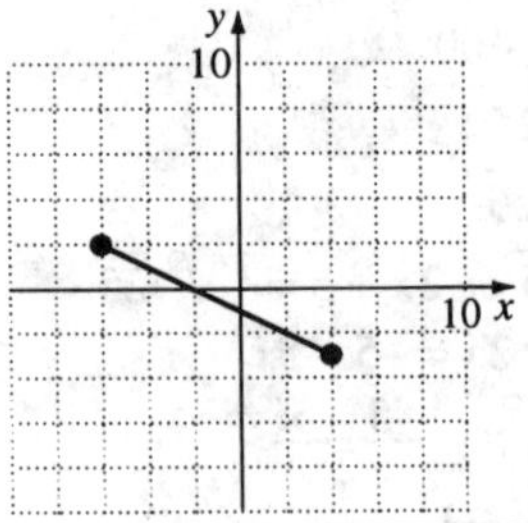

$d = \sqrt{(-6-4)^2 + (2+3)^2}$
$= \sqrt{(-10)^2 + (5)^2}$
$= \sqrt{100+25} = \sqrt{125} = \sqrt{25(5)} = 5\sqrt{5}$

63. (3, 2) and (6, 7)

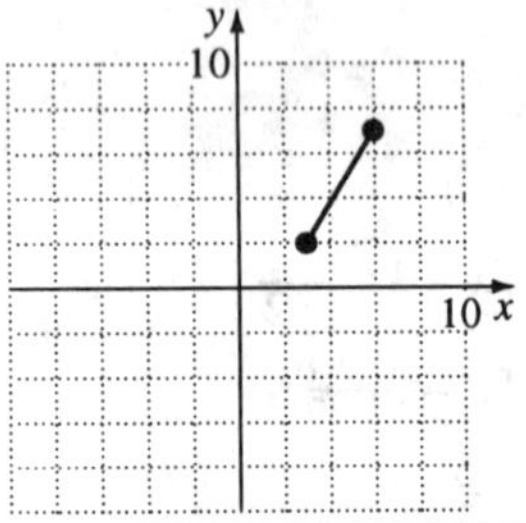

$d = \sqrt{(6-3)^2 + (7-2)^2} = \sqrt{9+25} = \sqrt{34}$

65. (0, 0) and (5, –12)

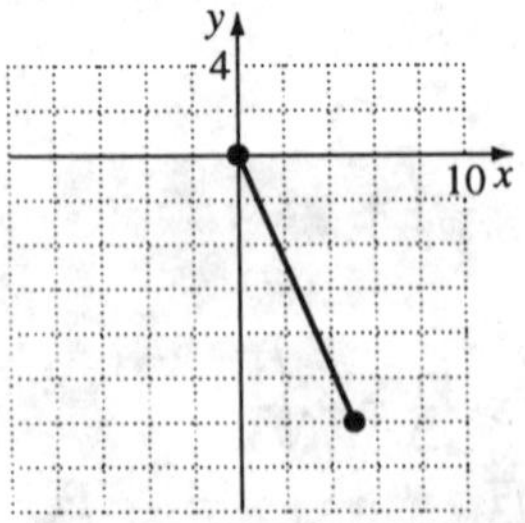

$d = \sqrt{(0-5)^2 + (0+12)^2} = \sqrt{25+144}$
$= \sqrt{169} = 13$

67. (0, –3) and (–3, 3)

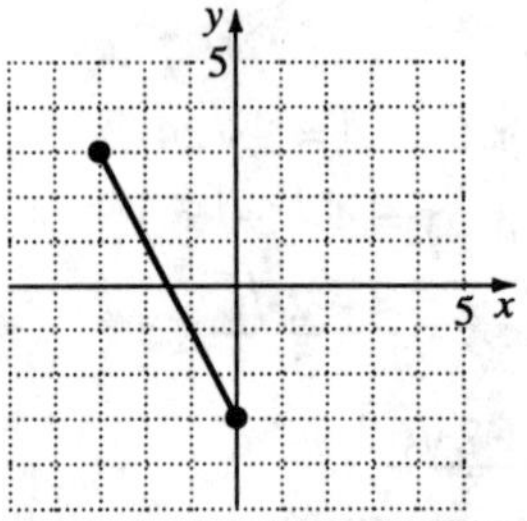

$d = \sqrt{(-3-0)^2 + (3+3)^2} = \sqrt{9+36}$
$= \sqrt{45} = 3\sqrt{5}$

69. (1, –2) and (–3, 6)

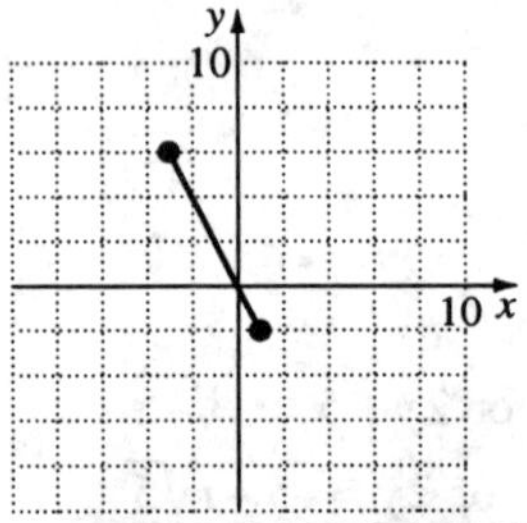

$d = \sqrt{(-3-1)^2 + (6+2)^2}$
$= \sqrt{16+64} = \sqrt{80} = 4\sqrt{5}$

71. $A(5, 7)$, $B(1, 10)$, $C(-3, -8)$

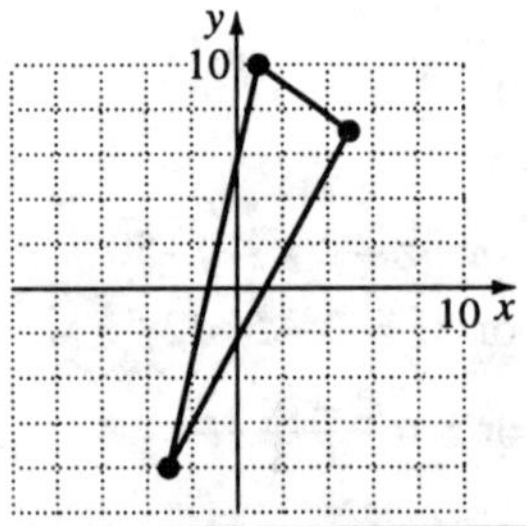

$AB = \sqrt{(1-5)^2 + (10-7)^2} = \sqrt{16+9}$
$= \sqrt{25} = 5$
$CB = \sqrt{(1+3)^2 + (10+8)^2} = \sqrt{16+324}$
$= \sqrt{340} = \sqrt{4 \cdot 85} = 2\sqrt{85}$
$CA = \sqrt{(5+3)^2 + (7+8)^2} = \sqrt{64+225}$
$= \sqrt{289} = 17$
Perimeter $= 5 + 2\sqrt{85} + 17 = 22 + 2\sqrt{85}$

73. $A(2, 3)$, $B(-1, -1)$, $C(3, -4)$

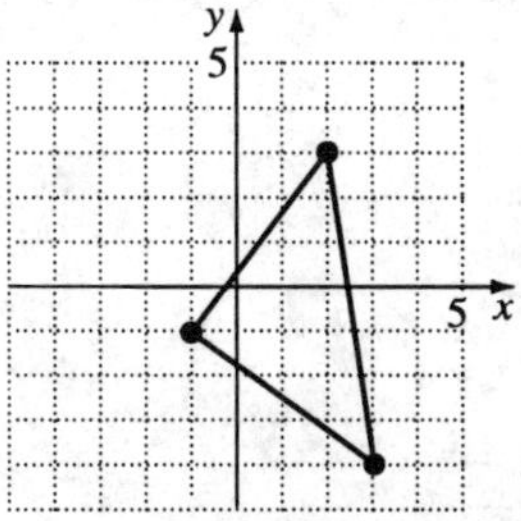

$AB = \sqrt{(2+1)^2 + (3+1)^2} = \sqrt{9+16}$
$= \sqrt{25} = 5$
$AC = \sqrt{(2-3)^2 + (3+4)^2} = \sqrt{1+49}$
$= \sqrt{50} = 5\sqrt{2}$
$BC = \sqrt{(-1-3)^2 + (-1+4)^2} = \sqrt{16+9}$
$= \sqrt{25} = 5$
Since the two sides have equal measure ($AB = BC$), the triangle is isosceles.

75. $AB = \sqrt{(1+4)^2 + (0+6)^2} = \sqrt{25+36}$
$= \sqrt{61}$
$BC = \sqrt{(11-1)^2 + (12-0)^2} = \sqrt{100+144}$
$= \sqrt{244} = \sqrt{4 \cdot 61} = 2\sqrt{61}$
$AC = \sqrt{(11+4)^2 + (12+6)^2} = \sqrt{225+324}$
$= \sqrt{549} = \sqrt{9 \cdot 61} = 3\sqrt{61}$
$AB + BC = \sqrt{61} + 2\sqrt{61} = 3\sqrt{61} = AC$
Thus A, B, and C are collinear.

77. $(-8, -2)$ and $(-2, -6)$
$\text{Midpoint} = \left(\frac{-8+(-2)}{2}, \frac{-2+(-6)}{2}\right)$
$= \left(-\frac{10}{2}, -\frac{8}{2}\right) = (-5, -4)$

79. $(4, -3)$ and $(-2, 8)$
$\text{Midpoint} = \left(\frac{4+(-2)}{2}, \frac{-3+8}{2}\right)$
$= \left(\frac{2}{2}, \frac{5}{2}\right) = \left(1, \frac{5}{2}\right)$

81. $(2.8, -8.6)$ and $(-3.2, 12.4)$
$\text{Midpoint} = \left(\frac{2.8+(-3.2)}{2}, \frac{-8.6+12.4}{2}\right)$
$= \left(\frac{-0.4}{2}, \frac{3.8}{2}\right) = (-0.2, 1.9)$

83. $(3, 5)$ and $(-3, -5)$
$\text{Midpoint} = \left(\frac{3+(-3)}{2}, \frac{5+(-5)}{2}\right)$
$= \left(\frac{0}{2}, \frac{0}{2}\right) = (0, 0)$

85. $\left(\frac{5}{6}, -\frac{1}{3}\right)$ and $\left(-\frac{3}{4}, \frac{1}{2}\right)$
$\text{Midpoint} = \left(\frac{\frac{5}{6}+\left(-\frac{3}{4}\right)}{2}, \frac{-\frac{1}{3}+\frac{1}{2}}{2}\right)$
$= \left(\frac{\frac{1}{12}}{2}, \frac{\frac{1}{6}}{2}\right) = \left(\frac{1}{24}, \frac{1}{12}\right)$

87. $\left(7\sqrt{3}, -6\right)$ and $\left(3\sqrt{3}, -2\right)$
$\text{Midpoint} = \left(\frac{7\sqrt{3}+3\sqrt{3}}{2}, \frac{-6+(-2)}{2}\right)$
$= \left(\frac{10\sqrt{3}}{2}, \frac{-8}{2}\right) = \left(5\sqrt{3}, -4\right)$

89. $\left(-8, 2\sqrt{27}\right)$ and $\left(8, -4\sqrt{3}\right)$
$\text{Midpoint} = \left(\frac{-8+8}{2}, \frac{2\sqrt{27}+\left(-4\sqrt{3}\right)}{2}\right)$
$= \left(\frac{0}{2}, \frac{6\sqrt{3}-4\sqrt{3}}{2}\right) = \left(0, \frac{2\sqrt{3}}{2}\right)$
$= \left(0, \sqrt{3}\right)$

91. $B = 10^5(1+2t^2)$
($B = 33 \times 10^5$ bacteria):
$10^5(1+2t^2) = 33 \times 10^5$
$\frac{1}{10^5}[10^5(1+2t^2)] = (33 \times 10^5)\left(\frac{1}{10^5}\right)$
$1 + 2t^2 = 33$
$2t^2 = 32$

$t^2 = 16$
$t = 4$ or $t = -4$ (reject)
After 4 hours

93. $A = \pi r^2$
$100\pi = \pi r^2$
$100 = r^2$
$r = 10$ or $r = -10$ (reject)
Radius = 10 ft

95. Let x = width of original rectangle
$3x$ = length of original rectangle
$x - 1$ = width of new rectangle
$3x + 3$ = length of new rectangle
area of new rectangle = 72
$(x - 1)(3x + 3) = 72$
$3x^2 - 3 = 72$
$3x^2 = 75$
$x^2 = 25$
$x = 5$ or $x = -5$ (reject)
$3x = 3(5) = 15$
Width: 5 yd Length: 15 yd

97. $mc^2 = E$
$c^2 = \frac{E}{m}$
$c = \sqrt{\frac{E}{m}}$

99. $A = \pi r^2$, for r
$\frac{A}{\pi} = r^2$
$\sqrt{\frac{A}{\pi}} = r$
$r = \sqrt{\frac{A}{\pi}}$

101. $C = 12 + \frac{s^2}{60}$, for s
$C - 12 = \frac{s^2}{60}$
$60(C - 12) = s^2$
$\sqrt{60(C-12)} = s$
$\sqrt{4 \cdot 15(C-12)} = s$
$2\sqrt{15(C-12)} = s$
$s = 2\sqrt{15(C-12)}$

103. x = side of square
By Pythagorean Theorem:
$x^2 + x^2 = \left(3\sqrt{2}\right)^2$
$2x^2 = 18$
$x^2 = 9$
$x = 3$ or $x = -3$ (reject)
side of square: 3 feet
Dimensions: 3 ft by 3 ft

105. Let
x = distance from home plate to second base.
$x^2 = (90)^2 + (90)^2$
$x^2 = 2(90)^2$
$x = \sqrt{2(90)^2}$
$x = 90\sqrt{2}$
The distance is $90\sqrt{2} \approx 127.3$ ft.

107. Let x = distance up the pole.
$18^2 = 12^2 + x^2$
$x^2 = 18^2 - 12^2$
$x = \sqrt{180} = 6\sqrt{5}$
The wire is attached at $6\sqrt{5}$ ft.

109. Let x = length of wire.
Difference between heights = 49 – 42 = 7
$x^2 = 24^2 + 7^2$
$x = \sqrt{625}$
$x = 25$
The wire is 25 ft long.

111. $A(-3, 6)$, $B(2, -3)$, $C(11, 2)$, $D(6, 11)$

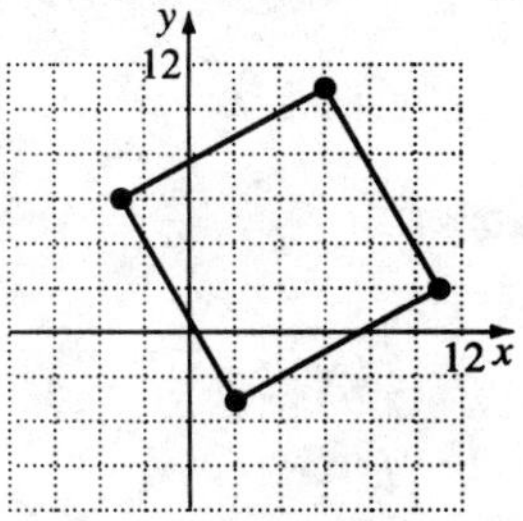

a. AD: $m = \frac{11-6}{6+3} = \frac{5}{9}$
BC: $m = \frac{2+3}{11-2} = \frac{5}{9}$
AD is parallel to BC.
AB: $m = \frac{-3-6}{2+3} = -\frac{9}{5}$
DC: $m = \frac{11-2}{6-11} = -\frac{9}{5}$
AB is parallel to DC.

b. AD:
$d = \sqrt{(6+3)^2 + (11-6)^2} = \sqrt{81+25}$
$= \sqrt{106}$
AB:
$d = \sqrt{(2+3)^2 + (-3-6)^2} = \sqrt{25+81}$
$= \sqrt{106}$
A parallelogram with two equal adjacent sides is a rhombus.

c. AB: $m = -\frac{9}{5}$
BC: $m = \frac{5}{9}$
$-\frac{9}{5}\left(\frac{5}{9}\right) = -1$
Since the slopes are negative reciprocals, AB and BC are perpendicular.

113. c is true.
If x = the person's age, we must find
$\sqrt{(x+1)(x-1)+1} = \sqrt{x^2-1+1}$
$= \sqrt{x^2} = x$ (Since $x > 0$)

115. $y = 10.675x^2 + 1007.775$
$1392 = 10.675x^2 + 1007.775$
$10.675x^2 = 384.225$
$x^2 \approx 35.9930$
$x \approx 6$
which corresponds to the year 1976.

117–119. Students should verify results.

121–125. Answers may vary.

127. $P(x_1, y_1)$; $(-1, 0)$ and $(-1, 5)$
$\sqrt{(x_1+1)^2 + (y_1-0)^2} = \sqrt{(x_1+1)^2 + (y_1-5)^2}$
$(x_1+1)^2 + y_1^2 = (x_1+1)^2 + (y_1-5)^2$
$y_1^2 = (y_1-5)^2$
$y_1^2 = y_1^2 - 10y_1 + 25$
$0 = -10y_1 + 25$
$y_1 = \frac{25}{10}$
$y_1 = \frac{5}{2}$

129. Let x = length of longer leg

$$x^2 + a^2 = (2a)^2$$
$$x^2 = 4a^2 - a^2$$
$$x = \sqrt{a^2(4-1)}$$
$$x = a\sqrt{3}$$

Thus, the length of the longer leg is $a\sqrt{3}$.

131. a. (x_1, y_1) and $\left(\frac{x_1+x_2}{2}, \frac{y_1+y_2}{2}\right)$

$$d = \sqrt{\left(\frac{x_1+x_2}{2} - x_1\right)^2 + \left(\frac{y_1+y_2}{2} - y_1\right)^2}$$
$$= \sqrt{\left(\frac{x_1+x_2-2x_1}{2}\right)^2 + \left(\frac{y_1+y_2-2y_1}{2}\right)^2}$$
$$= \sqrt{\left(\frac{x_2-x_1}{2}\right)^2 + \left(\frac{y_2-y_1}{2}\right)^2}$$
$$= \sqrt{\frac{(x_2-x_1)^2}{4} + \frac{(y_2-y_1)^2}{4}}$$
$$= \sqrt{\frac{(x_2-x_1)^2 + (y_2-y_1)^2}{4}}$$
$$= \frac{1}{2}\sqrt{(x_2-x_1)^2 + (y_2-y_1)^2}$$

(x_2, y_2) and $\left(\frac{x_1+x_2}{2}, \frac{y_1+y_2}{2}\right)$

$$d = \sqrt{\left(\frac{x_1+x_2}{2} - x_2\right)^2 + \left(\frac{y_1+y_2}{2} - y_2\right)^2}$$
$$= \sqrt{\left(\frac{x_1+x_2-2x_2}{2}\right)^2 + \left(\frac{y_1+y_2-2y_2}{2}\right)^2}$$
$$= \sqrt{\left(\frac{x_1-x_2}{2}\right)^2 + \left(\frac{y_1-y_2}{2}\right)^2}$$
$$= \sqrt{\frac{(x_2-x_1)^2}{4} + \frac{(y_2-y_1)^2}{4}}$$
$$= \sqrt{\frac{(x_2-x_1)^2 + (y_2-y_1)^2}{4}}$$
$$= \frac{1}{2}\sqrt{(x_2-x_1)^2 + (y_2-y_1)^2}$$

Both distances are equal to $\frac{1}{2}\sqrt{(x_2-x_1)^2 + (y_2-y_1)^2}$

b. points (x_1, y_1) and $\left(\frac{x_1+x_2}{2}, \frac{y_1+y_2}{2}\right)$

$$d = \sqrt{\left(\frac{x_1+x_2}{2} - x_1\right)^2 + \left(\frac{y_1+y_2}{2} - y_1\right)^2}$$
$$= \frac{1}{2}\sqrt{(x_2-x_1)^2 + (y_2-y_1)^2}$$

points $\left(\frac{x_1+x_2}{2}, \frac{y_1+y_2}{2}\right)$ and (x_2, y_2)

$$d_2 = \sqrt{\left(x_2 - \frac{x_1+x_2}{2}\right)^2 + \left(y_2 - \frac{y_1+y_2}{2}\right)^2}$$
$$= \frac{1}{2}\sqrt{(x_2-x_1)^2 + (y_2-y_1)^2}$$
$$d_1 + d_2 = 2 \cdot \frac{1}{2}\sqrt{(x_2-x_1)^2 + (y_2-y_1)^2}$$
$$= \sqrt{(x_2-x_1)^2 + (y_2-y_1)^2}$$

= distance between (x_1, y_1) and (x_2, y_2)

Review Problems

132. $$\frac{y^2-2y+1}{3y^2+7y-20} \cdot \frac{3y^2-2y-5}{y^2+3y-4} \div \frac{y^2-4y+3}{y+4} = \frac{(y-1)(y-1)}{(3y-5)(y+4)} \cdot \frac{(3y-5)(y+1)}{(y+4)(y-1)} \cdot \frac{(y+4)}{(y-3)(y-1)}$$
$$= \frac{y+1}{(y+4)(y-3)}$$

133. $$\frac{2}{\sqrt{3}+1} = \frac{2}{\sqrt{3}+1} \cdot \frac{\sqrt{3}-1}{\sqrt{3}-1} = \frac{2\left(\sqrt{3}-1\right)}{3-1}$$
$$= \frac{2\left(\sqrt{3}-1\right)}{2} = \sqrt{3}-1$$

134. Let x = width of rectangle,
$x + 4$ = length of rectangle, and
Area = 96 square yards.
$x(x + 4) = 96$
$x^2 + 4x - 96 = 0$
$(x + 12)(x - 8) = 0$
$x + 12 = 0$ or $x - 8 = 0$
$x = -12$ (reject) or $x = 8$
$x + 4 = 8 + 4 = 12$
Width: 8 yd   Length: 12 yd

Problem Set 7.2

1. $x^2 + 12x$
$$\left(\frac{12}{2}\right)^2 = 6^2 = 36$$
$$x^2 + 12x + 36 = (x+6)^2$$

3. $x^2 - 16x$
$$\left(-\frac{16}{2}\right)^2 = (-8)^2 = 64$$
$$x^2 - 16x + 64 = (x-8)^2$$

5. $x^2 + 7x$
$$\left(\frac{7}{2}\right)^2 = \frac{49}{4}$$
$$x^2 + 7x + \frac{49}{4} = \left(x + \frac{7}{2}\right)^2$$

7. $x^2 - 3x$

$\left(-\frac{3}{2}\right)^2 = \frac{9}{4}$

$x^2 - 3x + \frac{9}{4} = \left(x - \frac{3}{2}\right)^2$

9. $x^2 + \frac{1}{3}x$

$\left(\frac{\frac{1}{3}}{2}\right)^2 = \left(\frac{1}{6}\right)^2 = \frac{1}{36}$

$x^2 + \frac{1}{3}x + \frac{1}{36} = \left(x + \frac{1}{6}\right)^2$

11. $x^2 - \frac{2}{3}x$

$\left(\frac{-\frac{2}{3}}{2}\right)^2 = \left(-\frac{2}{6}\right)^2 = \left(-\frac{1}{3}\right)^2 = \frac{1}{9}$

$x^2 - \frac{2}{3}x + \frac{1}{9} = \left(x - \frac{1}{3}\right)^2$

For **13–35.**, the check is left to the student.

13. $x^2 - 4x = 21$

$x^2 - 4x + 4 = 21 + 4$

$(x-2)^2 = 25$

$x - 2 = \pm 5$

$x = 2 \pm 5$

$x = 7$ or $x = -3$

$\{-3, 7\}$

15. $x(x-6) = 16$

$x^2 - 6x + 9 = 16 + 9$

$(x-3)^2 = 25$

$x - 3 = \pm 5$

$x = 3 \pm 5$

$x = 8$ or $x = -2$

$\{-2, 8\}$

17. $y^2 - 6y + 2 = 0$

$y^2 - 6y = -2$

$y^2 - 6y + 9 = -2 + 9$

$\left[\left(\frac{1}{2}\right)(-6) = -3 \text{ and } (-3)^2 = 9\right]$

$y - 3 = \pm\sqrt{7}$

$y = 3 \pm \sqrt{7}$

$\{3+\sqrt{7}, 3-\sqrt{7}\}$

19. $x^2 + x - 1 = 0$

$x^2 + x + \frac{1}{4} = 1 + \frac{1}{4}$

$\left(x + \frac{1}{2}\right)^2 = \frac{5}{4}$

$\left[\frac{1}{2}(1) = \frac{1}{2} \text{ and } \left(\frac{1}{2}\right)^2 = \frac{1}{4}\right]$

$x + \frac{1}{2} = \pm\frac{\sqrt{5}}{2}$

$x = -\frac{1}{2} \pm \frac{\sqrt{5}}{2}$

$x = \frac{-1 \pm \sqrt{5}}{2}$

$\left\{\frac{-1+\sqrt{5}}{2}, \frac{-1-\sqrt{5}}{2}\right\}$

21. $2y^2 - 5y = 3$

$y^2 - \frac{5}{2}y = \frac{3}{2}$

$y^2 - \frac{5}{2}y + \frac{25}{16} = \frac{3}{2} + \frac{25}{16}$

$\left[\text{Note: } \left(\frac{1}{2}\right)\left(-\frac{5}{2}\right) = -\frac{5}{4} \text{ and } \left(-\frac{5}{4}\right)^2 = \frac{25}{16}\right]$

$\left(y - \frac{5}{4}\right)^2 = \frac{24}{16} + \frac{25}{16}$

$\left(y - \frac{5}{4}\right)^2 = \frac{49}{16}$

$y - \frac{5}{4} = \pm\frac{7}{4}$

$y = \frac{5}{4} \pm \frac{7}{4}$

$y = \frac{12}{4}$ or $y = -\frac{2}{4}$

$y = 3$ or $y = -\frac{1}{2}$

$\left\{-\frac{1}{2}, 3\right\}$

23. $9z^2 - 30z + 25 = 0$

$z^2 - \frac{30}{9}z + \frac{25}{9} = -\frac{25}{9} + \frac{25}{9}$

$\left[\frac{1}{2}\left(-\frac{10}{3}\right) = -\frac{5}{3} \text{ and } \left(-\frac{5}{3}\right)^2 = \frac{25}{9}\right]$

$\left(z-\frac{5}{3}\right)^2=0$

$z=\frac{5}{3}$

$\left\{\frac{5}{3}\right\}$

25. $2z^2+z=5$

$z^2+\frac{1}{2}z=\frac{5}{2}$

$z^2+\frac{1}{2}z+\frac{1}{16}=\frac{5}{2}+\frac{1}{16}$

$\left[\text{Note: } \frac{1}{2}\left(\frac{1}{2}\right)=\frac{1}{4} \text{ and } \left(\frac{1}{4}\right)^2=\frac{1}{16}\right]$

$\left(z+\frac{1}{4}\right)^2=\frac{41}{16}$

$z+\frac{1}{4}=\pm\sqrt{\frac{41}{16}}$

$z+\frac{1}{4}=\pm\frac{\sqrt{41}}{4}$

$z=-\frac{1}{4}\pm\frac{\sqrt{41}}{4}$

$z=\frac{-1\pm\sqrt{41}}{4}$

$\left\{\frac{-1+\sqrt{41}}{4}, \frac{-1-\sqrt{41}}{4}\right\}$

27. $2x^2-2x=3$

$x^2-x+\frac{1}{4}=\frac{3}{2}+\frac{1}{4}$

$\left[\frac{1}{2}(-1)=-\frac{1}{2} \text{ and } \left(-\frac{1}{2}\right)^2=\frac{1}{4}\right]$

$\left(x-\frac{1}{2}\right)^2=\frac{7}{4}$

$x-\frac{1}{2}=\pm\frac{\sqrt{7}}{2}$

$x=\frac{1\pm\sqrt{7}}{2}$

$\left\{\frac{1+\sqrt{7}}{2}, \frac{1-\sqrt{7}}{2}\right\}$

29. $y^2+2y+2=0$

$y^2+2y+1=-2+1$

$(y+1)^2=-1$

$y+1=\pm\sqrt{-1}$

$y+1=\pm i$

$y=-1\pm i$

(For practice) check $-1-i$:

$(-1-i)^2+2(-1-i)+2=0$

$1+2i+i^2-2-2i+2=0$

$1+i^2=0$

$1+(-1)=0$

$0=0$

$\{-1+i, -1-i\}$

31. $x^2-x+1=0$

$x^2-x+\frac{1}{4}=-1+\frac{1}{4}$

$\left[\frac{1}{2}(-1)=-\frac{1}{2} \text{ and } \left(-\frac{1}{2}\right)^2=\frac{1}{4}\right]$

$\left(x-\frac{1}{2}\right)^2=-\frac{3}{4}$

$x-\frac{1}{2}=\pm\sqrt{-\frac{3}{4}}$

$x-\frac{1}{2}=\pm\frac{i\sqrt{3}}{2}$

$x=\frac{1\pm i\sqrt{3}}{2}$

$\left\{\frac{1+i\sqrt{3}}{2}, \frac{1-i\sqrt{3}}{2}\right\}$

33. $8z^2-4z=-1$

$z^2-\frac{1}{2}z=-\frac{1}{8}$

$\left[\text{Note: } \frac{1}{2}\left(-\frac{1}{2}\right)=-\frac{1}{4};\ \left(-\frac{1}{4}\right)^2=\frac{1}{16}\right]$

$z^2-\frac{1}{2}z+\frac{1}{16}=-\frac{1}{8}+\frac{1}{16}$

$\left(z-\frac{1}{4}\right)^2=-\frac{1}{16}$

$z-\frac{1}{4}=\pm\sqrt{-\frac{1}{16}}$

$z-\frac{1}{4}=\pm\frac{i}{4}$

$z=\frac{1}{4}\pm\frac{i}{4}=\frac{1\pm i}{4}$

$\left\{\frac{1+i}{4}, \frac{1-i}{4}\right\}$

35. $3y^2 + 2y + 4 = 0$

$y^2 + \frac{2}{3}y + \frac{4}{3} = 0$

$y^2 + \frac{2}{3}y = -\frac{4}{3}$

$\left[\text{Note: } \frac{1}{2}\left(\frac{2}{3}\right) = \frac{1}{3};\ \left(\frac{1}{3}\right)^2 = \frac{1}{9}\right]$

$y^2 + \frac{2}{3}y + \frac{1}{9} = -\frac{4}{3} + \frac{1}{9}$

$\left(y + \frac{1}{3}\right)^2 = -\frac{12}{9} + \frac{1}{9}$

$\left(y + \frac{1}{3}\right)^2 = -\frac{11}{9}$

$y + \frac{1}{3} = \pm\sqrt{-\frac{11}{9}} = \pm\frac{i\sqrt{11}}{3}$

$y = -\frac{1}{3} \pm \frac{i\sqrt{11}}{3} = \frac{-1 \pm i\sqrt{11}}{3}$

$\left\{\frac{-1 + i\sqrt{11}}{3}, \frac{-1 - i\sqrt{11}}{3}\right\}$

37. **a.** vertex = $(h, k) = (2, 1)$

b. shifted 2 units right, shifted 1 unit up

c. $y = (x - 2)^2 + 1$

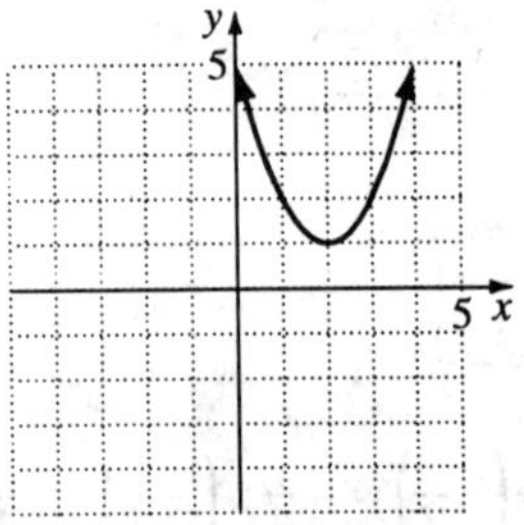

39. **a.** vertex = $(h, k) = (-1, -2)$

b. shifted 1 unit left, shifted 2 units down

c. $y = (x + 1)^2 - 2$

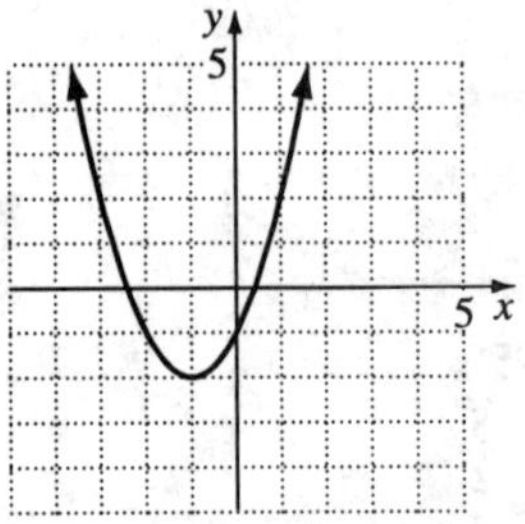

41. $f(x) = -(x - 3)^2 - 1$

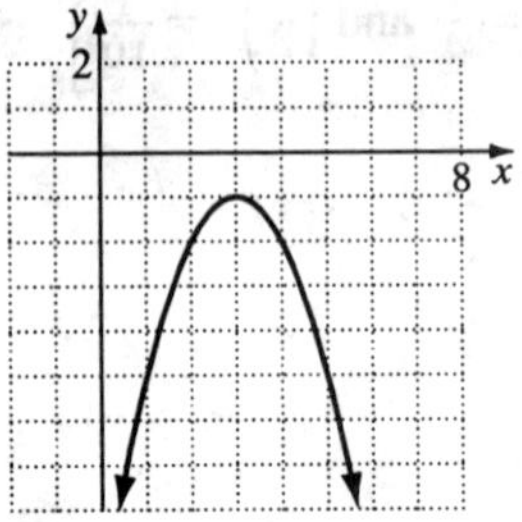

43. $y = 2(x + 1)^2 - 3$

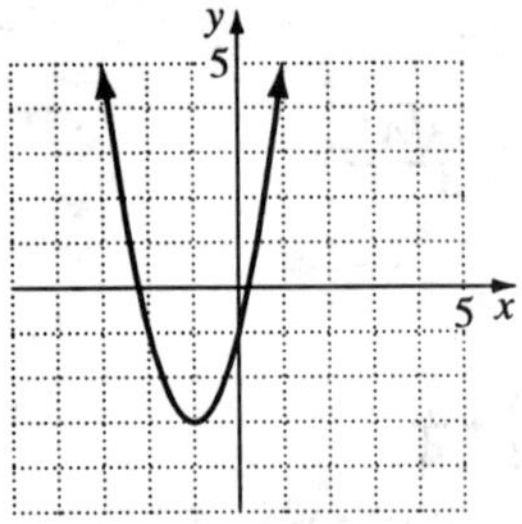

45. $g(x) = -2(x - 4)^2 + 3$

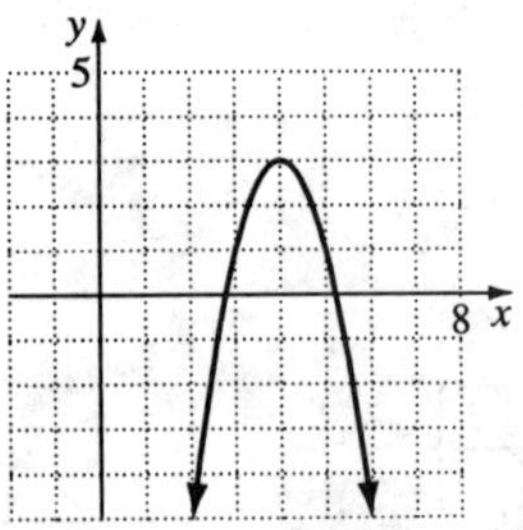

47. $y = -3(x+1)^2 + 4$

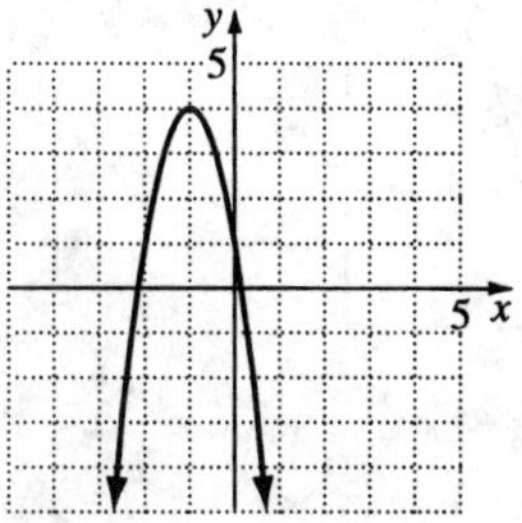

49. a.
$$y = x^2 + 6x + 5$$
$$y - 5 = x^2 + 6x$$
$$y - 5 + 9 = x^2 + 6x + 9$$
$$y + 4 = (x+3)^2$$
$$y = (x+3)^2 - 4$$

b.

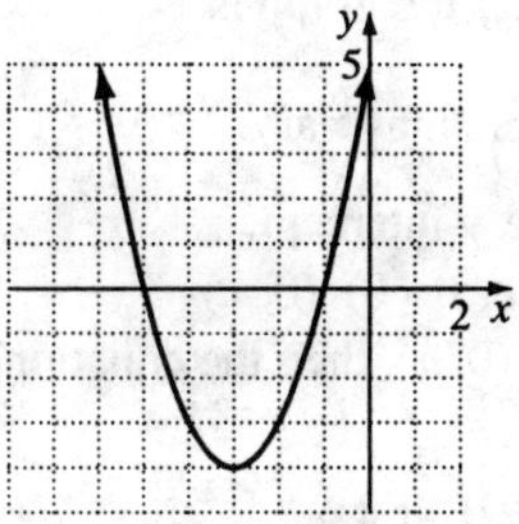

51. a.
$$y = -x^2 - 4x - 3$$
$$-y = x^2 + 4x + 3$$
$$-y - 3 = x^2 + 4x$$
$$-y - 3 + 4 = x^2 + 4x + 4$$
$$-y + 1 = (x+2)^2$$
$$-y = (x+2)^2 - 1$$
$$y = -(x+2)^2 + 1$$

b.

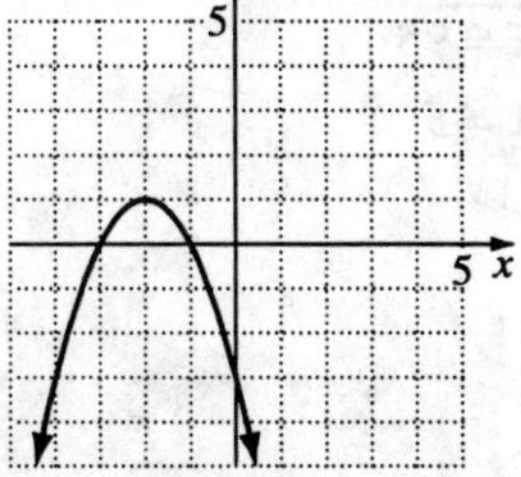

53. a.
$$y = 2x^2 + 6x + 8$$
$$\frac{y}{2} = x^2 + 3x + 4$$
$$\frac{y}{2} - 4 = x^2 + 3x$$
$$\frac{y}{2} - 4 + \frac{9}{4} = x^2 + 3x + \frac{9}{4}$$
$$\frac{2y-7}{4} = \left(x + \frac{3}{2}\right)^2$$
$$2y - 7 = 4\left(x + \frac{3}{2}\right)^2$$
$$2y = 4\left(x + \frac{3}{2}\right)^2 + 7$$
$$y = 2\left(x + \frac{3}{2}\right)^2 + \frac{7}{2}$$

b.

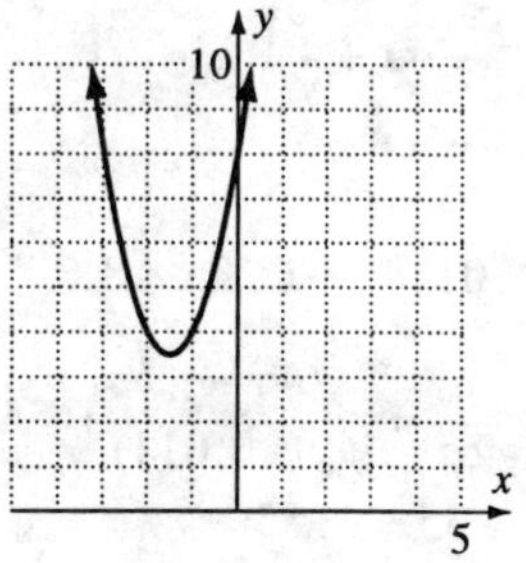

55. a.
$$y = 3x^2 - 12x + 13$$
$$y - 13 = 3x^2 - 12x$$
$$\frac{y-13}{3} = x^2 - 4x$$
$$\frac{y-13}{3} + 4 = x^2 - 4x + 4$$
$$\frac{y-1}{3} = (x-2)^2$$
$$y - 1 = 3(x-2)^2$$
$$y = 3(x-2)^2 + 1$$

b.

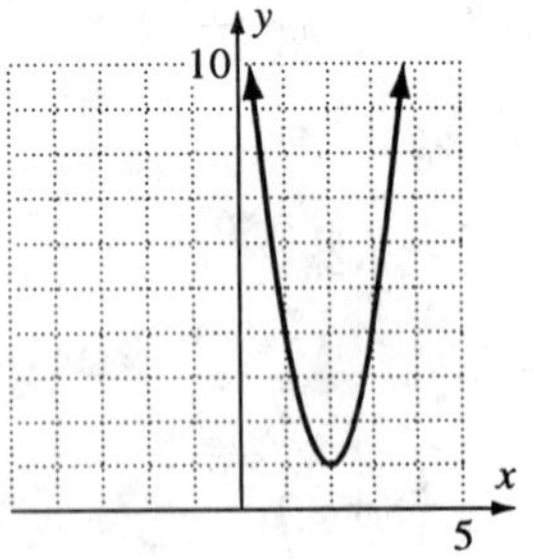

57. **a.** $y = (x+2)^2$

b. $y = x^2 + 4x + 4$

59. **a.** $y = (x-2)^2 + 2$

b. $y = x^2 - 4x + 4 + 2$
$y = x^2 - 4x + 6$

61. $f(x) = -0.011x^2 + 1.22x - 8.5$

a. $x = -\frac{b}{2a} = -\frac{1.22}{2(-0.011)} \approx 55$
$1930 + 55 = 1985$
In 1985, the number of union members was at a maximum.

b. $y = -0.011(55)^2 + 1.22(55) - 8.5$
$y = 25.325$
The maximum number of members was about 25 million.

63. c is true.

65–67. Students should verify results.

69. $y = x^2$, $y = 4x^2$, $y = \frac{1}{4}x^2$

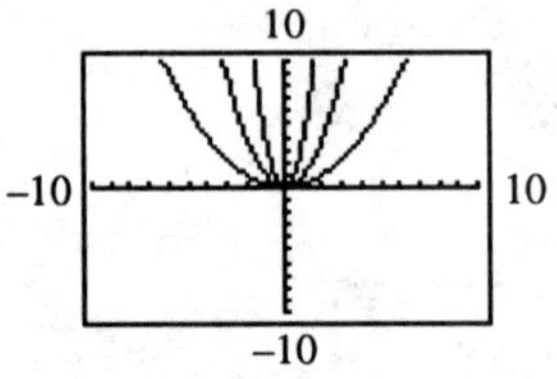

$y = x^2$, $y = 6x^2$, $y = \frac{1}{6}x^2$

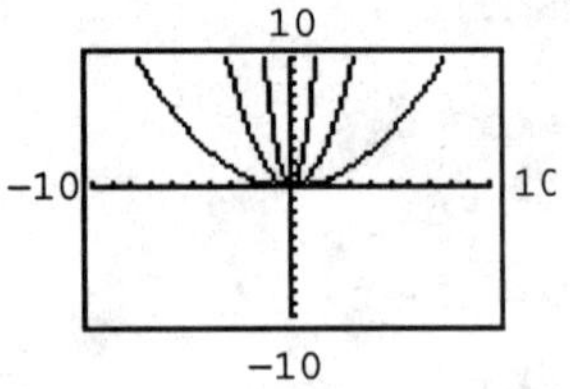

For $a > 1$, the larger a the more steeply the graph rises. For $0 < a < 1$, the closer a is to 0, the flatter the graph is.

71–73. Answers may vary.

75. Let c = the y-intercept.
y-intercept ($x = 0$): $(0, c)$
If $(h, k) = (0, c)$, then the equation of the parabola is
$y = a(x-0)^2 + c$
$y = ax^2 + c$

Review Problems

76. $3x - 2y = 1$ (1)
$5y + 3z = -7$ (2)
$2x + 5y = 45$ (3)
Multiply equation (1) by 5 and equation (3) by 2 and add.

$$\begin{array}{r} 15x - 10y = 5 \\ 4x + 10y = 90 \\ \hline 19x = 95 \\ x = 5 \end{array}$$

Equation 1:
$3x - 2y = 1$
$15 - 2y = 1$
$-2y = -14$
$y = 7$

Equation 2:

$5y + 3z = -7$

$5(7) + 3z = -7$

$3z = -42$

$z = -14$

$\{(5, 7, -14)\}$

77. $\sqrt{2x+3} - \sqrt{x+1} = 1$

$\sqrt{2x+3} = \sqrt{x+1} + 1$

$2x+3 = x+1+2\sqrt{x+1}+1$

$2x+3 = x+2+2\sqrt{x+1}$

$x+1 = 2\sqrt{x+1}$

$x^2 + 2x + 1 = 4x + 4$

$x^2 - 2x - 3 = 0$

$(x-3)(x+1) = 0$

$x = 3$ or $x = -1$ (both check)

$\{-1, 3\}$

78. $\dfrac{\frac{x}{y} - \frac{y}{x}}{\frac{x}{y} + 1} = \dfrac{\frac{x^2-y^2}{xy}}{\frac{x+y}{y}} = \dfrac{x^2 - y^2}{xy} \cdot \dfrac{y}{x+y}$

$= \dfrac{(x+y)(x-y)}{xy} \cdot \dfrac{y}{x+y} = \dfrac{x-y}{x}$

Problem Set 7.3

1. $x^2 - 9x + 20 = 0$

$x = \dfrac{9 \pm \sqrt{81 - 4(1)(20)}}{2}$

$= \dfrac{9 \pm \sqrt{1}}{2} = \dfrac{9 \pm 1}{2} = \dfrac{8}{2}, \dfrac{10}{2}$

$\{4, 5\}$

3. $x^2 - 4x - 60 = 0$

$x = \dfrac{4 \pm \sqrt{16 - 4(1)(-60)}}{2} = \dfrac{4 \pm \sqrt{256}}{2}$

$= \dfrac{4 \pm 16}{2} = -\dfrac{12}{2}, \dfrac{20}{2}$

$\{-6, 10\}$

5. $2x^2 - 7x = -5$

$2x^2 - 7x + 5 = 0$

$x = \dfrac{7 \pm \sqrt{49 - 4(2)(5)}}{4} = \dfrac{7 \pm \sqrt{9}}{4}$

$= \dfrac{7 \pm 3}{4} = \dfrac{4}{4}, \dfrac{10}{4}$

$\left\{1, \dfrac{5}{2}\right\}$

7. $2y^2 + y = 5$

$2y^2 + y - 5 = 0$

$y = \dfrac{-1 \pm \sqrt{1 - 4(2)(-5)}}{4} = \dfrac{-1 \pm \sqrt{41}}{4}$

$\left\{\dfrac{-1+\sqrt{41}}{4}, \dfrac{-1-\sqrt{41}}{4}\right\}$

9. $2 = 3y^2 + 4y$

$0 = 3y^2 + 4y - 2$

$y = \dfrac{-4 \pm \sqrt{16 - 4(3)(-2)}}{6}$

$= \dfrac{-4 \pm 2\sqrt{10}}{6} = \dfrac{-2 \pm \sqrt{10}}{3}$

$\left\{\dfrac{-2+\sqrt{10}}{3}, \dfrac{-2-\sqrt{10}}{3}\right\}$

11. $2z^2 = 2z - 1$

$2z^2 - 2z + 1 = 0$

$a = 2, b = -2, c = 1$

$z = \dfrac{-b \pm \sqrt{b^2 - 4ac}}{2a}$

$= \dfrac{-(-2) \pm \sqrt{4 - 4(2)(1)}}{4} = \dfrac{2 \pm \sqrt{-4}}{4}$

$= \dfrac{2 \pm 2i}{4} = \dfrac{2(1 \pm i)}{4} = \dfrac{1 \pm i}{2}$

$\left\{\dfrac{1+i}{2}, \dfrac{1-i}{2}\right\}$

13. $5y^2 = 2y - 3$

$5y^2 - 2y + 3 = 0$

$y = \frac{2 \pm \sqrt{4 - 4(5)(3)}}{10} = \frac{2 \pm \sqrt{-56}}{10}$

$= \frac{2 \pm 2i\sqrt{14}}{10} = \frac{1 \pm i\sqrt{14}}{5}$

$\left\{\frac{1 + i\sqrt{14}}{5}, \frac{1 - i\sqrt{14}}{5}\right\}$

15. $3 + \frac{7}{x} = \frac{6}{x^2}$

$x^2\left(3 + \frac{7}{x}\right) = x^2\left(\frac{6}{x^2}\right)$

$3x^2 + 7x - 6 = 0$

$a = 3, b = 7, c = -6$

$x = \frac{-b \pm \sqrt{b^2 - 4ac}}{2a} = \frac{-7 \pm \sqrt{49 - 4(3)(-6)}}{6}$

$= \frac{-7 \pm \sqrt{49 - (-72)}}{6} = \frac{-7 \pm \sqrt{121}}{6}$

$= \frac{-7 \pm 11}{6}$

$x = \frac{-7 + 11}{6}$ or $x = \frac{-7 - 11}{6}$

$= \frac{4}{6}$ $\quad = \frac{-18}{6}$

$= \frac{2}{3}$ $\quad = -3$

$\left\{-3, \frac{2}{3}\right\}$

17. $\frac{5x + 6}{2x + 3} = 3x$

$5x + 6 = 3x(2x + 3)$

$5x + 6 = 6x^2 + 9x$

$6x^2 + 4x - 6 = 0$

$3x^2 + 2x - 3 = 0$

$x = \frac{-2 \pm \sqrt{4 - 4(3)(-3)}}{6}$

$= \frac{-2 \pm \sqrt{40}}{6} = \frac{-2 \pm 2\sqrt{10}}{6}$

$\left\{\frac{-1 - \sqrt{10}}{3}, \frac{-1 + \sqrt{10}}{3}\right\}$

19. $\frac{1}{2}x^2 - \frac{1}{3}x + \frac{1}{4} = 0$

$12\left(\frac{1}{2}x^2 - \frac{1}{3}x + \frac{1}{4}\right) = 12 \cdot 0$

$6x^2 - 4x + 3 = 0$

$x = \frac{4 \pm \sqrt{16 - 4(6)(3)}}{12}$

$= \frac{4 \pm \sqrt{-56}}{12} = \frac{4 \pm 2i\sqrt{14}}{12} = \frac{2 \pm i\sqrt{14}}{6}$

$\left\{\frac{2 - i\sqrt{14}}{6}, \frac{2 + i\sqrt{14}}{6}\right\}$

21. $(x - 3)(x + 3) = 12$

$x^2 - 9 = 12$

$x^2 = 21$

$x = \pm\sqrt{21}$

$\left\{-\sqrt{21}, \sqrt{21}\right\}$

23. $(5x - 1)(2x + 3) = 3x - 3$

$10x^2 + 13x - 3 = 3x - 3$

$10x^2 + 10x = 0$

$10x(x + 1) = 0$

$x = 0$ or $x = -1$

$\{-1, 0\}$

25. $(3x - 4)^2 = 81$

$3x - 4 = \pm\sqrt{81}$

$3x - 4 = -9$ or $3x - 4 = 9$

$3x = -5$ or $3x = 13$

$x = -\frac{5}{3}$ or $x = \frac{13}{3}$

$\left\{-\frac{5}{3}, \frac{13}{3}\right\}$

27. $3x^2 - 4x + 2 = 0$

$x = \frac{4 \pm \sqrt{16 - 4(3)(2)}}{6} = \frac{4 \pm \sqrt{-8}}{6}$

$= \frac{4 \pm 2i\sqrt{2}}{6} = \frac{2 \pm i\sqrt{2}}{3}$

$\left\{\frac{2 - i\sqrt{2}}{3}, \frac{2 + i\sqrt{2}}{3}\right\}$

29. $2x^2 + 5x = 3$
$2x^2 + 5x - 3 = 0$
$(2x - 1)(x + 3) = 0$
$2x - 1 = 0 \text{ or } x + 3 = 0$
$x = \frac{1}{2} \text{ or } x = -3$
$\left\{-3, \frac{1}{2}\right\}$

31. $3x^2 = 11x - 10$
$3x^2 - 11x + 10 = 0$
$x = \frac{11 \pm \sqrt{121 - 4(3)(10)}}{6} = \frac{11 \pm \sqrt{1}}{6}$
$= \frac{11 \pm 1}{6} = \frac{10}{6}, \frac{12}{6}$
$\left\{\frac{5}{3}, 2\right\}$

33. a. $x^2 - 6x + 7 = 0$
$x = \frac{6 \pm \sqrt{36 - 4(1)(7)}}{2} = \frac{6 \pm \sqrt{8}}{2}$
$= \frac{6 \pm 2\sqrt{2}}{2} = 3 \pm \sqrt{2}$
$\left\{3 - \sqrt{2}, 3 + \sqrt{2}\right\}$

b. $y = x^2 - 6x + 7$

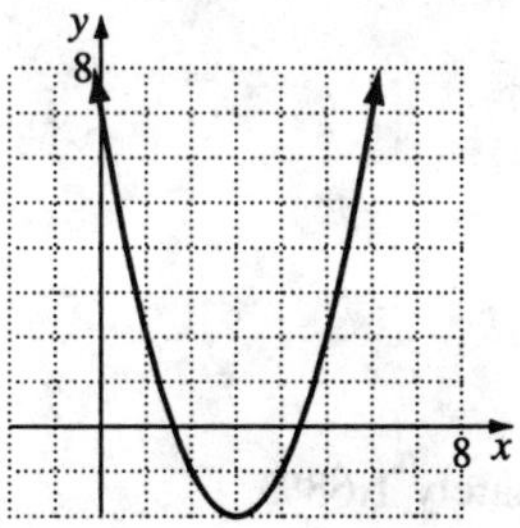

35. a. $2x^2 - 6x - 9 = 0$
$x = \frac{6 \pm \sqrt{36 - 4(2)(-9)}}{4} = \frac{6 \pm \sqrt{108}}{4}$
$= \frac{6 \pm 6\sqrt{3}}{4} = \frac{3 \pm 3\sqrt{3}}{2}$
$\left\{\frac{3 - 3\sqrt{3}}{2}, \frac{3 + 3\sqrt{3}}{2}\right\}$

b. $y = 2x^2 - 6x - 9$

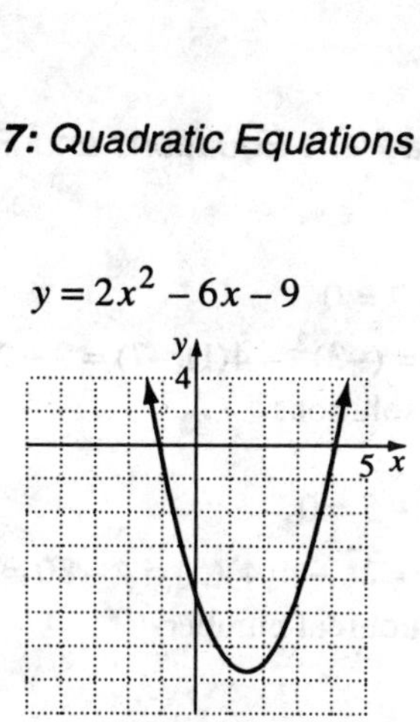

37. a. $x^2 + x + 5 = 0$
$x = \frac{-1 \pm \sqrt{1 - 4(1)(5)}}{2} = \frac{-1 \pm \sqrt{-19}}{2}$
$= \frac{-1 \pm i\sqrt{19}}{2}$
$\left\{\frac{-1 - i\sqrt{19}}{2}, \frac{-1 + i\sqrt{19}}{2}\right\}$

b. $y = x^2 + x + 5$

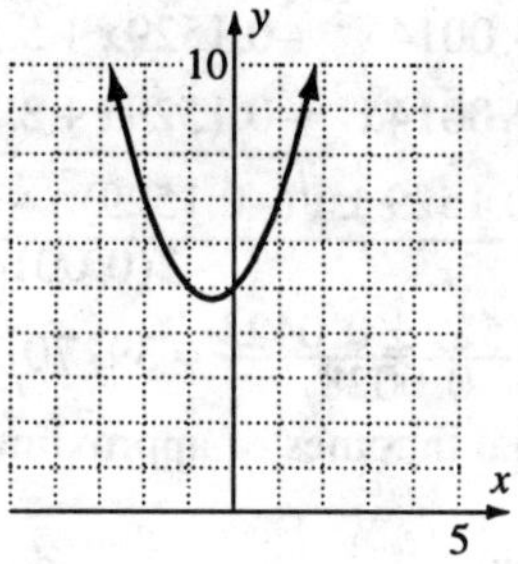

39. $y^2 - 4y - 5 = 0$
$b^2 - 4ac = 16 - 4 \cdot 1(-5) = 36$
rational solutions

41. $2x^2 - 11x + 3 = 0$
$b^2 - 4ac = (-11)^2 - 4(2)(3)$
$= 121 - 24 = 97$
irrational solutions

43. $x^2 - 2x + 1 = 0$
$b^2 - 4ac = (-2)^2 - 4(1)(1) = 4 - 4 = 0$
rational solutions

45. $x^2 - 3x - 7 = 0$
$b^2 - 4ac = (-3)^2 - 4(1)(-7) = 9 + 28 = 37$
irrational solutions

47. $4y^2 + 2y + 5 = 0$
$b^2 - 4ac = 2^2 - 4(4)(5) = 4 - 80 = -76$
solutions not real numbers

49. $P = 0.78t^2 + 76.7t + 4449$
$5000 = 0.78t^2 + 76.7t + 4449$
$0 = 0.78t^2 + 76.7t - 551$
$t = \dfrac{-76.7 \pm \sqrt{(76.7)^2 - 4(0.78)(-551)}}{2(0.78)}$
$= \dfrac{-76.7 \pm \sqrt{7602.01}}{1.56}$
$\approx -105, 7$
Disregard $t < 0$.
$1980 + 7 = 1987$
The year was 1987. The population starts to increase much more rapidly.

51. $P = 0.0014x^2 - 0.1529x + 5.855$

a. $2 = 0.0014x^2 - 0.1529x + 5.855$
$0 = 0.0014x^2 - 0.1529x + 3.855$
$x = \dfrac{0.1529 \pm \sqrt{(-0.1529)^2 - 4(0.0014)(3.855)}}{2(0.0014)}$
$\approx \dfrac{0.1529 \pm 0.0423}{0.0028} \approx 39, 70$
Annual incomes of approximately \$39,000 and \$70,000.

b. Vertex:
$x = -\dfrac{b}{2a} = \dfrac{0.1529}{2(0.0014)} \approx 54.6$
$y = 0.0014(54.6)^2 - 0.1529(54.6) + 5.855$
$y \approx 1.68$
The income is approximately \$54,600 and the percent is approximately 1.68%.

53. a. $y = ax^2 + bx + c$

$107,298 = a(0)^2 + b(0) + c$

$98,588 = a(1)^2 + b(1) + c$

$90,798 = a(2)^2 + b(2) + c$

or

$c = 107,298$

$a + b + c = 98,588$

$4a + 2b + c = 90,798$

Substitute c into equations

$a + b = -8710$

$4a + 2b = -16,500$

Add –2 times first equation to second.

$$\begin{array}{r} -2a - 2b = 17,420 \\ 4a + 2b = -16,500 \\ \hline 2a = 920 \\ a = 460 \end{array}$$

Substitute back.

$460 + b + 107,298 = 98,588$

$b = -9170$

$y = 460x^2 - 9170x + 107,298$

b. $165,548 = 460x^2 - 9170x + 107,298$

$0 = 460x^2 - 9170x - 58,250$

$$x = \frac{9170 \pm \sqrt{(-9170)^2 - 4(460)(-58,250)}}{2(460)}$$

$$= \frac{9170 \pm 13,830}{920} \approx -5.1,\ 25$$

Disregard a negative time.

$1975 + 25 = 2000$

In the year 2000, expenditures will be \$165,548.

c. Vertex:

$$x = -\frac{b}{2a} = \frac{-(-9170)}{2(460)} \approx 9.97$$

$y = 460(9.97)^2 - 9170(9.97) + 107,298$

$\approx 61,600$

$1975 + 10 = 1985$

In about 1985, the expenditures were at a minimum of approximately \$61,600 million.

55. $P = -5I^2 + 80I$
$340 = -5I^2 + 80I$
$0 = -5I^2 + 80I - 340$
$b^2 - 4ac = (80)^2 - 4(-5)(-340) = -400$
Since the discriminant is negative, there are no real solutions. No, there will never be enough current to generate 340 volts of power.

57. c is true.

59–61. Students should verify results.

63. $y = -0.163x + 40.5$
$y = -0.163(19) + 40.5$
$y \approx 37.4$
\$37.4 million in years 1999–2000

65. Answers may vary.

67. $ix^2 = 5x - 2i$
$ix^2 - 5x + 2i = 0$
$x = \frac{5 \pm \sqrt{25 - 8i^2}}{2i} = \frac{5 \pm \sqrt{25+8}}{2i}$
$= \frac{5 \pm \sqrt{33}}{2i} \cdot \frac{2i}{2i} = \frac{\left(10 \pm 2\sqrt{33}\right)i}{-4}$
$= \frac{\left(-5 \pm \sqrt{33}\right)i}{2}$
$\left\{\left(\frac{-5-\sqrt{33}}{2}\right)i, \left(\frac{-5+\sqrt{33}}{2}\right)i\right\}$

69. $f(x) = 5x^2 + hx + 3$
Set $b^2 - 4ac = 0.$
$h^2 - 4(5)(3) = 0$
$h^2 - 60 = 0$
$h^2 = 60$
$h = \pm\sqrt{60} = \pm\sqrt{4 \cdot 15}$
$h = \pm 2\sqrt{15}$

71. $3x^2 - 6x + a = 0$
Set $b^2 - 4ac > 0.$
$36 - 4(3)a > 0$
$36 - 12a > 0$
$-12a > -36$
$a < 3$

73. Roots are 3 and 7
Sum = 10; product = 21
$x^2 - 10x + 21 = 0$

75. Roots are $2 + \sqrt{3}$ and $2 - \sqrt{3}$
Sum = 4; product = 4 – 3 = 1
$x^2 - 4x + 1 = 0$

77. $(x + 2)(x - 5) = 0$
$x^2 - 3x - 10 = 0;$
$\left[x - \left(2 + \sqrt{3}\right)\right]\left[x - \left(2 - \sqrt{3}\right)\right] = 0$
$x^2 - 4x + 1;$
$[x - (3 + 2i)][x - (3 - 2i)] = 0$
$x^2 - 6x + 13 = 0$

79. Since $a = 1$, $b = p$, and $c = q$, the discriminant is
$b^2 - 4ac = p^2 - 4(1)q = p^2 - 4q.$
Since we want only real roots,
$p^2 - 4q \geq 0$
$1 - 4q \geq 0$
$\left(\text{Since } |p| = 1, \text{ then } p^2 = 1.\right)$
$-4q \geq -1$
$q \leq \frac{1}{4}$
We know that $|q| = 1$, so since $q \leq \frac{1}{4}$, then $q = -1$. The solution shows that $p = \pm 1$ and $q = -1$.

81. $x = \frac{-(a+c) \pm \sqrt{(a+c)^2 - 4ac}}{2a}$

$= \frac{-(a+c) \pm \sqrt{a^2 - 2ac + c^2}}{2a}$

$= \frac{-(a+c) \pm \sqrt{(a-c)^2}}{2a}$

$= \frac{-(a+c) \pm |a-c|}{2a}$

$= \frac{-(a+c) \pm (c-a)}{2a}$

$(|a-c| = c - a$ because $c > a.)$

$x = \frac{-(a+c)+(c-a)}{2a}$ or

$x = \frac{-(a+c)-(c-a)}{2a}$

$x = -1$ or $x = -\frac{c}{a}$

Since $c > a > 0$, then the larger root is -1.

Review Problems

83. $\frac{\frac{1}{x}+1}{\frac{1}{x^2}-1} = \frac{\frac{1+x}{x}}{\frac{1-x^2}{x^2}} = \frac{1+x}{x} \cdot \frac{x^2}{1-x^2}$

$= \frac{1+x}{x} \cdot \frac{x^2}{(1+x)(1-x)} = \frac{x}{1-x}$

84. Points $(1, -2)$ and $(5, 2)$

slope $= \frac{2-(-2)}{5-1} = \frac{4}{4} = 1$

point-slope:

$y - (-2) = 1(x - 1)$

$y + 2 = x - 1$ or $y - 2 = x - 5$

slope-intercept: $y = x - 3$

85. $5x - 2y = 10$

$-2y = -5x + 10$

$y = \frac{-5x}{-2} + \frac{10}{-2}$

$y = \frac{5}{2}x - 5$

Problem Set 7.4

1. $5(x^2 - 4) = 12(x-2) + 12(x+2)$

$5x^2 - 20 = 12x - 24 + 12x + 24$

$5x^2 - 24x - 20 = 0$

$x = \frac{24 \pm \sqrt{(-24)^2 - 4(5)(-20)}}{2(5)}$

$= \frac{24 \pm \sqrt{976}}{10} = \frac{24 \pm 4\sqrt{61}}{10}$

$= \frac{12 \pm 2\sqrt{61}}{5}$

$\left\{\frac{12 - 2\sqrt{61}}{5}, \frac{12 + 2\sqrt{61}}{5}\right\}$

3. $\frac{x}{x-2} + \frac{2}{x} = 4$

$x(x-2)\left[\frac{x}{x-2} + \frac{2}{x}\right] = 4x(x-2)$

$x^2 + 2x - 4 = 4x^2 - 8x$

$3x^2 - 10x + 4 = 0$

$x = \frac{10 \pm \sqrt{(-10)^2 - 4(3)(4)}}{2(3)}$

$= \frac{10 \pm \sqrt{52}}{6} = \frac{10 \pm 2\sqrt{13}}{6} = \frac{5 \pm \sqrt{13}}{3}$

$\left\{\frac{5 - \sqrt{13}}{3}, \frac{5 + \sqrt{13}}{3}\right\}$

5. $\frac{6}{x+12}+\frac{6}{12-x}=\frac{16}{15}$

$15(x+12)(12-x)\left[\frac{6}{x+12}+\frac{6}{12-x}\right]=\frac{16}{15}(15)(x+12)(12-x)$

$90(12-x)+90(x+12)=16(x+12)(12-x)$

$1080-90x+90x+1080=-16x^2+2304$

$16x^2-144=0$

$x^2-9=0$

$x^2=9$

$x=\pm 3$

$x=\{-3, 3\}$

7. $\frac{1}{x-2}+\frac{1}{x+2}=1$

$(x-2)(x+2)\left[\frac{1}{x-2}+\frac{1}{x+2}\right]=(x-2)(x+2)$

$x+2+x-2=x^2-4$

$x^2-2x-4=0$

$x=\frac{2\pm\sqrt{(-2)^2-4(1)(-4)}}{2(1)}=\frac{2\pm\sqrt{20}}{2}$

$=\frac{2\pm 2\sqrt{5}}{2}=1\pm\sqrt{5}$

$\left\{1-\sqrt{5}, 1+\sqrt{5}\right\}$

9. $\frac{1}{x+2}=\frac{1}{3}-\frac{1}{x}$

$3x(x+2)\left(\frac{1}{x+2}\right)=3x(x+2)\left[\frac{1}{3}-\frac{1}{x}\right]$

$3x=x^2+2x-3x-6$

$x^2-4x-6=0$

$x=\frac{4\pm\sqrt{(-4)^2-4(1)(-6)}}{2(1)}$

$=\frac{4\pm\sqrt{40}}{2}=\frac{4\pm 2\sqrt{10}}{2}=2\pm\sqrt{10}$

$\left\{2-\sqrt{10}, 2+\sqrt{10}\right\}$

11. $(x+5)\left(\frac{50}{x}-\frac{1}{2}\right)=50$

$2x(x+5)\left(\frac{50}{x}-\frac{1}{2}\right)=50(2x)$

$2(x+5)(50)-x(x+5)=50(2x)$

$100x+500-x^2-5x=100x$

$x^2+5x-500=0$

$x=\frac{-5\pm\sqrt{5^2-4(1)(-500)}}{2(1)}$

$=\frac{-5\pm 45}{2}=-\frac{50}{2}, \frac{40}{2}$

$\{-25, 20\}$

13. $x^4 - 5x^2 + 4 = 0$
Let $u = x^2$.
$u^2 - 5u + 4 = 0$
$(u-4)(u-1) = 0$
$u = 4$ or $u = 1$
$x^2 = 4$ or $x^2 = 1$
$x = \pm 2$ or $x = \pm 1$
$\{-2, -1, 1, 2\}$

15. $9x^4 = 25x^2 - 16$
$9x^4 - 25x^2 + 16 = 0$
Let $u = x^2$.
$9u^2 - 25u + 16 = 0$
$(u-1)(9u-16) = 0$
$u = 1$ or $u = \frac{16}{9}$
$x^2 = 1$ or $x^2 = \frac{16}{9}$
$x \pm 1$ or $x = \pm\frac{4}{3}$
$\left\{-\frac{4}{3}, -1, 1, \frac{4}{3}\right\}$

17. $y - 7\sqrt{y} + 10 = 0$
Let $u = \sqrt{y}$.
$u^2 - 7u + 10 = 0$
$(u-5)(u-2) = 0$
$u = 5$ or $u = 2$
$\sqrt{y} = 5$ or $\sqrt{y} = 2$
$y = 25$ or $y = 4$
$\{4, 25\}$

19. $x^{-2} - x^{-1} - 12 = 0$
Let $u = x^{-1}$.
$u^2 - u - 12 = 0$
$(u-4)(u+3) = 0$
$u = 4$ or $u = -3$
$x^{-1} = 4$ or $x^{-1} = -3$
$\frac{1}{x} = 4$ or $\frac{1}{x} = -3$
$x = \frac{1}{4}$ or $x = -\frac{1}{3}$
$\left\{-\frac{1}{3}, \frac{1}{4}\right\}$

21. $x - 13\sqrt{x} + 40 = 0$
Let $u = \sqrt{x}$.
$u^2 - 13u + 40 = 0$
$(u-5)(u-8) = 0$
$u = 5$ or $u = 8$
$\sqrt{x} = 5$ or $\sqrt{x} = 8$
$x = 25$ or $x = 64$
$\{25, 64\}$

23. $x^{-2} + 6x^{-1} = 16$
$x^{-2} + 6x^{-1} - 16 = 0$
Let $u = x^{-1}$.
$u^2 + 6u - 16 = 0$
$(u+8)(u-2) = 0$
$u = -8$ or $u = 2$
$x^{-1} = -8$ or $x^{-1} = 2$
$\frac{1}{x} = -8$ or $\frac{1}{x} = 2$
$x = -\frac{1}{8}$ or $x = \frac{1}{2}$
$\left\{-\frac{1}{8}, \frac{1}{2}\right\}$

25. $6(w-1)^{-2} + (w-1)^{-1} - 2 = 0$
Let $u = (w-1)^{-1}$.
$6u^2 + u - 2 = 0$
$(3u+2)(2u-1) = 0$
$u = -\frac{2}{3}$ or $u = \frac{1}{2}$
$(w-1)^{-1} = -\frac{2}{3}$ or $(w-1)^{-1} = \frac{1}{2}$
$\frac{1}{w-1} = -\frac{2}{3}$ or $\frac{1}{w-1} = \frac{1}{2}$
$w = -\frac{1}{2}$ or $w = 3$
$\left\{-\frac{1}{2}, 3\right\}$

27. $(x^2 + 3x)^2 - 8(x^2 + 3x) - 20 = 0$
Let $u = x^2 + 3x$.
$u^2 - 8u - 20 = 0$
$(u+2)(u-10) = 0$
$u = -2$ or $u = 10$
$x^2 + 3x = -2$ or $x^2 + 3x = 10$
$x^2 + 3x + 2 = 0$ or $x^2 + 3x - 10 = 0$
$(x+2)(x+1) = 0$ $(x+5)(x-2) = 0$

$x = -2, x = -1, x = -5, x = 2$
$\{-5, -2, -1, 2\}$

29. $(x^2 + 2x - 3)^2 + 6(x^2 + 2x - 3) + 8 = 0$
Let $u = x^2 + 2x - 3$.
$u^2 + 6u + 8 = 0$
$(u + 4)(u + 2) = 0$
$u = -4$ or $u = -2$
$x^2 + 2x - 3 = -4$ or $x^2 + 2x - 3 = -2$
$x^2 + 2x + 1 = 0$ or $x^2 + 2x - 1 = 0$
$(x+1)^2 = 0$ or $x = \frac{-2 \pm \sqrt{2^2 - 4(1)(-1)}}{2}$
$x = -1$ or $x = -1 \pm \sqrt{2}$
$\{-1 - \sqrt{2}, -1, -1 + \sqrt{2}\}$

31. $y^{-2} - 4y^{-1} - 3 = 0$
Let $u = y^{-1}$.
$u^2 - 4u - 3 = 0$
$u = \frac{4 \pm \sqrt{(-4)^2 - 4(1)(-3)}}{2(1)} = \frac{4 \pm \sqrt{28}}{2}$
$= 2 \pm \sqrt{7}$
$y^{-1} = 2 - \sqrt{7}$ or $y^{-1} = 2 + \sqrt{7}$
$y = \frac{1}{2 - \sqrt{7}}$ or $y = \frac{1}{2 + \sqrt{7}}$
$\left\{\frac{1}{2 - \sqrt{7}}, \frac{1}{2 + \sqrt{7}}\right\}$

33. $y^{2/3} - y^{1/3} - 6 = 0$
Let $u = y^{1/3}$.
$u^2 - u - 6 = 0$
$(u - 3)(u + 2) = 0$
$u = 3$ or $u = -2$
$y^{1/3} = 3$ or $y^{1/3} = -2$
$y = 27$ or $y = -8$
$\{-8, 27\}$

35. $2y - 3y^{1/2} + 1 = 0$
Let $u = y^{1/2}$.
$2u^2 - 3u + 1 = 0$
$(2u - 1)(u - 1) = 0$
$u = \frac{1}{2}$ or $u = 1$
$y^{1/2} = \frac{1}{2}$ or $y^{1/2} = 1$
$y = \frac{1}{4}$ or $y = 1$
$\left\{\frac{1}{4}, 1\right\}$

37. $x^{3/2} + 1 = 2x^{3/4}$
$x^{3/2} - 2x^{3/4} + 1 = 0$
Let $u = x^{3/4}$.
$u^2 - 2u + 1 = 0$
$(u - 1)^2 = 0$
$u = 1$
$x^{3/4} = 1$
$x = 1$
$\{1\}$

39. $x^{2/5} = -8x^{1/5} - 16$
$x^{2/5} + 8x^{1/5} + 16 = 0$
Let $u = x^{1/5}$.
$u^2 + 8u + 16 = 0$
$(u + 4)^2 = 0$
$u = -4$
$x^{1/5} = -4$
$x = -1024$
$\{-1024\}$

41. $\left(\frac{x+3}{x+1}\right)^2 - 12\left(\frac{x+3}{x+1}\right) + 27 = 0$
Let $u = \frac{x+3}{x+1}$.
$u^2 - 12u + 27 = 0$
$(u - 9)(u - 3) = 0$
$u = 9$ or $u = 3$
$\frac{x+3}{x+1} = 9$ or $\frac{x+3}{x+1} = 3$
$x + 3 = 9x + 9$ or $3x + 3 = x + 3$
$-8x = 6$ or $2x = 0$
$x = -\frac{3}{4}$ or $x = 0$
$\left\{-\frac{3}{4}, 0\right\}$

43. $2\left(\frac{x-1}{x-3}\right)^2 - 7\left(\frac{x-1}{x-3}\right) + 5 = 0$

Let $u = \frac{x-1}{x-3}$.

$2u^2 - 7u + 5 = 0$

$(2u-5)(u-1) = 0$

$u = \frac{5}{2}$ or $u = 1$

$\frac{x-1}{x-3} = \frac{5}{2}$ or $\frac{x-1}{x-3} = 1$

$2x - 2 = 5x - 15$ or $x - 1 = x - 3$

$-3x = -13$ or $-1 = -3$ (not true)

$x = \frac{13}{3}$

$\left\{\frac{13}{3}\right\}$

45.

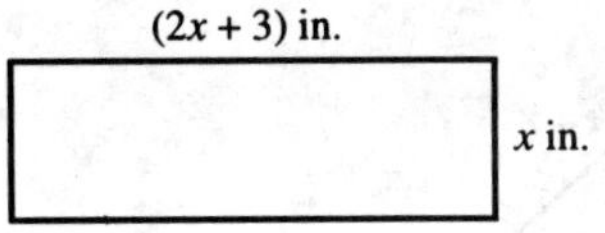

Let x = width, $2x + 3$ = length, and Area = 10 square inches.

$x(2x+3) = 10$

$2x^2 + 3x - 10 = 0$

$x = \frac{-3 \pm \sqrt{9 - 4(2)(-10)}}{4} = \frac{-3 \pm \sqrt{89}}{4}$

$\left(\text{reject } \frac{-3 - \sqrt{89}}{4}\right)$

width: $\frac{-3 + \sqrt{89}}{4}$ in. ≈ 1.6 in.

47.

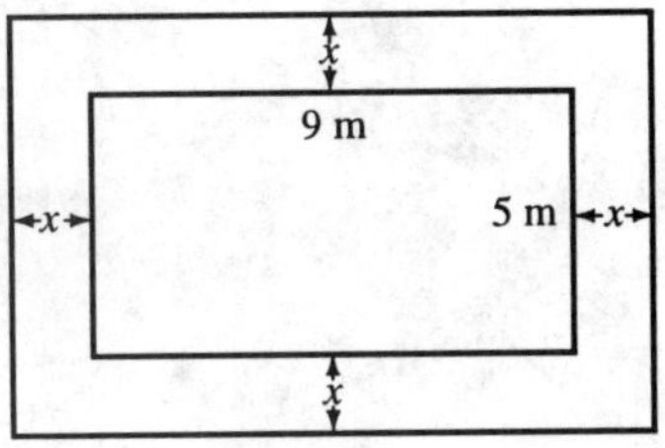

Let x = width of the border,
$5 + 2x$ = width of pool plus border, and
$9 + 2x$ = length of pool plus border.
(area of pool plus border) – (area of pool) = (area of border)

$(5 + 2x)(9 + 2x) - (5)(9) = 40$

$45 + 28x + 4x^2 - 45 = 0$

$4x^2 + 28x - 40 = 0$

$x^2 + 7x - 10 = 0$

$x = \frac{-7 \pm \sqrt{7^2 - 4(1)(-10)}}{2(1)}$

$x = \frac{-7 \pm \sqrt{89}}{2}$

$x = \frac{-7 + \sqrt{89}}{2}$ or $x = \frac{-7 - \sqrt{89}}{2}$ (reject)

$x \approx 1.2$

The border should be $\frac{-7 + \sqrt{89}}{2}$ m or approximately 1.2 m wide.

49.

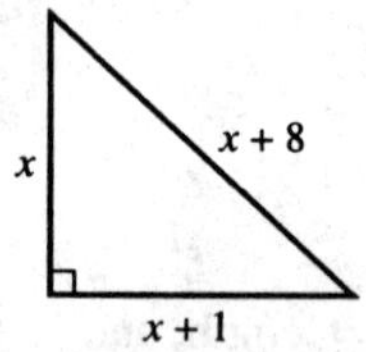

Let x = length of the shortest leg ofa right triangle,
$x + 1$ = length of other leg of right triangle, and
$(x + 1) + 7 = x + 8$ = length of hypotenuse.

$$x^2 + (x+1)^2 = (x+8)^2$$
$$x^2 + x^2 + 2x + 1 = x^2 + 16x + 64$$
$$x^2 - 14x - 63 = 0$$
$$x = \frac{-(-14) \pm \sqrt{(-14)^2 - 4(1)(-63)}}{2(1)}$$
$$x = \frac{14 \pm \sqrt{448}}{2}$$
$$x = 7 \pm 4\sqrt{7} \quad \left(\text{reject } 7 - 4\sqrt{7}\right)$$

The length of the shorter leg is $7 + 4\sqrt{7}$ in. ≈ 17.6 in.

51. Let x = length of original piece of cardboard,
$x - 10$ = width of open top box, and
$x - 10$ = length of open top box.
height: 5 inches
volume: 520 cubic inches

$$V = LWH$$
$$520 = (x - 10)(x - 10)5$$
$$104 = x^2 - 20x + 100$$
$$0 = x^2 - 20x - 4$$
$$x = \frac{20 \pm \sqrt{400 - 4(1)(-4)}}{2} = \frac{20 \pm \sqrt{416}}{2}$$
$$= \frac{20 \pm 4\sqrt{26}}{2} = 10 \pm 2\sqrt{26}$$
$$\left(\text{reject } 10 - 2\sqrt{26}\right)$$

Length of cardboard:
$10 + 2\sqrt{26}$ inches ≈ 20.2 inches

53. Let $PB = x$, $AB = AP + PB = 10 + x$,
and $\frac{AB}{AD} = \frac{BC}{QC}$.

$$\frac{10+x}{10} = \frac{10}{x}$$
$$10x + x^2 = 100$$
$$x^2 + 10x - 100 = 0$$
$$x = \frac{-10 \pm \sqrt{100 - 4(1)(-100)}}{2}$$
$$= \frac{-10 \pm \sqrt{500}}{2} = \frac{-10 \pm 10\sqrt{5}}{2} = -5 \pm 5\sqrt{5}$$
$$\left(\text{reject } -5 - 5\sqrt{5}\right)$$
$$AB = 10 + x = 10 + \left(-5 + 5\sqrt{5}\right) = 5 + 5\sqrt{5}$$

The length of side AB or L is
$5 + 5\sqrt{5}$ ft ≈ 16.2 ft.

55. Let x = number of original people.

$$\left(\frac{480}{x} - 4\right)(x + 4) = 480$$
$$(480 - 4x)(x + 4) = 480x$$
$$480x + 1920 - 4x^2 - 16x = 480x$$
$$4x^2 + 16x - 1920 = 0$$
$$x^2 + 4x - 480 = 0$$
$$(x - 20)(x + 24) = 0$$
$$x = 20 \text{ or } x = -24$$

Disregard the negative number.
There were 20 original people.

57. Let x = number of hours for slower person to complete job alone and
$x - 1$ = number of hours for faster person to complete job alone.

	Fractional part of job completed in 1 hour	Time spent working together	Fractional part of the job completed in 4 hours
slower person	$\frac{1}{x}$	4	$\frac{4}{x}$
faster person	$\frac{1}{x-1}$	4	$\frac{4}{x-1}$

(Fractional part by slower person in 4 hours) + (Fractional part by faster person in 4 hours)
= (one whole job)

$$\frac{4}{x} + \frac{4}{x-1} = 1$$
$$x(x-1)\left[\frac{4}{x} + \frac{4}{x-1}\right] = x(x-1)\cdot 1$$

$4(x-1)+4x=x(x-1)$
$4x-4+4x=x^2-x$
$0=x^2-9x+4$
$x=\dfrac{-(-9)\pm\sqrt{(-9)^2-4(1)(4)}}{2(1)}$
$x=\dfrac{9\pm\sqrt{65}}{2}\ \left(\text{reject } \dfrac{9-\sqrt{65}}{2} \text{ since } x-1=\dfrac{7-\sqrt{65}}{2}<0\right)$
$x=\dfrac{9+\sqrt{65}}{2}$ (slower)
$x-1=\dfrac{7+\sqrt{65}}{2}$ (faster)
slower: $\dfrac{9+\sqrt{65}}{2}$ hr ≈ 8.5 hr
faster: $\dfrac{7+\sqrt{65}}{2}$ hr ≈ 7.5 hr

59. Let x = number of cars dealer purchased.
$\left(1600+\dfrac{64,000}{x}\right)(x-2)=64,000$
$(1600x+64,000)(x-2)=64,000x$
$1600x^2-3200x+64,000x-128,000=64,000x$
$x^2-2x-80=0$
$(x-10)(x+8)=0$
$x=10$ or $x=-8$
Disregard a negative quantity.
The dealer purchased 10 cars.

61. Let x = speed of boat in still water, $x-2$ = speed of boat against current (going upstream), and $x+2$ = speed of boat with current (going downstream).

	Distance (miles)	Rate	Time = $\frac{\text{Distance}}{\text{Rate}}$
upstream	7	$x-2$	$\frac{7}{x-2}$
downstream	7	$x+2$	$\frac{7}{x+2}$

(time upstream) + (time downstream) = (3 hours)
$\dfrac{7}{x-2}+\dfrac{7}{x+2}=3$
$(x-2)(x+2)\left(\dfrac{7}{x-2}+\dfrac{7}{x+2}\right)=(x-2)(x+2)3$
$7(x+2)+7(x-2)=3(x^2-4)$
$7x+14+7x-14=3x^2-12$
$0=3x^2-14x-12$

$x = \frac{-(-14) \pm \sqrt{(-14)^2 - 4(3)(-12)}}{2(3)}$

$x = \frac{14 \pm \sqrt{340}}{6} = \frac{14 \pm 2\sqrt{85}}{6}$

$x = \frac{7+\sqrt{85}}{3}$ $\left(\text{reject } \frac{7-\sqrt{85}}{3}\right)$

The speed of the boat in still water is $\frac{7+\sqrt{85}}{3}$ mph ≈ 5.4 mph.

63. d is true. **65.** Students should verify results. **67.** Answers may vary.

69. Area of triangle – Area of rectangle = shaded region

$\frac{1}{2}[(y+5+y+1+3)(2y)] - y(y+1) = 10$

$\frac{1}{2}(2y+9)(2y) - y(y+1) = 10$

$2y^2 + 9y - y^2 - y = 10$

$y^2 + 8y - 10 = 0$

$y = \frac{-8 \pm \sqrt{64 - 4(1)(-10)}}{2} = \frac{-8 \pm \sqrt{104}}{2}$

$= \frac{-8 \pm 2\sqrt{26}}{2} = -4 \pm \sqrt{26}$ $\left(\text{reject } -4 - \sqrt{26}\right)$

$y = -4 + \sqrt{26}$ yd ≈ 1.1 yd

71. $\sqrt{\frac{x+4}{x-1}} + \sqrt{\frac{x-1}{x+4}} = \frac{5}{2}$

Let $u = \sqrt{\frac{x+4}{x-1}}$.

$u + \frac{1}{u} = \frac{5}{2}$

$2u\left(u + \frac{1}{u}\right) = 2u\left(\frac{5}{2}\right)$

$2u^2 + 2 = 5u$

$2u^2 - 5u + 2 = 0$

$(2u - 1)(u - 2) = 0$

$u = \frac{1}{2}$ or $u = 2$

$\sqrt{\frac{x+4}{x-1}} = \frac{1}{2}$ or $\sqrt{\frac{x+4}{x-1}} = 2$

$\frac{x+4}{x-1} = \frac{1}{4}$ or $\frac{x+4}{x-1} = 4$

$4x + 16 = x - 1$ or $x + 4 = 4x - 4$

$3x = -17$ or $-3x = -8$

$x = -\frac{17}{3}$ or $x = \frac{8}{3}$

$\left\{-\frac{17}{3}, \frac{8}{3}\right\}$

73. $u = \sqrt{x+y},\ v = \sqrt{x-y}$

$x + y + \sqrt{x+y} = 12$

$x - y + \sqrt{x-y} = 6$

$u^2 + u - 12 = 0$

$(u + 4)(u - 3) = 0$

$u = -4$ or $u = 3$

$\sqrt{x+y} = -4$ or $\sqrt{x+y} = 3$

Disregard the negative square root.

$x + y = 9$

$v^2 + v - 6 = 0$

$(v + 3)(v - 2) = 0$

$v = -3$ or $v = 2$

$\sqrt{x-y} = -3$ or $\sqrt{x-y} = 2$

Disregard the negative square root.

$x - y = 4$

$$\begin{aligned} x + y &= 9 \\ \underline{x - y} &\underline{= 4} \\ 2x &= 13 \\ x &= \frac{13}{2} \end{aligned}$$

$\frac{13}{2} + y = 9$

$y = \frac{5}{2}$

$\left\{\left(\frac{13}{2}, \frac{5}{2}\right)\right\}$

Review Problems

75. $2x + 3y = -2$

$x - 4y = 6$

$D = \begin{vmatrix} 2 & 3 \\ 1 & -4 \end{vmatrix} = 2(-4) - (1)(3) = -8 - 3 = -11$

$D_x = \begin{vmatrix} -2 & 3 \\ 6 & -4 \end{vmatrix} = -2(-4) - (3)(6) = 8 - 18$

$= -10$

$D_y = \begin{vmatrix} 2 & -2 \\ 1 & 6 \end{vmatrix} = 2(6) - (1)(-2) = 12 + 2 = 14$

$x = \frac{D_x}{D} = \frac{-10}{-11} = \frac{10}{11}$

$y = \frac{D_y}{D} = \frac{14}{-11} = -\frac{14}{11}$

$\left\{\left(\frac{10}{11}, -\frac{14}{11}\right)\right\}$

76. $3(x + 2) + 3x = 4(2x + 3) + 2$

$3x + 6 + 3x = 8x + 12 + 2$

$6x + 6 = 8x + 14$

$-2x = 8$

$x = -4$

$\{-4\}$

77. $\sqrt[3]{48x^7y^2} = \sqrt[3]{8x^6 \cdot 6xy^2}$

$= \sqrt[3]{(2x^2)^3 \cdot 6xy^2} = 2x^2\sqrt[3]{6xy^2}$

Problem Set 7.5

1. $(x - 4)(x + 2) > 0$

$(x - 4)(x + 2) = 0$

$x = 4$ or $x = -2$

T	F	T
	−2	4

Test −3:

$(-3 - 4)(-3 + 2) > 0$

$(-7)(-1) > 0$

$7 > 0$ True

Test 0:

$(0 - 4)(0 + 2) > 0$

$(-4)(2) > 0$

$-8 > 0$ False

Test 5:

$(5 - 4)(5 + 2) > 0$

$(1)(7) > 0$

$7 > 0$ True

$\{x | x < -2 \text{ or } x > 4\}$, $(-\infty, -2) \cup (4, \infty)$

3. $(x-7)(x+3) \le 0$
$(x-7)(x+3) = 0$
$x = 7,\ x = -3$

F | T | F
$-3 \quad 7$

Test -4:
$(-4-7)(-4+3) \le 0$
$11 \le 0$ False
Test 0:
$(0-7)(0+3) \le 0$
$-21 \le 0$ True
Test 8:
$(8-7)(8+3) \le 0$
$11 \le 0$ False
$\{x \mid -3 \le x \le 7\}$, $[-3, 7]$

5. $x^2 - 5x + 4 > 0$
$x^2 - 5x + 4 = 0$
$(x-4)(x-1) = 0$
$x = 4$ or $x = 1$

T | F | T
$1 \quad 4$

Test 0: $0^2 - 5 \cdot 0 + 4 > 0$
$4 > 0$ True
Test 2: $2^2 - 5 \cdot 2 + 4 > 0$
$-2 > 0$ False
Test 5: $5^2 - 5 \cdot 5 + 4 > 0$
$4 > 0$ True
$\{x \mid x < 1 \text{ or } x > 4\}$, $(-\infty, 1) \cup (4, \infty)$

7. $x^2 + 5x + 4 > 0$
$(x+1)(x+4) = 0$
$x = -1$ or $x = -4$

T | F | T
$-4 \quad -1$

Test -5: $(-5)^2 + 5(-5) + 4 > 0$
$4 > 0$ True
Test -2: $(-2)^2 + 5(-2) + 4 > 0$
$-2 > 0$ False
Test 0: $0 + 4 > 0$
$4 > 0$ True
$\{x \mid x < -4 \text{ or } x > -1\}$, $(-\infty, -4) \cup (-1, \infty)$

9. $x^2 - 6x + 9 < 0$
$(x-3)^2 < 0$
$(x-3)^2$ is not less than 0 for any value of x.

F | F
3

$\varnothing$

11. $x^2 - 6x + 8 \le 0$
$(x-2)(x-4) \le 0$
$x = 2$ or $x = 4$

F | T | F
$2 \quad 4$

Test 0: $0^2 - 6 \cdot 0 + 8 \le 0$
$8 \le 0$ False
Test 3: $3^2 - 6 \cdot 3 + 8 \le 0$
$-1 \le 0$ True
Test 5: $5^2 - 6 \cdot 5 + 8 \le 0$
$3 \le 0$ False
$\{x \mid 2 \le x \le 4\}$, $[2, 4]$

13. $3x^2 + 10x - 8 \le 0$
$(3x-2)(x+4) \le 0$
$x = \frac{2}{3}$ or $x = -4$

F | T | F
$-4 \quad \frac{2}{3}$

Test −5: $3(-5)^2 + 10(-5) - 8 \le 0$
$17 \le 0$ False
Test 0: $0 - 8 \le 0$
$-8 \le 0$ True
Test 1: $3(1) + 10(1) - 8 \le 0$
$5 \le 0$ False
$\left\{x \middle| -4 \le x \le \frac{2}{3}\right\}, \left[-4, \frac{2}{3}\right]$

15. $2x^2 + x < 15$
$2x^2 + x - 15 < 0$
$(2x - 5)(x + 3) < 0$
$x = \frac{5}{2}$ or $x = -3$

F	T	F
	−3	$\frac{5}{2}$

Test −4: $2(16) + (-4) - 15 < 0$
$13 < 0$ False
Test 0: $0 - 15 < 0$
$-15 < 0$ True
Test 3: $2(9) + 3 - 15 < 0$
$6 < 0$ False
$\left\{x \middle| -3 < x < \frac{5}{2}\right\}, \left(-3, \frac{5}{2}\right)$

17. $4x^2 + 7x < -3$
$4x^2 + 7x + 3 < 0$
$(4x + 3)(x + 1) < 0$
$x = -\frac{3}{4}$ or $x = -1$

F	T	F
	−1	$-\frac{3}{4}$

Test −2: $4(4) + 7(-2) + 3 < 0$
$5 < 0$ False
Test $-\frac{7}{8}$: $4\left(\frac{49}{64}\right) + 7\left(-\frac{7}{8}\right) + 3 < 0$
$-0.0625 < 0$ True
Test 0: $0 + 3 < 0$
$3 < 0$ False
$\left\{x \middle| -1 < x < -\frac{3}{4}\right\}, \left(-1, -\frac{3}{4}\right)$

19. $5x \le 2 - 3x^2$
$3x^2 + 5x - 2 \le 0$
$(3x - 1)(x + 2) \le 0$
$x = \frac{1}{3}$ or $x = -2$

F	T	F
	−2	$\frac{1}{3}$

Test 0 (middle interval):
$5 \cdot 0 \le 2 - 3 \cdot 0^2$
$0 \le 2$ True
$\left\{x \middle| -2 \le x \le \frac{1}{3}\right\}, \left[-2, \frac{1}{3}\right]$

21. $x^2 - 4x \ge 0$
$x(x - 4) \ge 0$
$x = 0$ or $x = 4$

T	F	T
	0	4

Test 1 (middle interval):
$1^2 - 4(1) \ge 0$
$-3 \ge 0$ False
$\{x | x \le 0 \text{ or } x \ge 4\}, (-\infty, 0] \cup [4, \infty)$

23. $2x^2 + 3x > 0$
$x(2x + 3) > 0$
$x = 0$ or $x = -\frac{3}{2}$

T	F	T
	$-\frac{3}{2}$	0

Test −1 (middle interval):
$2(1) + 3(-1) > 0$
$-1 > 0$ False
$\left\{x \middle| x < -\frac{3}{2} \text{ or } x > 0\right\}, \left(-\infty, -\frac{3}{2}\right) \cup (0, \infty)$

25. $-x^2 + x \geq 0$

$x^2 - x \leq 0$

$x(x-1) \leq 0$

$x = 0$ or $x = 1$

F | T | F

0   1

Test $\frac{1}{2}$ (middle interval):

$-\frac{1}{4} + \frac{1}{2} \geq 0$

$\frac{1}{4} \geq 0$ True

$\{x \mid 0 \leq x \leq 1\}$, $[0, 1]$

27. $\frac{x-4}{x+3} > 0$

points where the quotient is zero or undefined:

$x - 4 = 0$ or $x + 3 = 0$

$x = 4$ or $x = -3$

T | F | T

−3   4

Test −4: $\frac{-4-4}{-4+3} > 0$

$\frac{-8}{-1} > 0$

$8 > 0$ True

Test 0: $\frac{0-4}{0+3} > 0$

$\frac{-4}{3} > 0$ False

Test 5: $\frac{5-4}{5+3} > 0$

$\frac{1}{8} > 0$ True

$\{x \mid x < -3 \text{ or } x > 4\}$, $(-\infty, -3) \cup (4, \infty)$

29. $\frac{x+3}{x+4} < 0$

$x + 3 = 0$ or $x + 4 = 0$

$x = -3$ or $x = -4$

F | T | F

−4   −3

Test −3.5 (middle interval):

$\frac{-0.5}{0.5} < 0$

$-1 < 0$ True

$\{x \mid -4 < x < -3\}$, $(-4, -3)$

31. $\frac{-x+2}{x-4} \geq 0$

$-x + 2 = 0$ or $x - 4 = 0$

$x = 2$ or $x = 4$

F | T | F

2   4

Test 0: $\frac{0+2}{0-4} \geq 0$

$-\frac{1}{2} \geq 0$ False

Test 3: $\frac{-3+2}{3-4} \geq 0$

$1 \geq 0$ True

Test 5: $\frac{-5+2}{5-4} \geq 0$

$-3 \geq 0$ False

$\{x \mid 2 \leq x < 4\}$, $[2, 4)$

33. $\frac{4-2x}{3x+4} \leq 0$

$4 - 2x = 0$ or $3x + 4 = 0$

$x = 2$ or $x = -\frac{4}{3}$

T | F | T

$-\frac{4}{3}$ 2

Test −2: $\frac{4-2(-2)}{3(-2)+4} \leq 0$

$-4 \leq 0$ True

Test 0: $\frac{4-2(0)}{3(0)+4} \le 0$

$1 \le 0$ False

Test 3: $\frac{4-2(3)}{3(3)+4} \le 0$

$-\frac{2}{13} \le 0$ True

$\left\{x \middle| x < -\frac{4}{3} \text{ or } x \ge 2\right\}, \left(-\infty, -\frac{4}{3}\right) \cup [2, \infty)$

35. $\frac{x}{x-3} > 0$

$x = 0$ or $x = 3$

T	F	T
	0	3

Test −1: $\frac{-1}{-1-3} > 0$

$\frac{1}{4} > 0$ True

Test 1: $\frac{1}{1-3} > 0$

$-\frac{1}{2} > 0$ False

Test 4: $\frac{4}{4-3} > 0$

$4 > 0$ True

$\{x | x < 0 \text{ or } x > 3\}, (-\infty, 0) \cup (3, \infty)$

37. $\frac{x+1}{x+3} < 2$

$\frac{x+1}{x+3} - 2 < 0$

$\frac{x+1}{x+3} - 2 \cdot \frac{x+3}{x+3} < 0$

$\frac{x+1-2x-6}{x+3} < 0$

$\frac{-x-5}{x+3} < 0$

$-x - 5 = 0$ or $x + 3 = 0$

$x = -5$ or $x = -3$

T	F	T
	−5	−3

Test −6: $\frac{-6+1}{-6+3} < 2$

$\frac{-5}{-3} < 2$

$\frac{5}{3} < 2$ True

Test −4: $\frac{-4+1}{-4+3} < 2$

$\frac{-3}{-1} < 2$

$3 < 2$ False

Test 0: $\frac{0+1}{0+3} < 2$

$\frac{1}{3} < 2$ True

$\{x | x < -5 \text{ or } x > -3\}, (-\infty, -5) \cup (-3, \infty)$

39. $\frac{x+4}{2x-1} \le 3$

$\frac{x+4}{2x-1} - 3 \cdot \frac{(2x-1)}{(2x-1)} \le 0$

$\frac{-5x+7}{2x-1} \le 0$

$-5x + 7 = 0$ or $2x - 1 = 0$

$x = \frac{7}{5}$ or $x = \frac{1}{2}$

T	F	T
	$\frac{1}{2}$	$\frac{7}{5}$

Test 1 (middle interval):

$\frac{1+4}{2-1} \le 3$

$5 \le 3$ False

$\left\{x \middle| x < \frac{1}{2} \text{ or } x \ge \frac{7}{5}\right\}, \left(-\infty, \frac{1}{2}\right) \cup \left[\frac{7}{5}, \infty\right)$

41. $\frac{x-2}{x+2} \le 2$

$\frac{x-2}{x+2} - 2 \cdot \frac{x+2}{x+2} \le 0$

$\frac{x-2-2x-4}{x+2} \le 0$

$\frac{-x-6}{x+2} \le 0$

$-x-6=0$ or $x+2=0$

$x=-6$ or $x=-2$

T	F	T

$-6 \quad -2$

Test -4 (middle interval):

$\frac{-4-2}{-4+2} \le 2$

$\frac{-6}{-2} \le 2$

$3 \le 2$ False

$\{x|x \le -6 \text{ or } x > -2\}$, $(-\infty, -6] \cup (-2, \infty)$

43. $s = -16t^2 + 80t$

$-16t^2 + 80t > 64$

$-16t^2 + 80t - 64 > 0$

$t^2 - 5t + 4 < 0$

$(t-4)(t-1) < 0$

$t = 4$ or $t = 1$

F	T	F

$1 \quad 4$

Test 0: $-16(0)^2 + 80(0) > 64$

$0 > 64$ False

Test 2: $-16(2)^2 + 80(2) > 64$

$96 > 64$ True

Test 5: $-16(5)^2 + 80(5) > 64$

$0 > 64$ False

$\{x|1 < x < 4\}$

It will be more than 64 feet above the ground between 1 and 4 seconds.

45. $N = 0.4x^2 - 36x + 1000$

$0.4x^2 - 36x + 1000 > 250$

$0.4x^2 - 36x + 750 > 0$

$x^2 - 90x + 1875 > 0$

$x = 45 \pm 5\sqrt{6}$

T	F	T

$45-5\sqrt{6} \quad 45+5\sqrt{6}$

Test 30: $(30)^2 - 90(30) + 1875 > 0$

$75 > 0$ True

Test 50: $(50)^2 - 90(50) + 1875 > 0$

$-125 > 0$ False

Test 60: $(60)^2 - 90(60) + 1875 > 0$

$75 > 0$ True

$\{x|x < 45-5\sqrt{6} \text{ or } x > 45+5\sqrt{6}\}$

For this problem, $16 \le x \le 74$. The age range is from 16 to about $32\frac{1}{2}$ years and about $57\frac{1}{2}$ to 74 years.

47. c is true; $f(x) = \sqrt{x^2 - 6x + 10}$

Domain of $f(x) = x^2 - 6x + 10 \ge 0$

$b^2 - 4ac = 36 - 4(1)(10) = -4 < 0$

Thus $x^2 - 6x + 10$ is greater than zero for all real numbers.

Domain of $f(x) = (-\infty, \infty)$

49. Students should verify results.

51. $x^2 + 3x - 10 > 0$

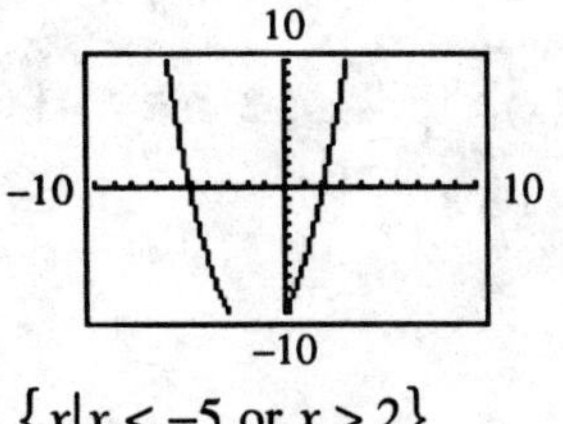

$\{x|x < -5 \text{ or } x > 2\}$

53. $\frac{x-4}{x-1} \le 0$

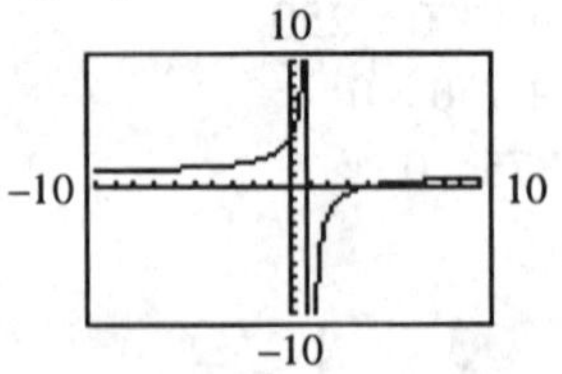

$\{x | 1 < x \le 4\}$

55. $\frac{1}{x+1} \le \frac{2}{x+4}$

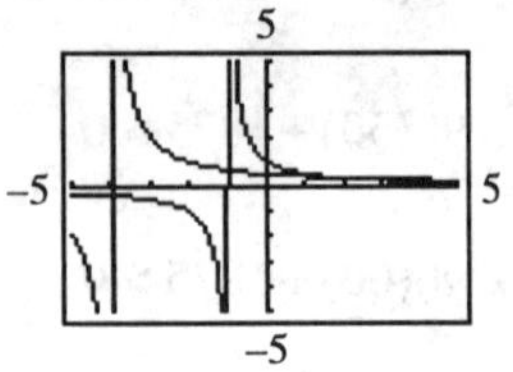

$\{x | -4 < x < -1 \text{ or } x \ge 2\}$

57. Answers may vary.

59. Answers may vary.

61. $f(x) = \sqrt{x^2 - x - 12}$

Domain:

$x^2 - x - 12 \ge 0$

$(x-4)(x+3) \ge 0$

$x = 4$ or $x = -3$

T	F	T
	−3	4

Test 0: $0 - 12 \ge 0$

$-12 \ge 0$ False

$x \le -3$ or $x \ge 4$

Domain of $f(x) = \{x | x \le -3 \text{ or } x \ge 4\}$

$= (-\infty, -3] \cup [4, \infty)$

63. $\frac{x^2 - x - 2}{x^2 - 4x + 3} > 0$

$\frac{(x-2)(x+1)}{(x-3)(x-1)} > 0$

$x = -1, x = 1, x = 2, x = 3$

T	F	T	F	T
	−1	1	2	3

Test −2: $\frac{(-2-2)(-2+1)}{(-2-3)(-2-1)} > 0$

$\frac{(-4)(-1)}{(-5)(-3)} > 0$

$\frac{4}{15} > 0$ True

Test 0: $\frac{(0-2)(0+1)}{(0-3)(0-1)}$

$-\frac{2}{3} > 0$ False

Test $1\frac{1}{2}$: $\frac{\left(1\frac{1}{2}-2\right)\left(1\frac{1}{2}+1\right)}{\left(1\frac{1}{2}-3\right)\left(1\frac{1}{2}-1\right)} > 0$

$\frac{\left(-\frac{1}{2}\right)\left(\frac{5}{2}\right)}{\left(-\frac{3}{2}\right)\left(\frac{1}{2}\right)} > 0$

$\frac{5}{3} > 0$ True

Test $2\frac{1}{2}$: $\frac{\left(\frac{1}{2}\right)\left(\frac{7}{2}\right)}{\left(-\frac{1}{2}\right)\left(\frac{3}{2}\right)} > 0$

$-\frac{7}{3} > 0$ False

Test $3\frac{1}{2}$: $\frac{\left(\frac{3}{2}\right)\left(\frac{9}{2}\right)}{\left(\frac{1}{2}\right)\left(\frac{5}{2}\right)} > 0$

$\frac{27}{5} > 0$ True

$x < -1$ or $1 < x < 2$ or $x > 3$

$\{x | x < -1 \text{ or } 1 < x < 2 \text{ or } x > 3\}$,

$(-\infty, -1) \cup (1, 2) \cup (3, \infty)$

65. $x^3 + 5x^2 - 4x - 20 \ge 0$
$x^2(x+5) - 4(x+5) \ge 0$
$(x+5)(x^2 - 4) \ge 0$
$(x+5)(x+2)(x-2) \ge 0$
$x = -5,\ x = -2,\ x = 2$

F	T	F	T
	−5	−2	2

using $(x+5)(x+2)(x-2) \ge 0$:
Test −6: $(-1)(-4)(-8) \ge 0$
$-32 \ge 0$ False
Test −3: $(2)(-1)(-5) \ge 0$
$10 \ge 0$ True
Test 0: $(5)(2)(-2) \ge 0$
$-20 \ge 0$ False
Test 3: $(8)(5)(1) \ge 0$
$40 \ge 0$ True
$-5 \le x \le -2$ or $x \ge 2$
$\{x \mid -5 \le x \le -2 \text{ or } x \ge 2\}$, $[-5, -2] \cup [2, \infty)$

Review Problems

67. $1 - 2x \ge 5 - x$
$-2x + x \ge 5 - 1$
$-x \ge 4$
$x \le -4$
$\{x \mid x \le -4\}$

68. $\sqrt{5x-1} - \sqrt{x+2} = 1$
$\sqrt{5x-1} = \sqrt{x+2} + 1$
$\left(\sqrt{5x-1}\right)^2 = \left(\sqrt{x+2} + 1\right)^2$
$5x - 1 = x + 2 + 2\sqrt{x+2} + 1$
$2x - 2 = \sqrt{x+2}$
$(2x-2)^2 = \left(\sqrt{x+2}\right)^2$
$4x^2 - 8x + 4 = x + 2$
$4x^2 - 9x + 2 = 0$
$(4x-1)(x-2) = 0$
$x = \frac{1}{4}$ or $x = 2$

Check $\frac{1}{4}$:
$\sqrt{5\left(\frac{1}{4}\right) - 1} - \sqrt{\frac{1}{4} + 2} = 1$
$\frac{1}{2} - \frac{3}{2} = 1$
$-1 = 1$ False
Check 2:
$\sqrt{5(2) - 1} - \sqrt{2+2} = 1$
$\sqrt{9} - \sqrt{4} = 1$
$3 - 2 = 1$
$1 = 1$ True
$\{2\}$

69. $A = \{3, 7, 8, 9\}$
$B = \{8, 9, 10\}$
$A \cup B = \{3, 7, 8, 9, 10\}$
$A \cap B = \{8, 9\}$

Chapter 7 Review Problems

1. $x^2 - 50 = 0$
$x^2 = 50$
$x = \pm\sqrt{50}$
$x = \pm 5i\sqrt{2}$
$\{-5i\sqrt{2}, 5i\sqrt{2}\}$

2. $2x^2 - 3 = 0$
$2x^2 = 3$
$x^2 = \frac{3}{2}$
$x = \pm\sqrt{\frac{3}{2}}$
$\left\{-\sqrt{\frac{3}{2}}, \sqrt{\frac{3}{2}}\right\}$

3. $(2x-3)^2 = 32$
$2x - 3 = \pm\sqrt{32}$
$2x = 3 \pm \sqrt{32}$
$x = \frac{3 \pm 4\sqrt{2}}{2}$
$\left\{\frac{3 - 4\sqrt{2}}{2}, \frac{3 + 4\sqrt{2}}{2}\right\}$

4. $(x-4)^2 = -36$
$x - 4 = \pm\sqrt{-36}$
$x = 4 \pm 6i$
$\{4-6i, 4+6i\}$

5. $a^2 + b^2 = c^2$
$a^2 = c^2 - b^2$
$a = \sqrt{c^2 - b^2}$

6. $A = P(1+r)^2$
$\frac{A}{P} = (1+r)^2$
$\sqrt{\frac{A}{P}} = 1 + r$
$r = \sqrt{\frac{A}{P}} - 1$

7. $s = 16t^2$
$1046 = 16t^2$
$\sqrt{\frac{1046}{16}} = t$
$t = \frac{\sqrt{1046}}{4} \approx 8.1$ seconds

8. $(12)^2 = x^2 + 4^2$
$x = \sqrt{144-16}$
$x = \sqrt{128}$
$x = 8\sqrt{2} \approx 11.3$ m

9. points (3, 7) and (–4, 6)

a. $d = \sqrt{(-4-3)^2 + (6-7)^2}$
$= \sqrt{(-7)^2 + (-1)^2} = \sqrt{49+1} = \sqrt{50}$
$= 5\sqrt{2}$

b. midpoint $= \left(\frac{3-4}{2}, \frac{7+6}{2}\right) = \left(-\frac{1}{2}, \frac{13}{2}\right)$

10. $x^2 + 20x$
$\left(\frac{20}{2}\right)^2 = (10)^2 = 100$
$x^2 + 20x + 100 = (x+10)^2$

11. $x^2 - 3x$
$\left(-\frac{3}{2}\right)^2 = \frac{9}{4}$
$x^2 - 3x + \frac{9}{4} = \left(x - \frac{3}{2}\right)^2$

12. $(x+3)^2 = x^2 + 6x + 9$
The area is 9.

13. $x^2 - 7x - 1 = 0$
$x^2 - 7x = 1$
$\left(-\frac{7}{2}\right)^2 = \frac{49}{4}$
$x^2 - 7x + \frac{49}{4} = 1 + \frac{49}{4}$
$\left(x - \frac{7}{2}\right)^2 = \frac{53}{4}$
$x - \frac{7}{2} = \pm\sqrt{\frac{53}{4}}$
$x = \frac{7}{2} \pm \frac{\sqrt{53}}{2}$
$\left\{\frac{7-\sqrt{53}}{2}, \frac{7+\sqrt{53}}{2}\right\}$

14. $2x^2 + 3x - 4 = 0$
$2x^2 + 3x = 4$
$x^2 + \frac{3}{2}x = 2$
$\left(\frac{\frac{3}{2}}{2}\right)^2 = \left(\frac{3}{4}\right)^2 = \frac{9}{16}$
$x^2 + \frac{3}{2}x + \frac{9}{16} = 2 + \frac{9}{16}$
$\left(x + \frac{3}{4}\right)^2 = \frac{41}{16}$
$x + \frac{3}{4} = \pm\sqrt{\frac{41}{16}}$
$x = -\frac{3}{4} \pm \frac{\sqrt{41}}{4}$
$\left\{\frac{-3-\sqrt{41}}{4}, \frac{-3+\sqrt{41}}{4}\right\}$

15. $y = x^2 - 2x - 8$

x-intercepts:

$x^2 - 2x - 8 = 0$

$(x - 4)(x + 2) = 0$

$x = 4$ or $x = -2$

$(4, 0), (-2, 0)$

y-intercept:

$y = (0)^2 - 2(0) - 8 = -8$

$(0, -8)$

Vertex:

$x = -\frac{b}{2a} = -\frac{(-2)}{2(1)} = 1$

$y = 1^2 - 2(1) - 8 = -9$

$(1, -9)$

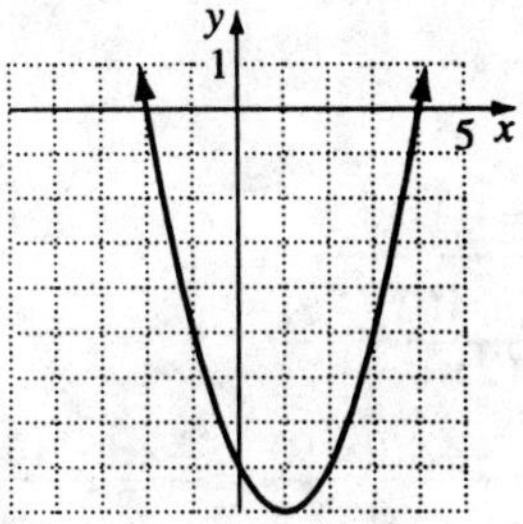

16. $f(x) = -2x^2 - 4x + 1$

x-intercepts:

$-2x^2 - 4x + 1 = 0$

$x = \frac{4 \pm \sqrt{(-4)^2 - 4(-2)(1)}}{2(-2)}$

$= \frac{4 \pm 2\sqrt{6}}{-4} = \frac{2 \pm \sqrt{6}}{-2}$

$\left(\frac{-2 - \sqrt{6}}{2}, 0\right), \left(\frac{-2 + \sqrt{6}}{2}, 0\right)$

y-intercept:

$y = -2(0)^2 - 4(0) + 1 = 1$

$(0, 1)$

Vertex:

$x = -\frac{b}{2a} = -\frac{(-4)}{2(-2)} = -1$

$y = (-2)(-1)^2 - 4(-1) + 1 = 3$

$(-1, 3)$

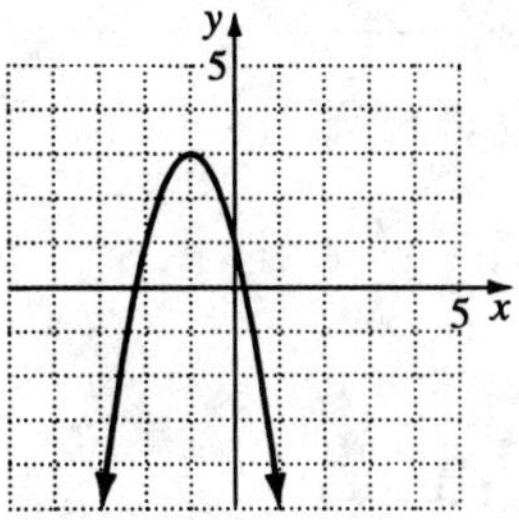

17. $y = (x - 1)^2 + 3$

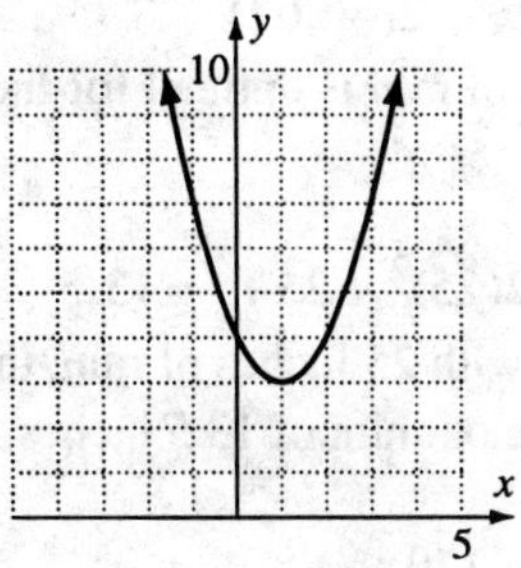

18. $f(x) = -(x + 1)^2 + 4$

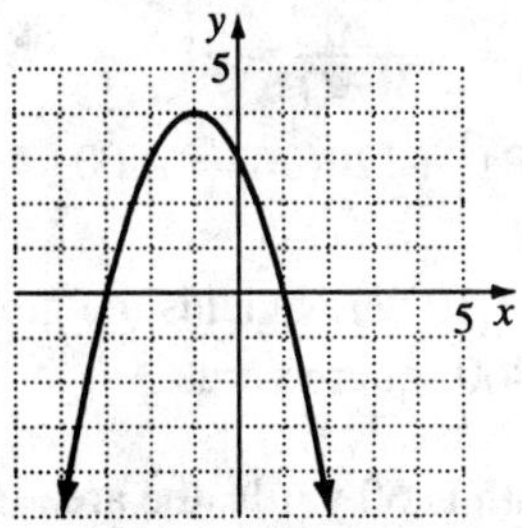

19. $y = x^2 - 2x - 2$

$y + 2 = x^2 - 2x$

$\left(-\frac{2}{2}\right)^2 = (-1)^2 = 1$

$y + 2 + 1 = x^2 - 2x + 1$

$y + 3 = (x - 1)^2$

$y = (x-1)^2 - 3$

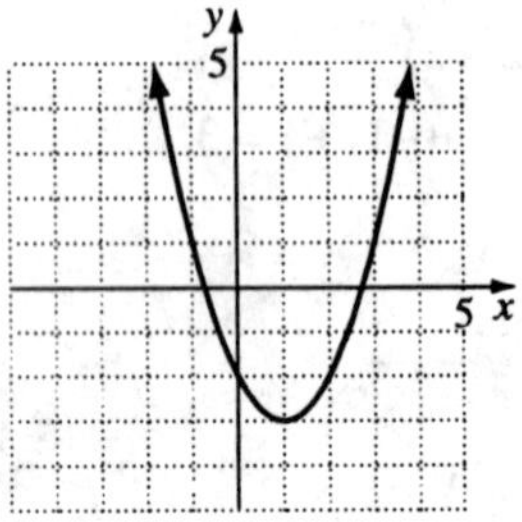

20. $y = 2(x+2)^2 - 1$

21. $f(x) = -0.02x^2 + x + 1$

a. $x = -\frac{b}{2a} = -\frac{1}{2(-0.02)} = 25$
25 inches of rain is optimal for the growth.

b. $y = -0.02(25)^2 + 25 + 1 = 13.5$
In a year with 25 inches of rain, the tree grows a maximum of 13.5 inches.

22. $A(x) = -2x^2 + 120x$

a. Vertex:
$x = -\frac{b}{2a} = -\frac{120}{2(-2)} = 30$
$y = -2(30)^2 + 120(30) = 1800$
(30, 1800)
A width of 30 yards yields the greatest area of 1800 square yards.

b. If the width is 60 yards, the area is 0 square yards. The fencing runs in two parallel lines right next to each other in a 60 yard line perpendicular to the river.

23. $x^2 = 2x + 4$
$x^2 - 2x - 4 = 0$
$$x = \frac{2 \pm \sqrt{(-2)^2 - 4(1)(-4)}}{2}$$
$$= \frac{2 \pm \sqrt{20}}{2} = \frac{2 \pm 2\sqrt{5}}{2} = 1 \pm \sqrt{5}$$
$\{1-\sqrt{5}, 1+\sqrt{5}\}$

24. $x^2 - 2x + 19 = 0$
$$x = \frac{2 \pm \sqrt{(-2)^2 - 4(1)(19)}}{2}$$
$$= \frac{2 \pm \sqrt{-72}}{2} = \frac{2 \pm 6i\sqrt{2}}{2} = 1 \pm 3i\sqrt{2}$$
$\{1-3i\sqrt{2}, 1+3i\sqrt{2}\}$

25. $2x^2 = 3 - 4x$
$2x^2 + 4x - 3 = 0$
$$x = \frac{-4 \pm \sqrt{4^2 - 4(2)(-3)}}{2(2)}$$
$$= \frac{-4 \pm \sqrt{40}}{4} = \frac{-4 \pm 2\sqrt{10}}{4} = \frac{-2 \pm \sqrt{10}}{2}$$
$\left\{\frac{-2-\sqrt{10}}{2}, \frac{-2+\sqrt{10}}{2}\right\}$

26. $2x^2 - 3x - 1 = 0$
$$x = \frac{3 \pm \sqrt{(-3)^2 - 4(2)(-1)}}{2(2)}$$
$$= \frac{3 \pm \sqrt{17}}{4}$$
$\left\{\frac{3-\sqrt{17}}{4}, \frac{3+\sqrt{17}}{4}\right\}$

27. $\frac{3}{x}+\frac{10}{x+6}=1$

$x(x+6)\left(\frac{3}{x}+\frac{10}{x+6}\right)=x(x+6)(1)$

$3x+18+10x=x^2+6x$

$x^2-7x-18=0$

$(x-9)(x+2)=0$

$x-9=0$ or $x+2=0$

$x=9$ or $x=-2$

$\{-2, 9\}$

28. $(5x-2)^2-9=0$

$[(5x-2)-3][(5x-2)+3]=0$

$(5x-5)(5x+1)=0$

$5x-5=0$ or $5x+1=0$

$x=1$ or $x=-\frac{1}{5}$

$\left\{-\frac{1}{5}, 1\right\}$

29. $3x^2+2x=4$

$3x^2+2x-4=0$

$x=\frac{-2\pm\sqrt{2^2-4(3)(-4)}}{2(3)}$

$=\frac{-2\pm\sqrt{52}}{6}=\frac{-2\pm2\sqrt{13}}{6}=\frac{-1\pm\sqrt{13}}{3}$

$\left\{\frac{-1-\sqrt{13}}{3}, \frac{-1+\sqrt{13}}{3}\right\}$

30. $\frac{5}{x+1}+\frac{x-1}{4}=2$

$4(x+1)\left(\frac{5}{x+1}+\frac{x-1}{4}\right)=2(4)(x+1)$

$20+(x+1)(x-1)=8(x+1)$

$20+x^2-1=8x+8$

$x^2-8x+11=0$

$x=\frac{8\pm\sqrt{(-8)^2-4(1)(11)}}{2}=\frac{8\pm\sqrt{20}}{2}$

$=\frac{8\pm2\sqrt{5}}{2}=4\pm\sqrt{5}$

$\{4-\sqrt{5}, 4+\sqrt{5}\}$

31. $x(x-2)=-5$

$x^2-2x+5=0$

$x=\frac{2\pm\sqrt{(-2)^2-4(1)(5)}}{2}=\frac{2\pm\sqrt{-16}}{2}$

$=\frac{2\pm4i}{2}=1\pm2i$

$\{1-2i, 1+2i\}$

32. $x^2-3x+7=0$

$b^2-4ac=(-3)^2-4(1)(7)=-19$

Solutions not real, imaginary

33. $2x^2+5x-3=0$

$b^2-4ac=5^2-4(2)(-3)=49$

Solutions rational

34. $2x^2+4x-3=0$

$b^2-4ac=4^2-4(2)(-3)=40$

Solutions irrational

35. $R=-2x^2+36x$

$19=-2x^2+36x$

$2x^2-36x+19=0$

$b^2-4ac=(-36)^2-4(2)(19)$

$=1144$

Since it is a positive number, there are real solutions, so the situation is possible and the applicant will be hired.

36. $N = 0.337x^2 - 2.265x + 3.962$

$10.9 = 0.337x^2 - 2.265x + 3.962$

$0 = 0.337x^2 - 2.265x - 6.938$

$$x = \frac{2.265 \pm \sqrt{(-2.265)^2 - 4(0.337)(-6.938)}}{2(0.337)}$$

$$\approx \frac{2.265 \pm 3.806}{0.674} \approx -2.3,\ 9.0$$

$1980 + 9 = 1989$

In year 1989

37. The graph is bell-shaped with one peak.

38. a. $s = at^2 + bt + c$

$80 = a(0)^2 + b(0) + c$

$128 = a(1)^2 + b(1) + c$

$144 = a(2)^2 + b(2) + c$

or

$c = 80$

$a + b + c = 128$

$4a + 2b + c = 144$

Substitute $c = 80$ into equations.

$a + b = 48$

$4a + 2b = 64$

Add first equation to $-\frac{1}{2}$ times the second equation.

$$\begin{aligned} a + b &= 48 \\ \underline{-2a - b} &\underline{= -32} \\ -a &= 16 \\ a &= -16 \end{aligned}$$

Substitute back.

$-16 + b + 80 = 128$

$b = 64$

$s = -16t^2 + 64t + 80$

b. $75 = -16t^2 + 64t + 80$

$16t^2 - 64t - 5 = 0$

$t = \frac{64 \pm \sqrt{(-64)^2 - 4(16)(-5)}}{2(16)}$

$= \frac{64 \pm \sqrt{4416}}{32}$

$\approx -0.08, 4.1$

Disregard a negative time.

After about 4.1 seconds

c. $0 = -16t^2 + 64t + 80$

$t^2 - 4t - 5 = 0$

$(t - 5)(t + 1) = 0$

$t = 5$ or $t = -1$

Disregard a negative time.

After 5 seconds

d. $s = -16t^2 + 64t + 80$

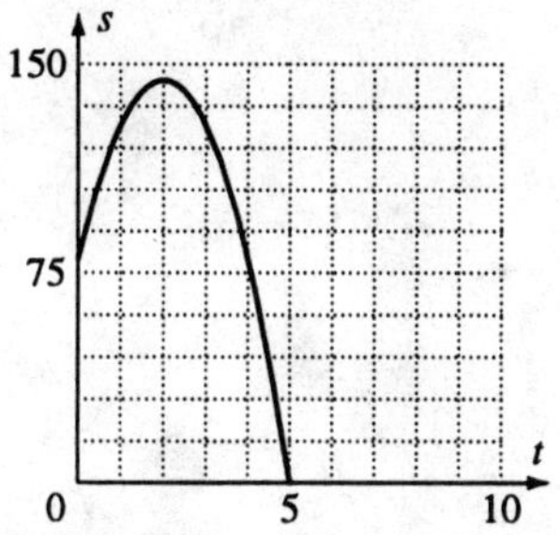

39. Let x = width of border, $12 + 2x$ = width of pool and border, and $20 + 2x$ = length of pool and border.

area of (pool + border) = area of pool and area of border

$(12 + 2x)(20 + 2x) = 12(20) + 160$

$240 + 64x + 4x^2 = 240 + 160$

$4x^2 + 64x - 160 = 0$

$x^2 + 16x - 40 = 0$

$x = \frac{-16 \pm \sqrt{16^2 - 4(-40)}}{2}$

$x = \frac{-16 \pm \sqrt{416}}{2}$

$x = \frac{-16 \pm 4\sqrt{26}}{2} = -8 \pm 2\sqrt{26}$

$x = -8 + 2\sqrt{26}$ $\left(\text{reject } -8 - 2\sqrt{26}\right)$

The width of the border is $-8 + 2\sqrt{26}$ m ≈ 2.2 m

40.

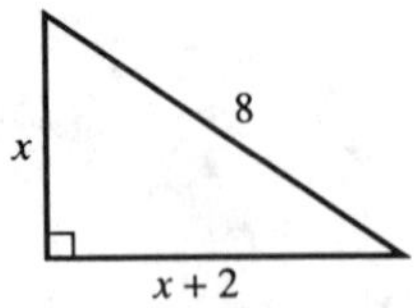

Let x = length of shorter leg of right triangle and $x + 2$ = length of longer leg.

$x^2 + (x+2)^2 = 8^2$

$x^2 + x^2 + 4x + 4 = 64$

$2x^2 + 4x - 60 = 0$

$x^2 + 2x - 30 = 0$

$x = \dfrac{-2 \pm \sqrt{4 - 4(-30)}}{2} = \dfrac{-2 \pm \sqrt{124}}{2}$

$x = \dfrac{-2 \pm 2\sqrt{31}}{2} = -1 \pm \sqrt{31}$

$x = -1 + \sqrt{31}$ $\left(\text{reject } -1 - \sqrt{31}\right)$

The length of the shorter leg is $-1 + \sqrt{31}$ cm ≈ 4.6 cm.

41.

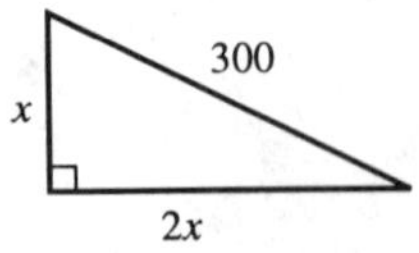

Let x = height of building and
$2x$ = length of shadow.

$x^2 + (2x)^2 = 300^2$

$5x^2 = 90{,}000$

$x^2 = 18{,}000$

$x = \pm 60\sqrt{5}$

$x = 60\sqrt{5}$ $\left(\text{reject } -60\sqrt{5}\right)$

The building is $60\sqrt{5}$ m ≈ 134.2 m high.

42. Let x = number of hours for slower person to complete the job alone and
$x - 1$ = number of hours for faster person to complete the job alone.

	Fractional part of job completed in 1 hour	Time spent working together	Fractional part of job completed in 2 hours
slower person	$\frac{1}{x}$	2	$\frac{2}{x}$
faster person	$\frac{1}{x-1}$	2	$\frac{2}{x-1}$

(Fractional part for slower person in 2 hours) + (Fractional part for faster person in 2 hours)
= (one complete job)

$\frac{2}{x}+\frac{2}{x-1}=1$

$x(x-1)\left[\frac{2}{x}+\frac{2}{x-1}\right]=x(x-1)\cdot 1$

$2(x-1)+2x=x(x-1)$

$2x-2+2x=x^2-x$

$0=x^2-5x+2$

$x=\frac{-(-5)\pm\sqrt{25-8}}{2}=\frac{5\pm\sqrt{17}}{2}$

$x=\frac{5+\sqrt{17}}{2}\quad\left(\text{reject } \frac{5-\sqrt{17}}{2}\right)$

$x-1=\frac{5+\sqrt{17}}{2}-1=\frac{3+\sqrt{17}}{2}$

slower person: $\frac{5+\sqrt{17}}{2}\approx 4.6$ hr

faster person: $\frac{3+\sqrt{17}}{2}\approx 3.6$ hr

43. Let x = width and $3x$ = length.
$(x-1)(3x+3)=72$
$3x^2-3=72$
$3x^2=75$
$x^2=25$
$x=\pm\sqrt{25}$
$x=5$ or $x=-5$ (reject)
$3x=3(5)=15$
Width: 5 yd; Length: 15 yd

44. Let x = original number of people.
$\left(\frac{12{,}000}{x}-1600\right)(x+2)=12{,}000$
$(12{,}000-1600x)(x+2)=12{,}000x$
$12{,}000x+24{,}000-1600x^2-3200x=12{,}000x$
$-1600x^2-3200x+24{,}000=0$
$x^2+2x-15=0$
$(x-3)(x+5)=0$
$x=3$ or $x=-5$
Disregard a negative number of people. There were 3 original people.

45. $3x(3x+10)-x(x+4)=150$
$9x^2+30x-x^2-4x=150$
$8x^2+26x-150=0$
$4x^2+13x-75=0$
$(4x+25)(x-3)=0$
$x=-\frac{25}{4}$ or $x=3$
Disregard a negative length.
$3x=3(3)=9$, $3x+10=3(3)+10=19$
$x=3$, $x+4=3+4=7$
The areas of the rectagles are 9 cm by 19 cm and 3 cm by 7 cm.

46. $AB=\sqrt{1^2+1^2}=\sqrt{2}$
$AD=\sqrt{1^2+(\sqrt{2})^2}=\sqrt{3}$
$AE=\sqrt{1^2+(\sqrt{3})^2}=2$
$AF=\sqrt{1^2+2^2}=\sqrt{5}$
$AG=\sqrt{1^2+(\sqrt{5})^2}=\sqrt{6}$

47. $x^4-5x^2+4=0$
Let $u=x^2$.
$u^2-5u+4=0$
$(u-4)(u-1)=0$
$u=4$ or $u=1$
$x^2=4$ or $x^2=1$
$x=\pm\sqrt{4}$ or $x=\pm\sqrt{1}$
$x=\pm 2$ or $x=\pm 1$
$\{-2, -1, 1, 2\}$

48. $(x^2+2x)^2-14(x^2+2x)=15$
$(x^2+2x)^2-14(x^2+2x)-15=0$
Let $u=x^2+2x$.
$u^2-14u-15=0$
$(u-15)(u+1)=0$
$u=15$ or $u=-1$
$x^2+2x=15$ or $x^2+2x=-1$
$x^2+2x-15=0$ or $x^2+2x+1=0$
$(x+5)(x-3)=0$ or $(x+1)^2=0$
$x=-5$, $x=3$ or $x=-1$
$\{-5, -1, 3\}$

49. $x^{2/3}-x^{1/3}-12=0$
Let $u=x^{1/3}$.
$u^2-u-12=0$
$(u-4)(u+3)=0$
$u=4$ or $u=-3$
$x^{1/3}=4$ or $x^{1/3}=-3$
$(x^{1/3})^3=4^3$ or $(x^{1/3})^3=(-3)^3$
$x=64$ or $x=-27$
$\{-27, 64\}$

50. $x+7\sqrt{x}=8$
$x+7\sqrt{x}-8=0$
Let $u=\sqrt{x}$.
$u^2+7u-8=0$
$(u+8)(u-1)=0$
$u=-8$ or $u=1$
$\sqrt{x}=-8$ or $\sqrt{x}=1$
$x=64$ (extraneous) or $x=1$ (checks)
$\{1\}$

51. $x^{-2}+x^{-1}-56=0$
Let $u=x^{-1}$.
$u^2+u-56=0$
$(u+8)(u-7)=0$
$u=-8$ or $u=7$
$x^{-1}=-8$ or $x^{-1}=7$
$\frac{1}{x}=-8$ or $\frac{1}{x}=7$
$x=-\frac{1}{8}$ or $x=\frac{1}{7}$ (both check)
$\left\{-\frac{1}{8}, \frac{1}{7}\right\}$

52. $2x^2 + 5x - 3 < 0$
$(2x - 1)(x + 3) < 0$
$x = \frac{1}{2}$ or $x = -3$

F	T	F
	-3	$\frac{1}{2}$

Test -4: $2(-4)^2 + 5(-4) - 3 < 0$
$9 < 0$ False
Test 0: $2(0)^2 + 5(0) - 3 < 0$
$-3 < 0$ True
Test 1: $2(1)^2 + 5(1) - 3 < 0$
$4 < 0$ False
$\left\{x \middle| -3 < x < \frac{1}{2}\right\}, \left(-3, \frac{1}{2}\right)$

53. $2x^2 + 9x + 4 \geq 0$
$(2x + 1)(x + 4) = 0$
$x = -\frac{1}{2}$ or $x = -4$

T	F	T
	-4	$-\frac{1}{2}$

Test -5: $2(-5)^2 + 9(-5) + 4 \geq 0$
$9 \geq 0$ True
Test -3: $2(-3)^2 + 9(-3) + 4 \geq 0$
$-5 \geq 0$ False
Test 0: $2(0)^2 + 9(0) + 4 \geq 0$
$4 \geq 0$ True
$\left\{x \middle| x \leq -4 \text{ or } x \geq -\frac{1}{2}\right\}, (-\infty, -4] \cup \left[-\frac{1}{2}, \infty\right)$

54. $\frac{x+7}{x-3} > 0$
$x + 7 = 0$ or $x - 3 = 0$
$x = -7$ or $x = 3$

T	F	T
	-7	3

Test -8: $\frac{-8+7}{-8-3} > 0$
$\frac{1}{11} > 0$ True
Test 0: $\frac{0+7}{0-3} > 0$
$-\frac{7}{3} > 0$ False
Test 4: $\frac{4+7}{4-3} > 0$
$11 > 0$ True
$\{x | x < -7 \text{ or } x > 3\}, (-\infty, -7) \cup (3, \infty)$

55. $\frac{x-3}{x+4} \leq 2$
$\frac{x-3}{x+4} - 2 \leq 0$
$\frac{x-3}{x+4} - \frac{2(x+4)}{x+4} \leq 0$
$\frac{x-3-2x-8}{x+4} \leq 0$
$\frac{-x-11}{x+4} \leq 0$
$-x - 11 = 0$ or $x + 4 = 0$
$x = -11$ or $x = -4$

T	F	T
	-11	-4

Test -12: $\frac{-12-3}{-12+4} \leq 2$
$\frac{15}{8} \leq 2$ True
Test -5: $\frac{-5-3}{-5+4} \leq 2$
$8 \leq 2$ False
Test 0: $\frac{0-3}{0+4} \leq 2$
$-\frac{3}{4} \leq 2$ True
$\{x | x \leq -11 \text{ or } x > -4\}, (-\infty, -11] \cup (-4, \infty)$

56. Let x = width of rectangle, and
$x + 2$ = length of rectangle.
(area) > (perimeter)
$x(x + 2) > 2x + 2(x + 2)$
$x^2 + 2x > 4x + 4$

$x^2 - 2x - 4 > 0$

$x = \frac{-(-2) \pm \sqrt{4+16}}{2} = \frac{2 \pm 2\sqrt{5}}{2} = 1 \pm \sqrt{5}$

T	F	T

$1-\sqrt{5}$ $\quad 1+\sqrt{5}$

Test 0 (middle interval):

$0 - 4 > 0$

$-4 > 0$ False

$\left(\text{Also reject } 1-\sqrt{5} \text{ since } 1-\sqrt{5} < 0.\right)$

Thus, the width must be greater than $1+\sqrt{5}$ m ≈ 3.24 m.

57. $x^2 - 2x - 3 = 0$

$x = -1,\ x = 3$

$\{-1, 3\}$

58. $x^2 - 3x - 28 > 0$

$\{x | x < -4 \text{ or } x > 7\}$

59. $x^4 - 13x^2 + 36 \le 0$

$\{x | -3 \le x \le -2 \text{ or } 2 \le x \le 3\}$

60. $x^2 - 4x + 4 \le 0$

$\{2\}$

Chapter 7 Test

1. $f(x) = x^2 - 6x + 7$

At $x = 2$, the graph shows $y = -1$, not $y = 4$.

2. $(3x-2)^2 = 50$

$3x - 2 = \pm\sqrt{50}$

$3x = 2 \pm 5\sqrt{2}$

$x = \frac{2 \pm 5\sqrt{2}}{3}$

$\left\{\frac{2-5\sqrt{2}}{3}, \frac{2+5\sqrt{2}}{3}\right\}$

3. $x(3x-8) = -4$

$3x^2 - 8x = -4$

$3x^2 - 8x + 4 = 0$

$(3x-2)(x-2) = 0$

$x = \frac{2}{3}$ or $x = 2$

$\left\{\frac{2}{3}, 2\right\}$

4. $(x+4)(x-8) = -42$

$x^2 - 4x - 32 = -42$

$x^2 - 4x + 10 = 0$

$x = \frac{4 \pm \sqrt{(-4)^2 - 4(1)(10)}}{2}$

$= \frac{4 \pm \sqrt{-24}}{2} = \frac{4 \pm 2i\sqrt{6}}{2} = 2 \pm i\sqrt{6}$

$\{2 - i\sqrt{6}, 2 + i\sqrt{6}\}$

5. $\frac{1}{x} + \frac{1}{x+2} = \frac{1}{3}$

$3x(x+2)\left(\frac{1}{x} + \frac{1}{x+2}\right) = 3x(x+2)\left(\frac{1}{3}\right)$

$3(x+2) + 3x = x(x+2)$

$3x + 6 + 3x = x^2 + 2x$

$x^2 - 4x - 6 = 0$

$x = \frac{4 \pm \sqrt{(-4)^2 - 4(1)(-6)}}{2} = \frac{4 \pm \sqrt{40}}{2}$

$= \frac{4 \pm 2\sqrt{10}}{2} = 2 \pm \sqrt{10}$

$\{2 - \sqrt{10}, 2 + \sqrt{10}\}$

6. $(x^2 + 2x)^2 - 11(x^2 + 2x) + 24 = 0$

Let $u = x^2 + 2x$.

$u^2 - 11u + 24 = 0$

$(u-3)(u-8) = 0$

$u = 3$ or $u = 8$

$x^2 + 2x = 3$ or $x^2 + 2x = 8$

$x^2 + 2x - 3 = 0$ or $x^2 + 2x - 8 = 0$

$(x+3)(x-1) = 0$ or $(x+4)(x-2) = 0$

$x = -3,\ x = 1,\ x = -4,\ x = 2$

$\{-4, -3, 1, 2\}$

7. $x^{1/6} - x^{1/3} + 2 = 0$
Let $u = x^{1/3}$.
$x^2 - x + 2 = 0$
$x = \frac{1 \pm \sqrt{(-1)^2 - 4(1)(2)}}{2} = \frac{1 \pm \sqrt{-7}}{2}$
$= \frac{1 \pm i\sqrt{7}}{2}$
$x^{1/3} = \frac{1 - i\sqrt{7}}{2}$ or $x^{1/3} = \frac{1 + i\sqrt{7}}{2}$
$x = \left(\frac{1 - i\sqrt{7}}{2}\right)^3$ or $x = \left(\frac{1 + i\sqrt{7}}{2}\right)^3$
$\left\{\frac{\left(1 - i\sqrt{7}\right)^3}{8}, \frac{\left(1 + i\sqrt{7}\right)^3}{8}\right\}$

8. $E = \frac{1}{2}mv^2$
$2E = mv^2$
$\frac{2E}{m} = v^2$
$\sqrt{\frac{2E}{m}} = v$
$v = \sqrt{\frac{2E}{m}}$

9. $x^2 - 6x + 7 = 0$
$x^2 - 6x = -7$
$\left(-\frac{6}{2}\right)^2 = (-3)^2 = 9$
$x^2 - 6x + 9 = -7 + 9$
$(x - 3)^2 = 2$
$x - 3 = \pm\sqrt{2}$
$x = 3 \pm \sqrt{2}$
$\{3 - \sqrt{2}, 3 + \sqrt{2}\}$

10. $d = \sqrt{(-5 - 5)^2 + (-3 - 3)^2}$
$= \sqrt{(-10)^2 + (-6)^2} = \sqrt{100 + 36}$
$= \sqrt{136} = 2\sqrt{34}$

11. $y = -x^2 + 2x + 3$

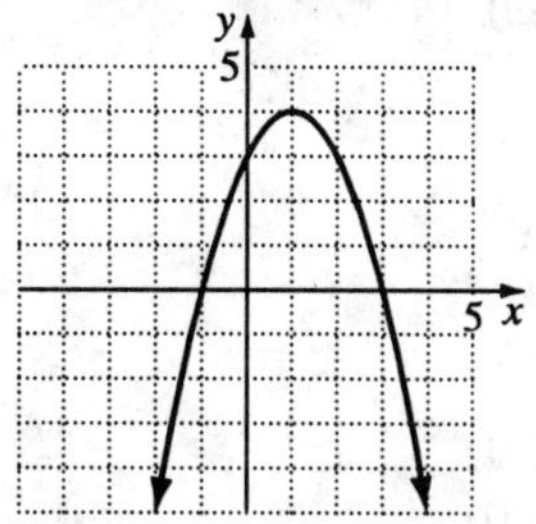

12. $f(x) = (x - 2)^2 + 3$

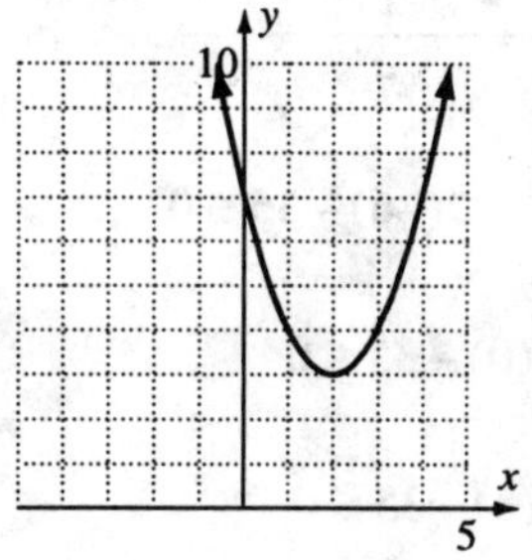

13. $2x^2 - x - 2 = 0$
$b^2 - 4ac = (-1)^2 - 4(2)(-2) = 17$
Solutions are irrational.

14. $f(x) = ax^2 + bx + c$
$0 = a(0)^2 + b(0) + c$
$0 = a(2)^2 + b(2) + c$
$3 = a(3)^2 + b(3) + c$
or
$c = 0$
$4a + 2b + c = 0$
$9a + 3b + c = 3$
Substitute c into equations.
$4a + 2b = 0$
$9a + 3b = 3$
Add –3 times 1st equation to 2 times second equation.
$-12a - 6b = 0$
$18a + 6b = 6$
$6a = 6$
$a = 1$

Substitute back.
$4(1)+2b+0=0$
$2b=-4$
$b=-2$
$f(x)=x^2-2x$

15. $x^2 \le 2x+35$
$x^2-2x-35 \le 0$
$(x-7)(x+5) \le 0$

F	T	F
	−5	7

Test −6: $(-6)^2-2(-6)-35 \le 0$
$13 \le 0$ False
Test 0: $0^2-2(0)-35 \le 0$
$-35 \le 0$ True
Test 8: $8^2-2(8)-35 \le 0$
$13 \le 0$ False
$\{x \mid -5 \le x \le 7\}$, $[-5, 7]$

16. $\frac{4}{x-2} \ge 2$
$\frac{4}{x-2}-2 \ge 0$
$\frac{4-2(x-2)}{x-2} \ge 0$
$\frac{4-2x+4}{x-2} \ge 0$
$\frac{8-2x}{x-2} \ge 0$
$8-2x=0$ or $x-2=0$
$x=4$ or $x=2$

F	T	F
	2	4

Test 0: $\frac{4}{0-2} \ge 2$, $-2 \ge 2$ False
Test 3: $\frac{4}{3-2} \ge 2$, $4 \ge 2$ True
Test 5: $\frac{4}{5-2} \ge 2$, $\frac{4}{3} \ge 2$ False
$\{x \mid 2 < x \le 4\}$, $(2, 4]$

17. Let x = width of pond.
$x^2=50^2+50^2$
$x=\pm\sqrt{5000}$
$x=\pm 50\sqrt{2}$
Disregard a negative distance.
The width is $50\sqrt{2}$ ft.

18. $s(t)=-16t^2+64t$
Maximum height at vertex
$t=-\frac{b}{2a}=-\frac{64}{2(-16)}=2$
$s(2)=-16(2)^2+64(2)=64$
(2, 64)
At 2 seconds it reaches a maximum height of 64 ft.

19. $(10-2x)(14-2x)=32$
$140-20x-28x+4x^2=32$
$4x^2-48x+108=0$
$x^2-12x+27=0$
$(x-9)(x-3)=0$
$x=9$ or $x=3$
Cutting a 9 in. by 9 in. square off is impossible. Therefore, the square is 3 in. by 3 in.

20. Pool + border area − pool area = 336
Let x = width of border.
$(20+2x)(30+2x)-20(30)=336$
$600+40x+60x+4x^2-600=336$
$4x^2+100x-336=0$
$x^2+25x-84=0$
$(x+28)(x-3)=0$
$x=-28$, $x=3$
Disregard a negative width.
The border is 3 ft wide.

Chapters 1–7 Cumulative Review Problems

1. $|x-1|>3$
$x-1>3$ or $x-1<-3$
$x>4$ or $x<-2$
$\{x|x<-2 \text{ or } x>4\}$, $(-\infty, -2) \cup (4, \infty)$

2. $x^{2/3}=2x^{1/3}+35$
$x^{2/3}-2x^{1/3}-35=0$
Let $t=x^{1/3}$.
$t^2-2t-35=0$
$(t-7)(t+5)=0$
$t=7$ or $t=-5$
$x^{1/3}=7$ or $x^{1/3}=-5$
$x=7^3$ or $x=(-5)^3$
$x=343$ or $x=-125$
$\{-125, 343\}$

3. $8(2x-1)+3(1-x)+2=5-2x(1+x)$
$16x-8+3-3x+2=5-2x-2x^2$
$13x-3=5-2x-2x^2$
$2x^2+15x-8=0$
$(2x-1)(x+8)=0$
$x=\frac{1}{2}$ or $x=-8$
$\left\{-8, \frac{1}{2}\right\}$

4. $(x-8)(x+7)=5$
$x^2-x-56=5$
$x^2-x-61=0$
$x=\frac{1\pm\sqrt{(-1)^2-4(1)(-61)}}{2}=\frac{1\pm\sqrt{245}}{2}$
$=\frac{1\pm 7\sqrt{5}}{2}$
$\left\{\frac{1-7\sqrt{5}}{2}, \frac{1+7\sqrt{5}}{2}\right\}$

5. $\sqrt{x+4}-\sqrt{x-4}=2$
$\sqrt{x+4}=2+\sqrt{x-4}$
$\left(\sqrt{x+4}\right)^2=\left(2+\sqrt{x-4}\right)^2$
$x+4=4+4\sqrt{x-4}+x-4$
$1=\sqrt{x-4}$
$1=x-4$
$5=x$
$\{5\}$

6. $-3\le\frac{2}{5}x-1\le 1$
$-2\le\frac{2}{5}x\le 2$
$-\frac{10}{2}\le x\le\frac{10}{2}$
$-5\le x\le 5$
$\{x|-5\le x\le 5\}$, $[-5, 5]$

7. $|6x-4|+6=1$
$|6x-4|=-5$
The absolute value is never negative.
No solution
$\varnothing$

8. $x-4\ge 0$ and $-3x\le -6$
$x\ge 4$ and $x\ge 2$
$\{x|x\ge 4\}$, $[4, \infty)$

9. $|5x+2|=|4-3x|$
$5x+2=4-3x$ or $5x+2=-(4-3x)$
$8x=2$ or $2x=-6$
$x=\frac{1}{4}$ or $x=-3$
$\left\{-3, \frac{1}{4}\right\}$

10. $-3x + 2y + 4z = 6$ (1)
$7x - y + 3z = 23$ (2)
$2x + 3y + z = 7$ (3)

Multiply equation (2) by 2 and add to equation (1).

$$\begin{array}{r} -3x + 2y + 4z = 6 \\ 14x - 2y + 6z = 46 \\ \hline 11x + 10z = 52 \quad (4) \end{array}$$

Multiply equation (2) by 3 and add to equation (3).

$$\begin{array}{r} 21x - 3y + 9z = 69 \\ 2x + 3y + z = 7 \\ \hline 23x + 10z = 76 \quad (5) \end{array}$$

Multiply equation (4) by -1 and add to equation (5).

$$\begin{array}{r} -11x - 10z = -52 \\ 23x + 10z = 76 \\ \hline 12x = 24 \\ x = 2 \end{array}$$

Substitute 2 for x in equation (4).

$11x + 10z = 52$
$22 + 10z = 52$
$10z = 30$
$z = 3$

Substitute 2 for x and 3 for z in equation (1).

$-3x + 2y + 4z = 6$
$-6 + 2y + 12 = 6$
$2y = 0$ so $y = 0$
$\{(2, 0, 3)\}$

11. $\dfrac{-5x^3y^7z^4}{15x^3y^9z^{-2}} = -\dfrac{x^{3-3}z^{4+2}}{3y^{9-7}} = -\dfrac{z^6}{3y^2}$

12. $(5x-2)(2x^2+3xy-y^2) = 5x(2x^2+3xy-y^2) - 2(2x^2+3xy-y^2)$
$= 10x^3 + 15x^2y - 5xy^2 - 4x^2 - 6xy + 2y^2$

13. $\dfrac{x+2}{x^2-6x+8} + \dfrac{3x-8}{x^2-5x+6} = \dfrac{x+2}{(x-4)(x-2)} + \dfrac{3x-8}{(x-3)(x-2)}$

$= \dfrac{(x+2)(x-3)}{(x-4)(x-2)(x-3)} + \dfrac{(3x-8)(x-4)}{(x-3)(x-2)(x-4)}$

$= \dfrac{x^2 - x - 6 + 3x^2 - 20x + 32}{(x-4)(x-2)(x-3)}$

$= \dfrac{4x^2 - 21x + 26}{(x-4)(x-2)(x-3)} = \dfrac{(4x-13)(x-2)}{(x-4)(x-2)(x-3)}$

$= \dfrac{4x-13}{(x-4)(x-3)}$

14. $(5x^3 - 24x^2 + 9) \div (5x+1) = x^2 - 5x + 1 + \dfrac{8}{5x+1}$

$$\begin{array}{r}
x^2 - 5x + 1 + \frac{8}{5x+1} \\
5x+1 \overline{) 5x^3 - 24x^2 + 0x + 9} \\
\underline{5x^3 + x^2} \quad\quad\quad \\
-25x^2 + 0x \quad \\
\underline{-25x^2 - 5x} \quad \\
5x + 9 \\
\underline{5x + 1} \\
8
\end{array}$$

15. $\dfrac{\sqrt[3]{32xy^{10}}}{\sqrt[3]{2xy^2}} = \sqrt[3]{\dfrac{32xy^{10}}{2xy^2}} = \sqrt[3]{16y^8}$

$= \sqrt[3]{8y^6 \cdot 2y^2} = \sqrt[3]{(2y^2)^3 \cdot 2y^2} = 2y^2\sqrt[3]{2y^2}$

16. $\dfrac{3-\sqrt{5}}{2+\sqrt{5}} = \dfrac{3-\sqrt{5}}{2+\sqrt{5}} \cdot \dfrac{2-\sqrt{5}}{2-\sqrt{5}}$

$= \dfrac{6 - 3\sqrt{5} - 2\sqrt{5} + 5}{2-5} = \dfrac{11 - 5\sqrt{5}}{-3}$

17. $x^2 - 8x + 16 - 25y^2 = (x^2 - 8x + 16) - 25y^2$

$= (x-4)^2 - 25y^2$

$= [(x-4) - 5y][(x-4) + 5y]$

$= (x - 4 - 5y)(x - 4 + 5y)$

18. **a.** $x^2 + y^2 = 4$

$y^2 = 4 - x^2$

$y = \pm\sqrt{4 - x^2}$

$y = \pm\sqrt{(2-x)(2+x)}$

b. No

c. No; it does not pass the vertical line test.

19. $s(t) = -16t^2 + 80t + 5$

$0 = -16t^2 + 80t + 5$

$t = \dfrac{-80 \pm \sqrt{(80)^2 - 4(-16)(5)}}{2(-16)}$

$= \dfrac{-80 \pm \sqrt{6720}}{-32} = \dfrac{-80 \pm 8\sqrt{105}}{-32}$

$= \dfrac{10 \pm \sqrt{105}}{4}$

Disregard the negative time.

The ball stries the ground at

$\dfrac{10 + \sqrt{105}}{4}$ seconds.

No; the graph indicates that it strikes the ground at slightly more than 5 seconds.

20. $V(x) = x(10 - 2x)(7 - 2x)$

$V(x) = x(70 - 20x - 14x + 4x^2)$

$V(x) = 4x^3 - 34x^2 + 70x$

21. $y = 1527x + 17{,}547$

where x = the number of years since 1982 and y = average annual salary.

$51{,}141 = 1527x + 17{,}547$

$33{,}594 = 1527x$

$22 = x$

$1982 + 22 = 2004$

In year 2004

22. slope = 1527
Each year, from 1982 on, the average salary increases by \$1527.

23. a. points $A(6, 32)$, $B(7, 24)$

$$\text{slope} = \frac{24-32}{7-6} = -8$$

The altitude decreased by an average of 8 m per second from 6 to 7 seconds.

b. points $P(22, 14)$, $Q(24, 22)$

$$\text{slope} = \frac{22-14}{24-22} = \frac{8}{2} = 4$$

The altitude increased by an average of 4 m per second from 22 to 24 seconds.

c. 0
The altitude did not change from 11 to 17 seconds.

24. All telephone area codes have the same sum of 2 + 5 + 2 = 9. One of the area codes is 252. A second area code begins with 6. The remaining 2 numbers must add to 9 – 6 = 3. The possibilities are 0 and 3 or 1 and 2. However this number cannot contain a 1 since the remaining area code ends in 1. Thus the second area code must be 603 or 630.
For the remaining area code, the possible remaining digits are 1, 4, 7, 9. The only combinations that sum to 9 and end in 1 are 711 and 171.

25. $z = 2x + 3y$
$(0, 6)$: $z = 2(0) + 3(6) = 18$
$(2, 2)$: $z = 2(2) + 3(2) = 10$
$(5, 0)$: $z = 2(5) + 3(0) = 10$
The minimum value is 10.

26. a. (20, 20,151) and (30, 18,301)

$$\text{slope} = \frac{18{,}301 - 20{,}151}{30-20} = -185$$

$y - 20{,}151 = -185(x - 20)$ or
$y - 18{,}301 = -185(x - 30)$

b. $y - 20{,}151 = -185x + 3700$
$y = -185x + 23{,}851$

c. $y = -185(55) + 23{,}851$
$y = 13{,}676$
The amount is \$13,676.

27. Let x = height of Library Tower.Then
$2x - 582$ = height of Sears Tower.

$$\frac{x + 2x - 582}{2} = 1236$$

$3x - 582 = 2472$
$3x = 3054$
$x = 1018$
$2x - 582 = 1454$
The heights of the buildings are as follows:
Library Tower, 1018 ft;
Sears Tower 1454 ft

28. Let x = width of strip,
$12 - 2x$ = width of room less strip, and
$20 - 2x$ = length of room less strip.
(area of room less strip) = (remaining area)
$(12 - 2x)(20 - 2x) = 180$
$240 - 64x + 4x^2 = 180$
$4x^2 - 64x + 60 = 0$
$x^2 - 16x + 15 = 0$
$(x - 15)(x - 1) = 0$
$x = 15$ or $x = 1$ (reject since $12 - 2x < 0$ and $20 - 2x < 0$)
The width of the strip is 1 ft.

29. Let x = speed of cyclist with no wind, $x + 4$ = speed of cyclist with headwind, and $x - 4$ = speed of cyclist against headwind.

	Rate	Time (hours)	Distance = Rate × Time
with headwind	$x+4$	2	$2(x+4)$
against headwind	$x-4$	3	$3(x-4)$

$2(x+4) = 3(x-4)$
$2x + 8 = 3x - 12$
$20 = x$
average speed of cyclist with no wind: 20 mph
distance = $2(x+4) = 2(20+4) = 48$ miles

30. $y = \frac{1}{9000}x^2 + 5$
$y = \frac{1}{9000}(2100)^2 + 5$
$y = 495$
They are 495 ft above the road.

Chapter 8

Problem Set 8.1

1. $y = 2^x$

x	-2	-1	0	1	2
y	$\frac{1}{4}$	$\frac{1}{2}$	1	2	4

$y = 2^{x+1}$

x	-3	-2	-1	0	1
y	$\frac{1}{4}$	$\frac{1}{2}$	1	2	4

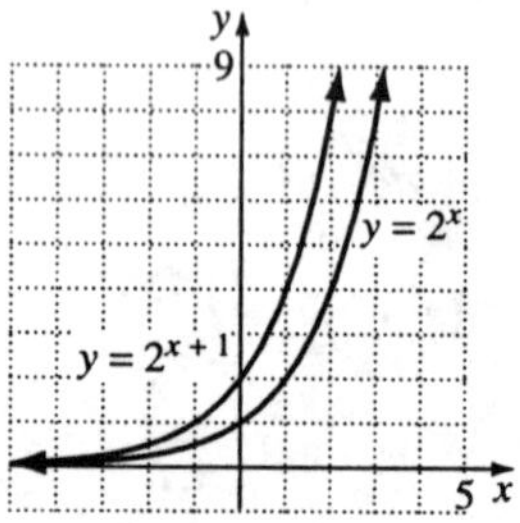

3. $y = 3^x$

x	-2	-1	0	1	2
y	$\frac{1}{9}$	$\frac{1}{3}$	1	3	9

$y = 3^{x-2}$

x	0	1	2	3	4
y	$\frac{1}{9}$	$\frac{1}{3}$	1	3	9

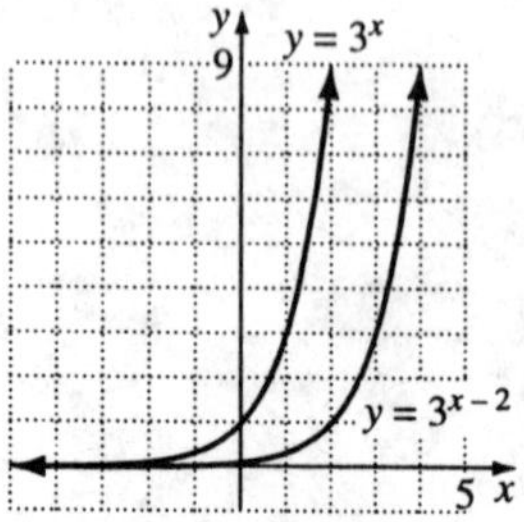

5. $y = b^{x+c}$ is $y = b^x$ shifted to the left if c is positive and shifted to the right if c is negative.

7. $f(x) = 2^x$, $g(x) = 2^x + 3$, $h(x) = 2^x - 1$

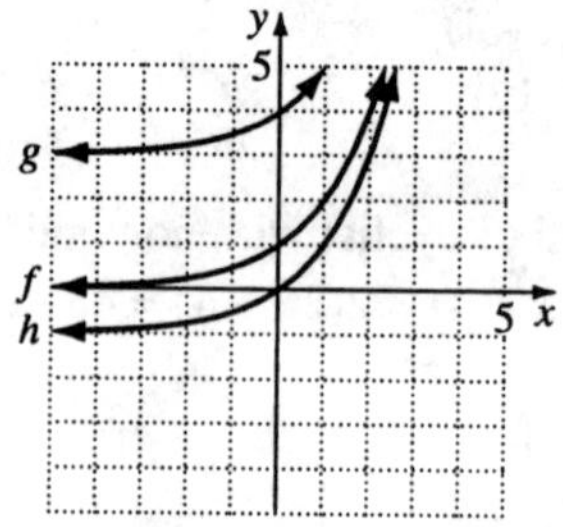

g is f shifted up 3 units and h is f shifted down 1 unit.

9. $f(x) = \left(\frac{3}{2}\right)^x$, $g(x) = \left(\frac{2}{3}\right)^x$

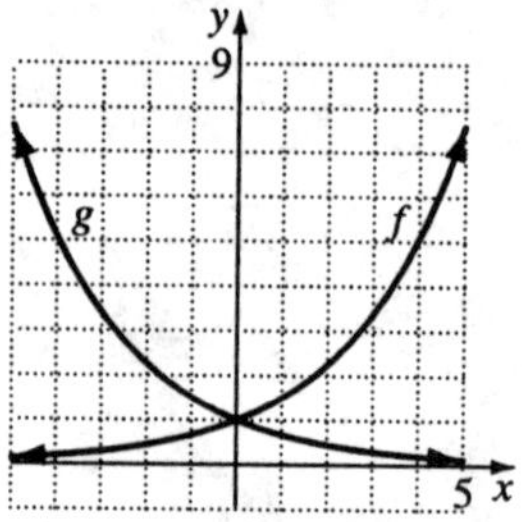

g is f reflected across the y-axis.

11. $y = 3^x$

x	-2	-1	0	1	2
y	$\frac{1}{9}$	$\frac{1}{3}$	1	3	9

$x = 3^y$

x	$\frac{1}{9}$	$\frac{1}{3}$	1	3	9
y	-2	-1	0	1	2

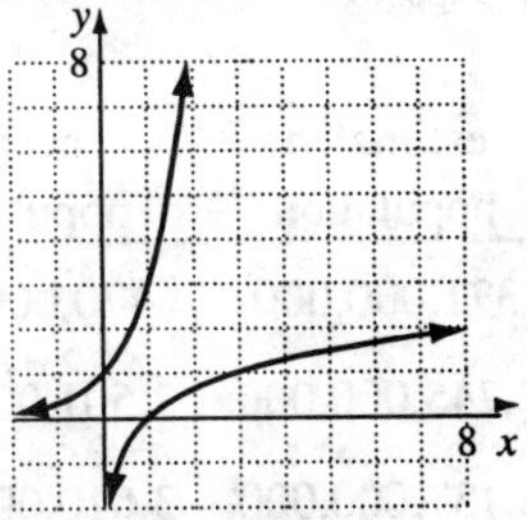

$x = 3^y$ is $y = 3^x$ reflected about the line $x = y$.

13. $y = 3^{-x}$

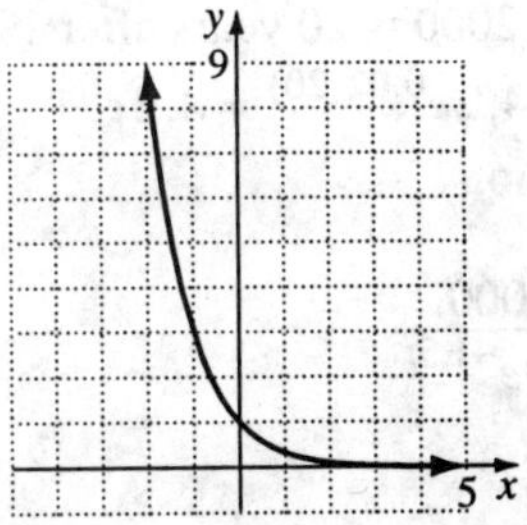

15. $g(x) = 2^{x/2}$

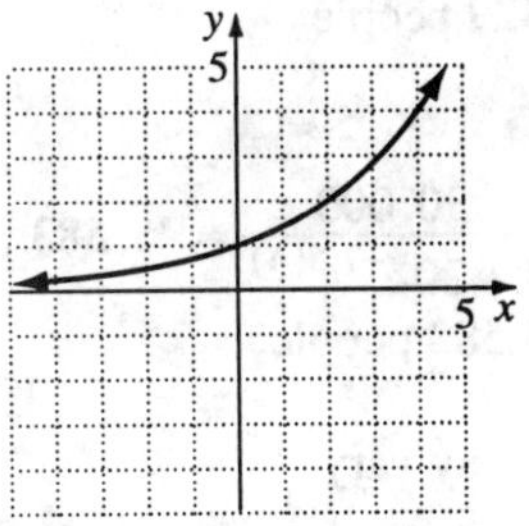

17. $y = 2^{x-1} - 1$

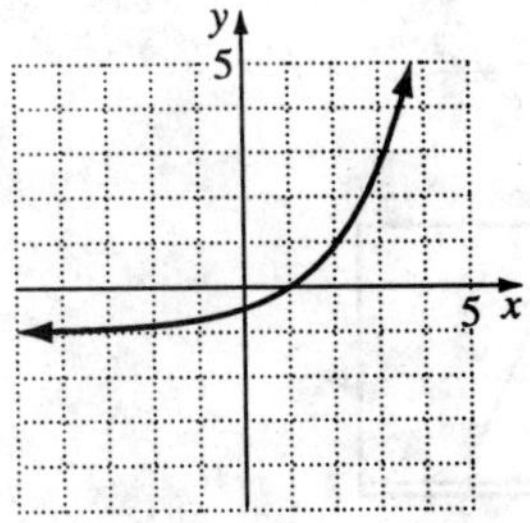

19. $y = 2^{x+1} + 1$

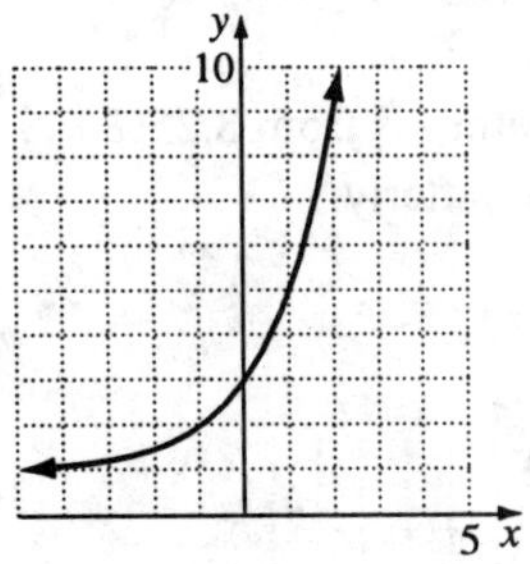

21. $f(x) = 3^{-\frac{x}{2}} - 1$

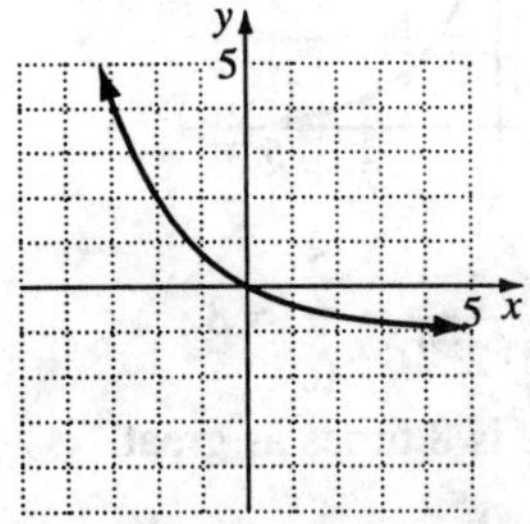

23. a. $A = P\left(1 + \frac{r}{n}\right)^{nt}$

$A = 5000\left(1 + \frac{0.065}{2}\right)^{2(10)}$

≈ 9479.19

It will be worth about \$9479.19.

b. $A = P\left(1 + \frac{r}{n}\right)^{nt}$

$A = 5000\left(1 + \frac{0.065}{12}\right)^{12(10)}$

≈ 9560.92

It will be worth about \$9560.92.

c. $A = Pe^{rt}$

$A = 5000e^{(0.065)(10)}$

≈ 9577.70

It will be worth about \$9577.70.

25. $A = P\left(1 + \frac{r}{n}\right)^{nt}$

$A = 6000\left(1 + \frac{0.0825}{4}\right)^{4(4)}$

≈ 8317.84

$A = 6000\left(1 + \frac{0.083}{2}\right)^{2(4)}$
≈ 8306.64
The greatest return is from 8.25% compounded quarterly.

27. **a.** $f(x) = 5 \cdot 2^{-x}$

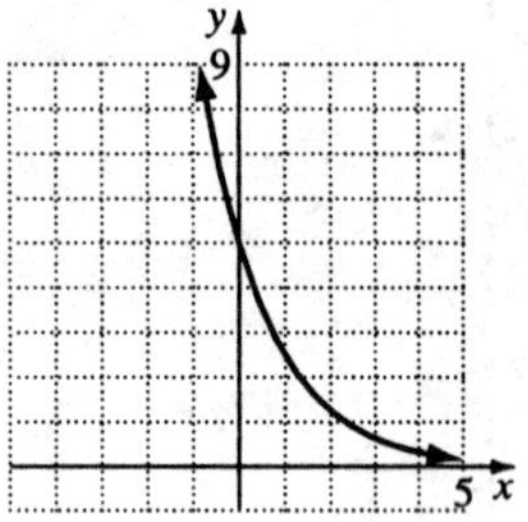

b. $\frac{f(1)}{f(4)} = \frac{5 \cdot 2^{-1}}{5 \cdot 2^{-4}} = \frac{\frac{5}{2}}{\frac{5}{16}} = 8$
The value is 8 times as great.

29. $y = Ae^{kt} \ (k > 0)$

a. $t = 11$
$y = 100,000e^{0.09(11)} = 269,123.45$
There will be about 269,123 cases.

b. $t = 31$
$y = 100,000e^{0.09(31)}$
=1,628,102
There will be 1,628,102 cases.

31. $y = 3.6e^{0.02t}$
$t = 2020 - 1969 = 51$
$y = 3.6e^{0.02(51)} \approx 10$

33. $f(x) = 6,164e^{0.00667x}$

a.

year	estimated population	actual population
1650	371,000,000	470,000,000
1950	2,745,000,000	2,501,000,000
1970	3,137,000,000	3,610,000,000

For 1650:
$f(1650) = 6164e^{0.00667(1650)}$
$\approx 371,000,000$

b. The year 2000 is 20 years after 1980.
$f(20) = 4.2e^{0.02(20)} = 4.2e^{0.4}$
≈ 6.3 billion

35. $N(t) = \frac{30,000}{1 + 20e^{-1.5t}}$

a. $t = 0$
$N(0) = \frac{30,000}{1 + 20e^{-1.5(0)}} \approx 1429$
About 1429 people

b. $t = 4$
$N(4) = \frac{30,000}{1 + 20e^{-1.5(4)}} \approx 28,583$
About 28,583 people

c. Answers may vary.

37. d is true.

39. $f(x) = \frac{1}{\sqrt{2\pi}} e^{-\frac{x^2}{2}}$

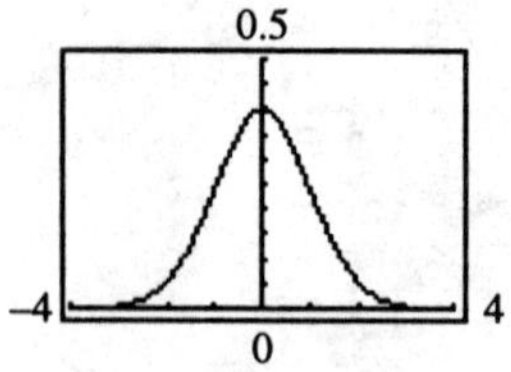

41. a. $y = e^x, y = 1 + x + \frac{x^2}{2}$

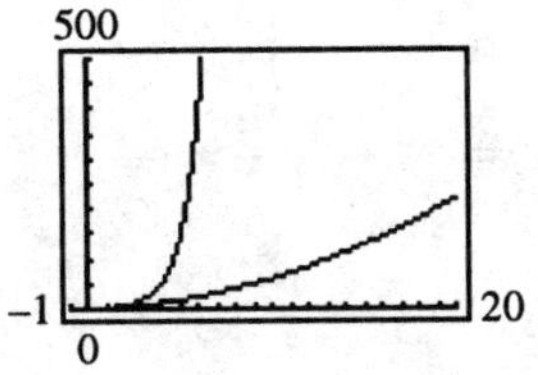

b. $y = e^x, y = 1 + x + \frac{x^2}{2} + \frac{x^3}{6}$

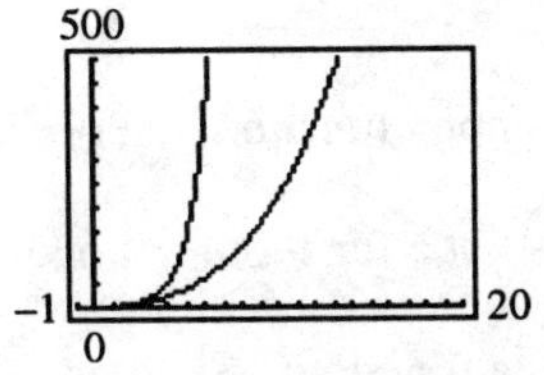

c. $y = e^x, y = 1 + x + \frac{x^2}{2} + \frac{x^3}{6} + \frac{x^4}{24}$

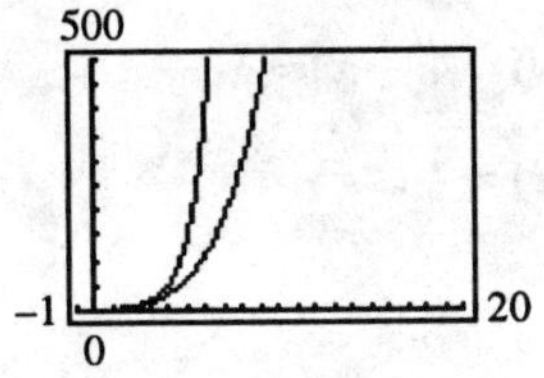

d. The second graph is approaching $y = e^x$ in a, then b, then c.

43–45. Students should verify results.

47. Answers may vary.

49. Determine the y-value at $x = \frac{1}{2}$.

51. a is $h(x)$
b is $g(x)$
c is $f(x)$
Answers may vary.

Review Problems

52. Let x = Rider's speed
Time against the wind + Time with the wind = 3 (the total time of the top)

$$\frac{40}{x-10} + \frac{40}{x+10} = 3$$
$$(x-10)(x+10)\left[\frac{40}{x-10} + \frac{40}{x+10}\right] = (x-10)(x+10)(3)$$
$$40(x+10) + 40(x-10) = 3x^2 - 300$$
$$40x + 400 + 40x - 400 = 3x^2 - 300$$
$$0 = 3x^2 - 80x - 300$$
$$0 = (x-30)(3x+10)$$
$$x - 30 = 0 \text{ or } 3x + 10 = 0$$
$$x = 30 \text{ or } x = -\frac{10}{3} \text{ (reject)}$$

Rider's speed: 30 mph

53. $D = \frac{ab}{a+b}$
$$D(a+b) = ab$$
$$Da + Db = ab$$
$$Db - ab = -Da$$
$$b(D-a) = -Da$$
$$b = -\frac{Da}{D-a} \text{ or } \frac{Da}{a-D}$$

54. $\begin{vmatrix} 3 & -2 \\ 7 & -5 \end{vmatrix} = 3(-5) - (-2)(7)$
$= -15 + 14 = -1$

Problem Set 8.2

1. $(f \circ g)(x) = 2(x^2 + 4x) - 5 = 2x^2 + 8x - 5$
$(f \circ g)(2) = 2(2)^2 + 8(2) - 5 = 19$

3. $(g \circ f)(x) = (2x-5)^2 + 4(2x-5)$
$= 4x^2 - 20x + 25 + 8x - 20$
$= 4x^2 - 12x + 5$
$(g \circ f)(2) = 4(2)^2 - 12(2) + 5 = -3$

5. $h(-1) = \frac{-1+5}{2} = 2$
$(g \circ h)(-1) = g(2) = 2^2 + 4(2) = 12$

7. $g(-1)=(-1)^2+4(-1)=-3$
$(h\circ g)(-1)=h(-3)=\frac{-3+5}{2}=1$

9. $h(11)=\frac{11+5}{2}=8$
$(f\circ h)(11)=f(8)=2(8)-5=11$

11. $f(11)=2(11)-5=17$
$(h\circ f)(11)=h(17)=\frac{17+5}{2}=11$

13. $f(6)=2(6)-5=7$
$(f\circ f)(6)=f(7)=2(7)-5=9$

15. $g(-2)=(-2)^2+4(-2)=-4$
$(g\circ g)(-2)=g(-4)=(-4)^2+4(-4)=0$

17. $f(x)=x^2+4,\ g(x)=2x+1$
$(f\circ g)(x)=(2x+1)^2+4$
$=4x^2+4x+1+4=4x^2+4x+5$
$(g\circ f)(x)=2(x^2+4)+1$
$=2x^2+8+1=2x^2+9$

19. $f(x)=5x+2,\ g(x)=3x^2-4$
$(f\circ g)(x)=5(3x^2-4)+2$
$=15x^2-20+2=15x^2-18$
$(g\circ f)(x)=3(5x+2)^2-4$
$=3(25x^2+20x+4)-4=75x^2+60x+8$

21. $f(x)=x^2+2,\ g(x)=x^2-2$
$(f\circ g)(x)=(x^2-2)^2+2$
$=x^4-4x^2+4+2=x^4-4x^2+6$
$(g\circ f)(x)=(x^2+2)^2-2$
$=x^4+4x^2+4-2=x^4+4x^2+2$

23. $f(x)=\sqrt{x},\ g(x)=x-1$
$(f\circ g)(x)=\sqrt{x-1}$
$(g\circ f)(x)=\sqrt{x}-1$

25. $f(x)=2x-3,\ g(x)=\frac{x+3}{2}$
$(f\circ g)(x)=2\left(\frac{x+3}{2}\right)-3=x+3-3=x$
$(g\circ f)(x)=\frac{2x-3+3}{2}=\frac{2x}{2}=x$

27. $f(x)=\frac{1}{x},\ g(x)=\frac{1}{x}$
$(f\circ g)(x)=\frac{1}{\frac{1}{x}}=x$
$(g\circ f)(x)=\frac{1}{\frac{1}{x}}=x$

29. Inverse is not a function; not one-to-one.

31. Inverse is not a function; not one-to-one.

33. Inverse is a function; one-to-one.

35. $f(x)=4x,\ g(x)=\frac{1}{4}x$
$(f\circ g)(x)=4\left(\frac{1}{4}x\right)=x$
$(g\circ f)(x)=\frac{1}{4}(4x)=x$
Yes

37. $f(x)=3x+8,\ g(x)=\frac{x-8}{3}$
$(f\circ g)(x)=3\left(\frac{x-8}{3}\right)+8=x-8+8=x$
$(g\circ f)(x)=\frac{3x+8-8}{3}=\frac{3x}{3}=x$
Yes

39. $f(x)=5x-9,\ g(x)=\frac{x+5}{9}$
$(f\circ g)(x)=5\left(\frac{x+5}{9}\right)-9=\frac{5}{9}x-\frac{56}{9}$
$(g\circ f)(x)=\frac{5x-9+5}{9}=\frac{5}{9}x-\frac{4}{9}$
No

41. $f(x)=\frac{3}{x-4}$, $g(x)=\frac{3}{x}+4$

$(f\circ g)(x)=\frac{3}{\frac{3}{x}+4-4}=x$

$(g\circ f)(x)=\frac{3}{\frac{3}{x-4}}+4=x$

Yes

43. $f(x)=-x$, $g(x)=-x$

$(f\circ g)(x)=-(-x)=x$

$(g\circ f)(x)=-(-x)=x$

Yes

45. **a.** $f(x)=x+3$

$y=x+3$

$x=y+3$

$x-3=y$

$f^{-1}(x)=x-3$

b. Verify:

$f(f^{-1}(x))=x-3+3=x$

$f^{-1}(f(x))=x+3-3=x$

c.

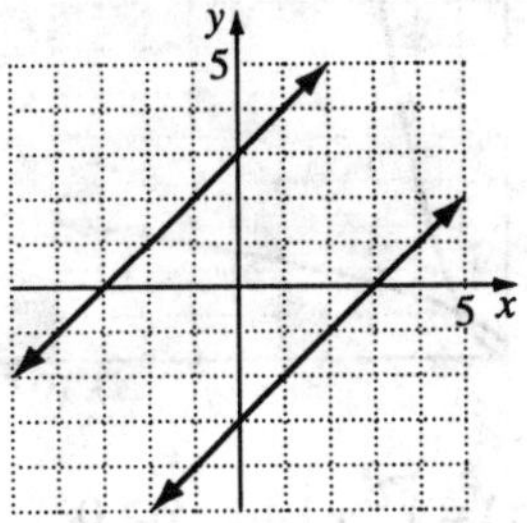

47. **a.** $f(x)=2x$

$y=2x$

$x=2y$

$\frac{x}{2}=y$

$f^{-1}(x)=\frac{x}{2}$

b. Verify:

$f(f^{-1}(x))=2\left(\frac{x}{2}\right)=x$

$f^{-1}(f(x))=\frac{2x}{2}=x$

c.

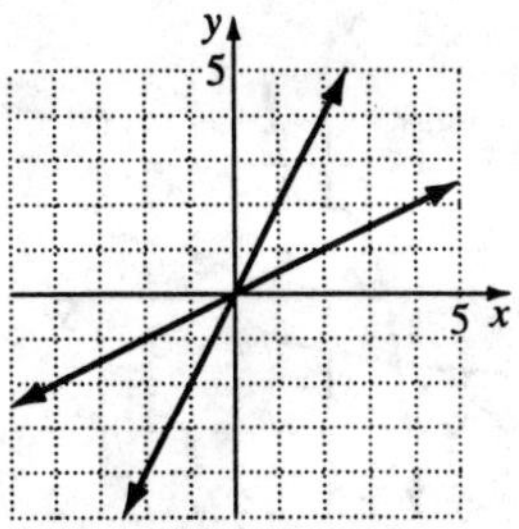

49. **a.** $f(x)=2x+3$

$y=2x+3$

$x=2y+3$

$\frac{x-3}{2}=y$

$f^{-1}(x)=\frac{x-3}{2}$

b. Verify:

$f\left(f^{-1}(x)\right)=2\left(\frac{x-3}{2}\right)+3=x$

$f^{-1}(f(x))=\frac{2x+3-3}{2}=x$

c.

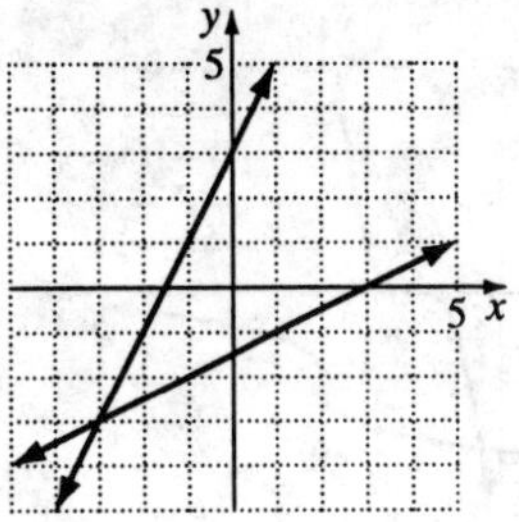

51. **a.** $f(x)=x^3+2$

$y=x^3+2$

$x=y^3+2$

$\sqrt[3]{x-2}=y$

$f^{-1}(x)=\sqrt[3]{x-2}$

b. Verify:

$f\left(f^{-1}(x)\right)=\left(\sqrt[3]{x-2}\right)^3+2=x$

$f^{-1}(f(x))=\sqrt[3]{x^3+2-2}=x$

c.

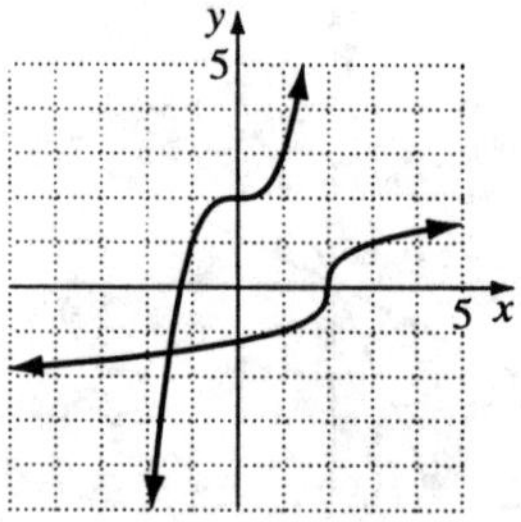

53. a. $f(x) = (x+2)^3$

$y = (x+2)^3$

$x = (y+2)^3$

$\sqrt[3]{x} - 2 = y$

$f^{-1}(x) = \sqrt[3]{x} - 2$

b. Verify:

$f(f^{-1}(x)) = \left(\sqrt[3]{x} - 2 + 2\right)^3 = x$

$f^{-1}(f(x)) = \sqrt[3]{(x+2)^3} - 2 = x$

c.

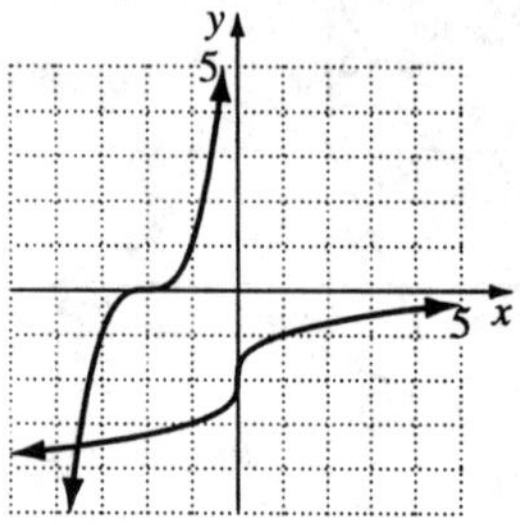

55. a. $f(x) = \frac{1}{x}$

$y = \frac{1}{x}$

$x = \frac{1}{y}$

$y = \frac{1}{x}$

$f^{-1}(x) = \frac{1}{x}$

b. Verify:

$f(f^{-1}(x)) = \frac{1}{\frac{1}{x}} = x$

$f^{-1}(f(x)) = \frac{1}{\frac{1}{x}} = x$

c.

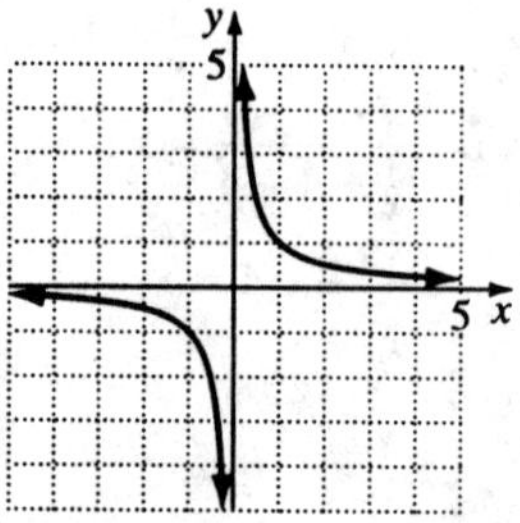

57. a. $f(x) = \sqrt{x}$

$y = \sqrt{x}$

$x = \sqrt{y}$

$x^2 = y$

$f^{-1}(x) = x^2$

b. Verify:

$f\left(f^{-1}(x)\right) = \sqrt{x^2} = x$

$f^{-1}(f(x)) = (\sqrt{x})^2 = x$

c.

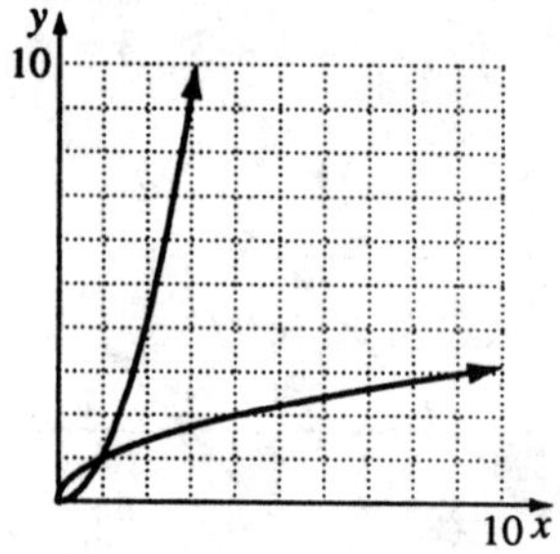

59. a. $f(x) = x^2 + 1$ for $x \geq 0$

$y = x^2 + 1$

$x = y^2 + 1$

$\sqrt{x-1} = y$

$f^{-1}(x) = \sqrt{x-1}$

b. Verify:

$f\left(f^{-1}(x)\right) = \left(\sqrt{x-1}\right)^2 + 1 = x$

$f^{-1}\left(f(x)\right) = \sqrt{x^2 + 1 - 1} = x$

c.

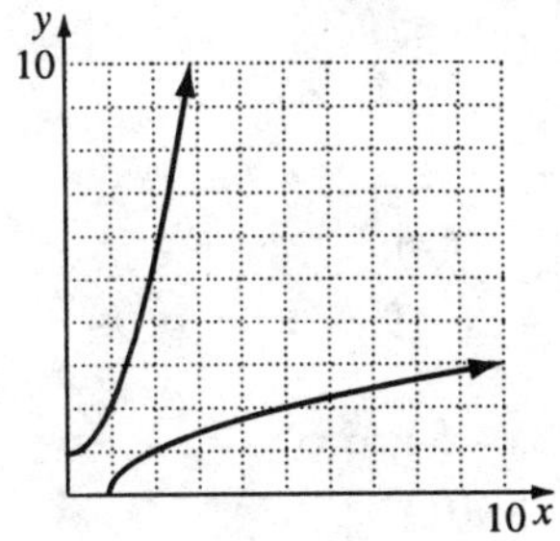

61. $g(1) = 0$
$(f \circ g)(1) = f(0) = -4$

63. Ordered pairs on original:
(0, –4), (2, 0), (4, 4)
Ordered pairs on inverse:
(–4, 0), (0, 2), (4, 4)

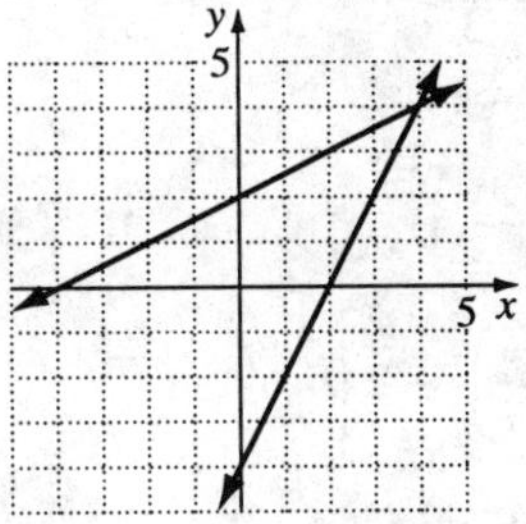

65. Ordered pairs on original function:
(0, 1), (1, 2), (2, 4)
Ordered pairs on inverse:
(1, 0), (2, 1), (4, 2)

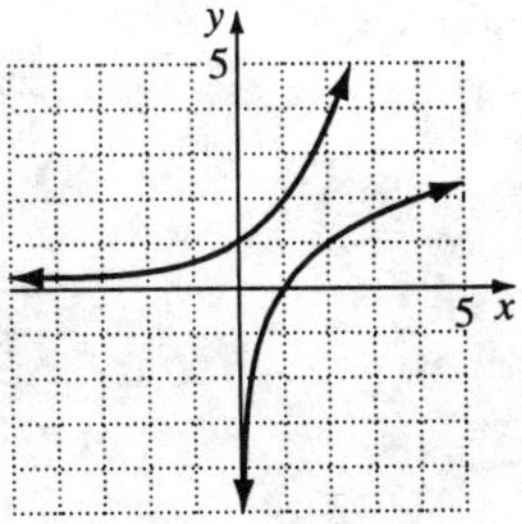

67. $f(x) = x - 5$, $g(x) = x - 0.4x$

a. f is \$5 off the regular price
g is 40% off the regular price

b. $(f \circ g)(x) = x - 0.4x - 5 = 0.6x - 5$
\$5 off the sale price of 40% off

c. $(g \circ f)(x) = (x - 5) - 0.4(x - 5)$
$= x - 5 - 0.4x + 2$
$= 0.6x - 3$
\$3 off the sale price of 40% off.

d. $(f \circ g)(x)$;
$0.6x - 5$ is less than $0.6x - 3$.

69. a. It passes the horiztonal line test.

b. (14, 20), so $f^{-1}(20) = 14$
There were 20 million women with AIDS in 1994.

c. Answers may vary.

71. d is true

73. Students should verify results.

75. $f(x) = \frac{1}{x} + 2$, $g(x) = \frac{1}{x - 2}$

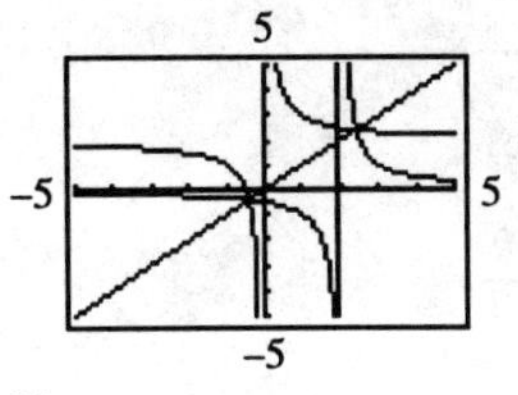

Yes

77. $f(x) = e^x$, $g(x) = \ln x$

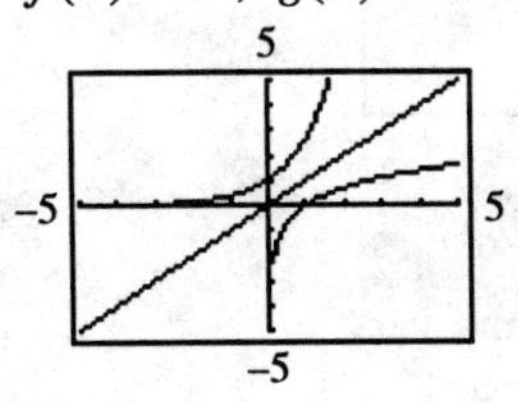

Yes

79. $f(x)=\sqrt[3]{2-x}$

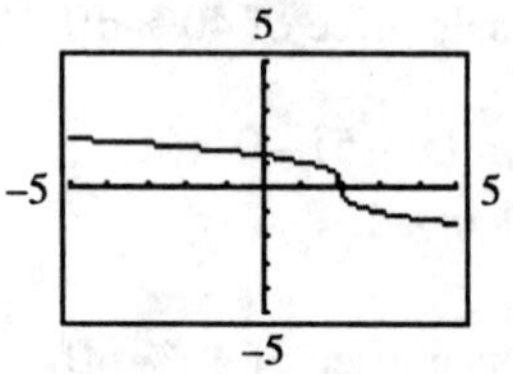

One-to-one

81. $f(x)=\frac{x^4}{4}$

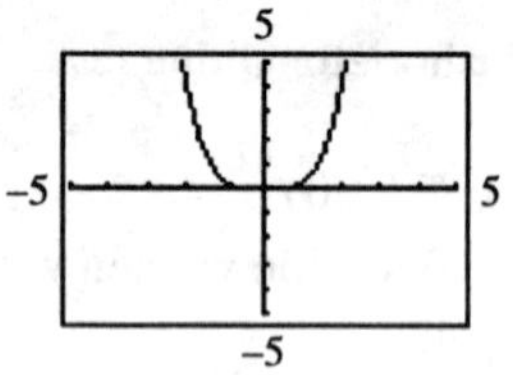

Not one-to-one

83. $f(x)=\frac{1}{x^2}$

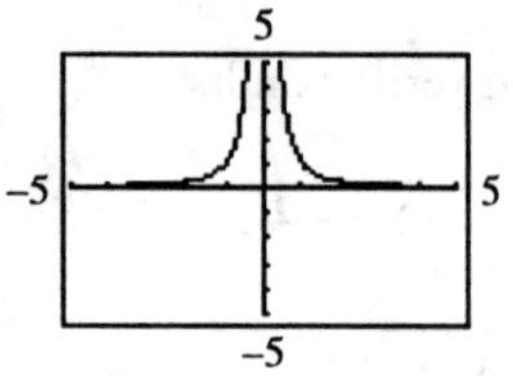

Not one-to-one

85. $f(x)=\frac{1}{(x-1)^2}$

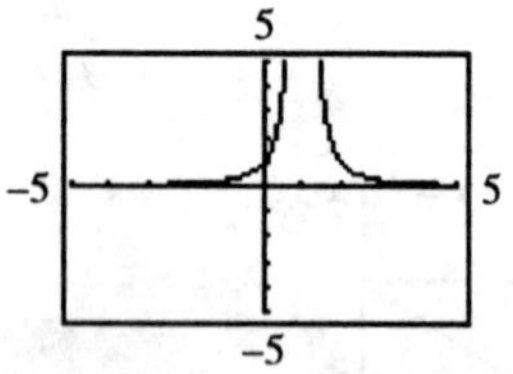

Not one-to-one

87–91. Answers may vary.

93. $f(x)=3x-2,\ g(x)=\ x^5$

$g(x)-f(x)=x^5-(3x-2)=x^5-3x+2$

$f(g(x)-f(x))\ =f(x^5-3x+2)$

$=3(x^5-3x+2)-2$

$=3x^5-9x+4$

$f(g(x))=3x^5-2$

$f(f(x))=3(3x-2)-2=9x-8$

$f(g(x))-f(f(x))$

$=3x^5-2-(9x-8)=3x^5-9x+6$

95. $f(x)=2x-5$ and $g(x)=3x+b$

$f(g(x))=g(f(x))$

$2(3x+b)-5=3(2x-5)+b$

$6x+2b-5=6x-15+b$

$b=-10$

97. $f(x)=6-5x$

$(f^{-1})^{-1}(x)=f(x)=6-5x$

99. $f(x)=3-\frac{2\sqrt[3]{3x+2}}{3}=y$

$2\sqrt[3]{3x+2}=9-3y$

$3x+2=\left(\frac{9-3y}{2}\right)^3$

$x=\frac{1}{3}\left[\left(\frac{9-3y}{2}\right)^3-2\right]$

$f^{-1}(x)=\frac{1}{3}\left[\left(\frac{9-3x}{2}\right)^3-2\right]$

101. $f(5)=13$

$1+f^{-1}(2x+3)=6$

$f^{-1}(2x+3)=6-1=5$

$2x+3=f(5)=13$

$2x=10$

$x=5$

103. $f(x)=\frac{x+1}{x-2}$

$g(x)=\frac{2x+1}{x-1}$

$g[f(x)]=\frac{2\left(\frac{x+1}{x-2}\right)+1}{\frac{x+1}{x-2}-1}$

$=\frac{2(x+1)+(x-2)}{(x+1)-(x-2)}$

$=\frac{3x}{3}=x$

Review Problems

105. $y = x^2 - 4x + 3$
x-intercepts (set $y = 0$):
$0 = x^2 - 4x + 3$
$0 = (x-3)(x-1)$
$x = 3$ and $x = 1$
y-intercept (set $x = 0$): $y = 3$
vertex: $x = -\frac{b}{2a} = -\frac{(-4)}{2(1)} = \frac{4}{2} = 2$
$y = 2^2 - 4(2) + 3 = -1$
Vertex: (2, −1)

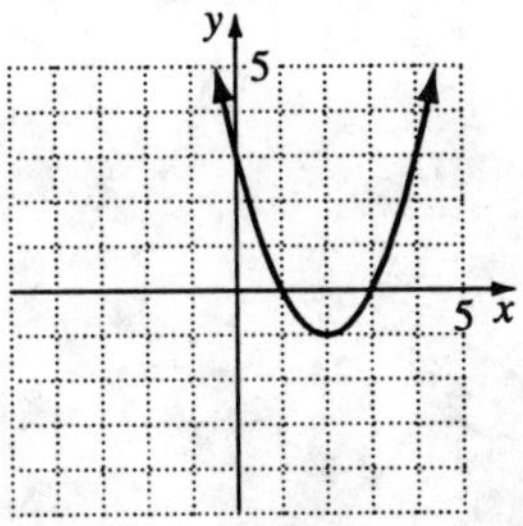

106. $a^4 - 16b^4 = (a^2 - 4b^2)(a^2 + 4b^2)$
$= (a-2b)(a+2b)(a^2 + 4b^2)$

107. $\frac{4.3 \times 10^5}{8.6 \times 10^{-4}} = \frac{4.3}{8.6} \times 10^{5+4}$
$= 0.5 \times 10^9$
$= 5 \times 10^8$

Problem Set 8.3

1. $\log_2 16 = 4$
$2^4 = 16$

3. $\log_5\left(\frac{1}{125}\right) = -3$
$5^{-3} = \frac{1}{125}$

5. $\log_{25} 5 = \frac{1}{2}$
$25^{1/2} = 5$

7. $y = \log_3 8$
$3^y = 8$

9. $\log_m P = c$
$m^c = P$

11. $\ln 5 = 1.6094$
$e^{1.6094} = 5$

13. $\log_b b^x = x$
$b^x = b^x$

15. $\log 10 = 1$
$10^1 = 10$

17. $2^3 = 8$
$\log_2 8 = 3$

19. $2^{-4} = \frac{1}{16}$
$\log_2 \frac{1}{16} = -4$

21. $\sqrt[3]{8} = 2$
$8^{1/3} = 2$
$\log_8 2 = \frac{1}{3}$

23. $8^{1/3} = 2$
$\log_8 2 = \frac{1}{3}$

25. $16^{3/4} = 8$
$\log_{16} 8 = \frac{3}{4}$

27. $10^2 = 100$
$\log_{10} 100 = 2$

29. $e^4 = 54.5982$
$\ln 54.5982 = 4$

31. $e^{-2} = 0.1353$
$\ln 0.1353 = -2$

33. $e^y = x$
$\ln x = y$

35. $P^a = m$
$\log_P m = a$

37. $\log_3 9 = x$
$3^x = 9$
$x = 2$

39. $\log_2 32 = x$
$2^x = 32$
$x = 5$

41. $\log_7 \sqrt{7} = x$
$7^x = \sqrt{7}$
$7^x = 7^{1/2}$
$x = \frac{1}{2}$

43. $\log_7\left(\frac{1}{7}\right) = x$
$7^x = \frac{1}{7}$
$x = -1$

45. $\log_2\left(\frac{1}{32}\right) = x$
$2^x = \frac{1}{32}$
$2^x = 2^{-5}$
$x = -5$

47. $\log_{81} 9 = x$
$81^x = 9$
$x = \frac{1}{2}$

49. $\log_8 8 = x$
$8^x = 8$
$x = 1$

51. $\log_4 1 = x$
$4^x = 1$
$x = 0$

53. $\log_5(-5) = x$
$5^x = -5$
Not possible, no value of x will make $5^x = -5$.

55. $\log_{16} 8 = x$
$16^x = 8$
$16^{3/4} = 8$
$x = \frac{3}{4}$

57. $\log 1000 = x$
$10^x = 1000$
$x = 3$

59. $\log 0.01 = x$
$10^x = 0.01$
$x = -2$

61. $\ln e^4 = x$
$e^x = e^4$
$x = 4$

63. $\ln(-1) = x$
$e^x = -1$
Not possible, no value of x makes $e^x = -1$.

65. $\log_5 5^6 = x$
$5^x = 5^6$
$x = 6$

67. $\log 10^{21} = x$
$10^x = 10^{21}$
$x = 21$

69. $\log_5(\log_2 32) = x$
$5^x = \log_2 32$
$2^{5^x} = 32$
$x = 1$

71. **a.** $f(x) = 3^x$

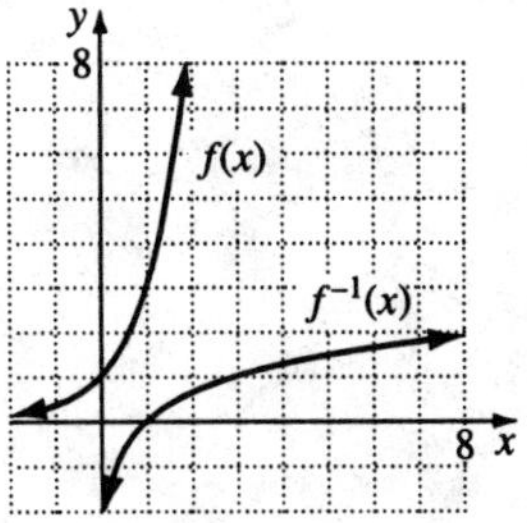

b. $f[\log_3 x] = 3^{\log_3 x} = x$

$f^{-1}[3^x] = \log_3 3^x = x$

Therefore, $\log_3 x = f^{-1}(x)$.

c. See part a.

73. **a.** $f(x) = \left(\frac{1}{2}\right)^x$

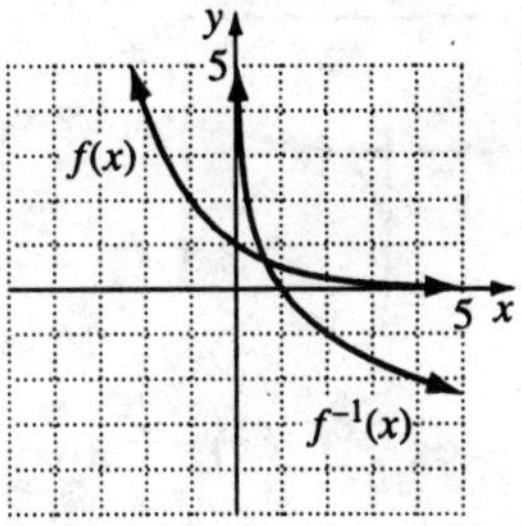

b. $f[\log_{1/2} x] = \frac{1}{2}\log_{1/2} x = x$

$f^{-1}\left[\left(\frac{1}{2}\right)^x\right] = \log_{1/2}\left(\frac{1}{2}\right)^x = x$

Therefore $\log_{1/2}(x) = f^{-1}(x)$

c. See part a.

75. $f(t) = 88 - 15\ln(t+1),\ 0 \le t \le 12$

a. $t = 0$

$f(0) = 88 - 15\ln(0 + 1)$

$= 88 - 15\ln 1$

$= 88 - 15(0)$

$= 88$

The original score was 88 points.

b. $f(2) = 88 - 15\ln(2 + 1) \approx 71.5$ points

$f(4) = 88 - 15\ln(4 + 1) \approx 63.9$ points

$f(6) = 88 - 15\ln(6 + 1) \approx 58.8$ points

$f(8) = 88 - 15\ln(8 + 1) \approx 55.0$ points

$f(12) = 88 - 15\ln(12 + 1) \approx 49.5$ points

c. $f(t) = 88 - 15\ln(t + 1)$

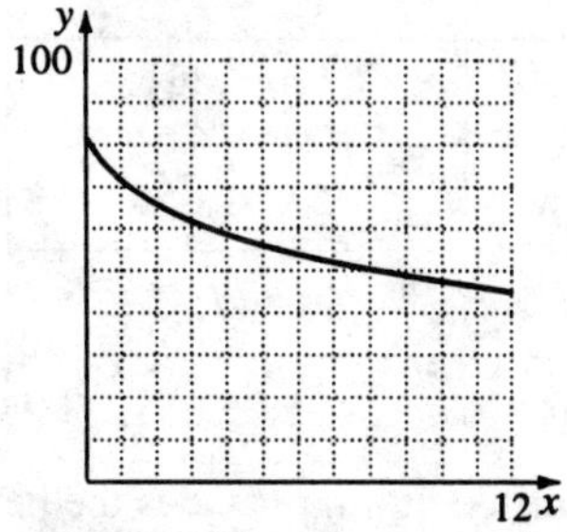

The students retained less material as time passed.

77. $f(R) = \dfrac{5600\log R}{\log 0.5}$

$f(0.25) = \dfrac{5600\log 0.25}{\log 0.5}$

$= \dfrac{5600(-0.60206)}{-0.30103} = 11{,}200$

The age is 11,200 years.

79.

x	Number of prime numbers that are less than x	$\frac{x}{\ln x}$	$\frac{x}{\ln x - 1.08366}$
100	25	22	28
10,000	1229	1086	1231
10^6	78,498	72,382	78,543
10^8	5,761,455	5,428,681	5,768,004
10^9	50,847,534	48,254,942	50,917,519
10^{10}	455,052,512	434,294,482	455,743,044

The fourth column provides the better estimate. They get further away.

81. d is true.

83. $f(x) = \ln x$, $g(x) = \ln(x + 3)$

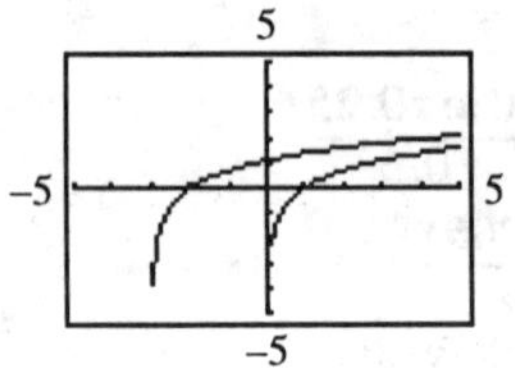

$g(x)$ is $f(x)$ shifted 3 units left.

85. $f(x) = \log x$, $g(x) = -\log x$

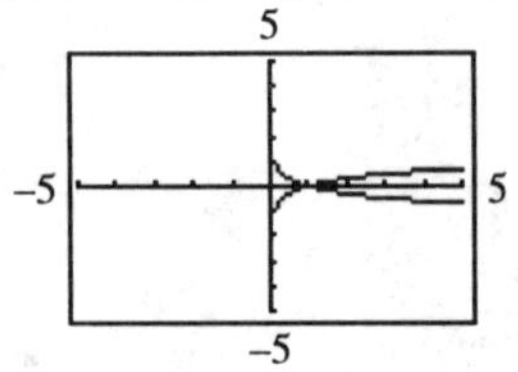

$g(x)$ is $f(x)$ reflected across the x-axis.

87. $f(t) = 75 - 10\log(t + 1), 0 \le t \le 12$

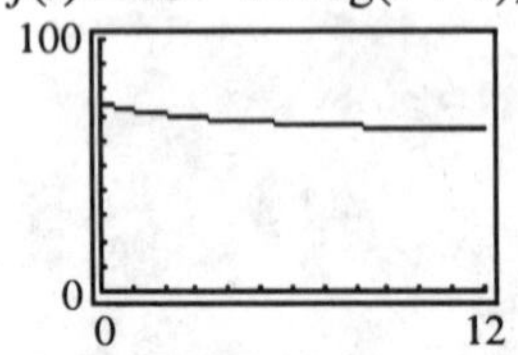

9 months must pass

89. a. $f(x) = \ln\frac{x}{2}$, $g(x) = \ln x - \ln 2$

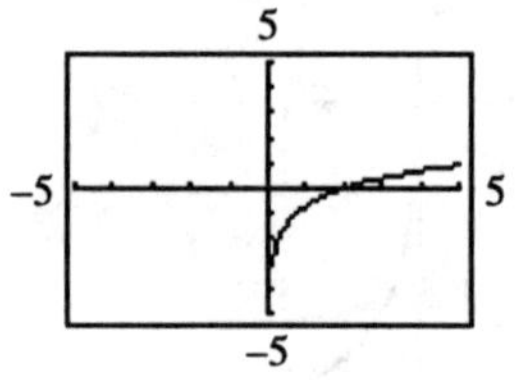

b. $f(x) = \log\frac{x}{5}$, $g(x) = \log x - \log 5$

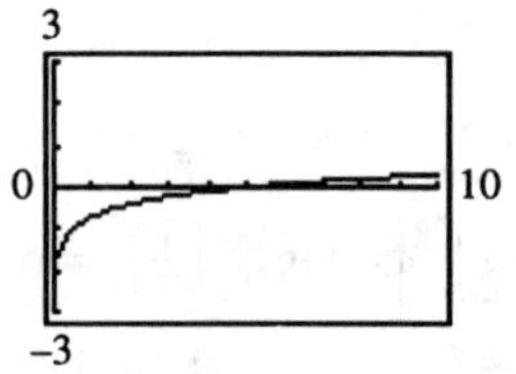

c. $f(x) = \ln\frac{x^2}{3}$, $g(x) = \ln x^2 - \ln 3$

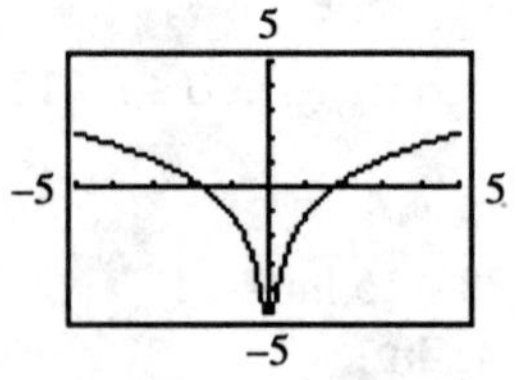

d. $f(x) = g(x)$;

$$\log_b\left(\frac{M}{N}\right) = \log_b M - \log_b N$$

e. the difference of the logs

91. In order from the one that increases most slowly to the one that increases most rapidly.

$y = \ln x,\ y = \sqrt{x},\ y = x,\ y = x^2,\ y = e^x,$

$y = x^x$

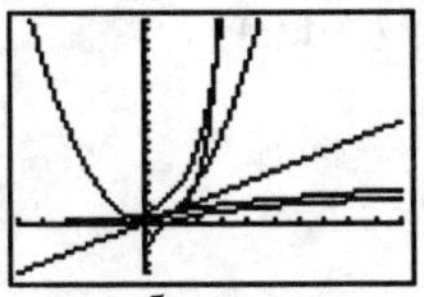

93. Answers may vary.

95. $\log_{25} 5 = x$

$25^x = 5$

$(5^2)^x = 5$

$5^{2x} = 5^1$

$2x = 1$

$x = \frac{1}{2}$

$\log_{1/16} 8 = y$

$\left(\frac{1}{16}\right)^y = 8$

$y = -\frac{3}{4}$

$\log_9 \frac{1}{27} = z$

$9^z = \frac{1}{27}$

$(3^2)^z = 3^{-3}$

$3^{2z} = 3^{-3}$

$2z = -3$

$z = -\frac{3}{2}$

$\log_4 1 = 0$

Thus,

$$\frac{\log_{25} 5 - \log_{1/16} 8}{\log_9 \frac{1}{27} + \log_4 1} = \frac{\frac{1}{2} - \left(-\frac{3}{4}\right)}{-\frac{3}{2} + 0}$$

$$= \frac{\frac{5}{4}}{-\frac{3}{2}} = -\frac{5}{6}$$

97. $\log_5 1 = x$

$5^x = 1$

$5^x = 5^0$

$x = 0$

$\log_8\left[4(\sqrt[5]{16})\right] = y$

$8^y = (4)(\sqrt[5]{16})$

$(2^3)^y = (2^2)(\sqrt[5]{2^4})$

$2^{3y} = 2^{2+4/5}$

$2^{3y} = 2^{14/5}$

$3y = \frac{14}{5}$

$y = \frac{14}{15}$

Thus,

$\log_5 1 + \log_8 (4)(\sqrt[5]{16}) = 0 + \frac{14}{15} = \frac{14}{15}$

99. $A(0, 1)$; $B(1, 0)$; $C(4, 2)$; $D(2, 4)$

101. $y = -\log_2 x$

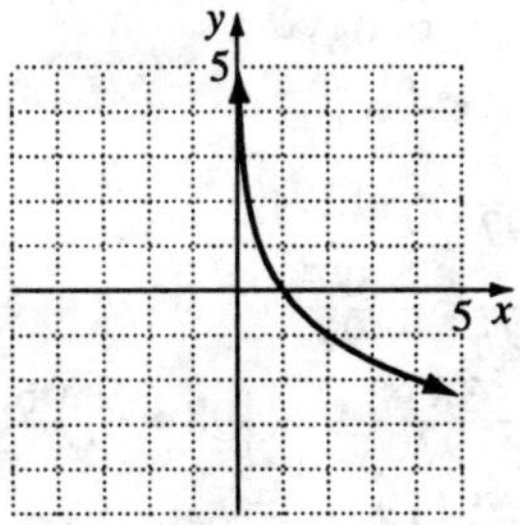

103. $\log_{1/25} 25\sqrt[3]{25} = x$

(Set the expression equal to x.)

$\log_{1/25} 25^{1+1/3} = x$

$\log_{1/25} 25^{4/3} = x$

$\left(\frac{1}{25}\right)^x = 25^{4/3}$ (Write in exponential form.)

$(25^{-1})^x = 25^{4/3}$

(Express with a common base.)

$25^{-x} = 25^{4/3}$

$-x = \frac{4}{3}$ (If $b^x = b^y$, then $x = y$.)

$x = -\frac{4}{3}$

Thus,

$\log_{1/25} 25\sqrt[3]{25} = -\frac{4}{3}$

Review Problems

104. $d = \sqrt{(x_2 - x_1)^2 + (y_2 - y_1)^2}$
$= \sqrt{(1+2)^2 + (-3+9)^2}$
$= \sqrt{3^2 + 6^2}$
$= \sqrt{9+36}$
$= \sqrt{45}$
$= \sqrt{(9)(5)}$
$= 3\sqrt{5}$

105. a. $m = \frac{y_2 - y_1}{x_2 - x_1}$
$= \frac{104-98}{323-129}$
$= \frac{6}{194}$
$= \frac{3}{97}$
$(x_1, y_1) = (129, 98)$
$y - y_1 = m(x - x_1)$
$y - 98 = \frac{3}{97}(x - 129)$
$y - 98 = \frac{3}{97}x - \frac{387}{97}$
$y = \frac{3}{97}x - \frac{387}{97} + 98 \quad \left[98 = \frac{9506}{97}\right]$
$y = \frac{3}{97}x + \frac{9119}{97}$

b. $y = \frac{3}{97}(800) + \frac{9119}{97}$
$y = \frac{11519}{97} \approx 119$
Approximately 119 people will die.

106. $2x = 11 - 5y$ (1)
$3x - 2y = -12$ (2)

Multiply equation (1) by 2.
Multiply equation (2) by 5.

$4x + 10y = 22$
$15x - 10y = -60$
$19x = -38$ (Add.)
$x = -2$

$2x + 5y = 11$
$2(-2) + 5y = 11$
$-4 + 5y = 11$
$5y = 15$
$y = 3$
$\{(-2, 3)\}$

Problem Set 8.4

1. $\log(10 \cdot 100) = \log 1000 = x$
$10^x = 1000$
$x = 3$

$\log 10 = y$
$10^y = 10$
$y = 1$
$\log 100 = z$
$10^z = 100$
$z = 2$

$\log(10)(100) = \log 10 + \log 100$
$3 = 1 + 2$
$3 = 3$ (verified)

3. $\log_3 9\left(\frac{1}{3}\right) = \log_3 3 = x$
$3^x = 3$
$x = 1$

$\log_3 9 = y$
$3^y = 9$
$y = 2$

$\log_3\left(\frac{1}{3}\right) = z$
$3^z = \frac{1}{3}$
$z = -1$

$\log_3(9)\left(\frac{1}{3}\right) = \log_3 9 + \log_3\left(\frac{1}{3}\right)$
$1 = 2 + (-1)$
$1 = 1$ (verified)

5. $\log_3\left(\frac{81}{3}\right) = \log_3 27 = 3$
$\log_3 81 - \log_3 3 = 4 - 1 = 3$
$3 = 3$ (verified)

7. $\ln\left(\frac{e^{17}}{e^4}\right) = \ln e^{13} = x$
$e^x = e^{13}$
$x = 13$

$\ln e^{17} = y$
$e^y = e^{17}$
$y = 17$

$\ln e^4 = z$
$e^z = e^4$
$z = 4$

$\ln\left(\frac{e^{17}}{e^4}\right) = \ln e^{17} - \ln e^4$
$13 = 17 - 4$
$13 = 13$ (verified)

9. $\log_5 5^3 = \log_5 125 = x$
$5^x = 125$
$x = 3$

$\log_5 5 = y$
$5^y = 5$
$y = 1$

$\log_5 5^3 = 3\log_5 5$
$3 = 3(1)$
$3 = 3$ (verified)

11. $\log_5 25^{1/2} = \log_5 5 = 1$
$\frac{1}{2}\log_5 25 = \frac{1}{2}(2) = 1$
$1 = 1$ (verified)

13. $\log_3 3x = \log_3 3 + \log_3 x = 1 + \log_3 x$

15. $\log 1000x$
$= \log 1000 + \log x$
$= \log 10^3 + \log x$
$= 3 + \log x$

17. $\log_4 Bx = \log_4 B + \log_4 x$

19. $\log_7\left(\frac{7}{x}\right) = \log_7 7 - \log_7 x = 1 - \log_7 x$

21. $\log\left(\frac{x}{100}\right)$
$= \log x - \log 100$
$= \log x - \log 10^2$
$= \log x - 2$

23. $\log_4\left(\frac{64}{y}\right)$
$= \log_4 64 - \log_4 y$
$= \log_4 4^3 - \log_4 y$
$= 3 - \log_4 y$

25. $\log_b x^3 = 3\log_b x$

27. $\log N^{-6} = -6\log N$

29. $\ln \sqrt[3]{x} = \ln x^{1/3} = \frac{1}{3}\ln x$

31. $\log_b x^2y = \log_b x^2 + \log_b y$
$= 2\log_b x + \log_b y$

33. $\log_4\left(\frac{\sqrt{x}}{16}\right)$
$= \log_4 \sqrt{x} - \log_4 16$
$= \log_4 x^{1/2} - \log_4 4^2$
$= \frac{1}{2}\log_4 x - 2$

35. $\log_b \frac{x^2 y}{z^4}$
$= \log_b x^2 y - \log_b z^4$
$= 2\log_b x + \log_b y - 4\log_b z$

37. $\log \sqrt[3]{100x}$
$= \frac{1}{3}\log 100x$
$= \frac{1}{3}(\log 100 + \log x)$
$= \frac{1}{3}(2 + \log x)$

39. $\ln \sqrt[4]{\frac{e^3 x}{yz^2}}$
$= \frac{1}{4}\ln \frac{e^3 x}{yz^2} = \frac{1}{4}(\ln e^3 x - \ln yz^2)$
$= \frac{1}{4}(3\ln e + \ln x - \ln y - 2\ln z)$
$= \frac{1}{4}(3 + \ln x - \ln y - 2\ln z)$

41. $\log_b \sqrt[5]{\frac{x^{10} y^{15}}{b^3 z^6}}$
$= \frac{1}{5}\log \frac{x^{10} y^{15}}{b^3 z^6}$
$= \frac{1}{5}(\log x^{10} y^{15} - \log b^3 z^6)$
$= \frac{1}{5}(10\log x + 15\log y - 3\log b - 6\log z)$

43. $\log 5 + \log 2 = \log(5 \cdot 2) = \log 10 = 1$

45. $\log_2 96 - \log_2 3$
$= \log_2\left(\frac{96}{3}\right)$
$= \log_2 32 = \log_2 2^5 = 5$

47. $\log 5 - 5\log x = \log 5 - \log x^5 = \log\left(\frac{5}{x^5}\right)$

49. $2\log_b x + 3\log_b y$
$= \log_b x^2 + \log_b y^3 = \log_b x^2 y^3$

51. $5\ln x - 2\ln y = \ln x^5 - \ln y^2 = \ln\left(\frac{x^5}{y^2}\right)$

53. $2\ln x - \frac{1}{2}\ln y$
$= \ln x^2 - \ln y^{1/2}$
$= \ln\left(\frac{x^2}{\sqrt{y}}\right)$

55. $\frac{1}{2}\log x + \frac{1}{3}\log y$
$= \log x^{1/2} + \log y^{1/3}$
$= \log(\sqrt{x})(\sqrt[3]{y})$

57. $\frac{1}{2}\log x + 3\log y - 2\log z$
$= \log x^{1/2} + \log y^3 - \log z^2$
$= \log\left(\frac{\sqrt{x} \cdot y^3}{z^2}\right)$

59. $\log_3(x^2 - 9) - \log_3(x - 3)$
$= \log_3\left(\frac{x^2 - 9}{x - 3}\right) = \log_3(x + 3)$

61. $\ln x + 2(\ln y - \ln z)$
$= \ln x + \ln\left(\frac{y}{z}\right)^2 = \ln x\left(\frac{y}{z}\right)^2$

63. $\log_b\left(\frac{b}{\sqrt{x}}\right) - \log_b \sqrt{bx}$
$= \log_b\left(\frac{\frac{b}{\sqrt{x}}}{\sqrt{bx}}\right)$

$= \log_b\left(\frac{b}{\sqrt{bx^2}}\right)$

$= \log_b\left(\frac{b}{x\sqrt{b}}\right)$

65. $\log_5 13 = \frac{\ln 13}{\ln 5} \approx 1.5937$

67. $\log_3 1.87 = \frac{\ln 1.87}{\ln 3} \approx 0.5698$

69. $\log_9 9.63 = \frac{\ln 9.63}{\ln 9} \approx 1.0308$

71. $\log_{14} 87.5 = \frac{\ln 87.5}{\ln 14} \approx 1.6944$

73. $\log_6 0.724 = \frac{\log 0.724}{\log 6} \approx -0.1802$

75. $\log_{0.1} 17 = \frac{\log 17}{\log 0.1} \approx -1.2304$

77. $\log_{200} 40 = \frac{\log 40}{\log 200} \approx 0.6962$

79. $\log_\pi 63 = \frac{\log 63}{\log \pi} \approx 3.6193$

81. $D = 10\log\frac{I}{I_0}$

$D = 10\log\frac{100I_0}{I_0}$

$D = 10 \log 100$

$D = 10(2)$

$D = 20$

20 decibels larger

83. d is true.

$\ln\sqrt{2} = \ln 2^{1/2}$

$= \frac{1}{2}\ln 2$

$= \frac{\ln 2}{2}$

85. b is true.

$\ln e^x = \log_e e^x = x$

87. b is true.

$\log_b \sqrt{\frac{xy}{z}} = \log_b\left(\frac{xy}{z}\right)^{1/2}$

$= \frac{1}{2}\log_b\left(\frac{xy}{z}\right)$

$= \frac{1}{2}(\log_b x + \log_b y - \log_b z)$

$(x > 0, y > 0, z > 0)$

89. $y = \log x$, $y = \log(10x)$, $y = \log(0.1x)$

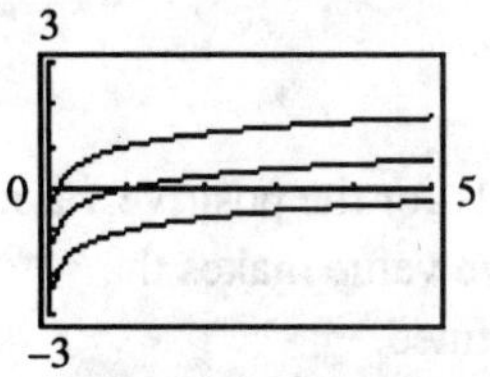

$y = \log(10x)$ is $y = \log x$ shifted 1 unit up.

$y = \log(0.1x)$ is $y = \log x$ shifted 1 unit down.

The Product Rule accounts for this relationship.

91–99. Answers may vary.

101. $\log_b M = R$ and $\log_b N = S$

$b^R = M$ and $b^s = N$

$\frac{M}{N} = \frac{b^R}{b^S} = b^{R-S}$

Rewrite $\frac{M}{N} = b^{R-S}$ in logarithmic form.

$\log_b\left(\frac{M}{N}\right) = R - S$

Substitute the original expressions for R and S.

$\log_b\left(\frac{M}{N}\right) = \log_b M - \log_b N$

103. $\log_7 9 = \frac{\log 9}{\log 7} = \frac{\log 3^2}{\log 7} = \frac{2\log 3}{\log 7} = \frac{2A}{B}$

105. $\log(\sqrt{13} - \sqrt{3})$

$= \log(\sqrt{13} - \sqrt{3}) \cdot \frac{(\sqrt{13} + \sqrt{3})}{(\sqrt{13} + \sqrt{3})}$

$= \log\frac{13 - 3}{\sqrt{13} + \sqrt{3}}$

$= \log\frac{10}{\sqrt{13}+\sqrt{3}}$
$= \log 10 - \log(\sqrt{13}+\sqrt{3})$
$= 1 - A$

107. $4^{\log_2 x} = 2^{2\log_2 x} = 2^{\log_2 x^2} = x^2$
$x^2 + 4^{\log_2 x} = 8$ (Given equation)
$x^2 + x^2 = 8$ (Substitute from above.)
$2x^2 = 8$
$x^2 = 4$
$x = 2$
The problem calls for the positive value of x since the negative value makes the logarithms undefined.

109. Since $\log_9 16 = x$ means that $9^x = 16$, in order to show that $\log_9 16 = \log_3 4$, we must show that
$9^{\log_3 4} \stackrel{?}{=} 16$
$(3^2)^{\log_3 4} \stackrel{?}{=} 16$
$3^{2\log_3 4} \stackrel{?}{=} 16$
$3^{\log_3 4^2} \stackrel{?}{=} 16$
$3^{\log_3 16} \stackrel{?}{=} 16$
$16 = 16$ (The equality is established since $b^{\log_b x} = x$.)

Review Problems

110. $f(x) = 3x + 17$
$y = 3x + 17$
$x = 3y + 17$ (Exchange x and y.)
$x - 17 = 3y$
$\frac{x-17}{3} = y$
$f^{-1}(x) = \frac{x-17}{3}$

$f\left(f^{-1}(x)\right) = 3\left[f^{-1}(x)\right] + 17$
$= 3\left(\frac{x-17}{3}\right) + 17$
$= x - 17 + 17 = x$

$f^{-1}(f(x)) = \frac{f(x)-17}{3}$
$= \frac{3x+17-17}{3}$
$= \frac{3x}{3} = x$ (verified)

111. $x - 5y - 2z = 6$ (1)
$2x - 3y + z = 13$ (2)
$3x - 2y + 4z = 22$ (3)

Equations 1 and 2:
(Multiply equation 2 by 2.)
$x - 5y - 2z = 6$
$4x - 6y + 2z = 26$
$5x - 11y = 32$
The result becomes equation 4.

Equations 1 and 3:
(Multiply equation 1 by 2.)
$2x - 10y - 4z = 12$
$3x - 2y + 4z = 22$
$5x - 12y = 34$
The result becomes equation 5.

Equations 4 and 5:
(Multiply equation 4 by –1.)
$-5x + 11y = -32$
$5x - 12y = 34$
$-y = 2$
$y = -2$

Equation 4:
$5x - 11(-2) = 32$
$5x = 10$
$x = 2$

Equation 2:
$2(2) - 3(-2) + z = 13$
$10 + z = 13$
$z = 3$
$\{(2, -2, 3)\}$

112. $5x - 2y > 10$

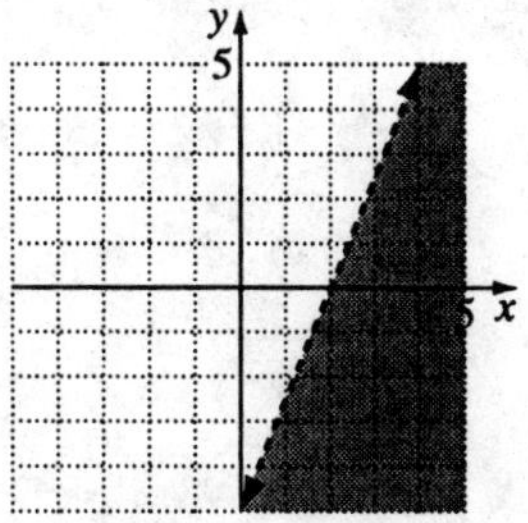

Problem Set 8.5

1. $2^x = 64$
$2^x = 2^6$
$x = 6$
$\{6\}$

3. $5^x = 125$
$5^x = 5^3$
$x = 3$
$\{3\}$

5. $2^{2x-1} = 32$
$2^{2x-1} = 2^5$
$2x - 1 = 5$
$2x = 6$
$x = 3$
$\{3\}$

7. $4^{2x-1} = 64$
$4^{2x-1} = 4^3$
$2x - 1 = 3$
$2x = 4$
$x = 2$
$\{2\}$

9. $32^x = 8$
$2^{5x} = 2^3$
$5x = 3$
$x = \frac{3}{5}$
$\left\{\frac{3}{5}\right\}$

11. $9^x = 27$
$3^{2x} = 3^3$
$2x = 3$
$x = \frac{3}{2}$
$\left\{\frac{3}{2}\right\}$

13. $9^x = \frac{1}{81}$
$3^{2x} = 3^{-4}$
$2x = -4$
$x = -2$
$\{-2\}$

15. $4^{2x-7} = \frac{1}{128}$
$(2^2)^{2x-7} = 2^{-7}$
$2(2x - 7) = -7$
$4x - 14 = -7$
$4x = 7$
$x = \frac{7}{4}$
$\left\{\frac{7}{4}\right\}$

17. $2^x = 7$
$\ln 2^x = \ln 7$
$x \ln 2 = \ln 7$
$x = \frac{\ln 7}{\ln 2} \approx 2.807$

19. $e^x = 5$
$\ln e^x = \ln 5$
$x = \ln 5 \approx 1.609$

21. $(2.7)^x = 31$
$\ln(2.7)^x = \ln 31$
$x \ln(2.7) = \ln 31$
$x = \frac{\ln(2.7)}{\ln 31} \approx 0.289$

23. $e^{0.5x} = 9$
$\ln e^{0.5x} = \ln 9$
$0.5x = \ln 9$
$x = \frac{\ln 9}{0.5} \approx 4.394$

25. $5^{-0.03t} = 0.07$
$\ln 5^{-0.03t} = \ln 0.07$
$-0.03t \ln 5 = \ln 0.07$
$t = \frac{\ln 0.07}{-0.03 \ln 5} \approx 55.076$

27. $e^{-0.03t} = 0.09$
$\ln e^{-0.03t} = \ln 0.09$
$-0.03t = \ln 0.09$
$t = \frac{\ln 0.09}{-0.03} \approx 80.265$

29. $30 - (1.4)^x = 0$
$30 = (1.4)^x$
$\ln 30 = \ln(1.4)^x$
$\ln 30 = x \ln(1.4)$
$x = \frac{\ln 30}{\ln(1.4)} \approx 10.108$

31. $1250e^{0.055t} = 3750$
$e^{0.055t} = 3$
$\ln e^{0.055t} = \ln 3$
$0.055t = \ln 3$
$t = \frac{\ln 3}{0.055} \approx 19.975$

33. $800 - 500 \cdot 2^{-0.5t} = 733$
$-500 \cdot 2^{-0.5t} = -67$
$2^{-0.5t} = 0.134$
$\ln 2^{-0.5t} = \ln 0.134$
$-0.5t \ln 2 = \ln 0.134$
$t = \frac{\ln 0.134}{-0.5 \ln 2} \approx 5.799$

35. $\log_3 x = 4$
$x = 3^4$
$x = 81$
$\{81\}$

37. $\log_2 x = -4$
$x = 2^{-4}$
$x = \frac{1}{16}$
$\left\{\frac{1}{16}\right\}$

39. $\log x = 2$
$x = 10^2$
$x = 100$
$\{100\}$

41. $\ln x = 3$
$x = e^3$
$\left\{e^3\right\}$

43. $\log_4(x+5) = 3$
$x + 5 = 4^3$
$x = 64 - 5$
$x = 59$
$\{59\}$

45. $\log_3(x-2) = -3$
$x - 2 = 3^{-3}$
$x = \frac{1}{27} + 2$
$x = \frac{55}{27}$
$\left\{\frac{55}{27}\right\}$

47. $\log_4(2x-1) = 3$
$2x - 1 = 4^3$
$2x = 64 + 1$
$x = \frac{65}{2}$
$\left\{\frac{65}{2}\right\}$

49. $\log_5 x + \log_5(4x-1) = 1$
$\log_5 x(4x-1) = 1$
$x(4x-1) = 5^1$
$4x^2 - x = 5$
$4x^2 - x - 5 = 0$
$(4x - 5)(x + 1) = 0$

$x = \frac{5}{4}$ or $x = -1$ (reject)

$\left\{\frac{5}{4}\right\}$

51. $\log_3(x-5)+\log_3(x+3)=2$

$\log_3(x-5)(x+3)=2$

$(x-5)(x+3)=3^2$

$x^2-2x-15=9$

$x^2-2x-24=0$

$(x-6)(x+4)=0$

$x=6,\ x=-4$ (reject)

$\{6\}$

53. $\log_4(x+2)-\log_4(x-1)=1$

$\log_4\left(\frac{x+2}{x-1}\right)=1$

$\frac{x+2}{x-1}=4^1$

$x+2=4x-4$

$-3x=-6$

$x=2$

$\{2\}$

55. $\log(3x-5)-\log 5x=2$

$\log\left(\frac{3x-5}{5x}\right)=2$

$\frac{3x-5}{5x}=10^2$

$3x-5=500x$

$-497x=5$

$x=-\frac{5}{497}$ (reject)

No solution

57. $\log x-\log(x+5)=-1$

$\log\left(\frac{x}{x+5}\right)=-1$

$\frac{x}{x+5}=10^{-1}$

$10x=x+5$

$9x=5$

$x=\frac{5}{9}$

$\left\{\frac{5}{9}\right\}$

59. $A=P\left(1+\frac{r}{n}\right)^{nt}$

$(A=2600,\ P=1000,\ n=4,\ r=6\%=0.06)$:

$2600=1000\left(1+\frac{0.06}{4}\right)^{4t}$

$(1+0.015)^{4t}=2.6$

$(1.015)^{4t}=2.6$

$\log(1.015)^{4t}=\log 2.6$

$4t\log(1.015)=\log 2.6$

$t=\frac{\log 2.6}{4\log(1.015)}\approx 16.0$

16 years

61. $A=Pe^{rt}$

$12{,}000=8{,}000e^{5r}$

$e^{5r}=1.5$

$\ln e^{5r}=\ln 1.5$

$5r=\ln 1.5$

$r=\frac{\ln 1.5}{5}\approx 0.0811\approx 8.1\%$

Required interest 8.1%

63. $f(x)=80e^{-0.5x}+20$

$80e^{-0.5x}+20=50$

$80e^{-0.5x}=30$

$e^{-0.5x}=0.375$

$\ln e^{-0.5x}=\ln 0.375$

$-0.5x=\ln 0.375$

$x=\frac{\ln 0.375}{-0.5}\approx 2.0$

After 2 weeks

65. $y=Ae^{kt}$

$6.37=5.702e^{0.015t}$

$t=\frac{\ln\left(\frac{6.37}{5.702}\right)}{0.015}\approx 7$

$1995+7=2002$

For less developed regions,

population $=4.533e^{0.019(7)}\approx 5.178$ billion

Percent $=\frac{5.178}{6.37}\cdot 100\approx 81.3\%$

67. $P(x) = 95 - 30\log_2 x \quad P(x) = 50$
$50 = 95 - 30\log_2 x$
$30\log_2 x = 95 - 50 = 45$
$\log_2 x = \frac{45}{30} = 1.5$
$x = 2^{1.5} \approx 2.83$ days
After approximately 2.83 days only half the students will recall the important features. The value can be located approximately on the graph.

69. $\text{pH} = -\log[H^+]$
$7.37 = -\log[H^+]$
$\log[H^+] = -7.37$
$H^+ = 10^{-7.37} \approx 4.266 \times 10^{-8}$
$7.44 = -\log[H^+]$
$H^+ = 10^{-7.44} \approx 3.631 \times 10^{-8}$
The range is from
3.631×10^{-8} to 4.266×10^{-8}

71. d is true.
$\log(x + 1) + \log(x - 1) = 5$
$\log(x + 1)(x - 1) = 5$
$10^5 = x^2 - 1$

73. Students should verify results.

75. $P = 145e^{-0.092t}$

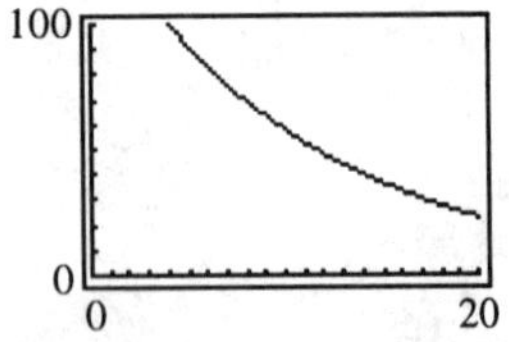

About 7.9 minutes
$70 = 145e^{-0.092t}$
$\frac{70}{145} = e^{-0.092t}$
$\ln\left(\frac{70}{145}\right) = \ln e^{-0.092t}$
$\ln\left(\frac{70}{145}\right) = -0.092t$
$\frac{\ln\left(\frac{70}{145}\right)}{-0.092} = t$
$7.9 \approx t$

77. Answers may vary.

79. Iraq: 20.6 million (1995 population)
growth rate $(k) = 3.7\% = 0.037$
U.K. 58.6 million (1995 population)
growth rate $(k) = 0.2\% = 0.002$
Iraq: $P = 20.6e^{0.037x}$
U.K. $P = 58.6e^{0.002x}$
$20.6e^{0.037x} = 58.6e^{0.002x}$
$\frac{e^{0.037x}}{e^{0.002x}} = \frac{58.6}{20.6}$
$e^{(0.037-0.002)x} = \frac{58.6}{20.6}$
$0.035x = \ln\left(\frac{58.6}{20.6}\right)$
$x \approx 29.9$
$1995 + 30 = 2025$

81. $x + y = 25$
$\log x + \log y = 2$
$\log(xy) = 2$
$xy = 10^2 = 100$
$y = \frac{100}{x}$
$x + \frac{100}{x} = 25$
$x^2 - 25x + 100 = 0$
$(x - 20)(x - 5) = 0$
$x = 5$ or $x = 20$
$\{5, 20\}$

83. $\log_b(\log_b Ax) = 1 \quad (A > 0)$
$\log_b Ax = b$
$Ax = b^b$
$x = \frac{b^b}{A}$

85. $\log_x 25 - \log_x 4 = \log_x \sqrt{x}$

$\log_x \frac{25}{4} = \log_x \sqrt{x}$

$\frac{25}{4} = \sqrt{x}$

$\frac{625}{16} = x$

The solution set is $\left\{\frac{625}{16}\right\}$.

87. Group activity

Review Problems

88. Line passing through (4, 2) and (–2, 3) has slope $\frac{3-2}{-2-4} = -\frac{1}{6}$

slope of line perpendicular $= -\frac{1}{-\frac{1}{6}} = 6$

line passing through (1, 5) with slope 6 is

$y - 5 = 6(x - 1)$

$y - 5 = 6x - 6$

$y = 6x - 1$

89. $\sqrt{x+4} - \sqrt{x-1} = 1$

$\sqrt{x+4} = \sqrt{x-1} + 1$

$\left(\sqrt{x+4}\right)^2 = \left(\sqrt{x-1}+1\right)^2$

$x + 4 = x - 1 + 2\sqrt{x-1} + 1$

$2 = \sqrt{x-1}$

$2^2 = \left(\sqrt{x-1}\right)^2$

$4 = x - 1$

$5 = x$

$\{5\}$

90. $\frac{3}{y+1} - \frac{5}{y} = \frac{19}{y^2+y}$

$\frac{3}{y+1} - \frac{5}{y} = \frac{19}{y(y+1)}$

$y(y+1)\left[\frac{3}{y+1} - \frac{5}{y}\right] = y(y+1)\left[\frac{19}{y(y+1)}\right]$

$3y - 5(y + 1) = 19$

$3y - 5y - 5 = 19 \quad (y \neq 0, -1)$

$-2y = 24$

$y = -12$

$\{-12\}$

Problem Set 8.6

1. a. $y = Ae^{kt}$

$10,586,223 = 6,907,387e^{k(10)}$

$\ln\left(\frac{10,586,223}{6,907,387}\right) = \ln e^{10k}$

$k = \frac{\ln\left(\frac{10,586,223}{6,907,387}\right)}{10}$

$k \approx 0.0427$

b. $y = 6,907,387e^{0.0427t}$

c. $y = 6,907,387e^{0.0427(54)}$

$y = 69,296,295$

This is not very close to the actual.

d. Answers may vary.

3. a. $y = Ae^{kt}$

$56,299 = 40,637e^{k(3)}$

$\ln\left(\frac{56,299}{40,637}\right) = \ln e^{3k}$

$k = \frac{\ln\left(\frac{56,299}{40,637}\right)}{3}$

$k \approx 0.1087$

$y = 40,637e^{0.1087t}$

b. $500,000 = 40,637e^{0.1087t}$

$\ln\left(\frac{500,000}{40,637}\right) = \ln e^{0.1087t}$

$t = \frac{\ln\left(\frac{500,000}{40,637}\right)}{0.1087}$

$t \approx 23$

$1989 + 23 = 2012$

In year 2012

c.–d. Answers may vary.

5. a. $y = Ae^{kt}$

$\frac{A}{2} = Ae^{k(5730)}$

$\ln\left(\frac{\frac{A}{2}}{A}\right) = \ln e^{5730k}$

$k = \frac{\ln\left(\frac{1}{2}\right)}{5730} = -0.000121$

$y = Ae^{-0.000121t}$

b. $0.76A = Ae^{-0.000121t}$

$\ln\left(\frac{0.76A}{A}\right) = \ln e^{-0.000121t}$

$t = \frac{\ln 0.76}{-0.000121}$

$t \approx 2268$

About 2268 years

7. $y = Ae^{kx}$

$0.5A = Ae^{8k}$

$0.5 = e^{8k}$

$\ln 0.5 = 8k$

$k = \frac{\ln 0.5}{8} \approx -0.0866$

$y = Ae^{-0.0866x}$

$(A = 3000)$:

$y = 3000e^{-0.0866x}$

($y = 0.1$ gram):

$0.1 = 3000e^{-0.0866x}$

$\ln \frac{0.1}{3000} = -0.0866x$

$x = \frac{1}{-0.0866} \ln \frac{0.1}{3000} \approx 119$ days

9. a. $y = Ae^{kt}$

$\frac{A}{2} = Ae^{kt}$, for $k < 0$

$\ln\left(\frac{1}{2}\right) = \ln e^{kt}$, for $k < 0$

$t = \frac{\ln\left(\frac{1}{2}\right)}{-k}$, for $k > 0$

$t = \frac{-\ln 2}{-k} = \frac{\ln 2}{k}$

b. $t = \frac{\ln 2}{0.063}$

$t \approx 11$

About 11 years

11. $\ln y = -1.2074 + 0.873x$

$e^{\ln y} = e^{-1.2074+0.873x}$

$e^{\ln y} = e^{-1.2074} \cdot e^{0.873x}$

$y = 0.29897e^{0.873x}$

13. a. logarithmic model

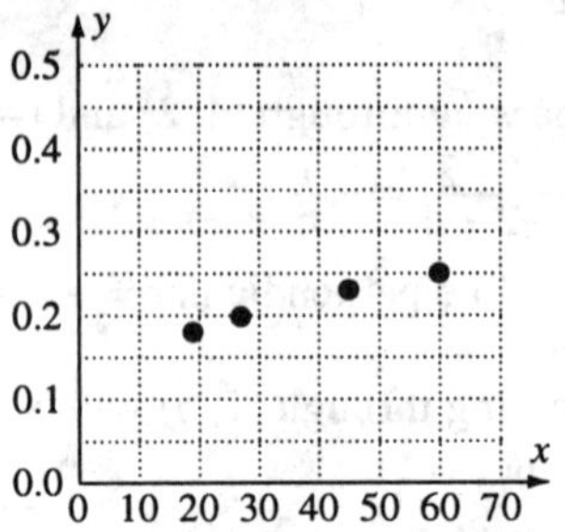

b. ≈(2.94, 0.18), (3.30, 0.20), (3.81, 0.23), (4.09, 0.25)

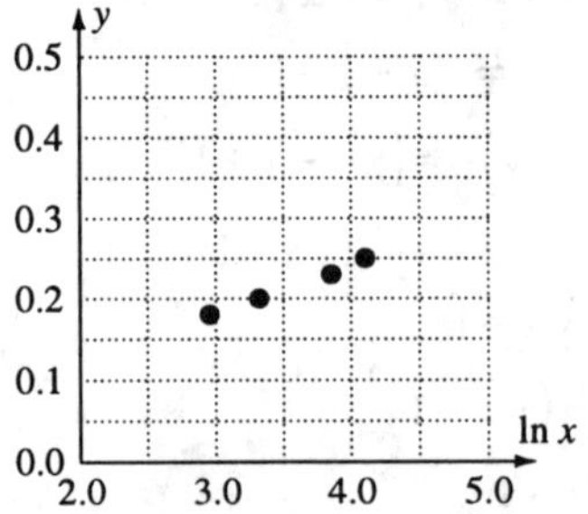

c. $y = 0.0604 \ln x + 0.00131$

d. $y = 0.0604 \ln 75 + 0.00131$

$y = 0.26$

reflex time = 0.26 second

15. Students should verify results.

17–19. Answers may vary.

Review Problems

21. $(3x^2 - 7y)^2$
$= (3x^2)^2 - 2(3x^2)(7y) + (-7y)^2$
$= 9x^4 - 42x^2y + 49y^2$

22. $\frac{63}{y^2 - 3y} - 11 = \frac{7y}{y-3} - \frac{21}{y}$
$y(y-3)\left(\frac{63}{y^2 - 3y} - 11\right)$
$= y(y-3)\left(\frac{7y}{y-3} - \frac{21}{y}\right)$
$63 - 11y(y-3) = 7y^2 - 21(y-3)$
$63 - 11y^2 + 33y = 7y^2 - 21y + 63$
$-18y^2 + 54y = 0$
$-18(y^2 - 3y) = 0$
$-18y(y-3) = 0$
$y = 0$ or $y = 3$
Both extraneous
No solution

23. $3x - 2y < -6$

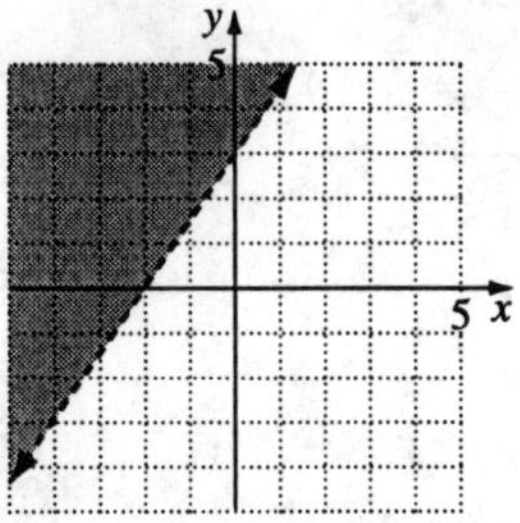

Chapter 8 Review Problems

1. $y = 2^x$

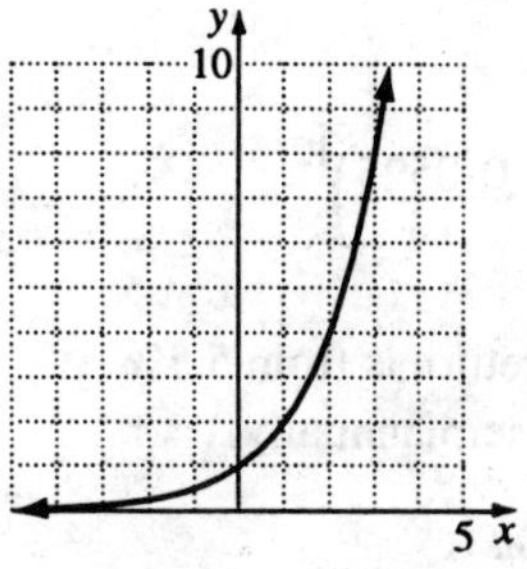

2. $y = 2^{x-2}$

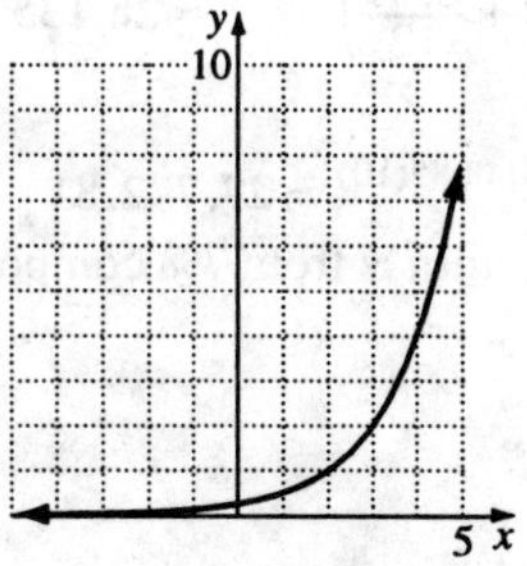

3. $y = 3^x - 1$

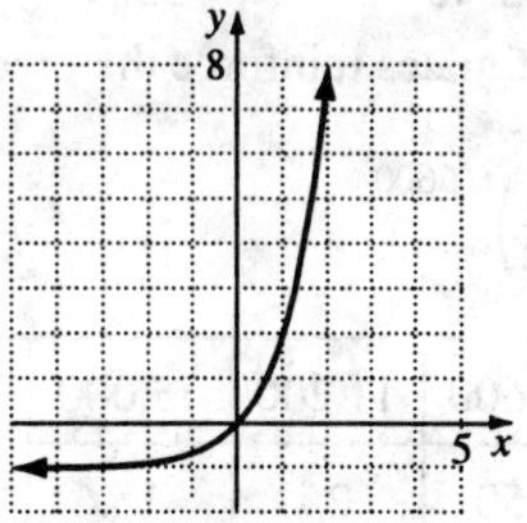

4. $f(x) = \left(\frac{1}{2}\right)^x$

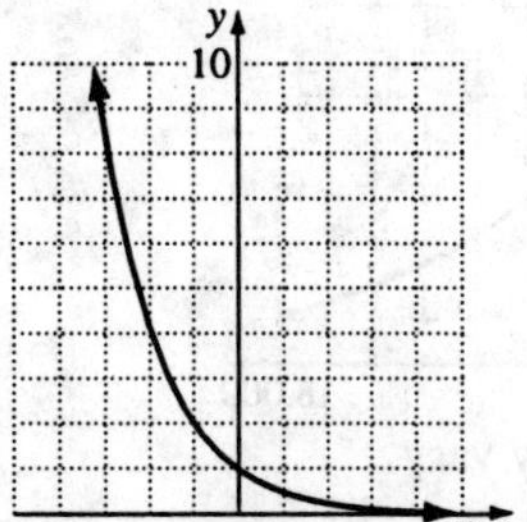

5. $A = P\left(1+\frac{r}{n}\right)^{nt}$

$A = 5000\left(1+\frac{0.055}{2}\right)^{2(5)}$

$A \approx 6558.26$

$A = 5000\left(1+\frac{0.0525}{12}\right)^{12(5)}$

$A = 6497.16$

The greatest return is from 5.5% compounded semiannually.

6. $A = P\left(1+\frac{r}{n}\right)^{nt}$

$A = 14,000\left(1+\frac{0.07}{12}\right)^{12(10)} \approx 28,135.26$

$A = Pe^{rt}$

$A = 14,000e^{0.0685(10)} = 27,772.81$

The greatest return is from 7% compounded monthly.

7. $y = Ae^{kt}$

$t = 60$

China: $y = 1119.9e^{0.013(60)} \approx 2443$

India: $y = 853.4e^{0.021(60)} \approx 3009$

Yes, these estimates reinforce the prediction.

8. $f(t) = 100\left(\frac{1}{2}\right)^{t/5600}$

t	0	2800	5600	11,200	15,000	16,800
$f(t)$	100	≈70.7	50	25	≈15.6	12.5

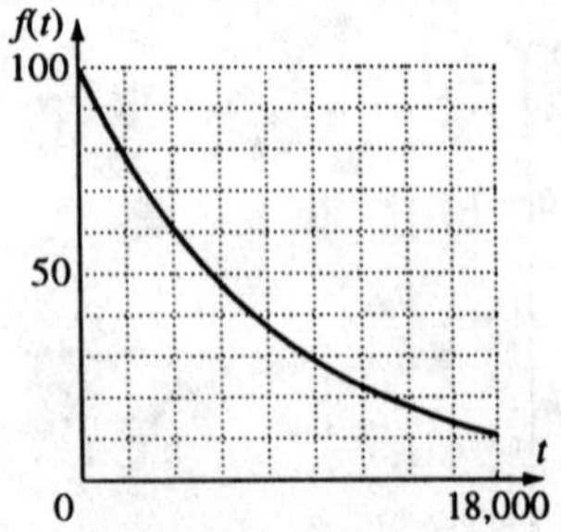

Answers may vary.

9. $T = 70 + 130e^{-0.04855t}$

a. 200°F

b. About 120°F

c. 70°F
Temperature of room is 70°F.

10. $f(x) = x^2 + 3,\ g(x) = 4x - 1$

$(f \circ g)(x) = (4x-1)^2 + 3$

$= 16x^2 - 8x + 1 + 3$

$= 16x^2 - 8x + 4$

$(g \circ f)(x) = 4(x^2+3) - 1$

$= 4x^2 + 12 - 1$

$= 4x^2 + 11$

11. $f(x) = \sqrt{x},\ g(x) = x + 1$

$(f \circ g)(x) = \sqrt{x+1}$

$(g \circ f)(x) = \sqrt{x} + 1$

12. Inverse of (a), (b), and (d) is not a function. Inverse of (c) is a function, so (c) is one-to-one.

13. a. $f(x) = 2x - 4$

$y = 2x - 4$

$x = 2y - 4$

$x + 4 = 2y$

$\frac{x+4}{2} = y$

$f^{-1}(x) = \frac{x+4}{2}$

b. Verify:

$f(f^{-1}(x)) = 2\left(\frac{x+4}{2}\right) - 4 = x$

$f^{-1}(f(x)) = \frac{2x-4+4}{2} = x$

c.

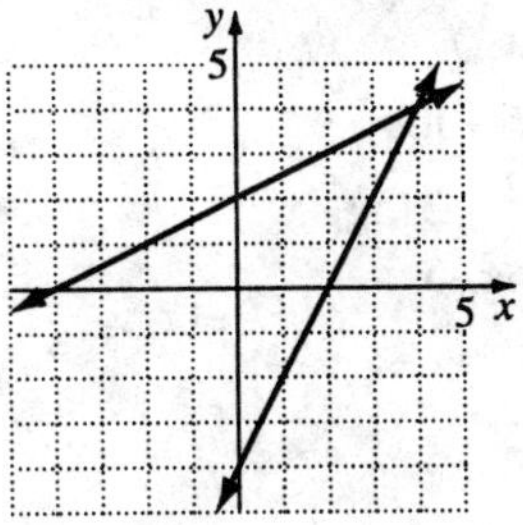

14. a. $f(x) = x^3 - 2$

$y = x^3 - 2$

$x = y^3 - 2$

$x + 2 = y^3$

$\sqrt[3]{x+2} = y$

$f^{-1}(x) = \sqrt[3]{x+2}$

b. Verify:

$f(f^{-1}(x)) = \left(\sqrt[3]{x+2}\right)^3 - 2 = x$

$f^{-1}(f(x)) = \sqrt[3]{x^3 - 2 + 2} = x$

c.

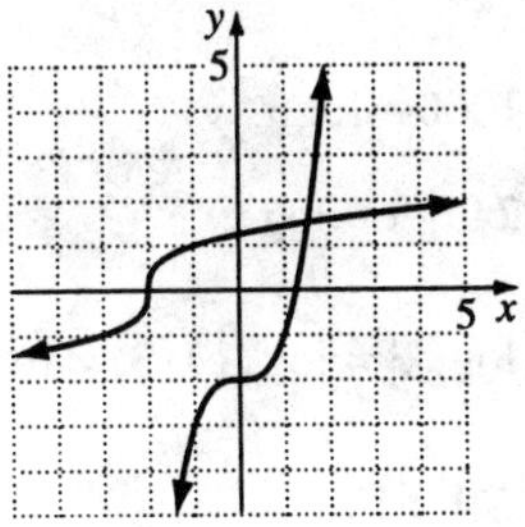

15. a. It passes the horiztonal line test.

b. $f^{-1}(60) = 1930$

The life expectancy was 60 years for someone born in 1930.

16. $\log_2 64 = 6$

$2^6 = 64$

17. $\log_{49} 7 = \frac{1}{2}$

$49^{1/2} = 7$

18. $6^3 = 216$

$\log_6 216 = 3$

19. $10^{-2} = \frac{1}{100}$

$\log_{10} \frac{1}{100} = -2$

$\log \frac{1}{100} = -2$

20. $\log_4 64 = x$

$4^x = 64$

$2^{2x} = 2^6$

$2x = 6$

$x = 3$

$\{3\}$

21. $\log_{125} 5 = x$

$125^x = 5$

$5^{3x} = 5^1$

$3x = 1$

$x = \frac{1}{3}$

$\left\{\frac{1}{3}\right\}$

22. $\log_3(-9) = x$

Not possible to take log of a negative number

No solution

23. $\log_3 \frac{1}{81} = x$

$3^x = \frac{1}{81}$

$3^x = 3^{-4}$

$x = -4$

$\{-4\}$

24. $f(x) = 2^x$, $f^{-1}(x) = \log_2 x$

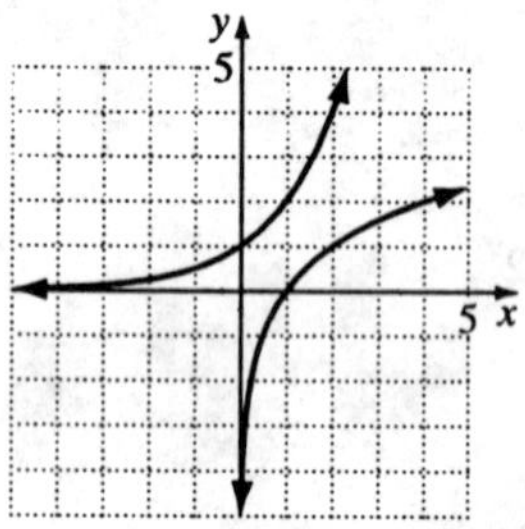

25. $f(t) = 76 - 18\log(t+1)$, $0 \le t \le 12$

a. $t = 0$
$76 - 18\log(0+1) = 76$
The score was 76.

b. $f(2) = 76 - 18\log(2+1) \approx 67$
$f(4) = 76 - 18\log(4+1) \approx 63$
$f(6) = 76 - 18\log(6+1) \approx 61$
$f(8) = 76 - 18\log(8+1) \approx 59$
$f(12) = 76 - 18\log(12+1) \approx 56$

c. $f(t) = 76 - 18\log(t+1)$

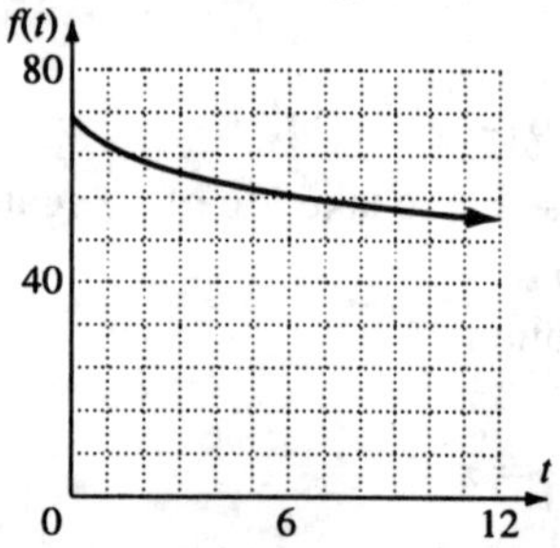

The material retained decreases over time.

26. $t = \frac{1}{c}\ln\frac{A}{A-N}$
($A = 12$ mph, $c = 0.06$, $N = 5$ mph):
$t = \frac{1}{0.06}\ln\frac{12}{12-5}$
$= \frac{1}{0.06}\ln\frac{12}{7} \approx 9$
Approximately 9 weeks

27. $\log_6(36x^3)$
$= \log_6 36 + \log_6 x^3$
$= \log_6 6^2 + 3\log_6 x$
$= 2 + 3\log_6 x$

28. $\log_2 \frac{xy^2}{64}$
$= \log_2 xy^2 - \log_2 64$
$= \log_2 x + \log_2 y^2 - \log_2 2^6$
$= \log_2 x + 2\log_2 y - 6$

29. $\log_4 \frac{\sqrt{x}}{64}$
$= \log_4 \sqrt{x} - \log_4 64$
$= \log_4 x^{1/2} - \log_4 4^3$
$= \frac{1}{2}\log_4 x - 3$

30. $\log \sqrt[3]{\frac{1000}{x^2 y}}$
$= \frac{1}{3}\log\frac{1000}{x^2 y}$
$= \frac{1}{3}(\log 1000 - \log x^2 y)$
$= \frac{1}{3}(3 - 2\log x - \log y)$

31. $\log_b 7 + \log_b 3 = \log_b(7\cdot 3) = \log_b 21$

32. $\log_5 250 - \log_5 2$
$= \log_5\left(\frac{250}{2}\right)$
$= \log_5 125 = \log_5 5^3 = 3$

33. $\log 3 - 3\log x = \log 3 - \log x^3 = \log\frac{3}{x^3}$

34. $3\log_b x + 4\log_b y$
$= \log_b x^3 + \log_b y^4$
$= \log_b x^3 y^4$

35. $\frac{1}{3}\ln x - \frac{1}{2}\ln y$

$= \ln x^{1/3} - \ln y^{1/2} = \ln\left(\frac{\sqrt[3]{x}}{\sqrt{y}}\right)$

36. a. $\log_3 7 = \frac{\log 7}{\log 3}$

b. $\log_3 7 = \frac{\ln 7}{\ln 3}$

37. b is true;

$\ln(\log_5 60 - \log_5 12) = \ln\left(\log_5 \frac{60}{12}\right)$

$= \ln(\log_5 5)$

$= \ln 1 = 0$

38. $2^{4x-2} = 64$

$2^{4x-2} = 2^6$

$4x - 2 = 6$

$4x = 8$

$x = 2$

$\{2\}$

39. $125^x = 25$

$5^{3x} = 5^2$

$3x = 2$

$x = \frac{2}{3}$

$\left\{\frac{2}{3}\right\}$

40. $9^x = \frac{1}{27}$

$3^{2x} = 3^{-3}$

$2x = -3$

$x = -\frac{3}{2}$

$\left\{-\frac{3}{2}\right\}$

41. $5^x = 119.4$

$\ln 5^x = \ln 119.4$

$x \ln 5 = \ln 119.4$

$x = \frac{\ln 119.4}{\ln 5} \approx 2.972$

42. $20,000e^{0.08t} = 40,000$

$e^{0.08t} = 2$

$\ln e^{0.08t} = \ln 2$

$0.08t = \ln 2$

$t = \frac{\ln 2}{0.08} \approx 8.664$

43. $8^{-0.04x} = 0.06$

$\ln 8^{-0.04x} = \ln 0.06$

$-0.04x \ln 8 = \ln 0.06$

$x = \frac{\ln 0.06}{-0.04 \ln 8} \approx 33.824$

44. $\log_5 x = -3$

$x = 5^{-3}$

$x = \frac{1}{125}$

$\left\{\frac{1}{125}\right\}$

45. $\log_x 32 = 6$

$x^6 = 32$

$x = \sqrt[6]{32} \approx 1.782$

$\left\{\sqrt[6]{32} \approx 1.782\right\}$

46. $\log x = -3$

$x = 10^{-3}$

$x = \frac{1}{1000}$

$\left\{\frac{1}{1000}\right\}$

47. $\log_5(2x-1) = 3$

$2x - 1 = 5^3$

$2x = 125 + 1$

$x = \frac{126}{2}$

$x = 63$

$\{63\}$

48. $\log x + \log(x-21) = 2$

$\log x(x-21) = 2$

$x(x-21) = 10^2$

$x^2 - 21x - 100 = 0$

$(x-25)(x+4)=0$
$x=25,\ x=-4$ (reject)
$\{25\}$

49. $\log_2(x+3)=4-\log_2(x-3)$
$\log_2(x+3)+\log_2(x-3)=4$
$\log_2(x+3)(x-3)=4$
$x^2-9=2^4$
$x^2=25$
$x=5,\ x=-5$ (reject)
$\{5\}$

50. $\log_3(x-1)-\log_3(x+2)=2$
$\log_3\left(\frac{x-1}{x+2}\right)=2$
$\frac{x-1}{x+2}=3^2$
$x-1=9x+18$
$-8x=19$
$x=-\frac{19}{8}$ (reject)
No solution

51. $A=P\left(1+\frac{r}{n}\right)^{nt}$
$20{,}000=12{,}500\left(1+\frac{0.065}{4}\right)^{4t}$
$1.6=\left(1+\frac{0.065}{4}\right)^{4t}$
$\ln 1.6=4t\ln\left(1+\frac{0.065}{4}\right)$
$t=\frac{\ln 1.6}{4\ln\left(1+\frac{0.065}{4}\right)}\approx 7.3$
About 7.3 years

52. $A=Pe^{rt}$

a. $3P=Pe^{rt}$
$3=e^{rt}$
$\ln 3=\ln e^{rt}$
$\ln 3=rt$
$t=\frac{\ln 3}{r}$
It will take $\frac{\ln 3}{r}$ yr.

b. $8=\frac{\ln 3}{r}$
$r=\frac{\ln 3}{8}\approx 0.137$
About 13.7%

53. $y=Ae^{kt}$

a. $10=4.043e^{0.015t}$
$\frac{10}{4.043}=e^{0.015t}$
$\ln\left(\frac{10}{4.043}\right)=\ln e^{0.015t}$
$\ln\left(\frac{10}{4.043}\right)=0.015t$
$t=\frac{\ln\left(\frac{10}{4.043}\right)}{0.015}\approx 60.4$
$1975+60=2035$
In year 2035

b. $10=4.134e^{0.02t}$
$t=\frac{\ln\left(\frac{10}{4.134}\right)}{0.02}\approx 44.2$
$1975+44=2019$
In year 2019

54. $W=0.35\ln P+2.74$
$W=0.35\ln 7323+2.74$
$W\approx 5.9$
About 5.9 feet/second

55. $R=0.67\log E-2.9$
$8.6=0.67\log E-2.9$
$11.5=0.67\log E$
$\frac{11.5}{0.67}=\log E$
$E=10^{\left(\frac{11.5}{0.67}\right)}$
$E=1.459\times 10^{17}$
Columbia, 1.459×10^{17} joules
$7.1=0.67\log E-2.9$
$\frac{10}{0.67}=\log E$
$E=10^{\left(\frac{10}{0.67}\right)}=8.421\times 10^{14}$
California, 8.421×10^{14} joules

56. a. $y = Ae^{kt}$

$26{,}077 = 14{,}609e^{k(14)}$

$\ln\left(\frac{26{,}077}{14{,}609}\right) = \ln e^{14k}$

$k = \frac{\ln\left(\frac{26{,}077}{14{,}609}\right)}{14}$

$k \approx 0.04139$

b. $y = 14{,}609e^{0.04139t}$

c. $t = 2005 - 1980 = 25$

$y = 14{,}609e^{0.04139(25)}$

$= 41{,}116$ thousand

$t = 2010 - 1980 = 30$

$y = 14{,}609e^{0.04139(30)}$

$\approx 50{,}569$ thousand

$t = 2025 - 1980 = 45$

$y = 14{,}609e^{0.04139(45)}$

$\approx 94{,}084$ thousand

57. a. $y = Ae^{kt}$

$\frac{A}{2} = Ae^{k(5730)}$

$\ln\frac{1}{2} = \ln e^{5730k}$

$k = \frac{\ln\frac{1}{2}}{5730} = -0.000121$

$y = Ae^{-0.000121t}$

b. $0.15A = Ae^{-0.000121t}$

$\ln 0.15 = \ln e^{-0.000121t}$

$t = \frac{\ln 0.15}{-0.000121}$

$t \approx 15{,}680$

About 15,680 years

58. $3A = Ae^{kt}$

$\ln\left(\frac{3A}{A}\right) = \ln e^{kt}$

$kt = \ln 3$

$t = \frac{\ln 3}{k}$

59. High: exponential

Medium: linear

Low: quadratic, negative

Explanations may vary.

There is not equal spacing representing the same time period.

60. a.

x	0	5	10	15	20	25	30	33
y	3.3	3.7	4.3	4.9	5.5	6.1	6.5	6.8

b.

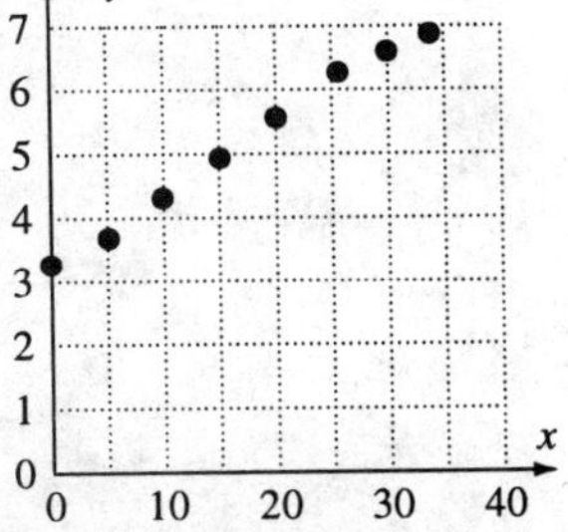

c. $\ln y = 0.110x + 3.25$

d. $e^{\ln y} = e^{0.110x+3.25}$
$y = e^{0.110x} \cdot e^{3.25}$
$y = 25.79e^{0.110x}$

e. $x = 2000 - 1960 = 40$
$y = 25.79e^{0.110(40)} = 2101$
The expenditures will be \$2101 billion.

61. a. (0, 0.30), (10, 0.95), (20, 1.36), (28, 1.40)

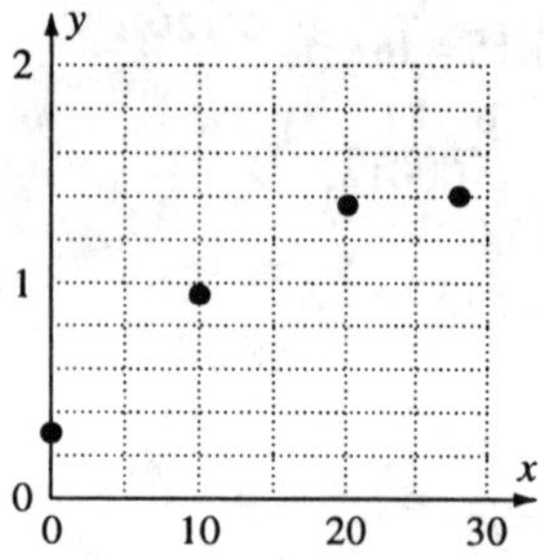

Logarithmic function

b. (2.3, 0.95), (3.0, 1.36), (3.3, 1.40)

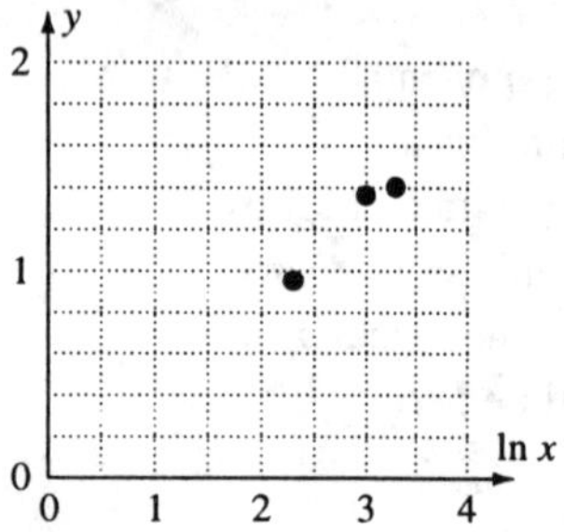

c. $y = 0.460 \ln x - 0.087$

d. $4 = 0.460 \ln x - 0.087$
$4.087 = 0.460 \ln x$
$x = e^{\left(\frac{4.087}{0.460}\right)}$
$x \approx 7221$
$1960 + 7221 = 9181$

Chapter 8 Test

1. Possible answer:
$f(1)=\frac{1}{2}$ on the graph, not $f(1)=2^1=2$

2. $f(x)=2^{x-3}$

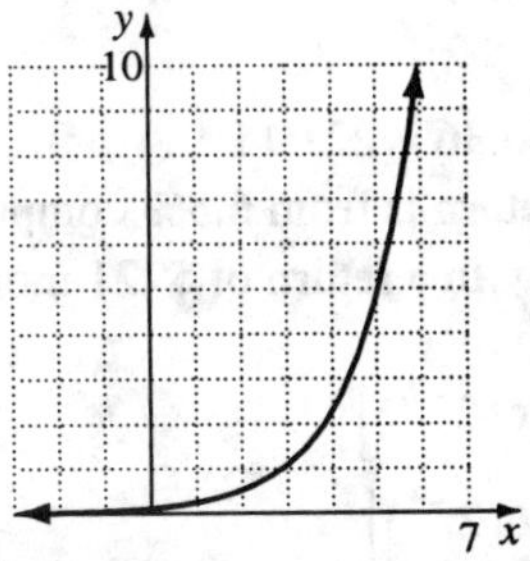

3. $f(x)=x^2+x,\ g(x)=3x-1$
$(f\circ g)(x)=(3x-1)^2+(3x-1)$
$=9x^2-6x+1+3x-1$
$=9x^2-3x$
$(g\circ f)(x)=3(x^2+x)-1=3x^2+3x-1$

4. $f(x)=5x-7$
$y=5x-7$
$x=5y-7$
$x+7=5y$
$\frac{x+7}{5}=y$
$f^{-1}(x)=\frac{x+7}{5}$

5. a. Yes; It passes the horizontal line test.

b. $f^{-1}(2000)=80$
A person who gives \$2000 to charity earns \$80,000.

6. $\log_5 125=3$
$5^3=125$

7. $\sqrt{36}=6$
$36^{1/2}=6$
$\log_{36} 6=\frac{1}{2}$

8. $f(x)=3^x,\ f^{-1}(x)=\log_3 x$

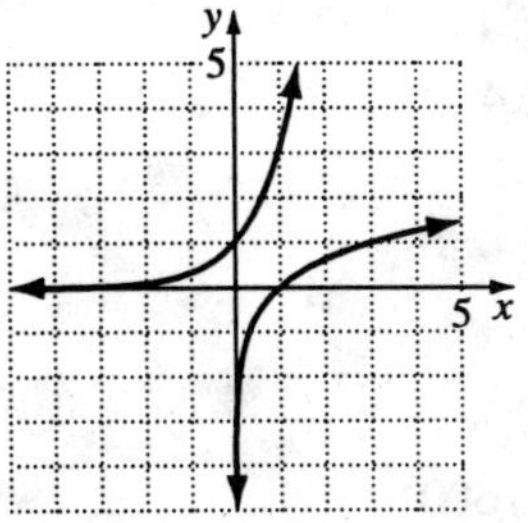

9. $\log_4(64x^5)$
$=\log_4 64+\log_4 x^5$
$=\log_4 4^3+5\log_4 x$
$=3+5\log_4 x$

10. $\log_3 \frac{\sqrt[3]{x}}{81}$
$=\log_3 \sqrt[3]{x}-\log_3 81$
$=\log_3 x^{1/3}-\log_3 3^4$
$=\frac{1}{3}\log_3 x-4$

11. $3\ln x+5\ln y$
$=\ln x^3+\ln y^5$
$=\ln x^3y^5$

12. $\log_3 405-\log_3 5$
$=\log_3 \frac{405}{5}$
$=\log_3 81$
$=\log_3 3^4=4$

13. $\log_{15} 71=\frac{\log 71}{\log 15}\approx 1.5741$

14. $3^{5x-2}=27$
$3^{5x-2}=3^3$
$5x-2=3$
$5x=5$
$x=1$
$\{1\}$

15. $5^x = 1.4$
$\log 5^x = \log 1.4$
$x \log 5 = \log 1.4$
$x = \frac{\log 1.4}{\log 5}$
$x = 0.209062$
$\{0.209062\}$

16. $400e^{0.005x} = 1600$
$e^{0.005x} = \frac{1600}{400}$
$\ln e^{0.005x} = 4$
$0.005x = 4$
$x = 800$
$\{800\}$

17. $\log_{25} x = \frac{1}{2}$
$25^{1/2} = x$
$\sqrt{25} = x$
$5 = x$
$\{5\}$

18. $\log_6(4x-1) = 3$
$6^3 = 4x-1$
$216 = 4x-1$
$217 = 4x$
$\frac{217}{4} = x$
$\left\{\frac{217}{4}\right\}$

19. $\log x + \log(x+15) = 2$
$\log x(x+15) = 2$
$10^2 = x(x+15)$
$100 = x^2 + 15x$
$x^2 + 15x - 100 = 0$
$(x+20)(x-5) = 0$
$x = -20$ (extraneous), $x = 5$
$\{5\}$

20. $A = P\left(1+\frac{r}{n}\right)^{nt}$
$A = 3000\left(1+\frac{0.065}{2}\right)^{2(10)}$
$A \approx 5687.51$
$A = Pe^{rt}$
$A = 3000e^{0.06(10)}$
$A \approx 5466.36$
$5687.51 - 5466.36 = 221.15$
The greatest return is from 6.5% computed semiannually with a return of $221 more.

21. $f(h) = 20\left[1+6(2^{-h})\right]$
$f(3) = 20\left[1+6(2^{-3})\right]$
$= 20\left[1+6\left(\frac{1}{8}\right)\right] = 20(1.75) = 35$
After 3 hours, the object's temperature is 35°C.

22. $f(t) = 75 - 16\log(t+1)$

a. $f(0) = 75 - 16\log(0+1) = 75$
The score was 75.

b. $f(9) = 75 - 16\log(9+1) = 59$
The score was 59.

23. $A = P\left(1+\frac{r}{n}\right)^{nt}$
$8000 = 4000\left(1+\frac{0.05}{4}\right)^{4t}$
$2 = \left(1+\frac{0.05}{4}\right)^{4t}$
$\log 2 = 4t\log\left(1+\frac{0.05}{4}\right)$
$\frac{\log 2}{4\log(1.0125)} = t$
$13.9 \approx t$
It will take 14 years.

24. $P = 14.7e^{-0.21x}$
$5.15 = 14.7e^{-0.21x}$
$\frac{5.15}{14.7} = e^{-0.21x}$
$\ln \frac{5.15}{14.7} = -0.21x$
$\frac{\ln \frac{5.15}{14.7}}{-0.21} = x$
$x \approx 4.99$
The altitude is 5.0 miles.

25. $R = 6e^{kx}$

a. $13 = 6e^{k(0.06)}$
$\frac{13}{6} = e^{k(0.06)}$
$\ln \frac{13}{6} = 0.06k$
$k = \frac{\ln \frac{13}{6}}{0.06}$
$k \approx 12.8865$

b. $100 = 6e^{12.8865x}$
$x = \frac{\ln \frac{100}{6}}{12.8865}$
$x \approx 0.218$
The concentration is about 0.218.

26. $R = \log \frac{I}{I_0}$
$R = \log \frac{10^{7.4} I_0}{I_0}$
$= \log 10^{7.4} = 7.4$
The magnitude is 7.4.

Chapters 1–8 Cumulative Review Problems

1. $\sqrt{2x+5} - \sqrt{x+3} = 2$
$\sqrt{2x+5} = \sqrt{x+3} + 2$
$\left(\sqrt{2x+5}\right)^2 = \left(\sqrt{x+3} + 2\right)^2$
$2x + 5 = x + 3 + 4\sqrt{x+3} + 4$
$\frac{x-2}{4} = \sqrt{x+3}$
$\frac{x^2 - 4x + 4}{16} = x + 3$
$\frac{x^2}{16} - \frac{5}{4}x - \frac{11}{4} = 0$
$x^2 - 20x - 44 = 0$
$(x - 22)(x + 2) = 0$
$x = 22$, $x = -2$ (reject)
$\{22\}$

2. $(x-5)^2 = -49$
$x - 5 = \pm\sqrt{-49}$
$x = 5 \pm 7i$
$\{5 - 7i, 5 + 7i\}$

3. $(9x + 3)(x - 1) = -8$
$9x^2 - 6x - 3 = -8$
$9x^2 - 6x + 5 = 0$
$x = \frac{6 \pm \sqrt{(-6)^2 - 4(9)(5)}}{2(9)}$
$= \frac{6 \pm \sqrt{-144}}{18}$
$= \frac{6 \pm 12i}{18}$
$\left\{\frac{1}{3} - \frac{2}{3}i, \frac{1}{3} + \frac{2}{3}i\right\}$

4. $x^2 + x > 6$
$x^2 + x - 6 > 0$
$(x + 3)(x - 2) > 0$

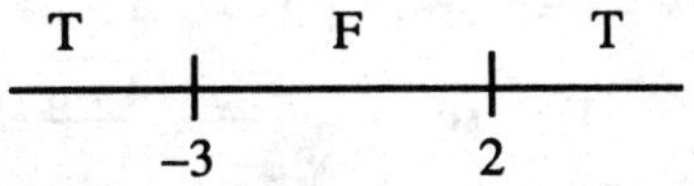

Test –4: $(-4)^2 + (-4) > 6$
$12 > 6$ True
Test 0: $0^2 + 0 > 6$
$0 > 6$ False
Test 3: $3^2 + 3 > 6$
$12 > 6$ True
$\{x \mid x < -3 \text{ or } x > 2\}$, $(-\infty, -3) \cup (2, \infty)$

5. $|2x-3| = |x+6|$
$2x-3 = x+6$ or $2x-3 = -x-6$
$x = 9$ or $3x = -3$
$x = -1$
$\{-1, 9\}$

6. $6x - 3(5x+2) = 4(1-x)$
$6x - 15x - 6 = 4 - 4x$
$-9x - 6 = 4 - 4x$
$-5x = 10$
$x = -2$
$\{-2\}$

7. $\frac{2}{x-3} - \frac{3}{x+3} = \frac{12}{x^2-9}$
$(x^2-9)\left(\frac{2}{x-3} - \frac{3}{x+3}\right) = \frac{12}{x^2-9}(x^2-9)$
$2(x+3) - 3(x-3) = 12$
$2x + 6 - 3x + 9 = 12$
$-x = -3$
$x = 3$, extraneous
No solution

8. $(x^2+2x)^2 - 5(x^2+2x) + 6 = 0$
$t^2 - 5t + 6 = 0$ (Let $t = x^2 + 2x$.)
$(t-3)(t-2) = 0$
$t = 3$ or $t = 2$
$x^2 + 2x = 3$ or $x^2 + 2x = 2$
$x^2 + 2x - 3 = 0$ or $x^2 + 2x - 2 = 0$
$(x+3)(x-1) = 0$ or $x = \frac{-2 \pm \sqrt{4+8}}{2}$
$x = -3$ or $x = 1$ or $x = -1 \pm \sqrt{3}$
$\left\{-3,\ 1,\ -1-\sqrt{3},\ -1+\sqrt{3}\right\}$

9. $3x + 2 < 4$ and $4 - x > 1$
$3x < 2$ and $-x > -3$
$x < \frac{2}{3}$ and $x < 3$
$\left\{x \middle| x < \frac{2}{3}\right\}$, $\left(-\infty, \frac{2}{3}\right)$

10. $\frac{x+1}{x-5} > 0$

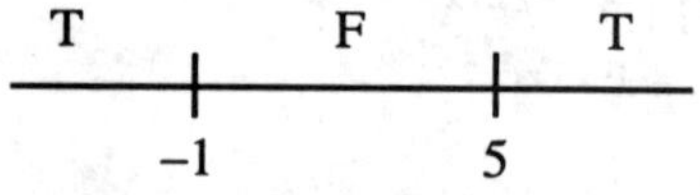

Test −2: $\frac{-2+1}{-2-5} = \frac{1}{7} > 0$ True
Test 0: $\frac{0+1}{0-5} = -\frac{1}{5} > 0$ False
Test 6: $\frac{6+1}{6-5} = 7 > 0$ True
$\{x | x < -1 \text{ or } x > 5\}$, $(-\infty, -1) \cup (5, \infty)$

11. $4x - 3y = 7$
$3x - 2y = 6$
Add −2 times first equation to 3 times second equation.

$$\begin{array}{r} -8x + 6y = -14 \\ 9x - 6y = \ \ 18 \\ \hline x = \ \ \ 4 \end{array}$$

Substitute back.
$4(4) - 3y = 7$
$-3y = -9$
$y = 3$
$\{(4, 3)\}$

12. $3x - 2y + z = 7$
$2x + 3y - z = 13$
$x - y + 2z = -6$

$$\left[\begin{array}{ccc|c} 3 & -2 & 1 & 7 \\ 2 & 3 & -1 & 13 \\ 1 & -1 & 2 & -6 \end{array}\right] \begin{array}{l} \\ R_2 - 2R_3 \\ R_1 - 3R_3 \end{array} \left[\begin{array}{ccc|c} 3 & -2 & 1 & 7 \\ 0 & 5 & -5 & 25 \\ 0 & 1 & -5 & 25 \end{array}\right]$$

$$\tfrac{1}{5}R_2 \to \left[\begin{array}{ccc|c} 3 & -2 & 1 & 7 \\ 0 & 1 & -1 & 5 \\ 0 & 1 & -5 & 25 \end{array}\right]$$

$$\begin{array}{l} \\ \\ R_2 - R_3 \to \end{array} \left[\begin{array}{ccc|c} 3 & -2 & 1 & 7 \\ 0 & 1 & -1 & 5 \\ 0 & 0 & 4 & -20 \end{array}\right]$$

$$\begin{array}{l} \frac{1}{3}R_1 \rightarrow \\ \\ \frac{1}{4}R_3 \rightarrow \end{array} \left[\begin{array}{ccc|c} 1 & -\frac{2}{3} & \frac{1}{3} & \frac{7}{3} \\ 0 & 1 & -1 & 5 \\ 0 & 0 & 1 & -5 \end{array}\right]$$

$$x - \frac{2}{3}y + \frac{1}{3}z = \frac{7}{3}$$
$$y - z = 5$$
$$z = -5$$

$y - (-5) = 5$
$y = 0$

$x - \frac{2}{3}(0) + \frac{1}{3}(-5) = \frac{7}{3}$
$x = \frac{12}{3} = 4$
$\{(4, 0, -5)\}$

13. $f(x) = \sqrt{x-2}$

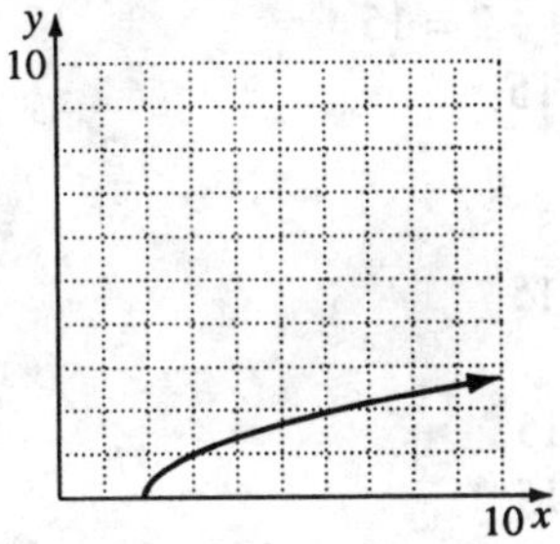

14. $y = (x+2)^2 - 4$

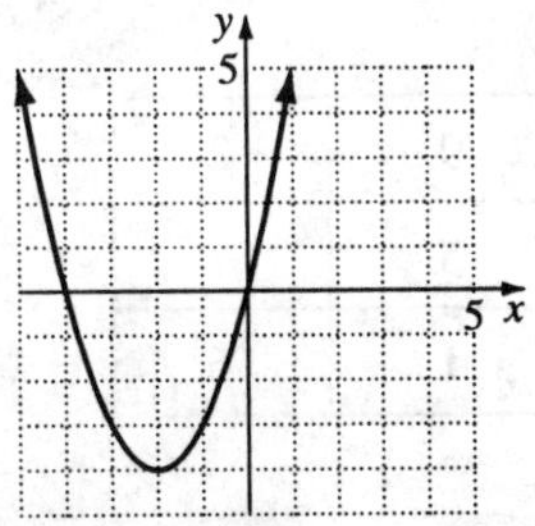

15. $y < -3x + 5$

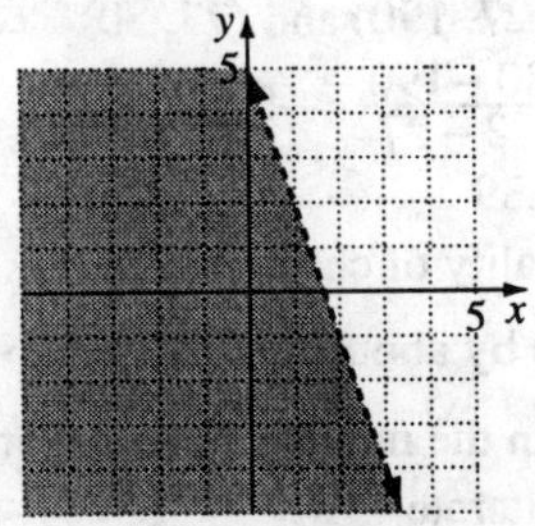

16. $2x - 3y \le 6$

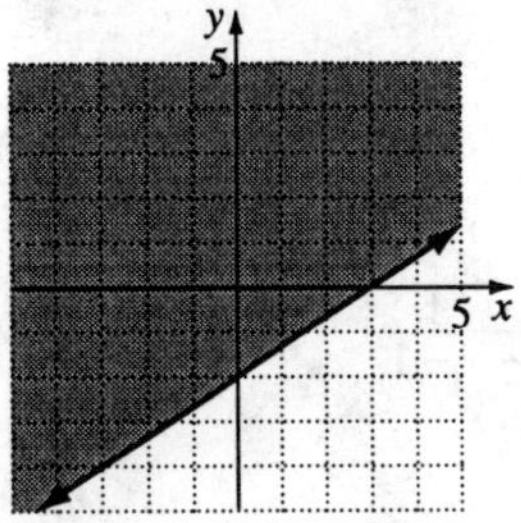

17. $y = 3^{x-2}$

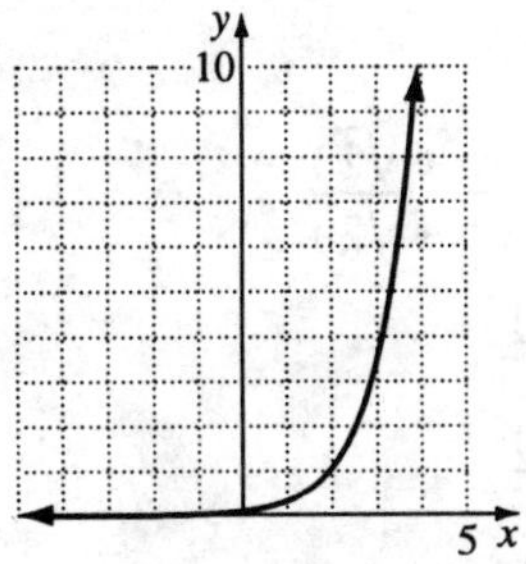

18. $y = \frac{1}{x-2}$

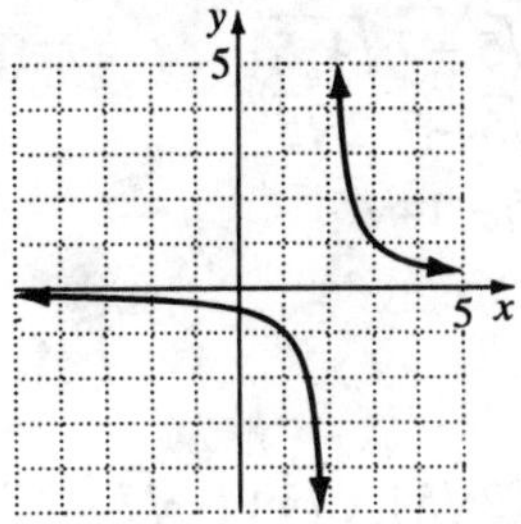

19. $\frac{2x+1}{x-5} - \frac{4}{x^2 - 3x - 10}$

$= \frac{2x+1}{x-5} - \frac{4}{(x-5)(x+2)}$

$= \frac{(2x+1)(x+2) - 4}{(x-5)(x+2)}$

$= \frac{2x^2 + 5x - 2}{(x-5)(x+2)}$

20. $\dfrac{\frac{1}{x-1}+1}{\frac{1}{x+1}-1}$

$=\dfrac{\frac{1+(x-1)}{x-1}}{\frac{1-(x+1)}{x+1}}$

$=\dfrac{1+x-1}{x-1}\cdot\dfrac{x+1}{1-x-1}$

$=\dfrac{x}{x-1}\cdot\dfrac{x+1}{-x}$

$=\dfrac{x+1}{1-x}$

21. $\dfrac{6}{\sqrt{5}-\sqrt{2}}$

$=\dfrac{6}{\sqrt{5}-\sqrt{2}}\cdot\dfrac{(\sqrt{5}+\sqrt{2})}{(\sqrt{5}+\sqrt{2})}$

$=\dfrac{6(\sqrt{5}+\sqrt{2})}{5-2}$

$=\dfrac{6(\sqrt{5}+\sqrt{2})}{3}$

$=2(\sqrt{5}+\sqrt{2})$

22. $8\sqrt{45}+2\sqrt{5}-7\sqrt{20}$

$=8\sqrt{9\cdot 5}+2\sqrt{5}-7\sqrt{4\cdot 5}$

$=8\cdot 3\sqrt{5}+2\sqrt{5}-7\cdot 2\sqrt{5}$

$=24\sqrt{5}+2\sqrt{5}-14\sqrt{5}$

$=12\sqrt{5}$

23. $(2\sqrt{5}-\sqrt{7})^2$

$=(2\sqrt{5})^2-2(2\sqrt{5})(\sqrt{7})+(-\sqrt{7})^2$

$=4\cdot 5-4\sqrt{35}+7$

$=27-4\sqrt{35}$

24. $\log_5(x+2)-\log_5 x=1$

$\log_5\dfrac{x+2}{x}=1$

$\dfrac{x+2}{x}=5^1$

$x+2=5x$

$2=4x$

$\dfrac{1}{2}=x$

x in the box $=\dfrac{1}{\frac{1}{2}}=2$

4	9	2
a	*b*	7
8	*c*	*d*

Sum $=4+9+2=15$

$4+a+8=15$

$a=3$

$a+b+7=3$

$3+b+7=15$

$b=5$

$9+b+c=15$

$9+5+c=15$

$c=1$

$2+7+d=15$

$d=6$

4	9	2
3	5	7
8	1	6

25. $2^{\sqrt{3}}\approx 3.3$

26. a. points ≈ (27, 190) and (73, 80)

slope $\approx\dfrac{80-190}{73-27}\approx -2.39$

About –2.39

The mortality of children under 5 decreases by about $2\frac{1}{3}$ for each percent increase in the number of adult females who are literate.

b. Answers may vary.

c. $y - 80 = -2.39(x - 73)$
$y = -2.39x + 254.47$

27. $f = \frac{Kmv^2}{r}$

Given: $m = 2500$ pounds, $v = 40$ mph, $r = 800$ ft, $f = 1500$ pounds

$1500 = \frac{K(2500)(40)^2}{800}$

$\frac{1500(800)}{2500(1600)} = K$

$\frac{3}{10} = K$

$f = \frac{3mv^2}{10r}$

If $m = 4000$ pounds, $r = 600$ ft, $v = 30$ mph, $f = ?$

$f = \frac{3(4000)(30)^2}{10(600)}$

$f = 1800$

1800 lb

28. $f(t) = 29{,}035t^2 + 429{,}200$

$690{,}515 = 29{,}035t^2 + 429{,}200$

$261{,}315 = 29{,}035t^2$

$9 = t^2$

$t = 3, t = -3$

Disregard a negative time.

$1983 + 3 = 1986$

The year was 1986.

29. $A(x)$ = Area of large triangle – Area of small triangle

$A(x) = \frac{1}{2}(4x)(2x+1) - \frac{1}{2}(2x)(x)$

$A(x) = 2x(2x+1) - x(x)$

$A(x) = 4x^2 + 2x - x^2$

$A(x) = 3x^2 + 2x$

30.

teams	lines
2	1
3	3
4	6
5	10
6	15
n	$\frac{n^2-n}{2}$

Chapter 9

Problem Set 9.1

1. Center: (3, 2)
$(h, k) = (3, 2), r = 5$
$(x-h)^2 + (y-k)^2 = r^2$
$(x-3)^2 + (y-2)^2 = 5^2$
$(x-3)^2 + (y-2)^2 = 25$

3. Center: (–1, 4)
$(h, k) = (-1, 4), r = 2$
$(x-h)^2 + (y-k)^2 = r^2$
$[x-(-1)]^2 + (y-4)^2 = 2^2$
$(x+1)^2 + (y-4)^2 = 4$

5. Center: (–3, –1)
$(h, k) = (-3, -1),\ r = \sqrt{3}$
$(x-h)^2 + (y-k)^2 = r^2$
$[x-(-3)]^2 + [y-(-1)]^2 = (\sqrt{3})^2$
$(x+3)^2 + (y+1)^2 = 3$

7. Center: (–4, 0):
$(h, k) = (-4, 0), r = 2$
$(x-h)^2 + (y-k)^2 = r^2$
$[x-(-4)]^2 + (y-0)^2 = 2^2$
$(x+4)^2 + y^2 = 4$

9. Center: (0, 0):
$(h, k) = (0, 0), r = 7$
$(x-h)^2 + (y-k)^2 = r^2$
$(x-0)^2 + (y-0)^2 = 7^2$
$x^2 + y^2 = 49$

11. $x^2 + y^2 = 16$
$(x-0)^2 + (y-0)^2 = 4^2$
$(x-h)^2 + (y-k)^2 = r^2$
Center: $(h, k) = (0, 0)$
radius: $r = 4$

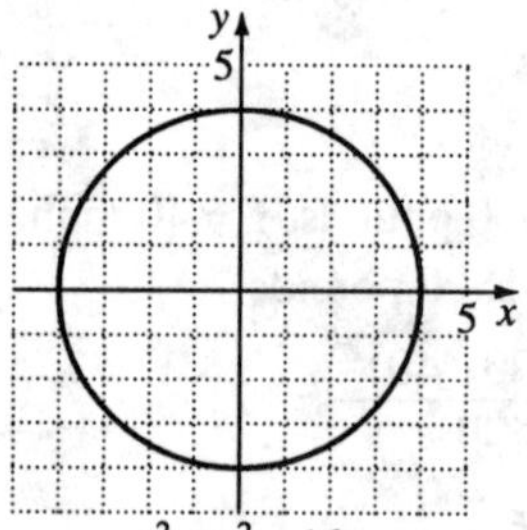

$x^2 + y^2 = 16$

13. $(x-3)^2 + (y-1)^2 = 36$
$(x-3)^2 + (y-1)^2 = 6^2$
$(x-h)^2 + (y-k)^2 = r^2$
Center: $(h, k) = (3, 1)$
radius: $r = 6$

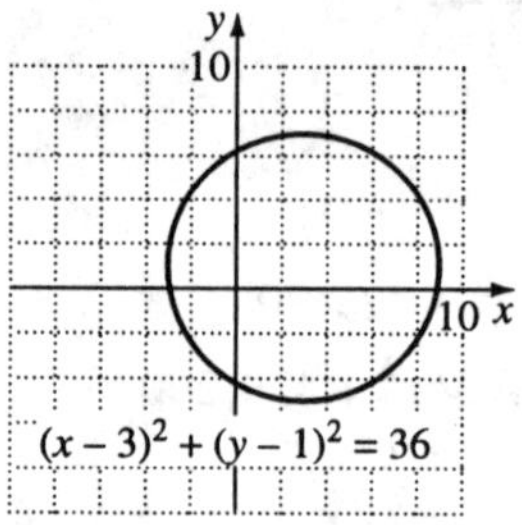

15. $(x+3)^2 + (y-2)^2 = 4$
$[x-(-3)]^2 + (y-2)^2 = 2^2$
$(x-h)^2 + (y-k)^2 = r^2$
Center: $(h, k) = (-3, 2)$
radius: $r = 2$

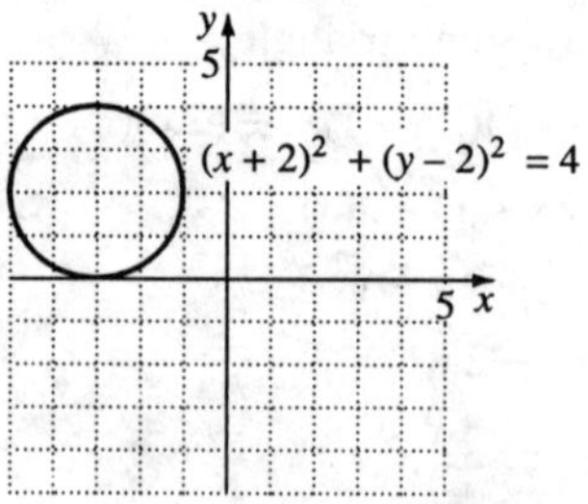

17. $(x+2)^2+(y+2)^2=4$

$[x-(-2)]^2+[y-(-2)]^2=2^2$

$(x-h)^2+(y-k)^2=r^2$

Center: $(h, k)=(-2,-2)$

radius: $r=2$

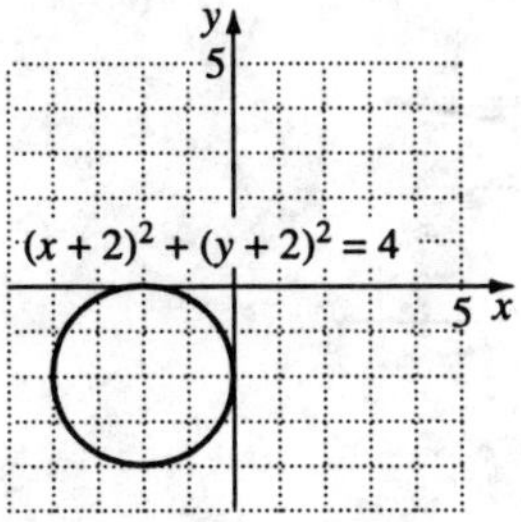

19. $x^2+y^2+6x+2y+6=0$

$x^2+6x+\_\_\_+y^2+2y+\_\_\_=-6$

$\uparrow \qquad\qquad \uparrow$

$\left(\frac{1}{2}\right)(6)=3 \qquad \left(\frac{1}{2}\right)(2)=1$

$3^2=9 \qquad 1^2=1$

$x^2+6x+9+y^2+2y+1=-6+9+1$

$(x+3)^2+(y+1)^2=4=2^2$

Center: $(h, k)=(-3,-1)$

radius: $r=2$

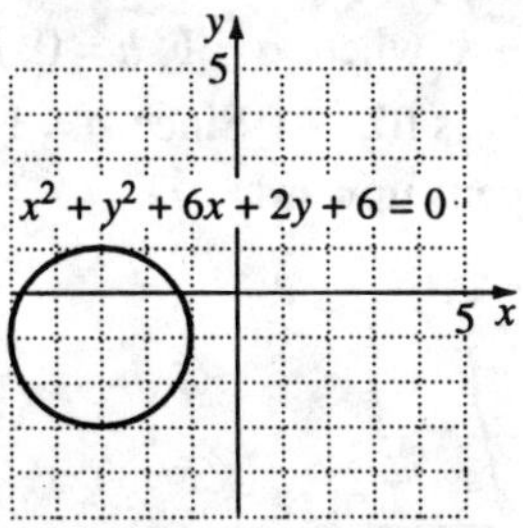

21. $x^2+y^2-10x-6y-30=0$

$x^2-10x+\_\_\_+y^2-6y+\_\_\_=30$

$\uparrow \qquad\qquad \uparrow$

$\left(\frac{1}{2}\right)(-10)=-5 \qquad \left(\frac{1}{2}\right)(-6)=-3$

$(-5)^2=25 \qquad (-3)^2=9$

$x^2-10x+25+y^2-6y+9=30+25+9$

$(x-5)^2+(y-3)^2=64=8^2$

Center: $(h, k)=(5, 3)$

radius: $r=8$

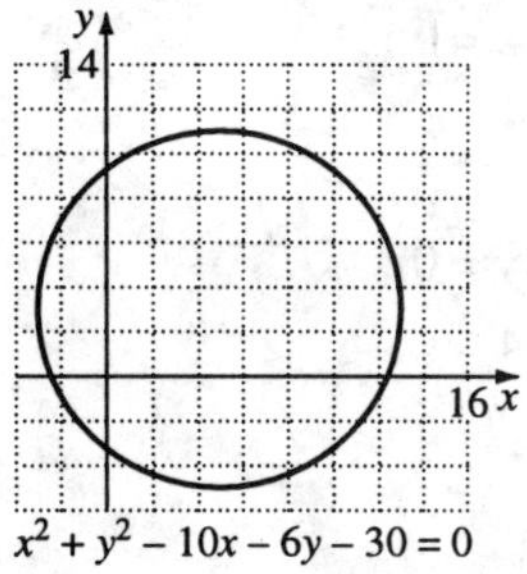

23. $x^2+y^2+8x-2y-8=0$

$x^2+8x+16+y^2-2y+1=8+16+1$

$(x+4)^2+(y-1)^2=25=5^2$

Center: $(h, k)=(-4, 1)$

radius: $r=5$

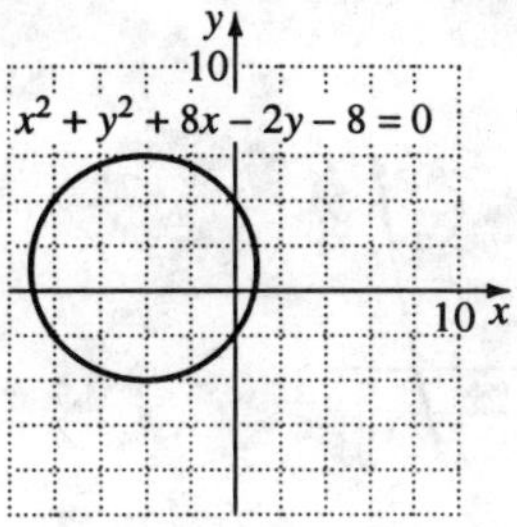

25. $x^2-2x+y^2-15=0$

$x^2-2x+\_\_\_+y^2=15$

$\uparrow$

$\left(\frac{1}{2}\right)(-2)=-1$

$(-1)^2=-1$

$x^2-2x+1+y^2=15+1$

$(x-1)^2+y^2=16=4^2$

Center: $(h, k)=(1, 0)$

radius: $r=4$

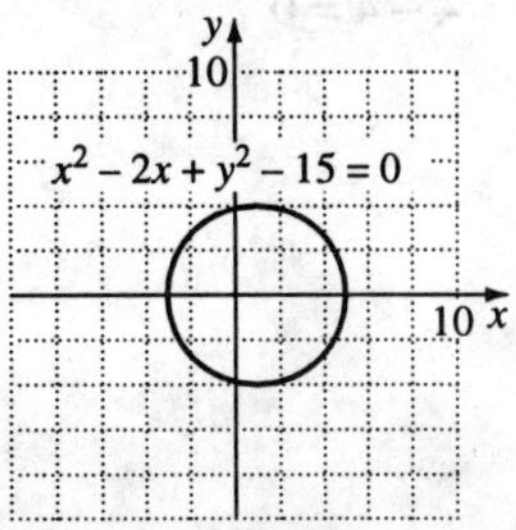

27. $y = (x-2)^2 - 4$

y-intercept (let $x = 0$):

$y = (0-2)^2 - 4$

$= 4 - 4 = 0$

x-intercept (let $y = 0$):

$0 = (x-2)^2 - 4$

$4 = (x-2)^2$

$\pm 2 = x - 2$

$x = 2 \pm 2$

$x = 4$ or $x = 0$

Vertex:

The form of this equation is

$y = a(x-h)^2 + k$ where $a = 1$, $h = 2$, $k = -4$.

Thus, the vertex is (2, –4).

Since a is positive, the parabola will open upward.

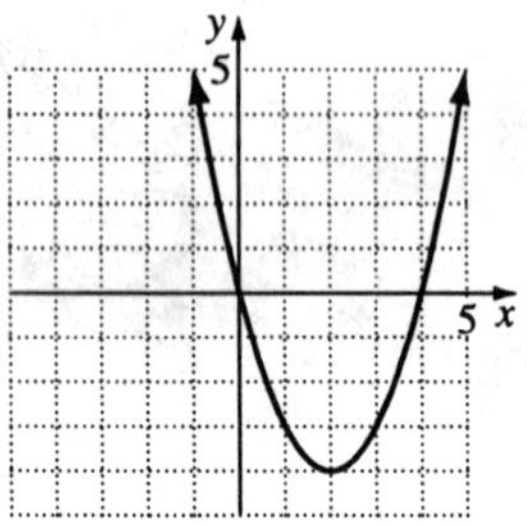

29. $x = (y-2)^2 - 4$

y-intercept (let $x = 0$):

$0 = (y-2)^2 - 4$

$4 = (y-2)^2$

$\pm 2 = y - 2$

$y = 2 \pm 2$

$y = 4$ or $y = 0$

x-intercept (let $y = 0$)

$x = (-2)^2 - 4 = 4 - 4 = 0$

Vertex:

The form of this equation is

$x = a(y-k)^2 + h$ where $a = 1$, $h = -4$, $k = 2$.

Thus, the vertex is (–4, 2).

Since $a = 1 > 0$, the parabola will open to the right.

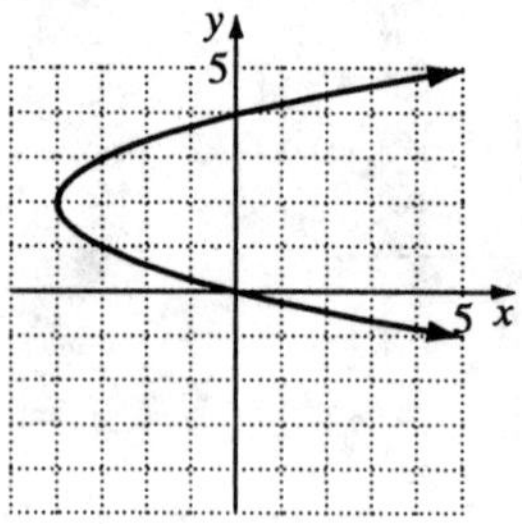

31. $y = x^2 - 4$

y-intercept (let $x = 0$):

$y = 0 - 4 = -4$

x-intercept (let $y = 0$):

$0 = x^2 - 4$

$4 = x^2$

$\pm 2 = x$

$x = 2$ or $x = -2$

Vertex:

The form of this equation is

$y = a(x-h)^2 + k$ where $a = 1$, $h = 0$, $k = -4$.

Thus, the vertex is (0, –4). Since $a = 1 > 0$, the parabola opens upward.

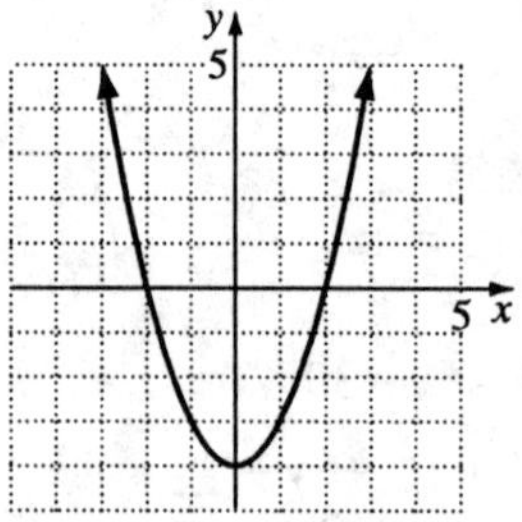

33. $x = y^2 + 4$

y-intercept (let $x = 0$):

$0 = y^2 + 4$

$-4 = y^2$

No y-intercept.

x-intercept ($y = 0$):

$x = 0 + 4 = 4$

Vertex:

The form of this equation is

$x = a(y-k)^2 + h$ where $a = 1$, $h = 4$, $k = 0$.

Thus, the vertex is (4, 0).

Since $a = 1 > 0$, the parabola will open to the right.

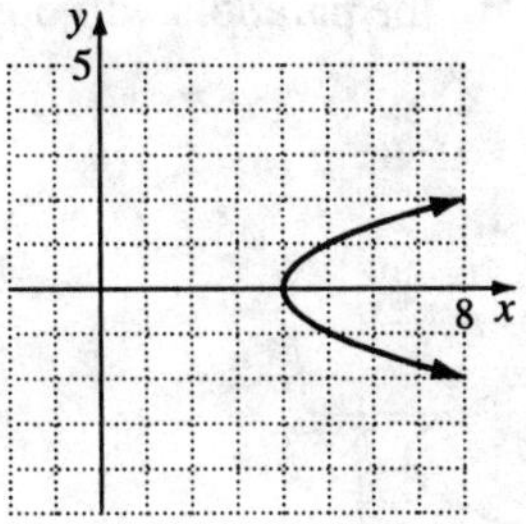

35. $y = -2(x-1)^2 - 1$

y-intercept ($x = 0$):

$y = -2(0-1)^2 - 1 = -2(1) - 1 = -3$

x-intercept ($y = 0$):

$0 = -2(x-1)^2 - 1$

$-\frac{1}{2} = (x-1)^2$

No x-intercept.

Vertex:

The form of this equation is

$y = a(x-h)^2 + k$ where $a = -2$, $h = 1$, $k = -1$. Thus, the vertex is (1, –1).

Since $a = -2 < 0$, the parabola opens downward.

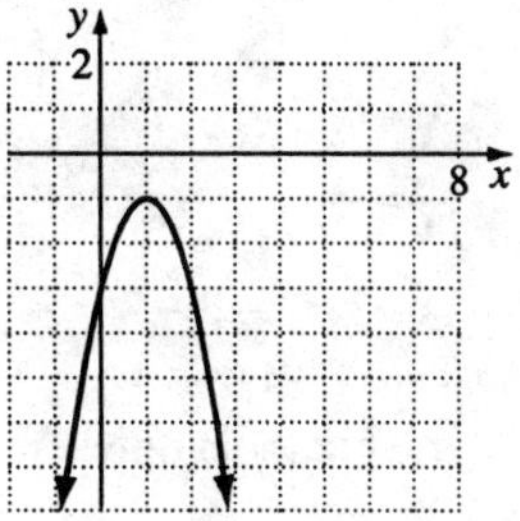

37. $x = -2(y-1)^2 - 1$

y-intercept ($x = 0$):

$0 = -2(y-1)^2 - 1$

$-\frac{1}{2} = (y-1)^2$

No y-intercept.

x-intercept ($y = 0$):

$x = -2(0-1)^2 - 1 = -2(1) = -3$

Vertex:

The form of this equation is

$x = a(y-k)^2 + h$ where $a = -2$, $h = -1$, $k = 1$. Thus, the vertex is (–1, 1).

Since $a = -2 < 0$, the parabola will open to the left.

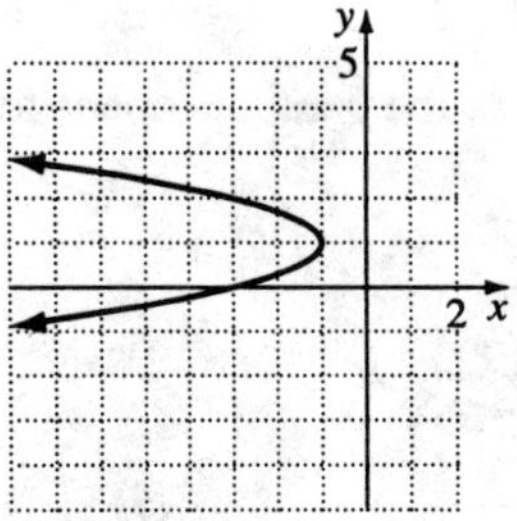

39. $x = y^2 + 2y - 3$
y-intercept ($x = 0$):
$0 = y^2 + 2y - 3$
$0 = (y + 3)(y - 1)$
$y = -3$ or $y = 1$
x-intercept ($y = 0$):
$x = 0 + 0 - 3 = -3$
Vertex:
$y = -\frac{b}{2a} = -\frac{2}{2(1)} = -1$
$x = (-1)^2 + 2(-1) - 3 = 1 - 2 - 3 = -4$
Thus, the vertex is (–4, –1).
Since $a = 1 > 0$, the parabola opens to the right.

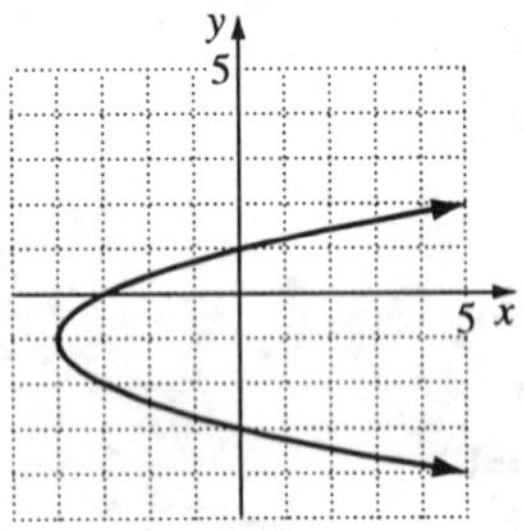

41. $x = y^2$
y-intercept ($x = 0$):
$0 = y^2$
$y = 0$
x-intercept ($y = 0$):
$x = 0$
Vertex: (0, 0)
Since $a = 1 > 0$, the parabola opens to the right.

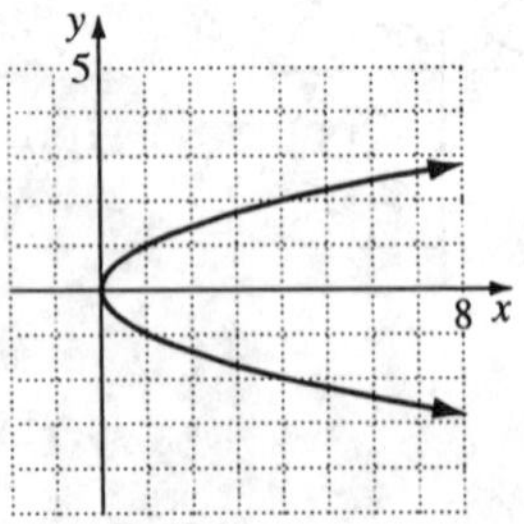

43. $x = -y^2 - 2y + 3$
y-intercept ($x = 0$):
$0 = -y^2 - 2y + 3$
$y^2 + 2y - 3 = 0$
$(y + 3)(y - 1) = 0$
$y = -3$ or $y = 1$
x-intercept ($y = 0$):
$x = -0 - 0 + 3 = 3$
Vertex:
$y = -\frac{b}{2a} = \frac{-(-2)}{2(-1)} = -1$
$x = -(-1)^2 - 2(-1) + 3 = -1 + 2 + 3 = 4$
Thus, the vertex is (4, –1).
Since $a = -1 < 0$, the parabola will open to the left.

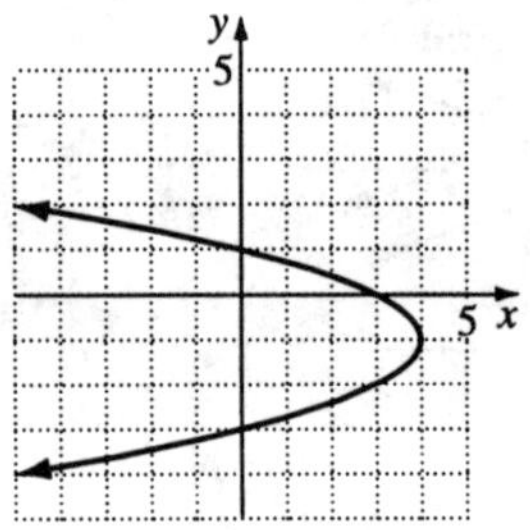

45. The vertex of the parabola lies at (0, 0).
At $x = 400$, $y = 160$
$y = ax^2$
$160 = a(400)^2$
$a = \frac{160}{160,000} = 0.001$
$y = 0.001x^2$
$x = 300$
$y = 0.001(300)^2$
$y = 90$
Height 100 ft from a tower: 90 ft

47. d is true:

A is *not* true: $x^2 + y^2 = 16 = 4^2$

Radius is 4 *not* 16.

B is *not* true: $(x-3)^2 + (y+5)^2 = 36 = 6^2$

$h = 3,\ k = -5,\ r = 6$

Center is at (3, –5), *not* (–3, 5)

C is *not* true: $(x-4)^2 + (y+6)^2 = 25 = 5^2$ is a circle of radius 5, centered at (4, –6) *not* $(x-4)+(y+6)=25$

d is true.

49. b is true;

$$x = ay^2 + by + c$$

$$x = a\left(y^2 + \frac{by}{a} + \frac{b^2}{4a^2}\right) + \left(c - \frac{b^2}{4a}\right)$$

$$x = a\left(y + \frac{b}{2a}\right)^2 + \left(c - \frac{b^2}{4a}\right)$$

If the vertex is (3, 2), then

$k = -\frac{b}{2a} = 2$ and $h = c - \frac{b^2}{4a} = 3$

Thus, $x = a(y-2)^2 + 3$

y-intercept ($x = 0$):

$$0 = a(y-2)^2 + 3$$

$$-\frac{3}{a} = (y-2)^2$$

If $a > 0$, then $(y-2)^2 < 0$ and the equation has no y-intercepts.

51. $16x^2 - 24xy + 9y^2 - 60x - 80y + 100 = 0$

$$9y^2 + (-24xy - 80y) + 16x^2 - 60x + 100 = 0$$

$$9y^2 - 8(3x+10)y + (16x^2 - 60x + 100) = 0$$

$$y = \frac{-b \pm \sqrt{b^2 - 4ac}}{2a}$$

$$= \frac{8(3x+10) \pm \sqrt{[-8(3x+10)]^2 - 4(9)(16x^2 - 60x + 100)}}{2(9)}$$

$$= \frac{(24x+80) \pm \sqrt{64(9x^2 + 60x + 100) - 36(16x^2 - 60x + 100)}}{18}$$

$$= \frac{(24x+80) \pm \sqrt{576x^2 + 3840x + 6400 - 576x^2 + 2160x - 3600}}{18}$$

$$= \frac{(24x+80) \pm \sqrt{6000x + 2800}}{18}$$

$$= \frac{(24x+80) \pm 20\sqrt{15x+7}}{18}$$

$$= \frac{2(12x+40) \pm 2(10)\sqrt{15x+7}}{2(9)}$$

$$= \frac{(12x+40) \pm 10\sqrt{15x+7}}{9}$$

$$y_1 = \frac{12x + 40 + 10\sqrt{15x+7}}{9}$$

$$y_2 = \frac{12x + 40 - 10\sqrt{15x+7}}{9}$$

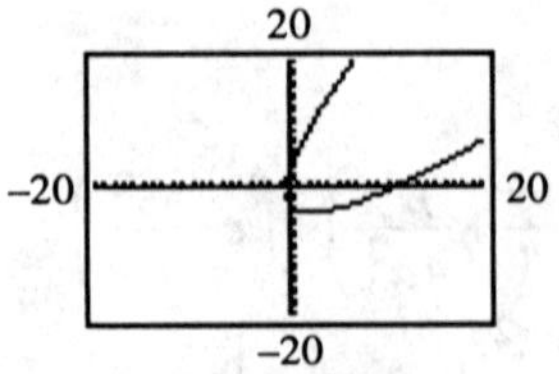

Answers may vary.

53–57. Answers may vary.

59. $(x-3)^2+(y+2)^2 \le 9 = 3^2$
Area enclosed by a circle with center (3, –2) and radius 3 is:
$\pi r^2 = \pi 3^2 = 9\pi$

61. Center: (0, 0)
radius:
$r = \sqrt{(4-0)^2+(3-0)^2}$
$= \sqrt{25} = 5$
$(x-0)^2+(y-0)^2 = 5^2$
$x^2+y^2 = 25$

63. Let $2x$ = length of rectangle
$2y$ = width of rectangle
$x^2+y^2 = 25$
$y = \pm\sqrt{25-x^2}$
$y = \sqrt{25-x^2}$
Reject $y = -\sqrt{25-x^2}$.
Area = $(2x)(2y) = 4xy$
$= 4x\sqrt{25-x^2}$

65. **a.** distance between (x, y) and the directrix is $\sqrt{(x-x)^2+[y-(-p)]^2} = |y+p|$.

b. distance from (x, y) to $(0, p)$
$= \sqrt{x^2+(y-p)^2}$.

c. equate the distances to get
$|y+p| = \sqrt{x^2+(y-p)^2}$
$(y+p)^2 = x^2+(y-p)^2$
$y^2+2py+p^2 = x^2+y^2-2py+p^2$
$x^2 = 4py$

Review Problems

67. $3x-2y=1$
$5y+3z=-7$
$2x+5y=45$

Equations 1 and 3:
Multiply equation 1 by 5.
Multiply equation 3 by 2.
$15x-10y = 5$
$4x+10y = 90$
$19x = 95$
$x = 5$
Equation 1:
$3x-2y=1$
$15-2y=1$
$-2y=-14$
$y=7$

Equation 2:
$5y+3z=-7$
$5(7)+3z=-7$
$3z=-42$
$z=-14$
$\{(5, 7, -14)\}$

68. $\sqrt{2x+3}-\sqrt{x+1}=1$
$\sqrt{2x+3}=\sqrt{x+1}+1$
$2x+3=x+1+2\sqrt{x+1}+1$
$2x+3=x+2+2\sqrt{x+1}$
$x+1=2\sqrt{x+1}$
$x^2+2x+1=4x+4$
$x^2-2x-3=0$
$(x-3)(x+1)=0$
$x=3$ or $x=-1$ (both check)
$\{-1, 3\}$

69. $(x^2-2x)^2-14(x^2-2x)=15$
$(x^2-2x)^2-14(x^2-2x)-15=0$
Let $t=x^2-2x$.
$t^2-14t-15=0$
$(t-15)(t+1)=0$
$t=15$ or $t=-1$
$x^2-2x=15$ or $x^2-2x=-1$
$x^2-2x-15=0$ or $x^2-2x+1=0$
$(x-5)(x+3)=0$ or $(x-1)^2=0$
$x=5$ or $x=-3$ or $x=1$
$\{-3, 1, 5\}$

Problem Set 9.2

1. $\frac{x^2}{9}+\frac{y^2}{4}=1$
y-intercepts (Let $x=0$):
$\frac{y^2}{4}=1$
$y^2=4$
$y=\pm 2$
x-intercepts (Let $y=0$):
$\frac{x^2}{9}=1$
$x^2=9$
$x=\pm 3$

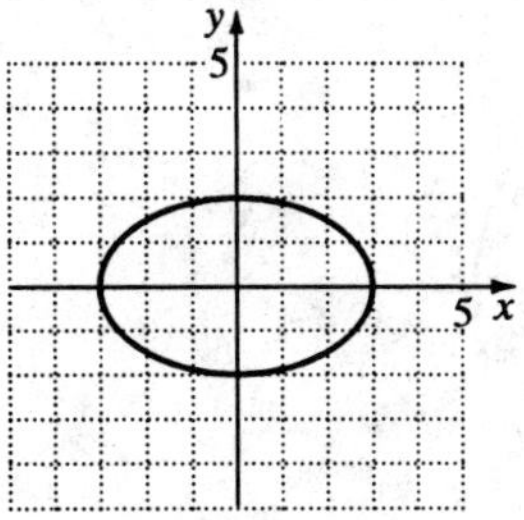

3. $\frac{x^2}{9}+\frac{y^2}{36}=1$
y-intercepts (Let $x=0$):
$\frac{y^2}{36}=1$
$y^2=36$
$y=\pm 6$
x-intercepts (Let $y=0$):
$\frac{x^2}{9}=1$
$x^2=9$
$x=\pm 3$

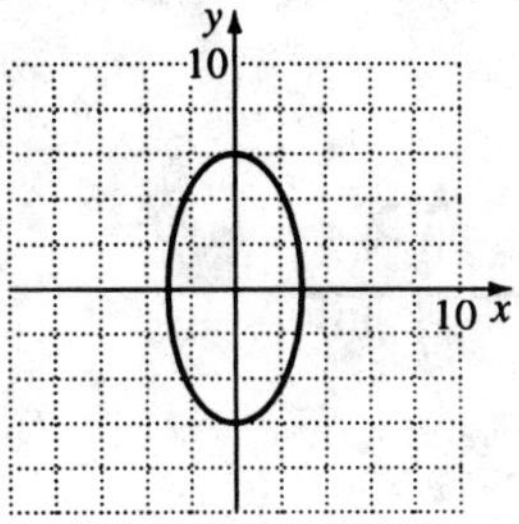

5. $\frac{x^2}{25}+\frac{y^2}{64}=1$
y-intercepts:
$\frac{y^2}{64}=1$
$y^2=64$
$y=\pm 8$
x-intercepts:
$\frac{x^2}{25}=1$
$x^2=25$

$x = \pm 5$

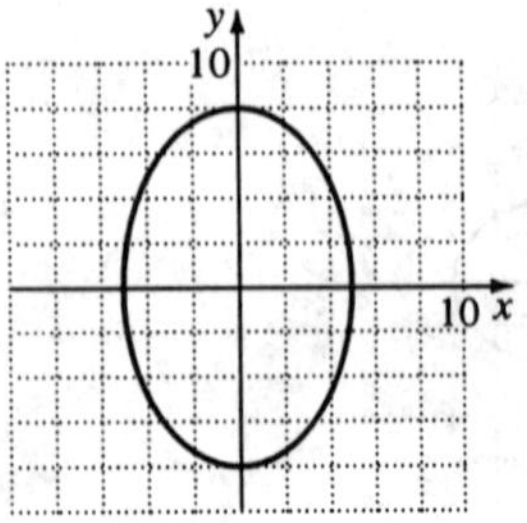

7. $\frac{x^2}{49} + \frac{y^2}{81} = 1$

y-intercepts:

$\frac{y^2}{81} = 1$

$y^2 = 81$

$y = \pm 9$

x-intercepts:

$\frac{x^2}{49} = 1$

$x^2 = 49$

$x = \pm 7$

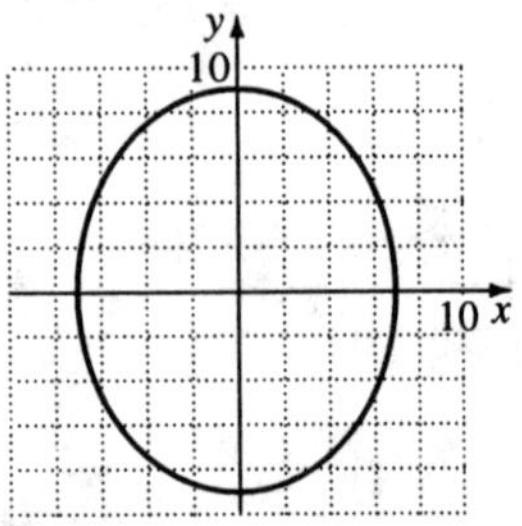

9. $25x^2 + 4y^2 = 100$

$\frac{25x^2}{100} + \frac{4y^2}{100} = 1$

$\frac{x^2}{4} + \frac{y^2}{25} = 1$

y-intercepts:

$\frac{y^2}{25} = 1$

$y^2 = 25$

$y = \pm 5$

x-intercepts:

$\frac{x^2}{4} = 1$

$x^2 = 4$

$x = \pm 2$

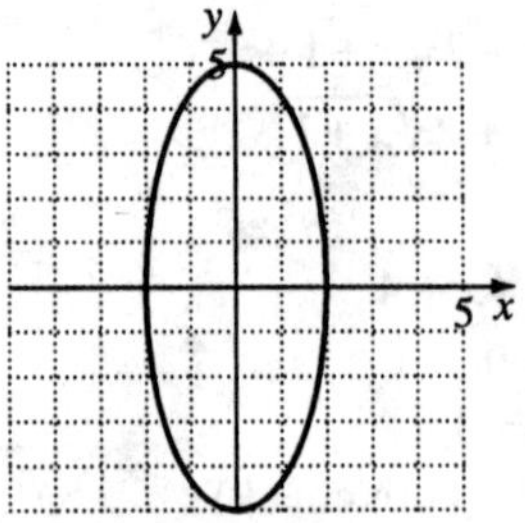

11. $4x^2 + 16y^2 = 64$

$\frac{x^2}{16} + \frac{y^2}{4} = 1$

y-intercepts:

$\frac{y^2}{4} = 1$

$y^2 = 4$

$y = \pm 2$

x-intercepts:

$\frac{x^2}{16} = 1$

$x^2 = 16$

$x = \pm 4$

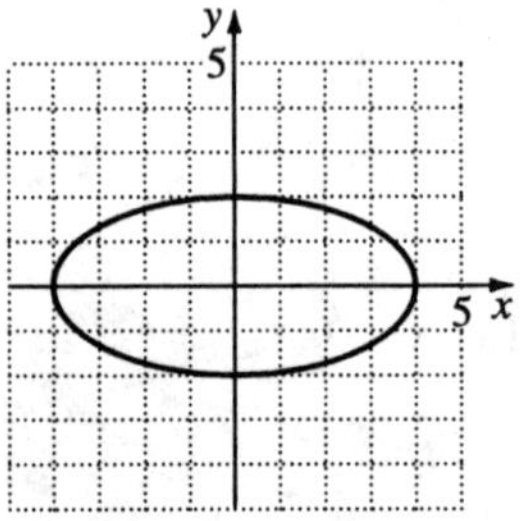

13. $25x^2 + 9y^2 = 225$

$\frac{x^2}{9} + \frac{y^2}{25} = 1$

y-intercepts:

$\frac{y^2}{25} = 1$

$y^2 = 25$

$y = \pm 5$

x-intercepts:

$\frac{x^2}{9}=1$

$x^2=9$

$x=\pm 3$

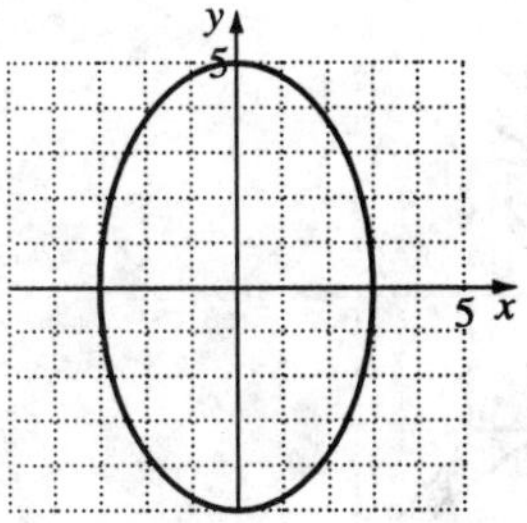

15. $x^2+2y^2=8$

$\frac{x^2}{8}+\frac{2y^2}{8}=1$

$\frac{x^2}{8}+\frac{y^2}{4}=1$

y-intercepts:

$\frac{y^2}{4}=1$

$y^2=4$

$y=\pm 2$

x-intercepts:

$\frac{x^2}{8}=1$

$x^2=8$

$x=\pm\sqrt{8}=\pm 2\sqrt{2}\approx \pm 2.8$

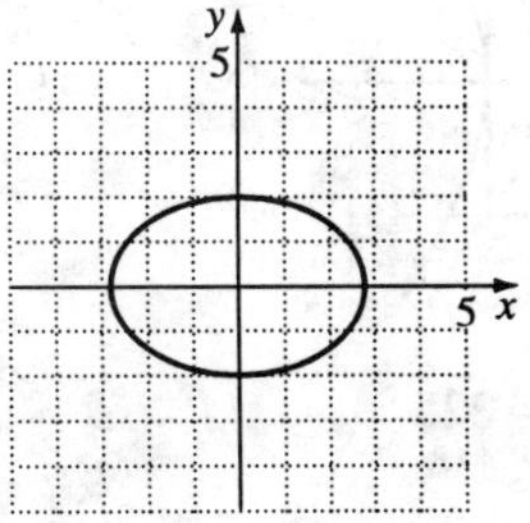

17. $\frac{x^2}{9}-\frac{y^2}{25}=1$

$\frac{x^2}{a^2}-\frac{y^2}{b^2}=1$

$a^2=9 \qquad b^2=25$

$a=\sqrt{9}=3 \qquad b=\sqrt{25}=5$

The sides of the rectangle used to draw the asymptotes pass through 3 and –3 on the *x*-axis and 5 and –5 on the *y*-axis.

x-intercepts (set $y = 0$):

$\frac{x^2}{9}=1$

$x^2=9$

$x=\pm 3$

y-intercepts (set $x = 0$):

$-\frac{y^2}{25}=1$

$y^2=-25$

$y=\pm\sqrt{-25}$

y is not a real number. No *y*-intercepts.

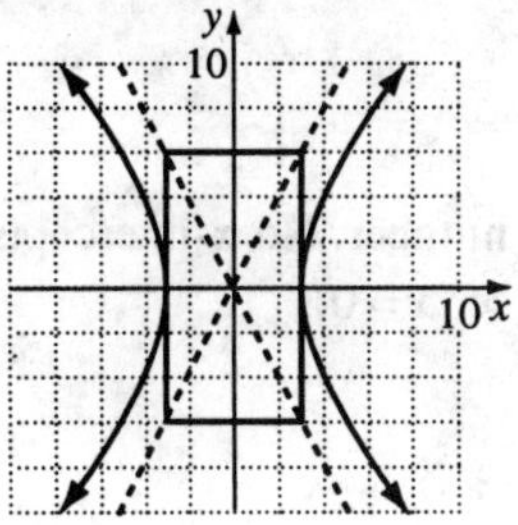

19. $\frac{x^2}{100}-\frac{y^2}{64}=1$

$a^2=100 \qquad a=10$

$b^2=64 \qquad b=8$

The sides of the rectangle used to draw the asymptotes pass through 10 and –10 on the *x*-axis and 8 and –8 on the *y*-axis. *x*-intercepts: ± 10

No y-intercepts.

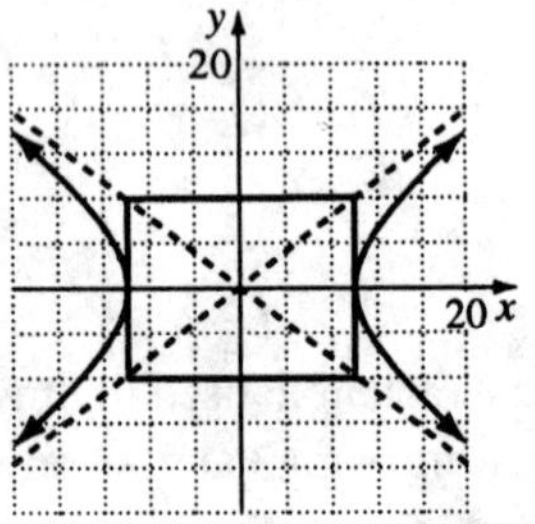

21. $\frac{y^2}{16}-\frac{x^2}{36}=1$

$\frac{y^2}{b^2}-\frac{x^2}{a^2}=1$

$b^2=16$ $\quad a^2=36$

$b=4$ $\quad a=6$

The sides of the rectangle used to draw the asymptotes pass through 4 and –4 on the y-axis and 6 and –6 on the x-axis.

x-intercepts (set $y=0$):

$-\frac{x^2}{36}=1$

$x^2=-36$

$x=\pm\sqrt{-36}$

x is not a real number. No x-intercepts.

y-intercepts (set $x=0$):

$\frac{y^2}{16}=1$

$y^2=16$

$y=\pm 4$

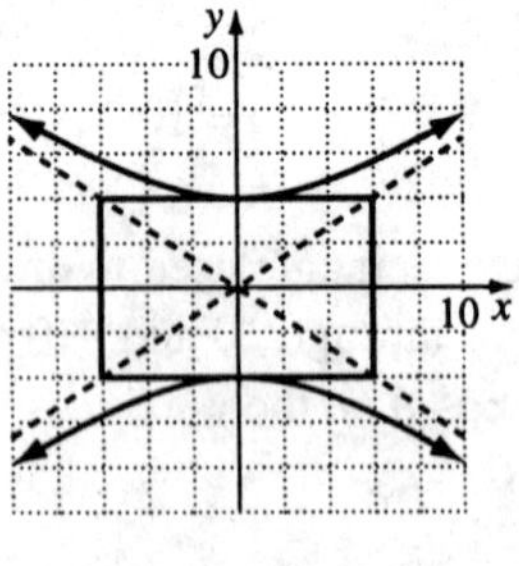

23. $\frac{y^2}{36}-\frac{x^2}{25}=1$

$b^2=36$ $\quad a^2=25$

$b=6$ $\quad a=5$

The sides of the rectangle used to draw the asymptotes pass through 5 and –5 on the x-axis and 6 and –6 on the y-axis.

y-intercepts: ± 6

No x-intercepts

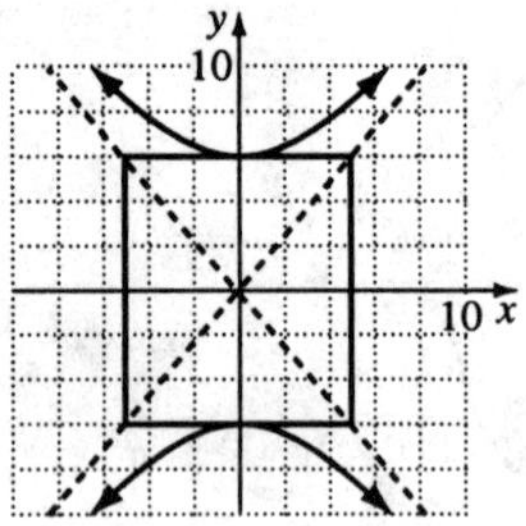

25. $9x^2-4y^2=36$

$\frac{9x^2}{36}-\frac{4y^2}{36}=1$

$\frac{x^2}{4}-\frac{y^2}{9}=1$

$a^2=4$ $\quad b^2=9$

$a=2$ $\quad b=3$

The sides of the rectangle used to draw the asymptotes pass through 2 and –2 on the x-axis and 3 and –3 on the y-axis.

x-intercepts: ± 2

No y-intercepts

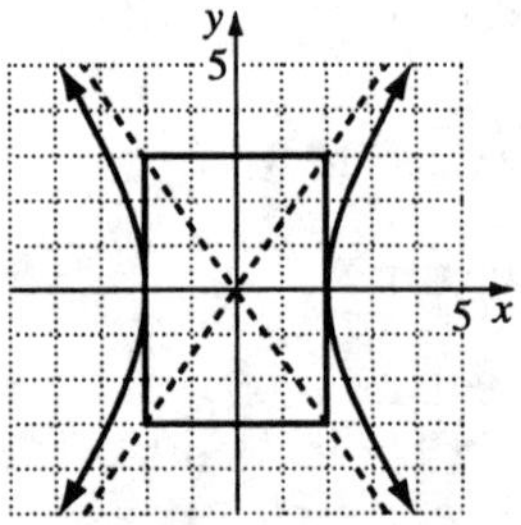

27. $9y^2-25x^2=225$

$\frac{y^2}{25}-\frac{x^2}{9}=1$

$b^2=25$ $\quad a^2=9$

$b=5$ $\quad a=3$

The sides of the rectangle used to draw the asymptotes pass through 3 and –3 on the x-axis and 5 and –5 on the y-axis.

y-intercepts: ±5
No x-intercepts.

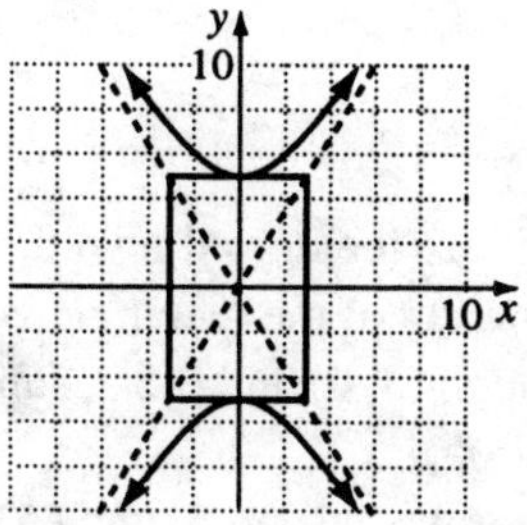

29. $4x^2 = 4 + y^2$

$4x^2 - y^2 = 4$

$\frac{4x^2}{4} - \frac{y^2}{4} = 1$

$\frac{x^2}{1} - \frac{y^2}{4} = 1$

$a^2 = 1$ $\quad$ $b^2 = 4$

$a = 1$ $\quad$ $b = 2$

The sides of the rectangle used to draw the asymptotes pass through 1 and −1 on the x-axis and 2 and −2 on the y-axis.
x-intercepts: ±1
No y-intercepts

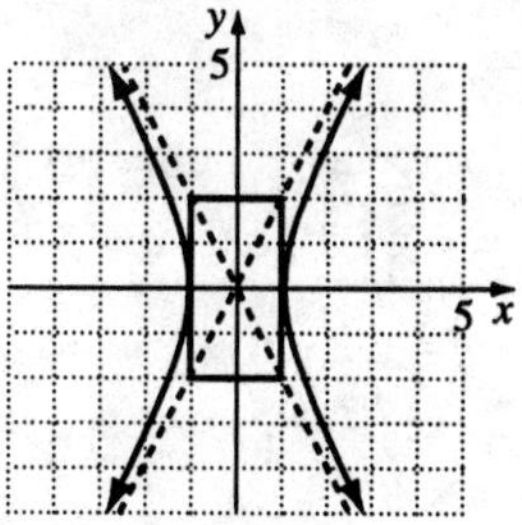

31. $xy = 4$

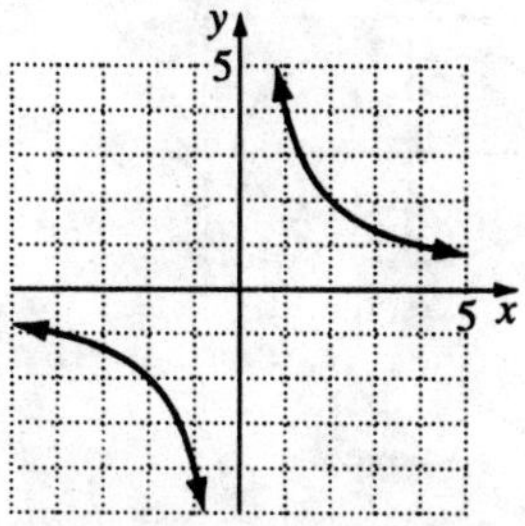

33. $xy = 2$

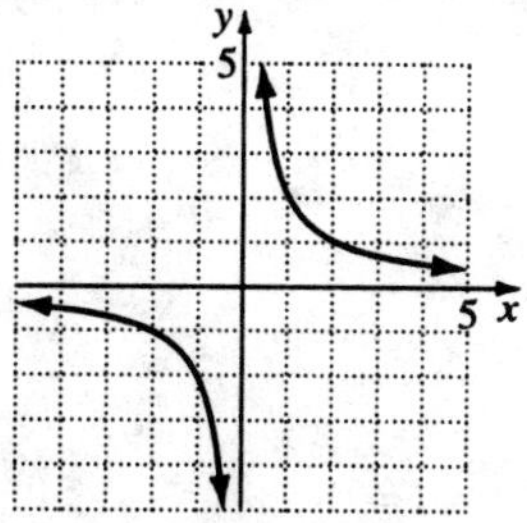

35. $xy = -4$

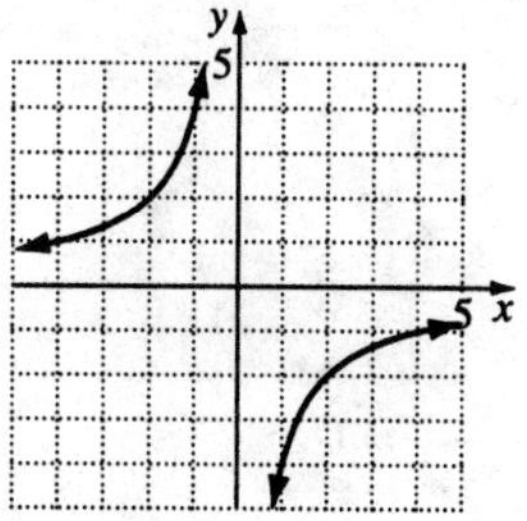

37. $xy = -2$

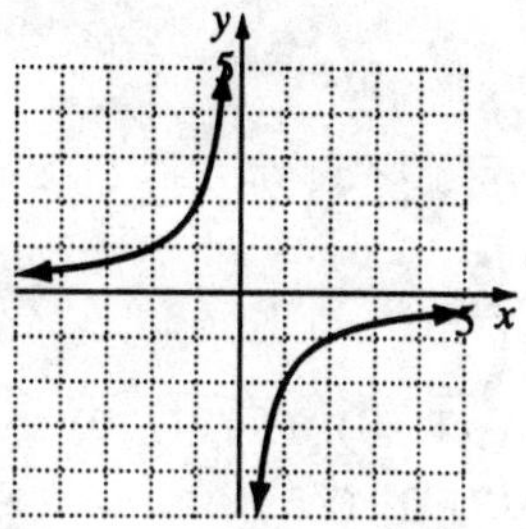

39. $x = 6y^2 + 4y + 3$
Parabola

41. $5x^2 - 5y^2 = 45$
Hyperbola

43. $x - 7 - 8y = y^2$
Parabola

45. $5x^2 = 12 - 10y^2$
Ellipse

47. $-3y^2 - 3x^2 = -27$
Circle

49. $x - \frac{5}{y} = 0$

Hyperbola

51. $x + y^2 = 4y^2 - y + 7$

Parabola

53. $\frac{x^2}{a^2} + \frac{y^2}{b^2} = 1$

$\frac{x^2}{(50)^2} + \frac{y^2}{(40)^2} = 1$

$\frac{x^2}{2500} + \frac{y^2}{1600} = 1$

55. a. $\frac{x^2}{a^2} + \frac{y^2}{b^2} = 1$

$\frac{x^2}{(48)^2} + \frac{y^2}{(23)^2} = 1$

$\frac{x^2}{2304} + \frac{y^2}{529} = 1$

b. $c^2 = a^2 - b^2$

$c = \pm\sqrt{2304 - 529}$

$c = \pm\sqrt{1775}$

$c = \pm 5\sqrt{71}$

Foci: $(-5\sqrt{71},\ 0)$, $(5\sqrt{71},\ 0)$

Desk: $(5\sqrt{71}$ ft, 0 ft$)$

57. $\frac{x^2}{a^2} - \frac{y^2}{b^2} = 1$

$y = \frac{b}{a}x$

$y = \frac{1}{2}x$

$y = \frac{\frac{3}{2}}{3}x$

$b = \frac{3}{2}, a = 3$

$\frac{x^2}{3^2} - \frac{y^2}{\left(\frac{3}{2}\right)^2} = 1$

$\frac{x^2}{9} - \frac{4y^2}{9} = 1$

59. d is true.

$\frac{x^2}{9} - \frac{y^2}{25} = 1$

$a^2 = 9,\ a = 3$

$b^2 = 25,\ b = 5$

The asymptotes are the lines drawn through the opposite corners of the rectangle whose sides pass through (3, 0) and (–3, 0) and (0, 5) and (0, –5).

The opposite corners are (3, 5) and (–3, -5) and (–3, 5) and (3, –5).

Equation for the lines drawn through the opposite corners are:

$y - 5 = \left(\frac{-5-5}{-3-3}\right)(x - 3)$

$y - 5 = \frac{5}{3}(x - 3)$

$y - 5 = \frac{5}{3}x - 5$

$y = \frac{5}{3}x$ and

$y - 5 = \left(\frac{-5-5}{3+3}\right)(x + 3)$

$y - 5 = -\frac{5}{3}(x + 3)$

$y - 5 = -\frac{5}{3}x - 5$

$y = -\frac{5}{3}x$

The equations for the asymptotes are $y = \frac{5}{3}x$ and $y = -\frac{5}{3}x$. (Note: $a = 3$, $b = 5$, $y = \pm\frac{b}{a}x$, $y = \pm\frac{5}{3}x$ are the equations for the asymptotes)

61. a. $x^2 + y^2 = (3950)^2$

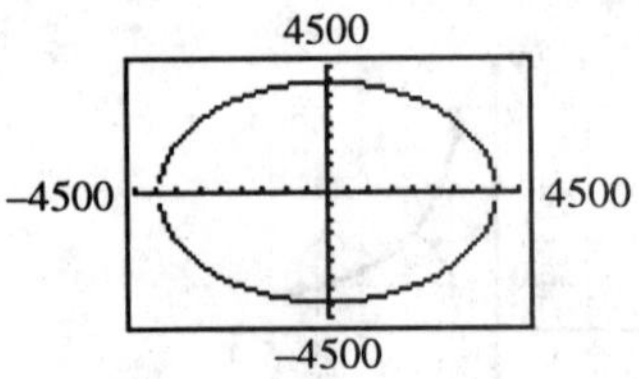

b. $\dfrac{(x+225.5)^2}{(4307.5)^2}+\dfrac{y^2}{(4301.6)^2}=1$

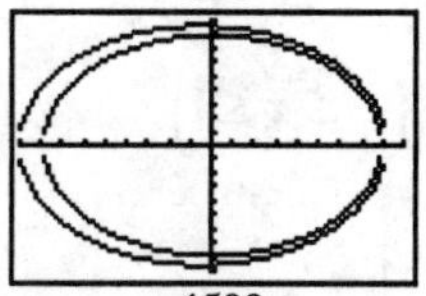

Altitude is higher

63. $\dfrac{x^2}{16}-\dfrac{y^2}{9}=1$ and $\dfrac{x|x|}{16}-\dfrac{y|y|}{9}=1$

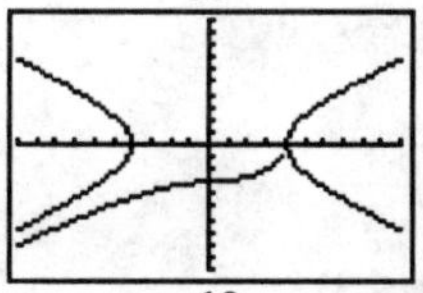

65. Answers may vary.

67. a. distance from (x, y) to $(-c, 0)$:

$\sqrt{(x+c)^2+(y-0)^2}=\sqrt{(x+c)^2+y^2}$

distance from (x, y) to $(c, 0)$:

$\sqrt{(x-c)^2+(y-0)^2}=\sqrt{(x-c)^2+y^2}$

sum of distances from (x, y) to foci $= 2a$

$\sqrt{(x+c)^2+y^2}+\sqrt{(x-c)^2+y^2}=2a$

$\sqrt{(x+c)^2+y^2}=2a-\sqrt{(x-c)^2+y^2}$

$(x+c)^2+y^2=4a^2-4a\sqrt{(x-c)^2+y^2}+(x-c)^2+y^2$

$x^2+2xc+c^2+y^2=4a^2-4a\sqrt{(x-c)^2+y^2}+x^2-2xc+c^2+y^2$

$4a\sqrt{(x-c)^2+y^2}=4a^2-4xc$

$a\sqrt{(x-c)^2+y^2}=a^2-xc$

$a^2(x-c)^2+a^2y^2=a^4-2a^2xc+x^2c^2$

$a^2x^2-2a^2xc+a^2c^2+a^2y^2=a^4-2a^2xc+x^2c^2$

$a^2x^2+a^2c^2+a^2y^2=a^4+x^2c^2$

$a^2x^2-x^2c^2+a^2y^2=a^4-a^2c^2$

$x^2(a^2-c^2)+a^2y^2=a^2(a^2-c^2)$

$$\frac{x^2(a^2-c^2)}{a^2(a^2-c^2)}+\frac{a^2y^2}{a^2(a^2-c^2)}$$
$$=\frac{a^2(a^2-c^2)}{a^2(a^2-c^2)}$$
$$\frac{x^2}{a^2}+\frac{y^2}{a^2-c^2}=1$$

b. Let $b^2=a^2-c^2$.

$$\frac{x^2}{a^2}+\frac{y^2}{b^2}=1$$

69. letter F

$y=2\ (0\le x\le 1)$

$y=1\ (0\le x\le 0.75)$

$x=0\ \ (0\le y\le 2)$

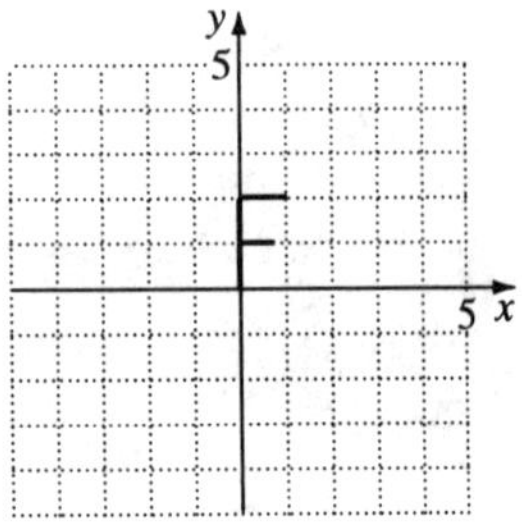

71. letter K:

$x=0\ \ (-1\le y\le 1)$

$y^2=x^3\rightarrow y=\pm x\sqrt{x}\ \ (0\le x\le 1)$

x	0	0.5	1
y	0	±0.35	±1

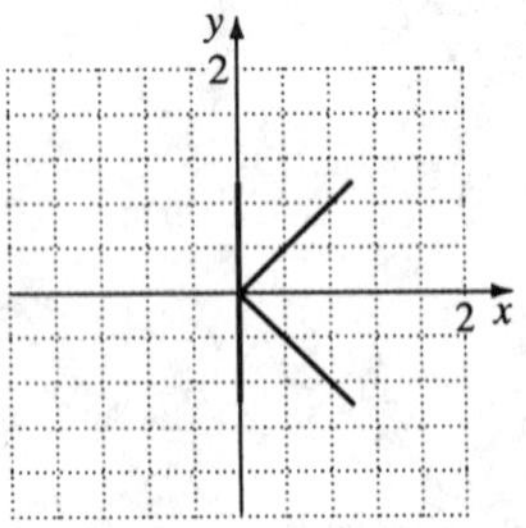

73. letter Q:

$x^2+y^2=1$ is a circle with $C(0, 0)$ and $r=1$

or $y=\pm\sqrt{1-x^2}$

$y=-x\ \ \left(\frac{1}{2}\le x\le 1\right)$

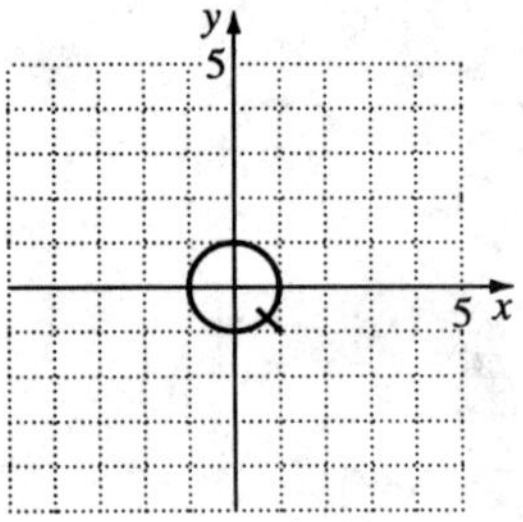

75. Letter Y:

$y=-x+3\ \ (1\le x\le 2)$

$y=x-1\ \ (2\le x\le 3)$

$x=2\ \ (0\le y\le 1)$

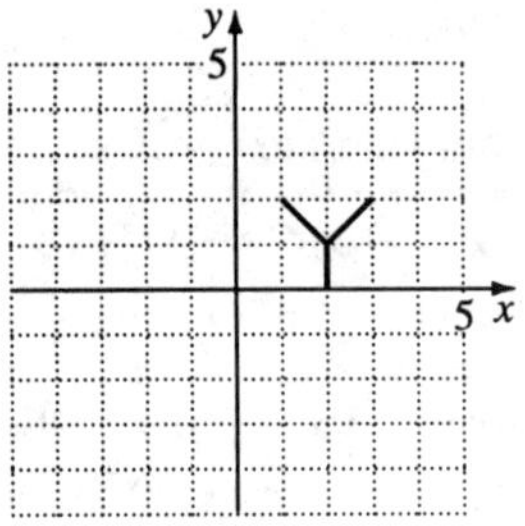

Review Problems

77. $x=2y+7$

$-5x+4y=-5$

Substitute $2y+7$ into the second equation.

$-5(2y+7)+4y=-5$

$-10y-35+4y=-5$

$-6y-35=-5$

$-6y=30$

$y=-5$

Substitute back.

$x=2(-5)+7$

$x=-10+7$

$x=-3$

$\{(-3, -5)\}$

78. $8x + 2y - 7 = 0$
$-2x - y + 2 = 0$
Solve the second equation for y.
$y = 2 - 2x$
Substitute $2 - 2x$ into the first equation for y.
$8x + 2(2 - 2x) - 7 = 0$
$8x + 4 - 4x - 7 = 0$
$4x = 3$
$x = \frac{3}{4}$
Substitute back.
$y = 2 - 2\left(\frac{3}{4}\right) = \frac{1}{2}$
$\left\{\left(\frac{3}{4}, \frac{1}{2}\right)\right\}$

79. $12x + 5y = -2$
$8x - 3y = -14$
Add 3 times the first equation to 5 times the second equation.
$36x + 15y = -6$
$40x - 15y = -70$
$76x = -76$
$x = -1$
Substitute back.
$12(-1) + 5y = -2$
$5y = 10$
$y = 2$
$\{(-1, 2)\}$

Problem Set 9.3

For problems **1–30.**, the solution confirmation is left to the student.

1. $x^2 + y^2 = 25$
$x - 3y = -5$
$x = 3y - 5$
$(3y - 5)^2 + y^2 = 25$
$9y^2 - 30y + 25 + y^2 = 25$
$10y^2 - 30y = 0$
$10y(y - 3) = 0$
$10y = 0$ or $y - 3 = 0$
$y = 0$ or $y = 3$
If $y = 0$: $x = 3(0) - 5 = -5$
If $y = 3$: $x = 3(3) - 5 = 4$
$\{(-5, 0), (4, 3)\}$

3. $y = x^2 - 1$
$4x + y + 5 = 0$
$4x + x^2 - 1 + 5 = 0$
$x^2 + 4x + 4 = 0$
$(x + 2)(x + 2) = 0$
$x + 2 = 0$
$x = -2$
If $x = -2$: $y = (-2)^2 - 1 = 3$
$\{(-2, 3)\}$

5. $\frac{x^2}{4} + \frac{y^2}{9} = 1$
$3x + 2y - 6 = 0$
$x = \frac{6 - 2y}{3}$
$\frac{\left(\frac{6-2y}{3}\right)^2}{4} + \frac{y^2}{9} = 1$
$\frac{36 - 24y + 4y^2}{9 \cdot 4} + \frac{y^2}{9} = 1$
$\frac{4y^2 - 24y + 36 + 4y^2 - 36}{36} = 0$
$\frac{8y^2 - 24y}{36} = 0$
$\frac{8}{36}y(y - 3) = 0$
$y = 0$ or $y = 3$
If $y = 0$: $x = \frac{6 - 2(0)}{3} = 2$
If $y = 3$: $x = \frac{6 - 2(3)}{3} = 0$
$\{(2, 0), (0, 3)\}$

7. $xy = 6$
$2x - y = 1$
$y = 2x - 1$
$x(2x - 1) = 6$
$2x^2 - x - 6 = 0$
$(2x + 3)(x - 2) = 0$

$x=-\frac{3}{2}$ or $x=2$

If $x=-\frac{3}{2}$, $y=2\left(-\frac{3}{2}\right)-1=-4$

If $x=2$, $y=2(2)-1=3$

$\left\{\left(-\frac{3}{2}, -4\right), (2, 3)\right\}$

9. $x^2+y^2=52$

$3x-2y=0$

$x=\frac{2y}{3}$

$\left(\frac{2y}{3}\right)^2+y^2=52$

$\frac{4y^2}{9}+y^2-52=0$

$\frac{13y^2-468}{9}=0$

$\frac{13}{9}(y^2-36)=0$

$\frac{13}{9}(y+6)(y-6)=0$

$y=-6$ or $y=6$

If $y=-6$: $x=\frac{2(-6)}{3}=-4$

If $y=6$: $x=\frac{2(6)}{3}=4$

$\{(-4, -6), (4, 6)\}$

11. $x-y^2=1$

$x=2y^2$

$2y^2-y^2=1$

$y^2=1$

$y=\pm 1$

If $y=1$: $x=2(1)^2=2$

If $y=-1$: $x=2(-1)^2=2$

$\{(2, 1), (2, -1)\}$

13. $y-4x=0$

$y=x^2+5$

$(x^2+5)-4x=0$

$x^2-4x+5=0$

$x=\frac{4\pm\sqrt{16-4(1)(5)}}{2}$

$=\frac{4\pm\sqrt{-4}}{2}=\frac{4\pm 2i}{2}=2\pm i$

If $x=2+i$: $y=4x=4(2+i)=8+4i$

If $x=2-i$: $y=4x=4(2-i)=8-4i$

$\{(2+i, 8+4i), (2-i, 8-4i)\}$

15. $y=4-x^2$

$2x+y=1$

$y=1-2x$

$1-2x=4-x^2$

$x^2-2x-3=0$

$(x-3)(x+1)=0$

$x=3$ or $x=-1$

If $x=3$: $y=4-(3)^2=-5$

If $x=-1$: $y=4-(-1)^2=3$

$\{(3, -5), (-1, 3)\}$

17. $x^2+y^2=25$

$x=y$

$y^2+y^2=25$

$2y^2=25$

$y=\pm\sqrt{\frac{25}{2}}$

$y=\pm\frac{5}{\sqrt{2}}$

If $y=-\frac{5}{\sqrt{2}}$: $x=-\frac{5}{\sqrt{2}}$

If $y=\frac{5}{\sqrt{2}}$: $x=\frac{5}{\sqrt{2}}$

$\left\{\left(-\frac{5\sqrt{2}}{2}, -\frac{5\sqrt{2}}{2}\right), \left(\frac{5\sqrt{2}}{2}, \frac{5\sqrt{2}}{2}\right)\right\}$

19. $x^2+y^2=13$

$\underline{x^2-y^2=5}$

$2x^2=18$

$x^2=9$

$x=\pm 3$

If $x=-3$:

$(-3)^2+y^2=13$

$y^2 = 4$
$y = \pm 2$

If $x = 3$:
$3^2 + y^2 = 13$
$y^2 = 4$
$y = \pm 2$
$\{(-3, -2), (-3, 2), (3, -2), (3, 2)\}$

21. $x^2 + y^2 = 13$
$2x^2 + 3y^2 = 30$

$-2x^2 - 2y^2 = -26$
$2x^2 + 3y^2 = 30$
$y^2 = 4$
$y = \pm 2$
If $y = -2$:
$x^2 + (-2)^2 = 13$
$x^2 = 9$
$x = \pm 3$
If $y = 2$:
$x^2 + (2)^2 = 13$
$x^2 = 9$
$x = \pm 3$
$\{(-3, -2), (3, -2), (-3, 2), (3, 2)\}$

23. $3x^2 - 7y^2 = -15$
$7x^2 + 9y^2 = 22$

$27x^2 - 63y^2 = -135$
$49x^2 + 63y^2 = 154$
$76x^2 = 19$
$x^2 = \frac{1}{4}$
$x = \pm\frac{1}{2}$

If $x = -\frac{1}{2}$:
$3\left(-\frac{1}{2}\right)^2 - 7y^2 = -15$
$y = \pm\frac{3}{2}$

If $x = \frac{1}{2}$:
$3\left(\frac{1}{2}\right)^2 - 7y^2 = -15$
$y = \pm\frac{3}{2}$
$\left\{\left(-\frac{1}{2}, -\frac{3}{2}\right), \left(-\frac{1}{2}, \frac{3}{2}\right), \left(\frac{1}{2}, -\frac{3}{2}\right), \left(\frac{1}{2}, \frac{3}{2}\right)\right\}$

25. $x^2 + y^2 = 4$
$\frac{x^2}{9} + \frac{y^2}{1} = 1$

$-x^2 - y^2 = -4$
$x^2 + 9y^2 = 9$
$8y^2 = 5$
$y = \pm\sqrt{\frac{5}{8}}$
If $y = -\sqrt{\frac{5}{8}}$:
$x^2 + \left(-\sqrt{\frac{5}{8}}\right)^2 = 4$
$x = \pm\sqrt{\frac{27}{8}} = \pm\frac{3}{2}\sqrt{\frac{3}{2}}$
If $y = \sqrt{\frac{5}{8}}$:
$x^2 + \left(\sqrt{\frac{5}{8}}\right)^2 = 4$
$x = \pm\sqrt{\frac{27}{8}} = \pm\frac{3}{2}\sqrt{\frac{3}{2}}$
$\left\{\left(-\frac{3}{2}\sqrt{\frac{3}{2}}, -\sqrt{\frac{5}{8}}\right), \left(\frac{3}{2}\sqrt{\frac{3}{2}}, -\sqrt{\frac{5}{8}}\right),\right.$
$\left.\left(-\frac{3}{2}\sqrt{\frac{3}{2}}, \sqrt{\frac{5}{8}}\right), \left(\frac{3}{2}\sqrt{\frac{3}{2}}, \sqrt{\frac{5}{8}}\right)\right\}$

27. $x^2 - y^2 = 12$

$x^2 - 7y = 2$

$$\begin{aligned} -x^2 + y^2 &= -12 \\ x^2 - 7y &= 2 \\ \hline y^2 - 7y &= -10 \end{aligned}$$

$y^2 - 7y + 10 = 0$

$(y-5)(y-2) = 0$

$y = 5$ or $y = 2$

If $y = 5$:

$x^2 - 35 = 2$

$x^2 = 37$

If $y = 2$:

$x^2 - 4 = 12$

$x^2 = 16$

$x = \pm 4$

$\{(\sqrt{37}, 5), (-\sqrt{37}, 5), (4, 2), (-4, 2)\}$

29. $y = x^2 + 5$

$x^2 + y^2 = 25$

$$\begin{aligned} -x^2 + y &= 5 \\ x^2 + y^2 &= 25 \\ \hline y^2 + y &= 30 \end{aligned}$$

$y^2 + y - 30 = 0$

$(y+6)(y-5) = 0$

$y = -6$ or $y = 5$

If $y = -6$:

$-6 = x^2 + 5$

$x^2 = -11$

$x = \pm\sqrt{11}i$

If $y = 5$:

$5 = x^2 + 5$

$x = 0$

$\left\{\left(\sqrt{11}i, -6\right), \left(-\sqrt{11}i, -6\right), (0, 5)\right\}$

31. $3x + y = 2$

$2x^2 - y^2 = 1$

$y = 2 - 3x$

$2x^2 - (2-3x)^2 = 1$

$2x^2 - 4 + 12x - 9x^2 - 1 = 0$

$7x^2 - 12x + 5 = 0$

$(7x-5)(x-1) = 0$

$x = \frac{5}{7}$ or $x = 1$

If $x = \frac{5}{7}$: $y = 2 - 3\left(\frac{5}{7}\right) = -\frac{1}{7}$

If $x = 1$: $y = 2 - 3(1) = -1$

$\left\{\left(\frac{5}{7}, -\frac{1}{7}\right), (1, -1)\right\}$

33. $2x^2 = 10 + 3y^2$

$x^2 + 4y^2 = -17$

$x^2 = -4y^2 - 17$

$2(-4y^2 - 17) = 10 + 3y^2$

$-8y^2 - 34 = 10 + 3y^2$

$-44 = 11y^2$

$y^2 = -4$

$y = \pm 2i$

When $y = -2i$, $x^2 = -4(-2i)^2 - 17$

$x^2 = -4(-4) - 17 = -1$

$x = \pm i$

When $y = 2i$, $x^2 = -4(2i)^2 - 17$

$x^2 = -4(4)(-1) - 17 = -1$

$x = \pm i$

$\{(-i, -2i), (+i, -2i), (-i, 2i), (+i, 2i)\}$

35. $x^2 + 4y^2 = 20$

$xy = 4$

$x = \frac{4}{y}$

$\left(\frac{4}{y}\right)^2 + 4y^2 = 20$

$\frac{16}{y^2} + 4y^2 = 20$

$16 + 4y^4 = 20y^2$

$4y^4 - 20y^2 + 16 = 0$

$y^4 - 5y^2 + 4 = 0$

$(y^2 - 4)(y^2 - 1) = 0$

$y = \pm 2$ or $y = \pm 1$

If $y=-2;\ x=\frac{4}{-2}=-2$

If $y=2,\ x=\frac{4}{2}=2$

If $y=-1,\ x=\frac{4}{-1}=-4$

If $y=1,\ x=\frac{4}{1}=4$

$\{(-2,-2),(2,2),(-4,-1),(4,1)\}$

37. $xy-2y^2=-6$

$xy-y^2=-4$

$$\begin{array}{r} xy-2y^2=-6 \\ -xy+\ y^2=\ 4 \\ \hline -y^2=-2 \\ y=\pm\sqrt{2} \end{array}$$

If $y=-\sqrt{2}$:

$x(-\sqrt{2})-2(-\sqrt{2})^2=-6$

$x=\frac{2}{\sqrt{2}}=\sqrt{2}$

If $y=\sqrt{2}$:

$x(\sqrt{2})-2(\sqrt{2})^2=-6$

$x=-\sqrt{2}$

$\left\{\left(-\sqrt{2},\ \sqrt{2}\right),\left(\sqrt{2},\ -\sqrt{2}\right)\right\}$

39. $a^2+b^2=25$

$ab=12$

$a=\frac{12}{b}$

$\left(\frac{12}{b}\right)^2+b^2=25$

$\frac{144}{b^2}=25-b^2$

$144=25b^2-b^4$

$b^4-25b^2+144=0$

$(b^2-9)(b^2-16)=0$

$b^2=9$ or $b^2=16$

If $b^2=9: b=\pm3;$

If $b=-3,\ a=\frac{12}{-3}=-4$

If $b=3,\ a=\frac{12}{3}=4$

If $b^2=16, b=\pm4,$

If $b=-4,\ a=\frac{12}{-4}=-3$

If $b=4,\ a=\frac{12}{4}=3$

$\{(-4,-3),(4,3),(-3,-4),(3,4)\}$

41. $\frac{x^2}{4}+\frac{y^2}{16}=1\to\frac{y+4}{4}+\frac{y^2}{16}=1$

$y=x^2-4\to x^2=y+4$

$y^2+4y=0$

$y(y+4)=0$

$y=0,-4$

$x^2=0+4\to x=\pm2$

$x^2=-4+4=0\to x=0$

$\{(-2,0),(2,0),(0,-4)\}$

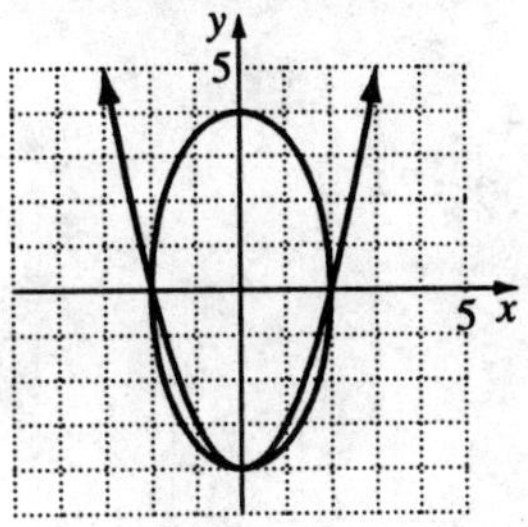

43. Let x = length, y = width

Perimeter is 20 feet.

$2x+2y=20$

Area is 21 ft^2.

$xy=21$

$(\div 2)\ x+y=10$ or $y=10-x$

$xy=21$

$x(10-x)=21$

$10x-x^2=21$

$0=x^2-10x+21$

$0=(x-7)(x-3)$

$x=7$ or $x=3$

If $x=7$: $y=10-x=10-7=3$

If $x=3$: $y=10-x=10-3=7$

Dimensions: 7 ft by 3 ft

45. $108=L\cdot W$

$(15)^2=L^2+W^2$

$L = \frac{108}{W}$

$225 = \left(\frac{108}{W}\right)^2 + W^2$

$225 = \frac{11,664}{W^2} + W^2$

$225 = \frac{11,664 + W^4}{W^2}$

$W^4 - 225W^2 + 11,664 = 0$

$(W^2 - 144)(W^2 - 81) = 0$

$(W - 12)(W + 12)(W - 9)(W + 9) = 0$

$W = 12,\ W = -12,\ W = 9,\ W = -9$

Disregard negative lengths.

The width is 9 ft and the length is 12 ft.

47. Area with fountain pool removed is 21.

$x^2 - y^2 = 21$

Perimeter is 24.

$3x + 3y + (x - y) = 24$

$x^2 - y^2 = 21$

$4x + 2y = 24$

$x^2 - y^2 = 21$

$2x + y = 12$

$y = 12 - 2x$

$x^2 - (12 - 2x)^2 = 21$

$x^2 - 144 + 48x - 4x^2 = 21$

$-3x^2 + 48x - 165 = 0$

$x^2 - 16x + 55 = 0$

$(x - 11)(x - 5) = 0$

$x = 11$ or $x = 5$

If $x = 5$, $y = 12 - 10 = 2$

If $x = 11$, $y = 12 - 22 = -10$

Disregard negative lengths.

Thus, $x = 5$ and $y = 2$.

$x = 5$ meters; $y = 2$ meters

49. $CD = 8$ inches

$x + y = 8$, so $y = 8 - x$

$y^2 = 4^2 + x^2$

$(8 - x)^2 = 16 + x^2$

$64 - 16x + x^2 = 16 + x^2$

$-16x = -48$

$x = 3$

$3 + y = 8$

$y = 5$

51. Let x = length, y = width, $xy = 480$

area = 480 square feet

$(x + 10)(y - 2) = 480 + 20 = 500$

$xy = 480 \rightarrow x = \frac{480}{y}$

$xy - 2x + 10y = 520$

$\left(\frac{480}{y}\right)y - 2\left(\frac{480}{y}\right) + 10y = 520$

$480y - 960 + 10y^2 = 520y$

$y^2 - 4y - 96 = 0$

$(y - 12)(y + 8) = 0$

$y = 12$ or $y = -8$ (reject)

If $y = 12$: $x = \frac{480}{y} = \frac{480}{12} = 40$

Length: 40 feet

Width: 12 feet

Dimensions: 40 ft by 12 ft

53. $LW = 216$

$2(L - 4)(W - 4) = 224$

$2 = \frac{216}{W}$

$2\left(\frac{216}{W} - 4\right)(W - 4) = 224$

$2\left(216 - \frac{864}{W} - 4W + 16\right) = 224$

$\frac{232W - 864 - 4W^2}{W} = 112$

$4W^2 - 120W + 864 = 0$

$W^2 - 30W + 216 = 0$

$(W - 18)(W - 12) = 0$

$W = 18$ or $W = 12$

The dimensions are width, 12 in., and length, 18 in.

55. b is true.

57. $\frac{x^2}{1.03}+\frac{y^2}{0.97}=1$

$x^2-2xy+y^2-2x-2y+1=0$

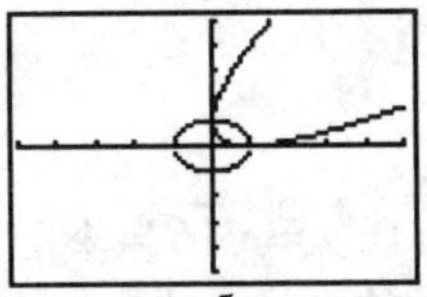

They will cross 2 times, at (0, 0.98488578) and (1.0148892, 0), or $\left(0, \sqrt{0.97}\right)$ and $\left(\sqrt{1.03}, 0\right)$.

59–61. Answers may vary.

63. The system's solution set is $\{(5, 4), (4, 5), (-5, -4), (-4, -5)\}$. Using the distance formula, the rectangle's dimensions are $\sqrt{2}$ and $\sqrt{162}$ and its area is $\sqrt{324}$ or 18 square units.

65. $(a+bi)^2=b+ai$

$a^2+2abi-b^2=b+ai$

$(a^2-b^2)+2abi=b+ai$

This means that

$a^2-b^2=b$ and

$2ab=a$

$b=\frac{1}{2}$

Since $a^2-b^2=b$ and $b=\frac{1}{2}$, we have

$a^2-\frac{1}{4}=\frac{1}{2}$

$a^2=\frac{3}{4}$

$a=\frac{\sqrt{3}}{2}$

we were told that $a>0$.

The ordered pair (a, b) is $\left(\frac{\sqrt{3}}{2}, \frac{1}{2}\right)$

67. $x^2-y^2=9$

$\sqrt{x+y}+\sqrt{x-y}=4$, so

$x+y+2\sqrt{x^2-y^2}+x-y=16$

$x+\sqrt{x^2-y^2}=8$

$x^2-y^2=64-16x+x^2$ (may introduce extraneous root)

$y^2=16x-64$

$x^2-16x+64=9$

$x^2-16x+55=0$

$(x-11)(x-5)=0$

$x=5, 11$

$y^2=x^2-9=16,\ 112$

$y=\pm 4,\ \pm 4\sqrt{7}$

$\{(5, -4), (5, 4)\}$

69. Let x = radius of one circle

y = radius of other circle

Perimeter = 12π

$2\pi x+2\pi y=12\pi$

$x+y=6$

$y=6-x$

$\pi x^2+\pi y^2=20\pi$

$x^2+y^2=20$

$x^2+(6-x)^2=20$

$x^2-6x+8=0$

$(x-4)(x-2)=0$

$x=2$ or $x=4$

$y=4$ or $y=2$

The radius are 2 m and 4 m.

71. Let r = average rate of travel
t = time
$t = 5r - 3$
$rt = 36$
$r(5r - 3) = 36$
$5r^2 - 3r - 36 = 0$
$(5r + 12)(r - 3) = 0$
$r = -\frac{12}{5}$ (reject) or $r = 3$
$t = 5r - 3 = 5(3) - 3 = 12$
No, Woody did not keep his promise.
His 36 mile hike took 12 hours > 8 hours.

Review Problems

73. $\log_x 36 = 2$
$x^2 = 36$
$x^2 = 6^2$
$x = 6$
$\{6\}$

74. $\dfrac{\frac{1}{x^2-9}}{\frac{1}{x+3}+\frac{1}{x-3}} = \dfrac{\frac{1}{x^2-9}}{\frac{x-3+x+3}{(x+3)(x-3)}}$
$= \dfrac{1}{x^2-9} \cdot \dfrac{x^2-9}{2x} = \dfrac{1}{2x}$

75. $7^x = 125$
$\ln 7^x = \ln 125$
$x \ln 7 = \ln 125$
$x = \dfrac{\ln 125}{\ln 7}$

Chapter 9 Review Problems

1. Center (–2, 4), radius of 6:
$(x-h)^2 + (y-k)^2 = r^2$
$h = -2,\ k = 4,\ r = 6$
$(x+2)^2 + (y-4)^2 = 36$

2. Center at origin, radius of 3:
$(x-h)^2 + (y-k)^2 = r^2$
$h = 0,\ k = 0,\ r = 3$
$x^2 + y^2 = 9$

3. $x^2 + y^2 = 16$
Circle: $h = 0, k = 0,\ r = \sqrt{16} = 4$
Center at (0, 0) and radius of 4

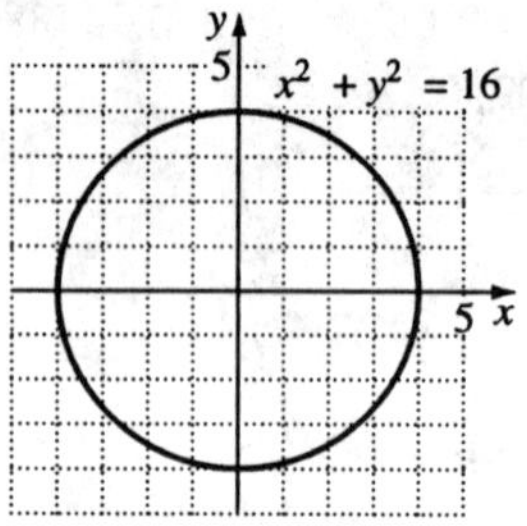

4. $(x+2)^2 + (y-3)^2 = 9$
$h = -2,\ k = 3$ and $r = \sqrt{9} = 3$
Circle: Center at (–2, 3) and radius of 3

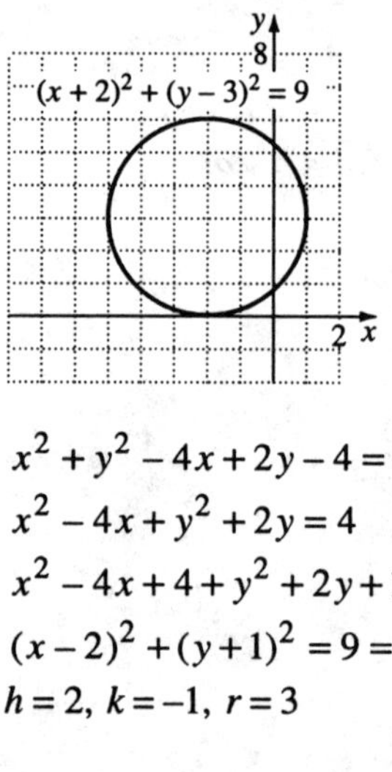

5. $x^2 + y^2 - 4x + 2y - 4 = 0$
$x^2 - 4x + y^2 + 2y = 4$
$x^2 - 4x + 4 + y^2 + 2y + 1 = 4 + 4 + 1$
$(x-2)^2 + (y+1)^2 = 9 = 3^2$
$h = 2,\ k = -1,\ r = 3$

Center at (2, –1) and radius of 3.

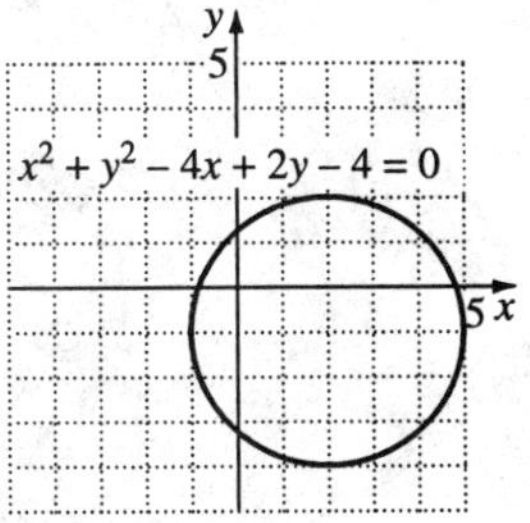

6. $y = x^2 - 2x + 3$

$y = x^2 - 2x + 1 + 3 - 1$

$y = (x - 1)^2 + 2$

$h = 1,\ k = 2,\ a = 1$

Parabola: vertex at (1, 2)

$a = 1 > 0$; opens up

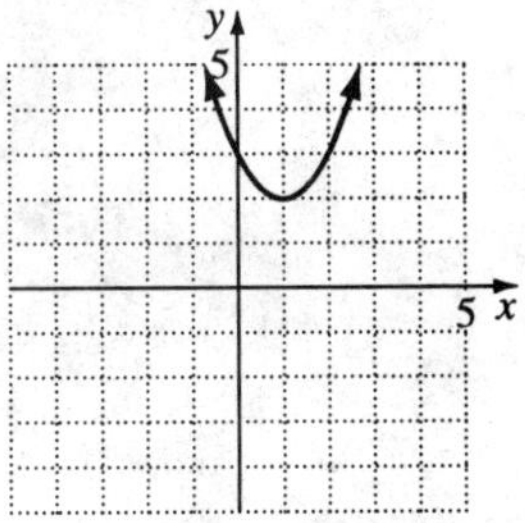

7. $y = -(x + 1)^2 + 4$

$h = -1,\ k = 4,\ a = -1$

parabola: vertex at (–1, 4)

$a = -1 < 1$; opens downward

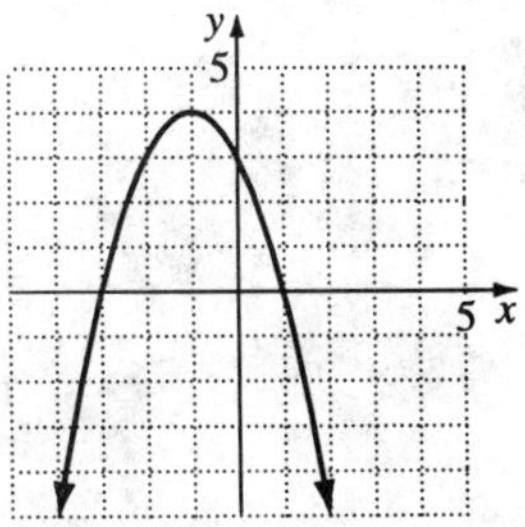

8. $x = y^2 - 8y + 12$

$x = y^2 - 8y + 16 + 12 - 16$

$x = (y - 4)^2 - 4$

$h = -4,\ k = 4,\ a = 1$

Parabola: vertex at (–4, 4)

$a = 1 > 0$; opens to the right

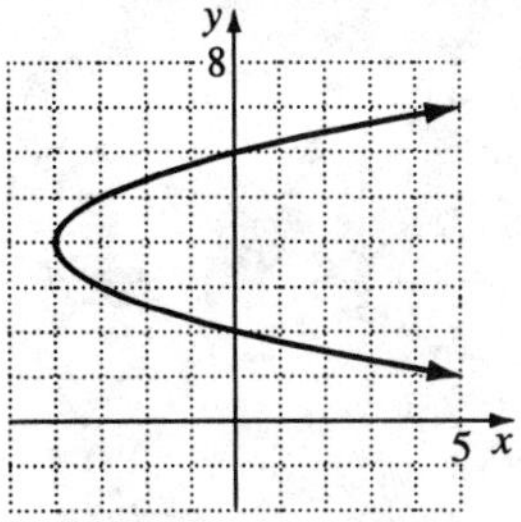

9. $x = (y - 3)^2 - 4$

$h = -4,\ k = 3,\ a = 1$

Parabola: vertex at (–4, 3)

$a = 1 > 0$; opens to right

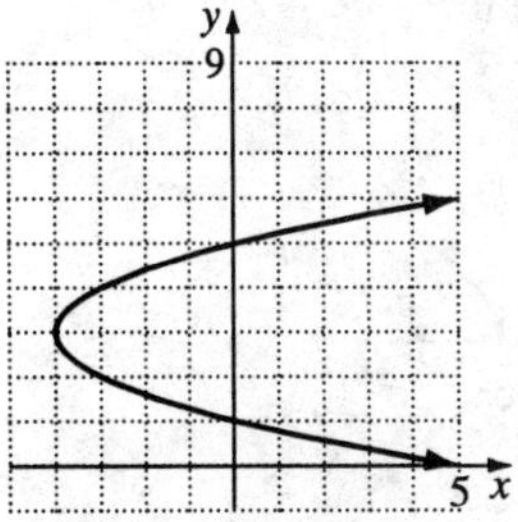

10. $\frac{x^2}{9} + \frac{y^2}{25} = 1$

Center (0, 0)

y-intercepts: ±5

x-intercepts: ±3

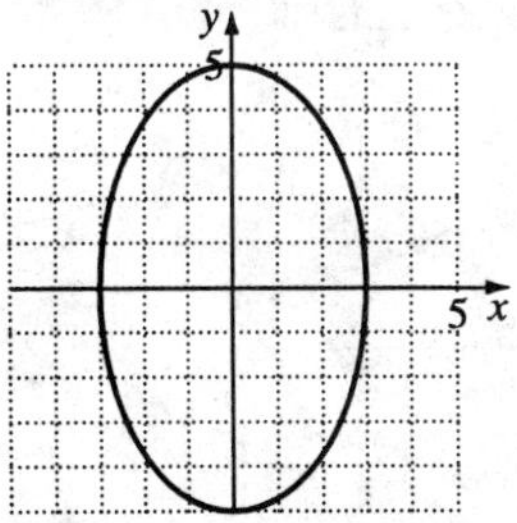

11. $4x^2 + 25y^2 = 100$

$\frac{x^2}{25} + \frac{y^2}{4} = 1$

Center (0, 0)

y-intercepts: ±2

x-intercepts: ± 5

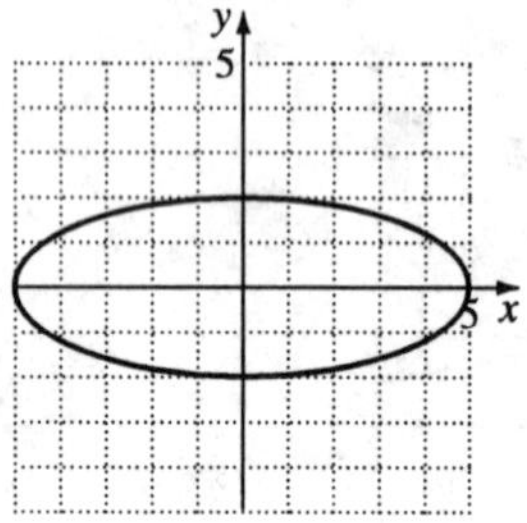

12. $\frac{x^2}{16} - \frac{y^2}{9} = 1$

$a^2 = 16,\ a = 4$

hyperbola: center at (0, 0) and vertices at $(\pm 4, 0)$; no y-intercepts

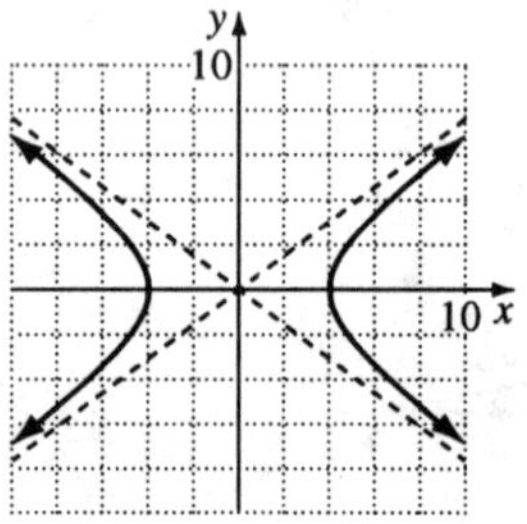

13. $\frac{y^2}{9} - \frac{x^2}{4} = 1$

$b^2 = 9,\ b = 3$

hyperbola: center (0, 0) and vertices at $(0, \pm 3)$; no x-intercepts

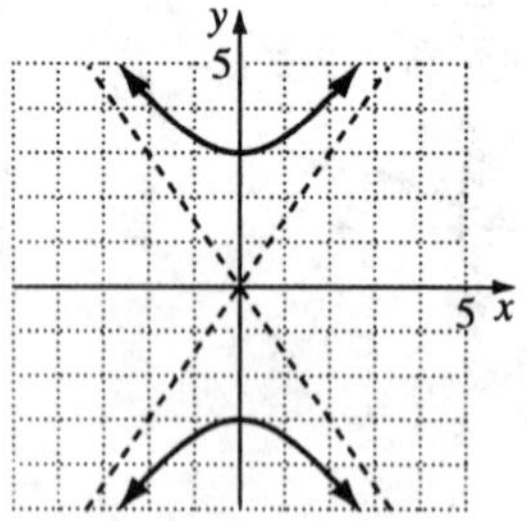

14. $x^2 - y^2 = 4$

$a^2 = 4,\ a = 2$

hyperbola: center at (0, 0) and vertices at $(\pm 2, 0)$; no y-intercept

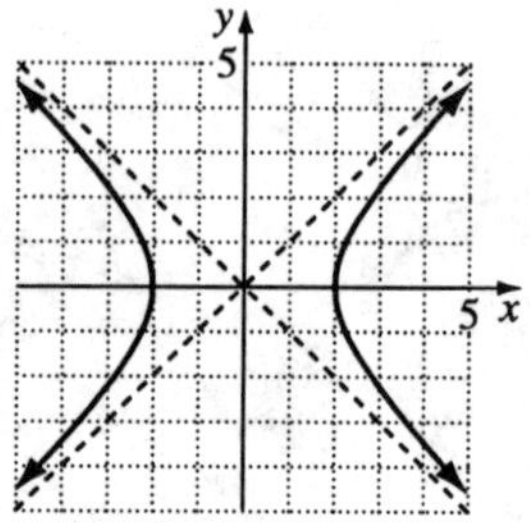

15. $xy = 8$

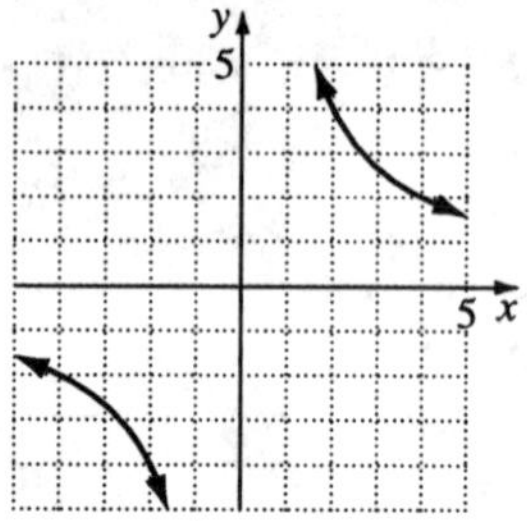

16. $xy = -4$

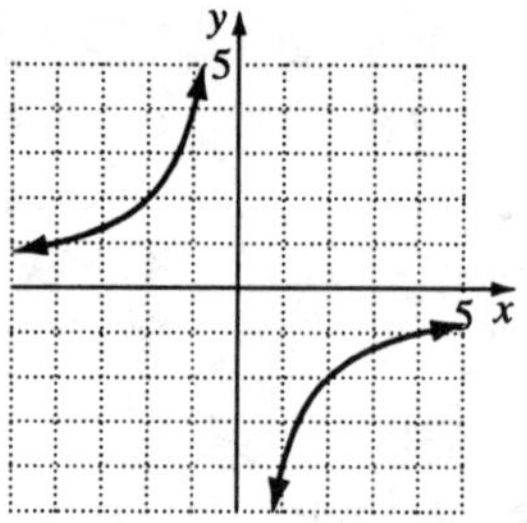

17. y-intercepts: ± 10

x-intercepts: ± 40

$$\frac{x^2}{(40)^2} + \frac{y^2}{(10)^2} = 1$$

$$\frac{x^2}{1600} + \frac{y^2}{100} = 1$$

18. Truck needs to clear point (14, 12)

ellipse: $\frac{x^2}{(25)^2}+\frac{y^2}{(15)^2}=1$

$\frac{x^2}{625}+\frac{y^2}{225}=1$

$\frac{(14)^2}{625}+\frac{(12)^2}{225}\ ?\ 1$

$0.9536<1$

Yes, the truck will clear the archway.

19. $5x^2+5y^2=180$

circle

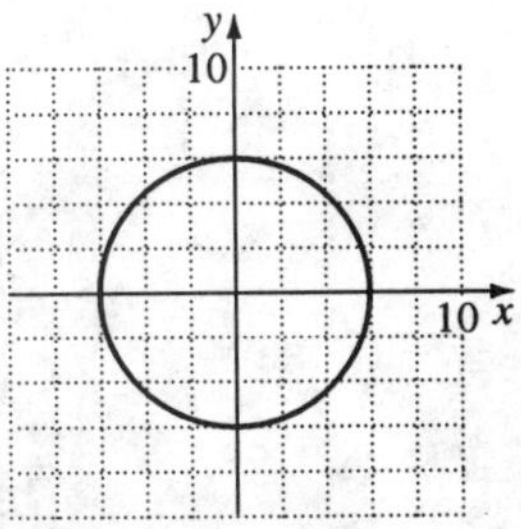

20. $4x^2+9y^2=36$

ellipse

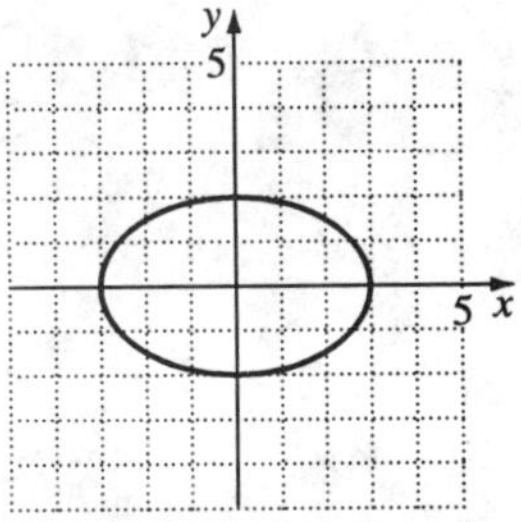

21. $x+3=-y^2+2y$

parabola

22. $4x^2-9y^2=36$

hyperbola

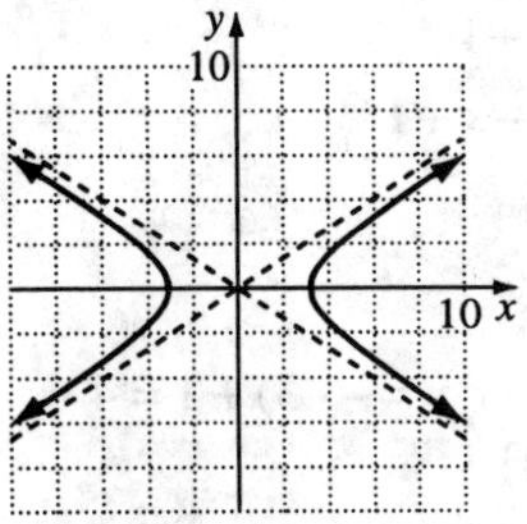

23. $x^2+y^2+6x-2y=-6$

circle

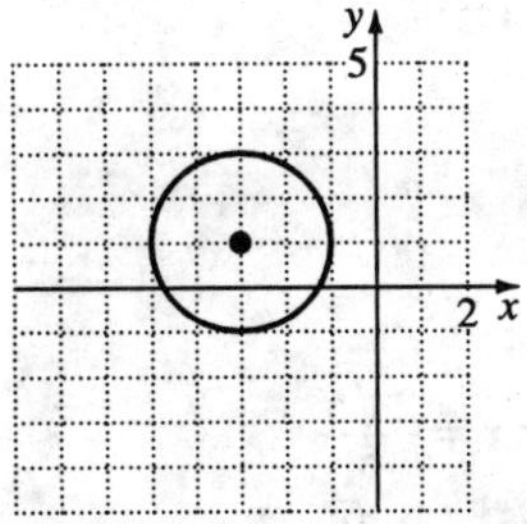

24. $xy=3$

hyperbola

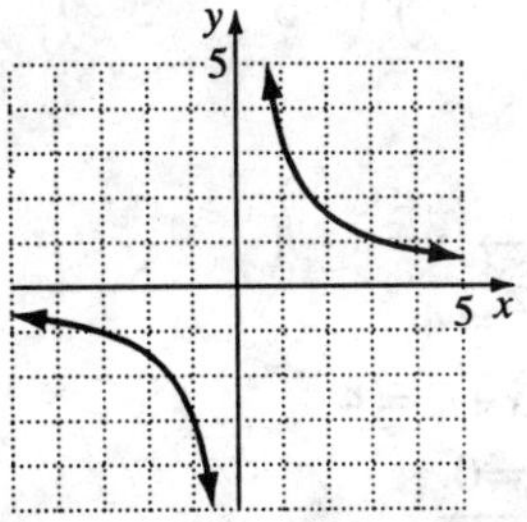

25–38. Check is left to the student.

25. $5y=x^2-1$

$x-y=1 \rightarrow y=x-1$

$5(x-1)=x^2-1$

$5x-5=x^2-1$

$0=x^2-5x+4$

$0=(x-4)(x-1)$

$x=4$ or $x=1$

$y=4-1=3$ or $y=1-1=0$

$\{(1, 0), (4, 3)\}$

26. $y = x^2 + 2x + 1 \rightarrow y = (x+1)^2$
$x + y - 1 = 0 \rightarrow y = -x + 1$
$(x+1)^2 = -x + 1$
$x^2 + 2x + 1 = -x + 1$
$x^2 + 3x = 0$
$x(x+3) = 0$
$x = 0$ or $x = -3$
$y = -0 + 1 = 1$ or $y = -(-3) + 1 = 4$
$\{(0, 1), (-3, 4)\}$

27. $x^2 + y^2 = 2$
$x + y = 0 \rightarrow y = -x$
$x^2 + (-x)^2 = 2$
$2x^2 = 2$
$x^2 = 1$
$x = \pm 1$
$x = -1$ or $x = 1$
$y = -(-1) = 1$ or $y = -1$
$\{(-1, 1), (1, -1)\}$

28. $x^2 + y^2 = 4$
$2x - y = 0 \rightarrow y = 2x$
$x^2 + (2x)^2 = 4$
$5x^2 = 4$
$x^2 = \frac{4}{5}$
$x = \pm\frac{2\sqrt{5}}{5}$
$x = -\frac{2\sqrt{5}}{5}$ or $x = \frac{2\sqrt{5}}{5}$
$y = -\frac{4\sqrt{5}}{5}$ or $y = \frac{4\sqrt{5}}{5}$
$\left\{\left(-\frac{2\sqrt{5}}{5}, -\frac{4\sqrt{5}}{5}\right), \left(\frac{2\sqrt{5}}{5}, \frac{4\sqrt{5}}{5}\right)\right\}$

29. $x^2 + 2y^2 = 4$
$x - y - 1 = 0 \rightarrow y = x - 1$
$x^2 + 2(x-1)^2 = 4$
$x^2 + 2x^2 - 4x + 2 = 4$
$3x^2 - 4x - 2 = 0$
$x = \frac{4 \pm \sqrt{16 + 24}}{6}$
$x = \frac{2 \pm \sqrt{10}}{3}$
$x = \frac{2 - \sqrt{10}}{3}$ or $x = \frac{2 + \sqrt{10}}{3}$
$y = \frac{2 - \sqrt{10}}{3} - 1 = \frac{-1 - \sqrt{10}}{3}$ or
$y = \frac{2 + \sqrt{10}}{3} - 1 = \frac{-1 + \sqrt{10}}{3}$
$\left\{\left(\frac{2 - \sqrt{10}}{3}, \frac{-1 - \sqrt{10}}{3}\right), \left(\frac{2 + \sqrt{10}}{3}, \frac{-1 + \sqrt{10}}{3}\right)\right\}$

30. $2x^2 + y^2 = 24 \rightarrow 2x^2 + y^2 = 24$

$x^2 + y^2 = 15 \rightarrow -x^2 - y^2 = -15$

$x^2 = 9$

$x = \pm 3$

$(\pm 3)^2 + y^2 = 15$

$9 + y^2 = 15$

$y^2 = 6$

$y = \pm\sqrt{6}$

If $x = -3,\ y = \pm\sqrt{6}.$

If $x = 3,\ y = \pm\sqrt{6}.$

$\left\{\left(-3, -\sqrt{6}\right), \left(-3, \sqrt{6}\right), \left(3, -\sqrt{6}\right), \left(3, \sqrt{6}\right)\right\}$

31. $xy - 4 = 0$

$y - x = 0 \rightarrow y = x$

$x(x) - 4 = 0$

$x^2 = 4$

$x = \pm 2$

$x = -2$ or $x = 2$

$y = -2$ or $y = 2$

$\{(-2, -2), (2, 2)\}$

32. $x^2 + y^2 = 2$

$y = x^2 \rightarrow x^2 = y$

$y + y^2 = 2$

$y^2 + y - 2 = 0$

$(y + 2)(y - 1) = 0$

$y = -2$ or $y = 1$

$x^2 = -2$ or $x^2 = 1$

$x = \pm\sqrt{2}i$ or $x = \pm 1$

$\left\{(1, 1), (-1, 1), \left(-\sqrt{2}i, -2\right), \left(\sqrt{2}i, -2\right)\right\}$

33. $y^2 = 4x$

$x - 2y + 3 = 0 \rightarrow x = 2y - 3$

$y^2 = 4(2y - 3)$

$y^2 = 8y - 12$

$y^2 - 8y + 12 = 0$

$(y - 6)(y - 2) = 0$

$y = 6$ or $y = 2$

$x = 2(6) - 3 = 9$ or $x = 2(2) - 3 = 1$

$\{(1, 2), (9, 6)\}$

34. $\frac{x^2}{4} + \frac{y^2}{9} = 1$

$2x - y = 0 \rightarrow y = 2x$

$\frac{x^2}{4} + \frac{4x^2}{9} = 1$

$\frac{25x^2}{36} = 1$

$x^2 = \frac{36}{25}$

$x = \pm\frac{6}{5}$

$x = -\frac{6}{5}$ or $x = \frac{6}{5}$

$y = 2\left(-\frac{6}{5}\right) = -\frac{12}{5}$ or $y = 2\left(\frac{6}{5}\right) = \frac{12}{5}$

$\left\{\left(-\frac{6}{5}, -\frac{12}{5}\right), \left(\frac{6}{5}, \frac{12}{5}\right)\right\}$

35. $4x^2 + y^2 = 1 \rightarrow -16x^2 - 4y^2 = -4$

$x^2 + 4y^2 = 1 \rightarrow x^2 + 4y^2 = 1$

$-15x^2 = -3$

$x^2 = \frac{1}{5}$

$x = \pm\frac{\sqrt{5}}{5}$

$4\left(\frac{1}{5}\right) + y^2 = 1$

$y^2 = \frac{1}{5}$

$y = \pm\frac{\sqrt{5}}{5}$

If $x = \pm\frac{\sqrt{5}}{5}$, $y = \pm\frac{\sqrt{5}}{5}$.

$\left\{\left(-\frac{\sqrt{5}}{5}, -\frac{\sqrt{5}}{5}\right), \left(-\frac{\sqrt{5}}{5}, \frac{\sqrt{5}}{5}\right), \left(\frac{\sqrt{5}}{5}, -\frac{\sqrt{5}}{5}\right), \left(\frac{\sqrt{5}}{5}, \frac{\sqrt{5}}{5}\right)\right\}$

36. $x^2 + y^2 = 9 \rightarrow -x^2 - y^2 = -9$

$(x-2)^2 + y^2 = 21 \rightarrow (x-2)^2 + y^2 = 21$

$(x-2)^2 - x^2 = 12$

$x^2 - 4x + 4 - x^2 = 12$

$-4x = 8$

$x = -2$

$(-2)^2 + y^2 = 9$

$y^2 = 5$

$y = \pm\sqrt{5}$

$\left\{\left(-2, -\sqrt{5}\right), \left(-2, \sqrt{5}\right)\right\}$

37. $x^2 + 2y^2 = 12$

$xy = 4 \rightarrow x = \frac{4}{y}$

$\left(\frac{4}{y}\right)^2 + 2y^2 = 12$

$\frac{16}{y^2} = 12 - 2y^2$

$16 = 12y^2 - 2y^4$

$y^4 - 6y^2 + 8 = 0$

$(y^2 - 4)(y^2 - 2) = 0$

$y^2 = 4$ or $y^2 = 2$

$y = \pm 2$ or $y = \pm\sqrt{2}$

If $y = -2$, $x = \frac{4}{-2} = -2$

If $y = 2$, $x = \frac{4}{2} = 2$

If $y = -\sqrt{2}$, $x = \frac{4}{-\sqrt{2}} \cdot \frac{\sqrt{2}}{\sqrt{2}} = -2\sqrt{2}$

If $y = \sqrt{2}$, $x = \frac{4}{\sqrt{2}} \cdot \frac{\sqrt{2}}{\sqrt{2}} = 2\sqrt{2}$

$\{(-2, -2), (2, 2),$

$(-2\sqrt{2}, -\sqrt{2}), (2\sqrt{2}, \sqrt{2})\}$

38. $y = x^2 - 2$
$x^2 + y^2 = 4$
$x^2 + (x^2 - 2)^2 = 4$
$x^2 + x^4 - 4x^2 + 4 = 4$
$x^4 - 3x^2 = 0$
$x^2(x^2 - 3) = 0$
$x^2 = 0$ or $x^2 = 3$
$x = 0$ or $x = \sqrt{3}$ or $x = -\sqrt{3}$
$y = 0^2 - 2 = -2$ or $y = \left(\sqrt{3}\right)^2 - 2 = 1$
or $y = \left(-\sqrt{3}\right)^2 - 2 = 1$
$\left\{(0, -2), \left(\sqrt{3}, 1\right), \left(-\sqrt{3}, 1\right)\right\}$

39. Let x and y equal the dimensions of a rectangle.
$2x + 2y = 26 \rightarrow y = 13 - x$
$xy = 40 \rightarrow x(13 - x) = 40$
$13x - x^2 = 40$
$0 = x^2 - 13x + 40$
$0 = (x - 8)(x - 5)$
$x = 8$ or $x = 5$
$y = 13 - 8 = 5$ or $y = 13 - 5 = 8$
The dimensions are 5 meters × 8 meters.

40. $x^2 + y^2 = 2900$
$3x + (x - y) + 3y = 240$
$4x + 2y = 240$
$2x + y = 120$
$y = 120 - 2x$
$x^2 + (120 - 2x)^2 = 2900$
$x^2 + 14,400 - 480x + 4x^2 = 2900$
$5x^2 - 480x + 11,500 = 0$
$x^2 - 96x + 2300 = 0$
$x = \dfrac{96 \pm \sqrt{(-96)^2 - 4(1)(2300)}}{2(1)}$
$x = \dfrac{96 \pm \sqrt{16}}{2} = \dfrac{96 \pm 4}{2}$
$x = 46$ or 50
If $x = 46$, $y = 120 - 2(46) = 28$
If $x = 50$, $y = 120 - 2(50) = 20$
{(46 ft, 28 ft), (50 ft, 20 ft)}

41. $\frac{1}{2}L_1L_2 = 54$
$L_1^2 + L_2^2 = (15)^2$
$L_1 = \dfrac{108}{L_2}$
$\left(\dfrac{108}{L_2}\right)^2 + L_2^2 = 225$
$\dfrac{11,664}{L_2^2} = 225 - L_2^2$
$11,664 = 225L_2^2 - L_2^4$
$L_2^4 - 225L_2^2 + 11,664 = 0$
$(L_2^2 - 81)(L_2^2 - 144) = 0$
$(L_2 - 9)(L_2 + 9)(L_2 - 12)(L_2 + 12) = 0$
$L_2 = 9, -9, 12, -12$
Disregard negative lengths.
The lengths are 9 yards and 12 yards.

42.
$$\begin{array}{r} -x^2 + 4y^2 = 1 \\ x^2 - 3y^2 = 1 \\ \hline y^2 = 2 \\ y = \pm\sqrt{2} \end{array}$$
Since the ship is in the first quadrant, disregard $y = -\sqrt{2}$. If $y = \sqrt{2}$, then
$x^2 - 3(\sqrt{2})^2 = 1$
$x^2 - 6 = 1$
$x^2 = 7$
$x = \pm\sqrt{7}$
Disregard $x = -\sqrt{7}$ since it is not in the first quadrant.
The ship is located at $\left(\sqrt{7}, \sqrt{2}\right)$.

43. Let r = average speed and t = time.
$rt = 2000$
$r = t - 10$
(Substitute for r) $\rightarrow (t-10)t = 2000$
$t^2 - 10t - 2000 = 0$
$(t-50)(t+40) = 0$
$t = 50$ or $t = -40$ (reject)
$r = 50 - 10 = 40$
rate: 40 kph; time: 50 hr

44. $xy = 6$
$2x + y = 8 \rightarrow y = 8 - 2x$
$x(8-2x) = 6$
$8x - 2x^2 = 6$
$-2x^2 + 8x - 6 = 0$
$x^2 - 4x + 3 = 0$
$(x-3)(x-1) = 0$
$x = 3$ or $x = 1$
$y = 8 - 2(3) = 2$ or $y = 8 - 2(1) = 6$
(1, 6) and (3, 2)

Chapter 9 Test

1. $(x-5)^2 + (y+3)^2 = 49$
$(x-5)^2 + (y-(-3))^2 = 7^2$
Center: (5, –3), radius: 7

2. $x^2 + y^2 + 4x - 6y - 3 = 0$
$(x^2 + 4x + 4) + (y^2 - 6y + 9) = 3 + 4 + 9$
$(x-2)^2 + (y-3)^2 = 4^2$
Center: (2, 3), radius: 4

3. $y = x^2 - 2x - 3$
parabola

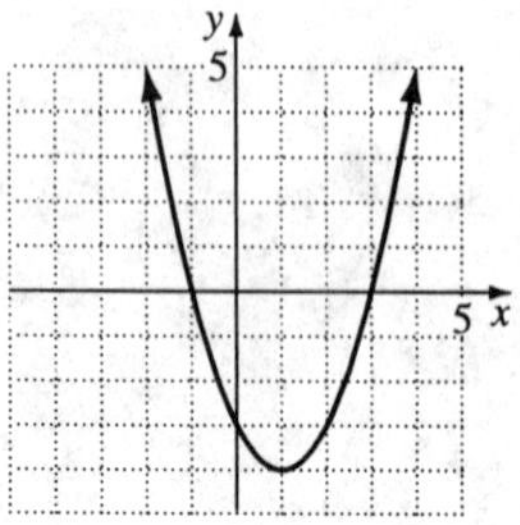

4. $\frac{x^2}{4} - \frac{y^2}{9} = 1$
hyperbola

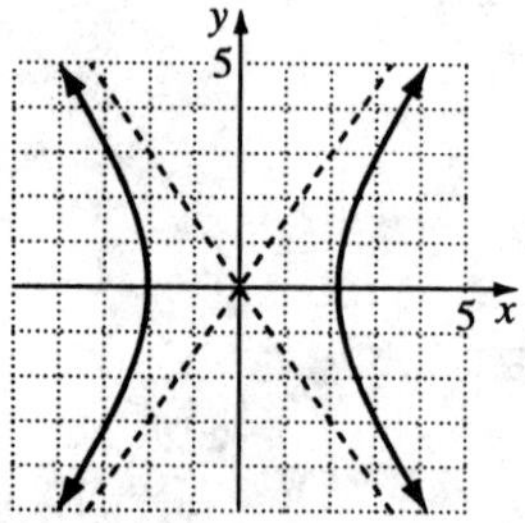

5. $4x^2 + 9y^2 = 36$
ellipse

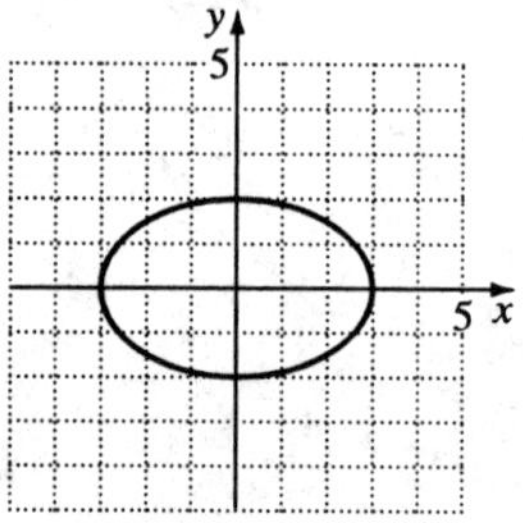

6. $xy = 4$
hyperbola

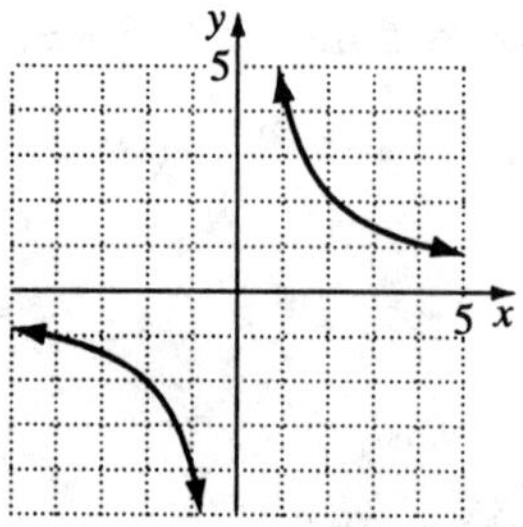

7. $x = (y-1)^2 + 4$
parabola

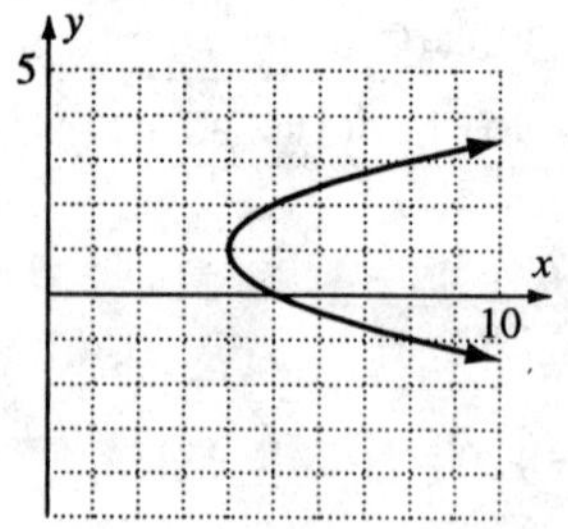

8. $16x^2 + y^2 = 16$

ellipse

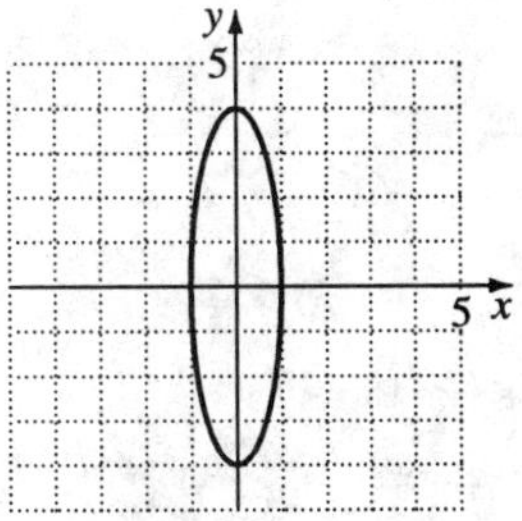

9. $\frac{x^2}{4} + \frac{y^2}{4} = 1$

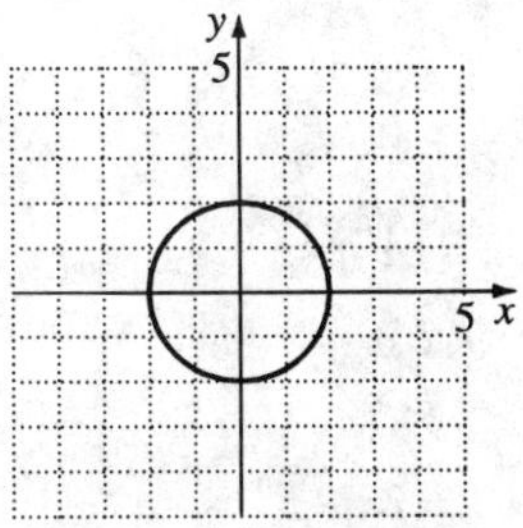

10. $25y^2 - 9x^2 = 225$

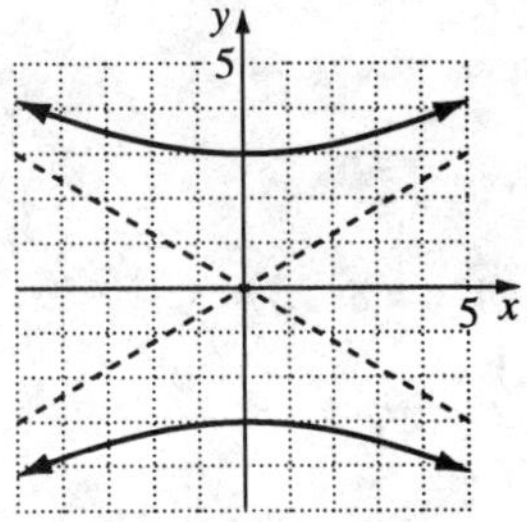

11. The graphs cross at (0, –1) and (3, 2).

12. $x^2 + y^2 = 25$

$x + y = 1$

$y = 1 - x$

$x^2 + (1-x)^2 = 25$

$x^2 + 1 - 2x + x^2 = 25$

$2x^2 - 2x - 24 = 0$

$x^2 - x - 12 = 0$

$(x-4)(x+3) = 0$

$x = 4$ or $x = -3$

$y = 1 - 4 = -3$ or $y = 1 - (-3) = 4$

$\{(4, -3), (-3, 4)\}$

13. $2x^2 - 5y^2 = -2$

$3x^2 + 2y^2 = 35$

$$\begin{array}{r} 4x^2 - 10y^2 = -4 \\ 15x^2 + 10y^2 = 175 \\ \hline 19x^2 = 171 \end{array}$$

$x^2 = \frac{171}{19} = 9$

$x = \pm 3$

If $x = -3$: $3(-3)^2 + 2y^2 = 35$

$y = \pm 2$

If $x = 3$: $3(3)^2 + 2y^2 = 35$

$y = \pm 2$

$\{(-3, -2), (-3, 2), (3, -2), (3, 2)\}$

14. Let l = length and w = width.

$l^2 + w^2 = (15)^2$

$2l + 2w = 42$

$l = 21 - w$

$(21 - w)^2 + w^2 = 225$

$441 - 42w + w^2 + w^2 - 225 = 0$

$2w^2 - 42w + 216 = 0$

$w^2 - 21w + 108 = 0$

$(w - 12)(w - 9) = 0$

$w = 12$ or $w = 9$

The width is 9 ft and the length is 12 ft.

15. $2x + y = 39$

$xy = 180$

$x = \frac{180}{y}$

$2\left(\frac{180}{y}\right) + y = 39$

$\frac{360}{y} = 39 - y$

$360 = 39y - y^2$

$y^2 - 39y + 360 = 0$
$(y - 15)(y - 24) = 0$
$y = 15$ or $y = 24$
$x = \frac{180}{15} = 12$ or $x = \frac{180}{24} = 7\frac{1}{2}$
There are two possible dimensions:
$y = 15$ ft, $x = 12$ ft or $y = 24$ ft, $x = 7\frac{1}{2}$ ft.

Chapters 1–9 Cumulative Review Problems

1. $2(2x + 3) - 10 \geq 6(x - 2)$
$4x + 6 - 10 \geq 6x - 12$
$4x - 4 \geq 6x - 12$
$4x - 6x \geq -12 + 4$
$-2x \geq -8$
$x \leq 4$
$\{x|x \leq 4\}, (-\infty, 4]$

2. $5x^{2/3} + 2x^{1/3} - 7 = 0$
$(5x^{1/3} + 7)(x^{1/3} - 1) = 0$
$x^{1/3} = -\frac{7}{5}$ or $x^{1/3} = 1$
$x = \left(-\frac{7}{5}\right)^3 = \frac{-343}{125}$ or $x = 1^3 = 1$
$\left\{-\frac{343}{125}, 1\right\}$

3. $(5x - 4)^2 + 6 = 8$
$(5x - 4)^2 = 2$
$5x - 4 = \pm\sqrt{2}$
$5x = 4 \pm \sqrt{2}$
$x = \frac{4 \pm \sqrt{2}}{5}$
$\left\{\frac{4 - \sqrt{2}}{5}, \frac{4 + \sqrt{2}}{5}\right\}$

4. $x^2 - 6x + 13 = 0$
$x = \frac{6 \pm \sqrt{36 - 52}}{2}$
$x = \frac{6 \pm \sqrt{-16}}{2}$
$x = \frac{6 \pm 4i}{2} = 3 \pm 2i$
$\{3 - 2i, 3 + 2i\}$

5. $2x^2 - 5x - 3 > 0$
$(2x + 1)(x - 3) > 0$

T	F	T
	$-\frac{1}{2}$	3

Test -1: $(-2 + 1)(-1 - 3) > 0$
$(-1)(-4) > 0$
$4 > 0$ True
Test 0: $(0 + 1)(0 - 3) = -3 > 0$
False
Test 4: $(8 + 1)(4 - 3) > 0$
$9(1) > 0$
$9 > 0$ True
$x < -\frac{1}{2}$ or $x > 3$
$\left\{x \middle| x < -\frac{1}{2} \text{ or } x > 3\right\}, \left(-\infty, -\frac{1}{2}\right) \cup (3, \infty)$

6. $x^2 + y^2 = 25$
$x - 2y = -5 \rightarrow x = 2y - 5$
$(2y - 5)^2 + y^2 = 25$
$4y^2 - 20y + 25 + y^2 = 25$
$5y^2 - 20y = 0$
$5y(y - 4) = 0$
$y = 0$ or $y = 4$
$x = 0 - 5 = -5$ or $x = 8 - 5 = 3$
$\{(-5, 0), (3, 4)\}$

7. $|2x - 1| < 3$
$2x - 1 < 3$ and $2x - 1 > -3$
$2x < 4$ and $2x > -2$
$x < 2$ and $x > -1$
$\{x|-1 < x < 2\}, (-1, 2)$

8. $x - 2y + 3z = 7$ (1)
$2x + y + z = 4$ (2)
$-3x + 2y + 3z = 19$ (3)
Equations 1 and 2:
$x - 2y + 3z = 7$
$\underline{4x + 2y + 2z = 8}$
$5x \quad + 5z = 15$
$x + z = 3$ (4)
Equations 1 and 3:
$x - 2y + 3z = 7$
$\underline{-3x + 2y + 3z = 19}$
$-2x \quad + 6z = 26$
$-x + 3z = 13$ (5)
Equations 4 and 5:
$x + z = 3$
$\underline{-x + 3z = 13}$
$4z = 16$
$z = 4$
Equation 4:
$x + 4 = 3$
$x = -1$
Equation 2:
$2(-1) + y + 4 = 4$
$y = 2$
$\{(-1, 2, 4)\}$

9. $\log(x + 3) + \log x = 1$
$\log x(x + 3) = 1$
$10^1 = x(x+3)$
$0 = x^2 + 3x - 10$
$0 = (x+5)(x-2)$
$x = -5$ (extraneous), $x = 2$
$\{2\}$

10. $\sqrt{2x+3} - \sqrt{x-2} = 2$
$\sqrt{2x+3} = 2 + \sqrt{x-2}$
$2x+3 = 4 + 4\sqrt{x-2} + x - 2$
$x + 1 = 4\sqrt{x-2}$
$x^2 + 2x + 1 = 16(x-2)$
$x^2 - 14x + 33 = 0$
$(x - 11)(x - 3) = 0$
$x = 11$ or $x = 3$
$\{3, 11\}$

11. $2x - 3y \geq 6$

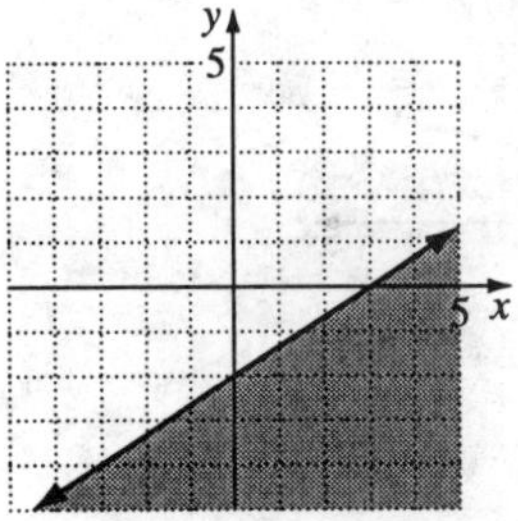

12. $y = \left(\frac{1}{2}\right)^x + 1$

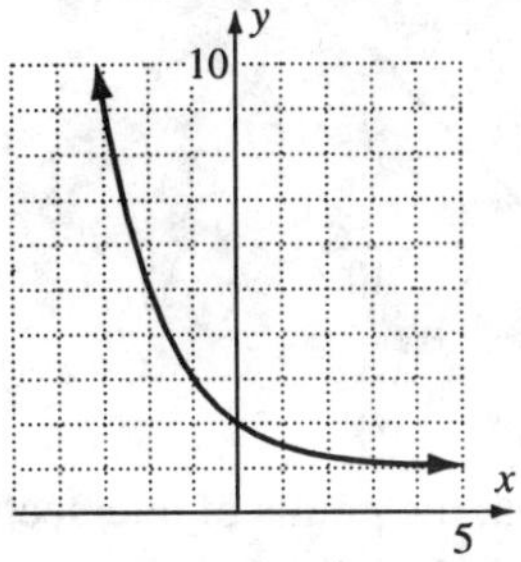

13. $9x^2 + 4y^2 = 36$

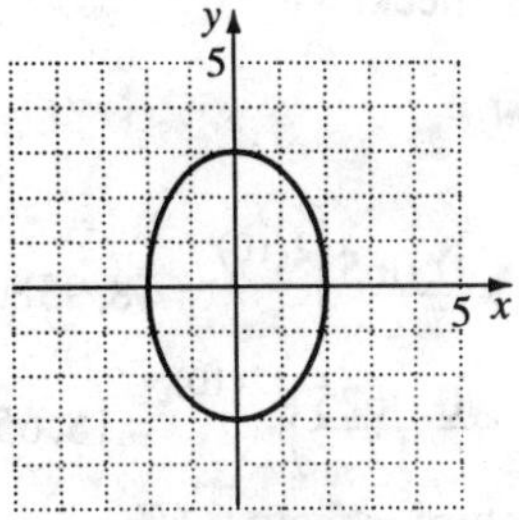

14. $y = \log_3 x$

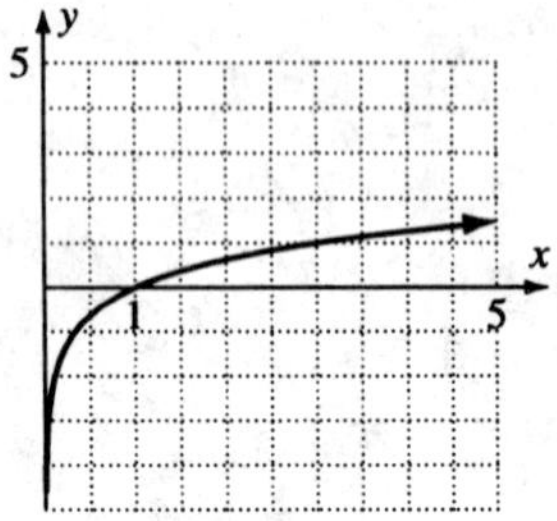

15. $f(x) = \sqrt{x+1}$

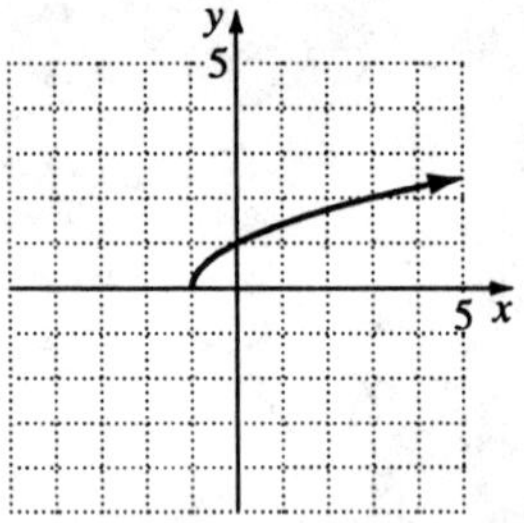

16. $x = y + 1$

$y = -(x+1)^2 + 4$

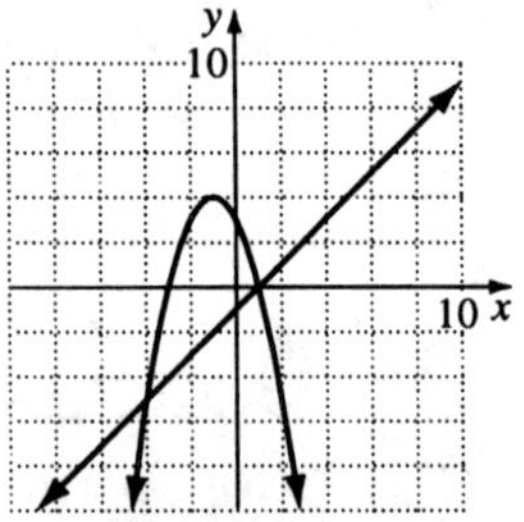

From the graph, the points of intersection are $x = -4$, $y = -5$ and $x = 1$, $y = 0$.

$\{(-4, -5), (1, 0)\}$

Both solutions check.

17. $A = P\left(1+\frac{r}{n}\right)^{nt}$

$A = 10{,}000\left(1+\frac{0.065}{2}\right)^{2(10)} = 18{,}958.38$

$A = 10{,}000\left(1+\frac{0.0625}{12}\right)^{12(10)} = 18{,}652.18$

The greatest return is from 6.5% compounded semiannually.

18. $f(x) = x^2 + 1,\ g(x) = 2x - 3$

$(f \circ g)(x) = (2x-3)^2 + 1 = 4x^2 - 12x + 10$

$(g \circ f)(x) = 2(x^2+1) - 3 = 2x^2 - 1$

19. $f(x) = 3x + 5$

$y = 3x + 5$

$x = 3y + 5$

$x - 5 = 3y$

$\frac{x-5}{3} = y$

$f^{-1}(x) = \frac{x-5}{3}$

20. $P = 14.7e^{-0.21x}$

a. $x = 0$

$P = 14.7e^{-0.21(0)} = 14.7$

The atmospheric pressure is 14.7 pounds per square inch.

b. $\frac{14.7}{2} = 14.7e^{-0.21x}$

$\frac{1}{2} = e^{-0.21x}$

$\ln\frac{1}{2} = \ln e^{-0.21x}$

$\ln\frac{1}{2} = -0.21x$

$\frac{\ln\frac{1}{2}}{-0.21} = x$

$x \approx 3.3$

The distance is 3.3 miles above sea level.

21. $\frac{3+2i}{2-i} \cdot \frac{2+i}{2+i} = \frac{6+7i+2i^2}{4-i^2}$

$= \frac{6+7i-2}{4+1} = \frac{4+7i}{5}$

22. $(x^3+2x^2-3x+4)\div(x-2)=x^2+4x+5+\frac{14}{x-2}$

$$\begin{array}{r|l} & x^2+4x+5+\frac{14}{x-2} \\ x-2 & x^3+2x^2-3x+4 \\ & \underline{x^3-2x^2} \\ & \quad 4x^2-3x \\ & \quad \underline{4x^2-8x} \\ & \qquad 5x+4 \\ & \qquad \underline{5x-10} \\ & \qquad\quad 14 \end{array}$$

23. $2\sqrt{12}-\sqrt{\frac{1}{3}}=2\sqrt{4\cdot3}-\frac{\sqrt{1}}{\sqrt{3}}$

$=2\cdot2\sqrt{3}-\frac{1}{\sqrt{3}}=4\sqrt{3}\cdot\frac{\sqrt{3}}{\sqrt{3}}-\frac{1}{\sqrt{3}}$

$=\frac{4\cdot3}{\sqrt{3}}-\frac{1}{\sqrt{3}}=\frac{12-1}{\sqrt{3}}=\frac{11}{\sqrt{3}}\cdot\frac{\sqrt{3}}{\sqrt{3}}=\frac{11\sqrt{3}}{3}$

24. $s(t)=-16t^2+16t+32$

a. maximum height occurs when
$t=-\frac{b}{2a}=\frac{-16}{2(-16)}=\frac{1}{2}$.
maximum height:
$s\left(\frac{1}{2}\right)=-16\left(\frac{1}{4}\right)+16\left(\frac{1}{2}\right)+32$
$=-4+8+32=36$
maximum height of 36 feet at 0.5 sec

b. The diver reaches the water when
$s(t)=0$.
$0=-16t^2+16t+32$
$16t^2-16t-32=0$
$t^2-t-2=0$
$(t-2)(t+1)=0$
$t=2$ or $t=-1$ (reject)
2 seconds

c. $s(t)=-16^2+16t+32$

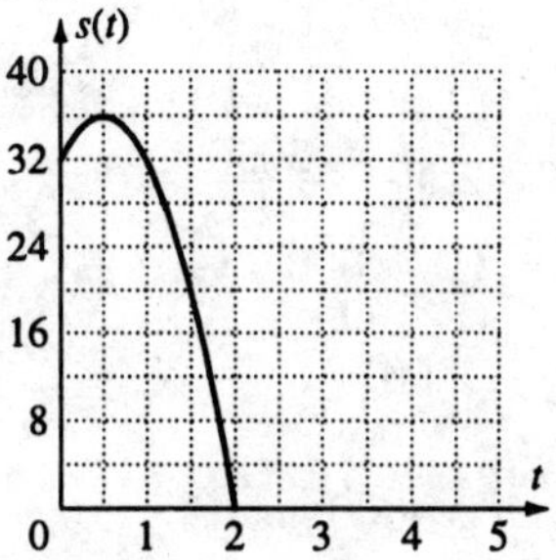

25. Let x = width of the border, $9 + 2x$ = width of pool plus border, and $12 + 2x$ = length of pool plus border.
area of pool plus border = area of pool + area of border
$(9+2x)(12+2x) = 9(12) + 162$
$108 + 42x + 4x^2 = 108 + 162$
$4x^2 + 42x - 162 = 0$
$2x^2 + 21x - 81 = 0$
$(2x+27)(x-3) = 0$
$x = -\frac{27}{2}$ (reject) or $x = 3$
width of border: 3 meters

26. $T = S + 2.5\left(\frac{x-3000}{100}\right)$
$45 = 20 + 2.5\left(\frac{x-3000}{100}\right)$
$\frac{45-20}{2.5} = \frac{x-3000}{100}$
$10(100) + 3000 = x$
$4000 = x$
The depth is 4000 meters.

27. $100 - 4 = 96$
$10^2 - 2^2 = 96$

28. $a^2 + 2ab + b^2 = (a+b)^2$

29. $N = 330.8t + 218.8$
$N = 330.8(4) + 218.8 = 1542$
$N = 330.8(7) + 218.8 = 2534.4$
$N = 62t^2 + 144.8t + 280.8$
$N = 62(4)^2 + 144.8(4) + 280.8 = 1852$
$N = 62(7)^2 + 144.8(7) + 280.8 = 4332.4$
The predictions for 1990 and 1993 are as follows:
first model: 1542 thousand, 2534.4 thousand
second model: 1852 thousand, 4332.4 thousand
The second model comes closer to describing reality.

30. $cx + dy = c^2$
$dx + cy = d^2$
$D = \begin{vmatrix} c & d \\ d & c \end{vmatrix} = c^2 - d^2$
$D_x = \begin{vmatrix} c^2 & d \\ d^2 & c \end{vmatrix} = c^3 - d^3$
$D_y = \begin{vmatrix} c & c^2 \\ d & d^2 \end{vmatrix} = cd^2 - c^2d$
$x = \frac{D_x}{D} = \frac{c^3 - d^3}{c^2 - d^2} = \frac{(c-d)(c^2+cd+d^2)}{(c-d)(c+d)}$
$x = \frac{c^2+cd+d^2}{c+d}$
$y = \frac{D_y}{D} = \frac{cd^2 - c^2d}{c^2 - d^2} = \frac{cd(d-c)}{(c-d)(c+d)}$
$= \frac{-cd(c-d)}{(c-d)(c+d)}$
$y = \frac{-cd}{c+d}$

Chapter 10

Problem Set 10.1

1. $f(x) = -x^4 + x^2$; degree 4, a_n negative. Graph (c)

3. $f(x) = (x-3)^2 = x^2 - 6x + 9$; degree 2, a_n positive; $f(0) = 9$. Graph (b)

5. $f(x) = x - 3$; degree 1, a_n positive; linear. Graph (a)

7. **a.** $a_n x^n = 1x^3$ so the graph of $f(x)$ falls to the left and rises to the right.

   b. $f(x) = x^3 + 2x^2 - x - 2$
   $= (x-1)(x+2)(x+1) = 0$
   The graph crosses the x-axis at –2, –1, and 1.

   c. $f(0) = (0)^3 + 2(0)^2 - (0) - 2 = -2$
   $f(x)$ crosses the y-axis at –2.

   d. y-axis symmetry:
   $f(-x) = -x^3 + 2x^2 + x - 2 \neq f(x)$;
   $f(x)$ has no y-axis symmetry.
   origin symmetry:
   $-f(x) = -x^3 - 2x^2 + x + 2 \neq f(-x)$;
   $f(x)$ does not have origin symmetry.

   e. $n - 1 = 2$ so there are no more than 2 turning points.

x	$f(x)$
–3	–8
2	12

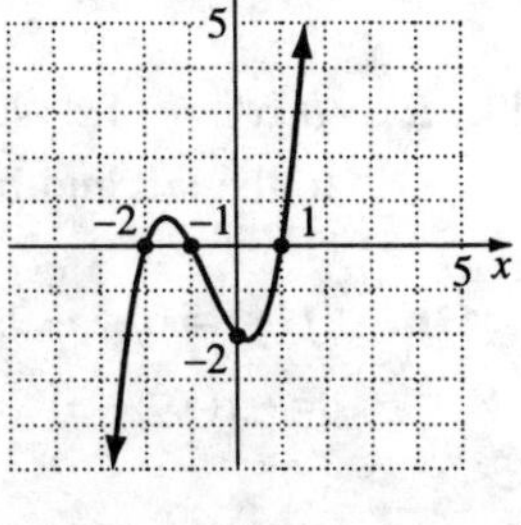

9. **a.** $a_n x^n = 1x^4$ so the graph of $f(x)$ rises to the left and to the right.

   b. $f(x) = x^4 - 9x^2 = x^2(x-3)(x+3) = 0$
   The graph crosses the x-axis at –3 and 3 and touches the x-axis at 0.

   c. $f(0) = (0)^4 - 9(0)^2 = 0$
   $f(x)$ crosses the y-axis at 0.

   d. y-axis symmetry:
   $f(-x) = x^4 - 9x^2 = f(x)$;
   $f(x)$ has y-axis symmetry.
   origin symmetry:
   $-f(x) = -x^4 + 9x^2 \neq f(-x)$
   $f(x)$ does not have origin symmetry.

   e. $n - 1 = 3$ so there are no more than 3 turning points.

x	$f(x)$
–2	–20
–1	–8

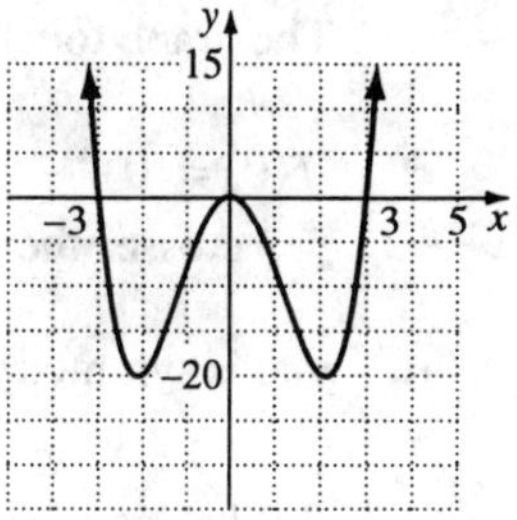

11. **a.** $a_n x^n = -1x^4$ so the graph of $f(x)$ falls to the left and to the right.

   b. $f(x) = -x^4 + 16x^2$
   $= -x^2(x-4)(x+4) = 0$
   The graph crosses the x-axis at –4 and 4 and touches the x-axis at 0.

   c. $f(0) = -(0)^4 + 16(0)^2 = 0$
   $f(x)$ crosses the y-axis at 0.

d. y-axis symmetry:

$f(-x) = -x^4 + 16x^2 = f(x)$;

$f(x)$ has y-axis symmetry.

Origin symmetry:

$-f(x) = x^4 - 16x^2 \neq f(-x)$;

$f(x)$ has no origin symmetry.

e. $n - 1 = 3$ so there are no more than 3 turning points.

x	$f(x)$
−3	63
−2	48
−1	15

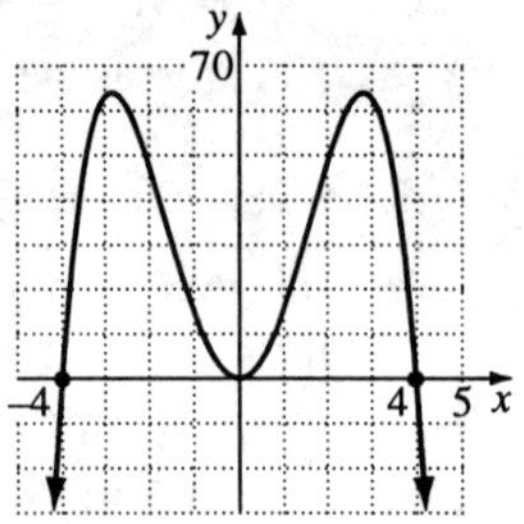

13. a. $a_n x^n = 1x^4$ so the graph of $f(x)$ rises to the left and to the right.

b. $f(x) = x^4 - 2x^3 + x^2 = x^2(x-1)^2 = 0$

The graph touches the x-axis at 0 and 1.

c. $f(0) = (0)^4 - 2(0)^3 + (0)^2 = 0$

$f(x)$ crosses the y-axis at 0.

d. y-axis symmetry:

$f(-x) = x^4 + 2x^3 + x^2 \neq f(x)$

$f(x)$ has no y-axis symmetry.

origin symmetry:

$-f(x) = -x^4 + 2x^3 - x^2 \neq f(-x)$

$f(x)$ has no origin symmetry.

e. $n - 1 = 3$ so there are no more than 3 turning points.

x	$f(x)$
−0.5	0.5625
0.5	0.0625

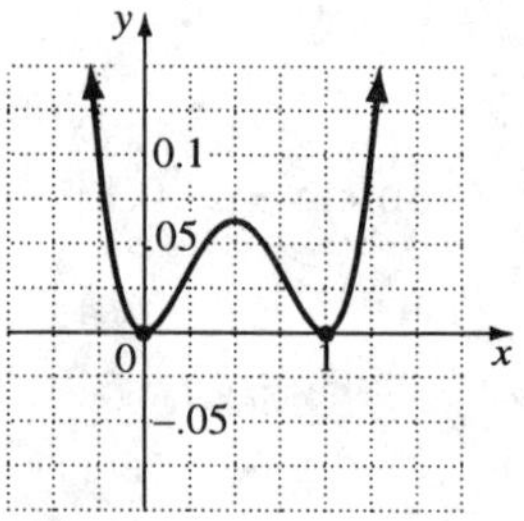

15. a. $a_n x^n = -2x^4$ so the graph of $f(x)$ falls to the left and to the right.

b. $f(x) = -2x^4 + 4x^3 = -2x^3(x-2) = 0$

The graph crosses the x-axis at 0 and 2.

c. $f(0) = -2(0)^4 + 4(0)^3 = 0$

$f(x)$ crosses the y-axis at 0.

d. y-axis symmetry;

$f(-x) = -2x^4 - 4x^3 \neq f(x)$

$f(x)$ does not have y-axis symmetry.

origin symmetry:

$-f(x) = 2x^4 - 4x^3 \neq f(-x)$

$f(x)$ has no origin symmetry.

e. $n - 1 = 3$ so there are no more than 3 turning points.

x	$f(x)$
−1	−6
1	2
1.5	3.375
3	−54

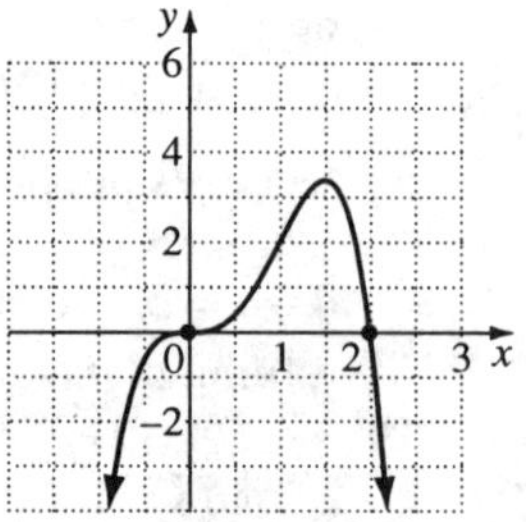

17. a. $a_n x^n = -1x^5$ so the graph of $f(x)$ rises to the left and falls to the right.

b. $f(x) = 6x^3 - 9x - x^5$

$= -x(x^2 - 3)^2 = 0$

$x = 0$ with multiplicity 1 and $x = \pm\sqrt{3}$ with multiplicity 2. The graph crosses the x-axis at 0 and touches the x-axis at $\pm\sqrt{3}$.

c. $f(0) = 6(0)^3 - 9(0) - (0)^5 = 0$
$f(x)$ crosses the y-axis at 0.

d. y-axis symmetry:
$f(-x) = -6x^3 + 9x + x^5 \neq f(x)$;
$f(x)$ does not have y-axis symmetry.
origin symmetry:
$-f(x) = -6x^3 + 9x + x^5 = f(-x)$;
$f(x)$ has origin symmetry.

e. $n - 1 = 4$ so there are no more than 4 turning points.

x	$f(x)$
–2	2
–1	4

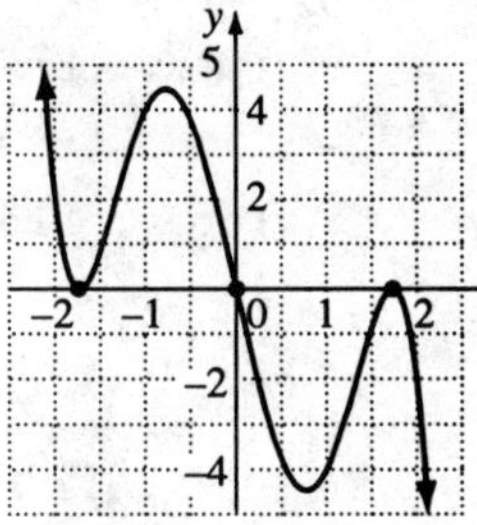

19. a. $a_nx^n = -1x^3$ so $f(x)$ rises to the left and falls to the right.

b. $f(x) = 3x^2 - x^3 = -x^2(x - 3) = 0$
The graph crosses the x-axis at 3 and touches the x-axis at 0.

c. $f(0) = 3(0)^2 - (0)^3 = 0$
$f(x)$ crosses the y-axis at 0.

d. y-axis symmetry:
$f(-x) = 3x^2 + x^3 \neq f(x)$;
$f(x)$ does not have y-axis symmetry.
origin symmetry:
$-f(x) = -3x^2 + x^3 \neq f(-x)$;
$f(x)$ has no origin symmetry.

e. $n - 1 = 2$ so there are no more than 2 turning points.

x	$f(x)$
–1	4
1	2
2	4
4	–16

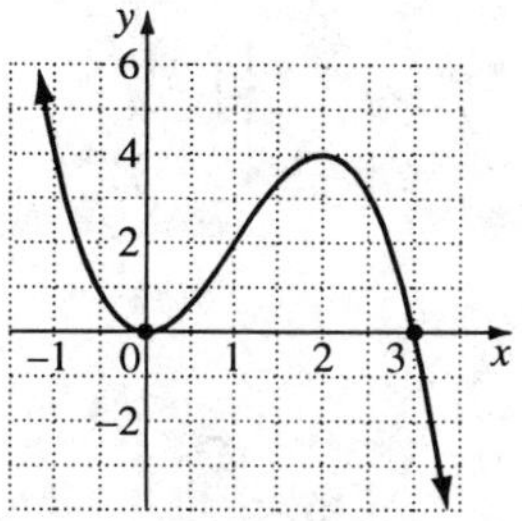

21. a. $f(x) = -3x^4 + 6x^3 + 9x^2 - 24x + 12$
$a_nx^n = -3x^4$ so $f(x)$ falls to the left and to the right.

b. $f(x) = -3(x-1)^2(x^2 - 4)$
$= -3(x-1)^2(x-2)(x+2) = 0$
The graph crosses the x-axis at –2 and 2 and touches the x-axis at 1.

c. $f(0) = -3(0)^4 + 6(0)^3 + 9(0)^2$
$- 24(0) + 12 = 12$
$f(x)$ crosses the y-axis at 12.

d. y-axis symmetry:
$f(-x) = -3x^4 - 6x^3 + 9x^2 + 24x + 12$
$\neq f(x)$
$f(x)$ does not have y-axis symmetry.
origin symmetry:
$-f(x) = 3x^4 - 6x^3 - 9x^2 + 24x - 12$
$\neq f(-x)$
$f(x)$ has no origin symmetry.

e. $n - 1 = 3$ so there are no more than 3 turning points.

x	$f(x)$
−3	−240
−1	36
1.5	1.3125
3	−60

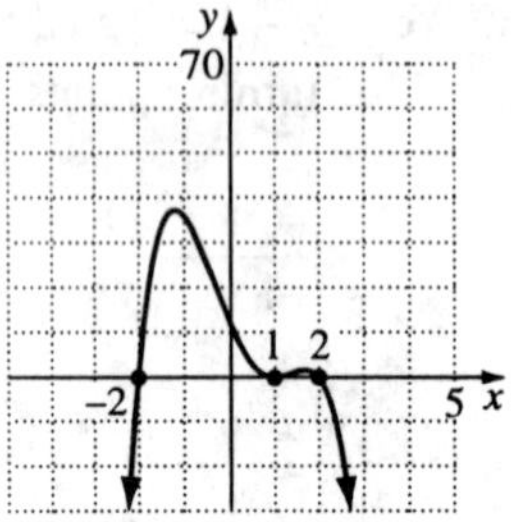

23. a. $a_n t^n = -1t^4$ so the graph of $N(t)$ falls to the right. Eventually $N(t) = 0$; that is, the elk population will die off.

b.

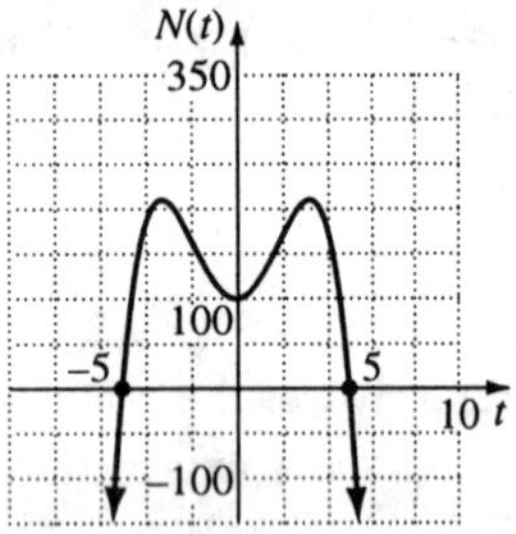

c. $N(t) = -t^4 + 21t^2 + 100 = 0$
$-(t-5)(t+5)(t^2+4) = 0$
$t = 5$ or $t = -5$ or $t = \pm\sqrt{-4}$
Since t must be real and positive, $t = 5$. The population becomes extinct in 5 years.

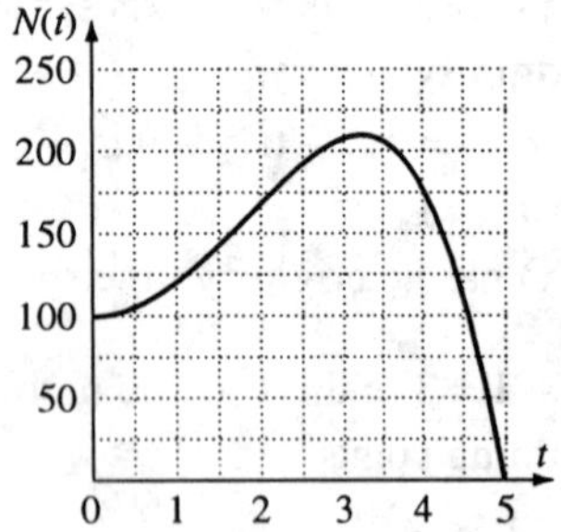

25. a. Volume = length · width · height
length = $18 - 2x$
width = $12 - 2x$
height = x
$V(x) = (18-2x)(12-2x)x$
$= 4x^3 - 60x^2 + 216x$

b. $a_n x^n = 4x^3$ so $V(x)$ falls to the left and rises to the right.
$V(x) = 4x^3 - 60x^2 + 216x = 0$
$4x(x^2 - 15x + 54) = 0$
$4x(x-6)(x-9) = 0$
$x = 0$ or $x = 6$ or $x = 9$
The graph crosses the x-axis at 0, 6, and 9.
$V(0) = 4(0)^3 - 60(0)^2 + 216(0) = 0$.
The graph crosses the y-axis at 0.
$f(-x) = -4x^3 - 60x^2 - 216 \neq f(x)$;
$V(x)$ does not have y-axis symmetry.
$-f(x) = -4x^3 + 60x^2 - 216x \neq f(-x)$;
$V(x)$ does not have origin symmetry.

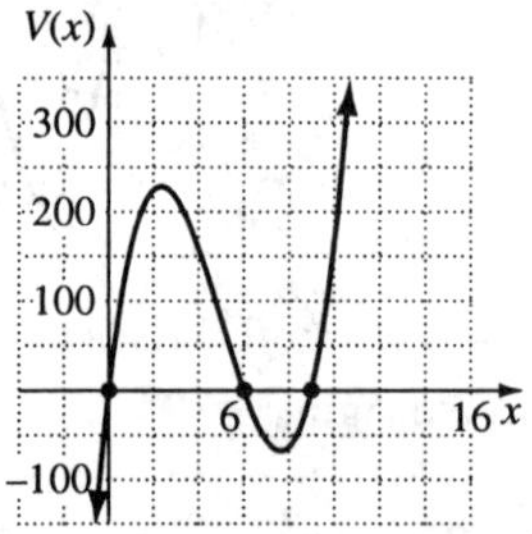

27. a. $f(x)$ crosses or touches the x-axis 3 times between −2 and 3 so there are 3 zeros between −2 and 3; false

b. $f(0) = 2$ so $f(f(0)) = f(2)$. At $x = 2$, $f(2) < 0$; false

c. If f were a function whose degree was even, either it would rise both on the left and on the right, or it would fall both on the left and on the right; true

d. $f(3) > 0$; false

29.

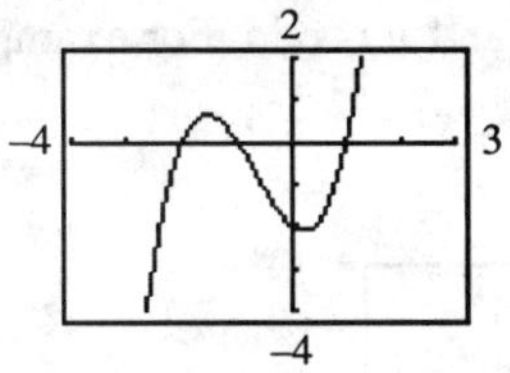

#7

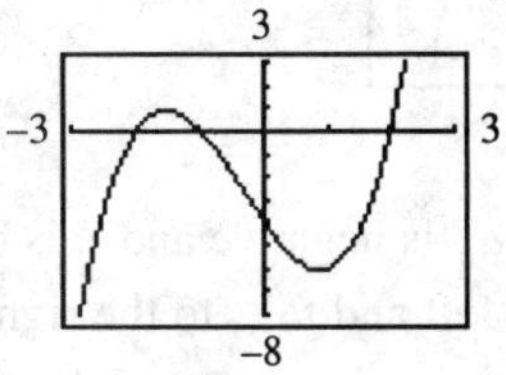

#8

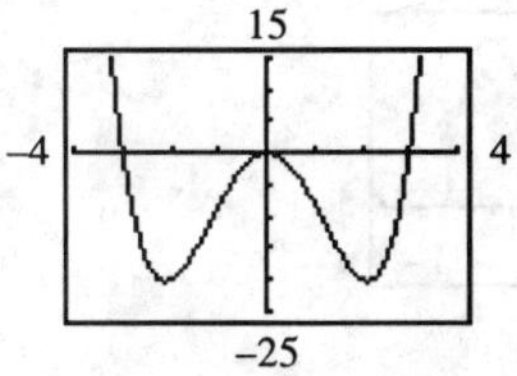

#9

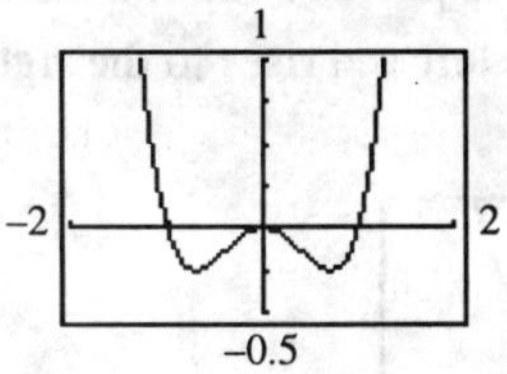

#10

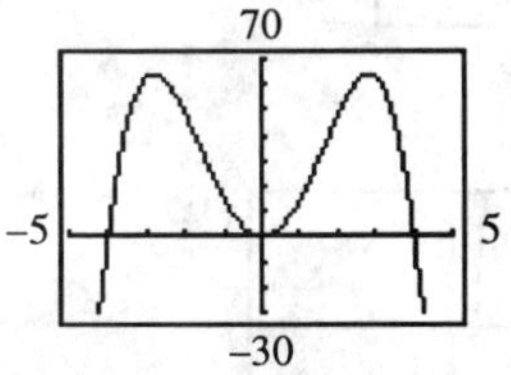

#11

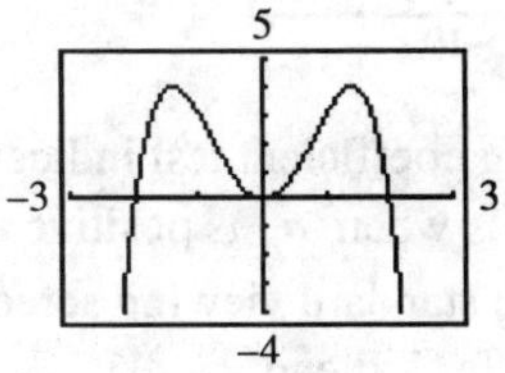

#12

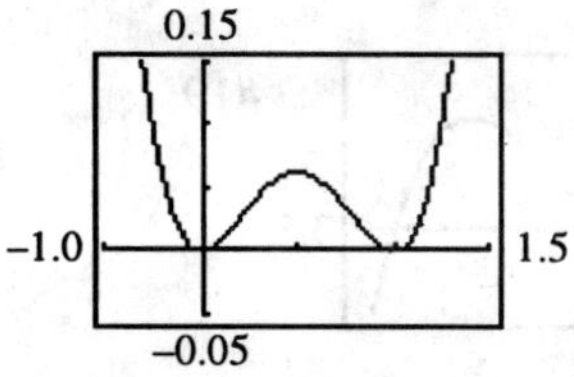

#13

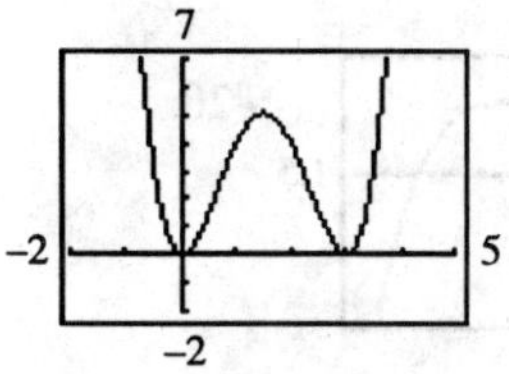

#14

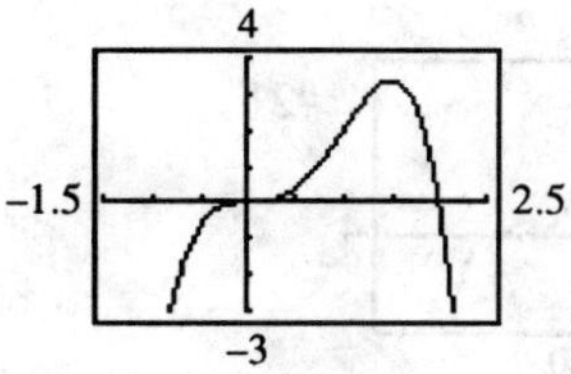

#15

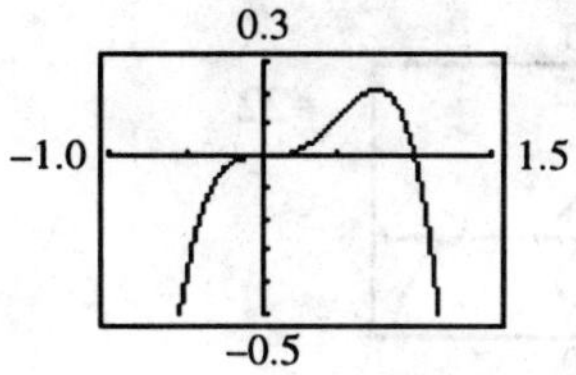

#16

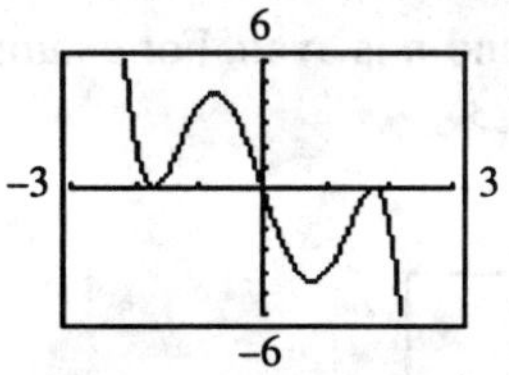

#17

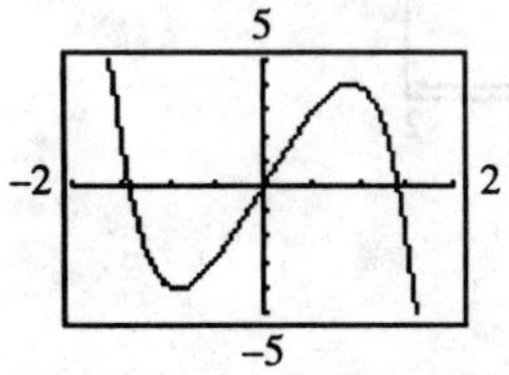

#18

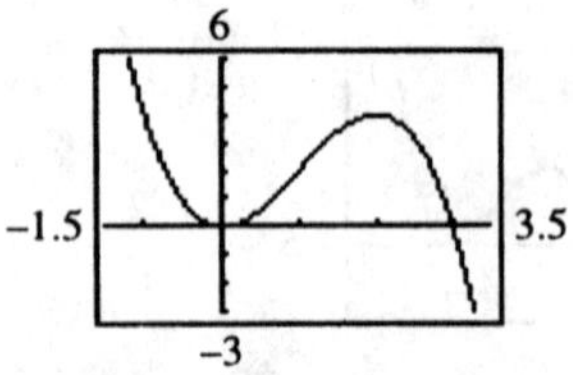

#19

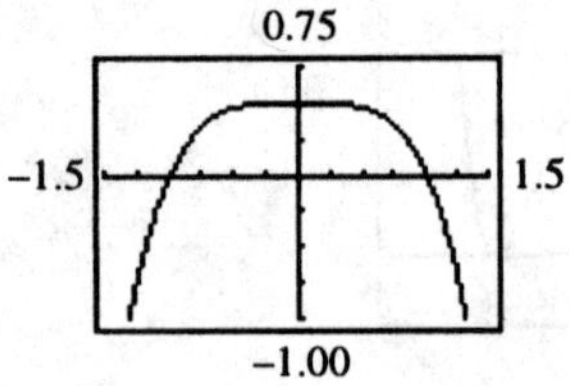

#20

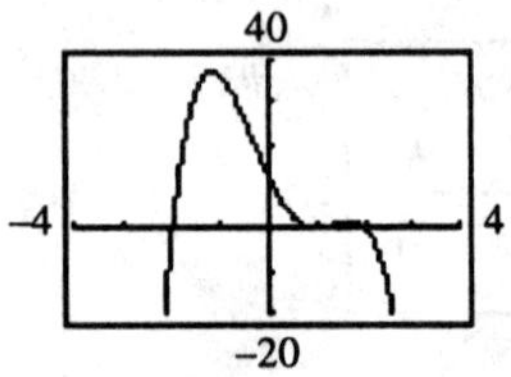

#21

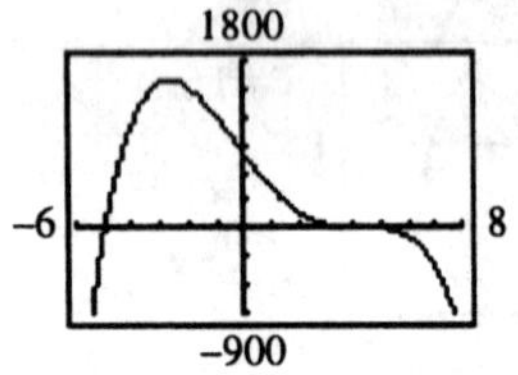

#22

31. a_n is positive and n is even. For example, $f(x) = 4x^4 - x^3 + 2$.

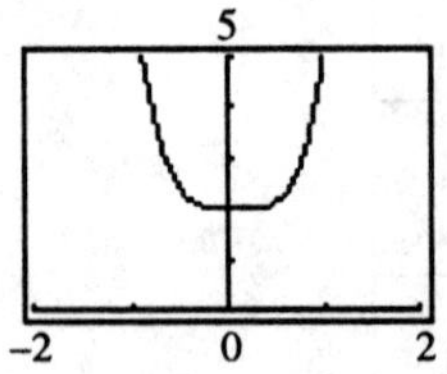

33. a_n is negative and n is odd. For example, $f(x) = -2x^5 + 3x^2 - x + 4$.

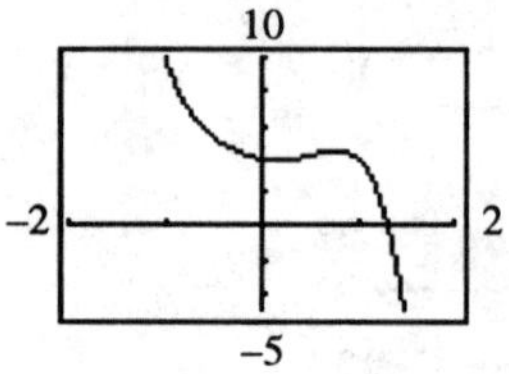

35. $a_n x^n = -2x^3$, a_n is negative and n is odd. $f(x)$ rises to the left and falls to the right.

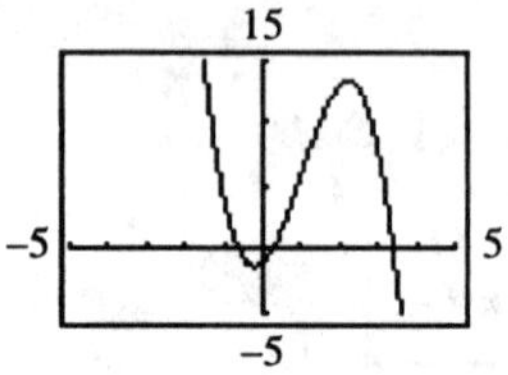

37. $a_n x^n = x^3$, a_n is positive and n is odd. $h(x)$ falls to the left and rises to the right.

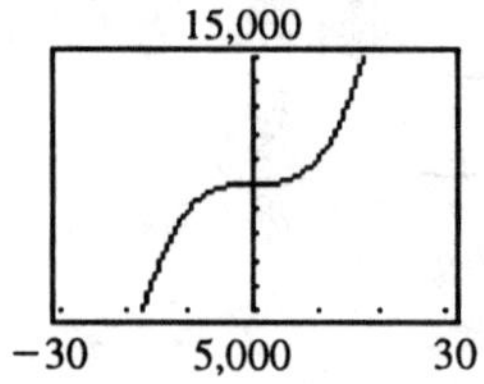

39. a.

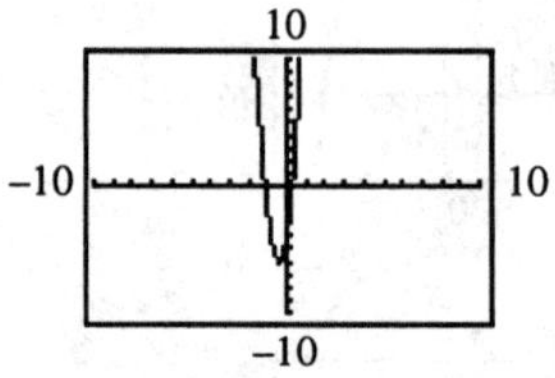

b. The leading coefficient test indicates the left end falls when a_n is positive and n is odd. The standard viewing screen misses the far left end.

c.

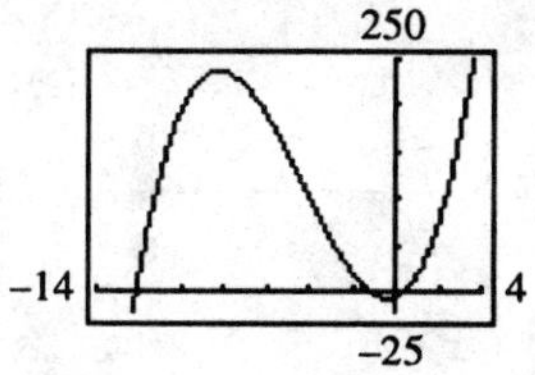

41. a.

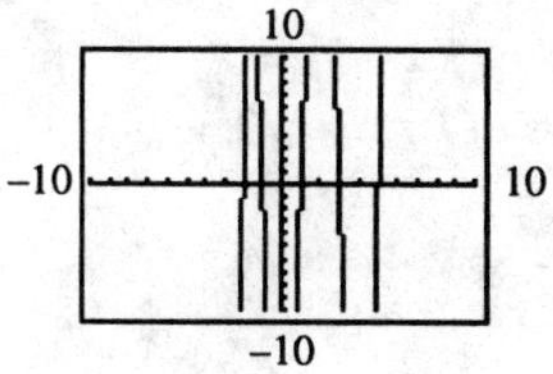

b. The peaks and troughs cannot be seen in the standard viewing screen. Without expanding $f(x)$, it is clear that $a_n x^n = x^5$. The left end of $f(x)$ falls and the right end of $f(x)$ rises with a maximum of $n - 1 = 4$ turning points. Extend the y-axis to complete the graph.

c.

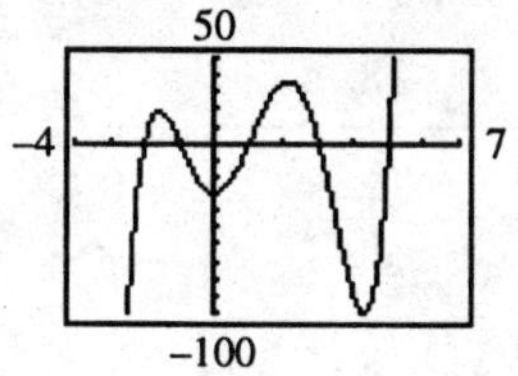

43.

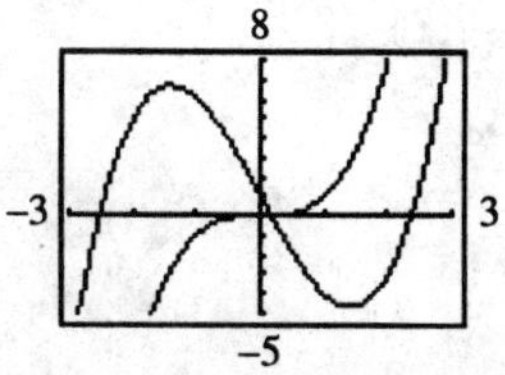

Zoom out to compare the behavior of $f(x)$ to $g(x)$ as $x \to \pm\infty$.

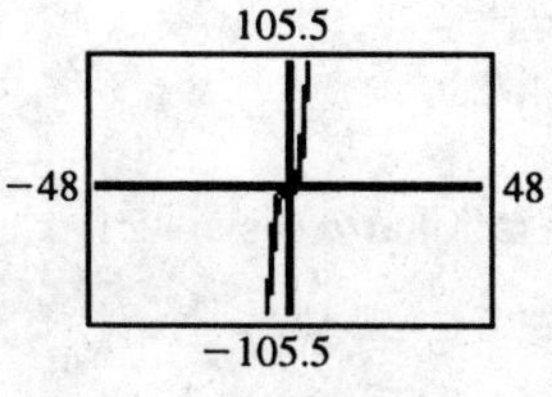

45.

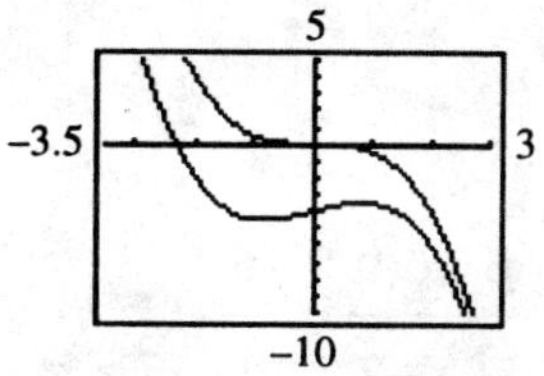

Zoom out to compare the behavior of $f(x)$ to $g(x)$ as $x \to \pm\infty$.

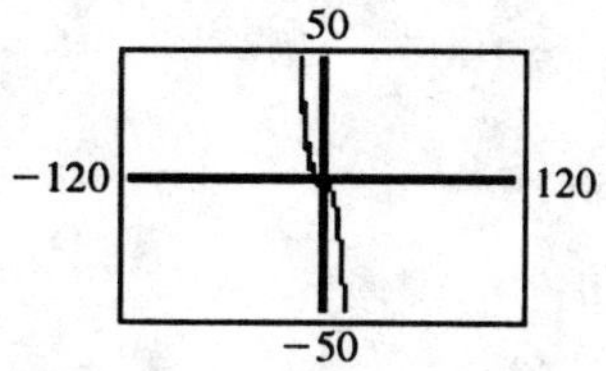

47. A function must pass the vertical line test and x-axis symmetry would violate that requirement.

49. Answers may vary.

51. The graph of $f(t)$ shows a growth period for the first 20 years followed by a quick decline over the next 5 years in which the reindeer population becomes extinct. Factors to consider include food and water sources, breeding habitat, disease, and climate.

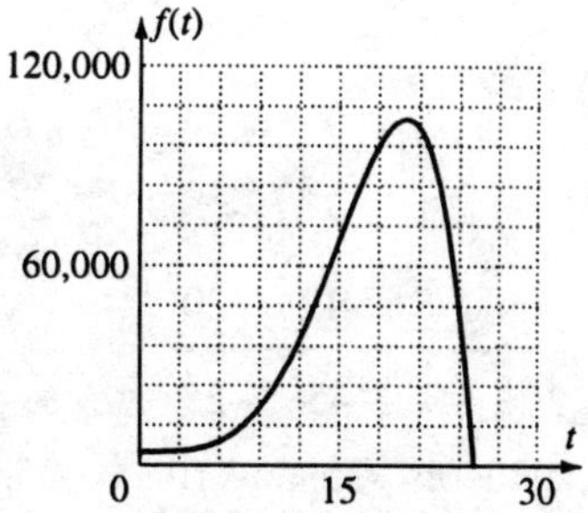

Review Problems

52.

3	2	−3	−11	6
		6	9	−6
	2	3	−2	0

$(2x^3-3x^2-11x+6)\div(x-3)$
$=2x^2+3x-2$

53. $4y^2-25x^2=100$

$\frac{4y^2}{100}-\frac{25x^2}{100}=1$

$\frac{y^2}{25}-\frac{x^2}{4}=1$

hyperbola

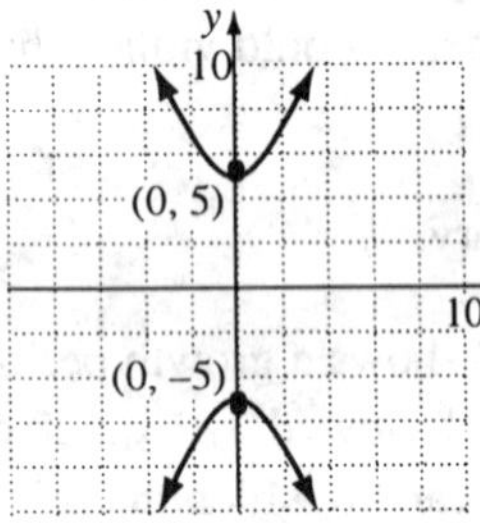

54. $3\log_4(x+2)-\log_4 x$

$=\log_4(x+2)^3-\log_4 x=\log_4\left[\frac{(x+2)^3}{x}\right]$

Problem Set 10.2

1.

4	2	−11	7	−5
		8	−12	−20
	2	−3	−5	−25

The remainder is −25, thus $f(4)=-25$.

3.

5	7	−3	0	6	9
		+35	160	800	4030
	7	32	160	806	4039

The remainder is 4039, thus $f(5)=4039$.

5.

−1	1	−4	1	6
		−1	5	−6
	1	−5	6	0

x^3-4x^2+x+6
$=(x+1)(x^2-5x+6)=0$
$(x+1)(x-2)(x-3)=0$
The zeros of f are −1, 2, and 3.

7.

2	2	−5	1	2
		4	−2	−2
	2	−1	−1	0

$2x^3-5x^2+x+2=0$
$(x-2)(2x^2-x-1)=0$
$(x-2)(2x+1)(x-1)=0$
The solution set is $\left\{-\frac{1}{2}, 1, 2\right\}$.

9.

$-\frac{3}{2}$	12	16	−5	−3
		−18	3	3
	12	−2	−2	0

$12x^3+16x^2-5x-3=0$
$\left(x+\frac{3}{2}\right)(12x^2-2x-2)=0$
$2\left(x+\frac{3}{2}\right)(3x+1)(2x-1)=0$
The solution set is $\left\{-\frac{3}{2}, -\frac{1}{3}, \frac{1}{2}\right\}$.

11. $a_0=15$ and $a_n=2$.
p: ±1, ±3, ±5, ±15
q: ±1, ±2
The possible rational zeros are ±1, ±3, ±5, ±15, $\pm\frac{1}{2}, \pm\frac{3}{2}, \pm\frac{5}{2}, \pm\frac{15}{2}$.

13. $a_0=8$ and $a_n=3$.
p: ± 1, ±2, ±4, ±8
q: ±1, ±3
The possible rational zeros are ±1, ±2, ±4, ±8, $\pm\frac{1}{3}, \pm\frac{2}{3}, \pm\frac{4}{3}, \pm\frac{8}{3}$.

15. There are 2 variations of sign, so there are 2 positive real solutions or 2 – 2 = 0 positive real solutions.
$f(-x) = 2(-x)^4 - 5(-x) + 2 = 2x^4 + 5x + 2$
There are no variations of sign for $f(-x)$, so there are no negative real solutions.

17. There are 3 variations of sign, so there are 3 positive real solutions or 3 – 2 = 1 positive real solution.
$f(-x) = 5(-x)^3 - 3(-x)^2 + 3(-x) - 1$
$= -5x^3 - 3x^2 - 3x - 1$
There are no variations of sign in $f(-x)$ so there are no negative real solutions.

19. There are 2 variations of sign so there are 2 positive real solutions or 2 – 2 = 0 positive real solutions.
$f(-x) = 2(-x)^4 - 5(-x)^3 - (-x)^2 - 6(-x) + 4$

$= 2x^4 + 5x^3 - x^2 + 6x + 4$
There are 2 variations of sign in $f(-x)$ so there are 2 negative real solutions or 2 – 2 = 0 negative real solutions.

21. a. $a_0 = 15$ and $a_n = 3$.
p: ±1, ±3, ±5, ±15
q: ±1, ±3
The possible rational zeros are ±1, ±3, ±5, ±15, $\pm\frac{1}{3}, \pm\frac{5}{3}$.

b.

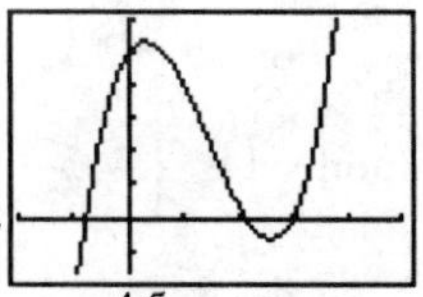

The graph suggests all possible rational zeros less than –1 and greater than 4 can be eliminated.

c. The graph from (b) shows one zero at $x = 3$. $3x^3 - 13x^2 + 7x + 15$
$= (x-3)(3x^2 - 4x - 5) = 0$
From the quadratic formula:
$3x^2 - 4x - 5 = 0$ when $x = \frac{2 \pm \sqrt{19}}{3}$.
The solution set for $f(x) = 0$ is
$\left\{3, \frac{2-\sqrt{19}}{3}, \frac{2+\sqrt{19}}{3}\right\}$.

23. a. $a_0 = -15$ and $a_n = 2$.
p: ±1, ±3, ±5, ±15
q: ±1, ±2
The possible rational zeros are ±1, ±3, ±5, ±15, $\pm\frac{1}{2}, \pm\frac{3}{2}, \pm\frac{5}{2}, \pm\frac{15}{2}$.

b.

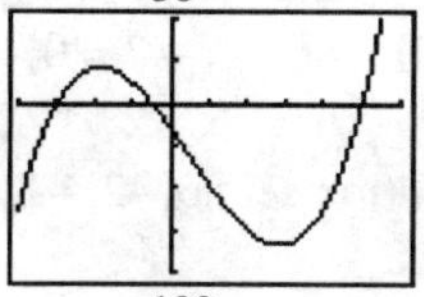

The graph suggests that –15, $-\frac{15}{2}$, –5, $-\frac{3}{2}, \frac{1}{2}, 1, \frac{3}{2}, \frac{5}{2}, 3, \frac{15}{2}$, 15 can be eliminated.

c. The graph from (b) shows one zero at $x = 5$.
$2x^3 - 3x^2 - 32x - 15$
$= (x-5)(2x^2 + 7x + 3)$
$= (x-5)(x+3)(2x+1) = 0$
The solution set for $f(x) = 0$ is
$\left\{-3, -\frac{1}{2}, 5\right\}$.

25. 1. $\frac{p}{q}$: ±1, ±2, ±4

2. 1 positive real solution and 2 or 0 negative real solutions.

3.

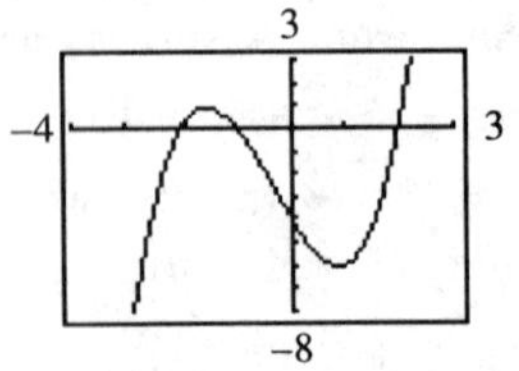

−2	1	1	−4	−4
		−2	2	4
	1	−1	−2	0

4. $x^3 + x^2 - 4x - 4 = (x+2)(x^2 - x - 2)$

5. $x + 2 = 0$ or $x^2 - x - 2 = 0$

6. $x = -2$
$x^2 - x - 2 = (x+1)(x-2)$
$x + 1 = 0$ or $x - 2 = 0$
$x = -1$ or $x = 2$
The solution set is $\{-2, -1, 2\}$.

27. 1. $\frac{p}{q}: \pm 1, \pm 2, \pm 3, \pm 6, \pm\frac{1}{2}, \pm\frac{3}{2}$

2. 2 or 0 positive real solutions and 1 negative real solution.

3.

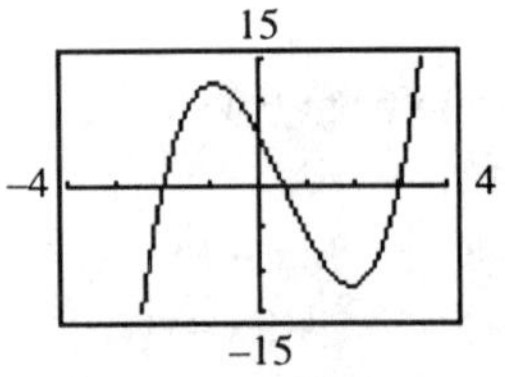

−2	2	−3	−11	6
		−4	14	−6
	2	−7	3	0

4. $2x^3 - 3x^2 - 11x + 6$
$= (x+2)(2x^2 - 7x + 3)$

5. $x + 2 = 0$ or $2x^2 - 7x + 3 = 0$

6. $x = -2$
$2x^2 - 7x + 3 = (2x-1)(x-3)$
$2x - 1 = 0$ or $x - 3 = 0$
$x = \frac{1}{2}$ or $x = 3$
The solution set is $\left\{-2, \frac{1}{2}, 3\right\}$.

29. 1. $\frac{p}{q}: \pm 1, \pm 2, \pm 4, \pm 8, \pm\frac{1}{3}, \pm\frac{2}{3}, \pm\frac{4}{3}, \pm\frac{8}{3}$

2. 1 positive real solution and 2 or 0 negative real solutions.

3.

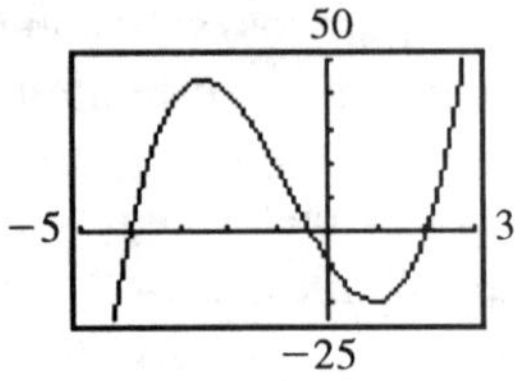

−4	3	7	−22	−8
		−12	20	8
	3	−5	−2	0

4. $3x^3 + 7x^2 - 22x - 8$
$= (x+4)(3x^2 - 5x - 2)$

5. $x + 4 = 0$ or $3x^2 - 5x - 2 = 0$

6. $x = -4$
$3x^2 - 5x - 2 = (3x+1)(x-2)$
$3x + 1 = 0$ or $x - 2 = 0$
$x = -\frac{1}{3}$ or $x = 2$
The solution set is $\left\{-4, -\frac{1}{3}, 2\right\}$.

31. 1. $\frac{p}{q}: \pm 1, \pm 2, \pm 4, \pm\frac{1}{2}$

2. 1 positive real solution and 2 or 0 negative real solutions.

3.

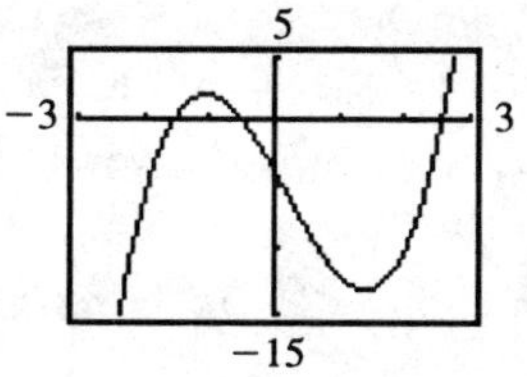

$-\frac{1}{2}$	2	−1	−9	−4
		−1	1	4
	2	−2	−8	0

4. $2x^3 - x^2 - 9x - 4$
$= \left(x + \frac{1}{2}\right)(2x^2 - 2x - 8)$

5. $x + \frac{1}{2} = 0$ or $2x^2 - 2x - 8 = 0$

6. $x = -\frac{1}{2}$
Use the quadratic formula for $2x^2 - 2x - 8 = 0$.
$x = \frac{2 \pm \sqrt{(-2)^2 - 4(2)(-8)}}{2(2)}$
$= \frac{2 \pm \sqrt{68}}{4} = \frac{1 \pm \sqrt{17}}{2}$
The solution set is
$\left\{-\frac{1}{2}, \frac{1-\sqrt{17}}{2}, \frac{1+\sqrt{17}}{2}\right\}$.

33. 1. $\frac{p}{q}$: ±1, ±2, ±4, ±8

2. 1 positive real solution and 3 or 1 negative real solutions.

3.

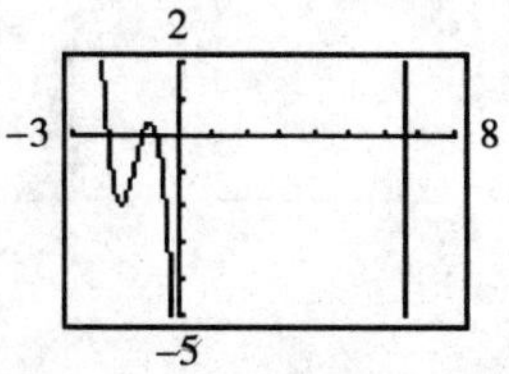

−2	1	−3	−20	−24	−8
		−2	10	20	8
	1	−5	−10	−4	0

4. $x^4 - 3x^3 - 20x^2 - 24x - 8$
$= (x+2)(x^3 - 5x^2 - 10x - 4)$

5. $x + 2 = 0$ or $x^3 - 5x^2 - 10x - 4 = 0$

6. $x = -2$

−1	1	−5	−10	−4
		−1	6	4
	1	−6	−4	0

$(x+1)(x^2 - 6x - 4) = 0$
$x + 1 = 0$ or $x^2 - 6x - 4 = 0$
$x = -1$; use the quadratic formula for $x^2 - 6x - 4 = 0$
$x = \frac{6 \pm \sqrt{(-6)^2 - 4(1)(-4)}}{2(1)}$
$= \frac{6 \pm \sqrt{52}}{2} = 3 \pm \sqrt{13}$
The solution set is
$\{-2, -1, 3-\sqrt{13}, 3+\sqrt{13}\}$.

35. 1. $\frac{p}{q}$: ±1, ±2, ±3, ±6, $\pm\frac{1}{3}, \pm\frac{2}{3}$

2. 2 or 0 positive real solutions and 2 or 0 negative real solutions.

3.

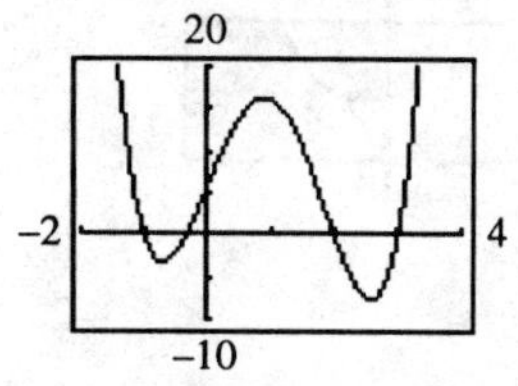

−1	3	−11	−1	19	6
		−3	14	−13	−6
	3	−14	13	6	0

4. $3x^4 - 11x^3 - x^2 + 19x + 6$
$= (x+1)(3x^3 - 14x^2 + 13x + 6)$

5. $x + 1 = 0$ or $3x^3 - 14x^2 + 13x + 6 = 0$

6. $x = -1$

3	3	−14	13	6
		9	−15	−6
	3	−5	−2	0

$(x-3)(3x^2 - 5x - 2) = 0$
$x - 3 = 0$ or $3x^2 - 5x - 2 = 0$
$x = 3$
$(x - 2)(3x + 1) = 0$
$x - 2 = 0$ or $3x + 1 = 0$
$x = 2$ or $x = -\frac{1}{3}$
The solution set is $\left\{-1, -\frac{1}{3}, 2, 3\right\}$.

37. 1. $\frac{p}{q}: \pm 1, \pm 2, \pm 3, \pm 6, \pm\frac{1}{4}, \pm\frac{1}{2}, \pm\frac{3}{4}, \pm\frac{3}{2}$

2. 3 or 1 positive real solutions and 1 negative real solution.

3.

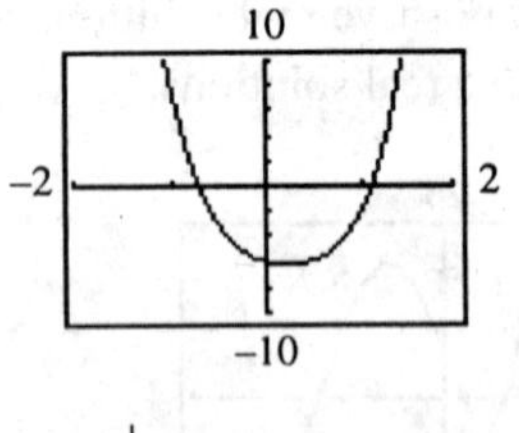

1	4	−1	5	−2	−6
		4	3	8	6
	4	3	8	6	0

4. $4x^4 - x^3 + 5x^2 - 2x - 6$
$= (x-1)(4x^3 + 3x^2 + 8x + 6)$

5. $x - 1 = 0$ or $4x^3 + 3x^2 + 8x + 6 = 0$

6. $x = 1$

$-\frac{3}{4}$	4	3	8	6
		−3	0	−6
	4	0	8	0

$\left(x + \frac{3}{4}\right)(4x^2 + 8) = 0$
$x + \frac{3}{4} = 0$ or $4x^2 + 8 = 0$
$x = -\frac{3}{4}$ or $4x^2 = -8$
$x^2 = -2$
$x = \pm\sqrt{-2} = \pm\sqrt{2}\,i$
The solution set is
$\left\{-\frac{3}{4}, 1, -\sqrt{2}\,i, \sqrt{2}\,i\right\}$.

39. 1. $\frac{p}{q}: \pm 1, \pm 2, \pm 4, \pm 8, \pm\frac{1}{2}$

2. 2 or 0 positive real solutions and 3 or 1 negative real solutions.

3.

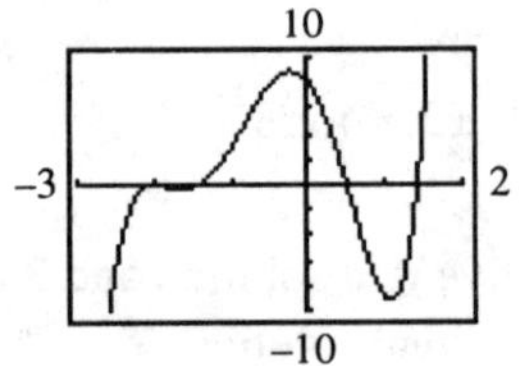

−2	2	7	0	−18	−8	8
		−4	−6	12	12	−8
	2	3	−6	−6	4	0

4. $2x^5 + 7x^4 - 18x^2 - 8x + 8$
$= (x+2)(2x^4 + 3x^3 - 6x^2 - 6x + 4)$

5. $x+2=0$ or
$2x^4+3x^3-6x^2-6x+4=0$

6. $x=-2$

$$\begin{array}{r|rrrrr} \frac{1}{2} & 2 & 3 & -6 & -6 & 4 \\ & & 1 & 2 & -2 & -4 \\ \hline & 2 & 4 & -4 & -8 & 0 \end{array}$$

$\left(x-\frac{1}{2}\right)(2x^3+4x^2-4x-8)=0$

$x=\frac{1}{2}$

$$\begin{array}{r|rrrr} -2 & 2 & 4 & -4 & -8 \\ & & -4 & 0 & 8 \\ \hline & 2 & 0 & -4 & 0 \end{array}$$

$(x+2)(2x^2-4)=0$
$x=-2$ or $2x^2=4$
$x^2=2$
$x=\pm\sqrt{2}$
The solution set is
$\left\{-2,-2,\frac{1}{2},-\sqrt{2},\sqrt{2}\right\}$.

41. 1. $\frac{p}{q}:\pm 1$

2. 1 positive real solution and 3 or 1 negative real solutions

3.
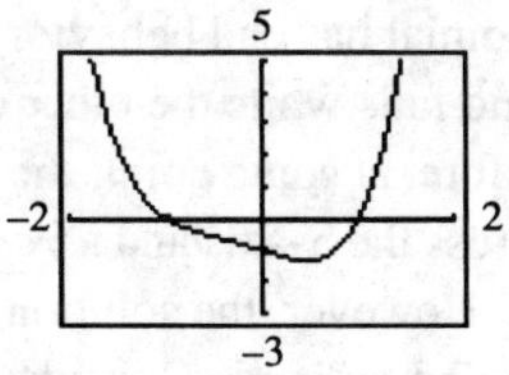

$$\begin{array}{r|rrrrr} -1 & 1 & 1 & 0 & -1 & -1 \\ & & -1 & 0 & 0 & 1 \\ \hline & 1 & 0 & 0 & -1 & 0 \end{array}$$

4. $x^4+x^3-x-1=(x+1)(x^3-1)$

5. $x+1=0$ or $x^3-1=0$

6. $x=-1$

$$\begin{array}{r|rrrr} 1 & 1 & 0 & 0 & -1 \\ & & 1 & 1 & 1 \\ \hline & 1 & 1 & 1 & 0 \end{array}$$

$(x-1)(x^2+x+1)=0$
$x=1$; use the quadratic formula for $x^2+x+1=0$.
$x=\frac{-1\pm\sqrt{(1)^2-4(1)(1)}}{2(1)}=\frac{-1\pm\sqrt{-3}}{2}$
$=\frac{-1\pm\sqrt{3}\,i}{2}$
The solution set is
$\left\{-1,1,\frac{-1-\sqrt{3}\,i}{2},\frac{-1+\sqrt{3}\,i}{2}\right\}$.

43.
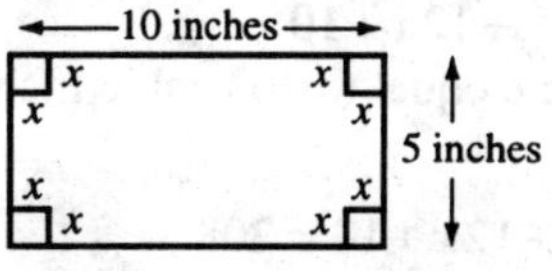

Volume = length · width · height
$V(x)=(10-2x)\cdot(5-2x)\cdot x$
$=4x^3-30x^2+50x$
If the volume is to be 24 cubic inches, then set $V(x)=24$ and solve for x.
$V(x)=4x^3-30x^2+50x=24$
$4x^3-30x^2+50x-24=0$

1	4	−30	50	−24
		4	−26	24
	4	−26	24	0

$(x-1)(4x^2-26x+24)$
$=2(x-1)(2x^2-13x+12)$
$x=1$; use the quadratic formula for
$2x^3-13x+12=0$.

$$x=\frac{13\pm\sqrt{(-13)^2-4(2)(12)}}{2(2)}=\frac{13\pm\sqrt{73}}{4}$$

The possible solution set is

$$\left\{1, \frac{13-\sqrt{73}}{4}, \frac{13+\sqrt{73}}{4}\right\}.$$

The possible solution $\frac{13+\sqrt{73}}{4}$ is greater than 5 (the width) so it cannot be a solution. The length of the square to be cut from each corner can be 1 inch or $\frac{13-\sqrt{73}}{4}$ inches for a volume of 24 cubic inches.

45. Volume
= length · width · height − open section
$=(x+5)\cdot(2x+1)\cdot(x+2)-(x+5)\cdot x\cdot 3$
$=2x^3+15x^2+27x+10-3x^2-15x$
$=2x^3+12x^2+12x+10$
Set the volume equal to 208 cubic inches and solve for x.
$2x^3+12x^2+12x+10=208$
$2x^3+12x^2+12x-198=0$

3	2	12	12	−198
		6	54	198
	2	18	66	0

$(x-3)(2x^2+18x+66)=0$
$x=3$; use the quadratic formula for
$x^2+9x+33=0$.

$$x=\frac{-9\pm\sqrt{(9)^2-4(1)(33)}}{2(1)}=\frac{-9\pm\sqrt{-51}}{2}$$
$$=\frac{-9\pm\sqrt{51}\,i}{2}$$

The possible solution set is

$$\left\{3, \frac{-9-\sqrt{51}\,i}{2}, \frac{-9+\sqrt{51}\,i}{2}\right\}.$$

The only real solution is $x=3$.

47. Cost per hour = 5200 + 2000x
Sales per hour $=(7070-0.4x^2)\cdot x$
Profit = sales − cost
$=(7070-0.4x^2)x-(5200+2000x)=214,500$

$-0.4x^3+5070x-219,700=0$
$4x^3-50,700x+2,197,000=0$
The possible solution set is $\{-130, 65, 65\}$.
The manufacturer should produce and sell 65 cars per hour for a $214,500 profit.

49. a. Using Descarte's Rule of Signs, there is no variation of sign in the equation so there are no positive real solutions. False

b. Descarte's Rule of Signs provides the *possible* number of positive and negative real roots for a polynomial equation. False

c. Using the leading coefficient test, when $n=3$, the graph of a third degree polynomial has end behavior such that one end falls while the other end rises. Therefore, at some point, the function will cross the x-axis and a *real* solution exists. However, the solution does not need to be *rational*. Consider
$x^3-3\sqrt{2}x^2+6x-2\sqrt{2}=\left(x-\sqrt{2}\right)^3$.
False

d. a., b., and c. are all false. True

51. **a.** False; consider how long it takes to solve $x^4 = 0$ and $x^2 = 0$.

b. By the properties of polynomial equations, if $3 - 5i$ is a solution, then $3 + 5i$ must also be a solution. Likewise, if $4 + 3i$ is a solution, then $4 - 3i$ is also a solution. Thus, if $3 - 5i$ and $4 + 3i$ are solutions, then there are at least 4 roots and $P(x)$ has at least degree 4. True

c. By the properties of polynomial equations, if $2 - 5i$ is a solution, then $2 + 5i$ must also be a solution. However, it is not true that $-2 + 5i$ must be a solution. False

d. Any polynomial with only one variation of sign will have *exactly* one positive or exactly one negative real root. False

53. See graphs in Problems 21–42 and use the zoom feature to verify the real zeros.

55. $f(x)$ has 5 variations of sign so there are 5, 3, or 1 positive real zeros. $f(-x) = -x^5 - x^4 - x^3 - x^2 - x - 8$ with no variations of sign so there are no negative real zeros.

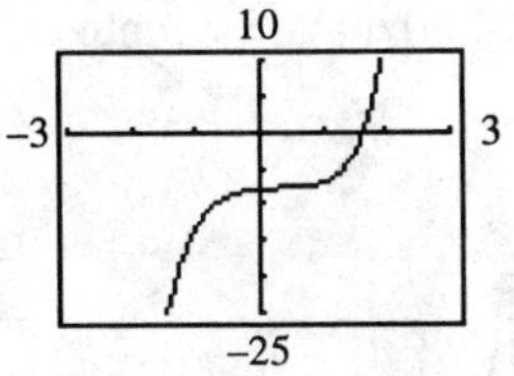

The graph of f suggests one positive real zero.

57. Answers may vary.

59. Answers may vary.

61. Answers may vary.

63. A polynomial function, $f(x)$, for which $f(-x)$ has no variation of sign and $f(x)$ has at least one variation of sign will have no negative rational zeros but the possibility of positive rational zeros. A polynomial function of the form

$$f(x) = \begin{cases} a_n x^n - a_{n-1}x^{n-1} + a_{n-2}x^{n-2} - \cdots - a_1 x + a_0 \text{ for } n \text{ is even, } a_i > 0 \\ -a_n x^n + a_{n-1}x^{n-1} - a_{n-2}x^{n-2} + \cdots - a_1 x + a_0 \text{ for } n \text{ is odd, } a_i > 0 \end{cases}$$

has the desired characteristics.

65. Use Descarte's Rule of Signs to find $x^5 + x + 1 = 0$ has no positive real solution and exactly one negative real solution.

67. Let $p = 3$. $x^3 + x^2 + x - 3 = 0$
$(x-1)(x^2 + 2x + 3)$
$x = 1$ or $x = \frac{-2 \pm \sqrt{-8}}{2} = -1 \pm \sqrt{2}\, i$
The solution set for $p = 3$ is
$\{1, -1-\sqrt{2}\, i, -1+\sqrt{2}\, i\}$.

69. $f(5) =$ remainder of $\frac{f(x)}{x-5}$

5	2	4	2	3	1
		10	70	360	1815
	2	14	72	363	1816

$24{,}231_5 = 1816$

Review Problems

71. $\left(\frac{x^{-3}y^{-5}}{xy^{-7}}\right)^{-6} = \left(\frac{y^2}{x^4}\right)^{-6} = \frac{y^{-12}}{x^{-24}} = \frac{x^{24}}{y^{12}}$

72. $\sqrt{8x^3} - 2x\sqrt{50x}$
$= \sqrt{(4x^2)(2x)} - 2x\sqrt{25(2x)}$
$= 2x\sqrt{2x} - 2x \cdot 5\sqrt{2x} = 2x\sqrt{2x} - 10x\sqrt{2x}$
$= -8x\sqrt{2x}$

73. $3^{2x-1} = 27^x$
$3^{2x-1} = (3^3)^x$
$3^{2x-1} = 3^{3x}$
$2x - 1 = 3x$
$x = -1$

Problem Set 10.3

1. No y-axis or origin symmetry, y-intercept is 0, x-intercept is 0, vertical asymptote is $x = 2$, and horizontal asymptote is $y = 4$.

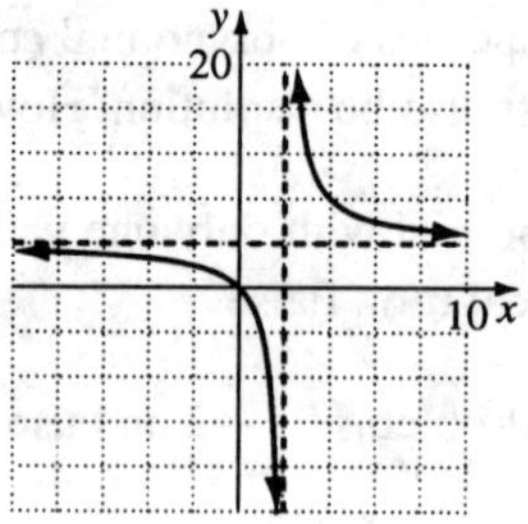

3. Origin symmetry, y-intercept is 0, x-intercept is 0, vertical asymptotes are $x = 2$ and $x = -2$, and horizontal asymptote is the x-axis.

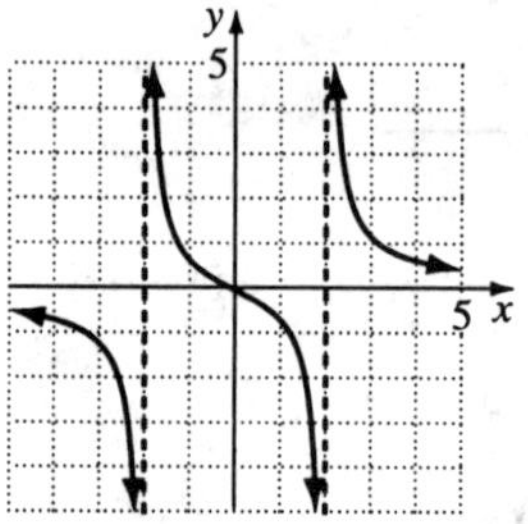

5. y-axis symmetry, y-intercept is 0, x-intercept is 0, vertical asymptotes are $x = -1$ and $x = 1$, and horizontal asymptote is $y = 2$.

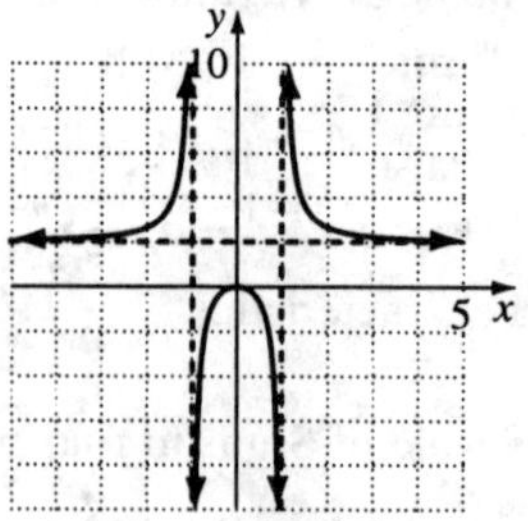

7. No y-axis or origin symmetry, y-intercept is 0, x-intercept is 0, vertical asymptote is $x = -1$, and horizontal asymptote is $y = -1$.

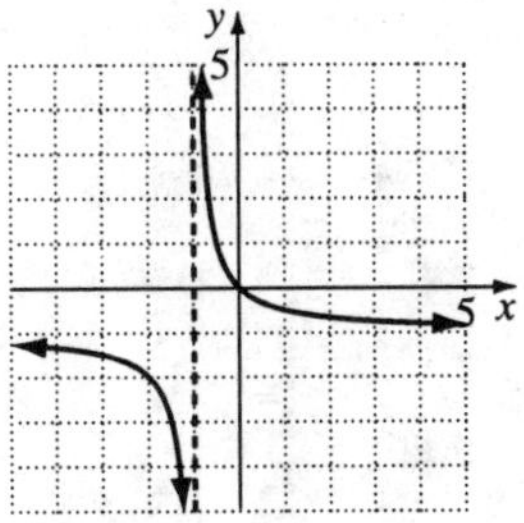

9. y-axis symmetry, y-intercept is $\frac{1}{4}$, no x-intercept, vertical asymptotes are $x = -2$ and $x = 2$, and horizontal asymptote is the x-axis.

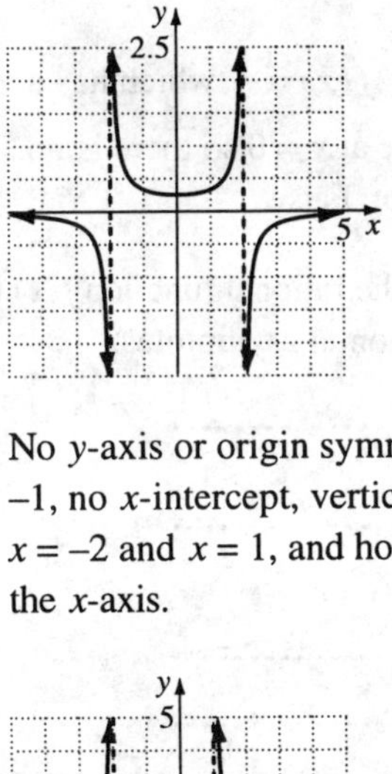

11. No y-axis or origin symmetry, y-intercept is -1, no x-intercept, vertical asymptotes are $x = -2$ and $x = 1$, and horizontal asymptote is the x-axis.

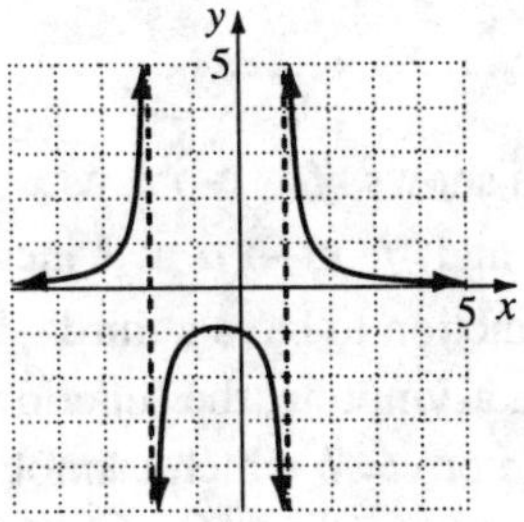

13. y-axis symmetry, y-intercept is 0, x-intercept is 0, no vertical asymptote, and horizontal asymptote is $y = 2$.

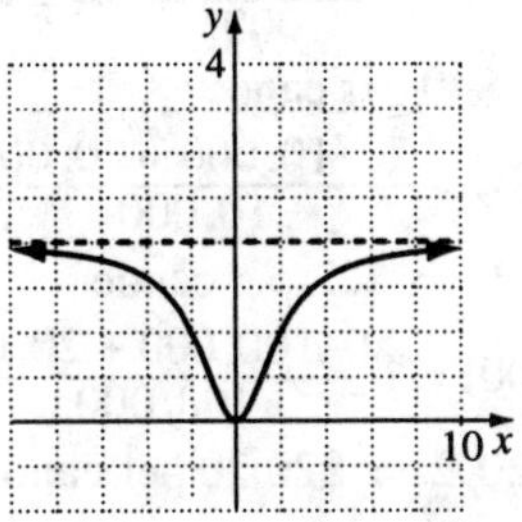

15. No y-axis or origin symmetry, y-intercept is $-\frac{1}{3}$, x-intercept is -2, vertical asymptotes are $x = -3$ and $x = 2$, and horizontal asymptote is the x-axis.

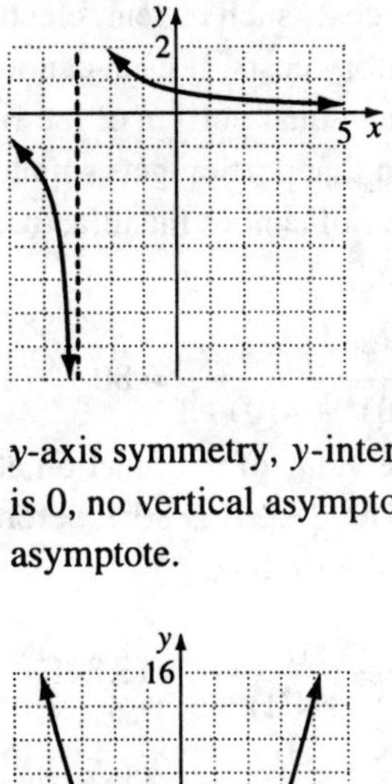

17. y-axis symmetry, y-intercept is 0, x-intercept is 0, no vertical asymptote, and no horizontal asymptote.

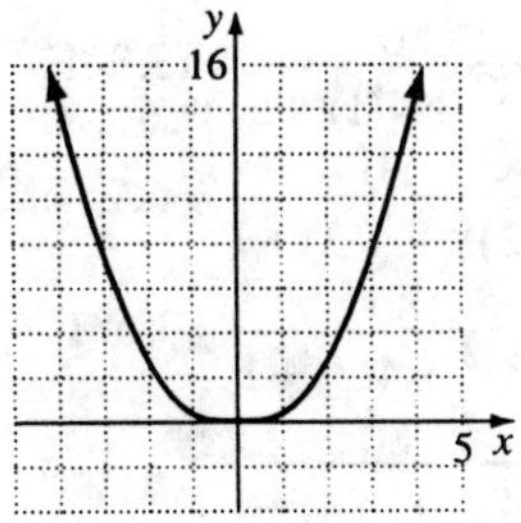

19. a. $\overline{C}(100) = \dfrac{20 \cdot 100 + 20{,}000}{100} = 220$

$\Rightarrow$ \$220 per canoe

$\overline{C}(1000) = \dfrac{20 \cdot 1000 + 20{,}000}{1000} = 40$

$\Rightarrow$ \$40 per canoe

$\overline{C}(10{,}000) = \dfrac{20 \cdot 10{,}000 + 20{,}000}{10{,}000}$

$= 22 \Rightarrow$ \$22 per canoe

$\overline{C}(100{,}000) = \dfrac{20 \cdot 100{,}000 + 20{,}000}{100{,}000}$

$= 20.2 \Rightarrow$ \$20.20 per canoe

b. The horizontal asymptote is \$20. Each canoe manufactured has a cost of at least \$20. The \$20 might be the cost of materials and labor for each canoe. The average cost above \$20 might be fixed operating costs such as rent, electricity, and inventory costs. If each canoe is assigned an equal portion of the fixed costs, then that portion gets smaller as the number of canoes manufactured increases.

21. a. $F(0) = \dfrac{80}{(0)^2 + 4(0) + 1} = 80$

$F(0)$ is the value of the function at time 0 hours. The dessert is 80°F before it is placed in the ice box.

b. $F(1) = \dfrac{80}{(1)^2 + 4(1) + 1} \approx 13.3°\text{F}$

$F(2) = \dfrac{80}{(2)^2 + 4(2) + 1} \approx 6.2°\text{F}$

$F(3) = \dfrac{80}{(3)^2 + 4(3) + 1} \approx 3.6°\text{F}$

$F(4) = \dfrac{80}{(4)^2 + 4(4) + 1} \approx 2.4°\text{F}$

$F(5) = \dfrac{80}{(5)^2 + 4(5) + 1} \approx 1.7°\text{F}$

c. The horizontal asymptote is the x-axis ($y = 0$). The temperature of the dessert approaches 0°F after several hours (t increases).

d.

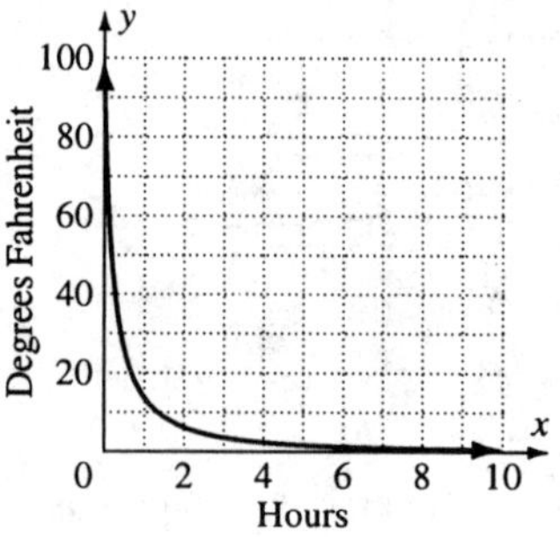

23. a. Consider $f(x) = \dfrac{4x}{x-2}$ which has a vertical asymptote at $x = 2$ and a horizontal asymptote at $y = 4$. False

b. Consider $f(x) = \dfrac{1}{x}$ which has a vertical asymptote at $x = 0$ so there is no y-intercept. False

c. Assume the rational function $f(x)$ has two horizontal asymptotes.

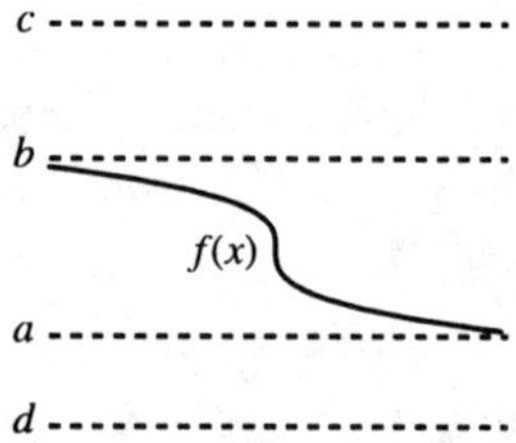

The graph shows $f(x) \to b$ as x decreases and $f(x) \to a$ as x increases. For the function to have a third horizontal asymptote, the function must cross $y = a$ or $y = b$ which cannot happen if $y = a$ and $y = b$ are in fact asymptotes. Therefore, there cannot be a third asymptote. False

d. A vertical asymptote at $x = a$ means $f(x)$ increases or decreases without bound as $x \to a$. $f(x)$ cannot cross a vertical asymptote. True

25.

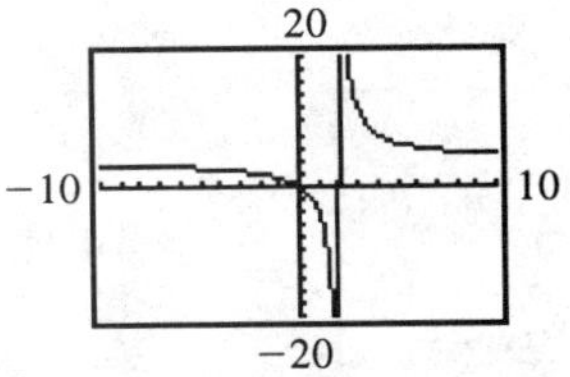

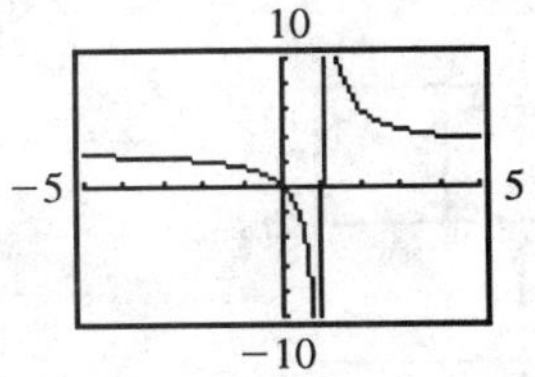

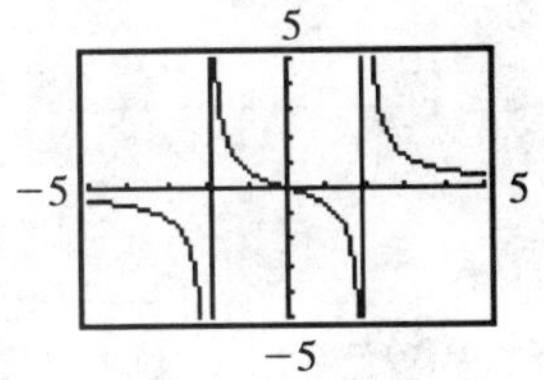

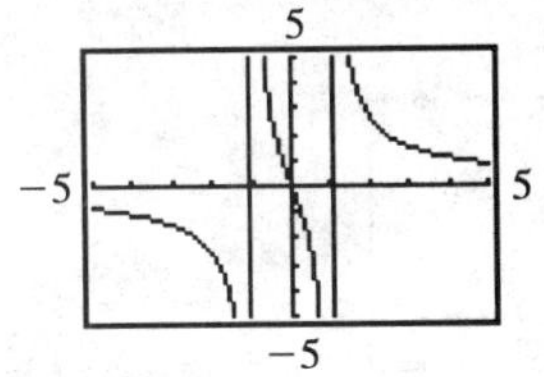

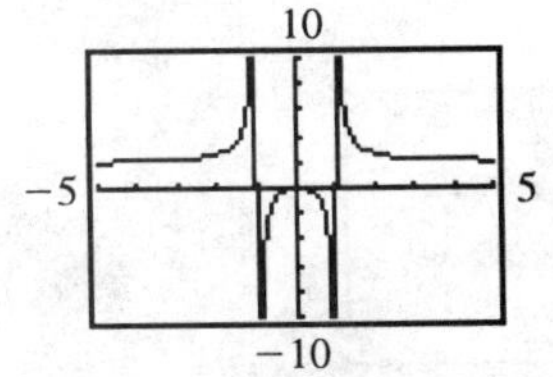

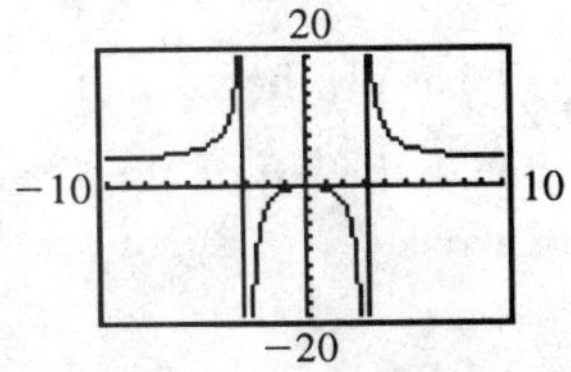

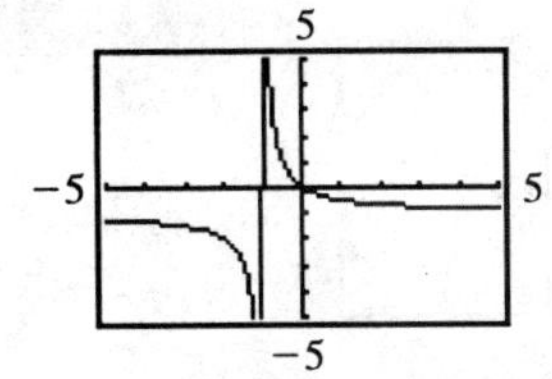

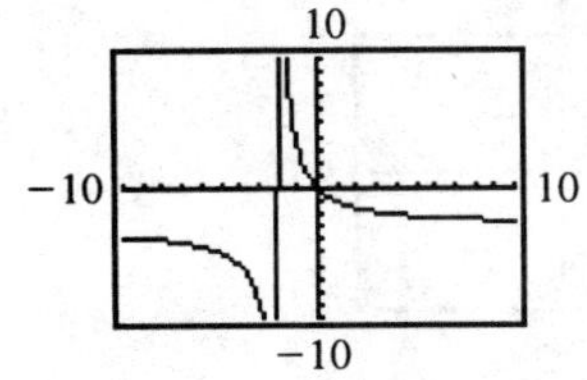

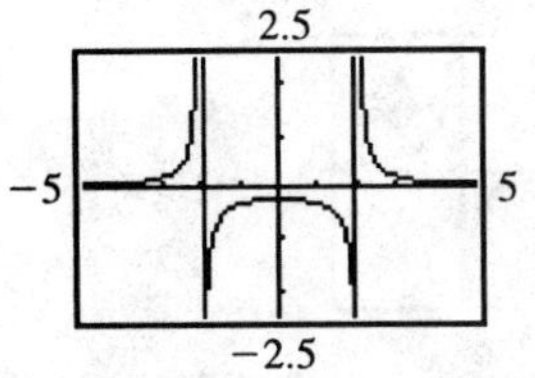

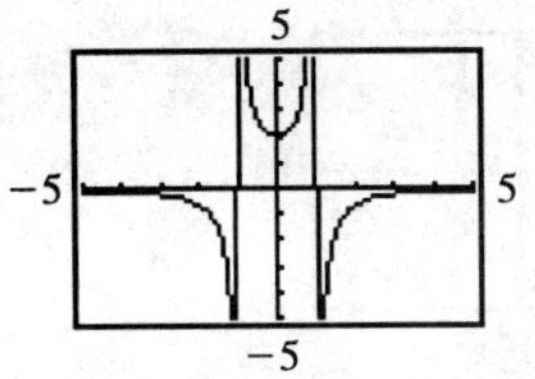

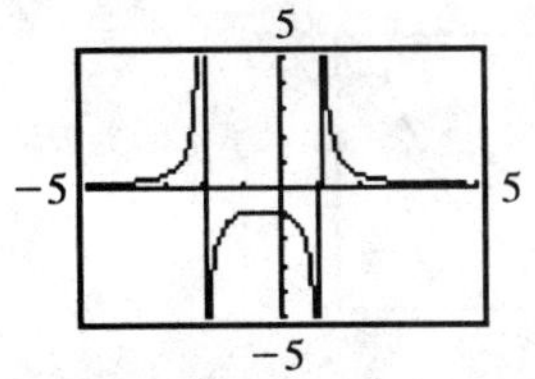

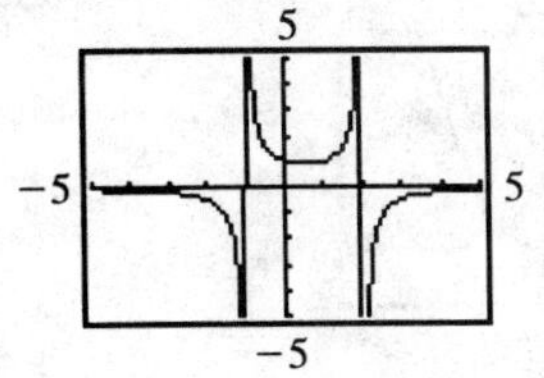

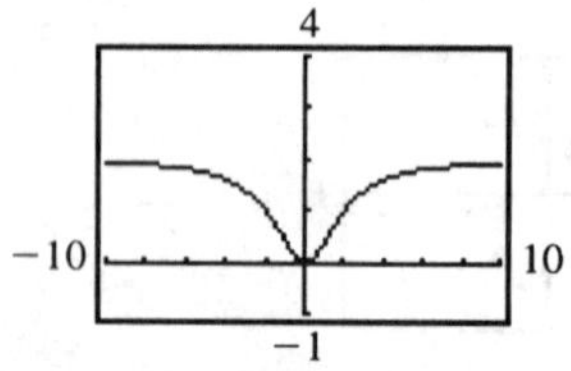

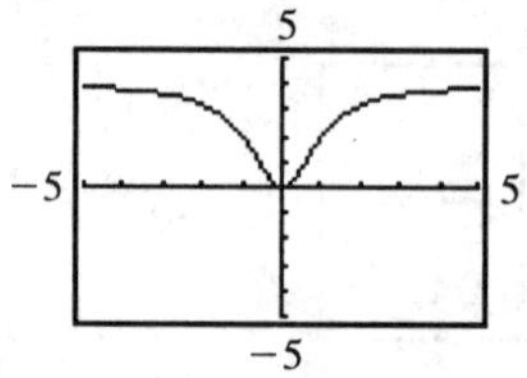

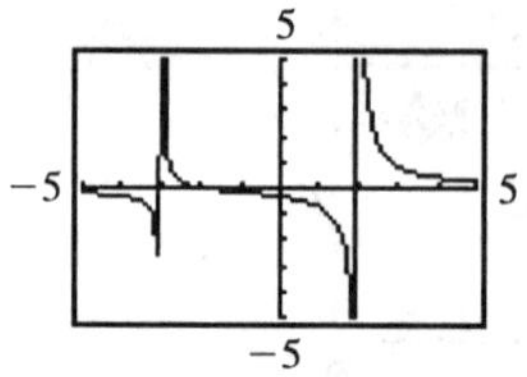

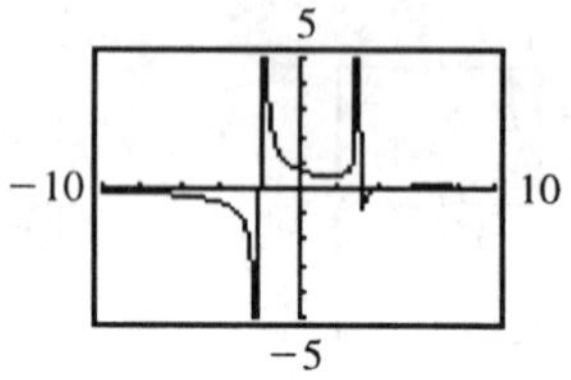

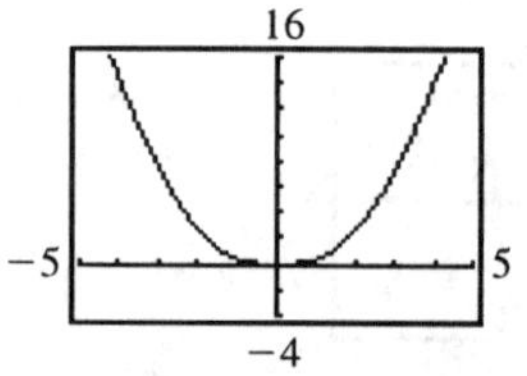

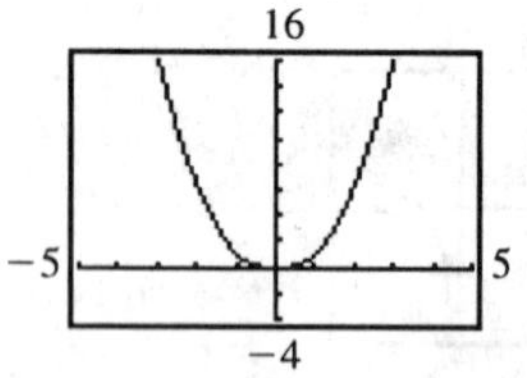

27.

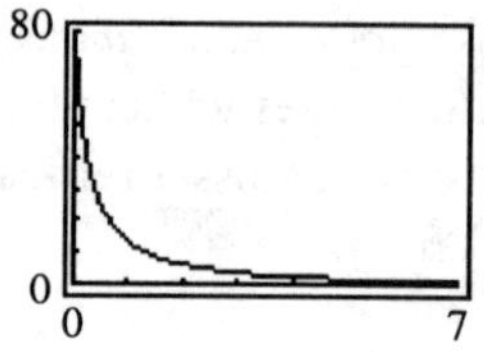

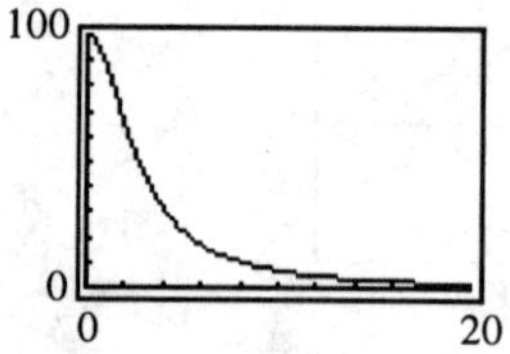

29.

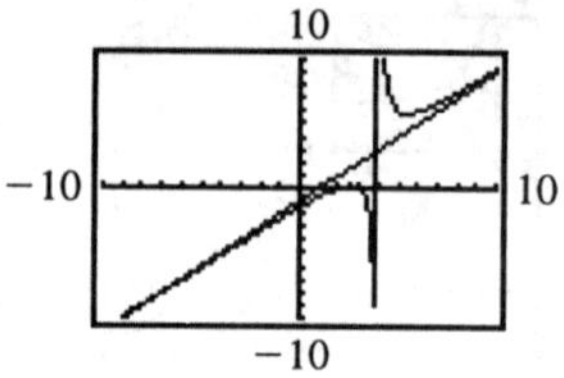

31. $f(x) = \dfrac{x^4}{x^2+2} = x^2 - 2 + \dfrac{4}{x^2+2}$

$= t(x) + \dfrac{r(x)}{q(x)}$

Conjecture: $y = t(x) = x^2 - 2$ is an equation for an asymptote for *f*(*x*).

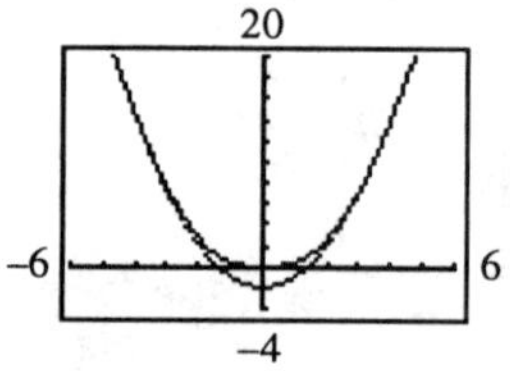

33.

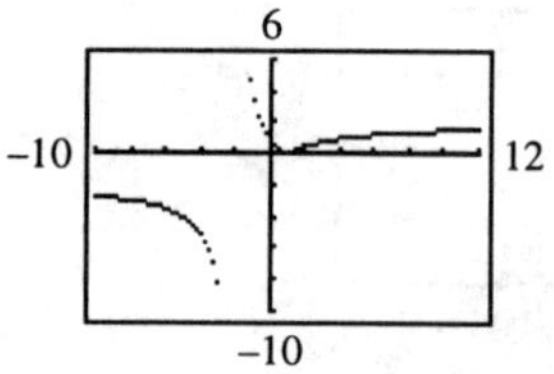

$f(x) = \dfrac{2|x-1|}{x+2}$ has *two* horizontal asymptotes; one to the left of $x = -2$ and one to the right of $x = 1$.

$$y = \begin{cases} -2; & -\infty < x < -2 \\ 2; & 1 < x < \infty \end{cases}$$

35. a.

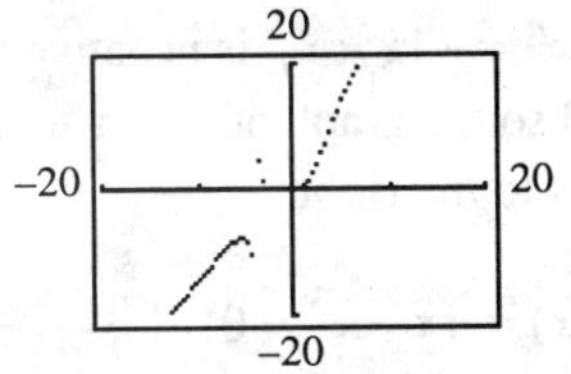

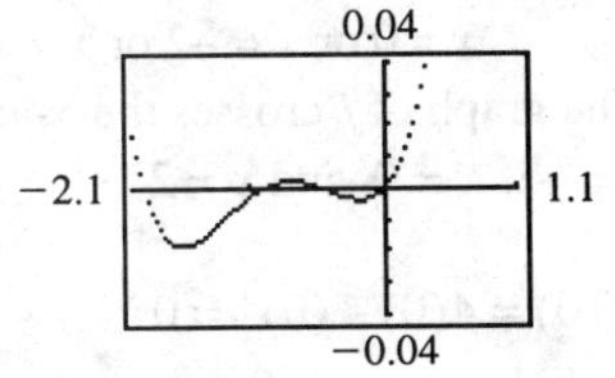

There are three turning points.

c. How to determine the maximum(s) and minimum(s) of the function.

37. Answers may vary.

39. Create $f(x)$ by considering each condition.

(1) f has a vertical asymptote given by $x = 3$

$\Rightarrow f(x) = \frac{1}{x-3}$

(2) f has a horizontal asymptote $y = 0$

$\Rightarrow f(x) = \frac{1}{x-3}$

(3) y-intercept $= -1$, ($f(0) = -1$)

$\Rightarrow f(x) = \frac{3}{x-3}$

(4) No x-intercept

$\Rightarrow f(x) = \frac{3}{x-3}$

41. Answers may vary; $y = c$ where $c \neq 0$.

Review Problems

43. $\left.\begin{array}{l} x^2 + xy - y^2 = 1 \\ x - y = 3 \end{array}\right\}$

$x = y + 3$

$(y+3)^2 + (y+3)y - y^2 = 1$
$(y^2 + 6y + 9) + (y^2 + 3y) - y^2 = 1$
$y^2 + 9y + 8 = 0$
$(y + 8)(y + 1) = 0$
$y = -8$ or $y = -1$

For $y = -8$, $x = y + 3 = -8 + 3 = -5$.
For $y = -1$, $x = y + 3 = -1 + 3 = 2$.
The solution set is $\{(-5, -8), (2, -1)\}$.

44.

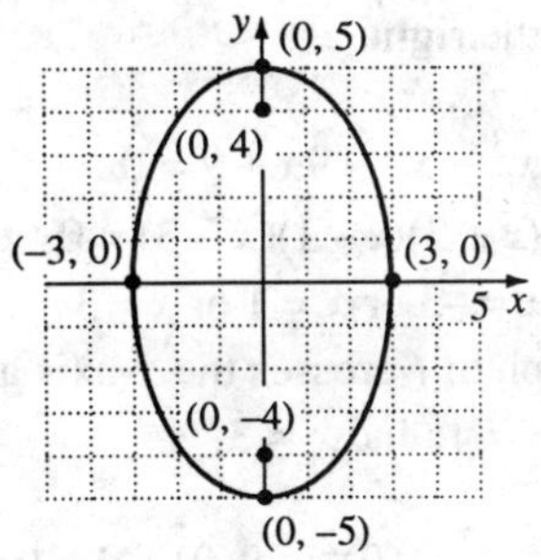

45. $\log_3 x + \log_3(2x+1) = 1$
$\log_3[x(2x+1)] = \log_3(2x^2 + x) = 1$
$2x^2 + x = 3^1$
$2x^2 + x - 3 = 0$
$(2x + 3)(x - 1) = 0$
$2x + 3 = 0$ or $x - 1 = 0$
$x = -\frac{3}{2}$ or $x = 1$

The solution set is $\left\{-\frac{3}{2}, 1\right\}$.

Chapter 10 Review Problems

1. $a_n x^n = -1x^3$; a_n is negative and n is odd so the graph of f rises to the left and falls to the right. $f(0) = -(0)^3 + 12(0)^2 - 0 = 0$ so the y-intercept is 0. Graph c.

2. $a_n x^n = 1x^6$; a_n is positive and n is even so the graph of g rises to the left and to the right. Graph b. (Note there are 5 turning points, so the degree of the function must be at least 6.)

3. $a_nx^n = 1x^5$; a_n is positive and n is odd so the graph of h falls to the left and rises to the right. $h(0) = 0$. Graph a.

4. $a_nx^n = 1x^3$; $a_n > 0$; degree 3 is odd; the curve falls to the left and rises to the right. $r(0) = 1$. Graph d.

5. a. $a_nx^n = 1x^3$; a_n is positive and n is odd so the graph of f falls to the left and rises to the right.

b. $f(x) = x^3 - x^2 - 9x + 9 = 0$
$(x+3)(x-1)(x-3) = 0$
$x = -3$ or $x = 1$ or $x = 3$
The graph of f crosses the x-axis at $x = -3$, $x = 1$, and $x = 3$.

c. $f(0) = (0)^3 - (0)^2 - 9(0) + 9 = 9$;
the y-intercept is 9.

d. $f(-x) = (-x)^3 - (-x)^2 - 9(-x) + 9$
$= -x^3 - x^2 + 9x + 9$;
There is no y-axis or origin symmetry.

e. $n - 1 = 2$; there are at most 2 turning points.

x	−4	−2	2	4
$f(x)$	−35	15	−5	21

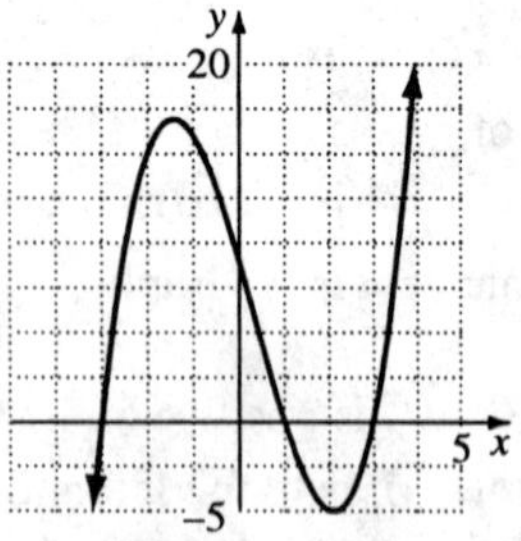

6. a. $a_nx^n = -1x^3$; a_n is negative and n is odd so the graph of f rises to the left and falls to the right.

b. $f(x) = 4x - x^3 = 0$
$-x(x+2)(x-2) = 0$
$x = 0$ or $x = -2$ or $x = 2$
The graph of f crosses the x-axis at $x = -2$, $x = 0$, and $x = 2$.

c. $f(0) = 4(0) - (0)^3 = 0$;
the y-intercept is 0.

d. $f(-x) = 4(-x) - (-x)^3 = -4x + x^3$
$= -f(x)$; the graph of f has origin symmetry.

e. $n - 1 = 2$; there are at most 2 turning points.

x	−3	−1	1	3
$f(x)$	15	−3	3	−15

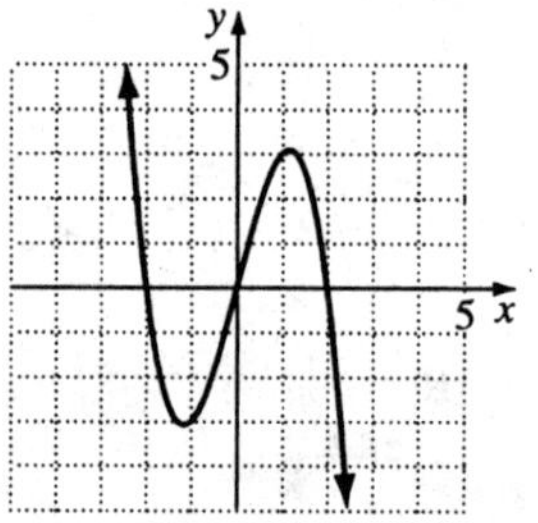

7. a. $a_nx^n = 2x^3$; a_n is positive and n is odd so f falls to the left and rises to the right.

b. $f(x) = 2x^3 + 3x^2 - 8x - 12 = 0$
$(2x+3)(x-2)(x+2) = 0$
$x = -\frac{3}{2}$ or $x = 2$ or $x = -2$
The graph of f crosses the x-axis at $x = -2$, $x = -\frac{3}{2}$, and $x = 2$.

c. $f(0) = 2(0)^3 + 3(0)^2 - 8(0) - 12$
$= -12$; the y-intercept is -12.

d. $f(-x)$
$= -2(-x)^3 + 3(-x)^2 - 8(-x) - 12$
$= -2x^3 + 3x^2 + 8x - 12$;
there is no y-axis or origin symmetry.

e. $n - 1 = 2$; there are at most 2 turning points.

x	−3	1	3
$f(x)$	−15	−15	45

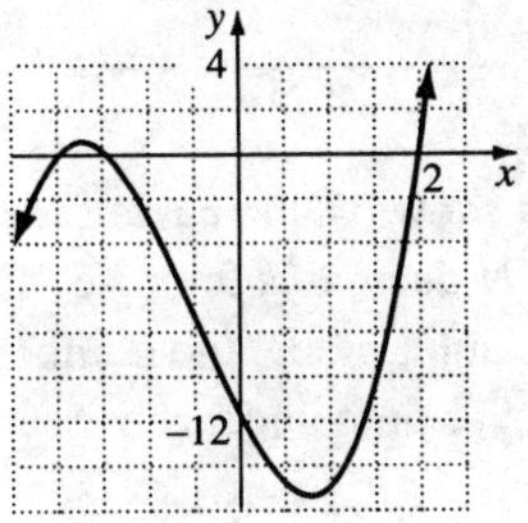

8. a. $a_n x^n = -1x^4$; a_n is negative and n is even so the graph of f falls to the left and to the right.

b. $f(x) = -x^4 + 25x^2 = 0$
$-x^2(x+5)(x-5) = 0$
$x = 0$ or $x = -5$ or $x = 5$
The graph of f crosses the x-axis at $x = -5$ and $x = 5$, and touches the x-axis at $x = 0$.

c. $f(0) = -(0)^4 + 25(0)^2 = 0$;
the y-intercept is 0.

d. $f(-x) = -(-x)^4 + 25(-x)^2$
$= -x^4 + 25x^2 = f(x)$;
there is y-axis symmetry.

e. $n - 1 = 3$; there are at most 3 turning points.

x	−6	−4	−3	−2	−1
$f(x)$	−396	144	144	84	24

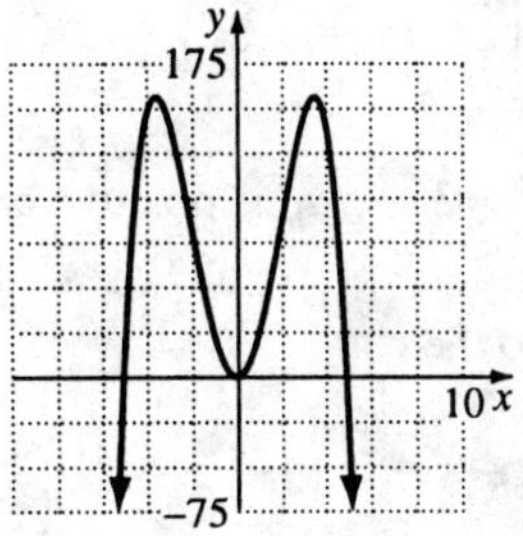

9. a. $a_n x^n = -1x^4$; a_n is negative and n is even so the graph of f falls to the left and to the right.

b. $f(x) = -x^4 + 6x^3 - 9x^2 = 0$
$-x^2(x-3)^2 = 0$
$x = 0$ or $x = 3$
The graph of f touches the x-axis at $x = 0$ and $x = 3$.

c. $f(0) = -(0)^4 + 6(0)^3 - 9(0)^2 = 0$
the y-intercept is 0.

d. $f(-x) = -(-x)^4 + 6(-x)^3 - 9(-x)^2$
$= -x^4 - 6x^3 - 9x^2$;
there is no y-axis or origin symmetry.

e. $n - 1 = 3$; there are at most 3 turning points.

x	−1	1.5
$f(x)$	−16	−5.0625

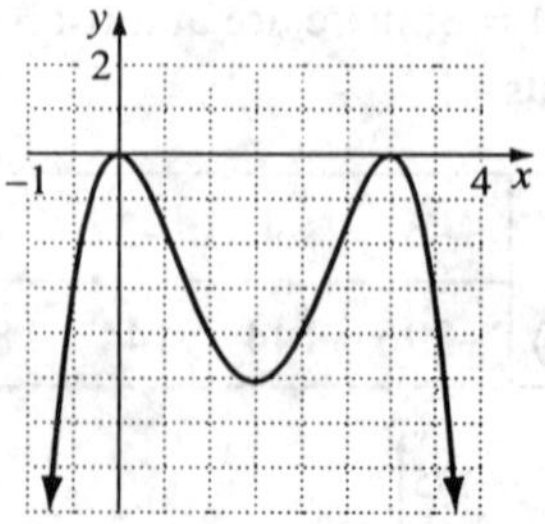

10. a. $a_n x^n = 3x^4$; a_n is positive and n is even so the graph of f rises to the left and to the right.

b. $f(x) = 3x^4 - 15x^3 = 0$

$3x^3(x-5) = 0$

$x = 0$ or $x = 5$

The graph of f crosses the x-axis at $x = 5$ and $x = 0$, and flattens out at $x = 0$.

c. $f(0) = 3(0)^4 - 15(0)^3 = 0$;

the y-intercept is 0.

d. $f(-x) = 3(-x)^4 - 15(-x)^3$

$= 3x^4 + 15x^3$;

there is no y-axis or origin symmetry.

e. $n - 1 = 3$; there are at most 3 turning points.

x	−1	1	2	3	4	6
$f(x)$	18	−12	−72	−162	−192	648

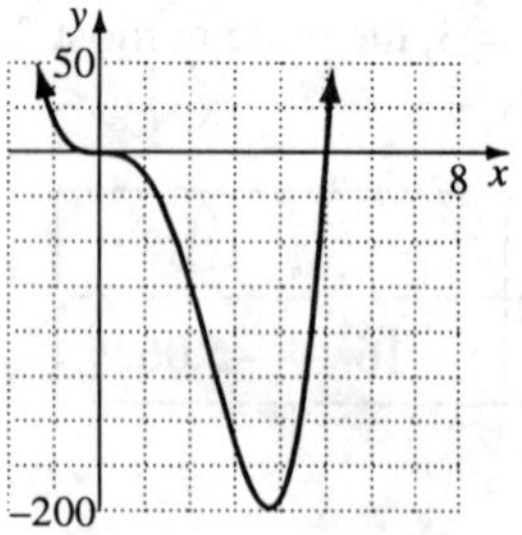

11. $a_n x^n = -0.0013x^3$; a_n is negative and n is odd so f falls to the right. At some point, f will fall below 0 as x increases. This model is only appropriate for values of x where f is between 0 and 100 meaning years after 1960 where the percentage of U.S. families below the poverty level is between 0% and 100%.

12.

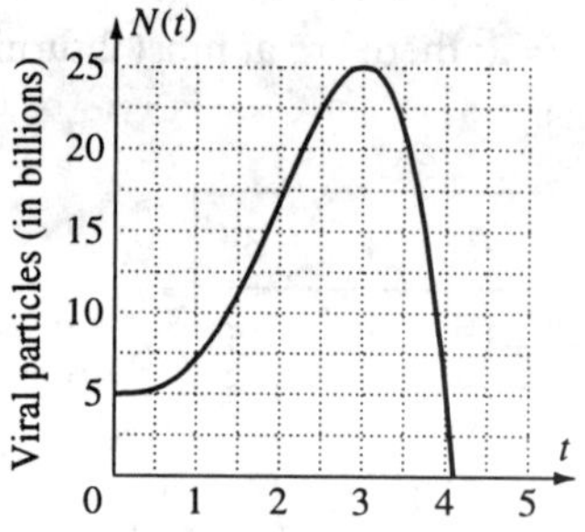

The graph shows N increases for t between 0 and 3 and N decreases for $t > 3$. This function cannot model the number of viral particles over time under the assumption that the number of viral particles is increasing until death.

13. a. $V(x) = (16 - 2x) \cdot (20 - 2x) \cdot x$

$= 4x^3 - 72x^2 + 320x$

b.

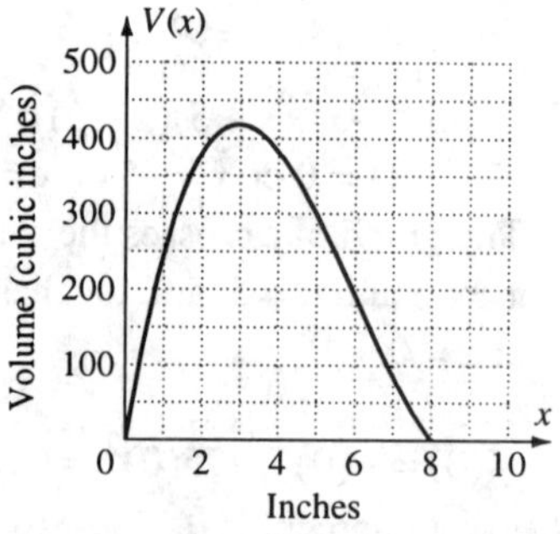

14.

−13	2	−7	9	−3
		−26	429	−5694
	2	−33	438	−5697

$f(-13) = -5697$

15.

2	2	1	−13	6
		4	10	−6
	2	5	−3	0

$f(x)=2x^3+x^2-13x+6$
$=(x-2)(2x^2+5x-3)$
$=(x-2)(x+3)(2x-1)$
$x=2$ or $x=-3$ or $x=\frac{1}{2}$
The solution set is $\left\{-3, \frac{1}{2}, 2\right\}$.

16.

4	1	0	−17	4
		4	16	−4
	1	4	−1	0

$x^3-17x+4=0$
$(x-4)(x^2+4x-1)=0$
$x=4$ or $x=\dfrac{-4\pm\sqrt{(4)^2-4(1)(-1)}}{2(1)}$
$=\dfrac{-4\pm\sqrt{20}}{2}=-2\pm\sqrt{5}$
The solution set is $\left\{4, -2-\sqrt{5}, -2+\sqrt{5}\right\}$.

17. $a_n=3$ and $a_0=-8$.
p: $\pm1, \pm2, \pm4, \pm8$
q: $\pm1, \pm3$
The possible rational zeros are $\frac{p}{q}$:
$\pm1, \pm2, \pm4, \pm8, \pm\frac{1}{3}, \pm\frac{2}{3}, \pm\frac{4}{3}, \pm\frac{8}{3}$.

18. There are two variations of sign, so there are 2 or 0 positive real solutions. Replace x with $-x$: $3(-x)^4-2(-x)^3-8(-x)+5$ $=3x^4+2x^3+8x+5=0$. There is no variation of sign, so there are no negative real solutions.

19. There are three variations of sign, so there are 3 or 1 positive real solutions. Replace x with $-x$:
$2(-x)^5-3(-x)^3-5(-x)^2+3(-x)-1$
$=-2x^5+3x^3-5x^2-3x-1=0$. There are two variations of sign, so there are 2 or 0 negative real solutions.

20. There is no variation of sign, so there are no positive real roots. Replace x with $-x$:
$2(-x)^4+6(-x)^2+8=2x^4+6x^2+8=0$.
There is no variation of sign, so there are no negative real roots. Therefore,
$2x^4+6x^2+8=0$ has no real roots.

21. a. $a_0=1$ and $a_n=6$.
p: ±1
q: $\pm1, \pm2, \pm3, \pm6$
The possible rational zeros are
$\pm1, \pm\frac{1}{2}, \pm\frac{1}{3}, \pm\frac{1}{6}$.

b.

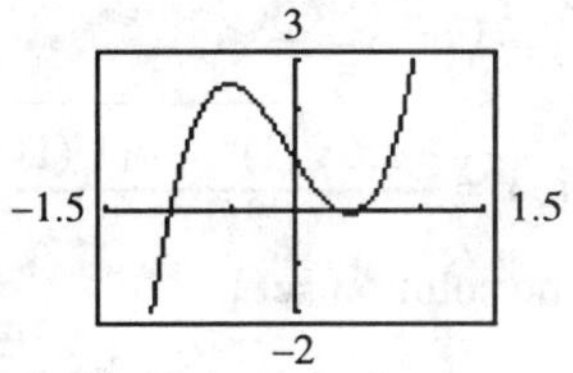

c.

−1	6	1	−4	1
		−6	5	−1
	6	−5	1	0

$f(x)=6x^3+x^2-4x+1$
$=(x+1)(6x^2-5x+1)$
$=(x+1)(3x-1)(2x-1)$
$x=-1$ or $x=\frac{1}{3}$ or $x=\frac{1}{2}$
The solution set is $\left\{-1, \frac{1}{3}, \frac{1}{2}\right\}$.

22. a. $a_0 = -3$ and $a_n = 2$.
$p: \pm 1, \pm 3$
$q: \pm 1, \pm 2$
The possible rational zeros are
$\pm 1, \pm 3, \pm \frac{1}{2}, \pm \frac{3}{2}$.

b.

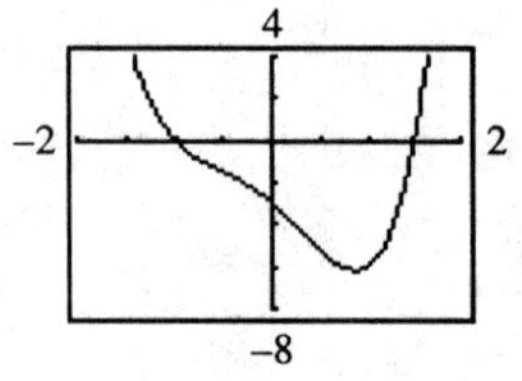

c.

−1	2	1	−2	−4	−3
		−2	1	1	3
	2	−1	−1	−3	0

$f(x) = 2x^4 + x^3 - 2x^2 - 4x - 3$
$= (x+1)(2x^3 - x^2 - x - 3)$
$= 2(x+1)\left(x - \frac{3}{2}\right)(x^2 + x + 1)$
$x = -1$ or $x = \frac{3}{2}$
or $x = \frac{-1 \pm \sqrt{(1)^2 - 4(1)(1)}}{2(1)} = \frac{-1 \pm \sqrt{3}\,i}{2}$
The solution set is
$\left\{-1, \frac{3}{2}, \frac{-1-\sqrt{3}\,i}{2}, \frac{-1+\sqrt{3}\,i}{2}\right\}$.

23. a. $\frac{p}{q}: \pm 1, \pm 2, \pm 4$

b. One positive and two or no negative roots

c.

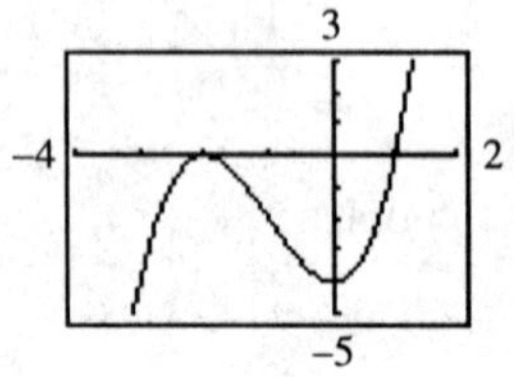

−2	1	3	0	−4
		−2	−2	4
	1	1	−2	0

d. $x^3 + 3x^2 - 4 = (x+2)(x^2 + x - 2) = 0$

e. $x + 2 = 0$ or $x^2 + x - 2 = 0$

f. $x = -2$
$(x + 2)(x - 1) = 0$
$x = -2$ or $x = 1$
The solution set is $\{-2, 1\}$.

24. a. $\frac{p}{q}: \pm 1, \pm 2, \pm 3, \pm 6, \pm \frac{1}{3}, \pm \frac{2}{3}$

b. Two or no positive and one negative real root

c.

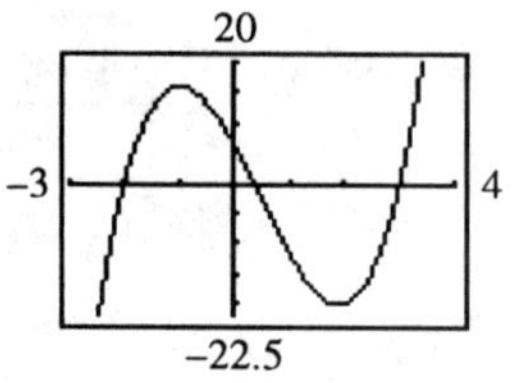

−2	3	−4	−17	−6
		−6	20	−6
	3	−10	3	0

d. $3x^3 - 4x^2 - 17x + 6$
$= (x+2)(3x^2 - 10x + 3) = 0$

e. $x + 2 = 0$ or $3x^2 - 10x + 3 = 0$

f. $x = -2$
$(3x - 1)(x - 3) = 0$
$x = \frac{1}{3}$ or $x = 3$
The solution set is $\left\{-2, \frac{1}{3}, 3\right\}$.

25. a. $\frac{p}{q}: \pm 1, \pm 2, \pm 4, \pm \frac{1}{2}$

b. Two or no positive and one negative real root

c.

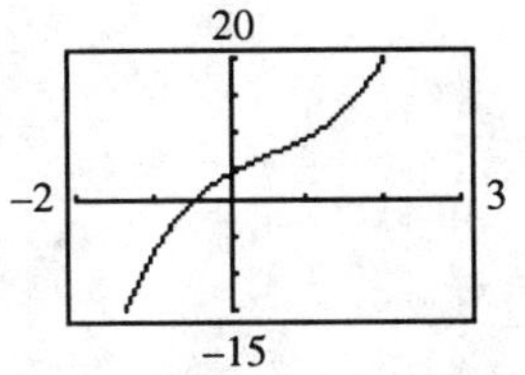

$-\frac{1}{2}$	2	–3	6	4
		–1	2	–4
	2	–4	8	0

d. $2x^3-3x^2+6x+4$
$=\left(x+\frac{1}{2}\right)(2x^2-4x+8)=0$

e. $x+\frac{1}{2}=0$ or $2x^2-4x+8=0$

f. $x=-\frac{1}{2}$
$2(x^2-2x+4)=0$
$x=\frac{2\pm\sqrt{(-2)^2-4(1)(4)}}{2(1)}=\frac{2\pm\sqrt{-12}}{2}$
$=1\pm\sqrt{3}\,i$
The solution set is
$\left\{-\frac{1}{2},\ 1-\sqrt{3}\,i,\ 1+\sqrt{3}\,i\right\}.$

26. a. $\frac{p}{q}$: ±1, ±2, ±4, ±5, ±8, ±10, ±20, ±40

b. Three or one positive and one negative real root

c.

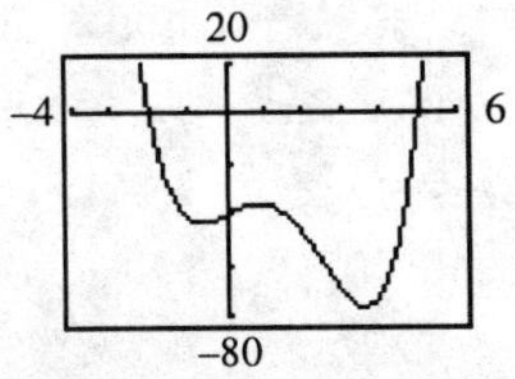

–2	1	–5	0	8	–40
		–2	14	–28	40
	1	–7	14	–20	0

d. $x^4-5x^3+8x-40$
$=(x+2)(x^3-7x^2+14x-20)$

e. $x+2=0$ or $x^3-7x^2+14x-20=0$

f. $x=-2$

5	1	–7	14	–20
		5	–10	20
	1	–2	4	0

$(x-5)(x^2-2x+4)=0$
$x=5$ or $x=\frac{2\pm\sqrt{(-2)^2-4(1)(4)}}{2(1)}$
$=\frac{2\pm\sqrt{-12}}{2}=1\pm\sqrt{3}\,i$
The solution set is
$\left\{-2,\ 5,\ 1-\sqrt{3}\,i,\ 1+\sqrt{3}\,i\right\}.$

27. a. $\frac{p}{q}$: ±1, ±2, $\pm\frac{1}{3}$, $\pm\frac{2}{3}$

b. Two or no positive and one negative real root

c.

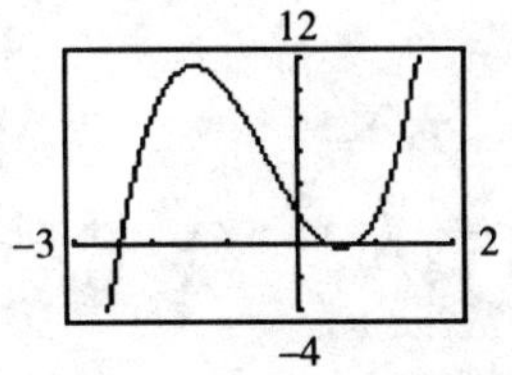

$\frac{2}{3}$	3	4	–7	2
		2	4	–2
	3	6	–3	0

d. $3x^3+4x^2-7x+2$
$=\left(x-\frac{2}{3}\right)(3x^2+6x-3)=0$

e. $x-\frac{2}{3}=0$ or $3x^2+6x-3=0$

f. $x=\frac{2}{3}$
$3(x^2+2x-1)=0$
$x=\frac{-2\pm\sqrt{(2)^2-4(1)(-1)}}{2(1)}=\frac{-2\pm\sqrt{8}}{2}$
$=-1\pm\sqrt{2}$
The solution set is
$\left\{\frac{2}{3},\ -1-\sqrt{2},\ -1+\sqrt{2}\right\}$.

28. a. $\frac{p}{q}: \pm 1, \pm 2, \pm\frac{1}{2}, \pm\frac{1}{4}$

b. One positive and one negative real root

c.

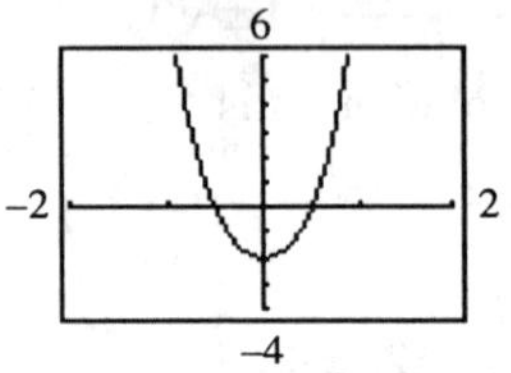

$\frac{1}{2}$	4	0	7	0	-2
		2	1	4	2
	4	2	8	4	0

d. $4x^4+7x^2-2$
$=\left(x-\frac{1}{2}\right)(4x^3+2x^2+8x+4)=0$

e. $x-\frac{1}{2}=0$ or $4x^3+2x^2+8x+4=0$

f. $x=\frac{1}{2}$

$-\frac{1}{2}$	4	2	8	4
		-2	0	-4
	4	0	8	0

$\left(x+\frac{1}{2}\right)(4x^2+8)=0$
$x=-\frac{1}{2}$
$4(x^2+2)=0$
$x=\frac{0\pm\sqrt{(0)^2-4(1)(2)}}{2(1)}=\frac{\pm\sqrt{-8}}{2}$
$=\pm\sqrt{2}\,i$
The solution set is
$\left\{-\frac{1}{2},\ \frac{1}{2},\ -\sqrt{2}\,i,\ \sqrt{2}\,i\right\}$.

29. a. $\frac{p}{q}: \pm 1, \pm 2, \pm 4, \pm\frac{1}{2}$

b. Two or no positive and two or no negative real roots

c.

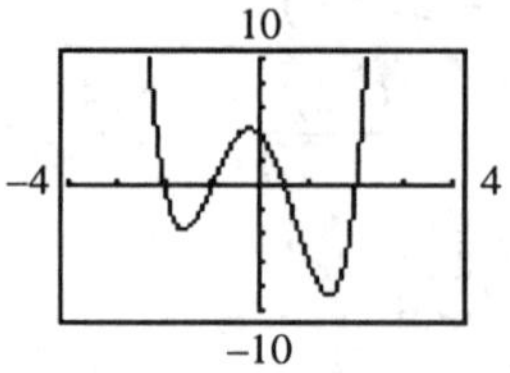

-2	2	1	-9	-4	4
		-4	6	6	-4
	2	-3	-3	2	0

d. $2x^4+x^3-9x^2-4x+4$
$=(x+2)(2x^3-3x^2-3x+2)=0$

e. $x+2=0$ or $2x^3-3x^2-3x+2=0$

f. $x=-2$

−1	2	−3	−3	2
		−2	5	−2
	2	−5	2	0

$(x+1)(2x^2-5x+2)=0$
$(x+1)(2x-1)(x-2)=0$
$x=-1$ or $x=\frac{1}{2}$ or $x=2$
The solution set is $\left\{-2, -1, \frac{1}{2}, 2\right\}$.

30. a. $\frac{p}{q}$: $\pm1, \pm2, \pm4, \pm8, \pm\frac{1}{3}, \pm\frac{2}{3}, \pm\frac{4}{3}, \pm\frac{8}{3}$

b. Three or one positive and two or no negative real roots

c.

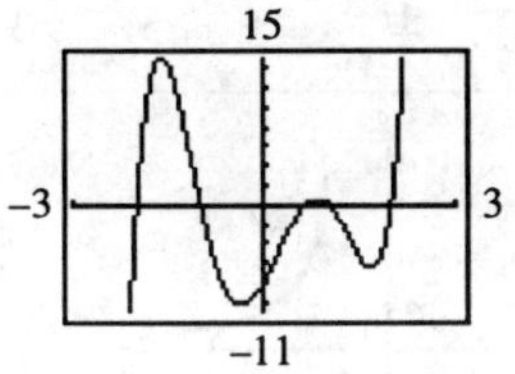

−2	3	−2	−15	10	12	−8
		−6	16	−2	−16	8
	3	−8	1	8	−4	0

d. $3x^5-2x^4-15x^3+10x^2+12x-8$
$=(x+2)(3x^4-8x^3+x^2+8x-4)=0$

e. $x+2=0$ or
$3x^4-8x^3+x^2+8x-4=0$

f. $x=-2$

−1	3	−8	1	8	−4
		−3	11	−12	4
	3	−11	12	−4	0

$x^4-8x^3+x^2+8x-4$
$=(x+1)(3x^3-11x^2+12x-4)$
$x+1=0$ or
$3x^3-11x^2+12-4=0$
$x=-1$

2	3	−11	12	−4
		6	−10	4
	3	−5	2	0

$(x-2)(3x^2-5x+2)=0$
$x=2$
$(3x-2)(x-1)=0$
$x=\frac{2}{3}$ or $x=1$
The solution set is $\left\{-2, -1, \frac{2}{3}, 1, 2\right\}$.

31. $V(h)=(h+7)(2h)(h)=2h^3+14h^2=72$
$2h^3+14h^2-72=0$
$2(h^3+7h^2-36)=0$

2	1	7	0	−36
		2	18	36
	1	9	18	0

$2(h-2)(h^2+9h+18)=0$
$2(h-2)(h+6)(h+3)=0$
$h=2$ or $h=-6$ or $h=-3$
h must be positive so $h=2$, $h+7=9$, and $2h=4$. The dimensions of the box are 2 inches by 9 inches by 4 inches.

32. Let $N=211$.
$211=14W^3-17W^2-16W+34$
$14W^3-17W^2-16W-177=0$

3	14	−17	−16	−177
		42	75	177
	14	25	59	0

$(W-3)(14W^2+25W+59)=0$

$W=3$ or $W=\dfrac{-25\pm\sqrt{(25)^2-4(14)(59)}}{2(14)}$

$=\dfrac{-25\pm\sqrt{2679}\,i}{28}$

W must be real such that $1.5 \le W \le 3.5$, so $W=3$.

The abdominal width is 3 millimeters when there are 211 eggs.

33. Volume of figure = volume of entire solid – volume of removed pieces

$164=(x+3)(x)(2x-1)-(x)(x)(x-2)$

$=(x^2+3x)(2x-1)-(x^3-2x^2)$

$=2x^3-x^2+6x^2-3x-x^3+2x^2$

$=x^3+7x^2-3x$

$x^3+7x^2-3x-164=0$

4	1	7	−3	−164
		4	44	164
	1	11	41	0

$x^3+7x^2-3x-164$

$=(x-4)(x^2+11x+41)$

$x=4$ or $x^2+11x+41=0$

$\dfrac{-11\pm\sqrt{(11)^2-4(1)(41)}}{2(1)}=\dfrac{-11\pm i\sqrt{43}}{2}$

Since x must be real, $x=4$ cm.

34. a. Since –6 is not in the interval (–6, –3], this statement is false.

b. $f(-6)=0$ and $f(-3)<0$ so $f(-6)-f(-3)>0$; false

c. $f(-2)=0$ so $f[f(-2)]=f(0)$ and $0<f(0)<2$ is true.

d. There are 4 turning points shown so f must be at least of degree 5; false

35. a. $f(-x)=\dfrac{-2x+1}{-x-1}$; there is no y-axis or origin symmetry.

b. $f(0)=\dfrac{2(0)+1}{(0)-1}=-1$; the y-intercept is –1.

c. $P(x)=2x+1=0$; $x=-\frac{1}{2}$ is the x-intercept.

d. $q(x)=x-1=0$; $x=1$ is a vertical asymptote.

e. $y=\frac{2}{1}=2$ is a horizontal asymptote.

f.

x	−1	$\frac{1}{2}$	2	4
$f(x)$	$\frac{1}{2}$	−4	5	3

g.

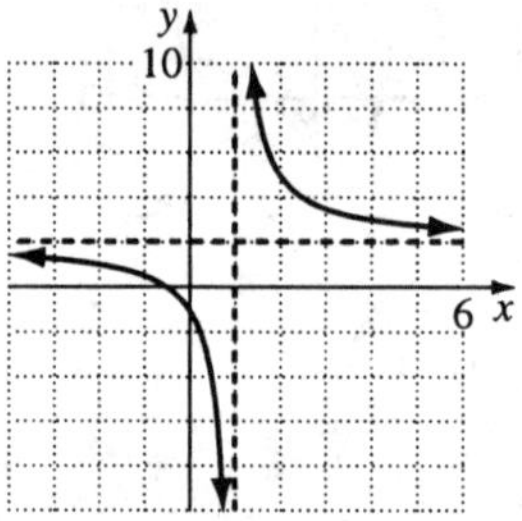

36. a. $f(-x)=\dfrac{-2x-3}{x^2-9}$; there is no y-axis or origin symmetry.

b. $f(0)=\dfrac{2(0)-3}{(0)^2-9}=\dfrac{1}{3}$; the y-intercept is $\frac{1}{3}$.

c. $p(x)=2x-3=0$; $x=\dfrac{3}{2}$ is the x-intercept.

d. $q(x)=x^2-9=0$; $x=\pm3$ are vertical asymptotes.

e. x-axis is a horizontal asymptote.

f.

x	-4	-2	2	4
$f(x)$	$-\frac{11}{7}$	$\frac{7}{5}$	$-\frac{1}{5}$	$\frac{5}{7}$

g.

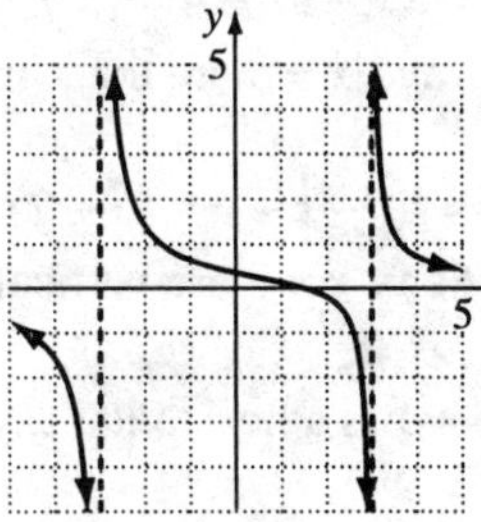

37. a. $f(-x)=\dfrac{4x^2}{x^2-1}=f(x);$
there is y-axis symmetry.

b. $f(0)=\dfrac{4(0)^2}{(0)^2-1}=0;$
the y-intercept is 0.

c. $p(x)=4x^2=0;\ x=0$ is the x-intercept.

d. $q(x)=x^2-1=0;$
$x=\pm1$ are vertical asymptotes.

e. $y=\frac{4}{1}=4$ is a horizontal asymptote.

f.

x	-4	-2	$-\frac{1}{2}$
$f(x)$	$\frac{64}{15}$	$\frac{16}{3}$	$-\frac{4}{3}$

g.

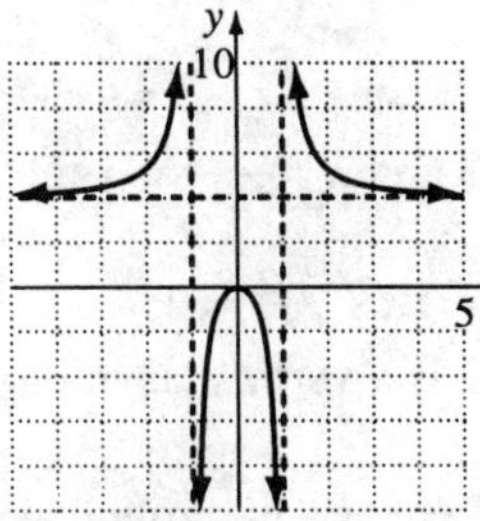

38. a. $f(-x)=\dfrac{x^2}{x^2+1}=f(x);$
there is y-axis symmetry.

b. $f(0)=\dfrac{(0)^2}{(0)^2+1}=0;$
the y-intercept is 0.

c. $p(x)=x^2=0;$
$x=0$ is the x-intercept.

d. $q(x)=x^2+1>0;$
there are no vertical asymptotes.

e. $y=\frac{1}{1}=1$ is a horizontal asymptote.

f.

x	-2	-1
$f(x)$	$\frac{4}{5}$	$\frac{1}{2}$

g.

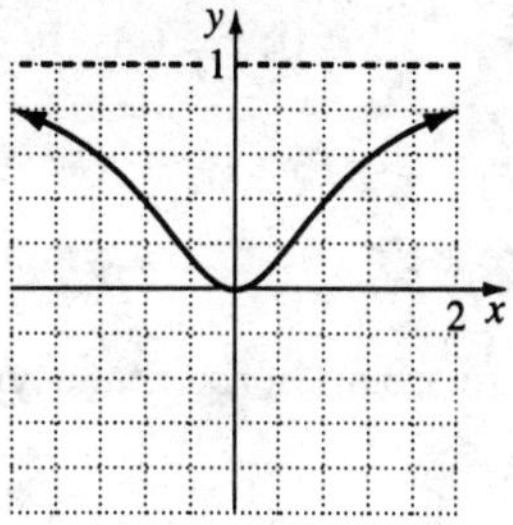

39. a. $f(-x)=\dfrac{-x-2}{x^2+2x+3};$
there is no y-axis or origin symmetry.

b. $f(0)=\dfrac{0-2}{0^2-2(0)+3}=-\dfrac{2}{3};$
the y-intercept is $-\frac{2}{3}$.

c. $p(x)=x-2=0;\ x=2$ is the x-intercept.

d. $q(x)=x^2-2x+3=0;\ x=1\pm\sqrt{2}\,i;$
there is no vertical asymptote.

e. The x-axis is a horizontal asymptote

f.

x	-4	-2	1	4	10
$f(x)$	$-\frac{2}{9}$	$-\frac{4}{11}$	$-\frac{1}{2}$	$\frac{2}{11}$	$\frac{8}{83}$

g.

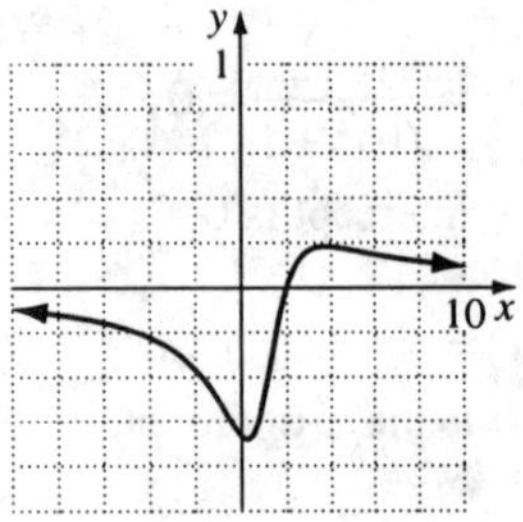

40. a. $f(-x)=\dfrac{x^4}{x^2+2}=f(x)$;
there is y-axis symmetry.

b. $f(0)=\dfrac{(0)^4}{(0)^2+2}=0$;
y-intercept is 0.

c. $p(x)=x^4=0$; $x=0$ is the x-intercept.

d. $q(x)=x^2+2>0$;
there are no vertical asymptotes.

e. There are no horizontal asymptotes.

f.

x	-4	-2
$f(x)$	$\frac{128}{9}$	$\frac{8}{3}$

g.

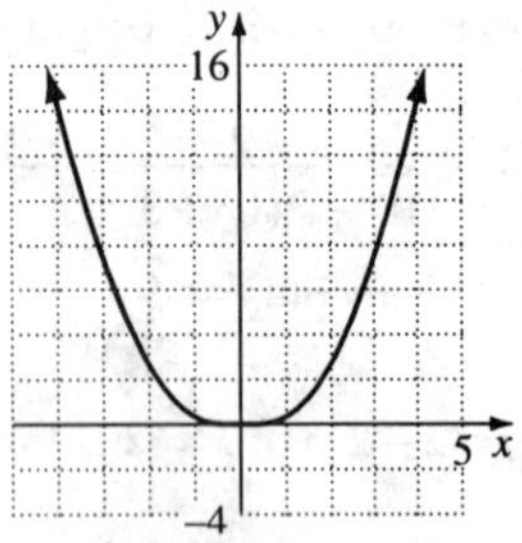

41. a. $f(-x)=\dfrac{x^2+3x-4}{x^2+x-6}$;
there is no y-axis or origin symmetry.

b. $f(0)=\dfrac{(0)^2-3(0)-4}{(0)^2-(0)-6}=\dfrac{2}{3}$;
the y-intercept is $\dfrac{2}{3}$.

c. $p(x)=x^2-3x-4=(x+1)(x-4)=0$;
$x=-1$ and $x=4$ are the x-intercepts.

d. $q(x)=x^2-x-6=(x+2)(x-3)=0$;
$x=-2$ and $x=3$ are vertical asymptotes.

e. $y=\frac{1}{1}=1$ is a horizontal asymptote.

f.

x	-4	-3	1	2	5
$f(x)$	$\frac{12}{7}$	$\frac{7}{3}$	1	$\frac{3}{2}$	$\frac{3}{7}$

g.

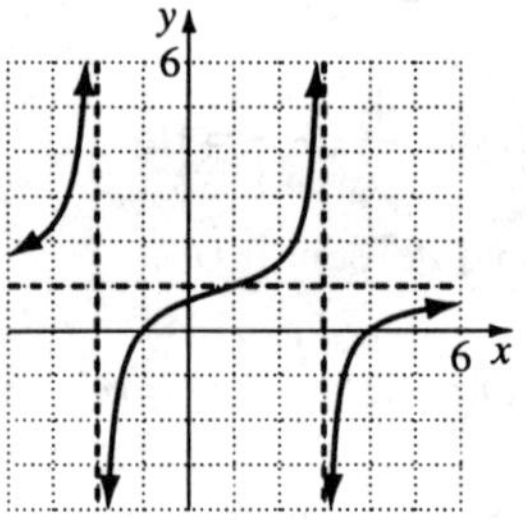

42. a. $f(-x)=\dfrac{x^2-4x+3}{(-x+2)^2}$;
there is no y-axis or origin symmetry.

b. $f(0)=\dfrac{(0)^2+4(0)+3}{(0+2)^2}=\dfrac{3}{4}$;
the y-intercept is $\dfrac{3}{4}$.

c. $p(x)=x^2+4x+3=(x+3)(x+1)=0$;
$x=-3$ and $x=-1$ are x-intercepts.

d. $q(x)=(x+2)^2=0$;
$x=-2$ is a vertical asymptote.

e. $y=\frac{1}{1}=1$ is a horizontal asymptote.

f.

x	-6	$-\frac{5}{2}$	$\frac{1}{2}$	2
$f(x)$	$\frac{15}{16}$	-3	$\frac{21}{25}$	$\frac{15}{16}$

g.

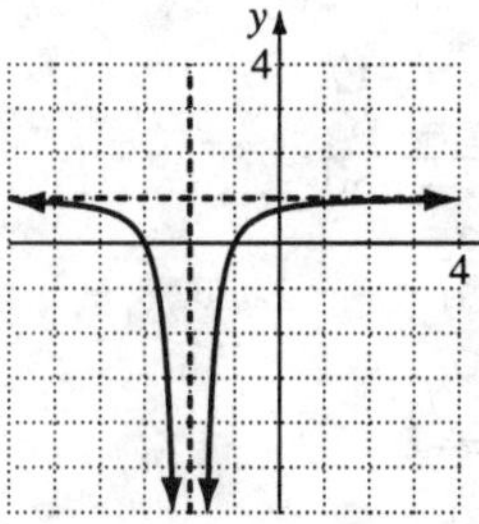

43. a. $\overline{C}(50)=\dfrac{25(50)+50,000}{50}=1025;$
the average cost per calculator is \$1025 when only 50 calculators are manufactured.

$\overline{C}(100)=\dfrac{25(100)+50,000}{100}=525;$
the average cost per calculator is \$525 when 100 calculators are manufactured.

$\overline{C}(1000)=\dfrac{25(1000)+50,000}{1000}=75;$
the average cost per calculator is \$75 when 1000 calculators are manufactured.

$\overline{C}(100,000)=\dfrac{25(100,000)+50,000}{100,000}$
$= 25.5$; the average cost per calculator is \$25.50 when 100,000 calculators are manufactured.

b. The horizontal asymptote is $y = 25$. The average cost per calculator approaches \$25 for large numbers of calculators manufactured. The cost per calculator is at least \$25 no matter how many are manufactured.

44. a. $C(90)=\dfrac{200(90)}{100-90}=1800$
$C(50)=\dfrac{200(50)}{100-50}=200$
$C(90)-C(50)=1800-200=1600$
The cost for removing an additional 40% of the contaminants after the first 50% are removed is \$16,000,000.

b. $y = -200$ is the horizontal asymptote. This asymptote is not relevant to this problem since costs (y-values) are positive when percentages (x-values) are between 0 and 100.

45. $N(x)=\dfrac{60(x-2)}{x}=\dfrac{60x-120}{x};$
$y = 60$ is the horizontal asymptote. This model suggests a maximum number of words per minute at 60, no matter how many weeks of practice. A person can approach 60 WPM after several weeks of practice.

46. $F(x)=\dfrac{30(4+5x)}{1+0.05x}=\dfrac{120+150x}{1+0.05x}:$
$y=\dfrac{150}{0.05}=3000$ is the horizontal asymptote. This model suggests the maximum number of fish the lake can support is 3,000,000. It will take a few years (many weeks) to reach this maximum.

47. The cost to seize *all* illegal drugs cannot be modeled with this cost function because $p \neq 100$. The cost for each additional percent seized increases dramatically as p increases (approaches 100%).

Chapters 1–10 Cumulative Review

1. $x^3 - 4x^2 - 10x + 4 = 0$

$$\begin{array}{r|rrrr} -2 & 1 & -4 & -10 & 4 \\ & & -2 & 12 & -4 \\ \hline & 1 & -6 & 2 & 0 \end{array}$$

$(x+2)(x^2 - 6x + 2) = 0$

$x = -2$ or $x = \dfrac{6 \pm \sqrt{(-6)^2 - 4(1)(2)}}{2}$

$x = \dfrac{6 \pm \sqrt{28}}{2}$

$x = \dfrac{6 \pm 2\sqrt{7}}{2}$

$x = 3 \pm \sqrt{7}$

$\left\{-2,\ 3-\sqrt{7},\ 3+\sqrt{7}\right\}$

2. $\dfrac{x-4}{x+6} \geq 0$

T	F	T

$-6 \quad 4$

Test –7: $\dfrac{-7-4}{-7+6} \geq 0$

$11 \geq 0$ True

Test 0: $\dfrac{0-4}{0+6} \geq 0$

$-\dfrac{2}{3} \geq 0$ False

Test 5: $\dfrac{5-4}{5+6} \geq 0$

$\dfrac{1}{11} \geq 0$ True

$\{x \mid x < -6 \text{ or } x \geq 4\}$, $(-\infty, -6) \cup [4, \infty)$

3. $x + y - z = -2$
$-3x + 2y + 5z = -7$
$2x - 3y + 4z = 17$

$$\begin{array}{r} 5x + 5y - 5z = -10 \\ -3x + 2y + 5z = -7 \\ \hline 2x + 7y = -17 \end{array}$$

$$\begin{array}{r} 4x + 4y - 4z = -8 \\ 2x - 3y + 4z = 17 \\ \hline 6x + y = 9 \end{array}$$

$$\begin{array}{r} 2x + 7y = -17 \\ -42x - 7y = -63 \\ \hline -40x = -80 \\ x = 2 \end{array}$$

$6(2) + y = 9$

$y = -3$

$2 - 3 - z = -2$

$z = 1$

$\{(2, -3, 1)\}$

4. $2x + y = 3$
$y = 2x^2 - 1$

$2x + 2x^2 - 1 = 3$

$2x^2 + 2x - 4 = 0$

$2(x-1)(x+2) = 0$

$x = 1$ or $x = -2$

$y = 2(1)^2 - 1 = 1$ or $y = 2(-2)^2 - 1 = 7$

$\{(1, 1), (-2, 7)\}$

5. $2 + \dfrac{4}{x-2} = \dfrac{8}{x^2 - 2x}$

$(x^2 - 2x)\left[2 + \dfrac{4}{x-2}\right] = \dfrac{8}{x^2 - 2x} \cdot (x^2 - 2x)$

$2(x^2 - 2x) + 4x = 8$

$2x^2 - 4x + 4x - 8 = 0$

$2x^2 - 8 = 0$

$2(x-2)(x+2) = 0$

$x = 2$ (extraneous) or $x = -2$

$\{-2\}$

6. $\sqrt{x} + 6 = x$

$\sqrt{x} = x - 6$

$x = x^2 - 12x + 36$

$0 = x^2 - 13x + 36$

$0 = (x-4)(x-9)$

$x = 4$ (extraneous) or $x = 9$

$\{9\}$

7. $x^{2/3}+3x^{1/3}-18=0$

Let $u=x^{1/3}$.

$u^2+3u-18=0$

$(u+6)(u-3)=0$

$u=-6$ or $u=3$

$x^{1/3}=-6$ or $x^{1/3}=3$

$x=-216$ or $x=27$

$\{-216, 27\}$

8. $3x^2+2y^2=7$

$5x^2-7y^2=-9$

$$\begin{aligned} 21x^2+14y^2&=49\\ 10x^2-14y^2&=-18\\ \hline 31x^2&=31\\ x^2&=1\\ x&=\pm1 \end{aligned}$$

If $x=-1$:

$3(-1)^2+2y^2=7$

$y^2=2$

$y=\pm\sqrt{2}$

If $x=1$:

$3(1)^2+2y^2=7$

$y=\pm\sqrt{2}$

$\left\{\left(-1, -\sqrt{2}\right), \left(-1, \sqrt{2}\right), \left(1, -\sqrt{2}\right), \left(1, \sqrt{2}\right)\right\}$

9. $27^x=3^{2x+5}$

$3^{3x}=3^{2x+5}$

$3x=2x+5$

$x=5$

$\{5\}$

10. $\log_2 x+\log_2 3=1$

$\log_2 3x=1$

$3x=2^1$

$x=\frac{2}{3}$

$\left\{\frac{2}{3}\right\}$

11. $\dfrac{1-\frac{8}{x^2-1}}{\frac{2}{x+1}-\frac{1}{x-1}}=\dfrac{\frac{x^2-1-8}{x^2-1}}{\frac{2(x-1)-(x+1)}{x^2-1}}$

$=\dfrac{(x-3)(x+3)}{x^2-1}\cdot\dfrac{x^2-1}{x-3}=x+3$

12. $\dfrac{x-5}{x^2-x-2}-\dfrac{2}{x+1}+\dfrac{5}{x-2}=\dfrac{x-5-2(x-2)+5(x+1)}{(x-2)(x+1)}$

$=\dfrac{x-5-2x+4+5x+5}{(x-2)(x+1)}=\dfrac{4x+4}{(x-2)(x+1)}$

$=\dfrac{4}{x-2}$

13. $\dfrac{5i}{3+4i}=\dfrac{5i}{3+4i}\cdot\dfrac{3-4i}{3-4i}$

$=\dfrac{15i-20i^2}{9-16i^2}=\dfrac{15i+20}{9+16}=\dfrac{5(3i+4)}{25}$

$=\dfrac{3i+4}{5}=\dfrac{4}{5}+\dfrac{3}{5}i$

14. $\frac{9x}{\sqrt[3]{3x^2}} = \frac{9x}{\sqrt[3]{3x^2}} \cdot \frac{\sqrt[3]{9x}}{\sqrt[3]{9x}}$

$= \frac{9x\sqrt[3]{9x}}{\sqrt[3]{(3x)^3}} = \frac{9x\sqrt[3]{9x}}{3x}$

$= 3\sqrt[3]{9x}$

15. $\left(\frac{-4x^5y^7}{2xy^9}\right)^{-3} = \left(\frac{2xy^9}{-4x^5y^7}\right)^3$

$= \frac{2^3x^3y^{27}}{(-4)^3x^{15}y^{21}} = \frac{8y^{27-21}}{-64x^{15-3}} = -\frac{y^6}{8x^{12}}$

16. $y = \log_2 x$

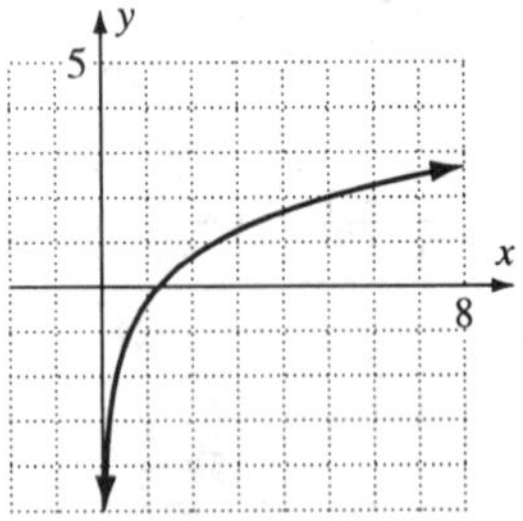

17. $\frac{x^2}{16} - \frac{y^2}{9} = 1$

hyperbola

x-intercepts at ± 4:

$\frac{x^2}{16} - \frac{0^2}{9} = 1,\ x^2 = 16,\ x = \pm 4$

no y-intercepts:

$\frac{0^2}{16} - \frac{y^2}{9} = 1,\ -\frac{y^2}{9} = 1,\ y^2 = -9$

asymptotes at $y = \pm\frac{3}{4}x$

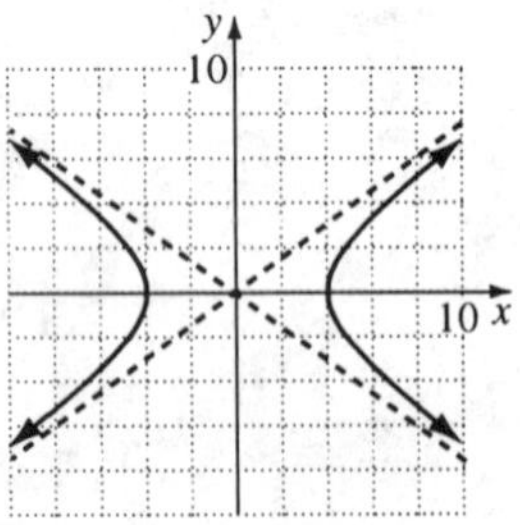

18. $3x - 2y > -6$

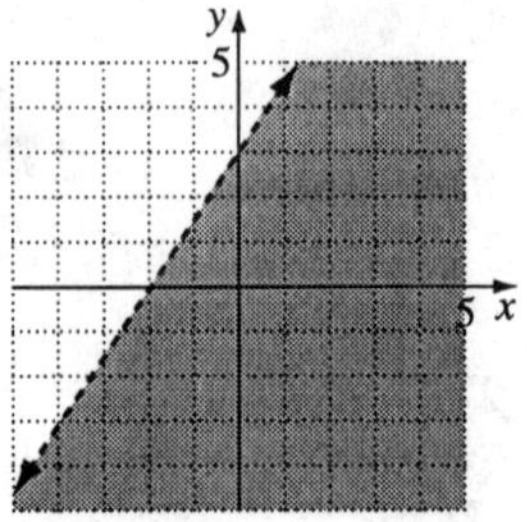

19. $a_nx^n = 1x^3$; a_n is positive and n is odd so the graph of f falls to the left and rises to the right.

$f(x) = x^3 - 8x^2 + 19x - 12 = 0$

$(x-1)(x-3)(x-4) = 0$

$x = 1$ or $x = 3$ or $x = 4$

The graph of f crosses the x-axis at $x = 1$, $x = 3$, and $x = 4$.

$f(0) = (0)^3 - 8(0)^2 + 19(0) - 12 = -12$; the y-intercept is at -12.

$f(-x) = -x^3 - 8x^2 - 19x - 12$; no y-axis or origin symmetry.

$n - 1 = 2$ so there are at most 2 turning points.

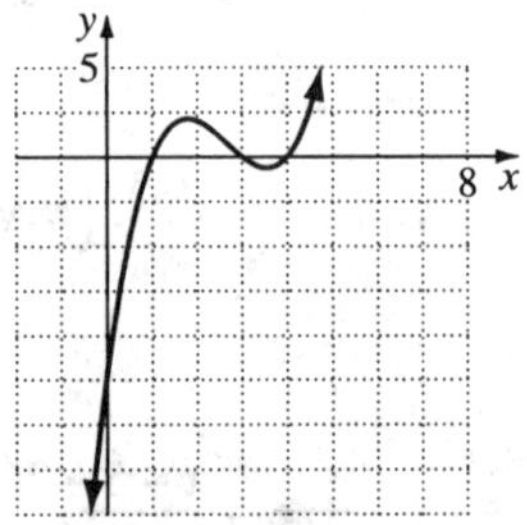

20. $f(x) = \frac{2x}{x+1}$

$f(-x) = \frac{-2x}{-x+1}$; no y-axis or origin symmetry.

$f(0) = \frac{2(0)}{0+1} = 0$; y-intercept is 0.

$p(x) = 2x = 0$; x-intercept is 0.

$q(x) = x + 1 = 0$; $x = -1$ is a vertical asymptote.

$y = 2$ is a horizontal asymptote.

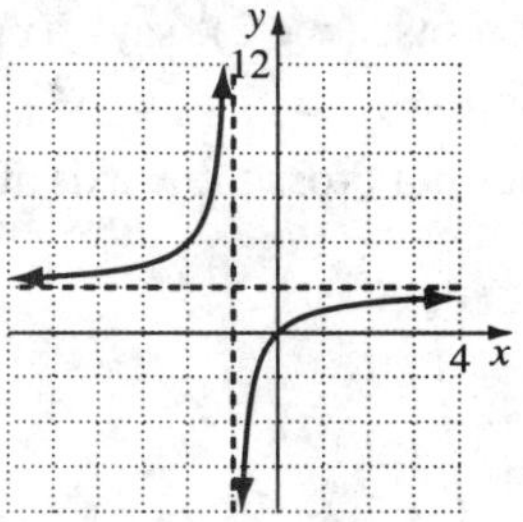

21. $y = Ae^{kt}$

$y = 24e^{0.08(372)}$

$y = 2.0175 \times 10^{14}$

The value is about $201,750,000,000,000.

22. Half-life means that 50% of its original carbon-14 is remaining. 50% of original carbon-14 remaining after 5750 years. The animal bone is 5750 years old.

23. Let x = side of square.
Area of square = x^2
Area of rectangle = $(x + 6)(x - 4)$

$x^2 = (x+6)(x-4)$

$x^2 = x^2 + 2x - 24$

$24 = 2x$

$12 = x$

$18 = x + 6$

$8 = x - 4$

The dimensions are as follows:
square: 12 cm by 12 cm
rectangle: 18 cm by 8 cm

24. Let x = length of one leg, then
$2x + 2$ = length of other leg.

$13^2 = x^2 + (2x+2)^2$

$169 = x^2 + 4x^2 + 8x + 4$

$0 = 5x^2 + 8x - 165$

$x = \dfrac{-8 \pm \sqrt{(8)^2 - 4(5)(-165)}}{2(5)}$

$x = \dfrac{-8 \pm 58}{10}$

$x = -6.6$ (disregard) or $x = 5$

$2x + 2 = 12$

The lengths of the legs are 5 in. and 12 in.

25. Let
x = length of each side of square painting.
Area of picture = Total area – Area of frame

$x^2 = (x+2)^2 - 28$

$x^2 = x^2 + 4x + 4 - 28$

$24 = 4x$

$6 = x$

The picture measures 6 in. by 6 in.

26. Let n = number of hours of sleep, c = cups of coffee, and k = constant.

$n = \dfrac{k}{c^2}$

$8 = \dfrac{k}{2^2}$

$k = 32$

$n = \dfrac{32}{c^2}$

$n = \dfrac{32}{4^2}$

$n = 2$

She should expect to get 2 hours of sleep.

27. $3x^3 + 6x^2 - 3x - 6 = 3x^2(x+2) - 3(x+2)$
$= (3x^2 - 3)(x+2) = 3(x^2 - 1)(x+2)$
$= 3(x + 1)(x - 1)(x + 2)$

28. $\begin{vmatrix} 4 & -3 \\ 5 & 7 \end{vmatrix} = 4(7) - 5(-3) = 28 + 15 = 43$

29. $x - 3y = 18$

$-3y = -x + 18$

$y = \frac{1}{3}x - 6$

Perpendicular slope = –3
point (–2, 5)
Point-slope form: $y - 5 = -3(x + 2)$
Slope-intercept form:

$y = -3x - 6 + 5$

$y = -3x - 1$

30. Let x = cost of one bath towel and y = cost of one hand towel.
$3x+2y=31$
$2x+5y=39$

$$\begin{aligned} -6x-4y&=-62\\ 6x+15y&=117\\ \hline 11y&=55\\ y&=5 \end{aligned}$$

$3x+2(5)=31$
$x=7$
$\{(7, 5)\}$

Chapter 10 Test

1. $a_nx^n=-x^7$. Because the coefficient is negative and the degree is odd, the curve rises to the left and falls to the right.

2. $x^3+5x^2-9x-45=0$
$x^2(x+5)-9(x+5)=0$
$(x^2-9)(x+5)=0$
$(x+3)(x-3)(x+5)=0$
$x=-3, 3, -5$

3. $f(x)=5x^7-x^3$
$f(-x)=-5x^7+x^3=-f(x)$
f is symmetric about the origin.

4. $f(x)=x^3-2x^2-4x+8$
$a_nx^n=x^3$, so curve rises to the right and falls to the left.
$f(x)=x^2(x-2)-4(x-2)$
$=(x^2-4)(x-2)=(x+2)(x-2)^2$
The x-axis intercepts are at 2 and –2.
$f(0)=8$ so the y-intercept is at 8.
Since $x^n=x^3$, there are at most $n-1=3-1=2$ turning points.

Multiplicity of roots: $(x+2)^1$ says $f(x)$ crosses the x axis at $x=-2$; $(x-2)^2$ says $f(x)$ touches but does not cross the x axis at $x=2$.

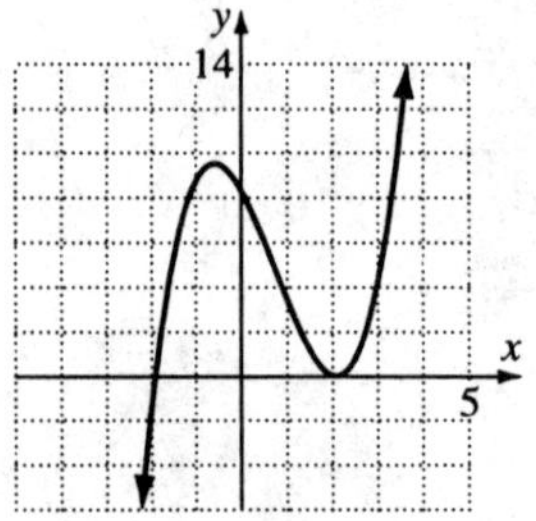

5. $f(x)=x^4-3x^3-4x^2+13x-8$

3	1	–3	–4	13	–8
		3	0	–12	3
	1	0	–4	1	–5

$f(3)=-5$

6. $f(x)=2x^3-11x^2+13x-4$

4	2	–11	13	–4
		8	–12	4
	2	–3	1	0

$f(x)=(x-4)(2x^2-3x+1)$
$=(x-4)(2x-1)(x-1)$
So $f(x)=0$ when $x=4, \frac{1}{2}$, or 1

7. $f(x)=2x^5+x^4+x^3+13x^2+5x-6$
$a_0=-6,\ a_n=2$
$p(x): \pm1, \pm2, \pm3, \pm6$
$q(x): \pm1, \pm2$
Possible rational zeros are
$p(x)/q(x)=\pm1, \pm2, \pm3, \pm6, \pm\frac{1}{2}, \pm\frac{3}{2}$

8. $f(x) = x^4 + 7x^3 + 17x^2 + 17x + 6$
No sign changes; no positive real roots
$f(-x) = x^4 - 7x^3 + 17x^2 - 17x + 6$
4 sign changes; there are 4, 2, or 0 negative real roots.

9. $6x^3 + x^2 - 9x - 4 = 0$

$p(x)/q(x) = \pm 1, \pm 2, \pm 4, \pm\frac{1}{2}, \pm\frac{1}{3}, \pm\frac{2}{3}, \pm\frac{4}{3}, \pm\frac{1}{6}$

$f(x)$ has one sign change, so there is exactly 1 positive real root. $f(-x) = -6x^3 + x^2 + 9x - 4$ has two sign changes, so there are 2 or 0 negative real solutions.

−1	6	1	−9	−4
		−6	5	+4
	6	−5	−4	0

$f(x) = (x+1)(6x^2 - 5x - 4)$

$x = \frac{+5 \pm \sqrt{(-5)^2 - 4(6)(-4)}}{2(6)} = \frac{5 \pm \sqrt{121}}{12} = \frac{5 \pm 11}{12} = \frac{-1}{2} \text{ or } \frac{4}{3}$

$f(x) = (x+1)\left(x + \frac{1}{2}\right)\left(x - \frac{4}{3}\right)$

$f(x) = 0$ when $x = -1, -\frac{1}{2}, \frac{4}{3}$

10. $v = lwh$
$x = \text{height}$
$40 = (2x+1)(2x)(x) = (2x+1)(2x^2) = 4x^3 + 2x^2$
$0 = 4x^3 + 2x^2 - 40 = 2(2x^3 + x^2 - 20)$

2	2	1	0	−20
		4	10	+20
	2	5	10	0

$0 = 2(x-2)(2x^2 + 5x + 10)$

$x = 2$ or $x = \frac{-5 \pm \sqrt{5^2 - 4(2)(10)}}{2(2)} = \frac{-5 \pm i\sqrt{55}}{4}$

Since x must be real, $x = 2$ and the box is 2 in. by 4 in. by 5 in.

11. $f(x)=\dfrac{4x-2}{x-1}$ $f(-x)=\dfrac{-4x-2}{-x-1}\neq f(x)$ or $-f(x)$

no y axis or origin symmetry

$f(0)=\dfrac{4(0)-2}{0-1}=2$ y-intercept is 2

$f(x)=0$ when $p(x)=4x-2=0$; $x=\dfrac{1}{2}$

$q(x)=0$ when $x-1=0$; $x=1$

$x=1$ is a vertical asymptote

horizontal asymptote is $y=\dfrac{4}{1}=4$

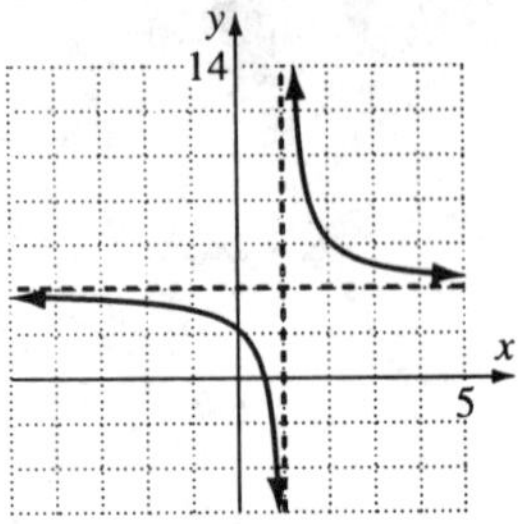

12. $f(x)=\dfrac{x^2-1}{x^2-4}$ $f(-x)=\dfrac{x^2-1}{x^2-4}=f(x)$ **y axis symmetry**

$f(0)=\dfrac{1}{4}$ y-intercept is $\dfrac{1}{4}$

$f(x)=0$ when $p(x)=x^2-1=(x+1)(x-1)$ is zero $x=1,-1$ are x-intercepts

$q(x)=0$ when $x^2-4=0$, i.e. when $x=\pm 2$

$x=-2, 2$ are vertical asymptotes

horizontal asymptote is $y=\dfrac{1}{1}=1$

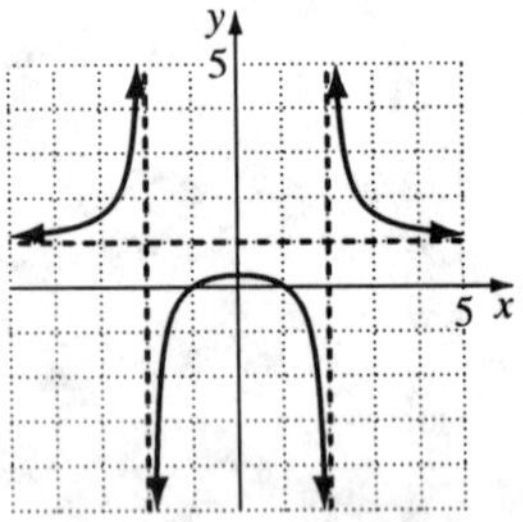

13. a. $\bar{c}(10{,}000)=\dfrac{400(10{,}000)+50{,}000}{10{,}000}=405$

The average cost per wheelchair is \$405 when 10,000 wheelchairs are produced. The overhead cost of \$50,000 is negligible (per chair) when 10,000 are produced.

b. The horizontal asymptote is 400, meaning that the variable cost of producing a wheelchair (labor, materials, overhead) averages $400 per chair regardless of production levels.

14. a. The function is not linear. False

b. At $500,000, approximately 85% of the pollution is eliminated. False

c. As the percentage approaches 100%, the cost increases without bound. False

d. True

15. a. 50 deer

b. $\approx$175 deer

c. The horizontal asymptote is approximately $y = 225$. This is the maximum population the habitat can support.

Chapter 11

Problem Set 11.1

1. $a_n = 2n - 1$
$a_1 = 2(1) - 1 = 1$
$a_2 = 2(2) - 1 = 3$
$a_3 = 2(3) - 1 = 5$
$a_4 = 2(4) - 1 = 7$
$a_{10} = 2(10) - 1 = 19$
$a_{15} = 2(15) - 1 = 29$
1, 3, 5, 7; 19; 29

3. $a_n = n^2 + 2$
$a_1 = 1 + 2 = 3$
$a_2 = 4 + 2 = 6$
$a_3 = 9 + 2 = 11$
$a_4 = 16 + 2 = 18$
$a_{10} = 10^2 + 2 = 102$
$a_{15} = 15^2 + 2 = 227$
3, 6, 11, 18; 102; 227

5. $a_n = \frac{n}{n+1}$
$a_1 = \frac{1}{1+1} = \frac{1}{2}$
$a_2 = \frac{2}{2+1} = \frac{2}{3}$
$a_3 = \frac{3}{4}$
$a_4 = \frac{4}{5}$
$a_{10} = \frac{10}{10+1} = \frac{10}{11}$
$a_{15} = \frac{15}{15+1} = \frac{15}{16}$
$\frac{1}{2}, \frac{2}{3}, \frac{3}{4}, \frac{4}{5}; \frac{10}{11}; \frac{15}{16}$

7. $a_n = 1 + \frac{1}{n}$
$a_1 = 1 + \frac{1}{1} = 2$
$a_2 = 1 + \frac{1}{2} = \frac{3}{2}$
$a_3 = 1 + \frac{1}{3} = \frac{4}{3}$
$a_4 = 1 + \frac{1}{4} = \frac{5}{4}$
$a_{10} = 1 + \frac{1}{10} = \frac{11}{10}$
$a_{15} = 1 + \frac{1}{15} = \frac{16}{15}$
$2, \frac{3}{2}, \frac{4}{3}, \frac{5}{4}; \frac{11}{10}; \frac{16}{15}$

9. $a_n = \frac{n-1}{n+2}$
$a_1 = \frac{1-1}{1+2} = 0$
$a_2 = \frac{1}{4}$
$a_3 = \frac{2}{5}$
$a_4 = \frac{3}{6} = \frac{1}{2}$
$a_{10} = \frac{10-1}{10+2} = \frac{9}{12} = \frac{3}{4}$
$a_{15} = \frac{15-1}{15+2} = \frac{14}{17}$
$0, \frac{1}{4}, \frac{2}{5}, \frac{1}{2}; \frac{3}{4}; \frac{14}{17}$

11. $a_n = (-1)^n n$
$a_1 = (-1)^1(1) = -1$
$a_2 = (-1)^2(2) = 2$
$a_3 = (-1)^3(3) = -3$
$a_4 = (-1)^4(4) = 4$
$a_{10} = (-1)^{10}(10) = 10$
$a_{15} = (-1)^{15}(15) = -15$
–1, 2, –3, 4; 10; –15

13. $a_n = (-1)^{n+1} n^2$
$a_1 = (-1)^{1+1}(1)^2 = 1$
$a_2 = (-1)^{2+1}(2)^2 = -4$
$a_3 = (-1)^{3+1}(3)^2 = 9$
$a_4 = (-1)^{4+1}(4)^2 = -16$
$a_{10} = (-1)^{10+1}(10)^2 = -100$
$a_{15} = (-1)^{15+1}(15)^2 = 225$
1, –4, 9, –16; –100; 225

15. $a_n = \left(-\frac{1}{2}\right)^n$

$a_1 = \left(-\frac{1}{2}\right)^1 = -\frac{1}{2}$

$a_2 = \left(-\frac{1}{2}\right)^2 = \frac{1}{4}$

$a_3 = \left(-\frac{1}{2}\right)^3 = -\frac{1}{8}$

$a_4 = \left(-\frac{1}{2}\right)^4 = \frac{1}{16}$

$a_{10} = \left(-\frac{1}{2}\right)^{10} = \frac{1}{1024}$

$a_{15} = \left(-\frac{1}{2}\right)^{15} = -\frac{1}{32,768}$

$-\frac{1}{2}, \frac{1}{4}, -\frac{1}{8}, \frac{1}{16}; \frac{1}{1024}; -\frac{1}{32,768}$

17. $a_n = 3n - 4$

$a_{12} = 3(12) - 4 = 32$

19. $a_n = (-2)^{n-2}(2n+1)$

$a_6 = (-2)^{6-2}(2 \cdot 6 + 1)$

$= (-2)^4(13) = 208$

21. $a_n = \ln e^n$

$a_{45} = \ln e^{45} = 45$

23-39. Possible answers:

23. 1, 3, 5, 7, 9, …

$a_n = 2n - 1$

25. 3, 6, 9, 12, 15, …

$a_n = 3n$

27. $\frac{2}{1}, \frac{3}{2}, \frac{4}{3}, \frac{5}{4}, \frac{6}{5}, \ldots$

$a_n = \frac{n+1}{n}$

29. 1, 4, 7, 10, 13, …

$a_n = 3n - 2$

31. –2, 4, –6, 8, –10, …

$a_n = (-1)^n 2n$

33. 2, –4, 6, –8, 10, …

$a_n = (-1)^{n+1} 2n$

35. –1, 1, –1, 1, –1, …

$a_n = (-1)^n$

37. –1, –4, –7, –10, –13, …

$a_n = 2 - 3n$

39. $\sqrt{5}, 5, 5\sqrt{5}, 25, 25\sqrt{5}, \ldots$

$a_n = (\sqrt{5})^n$

41. 3, 6, 9, 12, …

$S_6 = 3 + 6 + 9 + 12 + 15 + 18 = 63$

43. $\frac{1}{3}, \frac{2}{3}, \frac{3}{3}, \frac{4}{3}, \ldots$

$S_8 = \frac{1}{3} + \frac{2}{3} + \frac{3}{3} + \frac{4}{3} + \frac{5}{3} + \frac{6}{3} + \frac{7}{3} + \frac{8}{3} = 12$

45. $a_n = \frac{3n}{n+2}$

$a_1 = \frac{3(1)}{1+2} = 1$

$a_2 = \frac{3(2)}{2+2} = \frac{3}{2}$

$a_3 = \frac{3(3)}{3+2} = \frac{9}{5}$

$S_3 = 1 + \frac{3}{2} + \frac{9}{5} = \frac{43}{10}$

47. $a_n = \frac{(-1)^n}{2n-1}$

$a_1 = \frac{(-1)^1}{2(1)-1} = -1$

$a_2 = \frac{(-1)^2}{2(2)-1} = \frac{1}{3}$

$a_3 = \frac{(-1)^3}{2(3)-1} = -\frac{1}{5}$

$a_4 = \frac{(-1)^4}{2(4)-1} = \frac{1}{7}$

$S_4 = -1 + \frac{1}{3} - \frac{1}{5} + \frac{1}{7} = -\frac{76}{105}$

49. $\sum_{i=1}^{4} 3i = 3(1)+3(2)+3(3)+3(4)$
$= 3 + 6 + 9 + 12 = 30$

51. $\sum_{i=2}^{6} (i^2+3) = (2^2+3)+(3^2+3)+(4^2+3)$
$+ (5^2+3)+(6^2+3)$
$= 7 + 12 + 19 + 28 + 39$
$= 105$

53. $\sum_{i=1}^{5} i(i+4) = 1(1+4)+2(2+4)+3(3+4)$
$+ 4(4+4) + 5(5+4)$
$= 5 + 12 + 21 + 32 + 45$
$= 115$

55. $\sum_{j=1}^{4} (-1)^j = (-1)^1+(-1)^2+(-1)^3+(-1)^4$
$= -1 + 1 - 1 + 1$
$= 0$

57. $\sum_{k=1}^{4} \left(-\frac{1}{2}\right)^k = \left(-\frac{1}{2}\right)^1 + \left(-\frac{1}{2}\right)^2 + \left(-\frac{1}{2}\right)^3$
$+\left(-\frac{1}{2}\right)^4$
$= -\frac{1}{2}+\frac{1}{4}-\frac{1}{8}+\frac{1}{16}$
$= -\frac{8}{16}+\frac{4}{16}-\frac{2}{16}+\frac{1}{16}$
$= -\frac{5}{16}$

59. $\sum_{i=2}^{4} (-i)^i = (-2)^2+(-3)^3+(-4)^4$
$= 4 - 27 + 256$
$= 233$

61. $\sum_{i=3}^{5} \frac{2i-1}{i-1} = \frac{2(3)-1}{3-1}+\frac{2(4)-1}{4-1}+\frac{2(5)-1}{5-1}$
$= \frac{5}{2}+\frac{7}{3}+\frac{9}{4}$
$= \frac{30+28+27}{12}$
$= \frac{85}{12}$

63. $1 + 8 + 27 + 64 + 125 = \sum_{i=1}^{5} i^3$

65. $2+4+6+8+10 = \sum_{i=1}^{5} 2i$

67. $-2 + 4 - 6 + 8 - 10 = \sum_{i=1}^{5} (-1)^i 2i$

69. $\frac{1}{1}+\frac{1}{4}+\frac{1}{9}+\frac{1}{16}+\frac{1}{25} = \sum_{i=1}^{5} \frac{1}{i^2}$

71. $\frac{1}{1}-\frac{1}{4}+\frac{1}{9}-\frac{1}{16}+\frac{1}{25} = \sum_{i=1}^{5} \frac{(-1)^{i+1}}{i^2}$

73. $\frac{2}{3}+\frac{3}{4}+\frac{4}{5}+\frac{5}{6}+\frac{6}{7}+\frac{7}{8} = \sum_{i=1}^{6} \frac{i+1}{i+2}$

75. $6 + 12 + 18 + 24 + 30 + \ldots = \sum_{i=1}^{\infty} 6i$

77. $\frac{1}{1\cdot 2}+\frac{1}{2\cdot 3}+\frac{1}{3\cdot 4}+\frac{1}{4\cdot 5}+\ldots = \sum_{i=1}^{\infty} \frac{1}{i(i+1)}$

79. $a_n = 0.1\sqrt{9n^2 + 82}$

n	Year	$a_n = $ (Debt)
1	1981	1.0
2	1982	1.1
3	1983	1.3
4	1984	1.5
5	1985	1.8
6	1986	2.0
7	1987	2.3
8	1988	2.6
9	1989	2.8
10	1990	3.1
11	1991	3.4
12	1992	3.7
13	1993	4.0
14	1994	4.3
15	1995	4.6

Sequence = 1.0, 1.1, 1.3, 1.5, 1.8, 2.0, 2.3, 2.6, 2.8, 3.1, 3.4, 3.7, 4.0, 4.3, 4.6

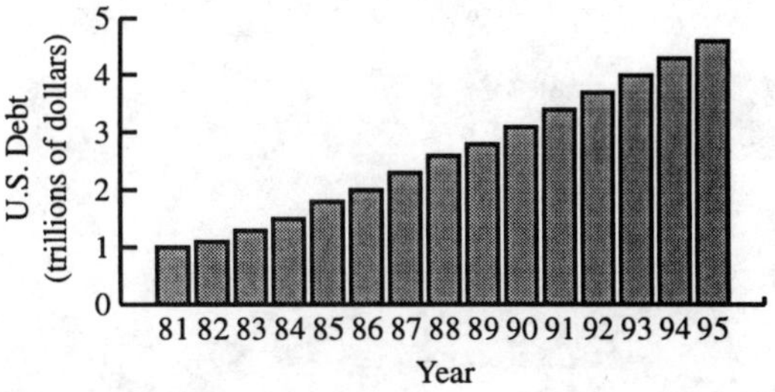

81. $a_n = -143n^3 + 1810n^2 - 187n + 2331$

$a_1 = -143(1)^3 + 1810(1)^2 - 187(1) + 2331$
$= 3811$

$a_2 = 8053$

$a_3 = 14,199$

$S_3 = 3811 + 8053 + 14,199 = 26,063$

The total number of AIDS cases reported for 1984, 1985, and 1986 approximated by this model is 26,063.

83. b is true.

85. d is true.

87. $a_n = \left(1 + \frac{1}{n}\right)^n$

$a_{10} = \left(1 + \frac{1}{10}\right)^{10} = \left(\frac{11}{10}\right)^{10} \approx 2.59374$

$a_{100} = \left(1 + \frac{1}{100}\right)^{100} = \left(\frac{101}{100}\right)^{100} \approx 2.70481$

$a_{1,000} = \left(1 + \frac{1}{1,000}\right)^{1,000} = \left(\frac{1,001}{1,000}\right)^{1000}$
≈ 2.71692

$a_{10,000} = \left(1 + \frac{1}{10,000}\right)^{10,000}$
$= \left(\frac{10,001}{10,000}\right)^{10,000} \approx 2.71815$

$a_{100,000} = \left(1 + \frac{1}{100,000}\right)^{100,000}$
$= \left(\frac{100,001}{100,000}\right)^{100,000} \approx 2.71827$

The terms approach the decimal approximation for *e*.

89–91. Students should verify results.

93–95. Answers may vary.

97.

1	1	1	1	1
2	2	2	2	2
3	4	6	4	6
4	7	15	7	15
5	12	40	12	40

Column 2 = Column 4
Column 3 = Column 5

99. $a_n = \frac{3 \cdot 5 \cdots (2n-1)(2n+1)}{2 \cdot 4 \cdots (2n-2)(2n)}$

a_1 is undefined due to division by zero

$a_2 = \frac{3 \cdot 5}{2 \cdot 4} = \frac{15}{8}$

$a_3 = \frac{3 \cdot 5 \cdot 7}{2 \cdot 4 \cdot 6} = \frac{35}{16}$

$a_4 = \frac{3 \cdot 5 \cdot 7 \cdot 9}{2 \cdot 4 \cdot 6 \cdot 8} = \frac{315}{128}$

$a_5 = \frac{3 \cdot 5 \cdot 7 \cdot 9 \cdot 11}{2 \cdot 4 \cdot 6 \cdot 8 \cdot 10} = \frac{693}{256}$

101. $\sum_{i=-3}^{2} \frac{i}{i+4} = \frac{-3}{1} + \frac{-2}{2} + \frac{-1}{3} + 0 + \frac{1}{5} + \frac{2}{6}$

$\sum_{j=-20}^{-15} \frac{j+17}{j+21} = \frac{-3}{1} + \frac{-2}{2} + \frac{-1}{3} + 0 + \frac{1}{5} + \frac{2}{6}$

$\sum_{k=14}^{19} \frac{k-17}{k-13} = \frac{-3}{1} + \frac{-2}{2} + \frac{-1}{3} + 0 + \frac{1}{5} + \frac{2}{6}$

Thus, the sums are the same:

$\frac{-3}{1} + \frac{-2}{2} + \frac{-1}{3} + 0 + \frac{1}{5} + \frac{2}{6}$

103. $\sum_{i=1}^{4} \log 2i = \log 2 + \log 4 + \log 6 + \log 8$

$= \log 2 \cdot 4 \cdot 6 \cdot 8 = \log 384$

105. $\sum_{i=2}^{4} 2i \log x = 4 \log x + 6 \log x + 8 \log x$

$= \log x^4 + \log x^6 + \log x^8$

$= \log x^4 \cdot x^6 \cdot x^8 = \log x^{18}$

107. $\sum_{i=1}^{n} a_i b_i = a_1 b_1 + a_2 b_2 + a_3 b_3 + a_4 b_4$
$+ \ldots + a_n b_n$

$\sum_{i=1}^{n} a_i = (a_1 + a_2 + a_3 + a_4 + \ldots + a_n)$

$\sum_{i=1}^{n} b_i = (b_1 + b_2 + b_3 + b_4 + \ldots + b_n)$

By the distributive property (distributing a_1 to each term in the second factor, a_2 to each term in the second factor, etc.), this is not equal to
$a_1 b_1 + a_2 b_2 + a_3 b_3 + a_4 b_4 + \ldots + a_n b_n$.
The given statement is not true.

109. $\sum_{i=1}^{6} a_i$ if $a_i = 1$, $a_2 = 2$, and

$a_n = a_{n-1}^2 + a_{n-2}^2$ for $n \geq 3$.

$\sum_{i=1}^{6} a_i = 1 + 2 + (4+1) + (25+4) + (29^2 + 25)$
$+ (866^2 + 29^2)$
$= 1 + 2 + 5 + 29 + 866 + 750{,}797$
$= 751{,}700$

Review Problems

111. $2y^6 + 16 = 2(y^6 + 8) = 2[(y^2)^3 + 2^3]$
$2(y^2 + 2)[(y^2)^2 - y^2 \cdot 2 + 2^2]$
$= 2(y^2 + 2)(y^4 - 2y^2 + 4)$

112. $\frac{x^2}{16} + \frac{y^2}{9} = 1$

Ellipse

x-intercepts: $\frac{x^2}{16} = 1$

$x^2 = 16$
$x = \pm 4$

y-intercepts: $\frac{y^2}{9} = 1$

$y^2 = 9$

$y = \pm 3$

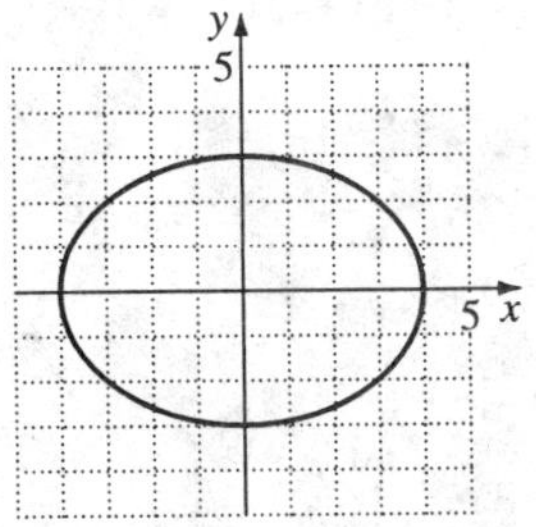

113. $\log_4 x + \log_4 (x-6) = 2$
$\log_4 x(x-6) = 2$
Exponential form: $4^2 = x(x-6)$
$16 = x^2 - 6x$
$0 = x^2 - 6x - 16$
$0 = (x-8)(x+2)$
$x - 8 = 0$ or $x + 2 = 0$
$x = 8$ or $x = -2$
Reject $x = -2$ because it causes the log of a negative number.
{8}

Problem Set 11.2

1. 2, 6, 10, 14, ...
$d = 6 - 2 = 4$

3. −7, −2, 3, 8, ...
$d = -2 - (-7) = -2 + 7 = 5$

5. 5.12, 5.25, 5.38, 5.51, ...
$d = 5.25 - 5.12 = 0.13$

7. $\frac{e}{6}, \frac{e}{3}, \frac{e}{2}, \frac{2e}{3}, \ldots$
$d = \frac{e}{3} - \frac{e}{6} = \frac{e}{6}$

9. 8, 10, 12, 14, 16, 18

11. −16, −20, −24, −28, −32, −36

13. $\frac{3}{2}, \frac{9}{4}, 3, \frac{15}{4}, \frac{9}{2}, \frac{21}{4}$

15. 4, 7, 10, 13, ...
$d = 7 - 4 = 3$
$a_n = a_1 + (n-1)d$
$a_{26} = 4 + (26-1)(3)$
$a_{26} = 79$

17. 7, 3, −1, −5, ...
$d = 3 - 7 = -4$
$a_n = a_1 + (n-1)d$
$a_{14} = 7 + (14-1)(-4)$
$a_{14} = -45$

19. 3.15, 3.10, 3.05, 3.00, ...
$d = 3.10 - 3.15 = -0.05$
$a_n = a_1 + (n-1)d$
$a_{22} = 3.15 + (22-1)(-0.05)$
$a_{22} = 2.10$

21. $\frac{3}{2}, \frac{9}{4}, 3, \frac{15}{4}, \ldots$
$d = \frac{9}{4} - \frac{3}{2} = \frac{3}{4}$
$a_n = a_1 + (n-1)d$
$a_{10} = \frac{3}{2} + (10-1)\left(\frac{3}{4}\right)$
$a_{10} = \frac{33}{4}$

23. $a_1 = 9, d = 2$
$a_n = a_1 + (n-1)d$
$a_{16} = 9 + (16-1)(2)$
$a_{16} = 39$

25. $a_1 = -\frac{1}{3}, d = \frac{1}{3}$
$a_n = a_1 + (n-1)d$
$a_{30} = -\frac{1}{3} + (30-1)\left(\frac{1}{3}\right)$
$a_{30} = \frac{28}{3}$

27. $a_1 = 4, d = -0.3$
$a_n = a_1 + (n-1)d$
$a_{12} = 4 + (12-1)(-0.3)$
$a_{12} = 0.7$

29. $d = 8 - 5 = 3$
$a_n = a_1 + (n-1)d$
$32 = 5 + (n-1)(3)$
$27 = 3n - 3$
$30 = 3n$
$10 = n$
The 10th term

31. $d = 1 - 4 = -3$
$a_n = a_1 + (n-1)d$
$-281 = 4 + (n-1)(-3)$
$-285 = -3n + 3$
$-288 = -3n$
$96 = n$
The 96th term

33. $d = -2 - \left(-\frac{5}{3}\right) = -\frac{6}{3} + \frac{5}{3} = -\frac{1}{3}$
$a_n = a_1 + (n-1)d$
$-9 = -\frac{5}{3} + (n-1)\left(-\frac{1}{3}\right)$
$-\frac{27}{3} + \frac{5}{3} = -\frac{1}{3}n + \frac{1}{3}$
$-\frac{22}{3} - \frac{1}{3} = -\frac{1}{3}n$
$23 = n$
The 23rd term

35. 4, 10, 16, 22, ...; $d = 6$
$a_n = a_1 + (n-1)d$
$a_{20} = 4 + (20-1)6 = 118$
$S_n = \frac{n}{2}(a_1 + a_n)$
$= \frac{20}{2}(4 + 118) = 1{,}220$

37. $-15, -7, 1, 9, \ldots; d = 8$
$a_n = a_1 + (n-1)d$
$a_{10} = -15 + (10-1)(8) = 57$
$S_n = \frac{n}{2}(a_1 + a_n)$
$S_{10} = \frac{10}{2}(-15 + 57) = 210$

39. $\sum_{i=1}^{17}(5i+3) = [5(1)+3] + [5(2)+3]$
$+ [5(3)+3] + \ldots + [5(17)+3]$
$= 8 + 13 + 18 + \ldots + 88$
$S_n = \frac{n}{2}(a_1 + a_n)$
$S_{17} = \frac{17}{2}(a_1 + a_{17})$
$S_{17} = \frac{17}{2}(8 + 88) = 816$

41. $\sum_{i=1}^{30}(-3i+5) = 2 + (-1) + (-4) + \ldots + (-85)$
$S_{30} = \frac{30}{2}(a_1 + a_{30})$
$S_{30} = \frac{30}{2}(2 - 85) = -1{,}245$

43. $\sum_{i=1}^{100} 4i = 4 + 8 + 12 + \ldots + 400$
$S_{100} = \frac{100}{2}(a_1 + a_{100})$
$S_{100} = \frac{100}{2}(4 + 400) = 20{,}200$

45. $1 + 2 + 3 + 4 + \ldots + 100$
$S_{100} = \frac{100}{2}(1 + 100) = 5050$

47. $22 + 24 + 26 + 28 + \ldots + 44$
$S_{12} = \frac{12}{2}(22 + 44) = 396$

49. $S_n = \frac{n}{2}(a_1 + a_n)$
$S_{25} = \frac{25}{2}(a_1 + a_{25})$
$a_1 = -9, d = 5$
We must find a_{25}.
$a_n = a_1 + (n-1)d$
$a_{25} = -9 + (25-1)(5) = 111$
$S_{25} = \frac{25}{2}(-9 + 111) = 1275$

51. $S_n = \frac{n}{2}(a_1 + a_n)$
$(n = 40,\ a_1 = 50,\ d = -3)$
$S_{40} = \frac{40}{2}(50 + a_{40})$
Find a_{40}.
$a_n = a_1 + (n-1)d$
$a_{40} = 50 + (40-1)(-3) = -67$
$S_{40} = \frac{40}{2}(50 - 67) = -340$

53. $S_n = \frac{n}{2}(a_1 + a_n)$
$\left(n = 12,\ a_1 = \frac{1}{2},\ d = -\frac{1}{2}\right)$
$S_{12} = \frac{12}{2}\left(\frac{1}{2} + a_{12}\right)$
Find a_{12}.
$a_n = a_1 + (n-1)d$
$a_{12} = \frac{1}{2} + (12-1)\left(-\frac{1}{2}\right) = \frac{1}{2} - \frac{11}{2} = -5$
$S_{12} = \frac{12}{2}\left(\frac{1}{2} - 5\right) = 6(-4.5) = -27$

55. **a.** $a_1 = 39.05,\ d = 0.45$
$a_n = a_1 + (n-1)d$
$a_n = 39.05 + (n-1)(0.45)$
$a_n = 0.45n + 38.6$

b. $n = 16$
$a_{16} = 0.45(16) + 38.6$
$a_{16} = 45.8$
The prediction is 45.8 million.

c. $50.75 = 0.45n + 38.6$
$12.15 = 0.45n$
$27 = n$
$1985 + (27 - 1) = 2011$
The year will be 2011.

57. Company A: $a_1 = 12{,}000,\ d = 800$
$a_n = a_1 + (n-1)d$
Year ten:
$a_{10} = 12{,}000 + (10-1)(800) = 19{,}200$
Company B: $a_1 = 14{,}000,\ d = 500$
Year ten:
$a_{10} = 14{,}000 + (10-1)(500) = 18{,}500$
Year ten difference: $19{,}200 - 18{,}500 = 700$
Company A pays \$700 more than Company B in year 10.

59. Company A: $a_1 = 20{,}000,\ d = 1{,}000$
$a_{12} = a_1 + (n-1)d$
$= 20{,}000 + (12-1)(1{,}000) = 31{,}000$
$S_{12} = \frac{n}{2}(a_1 + a_{12}) = \frac{12}{2}(20{,}000 + 31{,}000)$
$= 306{,}000$
Company B: $a_1 = 21{,}000,\ d = 800$
$a_{12} = 21{,}000 + (12-1)(800) = 29{,}800$
$S_{12} = \frac{12}{2}(21{,}000 + 29{,}800) = 304{,}800$
Over 12 years: Company A: \$306,000
Company B: \$304,800
Company A will pay the greater amount.

61. **a.** $a_1 = 87.1,\ d = 3.14$
$a_n = 87.1 + (n-1)(3.14)$
$a_n = 83.96 + 3.14n$

b. $n = 1, 2, 3, \ldots, 39$
$a_1 = 87.1$
$a_{39} = 83.96 + 3.14(39) = 206.42$
$S_{39} = \frac{39}{2}(87.1 + 206.42) = 5723.64$
The total amount is 5723.64 million tons.

63. Degree days: 23, 25, 27, …
$a_1 = 23,\ d = 2$
$S_{10} = \frac{10}{2}(a_1 + a_{10})$
$a_{10} = 23 + 9(2) = 41$
$S_{10} = \frac{10}{2}(23 + 41) = 320$
320 degree days

65. $a_1 = 3(2 \cdot 1) = 6 \cdot 1$
$a_2 = 3(2 \cdot 2) = 6 \cdot 2$
$a_3 = 3(2 \cdot 3) = 6 \cdot 3$
$a_{14} = 6 \cdot 14 = 84$

$$S_{14} = 6\sum_{i=1}^{14} i = 6 \cdot \frac{14}{2}(1+14) = 42(15) = 630$$

630 feet

67. d is true.

69. Students should verify results.

71-73. Answers may vary.

75. a. Since $\frac{1}{x}, \frac{1}{y}, \frac{1}{z}$ are consecutive terms:

$$\frac{1}{y} - \frac{1}{x} = \frac{1}{z} - \frac{1}{y}$$
$$\frac{x-y}{xy} = \frac{y-z}{zy}$$
$$zy(x-y) = xy(y-z)$$
$$\frac{x-y}{y-z} = \frac{xy}{zy} = \frac{x}{z}$$

b.
$$\frac{1}{z} - \frac{1}{x} = 2\left(\frac{1}{y} - \frac{1}{x}\right)$$
$$\frac{1}{z} - \frac{1}{x} = \frac{2}{y} - \frac{2}{x}$$
$$\frac{1}{z} + \frac{1}{x} = \frac{2}{y}$$
$$\frac{x+z}{zx} = \frac{2}{y}$$
$$(x+z)y = 2xz$$
$$y = \frac{2xz}{x+z}$$

77. h = hundred's digit
t = ten's digit
u = unit's digit

$100u + 10t + h = 100h + 10t + u + 396$
$h + t + u = 21$
$h - t = t - u$
$99u - 99h = 396$
$u + t + h = 21$
$u - 2t + h = 0$
$u - h = 4$ (1)
$u + t + h = 21$ (2)
$u - 2t + h = 0$ (3)
Eliminate t:
(Eq 1): $u - h = 4$
2(Eq 2) + (Eq 3): $3u + 3h = 42$

$$3u - 3h = 12$$
$$3u + 3h = 42$$
$$6u = 54$$
$$u = 9$$

$u - h = 4$
$9 - h = 4$
$h = 5$
$u + t + h = 21$
$9 + t + 5 + 21$
$t = 7$
Number:
$100h + 10t + u = 100(5) + 10(7) + 9 = 579$

79. Let d = the fixed sum.
700, 700 + d, 700 + 2d, ...

$$S_n = \frac{n}{2}(a_1 + a_n)$$
$$S_8 = \frac{8}{2}(a_1 + a_n) = 4(700 + a_8) = 6580$$
$$2800 + 4a_8 = 6580$$
$$4a_8 = 3780$$
$$a_8 = 945$$
$$a_n = a_1 + (n-1)d$$
$$a_8 = a_1 + (8-1)d$$
$$945 = 700 + 7d$$
$$7d = 245$$
$$d = 35$$

Fixed sum: \$35.

81.
$$\sum_{i=1}^{n}(ai + b) = (a+b) + (2a+b) + (3a+b) + \ldots + (na+b)$$
$$S_n = \frac{n}{2}(a_1 + a_n)$$
$$a_1 = a + b$$
$$a_n = na + b$$
$$n = n$$
$$S_n = \frac{n}{2}(a + b + na + b) = \frac{n}{2}(a + na + 2b)$$
$$= \frac{na}{2} + \frac{n^2 a}{2} + nb = \frac{n(n+1)a}{2} + nb$$

or $\frac{n}{2}(a + na + 2b)$

83. Group Activity

Review Problems

84. $(y+1)(2y+3)-3(y+2)(y+1)=-3(y+5)$

$2y^2+5y+3-3y^2-9y-6=-3y-15$

$-y^2-4y-3=-3y-15$

$0=y^2+y-12$

$0=(y+4)(y-3)$

$y+4=0$ or $y-3=0$

$y=-4$, $y=3$ (both check)

$\{-4, 3\}$

85. $x^2+4y^2=13 \quad (1)$

$x^2-y^2=8 \quad (2)$

Multiply equation 2 by –1 and add to equation 1.

$x^2+4y^2=13$

$\underline{-x^2+y^2=-8}$

$5y^2=5$

$y^2=1$

$y=\pm 1$

If $y=\pm 1$: $x^2-y^2=8$

$x^2-1=8$

$x^2=9$

$x=\pm 3$

(3, 1), (3, –1), (–3, 1), (–3, –1)

86. Resistance: R

Current: I

$R=\frac{k}{I^2}$

Given $I=0.8$ ampere, $R=50$ ohms

$50=\frac{k}{(0.8)^2}$

$50=\frac{k}{0.64}$

$(50)(0.64)=k$

$k=32$

$R=\frac{32}{I^2}$

If $I=0.5$ ampere, $R=?$

$R=\frac{32}{(0.5)^2}$

$R=\frac{32}{0.25}$

$R=128$

128 ohms

Problem Set 11.3

1. 5, 15, 45, 135, ...

$r=\frac{a_2}{a_1}=\frac{15}{5}=3$

3. –8, 8, –8, 8, ...

$r=\frac{8}{-8}=-1$

5. $5, -\frac{5}{2}, \frac{5}{4}, -\frac{5}{8}, \ldots$

$r=\frac{-\frac{5}{2}}{5}=-\frac{1}{2}$

7. $90, 30, 10, \frac{10}{3}, \ldots$

$r=\frac{30}{90}=\frac{1}{3}$

9. $\sqrt{5}, 5, 5\sqrt{5}, 25, \ldots$

$r=\frac{5}{\sqrt{5}}=\frac{5\sqrt{5}}{(\sqrt{5})^2}=\frac{5\sqrt{5}}{5}=\sqrt{5}$

11. $\frac{a}{b}, \frac{a}{b^2}, \frac{a}{b^3}, \frac{a}{b^4}, \ldots$

$r=\frac{\frac{a}{b^2}}{\frac{a}{b}}=\frac{a}{b^2}\cdot\frac{b}{a}=\frac{1}{b}$

13. $a_1=10$, $r=\frac{1}{2}$

$a_1=10$

$a_2=10\left(\frac{1}{2}\right)=5$

$a_3=5\left(\frac{1}{2}\right)=\frac{5}{2}$

$a_4=\frac{5}{2}\left(\frac{1}{2}\right)=\frac{5}{4}$

$a_5 = \frac{5}{4}\left(\frac{1}{2}\right) = \frac{5}{8}$

$10, 5, \frac{5}{2}, \frac{5}{4}, \frac{5}{8}$

15. $a_1 = -\frac{1}{4},\ r = -2$

$a_2 = -\frac{1}{4}(-2) = \frac{1}{2}$

$a_3 = \frac{1}{2}(-2) = -1$

$a_4 = -1(-2) = 2$

$a_5 = 2(-2) = -4$

$-\frac{1}{4}, \frac{1}{2}, -1, 2, -4$

17. $a_1 = 3,\ r = -3$

$a_2 = 3(-3) = -9$

$a_3 = -9(-3) = 27$

$a_4 = 27(-3) = -81$

$a_5 = -81(-3) = 243$

$3, -9, 27, -81, 243$

19. $a_1 = \frac{a^2}{b},\ r = \frac{2b}{a}$

$a_1 = \frac{a^2}{b}$

$a_2 = \frac{a^2}{b} \cdot \frac{2b}{a} = 2a$

$a_3 = 2a \cdot \frac{2b}{a} = 4b$

$a_4 = 4b \cdot \frac{2b}{a} = \frac{8b^2}{a}$

$a_5 = \frac{8b^2}{a} \cdot \frac{2b}{a} = \frac{16b^3}{a^2}$

$\frac{a^2}{b}, 2a, 4b, \frac{8b^2}{a}, \frac{16b^3}{a^2}$

21. $-3, -15, -75, -375, \ldots$

$r = \frac{-15}{-3} = 5$

$a_n = a_1 r^{n-1}$

$a_n = -3(5)^{n-1}$

$a_8 = -3(5)^{8-1} = -234{,}375$

23. $-12, -6, -3, -\frac{3}{2}, \ldots$

$r = \frac{-6}{-12} = \frac{1}{2}$

$a_n = a_1 r^{n-1}$

$a_n = -12\left(\frac{1}{2}\right)^{n-1}$

$a_{10} = -12\left(\frac{1}{2}\right)^{10-1} = -\frac{3}{128}$

25. $3, -1.5, 0.75, -0.375, \ldots$

$r = \frac{-1.5}{3} = -\frac{1}{2}$

$a_n = a_1 r^{n-1}$

$a_n = 3\left(-\frac{1}{2}\right)^{n-1}$

$a_6 = 3\left(-\frac{1}{2}\right)^{6-1} = -0.09375$

27. $\frac{2}{3}, \frac{2}{15}, \frac{2}{75}, \frac{2}{375}, \ldots$

$r = \frac{\frac{2}{15}}{\frac{2}{3}} = \frac{2}{15} \cdot \frac{3}{2} = \frac{1}{5}$

$a_n = a_1 r^{n-1}$

$a_n = \frac{2}{3}\left(\frac{1}{5}\right)^{n-1}$

$a_8 = \frac{2}{3}\left(\frac{1}{5}\right)^{8-1} = \frac{2}{234{,}375}$

29. $-4, -2\sqrt{2}, -2, -\sqrt{2}, \ldots$

$r = \frac{-2\sqrt{2}}{-4} = \frac{\sqrt{2}}{2}$

$a_n = a_1 r^{n-1}$

$a_n = -4\left(\frac{\sqrt{2}}{2}\right)^{n-1}$

$a_{11} = -4\left(\frac{\sqrt{2}}{2}\right)^{11-1} = -\frac{1}{8}$

31. $222\frac{2}{9}, 22\frac{2}{9}, 2\frac{2}{9}, \ldots$

$r = \frac{22\frac{2}{9}}{222\frac{2}{9}} = \frac{1}{10}$

$a_n = a_1 r^{n-1}$

$a_n = 222\frac{2}{9}\left(\frac{1}{10}\right)^{n-1}$

$a_6 = 222\frac{2}{9}\left(\frac{1}{10}\right)^{6-1} = 0.00\overline{2}$

33. $c^7d^6, c^6d^4, c^5d^2, \ldots$

$r = \frac{c^6d^4}{c^7d^6} = \frac{1}{cd^2}$

$a_n = a_1r^{n-1}$

$a_n = c^7d^6\left(\frac{1}{cd^2}\right)^{n-1}$

$a_{14} = c^7d^6\left(\frac{1}{cd^2}\right)^{14-1} = \frac{1}{c^6d^{20}}$

35. 2, 4, 8, 16, 32, ...

$r = \frac{4}{2} = 2$

$a_n = a_1r^{n-1} = 2(2)^{n-1} = 2^n$

37. 2, 6, 18, ...

$S_n = \frac{a_1 - a_1r^n}{1-r}$

$a_1 = 2,\ r = \frac{6}{2} = 3$

$S_6 = \frac{2-2(3)^6}{1-3} = \frac{2-1458}{-2} = \frac{-1456}{-2} = 728$

39. 3, –6, 12, ...

$a_1 = 3,\ r = \frac{-6}{3} = -2$

$S_n = \frac{a_1 - a_1r^n}{1-r}$

$S_5 = \frac{3-3(-2)^5}{1-(-2)} = \frac{99}{3} = 33$

41. $-\frac{3}{2}, 3, -6, \ldots$

$a_1 = -\frac{3}{2},\ r = \frac{3}{\left(-\frac{3}{2}\right)} = -2$

$S_n = \frac{a_1 - a_1r^n}{1-r}$

$S_7 = \frac{-\frac{3}{2}-\left(-\frac{3}{2}\right)(-2)^7}{1-(-2)} = \frac{-\frac{3}{2}+\frac{3}{2}(-128)}{3}$

$= \frac{-\frac{3}{2}-192}{3} = \frac{-3-384}{6} = \frac{-387}{6} = -\frac{129}{2}$

43. $\sum_{i=0}^{6} 3^i = 3^0 + 3^1 + 3^2 + \ldots + 3^6$

$= 1 + 3 + 9 + \ldots + 729$

$a_1 = 1,\ r = 3,\ n = 7$ (There are seven terms.)

$S_n = \frac{a_1 - a_1r^n}{1-r}$

$S_7 = \frac{1-1(3)^7}{1-3} = \frac{-2186}{-2} = 1093$

45. $\sum_{i=0}^{6} (-3)^i = 1 - 3 + 9 - 27 + \ldots + (-3)^6$

$a_1 = 1,\ r = -3,\ n = 7$

$S_n = \frac{a_1 - a_1r^n}{1-r}$

$S_7 = \frac{1-1(-3)^7}{1-(-3)} = \frac{2188}{4} = 547$

47. $\sum_{i=1}^{5} 2^{i-1} = 2^{1-1} + 2^{2-1} + 2^{3-1} + 2^{4-1} + 2^{5-1}$

$= 1 + 2 + 4 + 8 + 16$

$a_1 = 1,\ r = 2,\ n = 5$ (Since we begin at $i = 1$, not $i = 0$, we have five terms.)

$S_n = \frac{a_1 - a_1r^n}{1-r}$

$S_5 = \frac{1-1(2)^5}{1-2} = \frac{-31}{-1} = 31$

49. $\sum_{i=1}^{4}\left(-\frac{2}{3}\right)^i = \left(-\frac{2}{3}\right)^1 + \left(-\frac{2}{3}\right)^2 + \left(-\frac{2}{3}\right)^3 + \left(-\frac{2}{3}\right)^4$

$= -\frac{2}{3} + \frac{4}{9} + \left(-\frac{2}{3}\right)^3 + \left(-\frac{2}{3}\right)^4$

$a_1 = -\frac{2}{3},\ r = \frac{\frac{4}{9}}{-\frac{2}{3}} = -\frac{2}{3},\ n = 4$

$S_n = \frac{a_1 - a_1r^n}{1-r}$

$$S_4 = \frac{\left(-\frac{2}{3}\right)-\left(-\frac{2}{3}\right)\left(-\frac{2}{3}\right)^4}{1-\left(-\frac{2}{3}\right)} = \frac{-\frac{2}{3}+\frac{32}{243}}{\frac{5}{3}} \cdot \frac{243}{243}$$
$$= \frac{-162+32}{405} = \frac{-130}{405} = \frac{-26}{81}$$

51. $1+\frac{1}{4}+\frac{1}{16}+\ldots$

$a_1 = 1,\ r = \frac{1}{4}$

$S = \frac{a_1}{1-r} = \frac{1}{1-\frac{1}{4}} = \frac{1}{\frac{3}{4}} = \frac{4}{3}$

53. $12 + 6 + 3 + \ldots$

$a_1 = 12,\ r = \frac{6}{12} = \frac{1}{2}$

$S = \frac{a_1}{1-r} = \frac{12}{1-\frac{1}{2}} = \frac{12}{\frac{1}{2}} = 24$

55. $27 - 18 + 12 - \ldots$

$a_1 = 27,\ r = \frac{-18}{27} = -\frac{2}{3}$

$S = \frac{a_1}{1-r} = \frac{27}{1-\left(-\frac{2}{3}\right)} = \frac{27}{\frac{5}{3}} = \frac{81}{5}$

57. $5 + 10 + 20 + \ldots$

$a_1 = 5,\ r = \frac{10}{5} = 2$

Since r does not lie between –1 and 1 the infinite series has no finite sum.

59. $\frac{4}{3}+\frac{2}{9}+\frac{1}{27}+\ldots$

$a_1 = \frac{4}{3},\ r = \frac{\frac{2}{9}}{\frac{4}{3}} = \frac{2}{9}\cdot\frac{3}{4} = \frac{1}{6}$

$S = \frac{a_1}{1-r} = \frac{\frac{4}{3}}{1-\frac{1}{6}} = \frac{\frac{4}{3}}{\frac{5}{6}} = \frac{4}{3}\cdot\frac{6}{5} = \frac{8}{5}$

61. $1-\frac{1}{2}+\frac{1}{4}-\frac{1}{8}+\ldots$

$a_1 = 1,\ r = \frac{-\frac{1}{2}}{1} = -\frac{1}{2}$

$S = \frac{a_1}{1-r} = \frac{1}{1-\left(-\frac{1}{2}\right)} = \frac{1}{\frac{3}{2}} = \frac{2}{3}$

63. $0.\overline{5} = 0.5555\ldots = 0.5 + 0.05 + 0.005 + \ldots$

$a_1 = 0.5,\ r = 0.1$

$S = \frac{a_1}{1-r} = \frac{0.5}{1-0.1} = \frac{0.5}{0.9} = \frac{5}{9}$

65. $0.\overline{49} = 0.494949\ldots$

$= 0.49 + 0.0049 + 0.000049 + \ldots$

$a_1 = 0.49,\ r = 0.01$

$S = \frac{a_1}{1-r} = \frac{0.49}{1-0.01} = \frac{0.49}{0.99} = \frac{49}{99}$

67. $0.\overline{241} = 0.241241241\ldots$

$= 0.241 + 0.000241 + \ldots$

$a_1 = 0.241,\ r = 0.001$

$S = \frac{a_1}{1-r} = \frac{0.241}{1-0.001} = \frac{0.241}{0.999} = \frac{241}{999}$

69. $5.\overline{47}$

$0.\overline{47} = 0.474747\ldots$

$= 0.47 + 0.0047 + 0.000047 + \ldots$

$a_1 = 0.47,\ r = 0.01$

$S = \frac{a_1}{1-r} = \frac{0.47}{1-0.01} = \frac{0.47}{0.99} = \frac{47}{99}$

$5.\overline{47} = 5\frac{47}{99} = \frac{542}{99}$

71. $3.\overline{285}$

$0.\overline{285} = 0.285285\ldots$

$S = \frac{a_1}{1-r} = \frac{0.285}{1-0.001} = \frac{285}{999} = \frac{95}{333}$

$3.\overline{285} = 3\frac{95}{333} = \frac{1094}{333}$

73. $0.1\overline{2} = 0.12222\ldots$

Consider $0.0\overline{2} = 0.02222\ldots$

$= 0.02 + 0.002 + 0.0002 + \ldots$

$a_1 = 0.02,\ r = 0.1$

$S = \frac{a_1}{1-r}$

$S = \frac{0.02}{1-0.1} = \frac{0.02}{0.9} = \frac{2}{90} = \frac{1}{45}$

$0.1\overline{2} = 0.1 + \frac{1}{45} = \frac{1}{10} + \frac{1}{45} = \frac{9}{90} + \frac{2}{90} = \frac{11}{90}$

75. a. 7, 8.04, 9.23, 10.59, …, 110.77

$r \approx \frac{8.04}{7} \approx \frac{9.23}{8.04} \approx \frac{10.59}{9.23} \approx \ldots \approx 1.148$

b. $a_n = a_1 r^{n-1}$

$a_n = 7(1.148)^{n-1}$

c. $n = \frac{1998 - 1798}{5} = 40$

$a_{40} = 7(1.148)^{40-1} \approx 1523$

The population is predicted to be about 1523 million.

77. a. $w_1 = 200, r = 0.99$

$w_1 = 200$

$w_2 = 200(0.99) = 198$

$w_3 = 198(0.99) = 196.02$

$w_4 = 196.02(0.99) = 194.0598$

200, 198, 196.02, 194.0598

b. $w_n = w_1 r^{n-1}$

$w_n = 200(0.99)^{n-1}$

79. Company *A*:

20,000, 21,000, 22,000, ...

(Year 1, Year 2, Year 3, ...)

Arithmetic sequence: $a_n = a_1 + (n-1)d$

Year 6: $a_6 = 20,000 + (6-1)1000$

= 20,000 + 5000 = $25,000

Company *B*:

20,000, 21,000, 22,050, ...

(Year 1, Year 2, Year 3, ...)

Geometric Sequence: $a_n = a_1 r^{n-1}$

$r = \frac{21,000}{20,000} = 1.05$

Year 6: $a_6 = 20,000(1.05)^{6-1}$ = $25,525.63

Company *B* will pay more in the sixth year (approximately $526 more).

81. $S_n = \frac{a_1 - a_1 r^n}{1-r}$

$r = \frac{200\left(1 + \frac{0.09}{12}\right)^2}{200\left(1 + \frac{0.09}{12}\right)} = 1 + \frac{0.09}{12}$

$S_{72} = \frac{200\left(1 + \frac{0.09}{12}\right) - 200\left(1 + \frac{0.09}{12}\right)\left(1 + \frac{0.09}{12}\right)^{72}}{1 - \left(1 + \frac{0.09}{12}\right)}$

$= 19,143.916$

The balance is $19,143.92.

83. Company *A*:

Arithmetic sequence: $a_1 = 20,000$

$d = 1000$

$S_n = \frac{n}{2}(a_1 + a_n)$

$S_6 = \frac{6}{2}(a_1 + a_6)$

Find a_6: $a_n = a_1 + (n-1)d$

$a_6 = a_1 + (6-1)d$

$a_6 = 20,000 + 5(1,000)$

$a_6 = 25,000$

$S_6 = 3(a_1 + a_6)$

$S_6 = 3(20,000 + 25,000) = 135,000$

Company *B*:

Geometric sequence:

$a_1 = 20,000$

$a_2 = 20,000(0.05) + 20,000 = 21,000$

$r = \frac{21,000}{20,000} = 1.05$

$S_n = \frac{a_1 - a_1 r^n}{1-r}$

$S_6 = \frac{20,000 - 20,000(1.05)^6}{1 - 1.05}$

$S_6 = \frac{-6802}{-0.05} = 136,038$

Over six years:

Company *A*: $135,000

Company *B*: $136,040

Company *B* produces the better total income.

85. $r = 0.65$

$S_n = \frac{a_1 - a_1 r^n}{1-r}$

The bungee jumper travels the initial fall's distance plus seven "rebound then fall from the same height" cycles up through and including the 8th fall. The ith rebound-fall cycle covers

$2(200)(0.65)^i = 400(0.65)^i$ feet.

The total distance is

$200 + \sum_{i=1}^{7} 400(0.65)^i$

$= 200 + 400\sum_{i=1}^{7} 0.65^i$

where $\sum_{i=1}^{7} 0.65^i = 0.65 + 0.65^2 + \ldots + 0.65^7$

$\sum_{i=1}^{7} 0.65^i = S_7$ where

$a_1 = 0.65,\ r = 0.65$ and $n = 7$.

$S_n = \frac{a_1(1-r^n)}{1-r}$

$S_7 = \frac{0.65(1-0.65^7)}{1-0.65} \approx 1.766$

So total distance is

$200 + 400\sum_{i=1}^{7} 0.6s^i = 200 + 400S_7$

$= 200 + 400(1.766) = 906.44$

The bungee jumper travels 906.44 feet.

87. The amount spent is given by:

$0.9(40) + (0.9)^2(40) + (0.9)^3(40) + \ldots$

$a_1 = 0.9(40) = 36$

$r = 0.9$

$S = \frac{a_1}{1-r}$

$S = \frac{36}{1-0.9}$

$= \frac{36}{0.1} = 360$

Additional spending: \$360 billion

89. $a_1 = 12,\ r = \frac{1}{2}$

$S = \frac{a_1}{1-r} = \frac{12}{1-\frac{1}{2}} = 24$

24 inches

91. $10 + 1 + \frac{1}{10} + \frac{1}{100} + \ldots$

$a_1 = 10,\ r = \frac{1}{10}$

$S = \frac{a_1}{1-r} = \frac{10}{1-\frac{1}{10}} = \frac{10}{\frac{9}{10}} = \frac{100}{9}$

Achilles must run $\frac{100}{9} = 11\frac{1}{9}$ meters.

It will take him $\frac{100}{9}\left(\frac{1}{10}\right) = 1\frac{1}{9}$ seconds.

93. d is true.

95. d is true.

97. a. $f(x) = \frac{1-0.5^x}{0.5}$

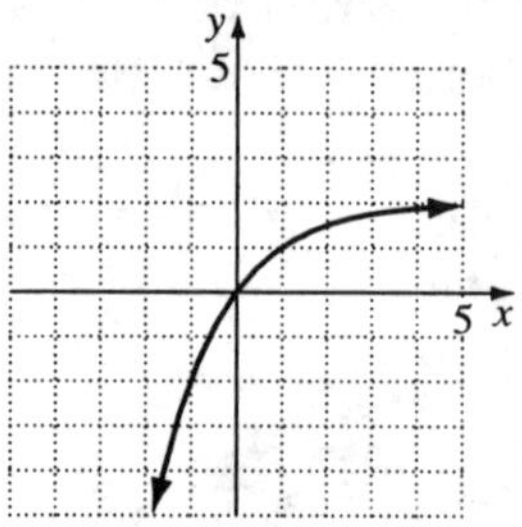

$f(x) = \frac{1-0.9^x}{0.1}$

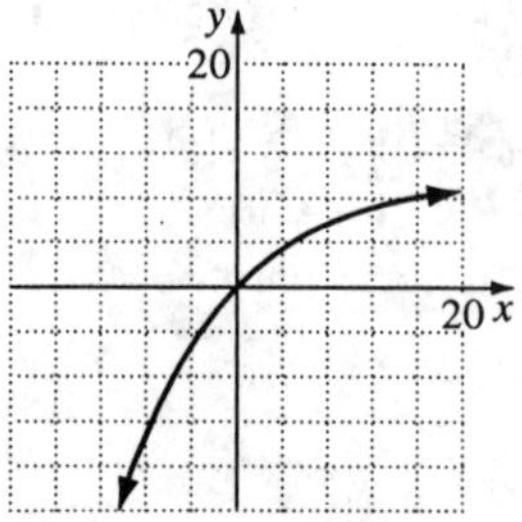

$f(x) = \frac{1-(1.1)^x}{-0.1}$

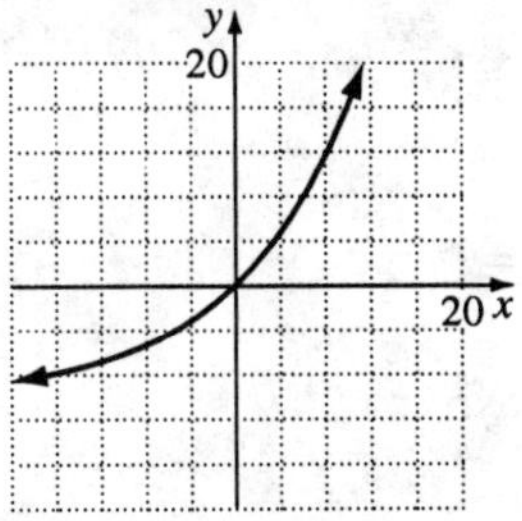

$f(x) = \frac{1-3^x}{-2}$

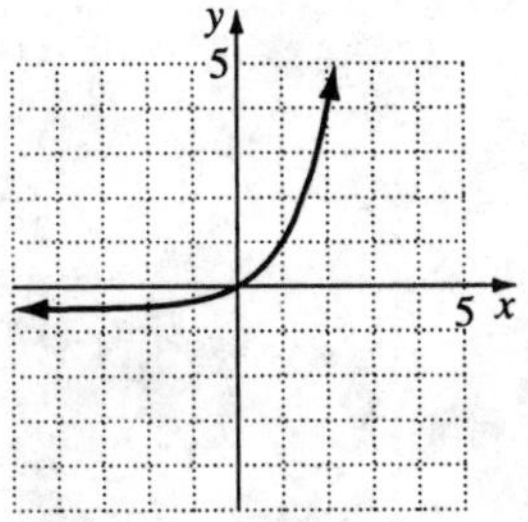

b. For $r = 0.5$, as $x \to \infty$, $f(x)$ has a horizontal asymptote of 2.
For $r = 0.9$, as $x \to \infty$, $f(x)$ has a horizontal asymptote of 10.
For $r = 1.1$, as $x \to \infty$, $f(x)$ has a horizontal asymptote of −10.
For $r = 3$, as $x \to \infty$, $f(x)$ has a horizontal asymptote of $-\frac{1}{2}$.

c. when $0 < r < 1$

99. $f(x) = \frac{4[1-(0.6)^x]}{1-0.6}$

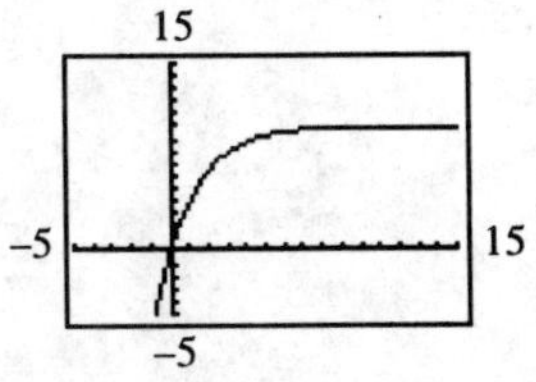

horizontal asymptote = 10

$\sum_{n=0}^{\infty} 4(0.6)^n = 10$

101-105. Answers may vary.

107. Given: $\frac{a^2}{a^1} = r$

Consider: $-a_1, -a_2, -a_3, \ldots, -a_n$

Common ratio $= \frac{-a_2}{-a_1} = \frac{a_2}{a_1} = r$

109. Given: $\frac{a_2}{a_1} = r$

Consider: $a_1, a_3, a_5, \ldots, a_{2n+1}$

Common ratio

$= \frac{a_3}{a_1} = \frac{a_2 r}{a_1} = \frac{a_2}{a_1} \cdot r = r \cdot r = r^2$

111. Given: $\frac{a_2}{a_1} = r$

Consider: $a_1, 4a_2, 16a_3, \ldots, 4^{n-1}a_n$

Common ratio $= \frac{4a_2}{a_1} = 4 \cdot \frac{a_2}{a_1} = 4r$

113. Since 6, a, b is arithmetic:

$b - a = a - 6$

$b = 2a - 6$

Since a, b, 16 is geometric:

$\frac{16}{b} = \frac{b}{a}$

$16a = b^2$

$16a = (2a-6)^2$

$16a = 4a^2 - 24a + 36$

$0 = 4a^2 - 40a + 36$

$0 = a^2 - 10a + 9$

$0 = (a-9)(a-1)$

$a = 9$ or $a = 1$

Since $b = 2a - 6$, the ordered pairs were (9, 12) and (1, −4).

115. $3 + 3^2 + 3^3 + \ldots + 3^n = 120$

$a_1 = 3, r = 3, S_n = 120$

Find n.

$S_n = \frac{a_1 - a_1 r^n}{1-r}$

$120 = \frac{3 - 3 \cdot 3^n}{1-3}$

$-240 = 3 - 3^{n+1}$

$3^{n+1} = 243$
$3^{n+1} = 3^5$
$n + 1 = 5$
$n = 4$

117. Let x = the sixth term of the arithmetic sequence. The sum of the first six terms of 1, 2, 4, 8, ... is

$\frac{1-1\cdot 2^6}{1-2} = \frac{1-64}{-1} = 63$

The sum of the first six terms of an arithmetic sequence whose first term is 1 and whose sixth term is x is

$S_6 = \frac{6}{2}(a_1 + a_n)$
$= 3(1+x) = 3+3x$

Thus,

$3 + 3x = 63$
$3x = 60$
$x = 20$

The sixth term of the arithmetic sequence is 20.

119. $\left(1+\frac{1}{2}+\frac{1}{4}+\frac{1}{8}+\ldots\right)\left(1+\frac{1}{3}+\frac{1}{9}+\frac{1}{27}+\cdots\right)$
$\cdot\left(1+\frac{1}{5}+\frac{1}{25}+\frac{1}{125}+\cdots\right)$
$= \left(\frac{1}{1-\frac{1}{2}}\right)\left(\frac{1}{1-\frac{1}{3}}\right)\left(\frac{1}{1-\frac{1}{5}}\right)$
$= (2)\left(\frac{3}{2}\right)\left(\frac{5}{4}\right)$
$= \frac{15}{4}$

121. $\frac{2}{1}+\frac{1}{3}+\frac{2}{9}+\frac{1}{27}+\frac{2}{81}+\frac{1}{243}+\ldots$
$= \frac{2}{1}+\frac{2}{9}+\frac{2}{81}+\ldots+\frac{1}{3}+\frac{1}{27}+\frac{1}{243}+\ldots$
$= \frac{a_1}{1-r}+\frac{a_1}{1-r}$
$= \frac{2}{1-\frac{1}{9}}+\frac{\frac{1}{3}}{1-\frac{1}{9}}$
$= \frac{2}{\frac{8}{9}}+\frac{\frac{1}{3}}{\frac{8}{9}}$
$= \frac{2}{1}\cdot\frac{9}{8}+\frac{1}{3}\cdot\frac{9}{8}$
$= \frac{9}{4}+\frac{3}{8}$
$= \frac{18+3}{8} = \frac{21}{8}$

Review Problems

122. $\sqrt[3]{54x^6y^7} = \sqrt[3]{27\cdot 2(x^2)^3(y^2)^3y}$
$= 3x^2y^2\sqrt[3]{2y}$

123. Algebraically:

$4x^2 + y^2 = 16$
$2x + y = 4$ (solve for y) $\rightarrow y = 4 - 2x$
$4x^2 + (4-2x)^2 = 16$
(substitute for y) $\rightarrow$
$4x^2 + 16 - 16x + 4x^2 = 16$
$8x^2 - 16x = 0$
$8x(x-2) = 0$
$8x = 0$ or $x - 2 = 0$
$x = 0$ or $x = 2$
If $x = 0$, $y = 4 - 2(0) = 4$
If $x = 2$, $y = 4 - 2(2) = 4 - 4 = 0$
$\{(0, 4), (2, 0)\}$

Graphically:

$4x^2 + y^2 = 16$
$\frac{4x^2}{16}+\frac{y^2}{16} = 1$
$\frac{x^2}{4}+\frac{y^2}{16} = 1$

Ellipse: x-intercepts

$\frac{x^2}{4} = 1$
$x^2 = 4$
$x = \pm 2$

y-intercepts:

$\frac{y^2}{16} = 1$
$y^2 = 16$
$y = \pm 4$

$2x + y = 4$

Line: x-intercept: $2x = 4$

$x = 2$
y–intercept: $y = 4$

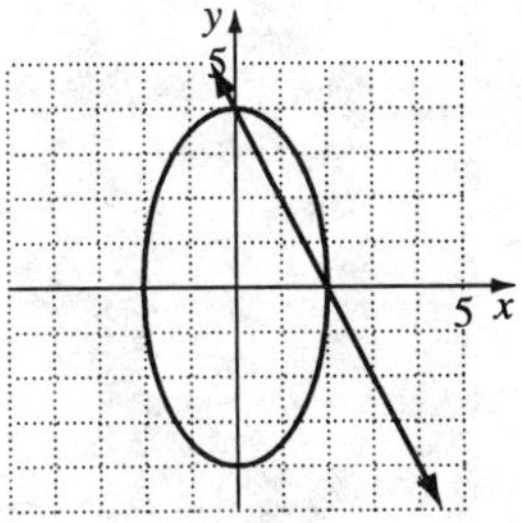

124. $\dfrac{\sqrt{5}+\sqrt{3}}{\sqrt{5}-\sqrt{3}} = \dfrac{\sqrt{5}+\sqrt{3}}{\sqrt{5}-\sqrt{3}} \cdot \dfrac{\sqrt{5}+\sqrt{3}}{\sqrt{5}+\sqrt{3}}$
$= \dfrac{5+2\sqrt{15}+3}{5-3} = \dfrac{8+2\sqrt{15}}{2} = 4+\sqrt{15}$

Problem Set 11.4

1. $3! = 3\cdot 2\cdot 1 = 6$

3. $2! = 2\cdot 1 = 2$

5. $\dfrac{10!}{8!2!} = \dfrac{10\cdot 9}{2\cdot 1} = 45$
(Factor 8! from 10!)

7. $\dfrac{7!}{6!1!} = \dfrac{7}{1} = 7$

9. $\binom{6}{3} = \dfrac{6!}{3!(6-3)!}$
$= \dfrac{6!}{3!3!}$
$= \dfrac{6\cdot 5\cdot 4}{3\cdot 2\cdot 1}$
$= 20$

11. $\binom{12}{1} = \dfrac{12!}{1!11!}$
$= \dfrac{12}{1}$
$= 12$

13. $\binom{6}{6} = \dfrac{6!}{6!(6-6)!}$
$= \dfrac{6!}{6!0!}$
$= \dfrac{1}{0!}$
$= 1$

15. $(c+2)^5 = \binom{5}{0}c^5 + \binom{5}{1}(2)c^4 + \binom{5}{2}(4)c^3 + \binom{5}{3}(8)c^2 + \binom{5}{4}(16)c + \binom{5}{5}(32)$
$= c^5 + 5c^4(2) + \dfrac{5\cdot 4}{2\cdot 1}(4)c^3 + \dfrac{5\cdot 4\cdot 3}{3\cdot 2\cdot 1}(8)c^2 + \dfrac{5\cdot 4\cdot 3\cdot 2}{4\cdot 3\cdot 2\cdot 1}(16)c + 32$
$= c^5 + 10c^4 + 10c^3(4) + 10c^2(8) + 5c(16) + 32$
$= c^5 + 10c^4 + 40c^3 + 80c^2 + 80c + 32$

17. $(x+y)^6 = \binom{6}{0}x^6 + \binom{6}{1}x^5y + \binom{6}{2}x^4y^2 + \binom{6}{3}x^3y^3 + \binom{6}{4}x^2y^4 + \binom{6}{5}xy^5 + \binom{6}{6}y^6$
$= x^6 + 6x^5y + \dfrac{6\cdot 5}{2\cdot 1}x^4y^2 + \dfrac{6\cdot 5\cdot 4}{3\cdot 2\cdot 1}x^3y^3 + \dfrac{6\cdot 5\cdot 4\cdot 3}{4\cdot 3\cdot 2\cdot 1}x^2y^4 + 6xy^5 + y^6$
$= x^6 + 6x^5y + 15x^4y^2 + 20x^3y^3 + 15x^2y^4 + 6xy^5 + y^6$

19. $(a-2)^4 = [a+(-2)]^4$
$= \binom{4}{0}a^4 + \binom{4}{1}a^3(-2) + \binom{4}{2}a^2(-2)^2 + \binom{4}{3}a(-2)^3 + \binom{4}{4}(-2)^4$
$= a^4 + 4a^3(-2) + 6a^2(4) + 4a(-8) + 16$
$= a^4 - 8a^3 + 24a^2 - 32a + 16$

21. $(x+2y)^6 = \binom{6}{0}x^6 + \binom{6}{1}x^5(2y) + \binom{6}{2}x^4(2y)^2 + \binom{6}{3}x^3(2y)^3 + \binom{6}{4}x^2(2y)^4 + \binom{6}{5}x(2y)^5 + \binom{6}{6}(2y)^6$

$x^6 + 6x^5(2y) + 15x^4(4y^2) + 20x^3(8y^3) + 15x^2(16y^4) + 6x(32y^5) + 64y^6$

$= x^6 + 12x^5y + 60x^4y^2 + 160x^3y^3 + 240x^2y^4 + 192xy^5 + 64y^6$

23. $\left(\frac{a}{2}+1\right)^4 = \binom{4}{0}\left(\frac{a}{2}\right)^4 + \binom{4}{1}\left(\frac{a}{2}\right)^3 \cdot 1 + \binom{4}{2}\left(\frac{a}{2}\right)^2 \cdot 1^2 + \binom{4}{3}\left(\frac{a}{2}\right)^1 \cdot 1^3 + \binom{4}{4} \cdot 1^4$

$= (1)\left(\frac{a^4}{16}\right) + 4\left(\frac{a^3}{8}\right) + 6\left(\frac{a^2}{4}\right) + 4\left(\frac{a}{2}\right) + 1$

$= \frac{a^4}{16} + \frac{a^3}{2} + \frac{3a^2}{2} + 2a + 1$

25. $(2x+3y)^4 = \binom{4}{0}(2x)^4 + \binom{4}{1}(2x)^3(3y) + \binom{4}{2}(2x)^2(3y)^2 + \binom{4}{3}(2x)^1(3y)^3 + \binom{4}{4}(3y)^4$

$= 1(16x^4) + 4(8x^3)(3y) + 6(4x^2)(9y^2) + 4(2x)(27y^3) + 1(81y^4)$

$= 16x^4 + 96x^3y + 216x^2y^2 + 162xy^3 + 81y^4$

27. $(2x^2 - y^2)^3 = [2x^2 + (-y^2)]^3$

$= \binom{3}{0}(2x^2)^3 + \binom{3}{1}(2x^2)^2(-y^2) + \binom{3}{2}(2x^2)(-y^2)^2 + \binom{3}{3}(-y^2)^3$

$= 1(8x^6) + 3(4x^4)(-y^2) + 3(2x^2)(y^4) + 1(-y^6)$

$= 8x^6 - 12x^4y^2 + 6x^2y^4 - y^6$

29. $\left(\frac{a}{3}+2\right)^3 = \binom{3}{0}\left(\frac{a}{3}\right)^3 + \binom{3}{1}\left(\frac{a}{3}\right)^2(2) + \binom{3}{2}\left(\frac{a}{3}\right)^1(2)^2 + \binom{3}{3}(2)^3$

$= 1\left(\frac{a^3}{27}\right) + 3\left(\frac{a^2}{9}\right)(2) + 3\left(\frac{a}{3}\right)(4) + 1(8)$

$= \frac{a^3}{27} + \frac{2a^2}{3} + 4a + 8$

31. $(\sqrt{2}-x)^6 = \binom{6}{0}(\sqrt{2})^6 + \binom{6}{1}(\sqrt{2})^5(-x) + \binom{6}{2}(\sqrt{2})^4(-x)^2 + \binom{6}{3}(\sqrt{2})^3(-x)^3$

$+ \binom{6}{4}\left(\sqrt{2}^2\right)(-x)^4 + \binom{6}{5}(\sqrt{2})^1(-x)^5 + \binom{6}{6}(-x)^6$

$= 8 + 6(4\sqrt{2})(-1)x + 15(4)(x^2) + 20(2\sqrt{2})(-1)x^3 + 15(2)x^4 + 6(\sqrt{2})(-1)x^5 + x^6$

$= 8 - 24\sqrt{2}x + 60x^2 - 40\sqrt{2}x^3 + 30x^4 - 6\sqrt{2}x^5 + x^6$

33. $(a^{1/2}+2)^4=\binom{4}{0}(a^{1/2})^4+\binom{4}{1}(a^{1/2})^3(2)+\binom{4}{2}(a^{1/2})^2(2)^2+\binom{4}{3}(a^{1/2})(2)^3+\binom{4}{4}(2)^4$
$=1(a^2)+4(a^{3/2})(2)+6a(4)+4a^{1/2}(8)+1(16)$
$=a^2+8a^{3/2}+24a+32a^{1/2}+16$

35. $(a^{-1}+b^{-1})^3=\binom{3}{0}(a^{-1})^3+\binom{3}{1}(a^{-1})^2(b^{-1})+\binom{3}{2}(a^{-1})(b^{-1})^2+\binom{3}{3}(b^{-1})^3$
$=1a^{-3}+3a^{-2}b^{-1}+3a^{-1}b^{-2}+1b^{-3}$
$=\frac{1}{a^3}+\frac{3}{a^2b}+\frac{3}{ab^2}+\frac{1}{b^3}$

37. $(x^2+x)^8=\binom{8}{0}(x^2)^8+\binom{8}{1}(x^2)^7x+\binom{8}{2}(x^2)^6x^2+\ldots$
$=1x^{16}+8x^{14}(x)+\frac{8\cdot7}{2\cdot1}x^{12}x^2+\ldots$
$=x^{16}+8(x^{14})x+28(x^{12})x^2+\ldots$
$=x^{16}+8x^{15}+28x^{14}+\ldots$

39. $(2a+b)^8=\binom{8}{0}(2a)^8+\binom{8}{1}(2a)^7b+\binom{8}{2}(2a)^6b^2+\ldots$
$=(1)(2a)^8+8(2a)^7b+\frac{8\cdot7}{2\cdot1}(2a)^6b^2+\ldots$
$=1(256a^8)+8(128a^7)b+28(64a^6)b^2+\ldots$
$=256a^8+1024a^7b+1729a^6b^2+\ldots$

41. $(a-2b)^8=[a+(-2b)]^8$
$=\binom{8}{0}a^8+\binom{8}{1}a^7(-2b)+\binom{8}{2}a^6(-2b)^2+\ldots$
$=1a^8+8a^7(-2b)+\frac{8\cdot7}{2\cdot1}a^6(-2b)^2+\ldots$
$=a^8+8a^7(-2b)+28a^6(4b^2)+\ldots$
$=a^8-16a^7b+112a^6b^2+\ldots$

43. $(a^2+b^2)^{10}=\binom{10}{0}(a^2)^{10}+\binom{10}{1}(a^2)^9(b^2)+\binom{10}{2}(a^2)^8(b^2)^2+\ldots$
$=a^{20}+10a^{18}b^2+\frac{10\cdot9}{2\cdot1}a^{16}b^4+\ldots$
$=a^{20}+10a^{18}b^2+45a^{16}b^4+\ldots$

45. $(a+b)^{42}=\binom{42}{0}a^{42}+\binom{42}{1}a^{41}b+\binom{42}{2}a^{40}b^2+\ldots$
$=a^{42}+42a^{41}b+\frac{42\cdot41}{2\cdot1}a^{40}b^2+\ldots$
$=a^{42}+42a^{41}b+861a^{40}b^2+\ldots$

47. $\left(y+\frac{1}{y}\right)^7 = \binom{7}{0}y^7 + \binom{7}{1}y^6\left(\frac{1}{y}\right) + \binom{7}{2}y^5\left(\frac{1}{y}\right)^2 + \ldots$

$= 1y^7 + 7y^6\left(\frac{1}{y}\right) + \frac{7\cdot 6}{2\cdot 1}y^5\left(\frac{1}{y}\right)^2 + \ldots$

$= y^7 + 7y^6\left(\frac{1}{y}\right) + 21y^5\left(\frac{1}{y^2}\right) + \ldots$

$= y^7 + 7y^5 + 21y^3 + \ldots$

49. $\binom{n}{k}p^k q^{n-k}$

$n = 10,\ k = 5,\ p = 0.5,\ q = 0.5$

$\binom{10}{5}(0.5)^5(0.5)^{10-5}$

$= \frac{10\cdot 9\cdot 8\cdot 7\cdot 6}{5\cdot 4\cdot 3\cdot 2\cdot 1}(0.5)^5(0.5)^5$

$= 252(0.03125)^2 \approx 0.246$

The probability is about 0.246.

51. $\binom{n}{k}p^k q^{n-k}$

$n = 5,\ k = 3,\ p = 0.3,\ q = 0.7$

$\binom{5}{3}(0.3)^3(0.7)^{5-3}$

$= 10(0.3)^3(0.7)^2 = 0.1323$

The probability is 0.1323.

53. c is true.

55. Students should verify results.

57. $f_1(x) = (x+1)^4$

$f_2(x) = x^4$

$f_3(x) = x^4 + 4x^3$

$f_4(x) = x^4 + 4x^3 + 6x^2$

$f_5(x) = x^4 + 4x^3 + 6x^2 + 4x$

$f_6(x) = x^4 + 4x^3 + 6x^2 + 4x + 1$

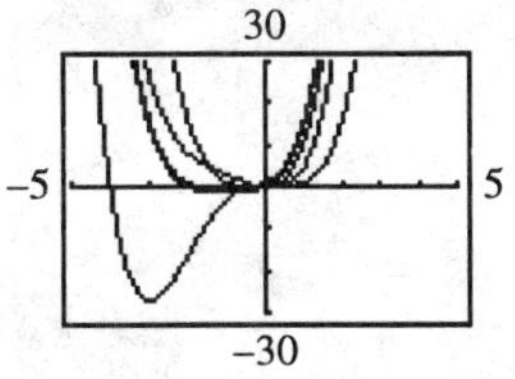

f_2, f_3, f_4, and f_5 are approaching $f_1 = f_6$.

59. $f_1(x) = (x-2)^4$

$= \binom{4}{0}x^4 + \binom{4}{1}x^3(-2) + \binom{4}{2}x^2(-2)^2 + \binom{4}{3}x(-2)^3 + \binom{4}{4}(-2)^4$

$= x^4 + 4x^3(-2) + 6x^2(4) + 4x(-8) + 16$

$= x^4 - 8x^3 + 24x^2 - 32x + 16$

Students should verify expansion with a graphing utility.

61. Answers may vary.

63. Answers may vary.

65. $\dfrac{(3!)^2 + (3!)!}{\binom{3!}{2!}} = \dfrac{(3\cdot 2\cdot 1)^2 + (3\cdot 2\cdot 1)!}{\binom{(3\cdot 2\cdot 1)}{(2\cdot 1)}}$

$= \dfrac{6^2 + 6!}{\binom{6}{2}} = \dfrac{36 + 720}{\frac{6!}{4!2!}} = \dfrac{756}{\frac{6\cdot 5}{2\cdot 1}}$

$= \dfrac{756}{15} = \dfrac{252}{5}$

67. $\dfrac{(n+1)!}{n!} = \dfrac{(n+1)n!}{n!} = n+1$

69. $\binom{n}{r} = \dfrac{n!}{r!(n-r)!} = \dfrac{n!}{(n-r)!r!}$

$= \dfrac{n!}{(n-r)![n-(n-r)]!} = \binom{n}{n-r}$

Review Problems

71. $4x^2 + 4y^2 = 65$

$6x - 2y = 5$

$x = \dfrac{5+2y}{6}$

$4\left(\dfrac{5+2y}{6}\right)^2 + 4y^2 = 65$

$4\left(\dfrac{25+20y+4y^2}{36}\right) = 65 - 4y^2$

$25 + 20y + 4y^2 = 585 - 36y^2$

$40y^2 + 20y - 560 = 0$

$2y^2 + y - 28 = 0$

$(2y - 7)(y + 4) = 0$

$y = \frac{7}{2}$ or $y = -4$

$x = \dfrac{5+2\left(\frac{7}{2}\right)}{6} = 2$ or $x = \dfrac{5+2(-4)}{6} = -\dfrac{1}{2}$

$\left\{\left(2, \frac{7}{2}\right), \left(-\frac{1}{2}, -4\right)\right\}$

72. $f(x) = 3x + 5$
$y = 3x + 5$
$x = 3y + 5$ (Exchange x and y.)
$x - 5 = 3y$
$\frac{x-5}{3} = y$
$f^{-1}(x) = \frac{x-5}{3}$
$f(f^{-1}(x)) = 3(f^{-1}(x)) + 5$
$= 3\left(\frac{x-5}{3}\right) + 5 = x - 5 + 5 = x$
$f^{-1}(f(x)) = \frac{f(x)-5}{3} = \frac{3x+5-5}{3}$
$= \frac{3x}{3} = x$

73. $\log 5x + 2\log x = \log 5x + \log x^2$
$= \log(5x)(x^2) = \log 5x^3$

Problem Set 11.5

1. $10 \cdot 9 \cdot 8 = 720$

3. $10 \cdot 10 \cdot 10 \cdot 10 = 10{,}000$

5. $2 \cdot 26 \cdot 26 \cdot 26 = 35{,}152$

7. $7 \cdot 10 \cdot 10 \cdot 5 = 3500$

9. 5##0 or 5##2 or 5##4
3(4)(3) = 36

11. $(2)^{10} = 1024$

13. W M W M W M W
$4 \cdot 3 \cdot 3 \cdot 2 \cdot 2 \cdot 1 \cdot 1 = 144$

15. ${}_6P_6 = \frac{6!}{0!} = 720$

17. ${}_9P_4 = \frac{9!}{5!} = 3024$

19. ${}_{10}P_2 = \frac{10!}{8!} = 90$

21. ${}_7P_6 = \frac{7!}{1!} = 5040$

23. ${}_{20}P_4 = \frac{20!}{16!} = 116{,}280$

25. ${}_6P_6 = \frac{6!}{0!} = 720$

27. ${}_7P_3 = \frac{7!}{4!} = 210$

29. $\binom{8}{1} = \frac{8!}{1!7!} = 8$

31. $\binom{6}{6} = \frac{6!}{6!0!} = 1$

33. $\binom{10}{6} = \frac{10!}{6!4!} = 210$

35. $\binom{30}{3} = \frac{30!}{3!27!} = 4060$

37. $\binom{6}{3} = \frac{6!}{3!3!} = 20$

39. $\binom{49}{6} = \frac{49!}{6!43!} = 13{,}983{,}816$

41. $10 \cdot 10 \cdot 10 = 1000$

43. $\binom{25}{5} = \frac{25!}{5!20!} = 53{,}130$

45. $\binom{25}{5} = \frac{25!}{5!20!} = 53{,}130$

47. $2 \cdot 2 \cdot 2 \cdot 3 \cdot 3 = 72$

49. ${}_8P_3 = \frac{8!}{5!} = 336$

51. $\binom{8}{3} = \frac{8!}{3!5!} = 56$

53. $\binom{3}{1}\binom{6}{2}\binom{5}{2} = \frac{3!}{1!2!} \cdot \frac{6!}{2!4!} \cdot \frac{5!}{2!3!} = 450$

55. $6 \cdot \binom{10}{4} = 6 \cdot \frac{10!}{4!6!} = 1260$

57. $\binom{12}{0}\binom{9}{5}+\binom{12}{1}\binom{9}{4}+\binom{12}{2}\binom{9}{3}+\binom{12}{3}\binom{9}{2}=1\cdot\frac{9!}{5!4!}+12\cdot\frac{9!}{4!5!}+\frac{12!}{2!10!}\cdot\frac{9!}{3!6!}+\frac{12!}{3!9!}\cdot\frac{9!}{2!7!}$
$= 126 + 1512 + 5544 + 7920 = 15{,}102$

59. $\binom{39}{4}=\frac{39!}{4!35!}=82{,}251$

61. $13(3)\cdot 12(3) = 1404$

63. $13(1)\cdot 48 = 624$

65. c is true.
$\binom{n}{r}=\frac{n!}{r!(n-4)!}$
${}_nP_r=\frac{n!}{(n-r)!}$
$\frac{{}_nP_r}{r!}=\frac{n!}{(n-r)!r!}=\binom{n}{r}$

67–75. Answers may vary.

77. ${}_4P_4({}_5P_5\cdot{}_3P_3\cdot{}_2P_2\cdot{}_1P_1)=\frac{4!}{0!}\left(\frac{5!}{0!}\cdot\frac{3!}{0!}\cdot\frac{2!}{0!}\cdot\frac{1!}{0!}\right)$
$= 4!(5!\cdot 3!\cdot 2!\cdot 1!) = 34{,}560$

79. Any combination of 5 heads and 5 tails
$\frac{10!}{5!5!}=252$

Review Problems

80. $\sqrt{x}+\sqrt{x-5}=5$
$\sqrt{x-5}=5-\sqrt{x}$
$x-5=25-10\sqrt{x}+x$
$3=\sqrt{x}$
$\pm 9=x$
Disregard the negative.
$\{9\}$

81. $\log(x^2-9)-\log(x+3)=2$
$\log\frac{x^2-9}{x+3}=2$
$\log(x-3)=2$
$10^2=x-3$
$103=x$
$\{103\}$

82. $x^2 + 3x > 4$
$x^2 + 3x - 4 > 0$
$(x + 4)(x - 1) > 0$

T	F	T
	-4	1

Test -5: $(-5)^2 + 3(-5) > 4$
$10 > 4$ True
Test 0: $0^2 + 3(0) > 4$
$0 > 4$ False
Test 4: $4^2 + 3(4) > 4$
$28 > 4$ True
$\{x \mid x < -4 \text{ or } x > 1\}$, $(-\infty, -4) \cup (1, \infty)$

Problem Set 11.6

1. $\frac{n(E)}{n(S)} = \frac{3}{8}$

3. $\frac{n(E)}{n(S)} = \frac{7}{8}$

5. $\frac{n(E)}{n(S)} = \frac{4}{36} = \frac{1}{9}$

7. $\frac{n(E)}{n(S)} = \frac{5}{36}$

9. $\frac{n(E)}{n(S)} = \frac{21}{36} = \frac{7}{12}$

11. a. $\frac{n(E)}{n(S)} = \frac{731}{3012}$

b. $\frac{n(E)}{n(S)} = \frac{864 + 607}{3012} = \frac{1471}{3012}$

13. $P(E) = \frac{n(E)}{n(S)} = \frac{1 \cdot 10^5 \cdot 1}{8 \cdot 10^6} = \frac{1}{80}$

15. $n(5) = 5! = 120$
event is Marilyn, x, x, x, Lee
$n(E) = 1 \cdot 3 \cdot 2 \cdot 1 \cdot 1 = 6$
so $P(E) = \frac{n(E)}{n(S)} = \frac{6}{120} = \frac{1}{20}$

17. $P(E) = \frac{n(E)}{n(S)} = \frac{1}{\binom{51}{6}} = \frac{1}{18,009,460}$
$P(E) = \frac{100}{18,009,460} = \frac{5}{900,473}$
$\frac{x}{18,009,460} = \frac{1}{2}$
$x = 9,004,730$
$\$2(9,004,730) = \$18,009,460$

19. $P(\text{En} \cup \text{Sp}) = P(\text{En}) + P(\text{Sp}) - P(\text{En} \cap \text{Sp})$
$= \frac{30}{100} + \frac{40}{100} - \frac{20}{100} = \frac{1}{2}$

21. $P(E) = \frac{n(E)}{n(S)} = \frac{75,791}{262,756} \approx 0.29$

23. $P(E) = \frac{n(E)}{n(S)} = \frac{28,805 + 11,145 + 3,628}{262,756}$
$= \frac{43,578}{262,756} \approx 0.16$

25. $P(E) = \frac{n(E)}{n(S)} = \frac{81,004 + 28,805}{262,756}$
$= \frac{109,809}{262,756} \approx 0.42$

27. Let A = 20–29 age group, B = 30–39 age group, and C = 20–39 age group.
$n(A) = n(C) - n(B) = 81,004 - 44,117$
$= 36,887$
$P(A) = \frac{n(A)}{n(S)} = \frac{36,887}{262,756} \approx 0.14$

29. $P(\text{AA or HA}) = P(\text{AA}) + P(\text{HA})$
$= \frac{8}{100} + \frac{24}{100} = \frac{32}{100} = \frac{8}{25} = 32\%$

31. $P(\text{HA and HA}) = P(\text{HA}) \cdot P(\text{HA})$
$= \frac{24}{100} \cdot \frac{24}{100} = \frac{756}{10,000} = \frac{189}{2500} = 5.76\%$
(The two draws can be considered independent on account of the large number of people involved.)

33. $P(\text{not HA and not HA})$
$= P(\text{not HA}) \cdot P(\text{not HA})$
$= [1 - P(\text{HA})] \cdot [1 - P(\text{HA})]$
$= \left(1 - \frac{24}{100}\right)^2 = \left(\frac{19}{25}\right)^2 = \frac{361}{625} = 0.5776$
$= 57.76\%$
(The draws may be considered independent.)

35. $P(E) = \frac{n(E)}{n(S)} \approx \frac{70,000}{203,000} = \frac{10}{29} \approx 0.34$

37. $P(E) = \frac{n(E)}{n(S)} \approx \frac{70,000 + 40,000}{203,000} = \frac{110}{203}$
≈ 0.54

39. $\frac{n(E)}{n(S)} = \frac{13}{52} = \frac{1}{4}$

41. $\frac{n(E)}{n(S)} = \frac{8}{52} = \frac{2}{13}$

43. $\frac{n(E)}{n(S)} = \frac{16}{52} = \frac{4}{13}$

45. $\frac{n(E)}{n(S)} = \frac{13 + 12 - 3}{52} = \frac{22}{52} = \frac{11}{26}$

47. $P(E) = \frac{\binom{39}{5}}{\binom{52}{5}} = \frac{\frac{39!}{5!34!}}{\frac{52!}{5!47!}}$
$= \frac{39 \cdot 38 \cdot 37 \cdot 36 \cdot 35}{5!(2,598,960)} = \frac{2109}{9520} \approx 0.22$

49. $P(E) = \frac{n(E)}{n(S)} = \frac{\binom{13}{5}}{\binom{52}{5}} = \frac{13!}{5!8!(2,598,960)}$
$\frac{13 \cdot 12 \cdot 11 \cdot 10 \cdot 9}{5!(2,598,960)} = \frac{154,440}{120(2,598,960)}$
$= \frac{1287}{2,598,960} = \frac{33}{66,640} \approx 0.0005$

51. c is true. The 2 balls selected can consist of 0, 1, or 2 defective balls.
E = at least one defective = 1 or 2 defective
E' = 0 defective = 2 good balls

53. Students should generate random numbers.

55. Answers may vary.

57. $P(E) = \left(\frac{1}{48}\right)^{18} = \frac{1}{1.8295 \times 10^{30}}$

Review Problems

58. $\frac{y - \frac{b^2}{y}}{1 + \frac{b}{y}} = \frac{\frac{y^2 - b^2}{y}}{\frac{y+b}{y}} = \frac{y^2 - b^2}{y + b} = \frac{(y - b)(y + b)}{y + b}$
$= y - b$

59. $\frac{42y}{x^2 - 49} + \frac{3y}{x + 7} + \frac{3y}{x - 7} = \frac{42y + 3y(x - 7) + 3y(x + 7)}{x^2 - 49}$
$= \frac{42y + 3xy - 21y + 3xy + 21y}{x^2 - 49} = \frac{42y + 6xy}{x^2 - 49}$
$= \frac{6y(7 + x)}{(x + 7)(x - 7)} = \frac{6y}{x - 7}$

60. $x^3 + 3x^2 - xy - 3y = x^2(x + 3) - y(x + 3)$
$= (x + 3)(x^2 - y)$

Problem Set 11.7

1. S_n: $2+4+6+\cdots+2n=n(n+1)$

S_1: $2=1(1+1)$

$2=2$

Assume S_k is true.

S_k: $2+4+6+\cdots+2k=k(k+1)$

This implies the truth of S_{k+1}.

S_{k+1}:
$2+4+6+\cdots+2(k+1)=(k+1)(k+2)$

Proof: (work with S_k)

$2+4+6+\cdots+2k+2(k+1)=k(k+1)+2(k+1)$

$2+4+6+\cdots+2(k+1)=(k+1)(k+2)$

This is the statement for $n=k+1$.

Therefore, S_n is true.

3. S_n: $6+10+14+\cdots+(4n+2)=2n(n+2)$

S_1: $6=2(1)(1+2)$

$6=6$

Assume S_k is true.

S_k: $6+10+14+\cdots+(4k+2)=2k(k+2)$

This implies the truth of S_{k+1}.

S_{k+1}: $6+10+14+\cdots+(4k+6)=2(k+1)(k+3)$

Proof: (Work with S_k.)

$6+10+14+\cdots+(4k+2)+(4k+6)=2k(k+2)+(4k+6)$

$6+10+14+\cdots+(4k+6)=2k^2+4k+4k+6$

$6+10+14+\cdots+(4k+6)=2(k^2+4k+3)$

$6+10+14+\cdots+(4k+6)=2(k+1)(k+3)$

This is the statement for $n=k+1$.

Therefore, S_n is true.

5. S_n: $2+4+8+\cdots+2^n=2^{n+1}-2$

S_1: $2=2^{1+1}-2$

$2=2$

Assume S_k is true.

S_k: $2+4+8+\cdots+2^k=2^{k+1}-2$

This implies the truth of S_{k+1}.

S_{k+1}: $2+4+8+\cdots+2^{k+1}=2^{k+2}-2$

Proof:

$2+4+8+\cdots+2^k+2^{k+1}=2^{k+1}-2+2^{k+1}$

$2+4+8+\cdots+2^{k+1}=2^{k+2}-2$

This is the statement for $n = k + 1$.

Therefore, S_n is true.

7. S_n: $1^3+2^3+3^3+\cdots+n^3=\dfrac{n^2(n+1)^2}{4}$

S_1: $1^3=\dfrac{1^2(1+1)^2}{4}$

$1=1$

Assume S_k is true.

S_k: $1^3+2^3+3^3+\cdots+k^3=\dfrac{k^2(k+1)^2}{4}$

This implies the truth of S_{k+1}.

S_{k+1}: $1^3+2^3+3^3+\cdots+(k+1)^3=\dfrac{(k+1)^2(k+2)^2}{4}$

Proof:

$1^3+2^3+3^3+\cdots+k^3+(k+1)^3=\dfrac{k^2(k+1)^2}{4}+(k+1)^3$

$1^3+2^3+3^3+\cdots+(k+1)^3=\dfrac{k^2(k+1)^2+4(k+1)^3}{4}$

$1^3+2^3+3^3+\cdots+(k+1)^3=\dfrac{[k^2+4(k+1)](k+1)^2}{4}$

$1^3+2^3+3^3+\cdots+(k+1)^3=\dfrac{(k+2)^2(k+1)^2}{4}$

This is the statement for $n = k + 1$.

Therefore, S_n is true.

9. S_n: $1\cdot2+2\cdot3+3\cdot4+\cdots+n(n+1)=\dfrac{n(n+1)(n+2)}{3}$

S_1: $1\cdot2=\dfrac{1(1+1)(1+2)}{3}$

$2=2$

Assume S_k is true.

S_k: $1\cdot2+2\cdot3+3\cdot4+\cdots+k(k+1)=\dfrac{k(k+1)(k+2)}{3}$

This implies the truth of S_{k+1}.

S_{k+1}: $1\cdot2+2\cdot3+3\cdot4+\cdots+(k+1)(k+2)=\dfrac{(k+1)(k+2)(k+3)}{3}$

Proof:

$1\cdot2+2\cdot3+3\cdot4+\cdots+k(k+1)+(k+1)(k+2)=\dfrac{k(k+1)(k+2)}{3}+(k+1)(k+2)$

$1\cdot2+2\cdot3+3\cdot4+\cdots+(k+1)(k+2)=\dfrac{k(k+1)(k+2)+3(k+1)(k+2)}{3}$

$1\cdot2+2\cdot3+3\cdot4+\cdots+(k+1)(k+2)=\dfrac{(k+1)(k+2)(k+3)}{3}$

This is the statement for $n = k + 1$.
Therefore, S_n is true.

11. S_n: $\frac{1}{1\cdot 2\cdot 3}+\frac{1}{2\cdot 3\cdot 4}+\frac{1}{3\cdot 4\cdot 5}+\cdots+\frac{1}{n(n+1)(n+2)}=\frac{n(n+3)}{4(n+1)(n+2)}$

S_1: $\frac{1}{1\cdot 2\cdot 3}=\frac{1(1+3)}{4(1+1)(1+2)}$

$\frac{1}{6}=\frac{1}{6}$

Assume S_k is true.

S_k: $\frac{1}{1\cdot 2\cdot 3}+\frac{1}{2\cdot 3\cdot 4}+\frac{1}{3\cdot 4\cdot 5}+\cdots+\frac{1}{k(k+1)(k+2)}=\frac{k(k+3)}{4(k+1)(k+2)}$

This implies the truth of S_{k+1}.

S_{k+1}: $\frac{1}{1\cdot 2\cdot 3}+\frac{1}{2\cdot 3\cdot 4}+\frac{1}{3\cdot 4\cdot 5}+\cdots+\frac{1}{(k+1)(k+2)(k+3)}=\frac{(k+1)(k+4)}{4(k+2)(k+3)}$

Proof:

$$\frac{1}{1\cdot 2\cdot 3}+\frac{1}{2\cdot 3\cdot 4}+\frac{1}{3\cdot 4\cdot 5}+\cdots+\frac{1}{k(k+1)(k+2)}+\frac{1}{(k+1)(k+2)(k+3)}$$
$$=\frac{k(k+3)}{4(k+1)(k+2)}+\frac{1}{(k+1)(k+2)(k+3)}$$

$$\frac{1}{1\cdot 2\cdot 3}+\frac{1}{2\cdot 3\cdot 4}+\frac{1}{3\cdot 4\cdot 5}+\cdots+\frac{1}{(k+1)(k+2)(k+3)}=\frac{k(k+3)^2+4}{4(k+1)(k+2)(k+3)}$$

$$\frac{1}{1\cdot 2\cdot 3}+\frac{1}{2\cdot 3\cdot 4}+\frac{1}{3\cdot 4\cdot 5}+\cdots+\frac{1}{(k+1)(k+2)(k+3)}=\frac{(k+1)^2(k+4)}{4(k+1)(k+2)(k+3)}$$

$$\frac{1}{1\cdot 2\cdot 3}+\frac{1}{2\cdot 3\cdot 4}+\frac{1}{3\cdot 4\cdot 5}+\cdots+\frac{1}{(k+1)(k+2)(k+3)}=\frac{(k+1)(k+4)}{4(k+2)(k+3)}$$

This is the statement for $n = k + 1$.
Therefore, S_n is true.

13. $\sum_{i=1}^{n} i^4=\frac{n(n+1)(2n+1)(3n^2+3n-1)}{30}$

S_n: $1^4+2^4+3^4+\cdots+n^4=\frac{n(n+1)(2n+1)(3n^2+3n-1)}{30}$

S_1: $1^4=\frac{1(2)(3)(5)}{30}$

$1 = 1$

Assume S_k is true.

S_k: $1^4+2^4+3^4+\cdots+k^4=\frac{k(k+1)(2k+1)(3k^2+3k-1)}{30}$

This implies the truth of S_{k+1}.

S_{k+1}: $1^4+2^4+3^4+\cdots+(k+1)^4=\frac{(k+1)(k+2)(2k+3)[3(k+1)^2+3(k+1)-1]}{30}$

S_{k+1}: $1^4+2^4+3^4+\cdots+(k+1)^4=\frac{(k+1)(k+2)(2k+3)(3k^2+9k+5)}{30}$

Proof:

$$1^4+2^4+3^4+\cdots+k^4+(k+1)^4=\frac{k(k+1)(2k+1)(3k^2+3k-1)}{30}+(k+1)^4$$

$$1^4+2^4+3^4+\cdots+(k+1)^4=\frac{k(k+1)(2k+1)(3k^2+3k-1)+30(k+1)^4}{30}$$

$$1^4+2^4+3^4+\cdots+(k+1)^4=\frac{(k+1)(k+2)(2k+3)(3k^2+9k+5)}{30}$$

This is the statement for $n = k + 1$.
Therefore S_n is true.

15. a. one disk: 1 move
2 disks: 3 moves = 1 move + 2 moves
3 disks: 7 moves = (1 move + 2 moves) + 4 moves
4 disks: 15 moves = (1 move + 2 moves + 4 moves) + 8 moves
For *n* disks, the game requires that the fewest number of moves necessary for moving the $n-1$ previous disks be increased by the moves necessary to move the *n*th disk at that stage of the game.
Two possible conjectures: for *n* disks
—the minimum number of moves is 2^n-1
—the minimum number of moves is $\sum 2^{(n-1)}$ (which takes into account the fact that the minimum number of moves for *n* disks depends on the minimum number of moves for $n-1$ disks)

b. Students should prove their conjecture.

17. d is true.

$$S_k=\sum_{i=1}^{k}a_i=a_1+a_2+\ldots+a_k$$

$$S_{k+1}=\sum_{i=1}^{k+1}a_i=a_1+a_2+\ldots+a_k+a_{k+1}$$

$$S_1=\sum_{i=1}^{1}a_i=a_1$$

Assume $S_k=\sum_{i=1}^{k}a_i=a_1+a_2+\ldots+a_k$

$$S_k+a_{k+1}=(a_1+a_2+\ldots+a_k)+a_{k+1}=\sum_{i=1}^{k+1}a_i$$

which is S_{k+1}

19–21. Answers may vary.

23. S_1: $\frac{1}{1\cdot 3}=\frac{1}{3}$

S_2: $\frac{1}{1\cdot 3}+\frac{1}{3\cdot 5}=\frac{1}{3}+\frac{1}{15}=\frac{2}{5}$

S_3: $\frac{1}{1\cdot 3}+\frac{1}{3\cdot 5}+\frac{1}{5\cdot 7}=\frac{1}{3}+\frac{1}{15}+\frac{1}{35}=\frac{3}{7}$

S_4: $\frac{1}{1\cdot 3}+\frac{1}{3\cdot 5}+\frac{1}{5\cdot 7}+\frac{1}{7\cdot 9}=\frac{4}{9}$

S_5: $\frac{1}{1\cdot 3}+\frac{1}{3\cdot 5}+\frac{1}{5\cdot 7}+\frac{1}{7\cdot 9}+\frac{1}{9\cdot 11}=\frac{45}{99}=\frac{5}{11}$

S_n: $\frac{1}{1\cdot 3}+\frac{1}{3\cdot 5}+\frac{1}{5\cdot 7}+\cdots+\frac{1}{(2n-1)(2n+1)}=\frac{n}{2n+1}$

Prove by mathematical induction.

25. S_1: $\left(1+\frac{1}{1}\right)=2$

S_2: $\left(1+\frac{1}{1}\right)\left(1+\frac{1}{2}\right)=2\left(\frac{3}{2}\right)=3$

$S_3=\left(1+\frac{1}{1}\right)\left(1+\frac{1}{2}\right)\left(1+\frac{1}{3}\right)=3\left(\frac{4}{3}\right)=4$

$S_4=\left(1+\frac{1}{1}\right)\left(1+\frac{1}{2}\right)\left(1+\frac{1}{3}\right)\left(1+\frac{1}{4}\right)=4\left(\frac{5}{4}\right)=5$

$S_5=\left(1+\frac{1}{1}\right)\left(1+\frac{1}{2}\right)\left(1+\frac{1}{3}\right)\left(1+\frac{1}{4}\right)\left(1+\frac{1}{5}\right)=5\left(\frac{6}{5}\right)=6$

$S_n=\left(1+\frac{1}{1}\right)\left(1+\frac{1}{2}\right)\left(1+\frac{1}{3}\right)\ldots\left(1+\frac{1}{n}\right)=(n+1)$

Prove by mathematical induction.

27. **a.** The sums are 4, 9, 16, 25, 36, or 2^2, 3^2, 4^2, 5^2, and 6^2. Count the number of "balls" on the diagonals as you move from a "square" of 1 ball to square of 2 balls on each side, to a square of 3 balls on each side, etc. The diagonal values for 2×2 are 1, 2, 1; the values for 3×3 are 1, 2, 3, 2, 1.

b. conjecture: $1+2+3+\ldots+(n-1)+n+(n-1)+\ldots+3+2+1=(n+1)^2$

c. proof by induction

29. Students should proceed step by step through the proof.

Review Problems

30. $5x(x-2)=x-5$

$5x^2-10x=x-5$

$5x^2-11x+5=0$

$x=\frac{-b\pm\sqrt{b^2-4ac}}{2a}$

$x=\frac{11\pm\sqrt{(-11)^2-4(5)(5)}}{2(5)}$

$x=\frac{11\pm\sqrt{21}}{10}$

$\left\{\frac{11-\sqrt{21}}{10},\ \frac{11+\sqrt{21}}{10}\right\}$

31. $|2x-1| \le 7$
$-7 \le 2x - 1 \le 7$
$-7 + 1 \le 2x \le 7 + 1$
$\frac{-6}{2} \le x \le \frac{8}{2}$
$-3 \le x \le 4$
$\{x \mid -3 \le x \le 4\}$, $[-3, 4]$

32. $3x^2 - y^2 = -1$
$2x^2 + 5y^2 = 22$

$15x^2 - 5y^2 = -5$
$2x^2 + 5y^2 = 22$
$17x^2 = 17$
$x^2 = 1$
$x = \pm 1$
If $x = -1$: $3(-1)^2 - y^2 = -1$
$y^2 = 4$
$y = \pm 2$
If $x = 1$: $3(1)^2 - y^2 = -1$
$y^2 = 4$
$y = \pm 2$
$\{(-1, -2), (-1, 2), (1, -2), (1, 2)\}$

Chapter 11 Review Problems

1. $a_n = 5n - 2$
$a_1 = 5(1) - 2 = 3$
$a_2 = 5(2) - 2 = 8$
$a_3 = 5(3) - 2 = 13$
$a_4 = 5(4) - 2 = 18$
$a_{10} = 5(10) - 2 = 48$
$a_{15} = 5(15) - 2 = 73$

2. $a_n = \frac{n-2}{n^2+1}$
$a_1 = \frac{1-2}{1^2+1} = -\frac{1}{2}$
$a_2 = \frac{2-2}{2^2+1} = 0$
$a_3 = \frac{3-2}{3^2+1} = \frac{1}{10}$
$a_4 = \frac{4-2}{4^2+1} = \frac{2}{17}$
$a_{10} = \frac{10-2}{10^2+1} = \frac{8}{101}$
$a_{15} = \frac{15-2}{15^2+1} = \frac{13}{226}$

3. $a_n = (-1)^{n+1}(n^2 - 1)$
$a_1 = (-1)^{1+1}(1^2 - 1) = 0$
$a_2 = (-1)^{2+1}(2^2 - 1) = -3$
$a_3 = (-1)^{3+1}(3^2 - 1) = 8$
$a_4 = (-1)^{4+1}(4^2 - 1) = -15$
$a_{10} = (-1)^{10+1}(10^2 - 1) = -99$
$a_{15} = (-1)^{15+1}(15^2 - 1) = 224$

4. Possible answer: $-3, -6, -9, -12, -15, \ldots$
$a_n = -3n$

5. Possible answer:
$\frac{1}{3}, -\frac{1}{4}, \frac{1}{5}, -\frac{1}{6}, \frac{1}{7}, -\frac{1}{8}, \ldots$
$a_n = (-1)^{n+1}\left(\frac{1}{n+2}\right)$

6. $1, 4, 9, 16, \ldots$
$S_7 = 1 + 4 + 9 + 16 + 25 + 36 + 49 = 140$

7. $a_n = \frac{n-1}{3n}$
$a_1 = \frac{1-1}{3(1)} = 0$
$a_2 = \frac{2-1}{3(2)} = \frac{1}{6}$
$a_3 = \frac{3-1}{3(3)} = \frac{2}{9}$
$a_4 = \frac{4-1}{3(4)} = \frac{1}{4}$
$a_5 = \frac{5-1}{3(5)} = \frac{4}{15}$
$S_5 = 0 + \frac{1}{6} + \frac{2}{9} + \frac{1}{4} + \frac{4}{15} = \frac{163}{180}$

8. $\sum_{i=2}^{6}(i^3-4)=(2^3-4)+(3^3-4)+(4^3-4)+(5^3-4)+(6^3-4)$
$=(8-4)+(27-4)+(64-4)+(125-4)+(216-4)$
$=4+23+60+121+212=420$

9. $\sum_{i=1}^{7} i^2 = 1+4+9+16+25+36+49$

10. $\sum_{i=1}^{6}\left(-\frac{1}{2}\right)^n = -\frac{1}{2}+\frac{1}{4}-\frac{1}{8}+\frac{1}{16}-\frac{1}{32}+\frac{1}{64}$

11. $a_n = 0.0012n^2 - 0.027n + 1.09$

a. $a_3 = 0.0012(3)^2 - 0.027(3) + 1.09$
$a_3 = 1.0198$
$a_4 = 0.0012(4)^2 - 0.027(4) + 1.09$
$a_4 = 1.0012$
$a_9 = 0.0012(9)^2 - 0.027(9) + 1.09$
$a_9 = 0.9442$
$a_{10} = 0.0012(10)^2 - 0.027(10) + 1.09$
$a_{10} = 0.94$
The ratio of men to women in the U.S. is 1.0198, 1.0012, 0.9442, and 0.94 in years 1930, 1940, 1990, and 2000, respectively.

b. Let x = number of women, then
$275 - x$ = number of men.
$\frac{275-x}{x} = 0.94$
$275 - x = 0.94x$
$\frac{275}{1.94} = x$
$141.75 \approx x$
$133.25 = 275 - x$
There will be approximately 141.75 million women and 133.25 million men.

12. $-7, -3, 1, 5, \ldots$
$d = -3 - (-7) = 4$
$a_n = a_1 + (n-1)d$
$a_n = -7 + (n-1)(4)$
$a_{15} = -7 + (15-1)(4)$
$a_{15} = 49$

13. $9, 4, -1, -6, \ldots$
$d = 4 - 9 = -5$
$a_n = a_1 + (n-1)d$
$a_n = 9 + (n-1)(-5)$
$a_{22} = 9 + (22-1)(-5)$
$a_{22} = -96$

14. $\frac{3}{5}, \frac{1}{10}, -\frac{2}{5}, \ldots$
$d = \frac{1}{10} - \frac{3}{5} = -\frac{1}{2}$
$a_n = a_1 + (n-1)d$
$a_n = \frac{3}{5} + (n-1)\left(-\frac{1}{2}\right)$
$a_{22} = \frac{3}{5} + (22-1)\left(-\frac{1}{2}\right)$
$a_{22} = 9.9 \text{ or } -\frac{99}{100}$

15. $a_1 = -12,\ d = -\frac{1}{2}$
$a_n = a_1 + (n-1)d$
$a_n = -12 + (n-1)\left(-\frac{1}{2}\right)$
$a_{19} = -12 + (19-1)\left(-\frac{1}{2}\right)$
$a_{19} = -21$

16. $\frac{5}{6}, \frac{9}{6}, \frac{13}{6}, \frac{17}{6}, \ldots$

$d = \frac{9}{6} - \frac{5}{6} = \frac{2}{3}$

$a_n = a_1 + (n-1)d$

$\frac{41}{6} = \frac{5}{6} + (n-1)\left(\frac{2}{3}\right)$

$6 = \frac{2}{3}n - \frac{2}{3}$

$\frac{20}{3} = \frac{2}{3}n$

$10 = n$

The 10th term

17. 200, 190, 180, 170, 160, 150

$d = 190 - 200 = -10$

$a_n = a_1 + (n-1)d$

$a_n = 200 + (n-1)(-10)$

$a_n = 200 - 10n + 10$

$a_n = 210 - 10n$

18. a. $a_n = a_1 + (n-1)d$

$a_n = 27{,}966 + (n-1)(553)$

$a_n = 27{,}966 + 553n - 553$

$a_n = 27{,}413 + 553n$

b. $n = 26$

$a_{26} = 27{,}413 + 553(26)$

$a_{26} = 41{,}791$

The average salary will be \$41,791.

19. 5, 12, 19, 26, …

$d = 12 - 5 = 7$

$S_n = \frac{n}{2}(a_1 + a_n)$

$S_{22} = \frac{22}{2}(5 + a_{22})$

Find a_{22}.

$a_n = a_1 + (n-1)d$

$a_{22} = 5 + (22-1)(7)$

$a_{22} = 5 + (21)(7)$

$a_{22} = 152$

$S_{22} = \frac{22}{2}(5 + 152)$

$S_{22} = 11(157) = 1{,}727$

20. Given: $n = 16,\ a_1 = 3,\ d = 5$

Find S_{16}.

$S_n = \frac{n}{2}(a_1 + a_n)$

$S_{16} = \frac{16}{2}(3 + a_{16})$

Find a_{16}.

$a_n = a_1 + (n-1)d$

$a_{16} = 3 + (16-1)(5)$

$a_{16} = 78$

$S_{16} = \frac{16}{2}(3 + 78)$

$S_{16} = 648$

21. $\sum_{i=1}^{16}(3i+2) = (3\cdot1+2) + (3\cdot2+2) + (3\cdot3+2) + \ldots + (3\cdot16+2)$

$= 5 + 8 + 11 + \ldots + 50$

$a_1 = 5,\ n = 16$ (There are 16 terms.)

$a_{16} = 50$

$S_n = \frac{n}{2}(a_1 + a_n)$

$S_{16} = \frac{16}{2}(a_1 + a_{16})$

$S_{16} = \frac{16}{2}(5 + 50)$

$S_{16} = 8(55) = 440$

22. $\sum_{i=1}^{25}(-2i+6)$

$a_1 = 4,\ d = -2,\ a_{25} = -44$

$S_{25} = \frac{25}{2}(4-44) = -500$

23. $\sum_{i=1}^{100} 3i = 3+6+9+12+\ldots+300$

$a_1 = 3,\ a_{100} = 300$

$S_{100} = \frac{100}{2}(3+300) = 15{,}150$

24. $\sum_{i=1}^{16} i = 1+2+3+4+\ldots+16$

$a_1 = 1,\ a_{16} = 16$

$S_{16} = \frac{16}{2}(1+16) = 136$

25. $\sum_{i=1}^{10}(21{,}800+1200i) = 23{,}000+\ldots+33{,}800$

$a_1 = 23{,}000,\ d = 1200,\ a_{10} = 33{,}800$

$a_n = 23{,}000+1200(n-1)$

$a_n = 21{,}800+1200n$

$S_{10} = \frac{10}{2}(23{,}000+33{,}800) = 284{,}000$

The total salary is $284,000.

26. $\sum_{i=1}^{15}(327i+2094) = 2421+2748+\ldots+6999$

$a_1 = 2421,\ a_{15} = 6999$

$S_{15} = \frac{15}{2}(2421+6999) = 70{,}650$

The total sales is $70,650 million.

27. $12,\ 4,\ \frac{4}{3},\ \frac{4}{9},\ \ldots$

$r = \frac{4}{12} = \frac{1}{3}$

$a_n = a_1 r^{n-1}$

$a_n = 12\left(\frac{1}{3}\right)^{n-1}$

$a_8 = 12\left(\frac{1}{3}\right)^{8-1}$

$a_8 = \frac{4}{729}$

28. $\frac{1}{3},\ \frac{1}{2},\ \frac{3}{4},\ \frac{9}{8},\ \ldots$

$r = \frac{\frac{1}{2}}{\frac{1}{3}} = \frac{1}{2}\cdot\frac{3}{1} = \frac{3}{2}$

$a_n = a_1 r^{n-1}$

$a_n = \frac{1}{3}\left(\frac{3}{2}\right)^{n-1}$

$a_{11} = \frac{1}{3}\left(\frac{3}{2}\right)^{11-1}$

$a_{11} = \frac{19{,}683}{1024}$

29. $3,\ 3\sqrt{3},\ 9,\ 9\sqrt{3},\ \ldots$

$r = \frac{3\sqrt{3}}{3} = \sqrt{3}$

$a_n = a_1 r^{n-1}$

$a_n = 3\left(\sqrt{3}\right)^{n-1}$

$a_{10} = 3\left(\sqrt{3}\right)^{10-1}$

$a_{10} = 243\sqrt{3}$

30. $3,\ -12,\ 48,\ -192,\ \ldots$

$r = \frac{-12}{3} = -4$

$a_n = a_1 r^{n-1}$

$a_n = 3(-4)^{n-1}$

31. $36,\ 12,\ 4,\ \frac{4}{3},\ \frac{4}{9},\ \ldots$

$r = \frac{12}{36} = \frac{1}{3}$

$a_n = a_1 r^{n-1}$

$a_n = 36\left(\frac{1}{3}\right)^{n-1}$

32. $100,\ 50,\ 25,\ 12.5,\ \ldots$

$r = \frac{50}{100} = \frac{1}{2}$

$a_n = a_1 r^{n-1}$

$a_n = 100\left(\frac{1}{2}\right)^{n-1}$

$n = 8$

$a_8 = 100\left(\frac{1}{2}\right)^{8-1} = \frac{100}{128} = \frac{25}{32}$

The remaining amount is $\frac{25}{32}$ mg.

33. $a_1 = 10{,}000,\ r = 0.8$

$a_n = a_1 r^{n-1}$

$a_7 = 10{,}000(0.8)^{7-1}$

$a_7 = 2621.44$

The value is \$2621.44.

34. $7, -14, 28, -56, \ldots$

$a_1 = 7,\ r = \frac{-14}{7} = -2$

$S_n = \frac{a_1 - a_1 r^n}{1-r}$

$S_6 = \frac{7 - 7(-2)^6}{1-(-2)} = \frac{-441}{3} = -147$

35. $\frac{3}{5}, 1, \frac{5}{3}, \frac{25}{9}, \ldots$

$a_1 = \frac{3}{5},\ r = \frac{1}{\frac{3}{5}} = \frac{5}{3}$

$S_n = \frac{a_1 - a_1 r^n}{1-r}$

$S_7 = \frac{\frac{3}{5} - \frac{3}{5}\left(\frac{5}{3}\right)^7}{1 - \frac{5}{3}}$

$S_7 \approx 31.25$

36. $a_1 = 84,\ r = -\frac{1}{4}$

$S_n = \frac{a_1 - a_1 r^n}{1-r}$

$S_6 = \frac{84 - 84\left(-\frac{1}{4}\right)^6}{1-\left(-\frac{1}{4}\right)}$

$S_6 \approx 67.18$

37. $\sum_{i=1}^{6} 3 \cdot 4^{i-1}$

$a_1 = 3,\ r = 4$

$S_n = \frac{a_1 - a_1 r^n}{1-r}$

$S_6 = \frac{3 - 3(4)^6}{1-4}$

$S_6 = 4095$

38. $a_1 = 30{,}000,\ r = 1.08$

$a_n = a_1 r^{n-1}$

$a_{20} = 30{,}000(1.08)^{19}$

$a_{20} = 129{,}471.03$

The salary is \$129,471.03.

$S_n = \frac{a_1 - a_1 r^n}{1-r}$

$S_{20} = \frac{30{,}000 - 30{,}000(1.08)^{20}}{1-1.08}$

$S_{20} = 1{,}372{,}858.9$

The total salary is \$1,372,858.90.

39. $a_1 = 100\left(1 + \frac{0.06}{12}\right)$

$r = 1 + \frac{0.06}{12}$

$S_n = \frac{a_1 - a_1 r^n}{1-r}$

$S_{48} = \frac{100\left(1+\frac{0.06}{12}\right) - 100\left(1+\frac{0.06}{12}\right)\left(1+\frac{0.06}{12}\right)^{48}}{1-\left(1+\frac{0.06}{12}\right)}$

$S_{48} = 5436.8329$

The balance is \$5436.83.

40. $a_1 = 12,\ r = \frac{3}{4},\ n = 8$

$$S_n = \frac{a_1 - a_1 r^n}{1-r}$$

$$S_8 = \frac{12 - 12\left(\frac{3}{4}\right)^8}{1-\frac{3}{4}}$$

$$S_8 = 43.19$$

The pendulum covers a distance of approximately 43.19 meters in the first 8 swings.

41. number of levels = 10 + 2 = 12

$a_1 = 2,\ r = 2$

$a_{12} = 2(2)^{12-1} = 4096$

42. $a_1 = 10,\ r = 3$

$$S_8 = \frac{10(1-3^8)}{1-3} = 5(6561-1) = 32,800$$

Remaining funds = 100,000 – 32,000 = \$67,200

43. $7 + \frac{7}{2} + \frac{7}{4} + \frac{7}{8} + \ldots$

$$r = \frac{\frac{7}{2}}{7} = \frac{1}{2}$$

$$S_\infty = \frac{a_1}{1-r}$$

$$S_\infty = \frac{7}{1-\frac{1}{2}} = 14$$

44. $5 - 1 + \frac{1}{5} - \frac{1}{25} + \ldots$

$$r = -\frac{1}{5}$$

$$S_\infty = \frac{a_1}{1-r}$$

$$S_\infty = \frac{5}{1+\frac{1}{5}} = \frac{25}{6}$$

45. $5 + \frac{10}{3} + \frac{20}{9} + \frac{40}{27} + \ldots$

$$r = \frac{\frac{10}{3}}{5} = \frac{2}{3}$$

$$S_\infty = \frac{5}{1-\frac{2}{3}} = 15$$

46. $0.75 + 1.5 + 3 + 6 + \ldots$

$$r = \frac{1.5}{0.75} = 2$$

Since $r > 1$, the infinite geometric series has no sum.

47. $6 + 3 + 1.5 + 0.75 + \cdots$

$$r = \frac{3}{6} = \frac{1}{2}$$

$$S_\infty = \frac{6}{1-\frac{1}{2}} = 12$$

48. $\frac{8}{10} + \frac{8}{100} + \frac{8}{1000} + \cdots$

$$r = \frac{\frac{8}{100}}{\frac{8}{10}} = \frac{8}{100} \cdot \frac{10}{8} = \frac{1}{10}$$

$$S_\infty = \frac{\frac{8}{10}}{1-\frac{1}{10}} = \frac{8}{10} \cdot \frac{10}{9} = \frac{8}{9}$$

49. $\frac{23}{100} + \frac{23}{10,000} + \frac{23}{1,000,000} + \cdots$

$$r = \frac{\frac{23}{10,000}}{\frac{23}{100}} = \frac{23}{10,000} \cdot \frac{100}{23} = \frac{1}{100}$$

$$S_\infty = \frac{\frac{23}{100}}{1-\frac{1}{100}} = \frac{23}{100} \cdot \frac{100}{99} = \frac{23}{99}$$

50. $a_1 = 3(6) = 18,\ r = \frac{1}{3}$

$$S = \frac{18}{1-\frac{1}{3}} = 27 \text{ m}$$

Sum of the perimeters: 27 m

51. 25, 20, 16, 12.8, …

$$r = \frac{20}{25} = \frac{4}{5}$$

$$S_\infty = \frac{25}{1-\frac{4}{5}} = 25 \cdot 5 = 125$$

It will travel 125 cm.

52. $\frac{1}{2}A+\frac{1}{4}A+\ldots$

$r=\frac{1}{2}$

$S_\infty=\frac{\frac{A}{2}}{1-\frac{1}{2}}=A$

It is the same size as the area of the full-size figure.

53. $4+4(0.7)+4(0.7)^2+\cdots$

$S_\infty=\frac{4}{1-0.7}=13\frac{1}{3}$

The total impact is $\$13\frac{1}{3}$ million.

54. $A_1=\pi r_1^2=\pi\cdot 1^2=\pi$

$A_2=\pi r_2^2=\pi\left(\frac{1}{2}\right)^2=\frac{\pi}{4}$

$r=\frac{1}{4}$

total area $=\frac{\pi}{1-\frac{1}{4}}=\frac{4\pi}{3}$

sum of the areas $=\frac{4\pi}{3}$ square meters

55. $6!=6\cdot 5\cdot 4\cdot 3\cdot 2\cdot 1=720$

56. $\binom{9}{2}=\frac{9!}{2!(9-2)!}=\frac{9!}{2!7!}=\frac{9\cdot 8}{2\cdot 1}=36$

57. $(3x+y)^4=\binom{4}{0}(3x)^4+\binom{4}{1}(3x)^3y+\binom{4}{2}(3x)^2y^2+\binom{4}{3}(3x)y^3+\binom{4}{4}y^4$

$=81x^4+4(27)x^3y+6(9)x^2y^2+4(3)xy^3+y^4$

$=81x^4+108x^3y+54x^2y^2+12xy^3+y^4$

58. $(x-2y)^5=\binom{5}{0}x^5+\binom{5}{1}x^4(-2y)+\binom{5}{2}x^3(-2y)^2+\binom{5}{3}x^2(-2y)^3+\binom{5}{4}x(-2y)^4+\binom{5}{5}(-2y)^5$

$=x^5+5(-2)x^4y+10(4)x^3y^2+10(-8)x^2y^3+5(16)xy^4-32y^5$

$=x^5-10x^4y+40x^3y^2-80x^2y^3+80xy^4-32y^5$

59. $(x^2+2y^3)^6=\binom{6}{0}(x^2)^6+\binom{6}{1}(x^2)^5(2y^3)+\binom{6}{2}(x^2)^4(2y^3)^2+\binom{6}{3}(x^2)^3(2y^3)^3$

$+\binom{6}{4}(x^2)^2(2y^3)^4+\binom{6}{5}(x^2)(2y^3)^5+\binom{6}{6}(2y^3)^6$

$= x^{12} + 6(2)x^{10}y^3 + 15(4)x^8y^6 + 20(8)x^6y^9 + 15(16)x^4y^{12} + 6(32)x^2y^{15} + 64y^{18}$
$= x^{12} + 12x^{10}y^3 + 60x^8y^6 + 160x^6y^9 + 240x^4y^{12} + 192x^2y^{15} + 64y^{18}$

60. $(2x - y^4)^7 = \binom{7}{0}(2x)^7 + \binom{7}{1}(2x)^6(-y^4) + \binom{7}{2}(2x)^5(-y^4)^2 + \binom{7}{3}(2x)^4(-y^4)^3 + \binom{7}{4}(2x)^3(-y^4)^4$
$+\binom{7}{5}(2x)^2(-y^4)^5 + \binom{7}{6}(2x)(-y^4)^6 + \binom{7}{7}(-y^4)^7$
$= 128x^7 + 7(64)(-1)x^6y^4 + 21(32)x^5y^8 + 35(16)(-1)x^4y^{12} + 35(8)x^3y^{16}$
$+21(4)(-1)x^2y^{20} + 7(2)xy^{24} + (-1)y^{28}$
$= 128x^7 - 448x^6y^4 + 672x^5y^8 - 560x^4y^{12} + 280x^3y^{16} - 84x^2y^{20} + 14xy^{24} - y^{28}$

61. $\binom{n}{k}p^kq^{n-k}$
$n = 6,\ k = 3,\ p = 0.6,\ q = 0.4$
$\binom{6}{3}(0.6)^3(0.4)^3 = 20(0.216)(0.064)$
$= 0.27648$

62. $(1)(1)(10)(10)(10)(10)(10) = 100{,}000$

63. $6! = 6 \cdot 5 \cdot 4 \cdot 3 \cdot 2 \cdot 1 = 720$

64. ${}_{15}P_4 = \frac{15!}{11!} = 15 \cdot 14 \cdot 13 \cdot 12 = 32{,}760$

65. ${}_7P_5 = \frac{7!}{2!} = 2520$

66. $(2)(10)(10)(5) = 1000$

67. **a.** $\binom{15}{4} = \frac{15!}{4!11!} = 1365$

b. $\binom{14}{3} = \frac{14!}{3!11!} = 364$

c. $\binom{15}{1} + \binom{15}{2} + \binom{15}{3} + \binom{15}{4} = \frac{15!}{1!14!} + \frac{15!}{2!13!} + \frac{15!}{3!12!} + \frac{15!}{4!11!}$
$= 15 + 105 + 455 + 1365 = 1940$

68. $\binom{12}{5}\binom{8}{4} = \frac{12!}{5!7!} \cdot \frac{8!}{4!4!} = 792(70) = 55,440$

69. $(4)^{10} = 1,048,576$

70. a. $P(E) = \frac{n(E)}{n(S)} = \frac{4}{6} = \frac{2}{3}$

b. $P(E) = \frac{n(E)}{n(S)} = \frac{3}{6} = \frac{1}{2}$

71. $P(E) = \frac{n(E)}{n(S)} = \frac{26}{36} = \frac{13}{18}$

72. $P = \frac{8258}{15,022} \approx 0.55$

73. $P = \frac{958+1766}{15,022} \approx 0.18$

74. $P = \frac{4152+1766}{15,022} \approx 0.39$

75. P(student is male or female) = 1

76. 4 O E O E O E

The numbers 1–9 include 5 odd and 4 even numbers $1 \cdot 5 \cdot 4 \cdot 5 \cdot 4 \cdot 5 \cdot 4 = 8,000$

(not 0 or 1) # # # # # #

$8 \cdot 10^6 = 8,000,000$

So, $P = \frac{8,000}{8,000,000} = \frac{1}{1000}$

77. a. $P(E) = \frac{n(E)}{n(S)} = \frac{1}{\binom{20}{5}} = \frac{1}{15,504}$

b. $P(E) = \frac{100}{15,504} = \frac{25}{3876}$

78. a. $A \cap B$ = outcome is 5

$P(A \cap B) = \frac{1}{6}$

b. $A \cup B$ = outcome is 1, 3, 5, or 6

$P(A \cup B) = \frac{4}{6} = \frac{2}{3}$

c. A' = outcome is even

$P(A') = \frac{3}{6} = \frac{1}{2}$

d. B' = outcome is 4 or less

$P(B') = \frac{4}{6} = \frac{2}{3}$

79. a. $P = \frac{15}{40} + \frac{11}{40} - \frac{5}{40} = \frac{21}{40}$

b. $P = \frac{5}{40} = \frac{1}{8}$

c. $P = 1 - \frac{15}{40} = \frac{25}{40} = \frac{5}{8}$

d. $P = 1 - \frac{11}{40} = \frac{29}{40}$

e. $P = 1 - \frac{21}{40} = \frac{19}{40}$

80. $P = \frac{13}{52} + \frac{16}{52} - \frac{4}{52} = \frac{25}{52}$

81. $P = \left(\frac{1}{2}\right)\left(\frac{1}{3}\right)\left(\frac{1}{6}\right) = \frac{1}{36}$

82. $P = \left(\frac{1}{2}\right)^{10} = \frac{1}{1024}$

83. $P = \frac{1}{6!} = \frac{1}{720}$

84. P("she" and "African American")

$= P(F \text{ and } AA)$

$= P(F) \cdot P(AA)$ since independent

$= (0.14)(0.17) = 0.0238$

85. Populations are not equal in each state. Therefore, the probability of a person being born in a state is not equal for each state.

86. $S_n\text{: } 1+3+5+\cdots+(2n-1)=n^2$

$S_1\text{: } 1=1^2$

$1=1$

Assume S_k is true.

$S_k\text{: } 1+3+5+\cdots+(2k-1)=k^2$

This implies the truth of S_{k+1}.

$S_{k+1}\text{: } 1+3+5+\cdots+(2k+1)=(k+1)^2$

Proof: (Start with S_k.)

$1+3+5+\cdots+(2k-1)+(2k+1)=k^2+(2k+1)$

$1+3+5+\cdots+(2k+1)=k^2+2k+1$

$1+3+5+\cdots+(2k+1)=(k+1)^2$

This is the statement for $n = k + 1$.

Therefore, S_n is true.

87. $S_n\text{: } \frac{1}{1\cdot 3}+\frac{1}{3\cdot 5}+\frac{1}{5\cdot 7}+\ldots+\frac{1}{(2n-1)(2n+1)}=\frac{n}{2n+1}$

$S_1\text{: } \frac{1}{1\cdot 3}=\frac{1}{2\cdot 1+1}$

$\frac{1}{3}=\frac{1}{3}$

Assume S_k is true.

$S_k\text{: } \frac{1}{1\cdot 3}+\frac{1}{3\cdot 5}+\frac{1}{5\cdot 7}+\ldots+\frac{1}{(2k-1)(2k+1)}=\frac{k}{2k+1}$

This implies the truth of S_{k+1}.

$S_{k+1}\text{: } \frac{1}{1\cdot 3}+\frac{1}{3\cdot 5}+\frac{1}{5\cdot 7}+\ldots+\frac{1}{(2k+1)(2k+3)}=\frac{k+1}{2k+3}$

Proof: $\frac{1}{1\cdot 3}+\frac{1}{3\cdot 5}+\frac{1}{5\cdot 7}+\ldots+\frac{1}{(2k-1)(2k+1)}+\frac{1}{(2k+1)(2k+3)}=\frac{k}{2k+1}+\frac{1}{(2k+1)(2k+3)}$

$\frac{1}{1\cdot 3}+\frac{1}{3\cdot 5}+\frac{1}{5\cdot 7}+\ldots+\frac{1}{(2k+1)(2k+3)}=\frac{k(2k+3)+1}{(2k+1)(2k+3)}$

$\frac{1}{1\cdot 3}+\frac{1}{3\cdot 5}+\frac{1}{5\cdot 7}+\ldots+\frac{1}{(2k+1)(2k+3)}=\frac{2k^2+3k+1}{(2k+1)(2k+3)}$

$\frac{1}{1\cdot 3}+\frac{1}{3\cdot 5}+\frac{1}{5\cdot 7}+\ldots+\frac{1}{(2k+1)(2k+3)}=\frac{(2k+1)(k+1)}{(2k+1)(2k+3)}$

$\frac{1}{1\cdot 3}+\frac{1}{3\cdot 5}+\frac{1}{5\cdot 7}+\ldots+\frac{1}{(2k+1)(2k+3)}=\frac{k+1}{2k+3}$

This is the statement for $n=k+1$

Therefore, S_n is true

88. $S_n\text{: } 1^2+3^2+5^2+\cdots+(2n-1)^2=\frac{n(2n-1)(2n+1)}{3}$

$S_1\text{: } 1^2=\frac{1(2\cdot 1-1)(2\cdot 1+1)}{3}$

$1=1$

Assume S_k is true.

S_k: $1^2+3^2+5^2+\cdots+(2k-1)^2=\dfrac{k(2k-1)(2k+1)}{3}$

This implies the truth of S_{k+1}.

S_{k+1}: $1^2+3^2+5^2+\cdots+(2k+1)^2=\dfrac{(k+1)(2k+1)(2k+3)}{3}$

Proof:

$$1^2+3^2+5^2+\cdots+(2k-1)^2+(2k+1)^2=\frac{k(2k-1)(2k+1)}{3}+(2k+1)^2$$

$$1^2+3^2+5^2+\cdots+(2k+1)^2=\frac{k(2k-1)(2k+1)+3(2k+1)^2}{3}$$

$$1^2+3^2+5^2+\cdots+(2k+1)^2=\frac{(2k+1)(2k^2-k+6k+3)}{3}$$

$$1^2+3^2+5^2+\cdots+(2k+1)^2=\frac{(2k+1)(2k+3)(k+1)}{3}$$

This is the statement for $n = k + 1$.

Therefore, S_n is true.

89. S_1: $\frac{2}{1}=2$

S_2: $\frac{2}{1}\cdot\frac{3}{2}=3$

S_3: $\frac{2}{1}\cdot\frac{3}{2}\cdot\frac{4}{3}=4$

S_4: $\frac{2}{1}\cdot\frac{3}{2}\cdot\frac{4}{3}\cdot\frac{5}{4}=5$

S_5: $\frac{2}{1}\cdot\frac{3}{2}\cdot\frac{4}{3}\cdot\frac{5}{4}\cdot\frac{6}{5}=6$

S_n: $\frac{2}{1}\cdot\frac{3}{2}\cdot\frac{4}{3}\cdot\ldots\cdot\frac{n+1}{n}=n+1$

S_1: $\frac{2}{1}=1+1$

$2=2$

Assume S_k is true.

S_k: $\frac{2}{1}\cdot\frac{3}{2}\cdot\frac{4}{3}\cdot\ldots\cdot\frac{k+1}{k}=k+1$

This implies the truth of S_{k+1}.

S_{k+1}: $\frac{2}{1}\cdot\frac{3}{2}\cdot\frac{4}{3}\cdot\ldots\cdot\frac{k+2}{k+1}=k+2$

Proof:

$$\frac{2}{1}\cdot\frac{3}{2}\cdot\frac{4}{3}\cdot\ldots\cdot\frac{k+1}{k}\cdot\frac{k+2}{k+1}=(k+1)\left(\frac{k+2}{k+1}\right)$$

$$\frac{2}{1}\cdot\frac{3}{2}\cdot\frac{4}{3}\cdot\ldots\cdot\frac{k+2}{k+1}=k+2$$

This is the statement for $n = k + 1$.

Therefore, S_n is true.

Chapter 11 Test

1. $a_n = \dfrac{n+1}{n^2}$

$a_1 = \dfrac{1+1}{1^2} = 2$

$a_2 = \dfrac{2+1}{2^2} = \dfrac{3}{4}$

$a_3 = \dfrac{3+1}{3^2} = \dfrac{4}{9}$

$a_4 = \dfrac{4+1}{4^2} = \dfrac{5}{16}$

$a_5 = \dfrac{5+1}{5^2} = \dfrac{6}{25}$

$a_{10} = \dfrac{10+1}{10^2} = \dfrac{11}{100}$

2. $\dfrac{1}{4}, \dfrac{4}{9}, \dfrac{9}{16}, \dfrac{16}{25}, \ldots$

$a_n = \dfrac{n^2}{(n+1)^2}$ or $\left(\dfrac{n}{n+1}\right)^2$

3. $\sum_{i=1}^{5}(2^i - 2) = (2^1 - 2) + (2^2 - 2) + (2^3 - 2) + (2^4 - 2) + (2^5 - 2)$

$= 0 + 2 + 6 + 14 + 30 = 52$

4. $\dfrac{5}{4(1)+1} + \dfrac{5}{4(2)+1} + \dfrac{5}{4(3)+1} + \cdots + \dfrac{5}{4(13)+1} = \sum_{i=1}^{13} \dfrac{5}{4i+1}$

5. $11, 7, 3, -1, \ldots$

$d = 7 - 11 = -4$

$a_n = a_1 + (n-1)d$

$a_n = 11 + (n-1)(-4)$

$a_n = 15 - 4n$

$a_{19} = 15 - 4(19)$

$a_{19} = -61$

6. $-16, -11, -6, -1, \ldots$

$d = -11 - (-16) = 5$

$a_{18} = -16 + (18-1)(5) = 69$

$S_{18} = \dfrac{18}{2}(-16 + 69) = 477$

7. 15, 30, …, 150

$S_{10} = \frac{10}{2}(15+150) = 825$

8. $\sum_{i=1}^{50}(3i-26)$

$a_1 = 3(1) - 26 = -23$

$a_{50} = 3(50) - 26 = 124$

$S_{50} = \frac{50}{2}(-23+124) = 2525$

9. $a_1 = 6,\ a_2 = 4,\ a_3 \approx 2.8$

$d = 4 - 6 = -2$

$d = 2.8 - 4 = -1.2$

The difference is not constant.

10. $2, \frac{2}{3}, \frac{2}{9}, \frac{2}{27}, \ldots$

$r = \frac{\frac{2}{3}}{2} = \frac{1}{3}$

$a_n = a_1 r^{n-1}$

$a_n = 2\left(\frac{1}{3}\right)^{n-1}$

$a_9 = 2\left(\frac{1}{3}\right)^{9-1}$

$a_9 = \frac{2}{6561}$

11. 5, −15, 45, −135, …

$S_n = \frac{a_1 - a_1 r^n}{1-r}$

$r = \frac{-15}{5} = -3$

$S_{15} = \frac{5 - 5(-3)^{15}}{1-(-3)}$

$= 17{,}936{,}135$

12. $\sum_{i=1}^{10} 12\left(\frac{3}{2}\right)^i = 18 + 27 + \cdots$

$S_n = \frac{a_1 - a_1 r^n}{1-r}$

$r = \frac{27}{18} = \frac{3}{2}$

$S_{10} = \frac{18 - 18\left(\frac{3}{2}\right)^{10}}{1 - \frac{3}{2}} \approx 2039.9$

13. $\frac{1}{4} + \frac{1}{12} + \frac{1}{36} + \frac{1}{108} + \cdots$

$r = \frac{\frac{1}{12}}{\frac{1}{4}} = \frac{1}{12} \cdot \frac{4}{1} = \frac{1}{3}$

$S_\infty = \frac{a_1}{1-r}$

$S_\infty = \frac{\frac{1}{4}}{1-\frac{1}{3}} = \frac{\frac{1}{4}}{\frac{2}{3}} = \frac{1}{4} \cdot \frac{3}{2} = \frac{3}{8}$

14. $0.\overline{73} = \frac{73}{100} + \frac{73}{10{,}000} + \frac{73}{1{,}000{,}000} + \cdots$

$r = \frac{\frac{73}{10{,}000}}{\frac{73}{100}} = \frac{73}{10{,}000} \cdot \frac{100}{73} = \frac{1}{100}$

$S_\infty = \frac{a_1}{1-r} = \frac{\frac{73}{100}}{1 - \frac{1}{100}} = \frac{73}{100} \cdot \frac{100}{99} = \frac{73}{99}$

15. $\binom{17}{14} = \frac{17!}{14!(17-14)!} = \frac{17!}{14!3!} = 680$

16. $(x^3 - 2y)^5 = \binom{5}{0}(x^3)^5 + \binom{5}{1}(x^3)^4(-2y) + \binom{5}{2}(x^3)^3(-2y)^2 + \binom{5}{3}(x^3)^2(-2y)^3$

$+\binom{5}{4}(x^3)(-2y)^4 + \binom{5}{5}(-2y)^5$

$= x^{15} + 5(-2)x^{12}y + 10(4)x^9y^2 + 10(-8)x^6y^3 + 5(16)x^3y^4 - 32y^5$

$= x^{15} - 10x^{12}y + 40x^9y^2 - 80x^6y^3 + 80x^3y^4 - 32y^5$

17. $a_n = 7.7n + 55$

$\sum_{n=0}^{20}(7.7n+55)$

$a_0 = 55,\ a_{20} = 209$

$S_{21} = \frac{21}{2}(55+209) = 2772$

The total annual advertising expenditures by U.S. companies from 1980 to 2000 is \$2772 billion.

18. 30, 38, 46, ...

$d = 46 - 38 = 8$

$a_n = a_1 + (n-1)d$

$a_{25} = 30 + (25-1)(8) = 222$

30 + 38 + 46 + ... + 222

$S_{25} = \frac{25}{2}(30+222) = 3150$

19. $22,000 + 22,000(1.06) + \cdots + 22,000(1.06)^9$

$r = 1.06$

$S_n = \frac{a_1 - a_1 r^n}{1-r}$

$S_{10} = \frac{22,000 - 22,000(1.06)^{10}}{1-1.06}$

$= 289,977.49$

The total salary is \$289,977.49.

20. 10 + 5 + 2.5 + ...

$S_\infty = \frac{a_1}{1-r}$

$S_\infty = \frac{10}{1-0.5} = 20$

The total impact is \$20 million.

21. (1)(1)(1)(10)(10)(10)(10) = 10,000

22. $\binom{11}{4} = \frac{11!}{4!7!} = 330$

23. $P(E) = \frac{n(E)}{n(S)} = \frac{(2)(3)(2)(1)}{4!} = \frac{12}{24} = \frac{1}{2}$

24. $P(E) = \frac{4}{120} = \frac{1}{30}$

25. $P = \frac{5}{10} + \frac{6}{10} - \frac{3}{5} = \frac{5}{10} = \frac{1}{2}$

26. $P = \frac{2}{8} \cdot \frac{2}{8} = \frac{1}{16}$

27. S_n: $1 + 4 + 7 + \ldots + (3n - 2) = \frac{n(3n-1)}{2}$

S_1: $1 = \frac{1(3 \cdot 1 - 1)}{2}$

$1 = 1$

Assume S_k is true.

S_k: $1 + 4 + 7 + \ldots + (3k - 2) = \frac{k(3k-1)}{2}$

This implies the truth of S_{k+1}.

S_{k+1}: $1 + 4 + 7 + \ldots + (3k - 2) + (3k + 1) = \frac{(k+1)(3k+2)}{2}$

Proof:

$1 + 4 + 7 + \ldots + (3k - 2) + (3k + 1) = \frac{k(3k-1)}{2} + 3k + 1$

$1 + 4 + 7 + \ldots + (3k - 2) + (3k + 1) = \frac{3k^2 - k + 6k + 2}{2}$

$1 + 4 + 7 + \ldots + (3k - 2) + (3k + 1) = \frac{3k^2 + 5k + 2}{2}$

$1 + 4 + 7 + \ldots + (3k - 2) + (3k + 1) = \frac{(k+1)(3k+2)}{2}$

This is the statement for $n = k + 1$.

Therefore, S_n is true.

Appendix

Review Problems Covering Entire Book

1. $2[3-2(x+4)]=3(4-x)$
$2(3-2x-8)=12-3x$
$6-4x-16=12-3x$
$-4x-10=12-3x$
$-x=22$
$x=-22$
$\{-22\}$

2. $\frac{4y-2}{3}-\frac{y+2}{4}=\frac{7y-2}{12}$
$\frac{12(4y-2)}{3}-\frac{12(y+2)}{4}=\frac{12(7y-2)}{12}$
$4(4y-2)-3(y+2)=7y-2$
$16y-8-3y-6=7y-2$
$6y=12$
$y=2$
$\{2\}$

3. $-4(x-2)\le 6x+4$
$-4x+8\le 6x+4$
$-10x\le -4$
$x\ge \frac{2}{5}$
$\left\{x \middle| x\ge \frac{2}{5}\right\}, \left[\frac{2}{5}, \infty\right)$

4. $3x+7>4$ or $6-x<1$
$3x>-3$ or $-x<-5$
$x>-1$ or $x>5$
$\{x|x>-1\}, (-1, \infty)$

5. $4x-3<13$ and $-3x-4\ge 8$
$4x<16$ and $-3x\ge 12$
$x<4$ and $x\le -4$
$\{x|x<-4\}, (-\infty, -4)$

6. $-7<\frac{4-2x}{3}\le \frac{1}{3}$
$-21<4-2x\le 1$
$-21-4<-2x\le 1-4$
$-25<-2x\le -3$
$\frac{25}{2}>x\ge \frac{3}{2}$
$\left\{x \middle| \frac{3}{2}\le x<\frac{25}{2}\right\}, \left[\frac{3}{2}, \frac{25}{2}\right)$

7. $\left|\frac{2x-3}{5}\right|=1$
$\frac{2x-3}{5}=1$ or $\frac{2x-3}{5}=-1$
$2x-3=5$ or $2x-3=-5$
$2x=8$ or $2x=-2$
$x=4$ or $x=-1$
$\{-1, 4\}$

8. $|6x|=|3x-9|$
$6x=3x-9$ or $6x=-3x+9$
$3x=-9$ or $9x=9$
$x=-3$ or $x=1$
$\{-3, 1\}$

9. $|3-2x|<7$
$3-2x<7$ and $3-2x>-7$
$-2x<4$ and $-2x>-10$
$x>-2$ and $x<5$
$\{x|-2<x<5\}, (-2, 5)$

10. $|2x-1|\ge 7$
$2x-1\ge 7$ or $2x-1\le -7$
$2x\ge 8$ or $2x\le -6$
$x\ge 4$ or $x\le -3$
$\{x|x\le -3 \text{ or } x\ge 4\}, (-\infty, -3]\cup[4, \infty)$

11. $3x+4y=2$
$2x+5y=-1$

$$\begin{array}{r} -6x-8y=-4 \\ 6x+15y=-3 \\ \hline 7y=-7 \\ y=-1 \end{array}$$

$3x+4(-1)=2$
$3x=6$
$x=2$
$\{(2, -1)\}$

12. $5x - 2y = -7$
$y = 3x + 5$

$5x - 2(3x + 5) = -7$
$5x - 6x - 10 = -7$
$-x = 3$
$x = -3$
$y = 3(-3) + 5 = -4$
$\{(-3, -4)\}$

13. $x - 2y + z = -4$ (1)
$2x + 4y - 3z = -1$ (2)
$-3x - 6y + 7z = 4$ (3)
Multiply equation (1) by 3 and add to (2).

$$\begin{array}{r} 3x - 6y + 3z = -12 \\ \underline{2x + 4y - 3z = -1} \\ 5x - 2y = -13 \quad (4) \end{array}$$

Multiply equation (1) by –7 and add to equation (3).

$$\begin{array}{r} -7x + 14y - 7z = 28 \\ \underline{-3x - 6y + 7z = 4} \\ -10x + 8y = 32 \quad (5) \end{array}$$

Multiply equation (4) by 2 and divide equation (5) by 2, then add.

$$\begin{array}{r} 10x - 4y = -26 \\ \underline{-5x + 4y = 16} \\ 5x = -10 \\ x = -2 \end{array}$$

Equation 4:
$5(-2) - 2y = -13$
$-2y = -3$
$y = \frac{3}{2}$
Equation 1:
$-2 - 2\left(\frac{3}{2}\right) + z = -4$
$-2 - 3 + z = -4$
$z = 1$
$\left\{\left(-2, \frac{3}{2}, 1\right)\right\}$

14. $x-2y+z=16$
$2x-y-z=14$
$3x+5y-4z=-10$

$$\left[\begin{array}{rrr|r} 1 & -2 & 1 & 16 \\ 2 & -1 & -1 & 14 \\ 3 & 5 & -4 & -10 \end{array}\right] \begin{array}{l} \\ R_2-2R_1 \to \\ R_3-3R_1 \to \end{array} \left[\begin{array}{rrr|r} 1 & -2 & 1 & 16 \\ 0 & 3 & -3 & -18 \\ 0 & 11 & -7 & -58 \end{array}\right]$$

$$\frac{1}{3}R_2 \to \left[\begin{array}{rrr|r} 1 & -2 & 1 & 16 \\ 0 & 1 & -1 & -6 \\ 0 & 11 & -7 & -58 \end{array}\right] R_3-11R_2 \to \left[\begin{array}{rrr|r} 1 & -2 & 1 & 16 \\ 0 & 1 & -1 & -6 \\ 0 & 0 & 4 & 8 \end{array}\right] \frac{1}{4}R_3 \to \left[\begin{array}{rrr|r} 1 & -2 & 1 & 16 \\ 0 & 1 & -1 & -6 \\ 0 & 0 & 1 & 2 \end{array}\right]$$

$x-2y+z=16$
$y-z=-6$
$z=2$

$y-2=-6$
$y=-4$

$x-2(-4)+2=16$
$x=6$
$\{(6,-4,2)\}$

15. $2x-z=1$
$3y+2z=0$
$x-y=-3$

$$D=\begin{vmatrix} 2 & 0 & -1 \\ 0 & 3 & 2 \\ 1 & -1 & 0 \end{vmatrix}$$

$$=2\begin{vmatrix} 3 & 2 \\ -1 & 0 \end{vmatrix}-0+1\begin{vmatrix} 0 & -1 \\ 3 & 2 \end{vmatrix}$$

$=2(0+2)-0+(0+3)=4+3=7$

$$D_x=\begin{vmatrix} 1 & 0 & -1 \\ 0 & 3 & 2 \\ -3 & -1 & 0 \end{vmatrix}$$

$$=1\begin{vmatrix} 3 & 2 \\ -1 & 0 \end{vmatrix}-0+(-3)\begin{vmatrix} 0 & -1 \\ 3 & 2 \end{vmatrix}$$

$=(0+2)-3(0+3)=2-9=-7$

$$x=\frac{D_x}{D}=\frac{-7}{7}=-1$$

$x=-1$

16. $x(2x-7)=4$

$2x^2-7x=4$

$2x^2-7x-4=0$

$(2x+1)(x-4)=0$

$x=-\frac{1}{2}$ or $x=4$

$\left\{-\frac{1}{2}, 4\right\}$

17. $x^3+3x^2-9x-27=0$

$x^2(x+3)-9(x+3)=0$

$(x^2-9)(x+3)=0$

$(x+3)(x-3)(x+3)=0$

$x=-3$ or $x=3$

$\{-3, 3\}$

18. $\frac{2}{3x+1}+\frac{4}{3x-1}=\frac{6x+8}{9x^2-1}$

$(9x^2-1)\left(\frac{2}{3x+1}+\frac{4}{3x-1}\right)=(9x^2-1)\left(\frac{6x+8}{9x^2-1}\right)$

$2(3x-1)+4(3x+1)=6x+8$

$6x-2+12x+4=6x+8$

$18x+2=6x+8$

$12x=6$

$x=\frac{1}{2}$

$\left\{\frac{1}{2}\right\}$

19. $\frac{5}{x-3}=1+\frac{30}{x^2-9}$

$(x^2-9)\left(\frac{5}{x-3}\right)=(x^2-9)\left(1+\frac{30}{x^2-9}\right)$

$5(x+3)=x^2-9+30$

$5x+15=x^2+21$

$0=x^2-5x+6$

$0=(x-6)(x+1)$

$x=6$ or $x=-1$

$\{-1, 6\}$

20. $\sqrt{2x+4}-\sqrt{x+3}-1=0$

$\sqrt{2x+4}=1+\sqrt{x+3}$

$2x+4=1+2\sqrt{x+3}+x+3$

$2x+4-1-x-3=2\sqrt{x+3}$

$x=2\sqrt{x+3}$

$x^2=4(x+3)$

$x^2-4x-12=0$

$(x-6)(x+2)=0$

$x=6$ or $x=-2$ (extraneous)

$\{6\}$

21. $2x^2=5-4x$

$2x^2+4x-5=0$

$x=\frac{-b\pm\sqrt{b^2-4ac}}{2a}$

$x=\frac{-4\pm\sqrt{4^2-4(2)(-5)}}{2(2)}=\frac{-4\pm\sqrt{56}}{4}$

$=\frac{-4\pm2\sqrt{14}}{4}=\frac{-2\pm\sqrt{14}}{2}$

$\left\{\frac{-2-\sqrt{14}}{2}, \frac{-2+\sqrt{14}}{2}\right\}$

22. $\frac{1}{y+2}-\frac{1}{3}=\frac{1}{y}$

$3y(y+2)\left[\frac{1}{y+2}-\frac{1}{3}\right]=3y(y+2)\cdot\frac{1}{y}$

$3y-y^2-2y=3y+6$

$-y^2-2y-6=0$

$y^2+2y+6=0$

$y=\frac{-2\pm\sqrt{4-24}}{2}=\frac{-2\pm2i\sqrt{5}}{2}$

$y=-1\pm i\sqrt{5}$

$\left\{-1-i\sqrt{5}, -1+i\sqrt{5}\right\}$

23. $x^{2/3}-x^{1/3}-6=0$

Let $u=x^{1/3}$.

$u^2-u-6=0$

$(u-3)(u+2)=0$

$u=3$ or $u=-2$

$x^{1/3}=3$ or $x^{1/3}=-2$

$x=3^3$ or $x=(-2)^3$

$x=27$ or $x=-8$

$\{-8, 27\}$

24. $(x^2+x)^2-5(x^2+x)=-6$

Let $u=x^2+x$.

$u^2-5u=-6$

$u^2-5u+6=0$

$(u-3)(u-2)=0$

$u=3$ or $u=2$

$x^2+x=3$ or $x^2+x=2$

$x^2+x-3=0$ or $x^2+x-2=0$

$x=\dfrac{-1\pm\sqrt{1^2-4(1)(-3)}}{2(1)}$ or $(x+2)(x-1)=$

0

$x=\dfrac{-1\pm\sqrt{13}}{2}$ or $x=-2$ or $x=1$

$\left\{\dfrac{-1-\sqrt{13}}{2},\ \dfrac{-1+\sqrt{13}}{2},\ -2,\ 1\right\}$

25. $3x^2+8x+5<0$

$(3x+5)(x+1)<0$

F	T	F
	$-\frac{5}{3}$	-1

Test –2: $3(-2)^2+8(-2)+5<0$

$1<0$ False

Test –1.5: $3(-1.5)^2+8(-1.5)+5<0$

$-0.25<0$ True

Test 0: $3(0)^2+8(0)+5<0$

$5<0$ False

$\left\{x\middle|-\frac{5}{3}<x<-1\right\},\ \left(-\frac{5}{3},-1\right)$

26. $\dfrac{x-1}{x+3}\le 0$

$x=1,\ x=-3$

F	T	F
	-3	1

Test –4: $\dfrac{-4-1}{-4+3}\le 0$

$5\le 0$ False

Test 0: $\dfrac{-1}{3}\le 0$ True

Test 2: $\dfrac{2-1}{2+3}\le 0$

$\frac{1}{5}\le 0$ False

$\{x|-3<x\le 1\}$, $(-3, 1]$

27. $\log_5(x-2)=3$

$5^3=x-2$

$125=x-2$

$127=x$

$\{127\}$

28. $\log_2 x+\log_2(2x-3)=1$

$\log_2 x(2x-3)=1$

$x(2x-3)=2^1$

$2x^2-3x=2$

$2x^2-3x-2=0$

$(2x+1)(x-2)=0$

$x=-\frac{1}{2}$ or $x=2$

$\left(\text{Reject } x=-\frac{1}{2} \text{ since } \log_2\left(-\frac{1}{2}\right) \text{ is not defined.}\right)$

$\{2\}$

29. $3^{2x-1}=81$

$3^{2x-1}=3^4$

$2x-1=4$

$2x=5$

$x=\frac{5}{2}$

$\left\{\frac{5}{2}\right\}$

30. $30e^{0.7x}=240$

$e^{0.7x}=8$

$\ln e^{0.7x}=\ln 8$

$0.7x=\ln 8$

$x=\dfrac{\ln 8}{0.7}\approx 2.971$

$\left\{\dfrac{\ln 8}{0.7}\approx 2.971\right\}$

31. $3x^2+4y^2=39$

$5x^2-2y^2=-13$

$$\begin{aligned} 3x^2 + 4y^2 &= 39 \\ 10x^2 - 4y^2 &= -26 \\ \hline 13x^2 &= 13 \\ x^2 &= 1 \\ x &= \pm 1 \end{aligned}$$

If $x = -1$: $3(-1)^2 + 4y^2 = 39$

$y^2 = 9$

$y = \pm 3$

If $x = 1$: $3(1)^2 + 4y^2 = 39$

$y = \pm 3$

$\{(-1, -3), (-1, 3), (1, -3), (1, 3)\}$

32. $2x^2 - y^2 = -8$

$x - y = 6$

$x = 6 + y$

$2(6 + y)^2 - y^2 = -8$

$2(36 + 12y + y^2) - y^2 = -8$

$72 + 24y + 2y^2 - y^2 = -8$

$y^2 + 24y + 80 = 0$

$(y + 4)(y + 20) = 0$

$y = -4$ or $y = -20$

$x = 6 - 4 = 2$ or $x = 6 - 20 = -14$

$\{(2, -4), (-14, -20)\}$

33. Possible rational roots of $3x^3 + 4x^2 - 7x + 2$ are $\pm\frac{1}{3}, \pm\frac{2}{3}, \pm 1, \pm 2$. Testing the roots gets $x = \frac{2}{3}$ as a solution:

$$3\left(\frac{2}{3}\right)^3 + 4\left(\frac{2}{3}\right)^2 - 7\left(\frac{2}{3}\right) + 2 = \frac{8}{9} + \frac{8}{9} - \frac{14}{3} + 2 = 0$$

Divide $3x - 2$ from $3x^3 + 4x^2 - 7x + 2$.

$$\begin{array}{r} x^2 + 2x - 1 \\ 3x - 2 \overline{)\, 3x^3 + 4x^2 - 7x + 2} \\ \underline{3x^3 + 2x^2} \quad\quad\quad \\ 6x^2 - 7x \quad\;\; \\ \underline{6x^2 - 4x} \quad\;\; \\ -3x + 2 \\ \underline{-3x + 2} \\ 0 \end{array}$$

Solve $x^2 + 2x - 1 = 0$.

$$x = \frac{-2 \pm \sqrt{(2)^2 - 4(1)(-1)}}{2}$$

$$= \frac{-2 \pm \sqrt{8}}{2}$$

$$= -1 \pm \sqrt{2}$$

$\left\{\frac{2}{3}, -1 - \sqrt{2}, -1 + \sqrt{2}\right\}$

34. a. Security guards, Gas station workers

b. Taxi drivers/chauffeurs

c. Gas station workers, Security guards

35. a. 1991, $2.1 B

b. It has been rising.

36. a. $x(x)(108 - 4x) = x^2(108 - 4x)$
$= 108x^2 - 4x^3 = -4x^3 + 108x^2$

b. $x = 18$ in., volume $\approx 11{,}700$ in^3

c. $-4(18)^3 + 108(18)^2 = 11{,}664$
The volume is $11{,}664$ in^3.

37. a. (1972, 450); in 1972, males and females had equal verbal SAT scores of 450.

b. Scores are decreasing over time.

38.

x	$f(x) = \lvert x \rvert$	$g(x) = \lvert x \rvert - 1$	$h(x) = \lvert x \rvert + 2$
−3	3	2	5
−2	2	1	4
−1	1	0	3
0	0	−1	2
1	1	0	3
2	2	1	4
3	3	2	5

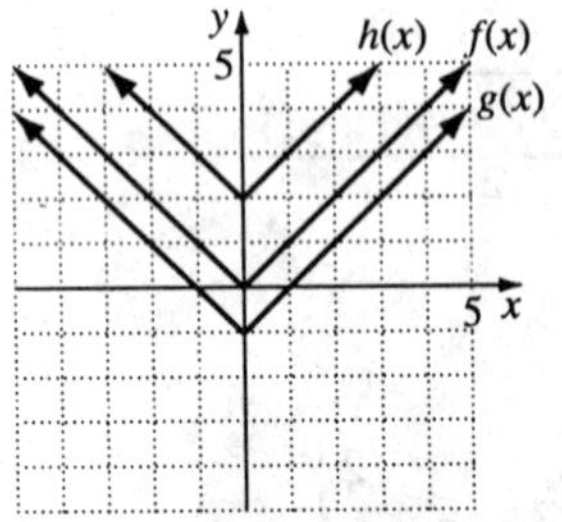

g is f with a vertical shift of –1.
h is f with a vertical shift of 2.

39. approximate points: (1980, 125), (1985, 135),
(1990, 90), (1995, 87)

$$\text{slope}_{80-85} = \frac{135-125}{1985-1980} = 2$$

$$\text{slope}_{85-90} = \frac{90-135}{1990-1985} = -9$$

$$\text{slope}_{90-95} = \frac{87-90}{1995-1990} = -\frac{3}{5}$$

The average rate of change per year in the average number of months imposed for violent offenses from 1980 to 1985 was 2, from 1985 to 1990 was –9, and from 1990 to 1995 was $-\frac{3}{5}$.

40. a. $-2y + 5 \le -75$
$-2y \le -80$
$y \ge 40$
Years 1970-2000

b. $85 \le 3y - 5 \le 115$
$90 \le 3y \le 120$
$30 \le y \le 40$
Years 1943, 1948-1970

41. a; all three planes intersect at one point.

42. Yes; they intersect at 4 points.

43. $y = -\frac{2}{3}x + 4$

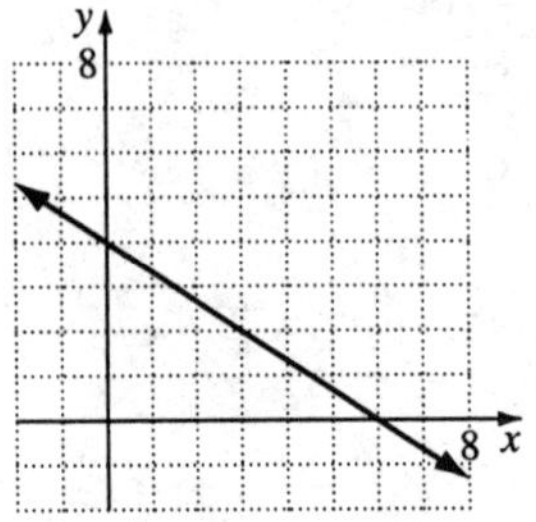

44. $3x - y > 6$

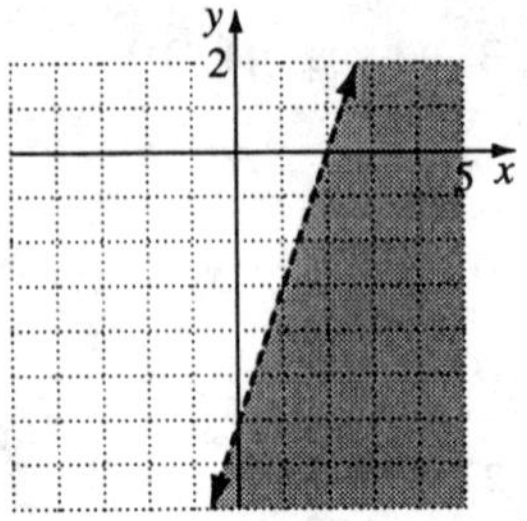

45. $y \ge \frac{3}{4}x - 5$

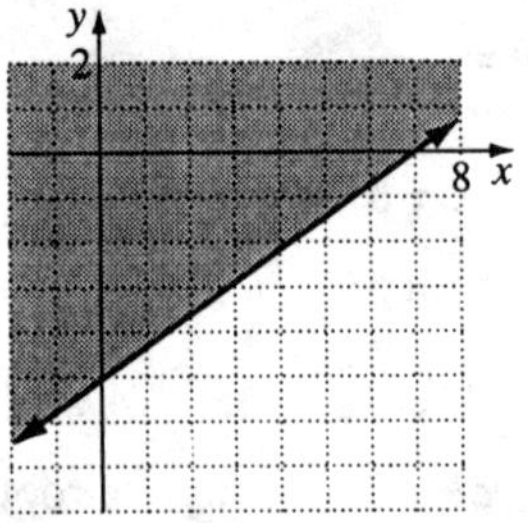

46. $2x - y \ge 4$
$x \le 2$

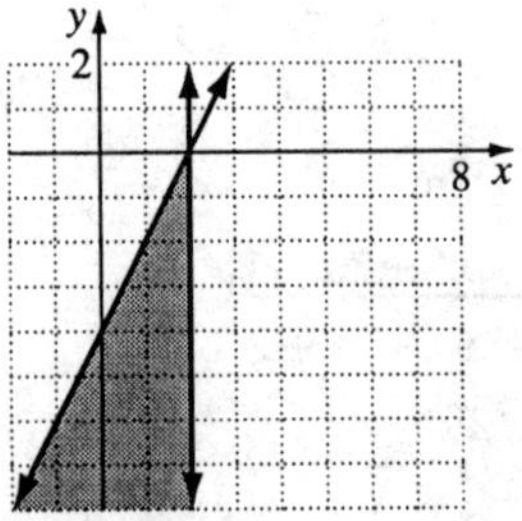

(2, 0)

47. $x \le 5$
$y \le 6$
$14x + 8y \ge 56$
$2x + 3y \ge 12$

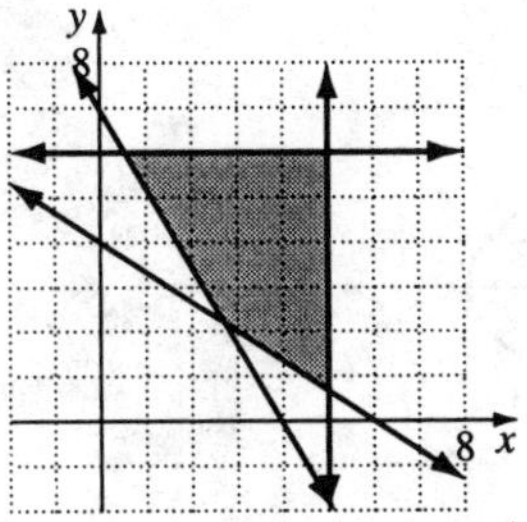

$(5, 6), \left(5, \frac{2}{3}\right), \left(\frac{36}{13}, \frac{28}{13}\right), \left(\frac{4}{7}, 6\right)$

48. $y = -2x - 1$
$x + 2y = 4$

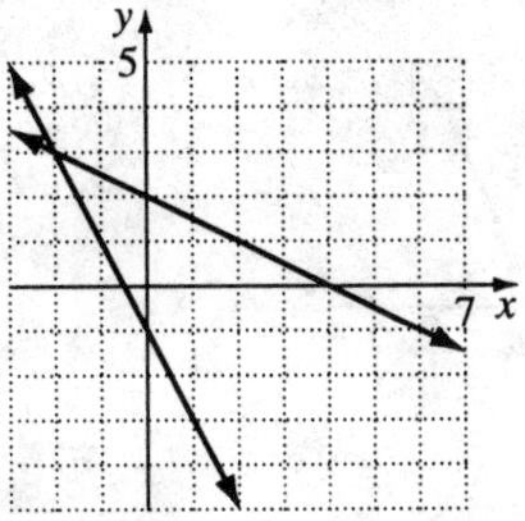

$\{(-2, 3)\}$

49. $f(x) = x^2 - 4x - 5$
x-intercepts:
$0 = x^2 - 4x - 5$
$0 = (x - 5)(x + 1)$
$x = 5$ or $x = -1$
$(5, 0), (-1, 0)$
y-intercept:
$y = 0^2 - 4(0) - 5 = -5$
$(0, -5)$
Vertex:
$x = -\frac{b}{2a} = -\frac{(-4)}{2(1)} = 2$
$y = 2^2 - 4(2) - 5 = -9$
$(2, -9)$

$y = x^2 - 4x - 5$

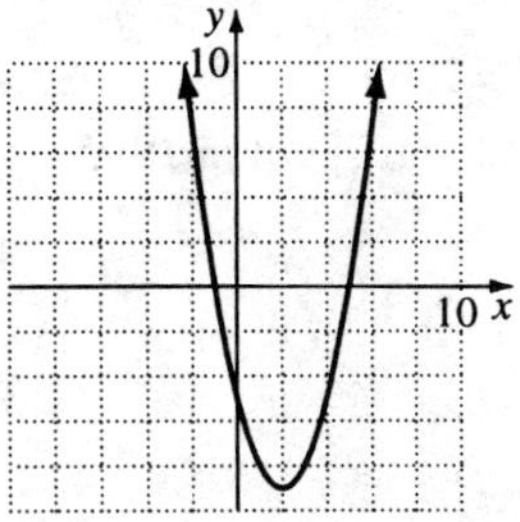

50. $f(x) = \sqrt{x},\ g(x) = \sqrt{x - 1}$

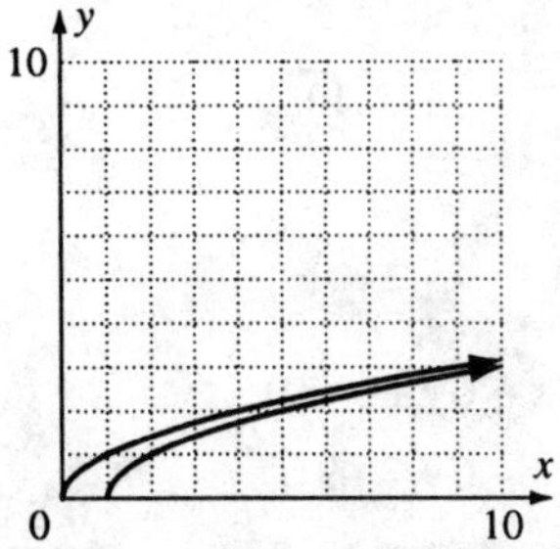

Domain of $f = [0, \infty)$, Range of $f = [0, \infty)$
Domain of $g = [1, \infty)$, Range of $g = [0, \infty)$
g is f shifted one unit to the right.

51. $y = (x - 1)^2 - 4$

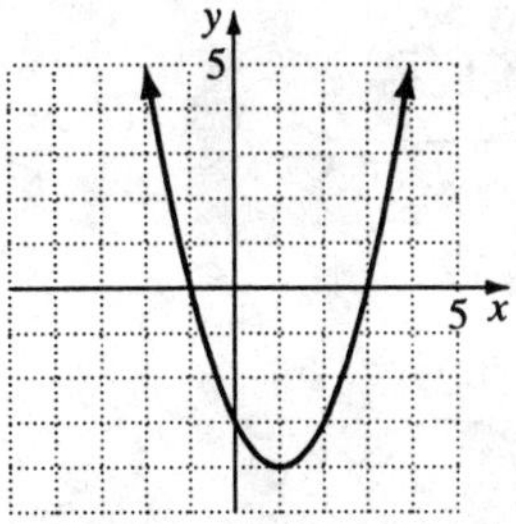

52. $y = 2^{x/2} + 1$

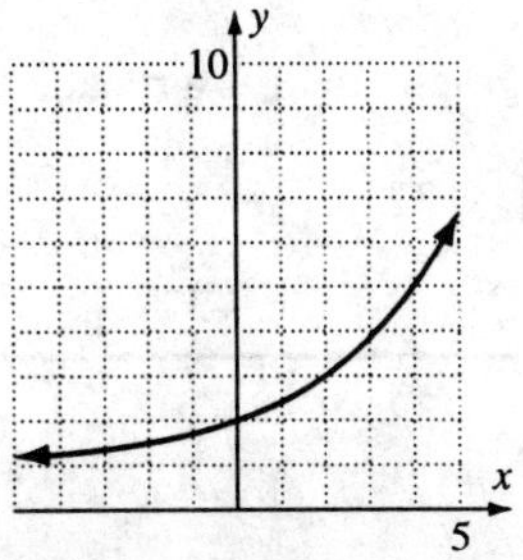

53. **a.** There is only one value of y for every x.

b. $f^{-1}(300) = 1995$
$300 billion was spent in 1995.

c. quadratic

54. $y = \log_2 x$

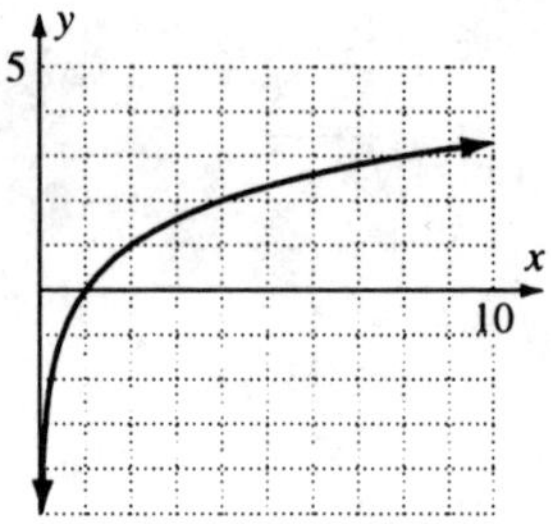

55. $x^2 + y^2 + 4x - 6y + 9 = 0$
$x^2 + 4x + y^2 - 6y = -9$
$x^2 + 4x + 4 + y^2 - 6y + 9 = -9 + 4 + 9$
$(x+2)^2 + (y-3)^2 = 4$
Circle with center (–2, 3) and radius 2

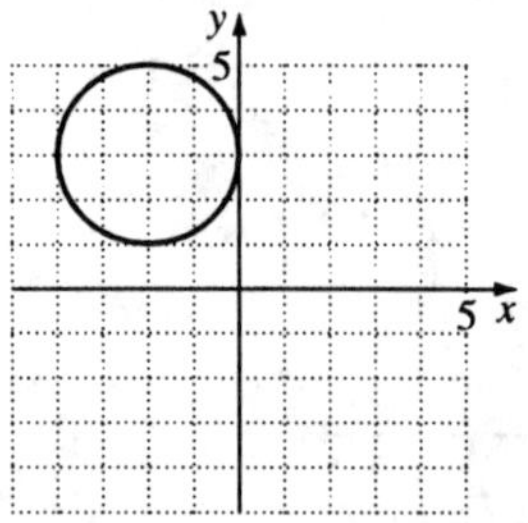

56. $\frac{x^2}{9} + \frac{y^2}{4} = 1$
Ellipse with center (0, 0) and vertices (±3, 0) and (0, ±2)

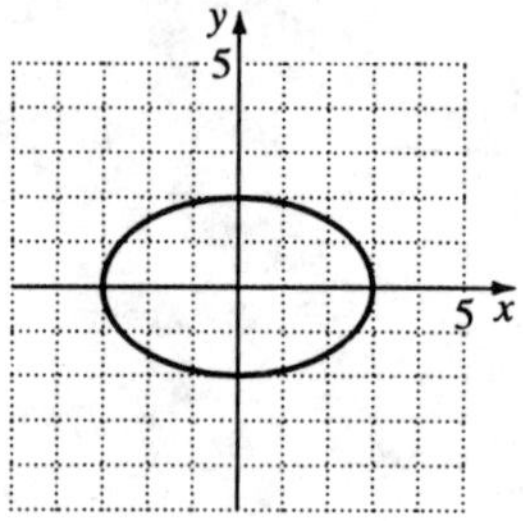

57. $\frac{x^2}{4} - \frac{y^2}{9} = 1$
Hyperbola with center (0, 0), vertices (±2, 0) and no y-intercept

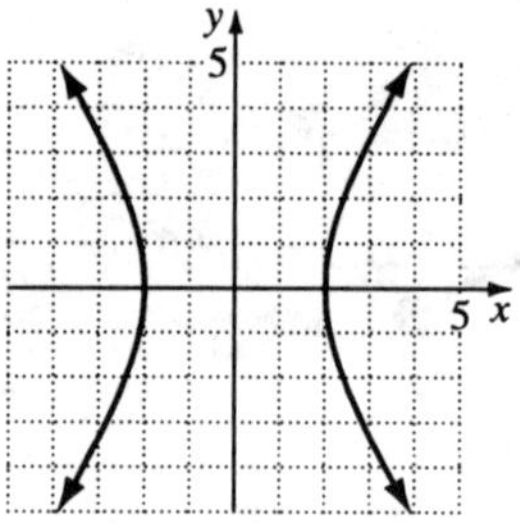

58. **a.** Possible rational roots: $\pm\frac{1}{32}, \pm\frac{1}{10}, \pm\frac{1}{8}, \pm\frac{1}{4}, \pm\frac{1}{2}, \pm 1, \pm\frac{3}{32}, \pm\frac{3}{16}, \pm\frac{3}{8}, \pm\frac{3}{4}, \pm\frac{3}{2}, \pm 3$

b. From the graph, $x = 1$ is a root.

–1	32	–52	17	3
		–32	20	3
	32	–20	–3	0

$32x^3 - 52x^2 + 17x + 3 = 0$
$(x-1)(32x^2 - 20x - 3) = 0$
$(x-1)(4x-3)(8x+1) = 0$
$x - 1 = 0$ or $4x - 3 = 0$ or $8x + 1 = 0$
$x = 1$ or $x = \frac{3}{4}$ or $x = -\frac{1}{8}$
The solution set is $\left\{-\frac{1}{8}, \frac{3}{4}, 1\right\}$.

59. $\frac{x^2+3x-40}{x^2-x-20} = \frac{(x-5)(x+8)}{(x-5)(x+4)} = \frac{x+8}{x+4} = 1 + \frac{4}{x+4} \quad (x \neq 5)$

Graph $y = 1 + \frac{4}{x+4}$ with a hole at $x = 5$. There is a vertical asymptote at $x = -4$ and a horizontal asymptote at $y = 1$.

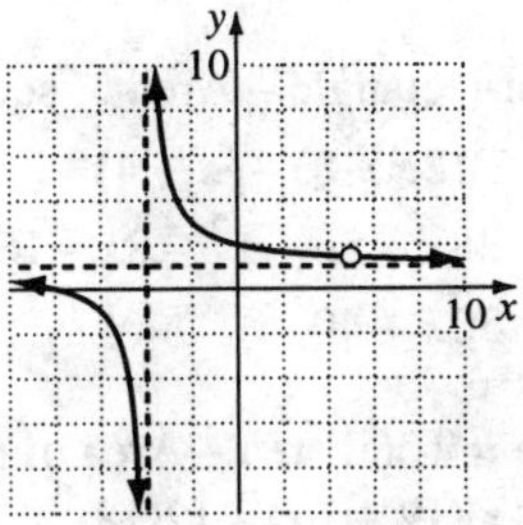

60. $t = \sqrt{\frac{2h}{g}}$
$t = \sqrt{\frac{2(64)}{32}}$
$t = \sqrt{4}$
$t = \pm 2$
Disregard a negative time.
$t = 2$ seconds

61. a. $t = 35d + 20$
$125 = 35d + 20$
$105 = 35d$
$3 = d$
$d = 3$ km

b. Point (3, 125)

62. $V_f = \dfrac{m_1 - m_2}{m_1 + m_2} \cdot v_i$, for m_1

$V_f(m_1 + m_2) = v_i(m_1 - m_2)$

$V_f m_1 + V_f m_2 = v_i m_1 - v_i m_2$

$V_f m_2 + v_i m_2 = v_i m_1 - V_f m_1$

$V_f m_2 + v_i m_2 = m_1(v_i - V_f)$

$\dfrac{V_f m_2 + v_i m_2}{v_i - V_f} = m_1$

$m_1 = \dfrac{m_2(V_f + v_i)}{v_i - V_f}$

63. $f(x) = 0.002x^2 + 4.05x + 71.78$

$f(10) = 0.002(10)^2 + 4.05(10) + 71.78$

$f(10) = 112.48$

$1960 + 10 = 1970$

In 1970, there were 112.48 million motor vehicle registrations in the U.S.

64. a. Negative; as x increases, y decreases.

b. Positive; as x increases, y increases.

65. a. slope $= \dfrac{171.1 - 119.4}{10 - 0} = 5.17$

$y - 119.4 = 5.17x$ or $y - 171.1 = 5.17(x - 10)$

b. $y = 5.17x + 119.4$

c. $y = 5.17(17) + 119.4 = 207.29$

There will be about 207 billion pieces.

66. a. $r = 0.25d + 30$

b. \$55

$r = 0.25(100) + 30 = 55$

c. Rent-a-Heap has a better deal.

67. a. $P = 8x + 12y$

b. $x + y \le 200$

$x \ge 10$

$y \ge 80$

c.

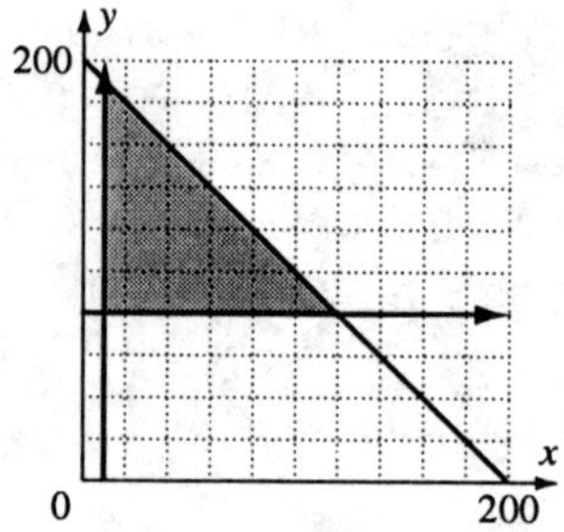

d. (10, 80), $P = 8(10) + 12(80) = 1040$

(10, 190), $P = 8(10) + 12(190) = 2360$

(120, 80), $P = 8(120) + 12(80) = 1920$

10, 190, \$2360

68. $A(x)$ = Area of rectangle – Area of square

$A(x) = (3x + 7)(2x + 8) - (x + 4)^2$

$A(x) = 6x^2 + 38x + 56 - x^2 - 8x - 16$

$A(x) = 5x^2 + 30x + 40$

69. Area of frame = Total area – Area of picture

$A(x) = (18 + 2x)(12 + 2x) - 12(18)$

$= 216 + 60x + 4x^2 - 216$

$A(x) = 4x^2 + 60x$

70. a. $s(t) = -16t^2 + 80t + 96$

$0 = -16t^2 + 80t + 96$

$0 = -16(t^2 - 5t - 6)$

$0 = -16(t - 6)(t + 1)$

$t = 6$ or $t = -1$

Disregard a negative time.

It will strike the ground in 6 seconds.

b. Vertex:

$x = -\dfrac{b}{2a} = -\dfrac{80}{2(-16)} = 2.5$

$s(2.5) = -16(2.5)^2 + 80(2.5) + 96 = 196$

In 2.5 seconds it will reach a maximum height of 196 ft.

c. $s(t) = -16t^2 + 80t + 96$

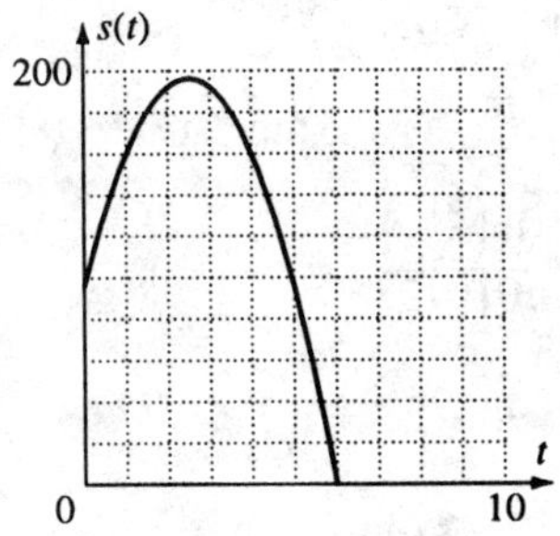

71. $C(x) = \frac{250x}{100-x}$

a. $C(90) - C(40) = \frac{250(90)}{100-90} - \frac{250(40)}{100-40}$

$= 2083\frac{1}{3}$

The difference in cost to remove 90% of the pollutants versus 40% of the pollutants is $\$2083\frac{1}{3}$ million.

b. 100; it is impossible to remove 100% of the pollutants.

72. $\overline{C}(x) = \frac{40x + 50,000}{x}$

$45 = \frac{40x + 50,000}{x}$

$45x = 40x + 50,000$

$5x = 50,000$

$x = 10,000$

10,000 pairs must be produced.

$\overline{C}(x) = \frac{40(100,000) + 50,000}{100,000}$

$\overline{C}(x) = 40.5$

The cost is \$40.50 per pair.

Cost decreases as the number produced increases.

73. $S = \frac{p}{1-nd}$, for d

$S(1 - nd) = p$

$S - Snd = p$

$S - p = Snd$

$\frac{S-p}{Sn} = d$

$d = \frac{S-p}{Sn}$

74. $\frac{1}{p} + \frac{1}{q} = \frac{1}{f}$, for q

$\frac{q+p}{pq} = \frac{1}{f}$

$f(q + p) = pq$

$fq + fp = pq$

$fq - pq = -fp$

$q(f - p) = -fp$

$q = \frac{-fp}{f-p}$

$q = \frac{fp}{p-f}$

75. $r = \sqrt[3]{\frac{3w}{4\pi d}}$, for d

$r^3 = \frac{3w}{4\pi d}$

$4\pi dr^3 = 3w$

$d = \frac{3w}{4\pi r^3}$

76. $N = 1220\sqrt[3]{t-42} + 4900$

$7340 = 1220\sqrt[3]{t-42} + 4900$

$2440 = 1220\sqrt[3]{t-42}$

$2 = \sqrt[3]{t-42}$

$2^3 = t - 42$

$8 + 42 = t$

$50 = t$

$1930 + 50 = 1980$

77. $f(x) = 2x^2 + 22x + 320$

$2120 = 2x^2 + 22x + 320$

$2120 = 2(x^2 + 11x + 160)$

$1060 = x^2 + 11x + 160$

$0 = x^2 + 11x - 900$

$0 = (x + 36)(x - 25)$

$x = -36$ or $x = 25$

Disregard a negative time.

$1980 + 25 = 2005$

78. a. $f(x) = x - 25$
$g(x) = x - 0.3x$
f is \$25 off the regular price while g is 30% off the regular price.

b. $(f \circ g)(x) = x - 0.3x - 25 = 0.7x - 25$
The discount price is \$25 off the sale price of 30% off.

c. $(g \circ f)(x) = (x-25) - 0.3(x-25)$
$= 0.7x - 17.5$
The discount price is 30% off the sale price of \$25 off.

d. $(f \circ g)(x)$; $0.7x - 25$ is less than $0.7x - 17.5$.

79. $y = Ae^{kt}$

a. $t = 2000 - 1995 = 5$
$y = 5.702e^{0.016(5)}$
$y = 6.176903$
It will be 6.176903 billion.

b. $10 = 5.702e^{0.016t}$
$\frac{10}{5.702} = e^{0.016t}$
$\ln\left(\frac{10}{5.702}\right) = \ln e^{0.016t}$
$\ln\left(\frac{10}{5.702}\right) = 0.016t$
$\frac{\ln\left(\frac{10}{5.702}\right)}{0.016} = t$
$35 \approx t$
$1995 + 35 = 2030$

80. $y = Ae^{kt}$

a. $263 = 250e^{k(5)}$
$\ln\left(\frac{263}{250}\right) = 5k$
$\frac{\ln\left(\frac{263}{250}\right)}{5} = k$
$k = 0.0101386$
$y = 250e^{0.0101386t}$

b. $317 = 250e^{0.0101386t}$
$\frac{\ln\left(\frac{317}{250}\right)}{0.0101386} = t$
$23 \approx t$
$1990 + 23 = 2013$

81. $f(t) = 80 - 27 \ln t$
$15 = 80 - 27 \ln t$
$\frac{-65}{-27} = \ln t$
$e^{65/27} = t$
$11 \approx t$
After about 11 minutes

82. a. $a_1 = 28{,}600,\ d = 2250$
$a_n = a_1 + (n-1)d$
$a_n = 28{,}600 + (n-1)2250$
$a_n = 26{,}350 + 2250n$

b. $a_8 = 26{,}350 + 2250(8)$
$a_8 = 44{,}350$
The salary will be \$44,350.

c. $S_8 = \frac{8}{2}(28{,}600 + 44{,}350) = 291{,}800$
The total salary will be \$291,800

83. a. $a_1 = 20{,}000,\quad r = 1.06$
$a_n = a_1 r^{n-1}$
$a_n = 20{,}000(1.06)^{n-1}$

b. $a_8 = 20{,}000(1.06)^{8-1}$
$a_8 = 30{,}072.606$
The salary will be \$30,072.61.

c. $S_n = \frac{a_1 - a_1 r^n}{1-r}$
$S_8 = \frac{20{,}000 - 20{,}000(1.06)^8}{1 - 1.06}$
$= 197{,}949.37$
The total salary will be \$197,949.37.

84. $12x^3 - 36x^2 + 27x = 3x(4x^2 - 12x + 9)$
$= 3x(2x - 3)(2x - 3)$
$= 3x(2x - 3)^2$

85. $27x^3 - 125 = (3x - 5)[(3x)^2 + (3x)(5) + 5^2]$
$= (3x - 5)(9x^2 + 15x + 25)$

86. $x^3 - 2x^2 - 9x + 18 = x^2(x - 2) - 9(x - 2)$
$= (x - 3)(x + 3)(x - 2)$

87. $15x^4 - 35x^2 - 100 = 5(3x^4 - 7x^2 - 20)$
$= 5(3x^2 + 5)(x^2 - 4) = 5(3x^2 + 5)(x + 2)(x - 2)$

88. $10x^{1/2} - 30x^{-1/2} = 10x^{-1/2}(x - 3)$

89. $-2(3^2 - 12)^3 - 45 \div 9 - 3 = -2(9 - 12)^3 - 5 - 3$
$= -2(-3)^3 - 5 - 3 = -2(-27) - 5 - 3$
$= 54 - 5 - 3 = 46$

90. $3x - [6 - 7(3 - 4x) - 5x] = 3x - (6 - 21 + 28x - 5x)$
$= 3x - (-15 + 23x) = 3x + 15 - 23x = 15 - 20x$

91. $(-4x^2y^{-5})(-3x^{-1}y^{-4}) = (-4)(-3)x^{2-1}y^{-5-4}$
$= 12xy^{-9} = \frac{12x}{y^9}$

92. $\frac{30x^2y^4}{-5x^{-3}y^6} = \left(\frac{30}{-5}\right)x^{2+3}y^{4-6} = -\frac{6x^5}{y^2}$

93. $(7.2 \times 10^{-3})(5.0 \times 10^{-5}) = (7.2)(5.0) \times 10^{-3-5}$
$= 36 \times 10^{-8} = 3.6 \times 10^{-7}$

94. $(8x^2 - 9xy - 11y^2) - (7x^2 - 4xy + 5y^2) = 8x^2 - 9xy - 11y^2 - 7x^2 + 4xy - 5y^2$
$= x^2 - 5xy - 16y^2$

95. $(2x - 7)(3x + 11) = 6x^2 + 22x - 21x - 77$
$= 6x^2 + x - 77$

96. $(5x^2 - 7y)^2 = (5x^2)^2 - 2(5x^2)(7y) + (-7y)^2$
$= 25x^4 - 70x^2y + 49y^2$

97. $\frac{x+2}{3x+9} \cdot \frac{x^2 - 9}{x^2 - x - 12} = \frac{x+2}{3(x+3)} \cdot \frac{(x+3)(x-3)}{(x-4)(x+3)}$
$= \frac{(x+2)(x-3)}{3(x+3)(x-4)}$

98. $\frac{3x^2+17xy+10y^2}{6x^2+13xy-5y^2} \div \frac{6x^2+xy-2y^2}{6x^2-5xy+y^2} = \frac{(3x+2y)(x+5y)}{(3x-y)(2x+5y)} \cdot \frac{(3x-y)(2x-y)}{(3x+2y)(2x-y)}$

$= \frac{x+5y}{2x+5y}$

99. $\frac{2y-6}{3y^2-14y-5} - \frac{y-3}{y^2-5y} = \frac{2y-6}{(3y+1)(y-5)} - \frac{y-3}{y(y-5)}$

(LCD is $y(3y+1)(y-5)$.)

$= \frac{2y-6}{(3y+1)(y-5)} \cdot \frac{y}{y} - \frac{y-3}{y(y-5)} \cdot \frac{(3y+1)}{(3y+1)}$

$= \frac{y(2y-6)-(y-3)(3y+1)}{y(3y+1)(y-5)}$

$= \frac{2y^2-6y-3y^2+8y+3}{y(3y+1)(y-5)} = \frac{-y^2+2y+3}{y(3y+1)(y-5)}$

100. $\frac{1-\frac{14y-45}{y^2}}{\frac{y}{9}-\frac{9}{y}} \cdot \frac{9y^2}{9y^2} = \frac{9y^2-9(14y-45)}{y^3-81y}$

$= \frac{9y^2-126y+405}{y^3-81y} = \frac{9(y^2-14y+45)}{y(y^2-81)}$

$= \frac{9(y-9)(y-5)}{y(y+9)(y-9)} = \frac{9(y-5)}{y(y+9)}$

101. $(3x^3-19x^2+17x+4) \div (3x-4) = x^2-5x-1$

$$
\begin{array}{r|l}
 & x^2-5x-1 \\
3x-4 & 3x^3-19x^2+17x+4 \\
 & \underline{3x^3-\ \ 4x^2} \\
 & -15x^2+17x \\
 & \underline{-15x^2+20x} \\
 & -3x+4 \\
 & \underline{-3x+4} \\
 & 0
\end{array}
$$

102. $\frac{3x^3-5x^2+2x-1}{x-2} = 3x^2+x+4+\frac{7}{x-2}$

2	3	–5	2	–1
		6	2	8
	3	1	4	7

103. $\sqrt[8]{16x^4y^2} = 16^{1/8}x^{4/8}y^{2/8}$
$= 16^{1/8}x^{1/2}y^{1/4} = \sqrt[8]{16}\sqrt{x}\,\sqrt[4]{y}$

104. $\sqrt{3}\cdot\sqrt[3]{3} = 3^{1/2}\cdot 3^{1/3} = 3^{\frac{1}{2}+\frac{1}{3}} = 3^{5/6} = \left(\sqrt[6]{3}\right)^5$

105. $\left(16x^{1/2}y^{4/3}\right)^{3/2} = 16^{3/2}x^{3/4}y^{12/6}$
$= 4^3x^{3/4}y^2 = 64y^2\left(\sqrt[4]{x}\right)^3$

106. $\sqrt[3]{16x^7y^{11}} = \sqrt[3]{8x^6y^9\cdot 2xy^2}$
$= \sqrt[3]{(2x^2y^3)^3\cdot 2xy^2} = 2x^2y^3\sqrt[3]{2xy^2}$

107. $\sqrt{5xy}\cdot\sqrt{10x^2y} = \sqrt{5xy\cdot 10x^2y}$
$= \sqrt{50x^3y^2} = \sqrt{25x^2y^2\cdot 2x} = \sqrt{(5xy)^2\cdot 2x}$
$= 5xy\sqrt{2x}$

108. $\sqrt[3]{4x^2y^5}\cdot\sqrt[3]{4xy^2z^2} = \sqrt[3]{16x^3y^7z^2}$
$= \sqrt[3]{8\cdot 2x^3(y^2)^3yz^2} = 2xy^2\sqrt[3]{2yz^2}$

109. $\dfrac{\sqrt[3]{54}}{\sqrt[3]{2y^2}} = \sqrt[3]{\dfrac{54}{2y^2}} = \sqrt[3]{\dfrac{27}{y^2}} = \dfrac{3}{\sqrt[3]{y^2}} = \dfrac{3\sqrt[3]{y}}{y}$

110. $7\sqrt{18x^5} - 3x\sqrt{2x^3}$
$= 7\sqrt{9\cdot 2(x^2)^2x} - 3x\sqrt{2x^2x}$
$= 7\cdot 3x^2\sqrt{2x} - 3x(x)\sqrt{2x}$
$= 21x^2\sqrt{2x} - 3x^2\sqrt{2x}$
$= 18x^2\sqrt{2x}$

111. $\dfrac{1-\sqrt{y}}{1+\sqrt{y}} = \dfrac{1-\sqrt{y}}{1+\sqrt{y}}\cdot\dfrac{1-\sqrt{y}}{1-\sqrt{y}}$
$= \dfrac{1-2\sqrt{y}+y}{1-y}$

112. $(7+3i)(9-4i) = 63 - 28i + 27i - 12i^2$
$= 63 - i - 12(-1) = 75 - i$

113. $\dfrac{6}{3+5i} = \dfrac{6}{3+5i}\cdot\dfrac{3-5i}{3-5i} = \dfrac{18-30i}{9+25}$
$= \dfrac{18-30i}{34} = \dfrac{9-15i}{17}$

114. $\dfrac{xy-(y-z)^2}{xz} = \dfrac{(3)(-2)-[-2-(-4)]^2}{(3)(-4)}$
$= \dfrac{-6-(2)^2}{-12} = \dfrac{-10}{-12} = \dfrac{5}{6}$

115. Passing through (1, –4) and (–5, 8):
slope $= \dfrac{8-(-4)}{-5-1} = \dfrac{12}{-6} = -2$
Point-slope:
$y + 4 = -2(x-1)$ or $y - 8 = -2(x+5)$
$y + 4 = -2x + 2$
Slope-intercept: $y = -2x - 2$

116. Passing through (3, –2) and perpendicular to the line whose equation is:
$-\frac{1}{4}x + y = 5$
$y = \frac{1}{4}x + 5$
slope $= \frac{1}{4}$
Slope of line perpendicular:
$\dfrac{-1}{\frac{1}{4}} = -4$
Point-slope: $y + 2 = -4(x-3)$
Slope-intercept:
$y + 2 = -4x + 12$
$y = -4x + 10$

117. $(f+g)(x) = x^2 - 4 + x + 2 = x^2 + x - 2$

118. $(f-g)(x) = x^2 - 4 - (x+2) = x^2 - x - 6$

119. $(fg)(x) = (x^2 - 4)(x+2) = x^3 + 2x^2 - 4x - 8$

120. $\left(\frac{f}{g}\right)(x) = \frac{x^2 - 4}{x+2} = \frac{(x-2)(x+2)}{x+2} = x - 2$

121. $(f \circ g)(x) = (x+2)^2 - 4 = x^2 + 4x + 4 - 4$
$= x^2 + 4x$

122. $(g \circ f)(x) = x^2 - 4 + 2 = x^2 - 2$

123. $f(x) = 3x^2 - 7x - 2$

$$\frac{f(a+h) - f(a)}{h} = \frac{3(a+h)^2 - 7(a+h) - 2 - (3a^2 - 7a - 2)}{h}$$
$$= \frac{3(a^2 + 2ah + h^2) - 7a - 7h - 2 - 3a^2 + 7a + 2}{h}$$
$$= \frac{3a^2 + 6ah + 3h^2 - 7a - 7h - 2 - 3a^2 + 7a + 2}{h}$$
$$= \frac{6ah + 3h^2 - 7h}{h} = 6a + 3h - 7$$

124. (4, 3) to (2, –1):

$$d = \sqrt{(4-2)^2 + (3+1)^2} = \sqrt{2^2 + 4^2}$$
$$= \sqrt{4 + 16} = 2\sqrt{5}$$

125. $f(x) = 4x - 3$

a. $y = 4x - 3$
$x = 4y - 3$
$x + 3 = 4y$
$\frac{x+3}{4} = y$
$f^{-1}(x) = \frac{x+3}{4}$

b. $f(f^{-1}(x)) = 4\left(\frac{x+3}{4}\right) - 3 = x$
$f^{-1}(f(x)) = \frac{4x - 3 + 3}{4} = x$

126. $\log_b x = 4$
$x = b^4$

127. $\log_5 \frac{x^3\sqrt{y}}{125} = \log_5 x^3 + \log_5 y^{1/2} - \log_5 5^3$
$= 3\log_5 x + \frac{1}{2}\log_5 y - 3$

128. $2\log_b x + 3\log_b y - \frac{1}{2}\log_b z = \log_b x^2 + \log_b y^3 - \log_b \sqrt{z}$

$= \log_b\left(\frac{x^2y^3}{\sqrt{z}}\right)$

129. $\sum_{i=2}^{5}(i^3 - 4) = (2^3 - 4) + (3^3 - 4) + (4^3 - 4) + (5^3 - 4)$

$= (8 - 4) + (27 - 4) + (64 - 4) + (125 - 4) = 4 + 23 + 60 + 121 = 208$

130. 2, 6, 10, ...

$a_1 = 2,\ n = 30,\ d = 6 - 2 = 4$

$S_n = \frac{n}{2}[2a_1 + (n-1)d]$

$= 15[4 + 29(4)] = 1800$

131. $\frac{1}{2}$, 2, 8, ...

$a_1 = \frac{1}{2},\ n = 8,\ r = \frac{a_2}{a_1} = \frac{2}{\frac{1}{2}} = 4$

$S_n = \frac{a_1 - a_1r^n}{1-r}$

$S_8 = \frac{\frac{1}{2} - \frac{1}{2}(4)^8}{1-4} = \frac{\frac{1}{2}(1 - 65{,}536)}{-3}$

$= \frac{-32{,}767.5}{-3} = 10{,}922.5$

132. $1 + \frac{1}{2} + \frac{1}{4} + \frac{1}{8} + \ldots$

$r = \frac{\frac{1}{2}}{1} = \frac{1}{2}$

$S_\infty = \frac{a_1}{1-r}$

$S_\infty = \frac{1}{1-\frac{1}{2}} = 2$

133. $\frac{34}{100} + \frac{34}{10{,}000} + \ldots$

$r = \frac{\frac{34}{10000}}{\frac{34}{100}} = \frac{1}{100}$

$S_\infty = \frac{a_1}{1-r}$

$S_\infty = \frac{\frac{34}{100}}{1-\frac{1}{100}} = \frac{34}{100}\cdot\frac{100}{99} = \frac{34}{99}$

134. $(2x-y^3)^5 = (2x)^5 + 5(2x)^4(-y^3) + 10(2x)^3(-y^3)^2 + 10(2x)^2(-y^3)^3 + 5(2x)(-y^3)^4 + (-y^3)^5$
$= 32x^5 - 80x^4y^3 + 80x^3y^6 - 40x^2y^9 + 10xy^{12} - y^{15}$

135. For each ? there are 10 possible digits, so there are $10 \cdot 10 = 100$ possibilities for the two digits. Thus, the probability is $\frac{1}{100}$.

136. The number of ways to select 3 people from the committee is $\binom{10}{3} = \frac{10!}{3!\,7!} = 120$. The number of ways to select 3 men from the committee is $\binom{3}{3} = \frac{3!}{3!\,0!} = 1$. The probability is $\frac{1}{120}$.

137. S_n: $3+7+11+15+\ldots+(4n-1) = n(2n+1)$
S_1: $3 = 1(2 \cdot 1 + 1)$
$3 = 3$
Assume S_k is true.
S_k: $3+7+11+15+\ldots+(4k-1) = k(2k+1)$
This implies the truth of S_{k+1}.
S_{k+1}: $3 + 7 + 11 + 15 \ldots + [4(k+1)-1] = (k+1)[2(k+1)+1]$
Proof:
$3 + 7 + 11 + 15 + \ldots + (4k-1) + [4(k+1)-1] = k(2k+1) + [4(k+1)-1]$
$3 + 7 + 11 + 15 + \ldots + [4(k+1)-1] = 2k^2 + k + 4k + 4 - 1$
$3 + 7 + 11 + 15 + \ldots + [4(k+1)-1] = 2k^2 + 5k + 3$
$3 + 7 + 11 + 15 + \ldots + [4(k+1)-1] = (k+1)(2k+3)$
$3 + 7 + 11 + 15 + \ldots + [4(k+1)-1] = (k+1)[2(k+1)+1]$
This is the statement for $n = k+1$. Therefore, S_n is true.

138. Let x = original price.
$784 = x - 0.3x$
$784 = 0.7x$
$1120 = x$
The price was $1120.

139. Let x = number of miles.
$39 + 0.16x = 25 + 0.24x$
$14 = 0.08x$
$175 = x$
$39 + 0.16(175) = 67$
175 miles must be driven for a cost of $67.

140. Let x = number of minutes.
$82.45 = 19.95 + 0.25(x-50)$
$62.5 = 0.25(x-50)$
$62.5 = 0.25x - 12.5$
$75 = 0.25x$
$300 = x$
300 minutes

141. $17 = x(x+4) - (x+1)(x-2)$
$17 = x^2 + 4x - (x^2 - x - 2)$
$17 = x^2 + 4x - x^2 + x + 2$
$17 = 5x + 2$
$15 = 5x$
$3 = x$

$7 = x + 4$
The rectangle is 3 m by 7 m.

142. $144 = 2(12 + x) + 2(x + 6 + 24)$
$144 = 24 + 2x + 2x + 60$
$60 = 4x$
$15 = x$
$27 = 12 + x$
$45 = x + 30$
$1215 = (27)(45)$
The plan is 27 ft by 45 ft, with an area of 1215 ft^2.

143. Let x = rate of the slower car.
$d = rt$
$342 = 3x + 3(x + 10)$
$342 = 3x + 3x + 30$
$312 = 6x$
$52 = x$
$62 = x + 10$
The cars are traveling at 52 mph and 62 mph.

144. Let x = first odd integer (Richard Nixon's age).
$x + x + 2 + x + 4 = 249$
$3x = 243$
$x = 81$
$x + 2 = 83$
$x + 4 = 85$
They lived 81 yr, 83 yr and 85 yr, respectively.

145. Let x = calories in one apple and y = calories in one banana.
$3x + 2y = 354$
$2x + 3y = 381$

$$\begin{array}{r} 6x + 4y = 708 \\ -6x - 9y = -1143 \\ \hline -5y = -435 \\ y = 87 \end{array}$$

$3x + 2(87) = 354$
$3x = 180$
$x = 60$
The apple has 60 calories and the banana has 87 calories.

146. Let x = length and y = width.
$2x + 2y = 320$
$9x + 5(2y) = 1510$
$9x + 10y = 1510$
$x + y = 160$
$x = 160 - y$
$9(160 - y) + 10y = 1510$
$1440 - 9y + 10y = 1510$
$y = 70$
$x = 160 - 70 = 90$
The lot is 90 ft × 70 ft

147. $x + y = 90$
$x + (3y - 20) = 180$
$y = 90 - x$
$x + 3(90 - x) = 200$
$x + 270 - 3x = 200$
$-2x = -70$
$x = 35$
$y = 90 - 35 = 55$
The measures are 35° and 55°.

148. Let x = liters of 15% acid solution. Then $6 - x$ = liters of 40% acid solution.
$0.15x + 0.4(6 - x) = 0.25(6)$
$0.15x + 2.4 - 0.4x = 1.5$
$-0.25x = -0.9$
$x = 3.6$
$6 - x = 2.4$
3.6 liters of 15% acid and 2.4 liters of 40% acid

149. Let x = speed of boat in calm water and y = speed of the current.
$d = rt$
$84 = (x + y)2$
$84 = (x - y)3$

$x + y = 42$
$x - y = 28$
$2x = 70$
$x = 35$
$y = 42 - 35 = 7$
The boat's speed is 35 mph and the current's speed is 7 mph.

150. $x = z + 35{,}600$
$x = 2y - 55{,}600$
$x + y + z = 168{,}200$

$z + 35{,}600 = 2y - 55{,}600$
$z + 35{,}600 = 168{,}200 - y - z$

$z - 2y = -91{,}200$
$4z + 2y = 265{,}200$
$5z = 174{,}000$
$z = 34{,}800$

$34{,}800 - 2y = -91{,}200$
$y = 63{,}000$
$x = 34{,}800 + 35{,}600 = 70{,}400$
The number of recruits are as follows:
Army: 70,400
Navy: 63,000
Marines: 34,800

151. Let x = smallest angle.
Then, $x + 29$ = largest angle and
$180 - x - (x + 29) = 151 - 2x$ = remaining angle.
$x + 29 = 2(151 - 2x) - 58$
$x + 29 = 302 - 4x - 58$
$5x = 215$
$x = 43$
$x + 29 = 72$
$151 - 2x = 65$
The angles are 43°, 72° and 65°.

152. $10(8) - 10x - x(8 - x) = 35$
$80 - 10x - 8x + x^2 = 35$
$x^2 - 18x + 45 = 0$
$(x - 3)(x - 15) = 0$
$x = 3$ or $x = 15$ (not possible)
The width of the path should be 3 m.

153. Let x = height, then $2x + 1$ = base.
$\frac{1}{2}[x(2x+1)] = 39$
$2x^2 + x - 78 = 0$
$(2x + 13)(x - 6) = 0$
$x = -\frac{13}{2}$ (disregard) or $x = 6$
$2x + 1 = 2(6) + 1 = 13$
The dimensions are:
height, 6 ft and base, 13 ft.

154. Let x = height up pole to where wire is attached.
$17^2 = (x - 7)^2 + x^2$
$289 = x^2 - 14x + 49 + x^2$
$0 = 2x^2 - 14x - 240$
$0 = 2(x + 8)(x - 15)$
$x = -8$ (disregard) or $x = 15$
The wire reaches 15 ft up the pole.

155. Let l = length of skid marks, and s = speed.
k = constant
$l = ks^2$
$40 = k(30)^2$
$k = \frac{40}{900} = \frac{2}{45}$
$250 = \frac{2}{45}s^2$
$s = 75$
No; they were traveling 75 mph.

156. $4(5x) + x^2 = 341$
$x^2 + 20x - 341 = 0$
$(x + 31)(x - 11) = 0$
$x = -31$ (disregard) or $x = 11$
The dimensions should be 11 in. by 11 in. by 5 in.

157. $2x(x + 1) = 40$
$2x^2 + 2x - 40 = 0$
$2(x + 5)(x - 4) = 0$
$x = -5$ (disregard) or $x = 4$
$x + 2 + 2 = 4 + 2 + 2 = 8$
$(x + 1) + 2 + 2 = 4 + 1 + 2 + 2 = 9$
The cardboard should be 8 in. by 9 in.

158. $2xy = 8400$
$3y + 4x = 450$
$x = \frac{4200}{y}$
$3y + 4\left(\frac{4200}{y}\right) = 450$
$\frac{16,800}{y} = 450 - 3y$
$16,800 = 450y - 3y^2$
$y^2 - 150y + 5600 = 0$
$(y - 80)(y - 70) = 0$
$y = 80$ or $y = 70$
$x = \frac{4200}{80} = 52.5$ or $x = \frac{4200}{70} = 60$
The dimensions are 52.5 ft by 80 ft or 60 ft by 70 ft.

159. Let x = length and y = width.
$2x + 2y = 14$
$x^2 + y^2 = 5^2$
$x = 7 - y$
$(7 - y)^2 + y^2 = 25$
$49 - 14y + y^2 + y^2 = 25$
$2y^2 - 14y + 24 = 0$
$y^2 - 7y + 12 = 0$
$(y - 4)(y - 3) = 0$
$y = 4$ or 3
If $y = 4$, $x = 7 - 4 = 3$
If $y = 3$, $x = 7 - 3 = 4$.
The book is 3 in. × 4 in.

160. Odd number less than 100: 1, 3, 5, 7, …, 97, 99
A multiple of 5: 5, 15, 25, 35, 45, 55, 65, 75, 85, 95
Divisible by 3: 15, 45, 75
Sum of digits is odd:

15: $1 + 5 = 6$
45: $4 + 5 = 9$
75: $7 + 5 = 12$
45 is the only number.

161.

	1st	2nd	3rd	4th	5th	6th	12th	nth
Pattern	$\frac{1+1}{2}$	$\frac{2(2+1)}{2}$	$\frac{3(3+1)}{2}$	$\frac{4(4+1)}{2}$	$\frac{5(5+1)}{2}$	$\frac{6(6+1)}{2}$	$\frac{12(12+1)}{2}$	$\frac{n(n+1)}{2}$
Triangular number	1	3	6	10	15	21	78	$\frac{n(n+1)}{2}$

15, 21, 78, $\frac{n(n+1)}{2}$

162. $\frac{10x+y}{w} = z$ with $w, x, y,$ and z different, non-zero, one-digit positive numbers.

a. $w = 3, y = 1$
$\frac{10x+1}{3} = z$
$x = 2$ and $z = 7$ since $\frac{20+1}{3} = 7.$

b. $x = y - 3$
$y = 4,\ x = 1,\ w = 7$ so $z = 2.$
$y = 4,\ x = 1,\ w = 2$ so $z = 7.$
since $\frac{10x+y}{w} = z$
$\frac{10+4}{7} = 2$ and $\frac{10+4}{2} = 7.$

c. $w = 2z$ and $x > y.$
$x = 3,\ y = 2,\ z = 4,\ w = 8$
$z = 4,\ w = 2(4) = 8$
$\frac{10(3)+2}{8} = 4,\ 3 > 2$

d. $w + z = x + y$
$x = 1,\ y = 8$ (with $w = 3$ and $z = 6$)
$3 + 6 = 1 + 8$
$\frac{10(1)+8}{3} = 6$

163. $2 \cdot (1 + 2 \cdot 3) - 2 \div (2 - 1) = 12$

164. $50 + 30 + 10 = 90$
$50 + 20 + 20 = 90$
$30 + 30 + 30 = 90$
3 ways

165.

Radius	Height	Area
1	2	$4(1 \times 2) = 8$
2	3	$4(2 \times 3) = 24$
3	4	$4(3 \times 4) = 48$
9	10	$4(9 \times 10) = 360$

The area is 360 square units.

166.

solid	1	2	3	4	5	6	7	8	9	10
cubes	1	6	11	16	21	26	31	36	41	46

It has 46 cubes.